Interference Analysis of Communication Systems

OTHER IEEE PRESS BOOKS

Integrated Injection Logic, *Edited by J. E. Smith*
Sensory Aids for the Hearing Impaired, *Edited by H. Levitt, J. M. Pickett, and R. A. Houde*
Data Conversion Integrated Circuits, *Edited by Daniel J. Dooley*
Semiconductor Injection Lasers, *Edited by J. K. Butler*
Satellite Communications, *Edited by H. L. Van Trees*
Frequency-Response Methods in Control Systems, *Edited by A.G.J. MacFarlane*
Programs for Digital Signal Processing, *Edited by the Digital Signal Processing Committee*
Automatic Speech & Speaker Recognition, *Edited by N. R. Dixon and T. B. Martin*
Speech Analysis, *Edited by R. W. Schafer and J. D. Markel*
The Engineer in Transition to Management, *I. Gray*
Multidimensional Systems: Theory & Applications, *Edited by N. K. Bose*
Analog Integrated Circuits, *Edited by A. B. Grebene*
Integrated-Circuit Operational Amplifiers, *Edited by R. G. Meyer*
Modern Spectrum Analysis, *Edited by D. G. Childers*
Digital Image Processing for Remote Sensing, *Edited by R. Bernstein*
Reflector Antennas, *Edited by A. W. Love*
Phase-Locked Loops & Their Application, *Edited by W. C. Lindsey and M. K. Simon*
Digital Signal Computers and Processors, *Edited by A. C. Salazar*
Systems Engineering: Methodology and Applications, *Edited by A. P. Sage*
Modern Crystal and Mechanical Filters, *Edited by D. F. Sheahan and R. A. Johnson*
Electrical Noise: Fundamentals and Sources, *Edited by M. S. Gupta*
Computer Methods in Image Analysis, *Edited by J. K. Aggarwal, R. O. Duda, and A. Rosenfeld*
Microprocessors: Fundamentals and Applications, *Edited by W. C. Lin*
Machine Recognition of Patterns, *Edited by A. K. Agrawala*
Turning Points in American Electrical History, *Edited by J. E. Brittain*
Charge-Coupled Devices: Technology and Applications, *Edited by R. Melen and D. Buss*
Spread Spectrum Techniques, *Edited by R. C. Dixon*
Electronic Switching: Central Office Systems of the World, *Edited by A. E. Joel, Jr.*
Electromagnetic Horn Antennas, *Edited by A. W. Love*
Waveform Quantization and Coding, *Edited by N. S. Jayant*
Communication Satellite Systems: An Overview of the Technology, *Edited by R. G. Gould and Y. F. Lum*
Literature Survey of Communication Satellite Systems and Technology, *Edited by J. H. W. Unger*
Solar Cells, *Edited by C. E. Backus*
Computer Networking, *Edited by R. P. Blanc and I. W. Cotton*
Communications Channels: Characterization and Behavior, *Edited by B. Goldberg*
Large-Scale Networks: Theory and Design, *Edited by F. T. Boesch*
Optical Fiber Technology, *Edited by D. Gloge*
Selected Papers in Digital Signal Processing, II, *Edited by the Digital Signal Processing Committee*
A Guide for Better Technical Presentations, *Edited by R. M. Woelfle*
Career Management: A Guide to Combating Obsolescence, *Edited by H. G. Kaufman*
Energy and Man: Technical and Social Aspects of Energy, *Edited by M. G. Morgan*
Magnetic Bubble Technology: Integrated-Circuit Magnetics for Digital Storage and Processing, *Edited by H. Chang*
Frequency Synthesis: Techniques and Applications, *Edited by J. Gorski-Popiel*
Literature in Digital Processing: Author and Permuted Title Index (Revised and Expanded Edition), *Edited by H. D. Helms, J. F. Kaiser, and L. R. Rabiner*
Data Communications via Fading Channels, *Edited by K. Brayer*
Nonlinear Networks: Theory and Analysis, *Edited by A. N. Willson, Jr.*
Computer Communications, *Edited by P. E. Green, Jr. and R. W. Lucky*
Stability of Large Electric Power Systems, *Edited by R. T. Byerly and E. W. Kimbark*
Automatic Test Equipment: Hardware, Software, and Management, *Edited by F. Liguori*
Key Papers in the Development of Coding Theory, *Edited by E. R. Berkekamp*
Technology and Social Institutions, *Edited by K. Chen*
Key Papers in the Development of Information Theory, *Edited by D. Slepian*
Computer-Aided Filter Design, *Edited by G. Szentirmai*
Integrated Optics, *Edited by D. Marcuse*
Digital Signal Processing, *Edited by L. R. Rabiner and C. M. Rader*
Minicomputers: Hardware, Software, and Applications, *Edited by J. D. Schoeffler and R. H. Temple*
Semiconductor Memories, *Edited by D. A. Hodges*

Interference Analysis of Communication Systems

Edited by

Peter Stavroulakis
Associate Professor of Engineering
Oakland University

A substantial portion of this book was prepared while the Editor was at Bell Laboratories.

A volume in the IEEE PRESS Selected Reprint Series, prepared under the sponsorship of the IEEE Electromagnetic Compatability Society.

IEEE PRESS

The Institute of Electrical and Electronics Engineers, Inc. New York

PRINTED IN THE UNITED STATES OF AMERICA

Sole Worldwide Distributor (Exclusive of IEEE):
JOHN WILEY & SONS, INC.
605 Third Ave.
New York, NY 10016

Wiley Order Numbers: Clothbound: 0-471-08674-6
Paperbound: 0-471-08673-8

IEEE Order Numbers: Clothbound: PC-01321
Paperbound: PP-01339

Library of Congress Cataloging in Publication Data

Main entry under title:

Interference analysis of communication systems.

(IEEE Press selected reprint series)
"Prepared under the sponsorship of the IEEE Electromagnetic Compatibility Society."
Includes index.
1. Electromagnetic interference--Addresses, essays, lectures. 2. Electromagnetic compatibility--Addresses, essays, lectures. 3. Telecommunication systems--Addresses, essays, lectures. I. Stavroulakis, Peter. II. IEEE Electromagnetic Compatibility Society.
TK155.I57 621.38'0436 80-18464
ISBN 0-87942-135-5
ISBN 0-87942-136-3 (pbk.)

Contents

Preface vii

Part I: Analog-Signal Interference 1

Interchannel Interference Considerations in Angle-Modulated Systems, *V. K. Prabhu and L. H. Enloe (Bell System Technical Journal,* September 1969) 3

Interference into Angle-Modulated Systems Carrying Multichannel Telephony Signals, *B. A. Pontano, J. C. Fuenzalida, and N. K. M. Chitre (IEEE Transactions on Communications,* June 1973) 16

Interchannel Interference in FM and PM Systems Under Noise Loading Conditions, *W. R. Bennett, H. E. Curtis, and S. O. Rice (Bell System Technical Journal,* May 1955) 29

An Extended Analysis of RF Interference in FDM-FM Radio Relay Systems, *G. J. Garrison (IEEE Transactions on Communication Technology,* October 1967) 47

Multipath Intermodulation Associated with Operation of FM-FDM Radio Relays in Heavily Built Areas, *D. Ongaro (IEEE Transactions on Communication Technology,* June 1971) 56

Intermodulation Distortion in Frequency-Division-Multiplex FM Systems—A Tutorial Summary, *G. J. Garrison (IEEE Transactions on Communication Technology,* April 1968) 65

Time-Domain Analysis of Intermodulation Effects Caused by Nonlinear Amplifiers, *J. C. Fuenzalida, O. Shimbo, and W. L. Cook (COMSAT Technical Review,* Spring 1973) 80

Expected Interference Levels due to Interactions Between Line-of-Sight Radio Relay Systems and Broadband Satellite Systems, *R. G. Medhurst and J. H. Roberts (Proceedings of the Institution of Electrical Engineers,* March 1964) 107

Interference between Satellite Communication Systems and Common Carrier Surface Systems, *H. E. Curtis (Bell System Technical Journal,* May 1962) 112

A Survey of Interference Problems and Applications to Geostationary Satellite Networks, *M. C. Jeruchim (Proceedings of the IEEE,* March 1977) 124

Part II: Digital-Signal Interference 139

Performance of Digital Phase-Modulation Communication Systems, *C. R. Cahn (IRE Transactions on Communications Systems,* May 1959) 142

PSK Error Performance with Gaussian Noise and Interference, *A. S. Rosenbaum (Bell System Technical Journal,* February 1969) 146

Error Rate Considerations for Coherent Phase-Shift Keyed Systems with Co-Channel Interference, *V. K. Prabhu (Bell System Technical Journal,* March 1969) 161

Binary PSK Error Probabilities with Multiple Cochannel Interferences, *A. S. Rosenbaum (IEEE Transactions on Communication Technology,* June 1970) 174

Multiple Error Performance of PSK Systems with Cochannel Interference and Noise, *J. Goldman (IEEE Transactions on Communication Technology,* August 1971) 187

Spacing Limitations of Geostationary Satellites Using Multilevel Coherent PSK Signals, *M. C. Jeruchim and F. E. Lilley (IEEE Transactions on Communications,* October 1972) 198

Binary and Quaternary PSK Radio Systems in a Multiple-Interference Environment, *C. Colavito and M. Sant'Agostino (IEEE Transactions on Communications,* September 1973) 204

Unified Analysis of a Class of Digital Systems in Additive Noise and Interference, *R. Fang and O. Shimbo (IEEE Transactions on Communications,* October 1973) 216

Effects of Cochannel Interference and Gaussian Noise in *M*-ary PSK Systems, *O. Shimbo and R. Fang (COMSAT Technical Review,* Spring 1973) 233

Combined Effects of Intersymbol, Interchannel, and Co-Channel Interferences in *M*-ary CPSK Systems, *S. Benedetto, E. Biglieri, and V. Castellani (IEEE Transactions on Communications,* September 1973) 245

An Error-Probability Upper Bound for Coherent Phase-Shift Keying with Peak-Limited Interference, *A. S. Rosenbaum and F. E. Glave (IEEE Transactions on Communications,* January 1974) 257

An Upper Bound Analysis for Coherent Phase-Shift Keying with Cochannel, Adjacent-Channel, and Intersymbol Interference, *F. E. Glave and A. S. Rosenbaum (IEEE Transactions on Communications,* June 1975) 268
TV Cochannel Interference on a PCM-PSK SCPC System, *D. Kurjan and M. Wachs (COMSAT Technical Review,* Fall 1976) 280
An Approximate Method to Estimate an Upper Bound on the Effect of Multipath Delay Distortion on Digital Transmission, *W. C. Jakes, Jr. (IEEE Transactions on Communications,* January 1979) 286
Imperfect Carrier Recovery Effect on Filtered PSK Signals, *V. K. Prabhu (IEEE Transactions on Aerospace and Electronic Systems,* July 1978) 292
Phase-Shift-Keyed Signal Detection with Noisy Reference Signals, *W. C. Lindsey (IEEE Transactions on Aerospace and Electronic Systems,* July 1966) 299
Effect of Noisy Phase Reference on Coherent Detection of Offset-QPSK Signals, *S. A. Rhodes (IEEE Transactions on Communications,* August 1974) 308
Co-Channel Inteference of Spread Spectrum Systems in a Multiple User Environment, *S. A. Musa and W. Wasylkiwskyj (IEEE Transactions on Communications,* October 1978) 318
Fading Effects on the Performance of a Spread Spectrum Multiple Access Communication System, *C. S. Gardner and J. A. Orr (IEEE Transactions on Communications,* January 1979) 326

Part III: Interference Reduction/Cancelling 333

Adaptive Antenna Systems, *B. Widrow, P. E. Mantey, L. J. Griffiths, and B. B. Goode (Proceedings of the IEEE,* December 1967) 334
Adaptive Filter for Interference Suppression, *A. M. Kowalski (National Telecommunications Conference Record,* December 1977) 351
Suppression of Co-Channel Interference with Adaptive Cancellation Devices at Communications Satellite Earth Stations, *P. D. Lubell and F. D. Rebhun (Proceedings of the International Conference on Communications,* June 1977) 357
An Adaptive Interference Cancellation System for Elimination of Co-Located Interference Signals, *W. F. Geist (Proceedings of the International Conference on Communications,* June 1977) 363
Echo Suppressor Design Considerations, *J. E. Unrue, Jr. (IEEE Transactions on Communication Technology,* August 1968) 367
Echo Canceller Utilizing Pseudo-Logarithmic Coding, *O. A. Horna (National Telecommunications Conference Record,* December 1977) 376
An Unusually Simple Technique for Sidelobe Reduction, *J. Perini (IEEE Transactions on Electromagnetic Compatability,* February 1969) 384
New Sidelobe Envelopes for Small Aperture Earth Stations, *J. M. Janky, B. B. Lusignan, L.-S. Lee, E. C. Ha, and E. F. Reinhart (IEEE Transactions on Broadcasting,* June 1976) 390

Part IV: Computer Simulation 397

Multi-channel Frequency Division Multiplex Simulation, *C. S. Lorens (Proceedings of the Fifth Annual Pittsburgh Conference on Modeling and Simulation,* April 1974) 398
Digital Computer Simulation of Satellite Quadrature Data Communication Systems, *M. C. Jeruchim (National Telecommunications Conference Record,* December 1975) 403
Computer Simulation of a Digital Satellite Communications Link, *D. L. Hedderly and L. Lundquist (IEEE Transactions on Communications,* April 1973) 409
Interactive Computer Simulation of Satellite Transmission Systems, *W. L. Cook (Proceedings of the Fifth Annual Pittsburgh Conference on Modeling and Simulation,* April 1974) 414

Author Index 420

Subject Index 421

Editor's Biography 424

Preface

ONE CAN SAY that the subject of interference is as old as the subject of communications. Over the last few years, however, the subject of interference as a self-contained discipline has been followed with special attention. The reason for this may be that the field of communications has experienced an unprecedented expansion and development both on the theoretical and applied fronts (radio, mobile, fiber, and satellite communications), and interference presents an unavoidable limitation to the capacity, efficiency, reliability, and cost of communication systems. Interference mechanisms, therefore, must be fully understood before any optimal communication systems can be designed.

Interference is a ubiquitous property of the environment in which communication systems exist, being caused not only by external means (neighboring systems) but also by nonideal mechanisms utilized in the process of communications. The following set of papers contains, in the opinion of the editor, the basic mathematical tools necessary for the understanding and analytical study of the subject. The reader may have already realized that such a subject is necessarily very broad, and any book of this size has to be very selective in order to provide adequate treatment of the subject and still present a balanced picture.

It was apparent that the subject can be divided naturally into two broad categories: analog and digital signal transmissions. Within each category, distortions of communications signals due to interfering signals caused by external sources (intersystem, cochannel, and adjacent channel interference) and by intrinsic sources (thermal noise, multipath, and noisy reference signals) are presented. Following this natural breakdown, the first two parts of the book are organized to cover analog system interference in Part I and digital system interference in Part II. It was felt that some examples of computer simulations used as tools for the study of interference and examples of interference reduction and cancelling should be included. These are presented in Parts III and IV, respectively. Preceding each part of the book, an introductory discussion is given to guide the reader through the papers and point out the relations of particular papers to the main subject matter.

The interference due to the spurious signals caused by the nonlinear mechanisms of the communications process will not be covered in this book through the inclusion of reprint papers. We shall discuss these concepts only as they relate to the main theme. Important papers that cover these concepts will be given as references in the discussion.

Part I
Analog-Signal Interference

THE SUBJECT of interference of analog signals covers cases in which a modulated analog signal experiences interference from an unwanted signal arising from within its own system, from another system, or from disturbances produced by the nonlinearities and frequency-dependent parameters of the transmission path of the wanted system. Basic papers that treat this type of interference are included in this part.

A detailed analysis of the interference of angle-modulated (PM or FM) signals is studied in the first paper of this part by V. K. Prabhu and L. H. Enloe. It is shown that, with band-limited white Gaussian noise modulation (simulating modulation by a frequency division multiplex signal), an explicit expression for the spectral density of the baseband interchannel interference can be given when two or more PM waves interfere with each other. This expression is given as the convolution of the spectral densities of the angle-modulated waves.[1] The dependence of interference on the modulation index is explicitly derived, and it is shown that expanding the bandwidth of the wanted signal can be used with advantage in combating interference. These results illustrate the well-known properties of frequency and phase modulation with respect to exchanging bandwidth for signal-to-noise ratio. The effects of adjacent channel interference when the interference is in the passband of the receiver are also examined in this paper.

Some of the restrictions of the first paper are relaxed in the second, by B. A. Pontano *et al.* It is shown that the low interference restriction (i.e., small interfering to desired carrier amplitude), the assumption of an angle-modulated interferer, and Gaussian modulation can be eliminated. The results are given in the form of graphic illustrations and can be used to study the advantages of interleaving for certain cases of interference involving two identical signals. The remaining papers on analog signal interference included in this part cover specialized cases of interference such as echo interference, FDM/FM interference, multipath, and intermodulation.

One of the principal sources of interchannel interference is the variation of attenuation and phase shift of the transmission path with frequency. An echo is a manifestation of such variation. In the paper by W. R. Bennett *et al.*, interference produced by echoes of relatively small amplitudes is analyzed. Several important special cases are studied in detail, and curves that can be used for the computation of interference power are presented.

A general investigation of the interference of FDM/FM from RF signals is presented in [1] by R. Hamer. Both modulated and unmodulated interference are considered, and the results are given in terms of noise power ratio (NPR). In this investigation the modulated FDM/FM signal is approximated by a random noise signal of uniform power density. The noise power ratio is used to relate S/I to C/I which facilitates the design of FDM/FM systems in an interference environment. The analogy of the interferer with a very long delay echo signal is also analyzed. In the fourth paper, by G. J. Garrison, particular cases are studied which include interfering signals in FDM/FM systems falling on or near all major "critical" frequencies of the receiver, which cause spurious IF responses.

When a frequency modulated wave is received by more than one path, considerable distortion can result from the interference of these signals. This distortion is liable to be encountered in the vicinity of large physical obstructions in the transmission path which reflect and absorb the waves and thus cause interference. The result obtained in measurements on FM-FDM relays suffering from multipath interference are reported and discussed in the paper by D. Ongaro.

It is a well-known phenomenon in communication systems employing frequency multiplexing that amplitude nonlinearity, such as amplifier saturation (clipping), causes crosstalk due to intermodulations between signals in the

[1] The original paper that contained this result was published by S. Hayashi, "On the interference characteristics of the phase modulation receiver for the multiplex transmission," *Journal of the Institute of Electrical Communications Engineers of Japan* (in Japanese), in 1952.

various channels. Such a situation frequently arises in connection with multicarrier use of satellite transponders. The results obtained in [2] indicate that the amount of crosstalk which results in the individual channel due to peak clipping of a frequency multiplexed signal is relatively low, even where clipping occurs for a large fraction of time. These results indicate that large amounts of clipping can be tolerated in a frequency multiplexed communication system when moderate output signal-to-interference ratios are sufficient. The case where the saturation is presented by a smoothly limiting error function curve is studied in [3]. The signal-to-crosstalk ratio is determined as function of the "degree" of limiting. A more general case, which considers the crosstalk arising from transmission deviations versus frequency followed by amplitude-to-phase AM/PM conversion, is studied in [4]. The effects of an arbitrary continuous preemphasis characteristic as well as an arbitrary gain and phase shape transmission medium are analyzed in [5]. A summary of the results of the effects of power spectra that are generated by amplitude, phase, and AM/PM conversion within an FM transmission path is given, in the sixth paper of this part, by G. J. Garrison.

A mathematically unified approach to the calculation of intermodulation effects of a memoryless nonlinear amplifier is presented in the paper by J. C. Fuenzalida *et al*. The power spectral density of the baseband signal caused by intermodulation has been derived by using a time-domain approach. The physical interpretations of the analytical results and the computational method used to implement these results are also given. A computer program is described, based on these results, which has been found to be a useful tool in the determination of intermodulation spectra and baseband distortion in multicarrier systems.

As mentioned before, interference can be caused by unwanted signals from another system entering the receiver of the wanted system. The widespread use of satellite communication systems has made the threat of interference between satellite and terrestrial systems a real worldwide problem. Two practical situations which involve this type of interference are analyzed in the eighth and ninth papers of this part. R. G. Medhurst and J. H. Roberts deal mainly with the physical separation between satellite, earth station, and terrestrial carrier stations, and H. E. Curtis deals with intermodulation distortion between satellite and terrestrial systems.

Finally, an overall, comprehensive survey of interference generated by satellite networks is presented by M. C. Jeruchim in the last paper of this part. Particular attention is given to those systems and signals of importance in space communications, namely, FDM/FM, TV/FM, and coherent PSK digital signals. Examples are given to show the limiting effects of interference on orbit utilization.

REFERENCES

[1] R. Hamer, "Radio-frequency interference in multichannel telephony FM radio systems," *Proc. of IEEE,* Dec. 1961.

[2] C. R. Chan, "Crosstalk due to finite limiting of frequency-multiplexed signals," *Proc. IRE*, Jan. 1960.

[3] J. S. Lee, "Signal to crosstalk power ratio in smoothly limited multichannel FDM signals," *IEEE Trans. Commun. Technol.,* vol. COM-16, Feb. 1968.

[4] R. C. Chapman, Jr., and J. B. Millard, "Intelligible crosstalk between frequency modulated carriers through AM-PM conversion," *IEEE Trans. Commun. Syst.,* June 1964.

[5] M. L. Liou, "Noise in an FM system due to an imperfect linear transducer," *Bell Syst. Tech. J.,* Nov. 1966.

Interchannel Interference Considerations in Angle-Modulated Systems

By V. K. PRABHU and L. H. ENLOE

(Manuscript received November 14, 1968)

This paper considers the deterioration in performance of angle-modulated systems resulting from interchannel interference. We show that with band-limited white gaussian noise modulation (simulating modulation by a frequency division multiplex signal), we can derive an explicit expression for the spectral density of the baseband interchannel interference when two or more PM waves interfere with each other.

We show that, if the interference is co-channel, maximum interference occurs at the lowest baseband frequency present in the system and we can derive upper and lower bounds to this minimum baseband signal-to-interference ratio. For high enough modulation index, we show that this minimum signal-to-interference ratio is proportional to the cube of the modulation index and that phase modulation can be used with advantage in interference limited systems. We do not consider the effects of linear filters on angle-modulated systems, but give some results about the effect of adjacent channel interference when the interference is in the passband of the receiver.

I. INTRODUCTION

The properties of frequency and phase modulation with respect to exchanging bandwidth for signal-to-noise ratio are well known,[1,2] but the type of noise considered is almost always limited to be random gaussian noise. In the design of any system, where the noise is likely to be interference limited, it is necessary to consider other kinds of disturbances such as co-channel and adjacent channel interference corrupting the desired received signal.

Consider the following situation. In the frequency bands above 10 GHz where the signal attenuation resulting from rain could be very severe, close spacings of the repeaters are almost always mandatory for reliable communication from point-to-point and for all periods of time.[3,4] If low noise receivers are used in the system, it is possible that the total interference power received by the system may be very much larger than the noise power in the system. For all practical purposes, the performance of such a system is determined by the interchannel interference.[3,4] It is therefore desirable to evaluate the effect of co-channel and adjacent channel interference on the performance of any modulation system like FM or PM (or PCM) so that its advantages in combating interference can be determined, and any system parameters (such as rms phase deviation, channel separation, and so on) can be properly chosen to keep the baseband interference below a certain desired level. (It is possible to reduce adjacent channel interference by using suitable receiving filters, but co-channel interference occupies the same band as the signal.)

The problem of interference in angle-modulated systems has been considered by many authors.[5–12] In the analysis, most of these authors have given an approximate expression (the first term in the power series expansion) for the baseband interchannel interference, and have shown that it can be expressed as the convolution of the spectral densities of the angle-modulated waves. The accuracy in this approximation has not been determined previously. Also, in the calculation of interchannel interference in high index FM and PM systems, most of these authors use the quasistatic approximation, the accuracy of which is unknown.

We first consider a general method of evaluating the baseband interchannel interference when two angle-modulated waves interfere with each other. We assume that an ideal angle (frequency or phase) demodulator is used in the system. (An ideal angle demodulator does not respond to any variations in the amplitude of the wave. This can be achieved in practice by using an ideal limiter at the front end of the receiver. If $A(t)e^{i\Phi(t)}$ is the input to an ideal limiter, its output is given by $A_o e^{i\Phi(t)}$ where A_o is a constant.)

We obtain a general expression for the baseband interference when the modulating wave is gaussian. This expression can be utilized even when the baseband signal is passed through a linear network (such as a pre-emphasis—de-emphasis network).

We are specifically interested in calculating the baseband interchannel interference between two or more waves phase modulated (without pre-emphasis) by band-limited white gaussian random processes. It has been found in practice that such a random gaussian noise of appropriate bandwidth and power spectral density adequately simulates (for some purposes) a variety of signals such as a frequency division multiplex (FDM) signal, a composite speech

Reprinted with permission from *Bell Syst. Tech. J.*, vol. 48, pp. 2333-2358, Sept. 1969.

signal, and so on.[13] Since the determination of the power spectrum is fundamental to the evaluation of baseband interference, first we review briefly the methods of obtaining this spectrum for a wave phase modulated by band-limited white gaussian noise.

In the case of band-limited white gaussian noise modulation, if the bandwidths of the modulating waveforms for the desired and interfering carriers are the same, we show that the determination of baseband interference power is relatively simple, and requires only the computation of the spectral density of a phase-modulated carrier for a variety of values of rms phase deviation. For small values of interference and for band-limited white gaussian noise modulation, we also show that the first term in the series gives most of the contribution to the baseband interference, and that this first term can be used as a good approximation.

For a co-channel interferer, we show that maximum interference occurs at the lowest baseband frequency present in the system (we assume that this lowest frequency is $f = 0$)* and that we can derive upper and lower bounds to this minimum signal-to-interference ratio. For sufficiently high modulation index, we show that these bounds are proportional to the cube of the modulation index, and that phase modulation can be used to advantage in combating interference.[14]

We show that maximum interference with an adjacent channel interferer occurs at the highest baseband frequency present in the system if the carrier frequency separation f_d between the two channels is relatively large compared with the baseband bandwidth W. For a set of values of f_d/W and for different modulation indexes of the two channels, we compute this minimum signal-to-interference ratio and give the results in graphic form.

We then consider the case in which more than one interferer may corrupt the desired received carrier and show that we can derive an expression for the spectral density of the resulting baseband interference. This expression is in the form of an infinite series and for its evaluation, in the case of band-limited white gaussian noise modulation and equal modulation bandwidths, it is only necessary to be able to compute the spectral density of a sinusoidal carrier phase modulated by gaussian noise. In case all these interferers are co-channel and all of them have the same (high) modulation index Φ, we show that we can derive upper and lower bounds to the minimum baseband signal-to-interference ratio.

* We do not imply that maximum baseband interchannel interference always occurs at $f = 0$ for any general system angle modulated by gaussian noise.

II. INTERFERENCE BETWEEN TWO ANGLE-MODULATED WAVES

We first assume that there is only one interfering wave corrupting the desired received signal, and that both of them are angle modulated by two independent gaussian random processes. Let the desired angle-modulated wave be given by

$$\begin{aligned} s(t) &= A \cos [\omega_o t + p(t) * \varphi(t)] \\ &= \operatorname{Re} A \exp \{j[\omega_o t + p(t) * \varphi(t)]\}, \end{aligned} \tag{1}$$

where A is the amplitude of the wave, $f_o = \omega_o/2\pi$ its carrier frequency, $p(t)$ the impulse response of the pre-emphasis network, and $\varphi(t)$ is a stationary gaussian random process with mean zero, and covariance function $R_\varphi(\tau)$. (We only assume that $p(t)$ is the impulse response of a linear network through which $\varphi(t)$ may be passed. Only for convenience, we refer to it as the impulse response of the pre-emphasis network.) The notation $A(x)*B(x)$ represents the convolution of function $A(x)$ with $B(x)$.

Let the interfering wave $i(t)$ be given by

$$\begin{aligned} i(t) &= R_i A \cos [\omega_i t + p_i(t) * \varphi_i(t) + \mu_i] \\ &= \operatorname{Re} AR_i \exp \{j[\omega_i t + p_i(t) * \varphi_i(t) + \mu_i]\}, \end{aligned} \tag{2}$$

where AR_i is its amplitude (R_i is the relative amplitude of the interfering wave with respect to the desired wave), ω_i is its angular frequency, $p_i(t)$ is the impulse response of its pre-emphasis network, and $\varphi_i(t)$ is a stationary gaussian random process with mean zero and covariance function $R_{\varphi_i}(\tau)$.

Since $s(t)$ and $i(t)$ usually originate from two different sources, it seems reasonable to assume that μ_i is a uniformly distributed random variable with probability density $\pi_{\mu_i}(\mu)$ where

$$\pi_{\mu_i}(\mu) = \begin{cases} \dfrac{1}{2\pi}, & 0 \leq \mu < 2\pi \\ 0, & \text{otherwise}. \end{cases} \tag{3}$$

Further, we assume that $\varphi(t)$ and $\varphi_i(t)$ are independent of each other and independent of μ_i. (Reference 13 treats of the case in which μ_i is a deterministic constant, and $\varphi(t)$ and $\varphi_i(t)$ are not independent of each other.)

If we assume that $s(t)$ and $i(t)$ are both in the passband of the receiver used in the system, the total signal $r(t)$ incident at the receiver is given by†

$$r(t) = \text{Re } A(\exp \{j[\omega_o t + p(t) * \varphi(t)]\} + R_i \exp \{j[\omega_i t + p_i(t) * \varphi_i(t) + \mu_i]\})$$
$$= \text{Re } A(1 + R_i \exp \{j[(\omega_i - \omega_o)t + p_i(t) * \varphi_i(t) - p(t) * \varphi(t) + \mu_i]\}) \cdot \exp \{j[\omega_o t + p(t) * \varphi(t)]\}$$
$$= \text{Re } Aa(t)e^{j\lambda(t)} \exp \{j[\omega_o t + p(t) * \varphi(t)]\}$$
$$= \text{Re } Aa(t) \exp \{j[\omega_o t + p(t) * \varphi(t) + \lambda(t)]\}, \quad (4)$$

where

$$a(t)e^{j\lambda(t)} = 1 + R_i \cdot \exp \{j[(\omega_i - \omega_o)t + p_i(t) * \varphi_i(t) - p(t) * \varphi(t) + \mu_i]\}. \quad (5)$$

Notice from equation (4) that the (excess) phase angle $\eta(t)$, as detected by an ideal angle demodulator, is given by

$$\eta(t) = \varphi(t) + \lambda(t). \quad (6)$$

(The gain—or proportionality factor—of the phase demodulator has been assumed to be unity.) Therefore, the spectral density of $\eta(t)$ can be written as

$$S_\eta(f) = \int_{-\infty}^{\infty} R_\eta(\tau)e^{-j2\pi f\tau}\, d\tau, \quad (7)$$

where $R\eta(\tau)$ is the covariance function of $\eta(t)$, and

$$R_\eta(\tau) = \langle \eta(t)\eta(t + \tau)\rangle. \quad (8)$$

(The notation $\langle x\rangle$ represents the ensemble average of random variable x.) If there is no interference, and if $q(t)$ is the impulse response of the de-emphasis network used in the system, the detected phase angle $\Omega(t)$ can be written as

$$[\Omega(t)]_{R_i=o} = q(t) * p(t) * \varphi(t). \quad (9)$$

If $R_i \neq 0$,

$$\Omega(t) = q(t) * p(t) * \varphi(t) + q(t) * \lambda(t). \quad (10)$$

Now if we assume that the de-emphasis network is the inverse of the pre-emphasis network, we have

$$q(t) * p(t) = \delta(t), \quad (11)$$

and

$$\Omega(t) = \varphi(t) + q(t) * \lambda(t), \quad (12)$$

where $\delta(t)$ is the Dirac delta function.

From equation (5), we have

$$\lambda(t) = \text{Im} \ln (1 + R_i \exp \{j[\omega_d t + p_i(t) * \varphi_i(t) - p(t) * \varphi(t) + \mu_i]\}), \quad (13)$$

where

$$\omega_d = \omega_i - \omega_o . \quad (14)$$

Notice that

$$\ln (1 + z) = \sum_{m=1}^{\infty} (-1)^{m+1} \frac{z^m}{m}, \qquad |z| < 1, \quad (15)$$

where z is any complex number.

Therefore, for $R_i < 1$, we have†

$$\lambda(t) = \sum_{m=1}^{\infty} \frac{(-1)^{m+1}}{m} \cdot R_i^m \left[\frac{\exp \{jm[\omega_d t + p_i(t) * \varphi_i(t) - p(t) * \varphi(t) + \mu_i]\}}{2j} - \frac{\exp \{-jm[\omega_d t + p_i(t) * \varphi_i(t) - p(t) * \varphi(t) + \mu_i]\}}{2j}\right]$$
$$= \sum_{m=1}^{\infty} \frac{(-1)^{m+1}}{m} R_i^m \sin \{m[\omega_d t + p_i(t) * \varphi_i(t) - p(t) * \varphi(t) + \mu_i]\}. \quad (16)$$

Since $\varphi(t)$, $\varphi_i(t)$, and μ_i are statistically independent random variables and since $\langle \exp (jk\mu_i)\rangle = 0$ with $k \neq 0$, we can show from equations (6), (8), (13), and (16) that

$$R_\eta(\tau) = R_p(\tau) * R_\varphi(\tau) + \sum_{m=1}^{\infty} \frac{R_i^{2m}}{2m^2} \cos m\omega_d\tau \cdot \exp (-m^2\{[R_{\varphi p}(0) - R_{\varphi p}(\tau)] + [R_{\varphi i p i}(0) - R_{\varphi i p i}(\tau)]\}), \quad (17)$$

† In this paper we do not consider the effects of linear filters usually used in receiving systems on the interchannel interference between two (or more) angle-modulated systems.

† For $R_i < 1$, notice that $a(t) > 0$.

where‡

$$R_p(\tau) = \int_{-\infty}^{\infty} p(t)p(t+\tau)\,dt, \tag{18}$$

$$R_{p_i}(\tau) = \int_{-\infty}^{\infty} p_i(t)p_i(t+\tau)\,dt, \tag{19}$$

$$R_{\varphi p}(\tau) = R_p(\tau) * R_\varphi(\tau), \tag{20}$$

and

$$R_{\varphi_i p_i}(\tau) = R_{p_i}(\tau) * R_{\varphi_i}(\tau). \tag{21}$$

Therefore, the spectral density of the output is given by

$$S_\Omega(f) = S_\varphi(f) + \frac{1}{|H_p(f)|^2} \sum_{m=1}^{\infty} \frac{R_i^{2m}}{4m^2} [T_m(f - mf_d) + T_m(f + mf_d)], \tag{22}$$

where $H_p(f)$ is the Fourier transform of $p(t)$, and

$$T_m(f) = \int_{-\infty}^{\infty} \exp(-m^2\{[R_{\varphi p}(0) - R_{\varphi p}(\tau)] + [R_{\varphi_i p_i}(0) - R_{\varphi_i p_i}(\tau)]\})e^{-i2\pi f\tau}\,d\tau. \tag{23}$$

From equation (23), we can show that

$$T_m(f) = U_m(f) * V_m(f) \tag{24}$$

where¶

$$U_m(f) = \int_{-\infty}^{\infty} \exp\{-m^2[R_{\varphi p}(0) - R_{\varphi p}(\tau)]\}e^{-i2\pi f\tau}\,d\tau, \tag{25}$$

and

$$V_m(f) = \int_{-\infty}^{\infty} \exp\{-m^2[R_{\varphi_i p_i}(0) - R_{\varphi_i p_i}(\tau)]\}e^{-i2\pi f\tau}\,d\tau. \tag{26}$$

Equation (22) gives a general expression for the baseband interchannel interference when two angle-modulated waves interfere with each other. To calculate this interchannel interference, equations (22) through (26) show that it is essential to determine the RF spectral density of a wave angle modulated by gaussian noise. Methods of

‡ Since $\varphi(t)$ and $\varphi_i(t)$ are assumed to be gaussian, $p(t) * \varphi(t)$, and $p_i(t) * \varphi_i(t)$ are also gaussian.[2,15] Notice also that the Fourier transform of $R_p(\tau)$ is equal to $|H_p(f)|^2$, if $H_p(f)$ is the Fourier transform of $p(t)$.

¶ Notice that $U_m(f)$ and $V_m(f)$ are the RF spectral densities of waves angle modulated by gaussian noise.

calculating this spectrum for low and medium index modulation are generally available, and the quasistatic approximation has been used for high index modulation.[2,13–16] Since the accuracy in the quasistatic approximation cannot often be determined, some rigorous methods of evaluating this spectrum for high index modulation have recently been developed.[2,16]

III. SPECTRAL DENSITY OF A PM WAVE

In this paper, we are specifically interested in determining the interchannel interference between two or more waves phase modulated by band-limited white gaussian random processes. Hence, we now review briefly the methods of obtaining the RF spectrum of such a wave. A sinusoidal wave of constant amplitude A phase modulated by a signal $n(t)$ can be written as

$$w(t) = A\cos[\omega_0 t + n(t) + \theta], \tag{27}$$

$$= \text{Re}\, A \exp\{j[\omega_0 t + n(t) + \theta]\}, \tag{28}$$

where $f_o = \omega_o/2\pi$ is the carrier frequency of the wave, and θ is a random variable with probability density function

$$\pi_\theta(\theta) = \begin{cases} 1/2\pi, & 0 \leqq \theta < 2\pi \\ 0, & \text{otherwise.} \end{cases} \tag{29}$$

If the modulating waveform is band-limited and white, its spectrum $S_n(f)$ is given by (see Fig. 1)

$$S_n(f) = \begin{cases} \Phi^2/2W, & |f| < W, \\ 0, & \text{otherwise.} \end{cases} \tag{30}$$

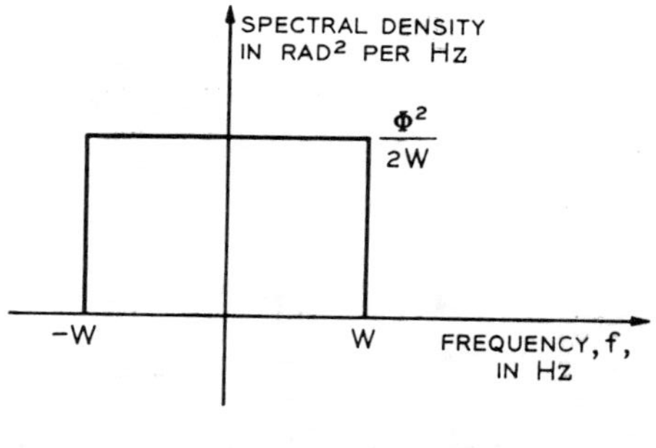

Fig. 1 — Spectral density of modulating wave.

Notice that Ref. 16 treats, in detail, the methods of obtaining the spectral characteristics of a sinusoidal carrier phase modulated by such a signal. From equation (30) we can show that (see Fig. 2)

$$R_n(\tau) = \Phi^2 \frac{\sin 2\pi W\tau}{2\pi W\tau}. \tag{31}$$

For $\Phi^2 \gg 1$ and for low frequencies, the quasistatic approximation yields[2,15]

$$S_V(f) \approx \exp(-\Phi^2)\,\delta(f) + \frac{1}{\Phi W}\left(\frac{3}{2\pi}\right)^{\frac{1}{2}} \exp\left[-\frac{3}{2}\frac{1}{\Phi^2}\left(\frac{f}{W}\right)^2\right]. \tag{32}$$

One can show that the approximation given by equation (32) is only good at low frequencies and that it is too small for large f.[16]

For large modulation indexes ($\Phi > 1.7432$ rad) and for all frequencies, we can show that[16]

$$S_V(f) = \exp(-\Phi^2)\left\{\delta(f) + \frac{\Phi^2}{2W}\left[u_{-1}(f+W) - u_{-1}(f-W)\right]\right\}$$
$$+ \frac{1}{2\pi W}\exp\left[-2\Phi^2\left(\cosh^2\frac{y_s}{2} - \frac{\sinh y_s}{y_s}\right)\right]\mu, \tag{33}$$

where

$$u_{-1}(x) = \begin{cases} 1, & x > 0, \\ 0, & \text{otherwise}, \end{cases} \tag{34}$$

$$\left(\frac{2\pi}{\Phi^2 A_2}\right)^{\frac{1}{2}}(1 - C) < \mu < \left(\frac{2\pi}{\Phi^2 A_2}\right)^{\frac{1}{2}}(1 + D), \tag{35}$$

$$\frac{\cosh y_s}{y_s} - \frac{\sinh y_s}{y_s^2} = \frac{f}{\Phi^2 W}, \tag{36}$$

and

$$A_2 = \frac{\sinh y_s}{y_s} - \frac{2}{y_s}\frac{f}{\Phi^2 W}. \tag{37}$$

We can also show that C and D, appearing in equation (35), are less than 8 per cent for $\Phi > (10)^{\frac{1}{2}}$ rad. Further, for all f, one can show that[16]

$$C < 2\% \quad \text{for} \quad \Phi > 5 \text{ rad}, \tag{38}$$

and

$$D < 2\% \quad \text{for} \quad \Phi > 5 \text{ rad}. \tag{39}$$

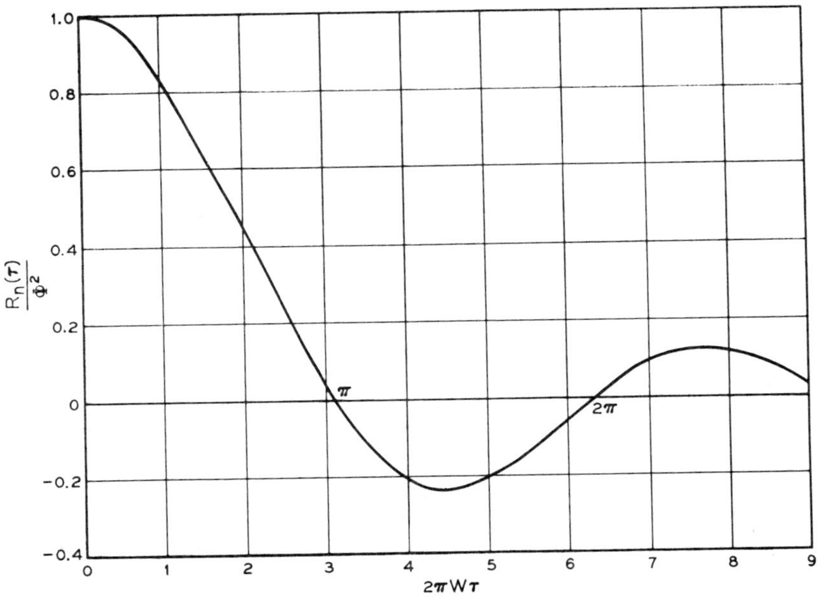

Fig. 2 — Covariance function $R_n(\tau)$. Since $R_n(\tau)$ is an even function of τ, we only show $R_n(\tau)$ for $\tau \geqq 0$.

Hence, we can say that

$$\mu \approx \left(\frac{2\pi}{\Phi^2 A_2}\right)^{\frac{1}{2}}, \tag{40}$$

and that the fractional error in this approximation is very much less than unity (less than 2 per cent, $\Phi > 5$ rad).

For $f = 0$, from equations (33) through (37) we can show that

$$0.92\left(\frac{3}{2\pi}\right)^{\frac{1}{2}}\frac{1}{\Phi W} < S_V(f) - \exp(-\Phi^2)\,\delta(f) < 1.08\left(\frac{3}{2\pi}\right)^{\frac{1}{2}}\frac{1}{\Phi W},$$
$$\Phi > (10)^{\frac{1}{2}} \text{ rad}. \tag{41}$$

For any f and Φ, the determination of the spectral density $S_V(f)$ from equations (33) through (40) is rather simple. For any given f, Φ^2, and W, we calculate y_s from equation (36), and A_2 from equation (37). The spectral density $S_V(f)$ is then calculated from equations (33) and (40).

IV. INTERFERENCE BETWEEN TWO PM WAVES

We now assume that $\varphi(t)$ and $\varphi_i(t)$ in Section II are band-limited white gaussian random processes with the same bandwidth W and rms phase deviations Φ and Φ_i. We also assume that $p(t) = p_i(t) = \delta(t)$, or that no pre-emphasis—de-emphasis networks are used in the system. Therefore, we have

$$R_\varphi(\tau) = \Phi^2 \frac{\sin 2\pi W\tau}{2\pi W\tau}, \tag{42}$$

and

$$R_{\varphi_i}(\tau) = \Phi_i^2 \frac{\sin 2\pi W\tau}{2\pi W\tau}. \tag{43}$$

From equations (22), (23), (42), and (43) we can write†

$$S_\Omega(f) = S_\varphi(f) + \sum_{m=1}^{\infty} \frac{R_i^{2m}}{m^2} G_m(f), \tag{44}$$

where

$$G_m(f) = \tfrac{1}{4}[H_m(f - mf_d) + H_m(f + mf_d)], \tag{45}$$

and

$$H_m(f) = \int_{-\infty}^{\infty} \exp\left[-m^2(\Phi^2 + \Phi_i^2)\left(1 - \frac{\sin 2\pi W\tau}{2\pi W\tau}\right)\right] e^{-i2\pi f\tau}\, d\tau. \tag{46}$$

Notice that $G_m(f)$ is the spectral density of a sinusoidal carrier (at carrier frequency mf_d, and having unit amplitude) phase modulated by a band-limited white gaussian random process having mean square phase deviation $m^2(\Phi^2 + \Phi^2)$. Section III gives methods of obtaining this spectrum for all values of f; hence, $S_\Omega(f)$ can easily be calculated. In order to evaluate $S_\Omega(f)$ from equation (44), we must be able to determine the spectral density of a carrier phase modulated by gaussian noise for any arbitrary modulation index. In the case of band-limited white gaussian noise modulation the technique presented in Ref. 16 is very convenient to calculate this spectrum. The series method of determining this spectral density can become rather tedious when Φ or Φ_i is large.

When there is no interference, the signal as detected by an ideal phase demodulator is given by $\varphi(t)$, and its spectral density by $S_\varphi(f)$. Therefore, from equation (44), the spectral density $S_I(f)$ of the baseband interchannel interference can be written as

$$S_I(f) = S_\Omega(f) - S_\varphi(f), \tag{47}$$

or

$$S_I(f) = \sum_{m=1}^{\infty} \frac{R_i^{2m}}{m^2} G_m(f). \tag{48}$$

Figure 3 is a graph of $S_I(f)$ for $f_d/W = 0$, 1, and 5; $\Phi = 3$ rad, and $\Phi_i = 2$ rad. Notice that, for $f_d/W = 1$, $S_I(f)$ is maximum at $f = 0$ or that maximum interchannel interference occurs at the lowest baseband frequency present in the system.

In practice the quantity of interest is usually the ratio of the average signal power to average interchannel interference power. In this case this signal-to-interference ratio $\sigma(f)$ can be written as

$$\sigma(f) = \frac{S_\varphi(f)\,\Delta f}{S_I(f)\,\Delta f} = \frac{S_\varphi(f)}{S_I(f)}, \tag{49}$$

where Δf is the spot frequency band of interest. Clearly, $\sigma(f)$ is a function of f and in designing an angle-modulated system one is usually

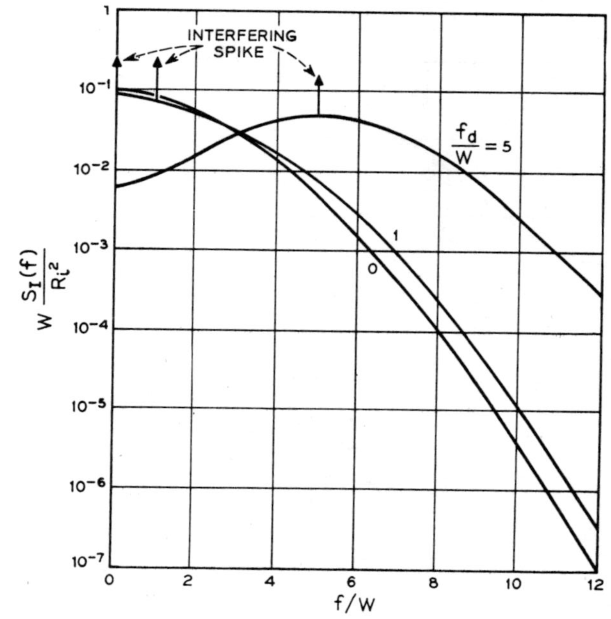

Fig. 3 — Spectral density $S_I(f)$ of baseband interference. $\Phi = 3$ rad; $\Phi_i = 2$ rad.

† Notice that in this case $\Omega(t) = \eta(t)$, since $p(t) = p_i(t) = \delta(t)$.

interested in the minimum value of $\sigma(f)$ for $0 < |f| \leqq W$. We denote the minimum value of this signal-to-interference ratio by S/I. In practice a phase demodulator is followed by a linear low pass filter. We assume that this filter is ideal and that it removes all the frequency components outside the desired signal frequency band $0 < |f| \leqq W$.

4.1 *Interference Between Two Co-Channel PM Waves*

In general, one can show (see Fig. 3) that $S_I(f)$ contains a (nonzero) Dirac delta function (corresponding to a line spectrum) at the frequency $\pm f_d$, and that the frequency division multiplex channel corresponding to this frequency may not be usable.‡ In case the interference is co-channel, $f_d = 0$, and the line spectrum lies at the frequency $f = 0$. In systems usually encountered in practice, there is no frequency division multiplex channel at dc even though the lowest frequency present in the baseband signal may approach a frequency arbitrarily close to zero.[14]

Notice from equation (48) and Fig. 3 that, in the case of co-channel interference between two PM waves, maximum baseband interference occurs at the lowest frequency present in the system; we assume that this lowest baseband frequency lies arbitrarily close to zero. In this case the minimum signal-to-interference ratio therefore occurs at $f = 0$ and

$$S/I = \frac{\Phi^2}{2W} \frac{1}{S_I'(0)}, \tag{50}$$

where

$$S_I'(0) = \sum_{m=1}^{\infty} \frac{R_i^{2m}}{2m^2} \{H_m(f) - \exp[-m^2(\Phi^2 + \Phi_i^2)]\, \delta(f)\}_{f=0}. \tag{51}$$

Since the interference is co-channel we further assume that $\Phi = \Phi_i$ so that the rms phase deviations in the two PM waves are the same. We can now write

$$S_I'(0) = \sum_{m=1}^{\infty} \frac{R_i^{2m}}{2m^2} [H_m(f) - \exp(-2m^2\Phi^2)\, \delta(f)]_{f=0}. \tag{52}$$

Consider the case $\Phi > (5)^{\frac{1}{2}}$ radians. In this case one can show that[16]

$$[H_m(f) - \exp(-2m^2\Phi^2)\, \delta(f)]_{f=0} \approx \frac{1}{2m\Phi W}\left(\frac{3}{\pi}\right)^{\frac{1}{2}}, \tag{53}$$

and that the error in this approximation is less than 8 per cent. Hence,

‡ We do not put any lower limit on the width of any frequency division multiplex channel present in the baseband signal.

we have

$$\frac{0.23}{\Phi W}\left(\frac{3}{\pi}\right)^{\frac{1}{2}} \sum_{m=1}^{\infty} \frac{R_i^{2m}}{m^3} < S_I'(0) < \frac{0.27}{\Phi W}\left(\frac{3}{\pi}\right)^{\frac{1}{2}} \sum_{m=1}^{\infty} \frac{R_i^{2m}}{m^3}, \qquad \Phi > (5)^{\frac{1}{2}} \text{ rad}. \tag{54}$$

It can be shown that

$$\sum_{m=1}^{\infty} \frac{R_i^{2m}}{m^3} = Q(R_i^2) = \int_0^{\infty} \frac{t^2\, dt}{e^t - R_i^2}. \tag{55}$$

Therefore, the signal-to-interference ratio at $f = 0$ is bounded by

$$\frac{1}{0.46}\left(\frac{\pi}{3}\right)^{\frac{1}{2}} \frac{\Phi^3}{Q(R_i^2)} > S/I > \frac{1}{0.54}\left(\frac{\pi}{3}\right)^{\frac{1}{2}} \frac{\Phi^3}{Q(R_i^2)}, \qquad \Phi > (5)^{\frac{1}{2}} \text{ rad}. \tag{56}$$

For any value of $R_i < 1$, equation (56) gives upper and lower bounds to S/I. We shall now investigate whether we can derive simpler upper and lower bounds to $Q(R_i^2)$.

From equation (55)

$$\sum_{m=1}^{\infty} \frac{R_i^{2m}}{m^3} = R_i^2 + \sum_{m=2}^{\infty} \frac{\exp[-m \ln(1/R_i^2)]}{m^3}. \tag{57}$$

Now one can show (see Fig. 4) that

$$0 < \sum_{m=2}^{\infty} \frac{\exp[-m \ln(1/R_i^2)]}{m^3} < \int_1^{\infty} \frac{\exp[-x \ln(1/R_i^2)]}{x^3}\, dx = E_3[\ln(1/R_i^2)], \tag{58}$$

where†

$$E_3(z) = \int_1^{\infty} \frac{e^{-zt}}{t^3}\, dt, \qquad z > 0. \tag{59}$$

We can show that for $R_i < 1$, $(\ln 1/R_i^2 > 0)$, (see Ref. 17)

$$0 < E_3(\ln 1/R_i^2) \leqq \frac{R_i^2}{2 + \ln(1/R_i^2)}, \tag{60}$$

or

$$Q(R_i^2) < R_i^2\left[1 + \frac{1}{2 + \ln(1/R_i^2)}\right]. \tag{61}$$

Since

† The function $E_3(z)$ is tabulated in Ref. 17 (see pp. 228–248). Notice also the inequality $E_n(z) \leqq e^{-z}/(z + n - 1)$ on p. 229.

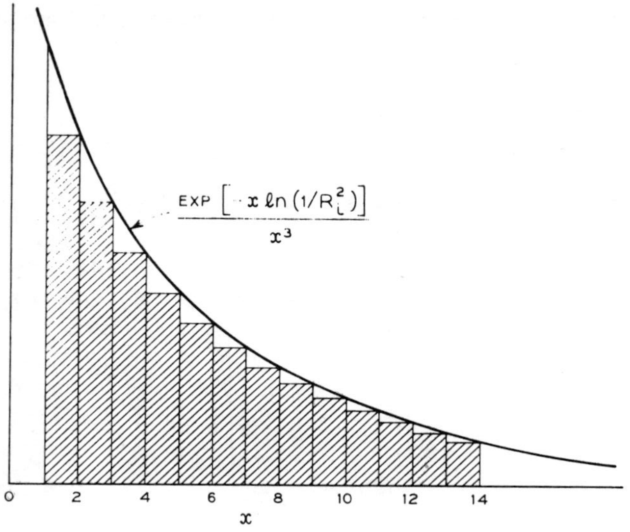

Fig. 4. — Function $\exp[-x \ln(1/R_i^2)]/x^3$ and $\sum_{m=2}^{\infty} R_i^{2m}/m^3$. The area in the shaded region is less than the area under the curve from $x = 1$.

$$\sum_{m=1}^{\infty} \frac{R_i^{2m}}{m^3} < \sum_{m=1}^{\infty} \frac{R_i^{2m}}{m} = -\ln(1 - R_i^2), \qquad Q(R_i^2) < -\ln(1 - R_i^2). \quad (62)$$

We are thankful to W. T. Barnett for having suggested another upper bound $R_i^2/(1 - R_i^2)$ to $Q(R_i^2)$.

One can show that the bound given in equation (62) is tighter than that given in equation (61) if

$$R_i < R_0 = 0.695573. \quad (63)$$

Let us write

$$U(R_i^2) = \begin{cases} 1 + \dfrac{1}{2 + \ln(1/R_i^2)}, & R_0 < R_i < 1, \\ -\dfrac{\ln(1 - R_i^2)}{R_i^2}, & 0 < R_i < R_0, \end{cases} \quad (64)$$

so that

$$R_i^2 < Q(R_i^2) < R_i^2 U(R_i^2), \qquad 0 < R_i < 1. \quad (65)$$

For carrier-to-interference ratio of 10 dB or for $R_i^2 = 0.1$

$$R_i^2 < \sum_{m=1}^{\infty} \frac{R_i^{2m}}{m^3} < 1.0536 R_i^2. \quad (66)$$

From equations (56) and (65), we next write

$$\frac{1}{0.46}\left(\frac{\pi}{3}\right)^{\frac{1}{2}} \frac{1}{R_i^2} \Phi^3 > S/I > \frac{1}{0.54}\left(\frac{\pi}{3}\right)^{\frac{1}{2}} \frac{1}{R_i^2} \frac{\Phi^3}{U(R_i^2)},$$

$$\Phi > (5)^{\frac{1}{2}} \text{ rad}, \quad R_i < 1. \quad (67)$$

Since the physical characteristics of elements used in a PM receiver are far from being ideal, and since thermal noise (which is always present) further deteriorates the performance of any PM receiver, we often find that $R_i^2 < 0.1$ in systems currently in use. Equations (66) and (67) show that the error introduced in truncating the series at $m = 1$ is less than 5.36 percent if $R_i^2 < 0.1$. For any $R_i \ll 1$, we therefore need take only the $m = 1$ term in equation (54) to estimate the baseband interference. Equation (67) gives upper and lower bounds to S/I for any $R_i < 1$. Also, note from equation (67) that co-channel interference can be suppressed in PM systems by using a large modulation index Φ.[14]

4.2 *Interference between Two Adjacent-Channel PM Waves*

As mentioned in Section II we do not consider the effects of linear filters on angle-modulated systems. We assume that the desired and interfering wave are both in the passband of the PM receiver used in the system, and that no filters are used to reduce the adjacent channel interference.

In any multichannel angle-modulated system generally encountered in practice there is usually both adjacent channel and co-channel interference. Protection against adjacent channel interference is often obtained by proper choice of the channel separation frequency and the required (linear) filters generally used in such systems. The assumptions made in this section are, therefore, a little unrealistic; hence, the results given may serve only as a guide in the actual calculation of adjacent channel interference.

For $0 < f_d/W < 1$, one can show that $S_I(f)$ contains a (nonzero) Dirac delta function (corresponding to a line spectrum) at the frequency $\pm f_d$ and that the frequency division multiplex channel corresponding to f_d/W may not be usable.

For $f_d \neq 0$ we can show, from equations (44) through (46), that

$$S_I(f) = \sum_{m=1}^{\infty} \frac{R_i^{2m}}{m^2} G_m(f), \quad (68)$$

where

$$G_m(f) = \tfrac{1}{4}[H_m(f - mf_d) + H_m(f + mf_d)], \quad (69)$$

and

$$H_m(f) = \int_{-\infty}^{\infty} \exp\left[-m^2(\Phi^2 + \Phi_i^2)\left(1 - \frac{\sin 2\pi W\tau}{2\pi W\tau}\right)\right] e^{-j2\pi f\tau}\, d\tau. \qquad (70)$$

For $0 \leqq |f| < W$, and $|f_d| \gg W$, one can show (by numerical methods) that $S_I(f)$ reaches its maximum at $f = W$, or that maximum baseband interchannel interference occurs at the highest frequency present in the baseband signal. For other values of channel separation frequency, this maximum is to be determined from equations (68) through (70).

For $(\Phi^2 + \Phi_i^2)^{\frac{1}{2}} > (30/\pi)^{\frac{1}{2}}$ rad, the saddle-point method of calculating $G_m(f)$ is very convenient;[16] and this method can be applied in a straightforward manner to estimate $S_I(f)$. (Since one can show that the saddle-point approximation reduces to the quasistatic approximation for $f_d/W \ll (\Phi^2 + \Phi_i^2)^{\frac{1}{2}}$, the quasistatic approximation may be used for convenience if this condition is satisfied. However, the error introduced as a result of the use of quasistatic approximation cannot often be estimated.) For $R_i \ll 1$, we can also show that we need take only the $m = 1$ term in equation (68) to estimate S/I with a very small fractional error (less than 5.36 percent for $R_i < 0.1$).

For $f_d/W = 2, 4, 6, 8$, and 10 and for a set of values of Φ and Φ_i, we have calculated this minimum signal-to-interference ratio; Figs. 5 through 9 give these results. For any value of f_d/W and for any S/I, the required values of Φ and Φ_i may be obtained from these figures. Since the effects of linear filters on adjacent channel interference has not been taken into account in this paper, these values of Φ and Φ_i may serve only as a guide in the design of any angle-modulated system.

V. INTERFERENCE BETWEEN L+1 PM WAVES

We now assume that there are L interfering waves, and that all of them are phase modulated by mutually independent gaussian random processes.† Let the desired PM wave be given by

$$s(t) = \text{Re } A \exp \{j[\omega_o t + \varphi(t)]\}. \qquad (71)$$

Let the kth interfering wave be represented as

$$i_k(t) = \text{Re } R_{ik} A \exp \{j[\omega_k t + \varphi_{ik}(t) + \mu_k]\}, \qquad 1 \leqq k \leqq L. \qquad (72)$$

† The analysis given in this section can suitably be modified for angle modulation by general gaussian random processes (see Section II).

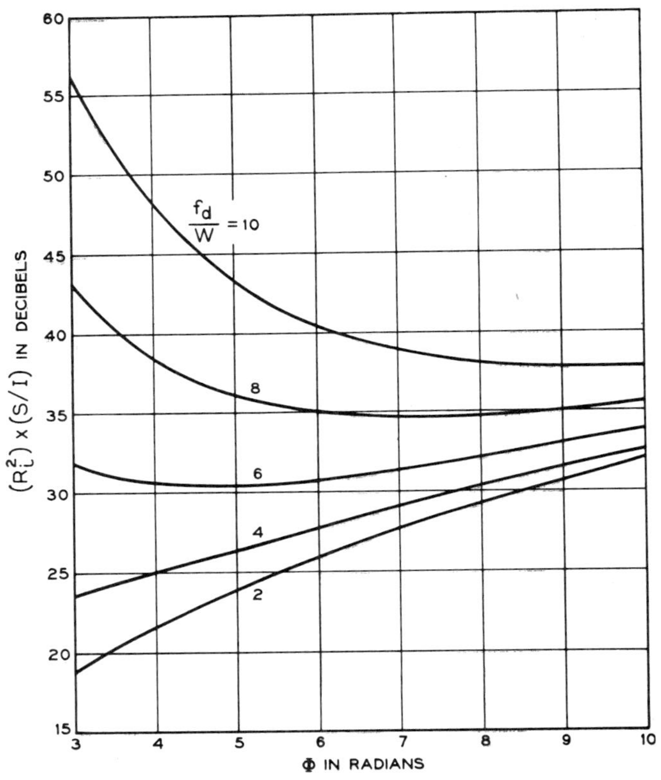

Fig. 5 — Signal-to-interference ratio as a function of rms phase deviations and channel separation for $\Phi_i = 2$ rad.

Since the L interfering waves are assumed to originate from L different sources, we assume that the μ_k's are independent of each other, and that μ_k, $1 \leqq k \leqq L$ has a uniform probability density function $\pi_{\mu_k}(\mu)$ where

$$\pi_{\mu_k}(\mu) = \begin{cases} 1/2\pi, & 0 \leqq \mu < 2\pi, \quad 1 \leqq k \leqq L, \\ 0, & \text{otherwise.} \end{cases} \qquad (73)$$

We further assume that $\varphi(t)$, the $\varphi_k(t)$'s, and the μ_k's (with $1 \leqq k \leqq L$) are mutually independent random variables.

If $s(t)$ and the $i_k(t)$'s are all in the passband of the PM receiver used in the system, the total signal incident at the receiver can be written as

$$r(t) = s(t) + \sum_{k=1}^{L} i_k(t)$$

$$= \text{Re } A\left(1 + \sum_{k=1}^{L} R_{ik} \exp \{j[\omega_{dk}t + \varphi_{ik}(t) - \varphi(t) + \mu_k]\}\right)$$

$$\cdot \exp \{j[\omega_0 t + \varphi(t)]\}, \tag{74}$$

where

$$\omega_{dk} = \omega_k - \omega_o = f_{dk}/2\pi. \tag{75}$$

From equation (74), we can show that the output $\theta(t)$ of an ideal phase demodulator can be represented as

$$\theta(t) = \varphi(t) + \text{Im ln}\left(1 + \sum_{k=1}^{L} R_{ik} \exp \{j[\omega_{dk}t + \varphi_{ik}(t) - \varphi(t) + \mu_k]\}\right). \tag{76}$$

Next we write

$$\ln\left(1 + \sum_{k=1}^{L} R_{ik} \exp \{j[\omega_{dk}t + \varphi_{ik}(t) - \varphi(t) + \mu_k]\}\right)$$

$$= \sum_{m=1}^{\infty} \frac{(-1)^{m+1}}{m}\left(\sum_{k=1}^{L} R_{ik} \exp \{j[\omega_{dk}t + \varphi_{ik}(t) - \varphi(t) + \mu_k]\}\right)^m$$

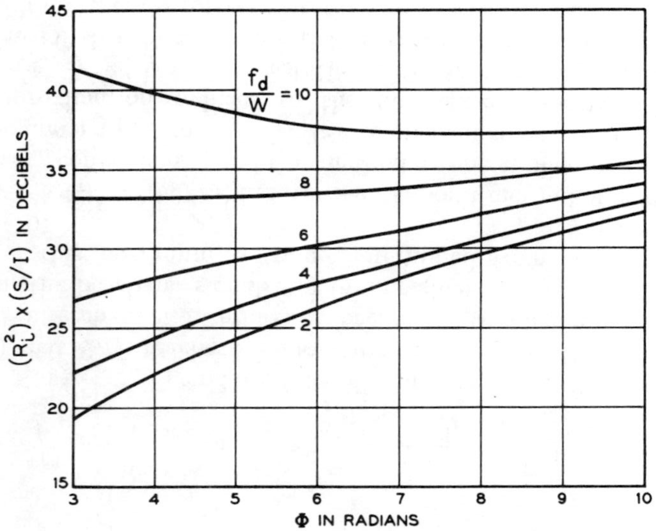

Fig. 6 — Signal-to-interference ratio as a function of rms phase deviations and channel separation for $\Phi_\iota = 4$ rad.

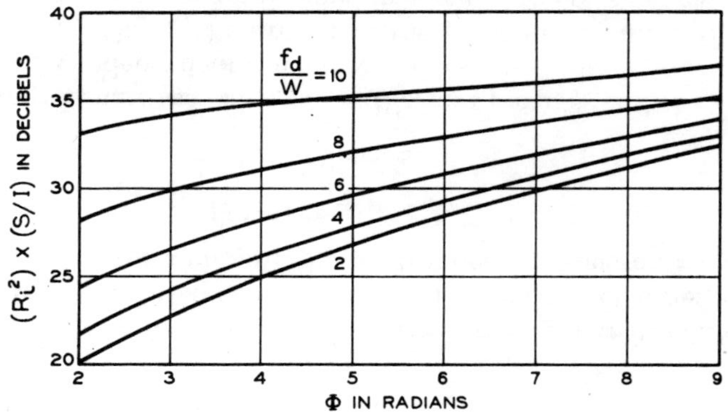

Fig. 7 — Signal-to-interference ratio as a function of rms phase deviations and channel separation for $\Phi_\iota = 6$ rad.

$$\sum_{k=1}^{L} R_{ik} < 1. \tag{77}$$

By the multinomial theorem, we have

$$\left(\sum_{k=1}^{L} R_{ik} \exp \{j[\omega_{dk}t + \varphi_{ik}(t) - \varphi(t)]\}\right)^m$$

$$= \sum \frac{m!}{\prod_{r=1}^{L} a_r!} \prod_{r=1}^{L} R_{ir}^{a_r} \exp \{ja_r[\omega_{dr}t + \varphi_{ir}(t) - \varphi(t) + \mu_r]\}, \tag{78}$$

where the a_r's are a set of nonnegative integers such that

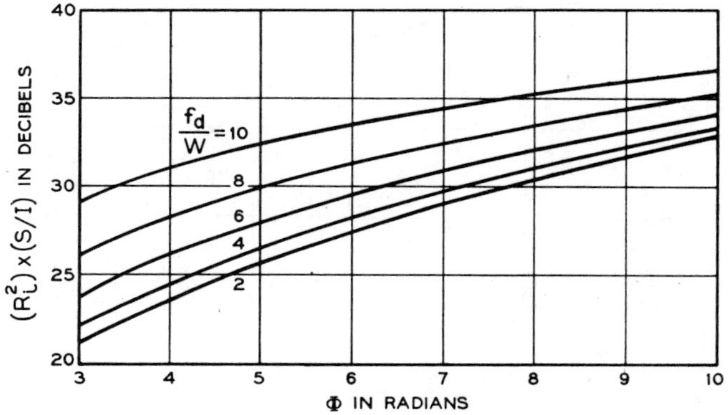

Fig. 8 — Signal-to-interference ratio as a function of rms phase deviations and channel separation for $\Phi_\iota = 8$ rad.

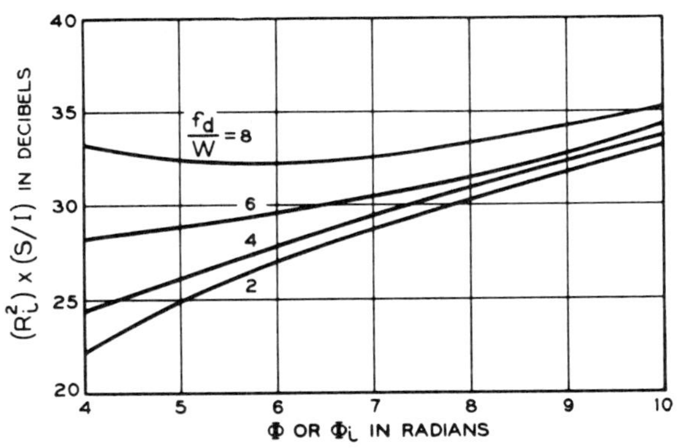

Fig. 9 — Signal-to-interference ratio as a function of rms phase deviations and channel separation for $\Phi = \Phi_i$.

$$\sum_{r=1}^{L} a_r = m. \tag{79}$$

From equations (73) and (76) through (79), one can show that the covariance function $R_\theta(\tau)$ of $\theta(t)$ can be written as

$$R_\theta(\tau) = \langle \theta(t)\theta(t+\tau) \rangle$$

$$= R_\varphi(\tau) + \sum_{m=1}^{\infty} \frac{1}{2m^2} \exp\{-m^2[R_\varphi(0) - R_\varphi(\tau)]\}$$

$$\cdot \left[\sum \left\{ \frac{m! \prod_{r=1}^{L} R_{ir}^{a_r}}{\prod_{r=1}^{L} a_r!} \right\}^2 \exp\left(-\sum_{r=1}^{L} a_r^2[R_{\varphi r}(0) - R_{\varphi r}(\tau)]\right) \right.$$

$$\left. \cdot \cos\left(\tau \sum_{r=1}^{L} a_r \omega_{dr}\right) \right]. \tag{80}$$

If the random gaussian noise is band-limited and white, and if all the modulating waveforms have the same bandwidth W, we have

$$R_\varphi(\tau) = \Phi^2 \cdot \frac{\sin 2\pi W\tau}{2\pi W\tau}, \tag{81}$$

and

$$R_{\varphi_k}(\tau) = \Phi_k^2 \frac{\sin 2\pi W\tau}{2\pi W\tau}, \qquad 1 \leqq k \leqq L. \tag{82}$$

In this case, equation (80) can be written as

$$R_\theta(\tau) = R_\varphi(\tau) + \sum_{m=1}^{\infty} \frac{1}{2m^2} \exp\left[-m^2\Phi^2\left(1 - \frac{\sin 2\pi W\tau}{2\pi W\tau}\right)\right]$$

$$\cdot \left\{ \sum \frac{(m!)^2 \prod_{r=1}^{L} R_{ir}^{2a_r}}{\prod_{r=1}^{L} (a_r!)^2} \exp\left[-\left(1 - \frac{\sin 2\pi W\tau}{2\pi W\tau}\right) \sum_{r=1}^{L} a_r^2 \Phi_r^2\right] \right.$$

$$\left. \cdot \cos\left(\tau \sum_{r=1}^{L} a_r \omega_{dr}\right) \right\}. \tag{83}$$

Therefore, the spectral density of baseband interchannel interference is given by

$$S_I(f) = \sum_{m=1}^{\infty} \frac{1}{4m^2} \left\{ \sum_{s} \frac{(m!)^2 \prod_{r=1}^{L} R_{ir}^{2a_r}}{\prod_{r=1}^{L} (a_r!)^2} \right.$$

$$\left. \cdot \left[T_{ms}\left(f - \sum_{r=1}^{L} a_r f_{dr}\right) + T_{ms}\left(f + \sum_{r=1}^{L} a_r f_{dr}\right) \right] \right\}, \tag{84}$$

where

$$T_{ms}(f) = \int_{-\infty}^{\infty} \exp\left[-\left(1 - \frac{\sin 2\pi W\tau}{2\pi W\tau}\right)\left(m^2\Phi^2 + \sum_{r=1}^{L} a_r^2 \Phi_r^2\right)\right] e^{-j2\pi f\tau} d\tau. \tag{85}$$

Next notice that the methods given in Section III can be used to calculate $T(f)$ for all values of Φ, and Φ_{ik}'s (with $1 \leqq k \leqq L$); hence, we can calculate $S_I(f)$ for all values of R_{ik}'s such that $\sum_{k=1}^{L} R_{ik} < 1$. The minimum signal-to-interference ratio S/I can then be obtained from equation (49).

Now assume that we have L co-channel interferers and that all have the same rms phase deviation Φ, or

$$\Phi_r = \Phi, \qquad 1 \leqq r \leqq L. \tag{86}$$

In this case equation (84) yields

$$S_I(f) = \sum_{m=1}^{\infty} \frac{1}{2m^2} \left\{ \sum_{s} \frac{(m!)^2 \prod_{r=1}^{L} R_{ir}^{2a_r}}{\prod_{r=1}^{L} (a_r!)^2} G_{ms}(f) \right\}, \tag{87}$$

where

$$G_{ms}(f) = \int_{-\infty}^{\infty} \exp\left[-\left(1 - \frac{\sin 2\pi W\tau}{2\pi W\tau}\right)\Phi^2\left(m^2 + \sum_{r=1}^{L} a_r^2\right)\right]e^{-j2\pi f\tau}\, d\tau. \tag{88}$$

From equations (87) and (88) and Refs. 2 and 16, one can show that the continuous part of $S_I(f)$ reaches its maximum at $f = 0$, and that†

$$0.92\left(\frac{3}{2\pi}\right)^{\frac{1}{2}} \frac{1}{\left(m^2 + \sum_{r=1}^{L} a_r^2\right)^{\frac{1}{2}}} \frac{1}{W\Phi} < G_{ms}(0)$$

$$< 1.08\left(\frac{3}{2\pi}\right)^{\frac{1}{2}} \frac{1}{\left(m^2 + \sum_{r=1}^{L} a_r^2\right)^{\frac{1}{2}}} \frac{1}{W\Phi}, \qquad \Phi > (5)^{\frac{1}{2}}\text{ rad}. \tag{89}$$

The expression $G_{ms}(0)$ in equation (89) does not include the delta function contained in $G_{ms}(f)$ at $f = 0$.

Since $\sum_{r=1}^{L} a_r = m$, one can prove that

$$\frac{m^2}{L} \leqq \sum_{r=1}^{L} a_r^2 \leqq m^2. \tag{90}$$

From equations (89) and (90) we have

$$0.46\left(\frac{3}{\pi}\right)^{\frac{1}{2}} \frac{1}{m\Phi W} < G_{ms}(0) < 0.54\left(\frac{3}{\pi}\right)^{\frac{1}{2}}\left(\frac{2L}{L+1}\right)^{\frac{1}{2}} \frac{1}{m\Phi W}. \tag{91}$$

Next

$$S_I(0) < \sum_{m=1}^{\infty} \frac{1}{2m^2}\left[\sum \frac{(m!)^2 \prod_{r=1}^{L} R_{ir}^{2a_r}}{\prod_{r=1}^{L} (a_r!)^2}\, 0.54\left(\frac{3}{\pi}\right)^{\frac{1}{2}}\left(\frac{2L}{L+1}\right)^{\frac{1}{2}} \frac{1}{m\Phi W}\right]. \tag{92}$$

If all x_i's are nonnegative, one can show that

$$\sum x_i^2 \leqq \left(\sum x_i\right)^2. \tag{93}$$

Using equation (93), equation (92) yields

$$S_I(0) < \sum_{m=1}^{\infty} 0.27\left(\frac{3}{\pi}\right)^{\frac{1}{2}}\left(\frac{2L}{L+1}\right)^{\frac{1}{2}} \frac{1}{\Phi W} \frac{1}{m^3}\left[\left[\sum \frac{m! \prod_{r=1}^{L} R_{ir}^{a_r}}{\prod_{r=1}^{L} a_r!}\right]^2\right]$$

† We consider only the continuous part of $S_I(f)$.

$$= 0.27\left(\frac{3}{\pi}\right)^{\frac{1}{2}}\left(\frac{2L}{L+1}\right)^{\frac{1}{2}} \frac{1}{\Phi W} \sum_{m=1}^{\infty} \frac{\left(\sum_{r=1}^{L} R_{ir}\right)^{2m}}{m^3} \tag{94}$$

or

$$S_I(0) < 0.27\left(\frac{3}{\pi}\right)^{\frac{1}{2}}\left(\frac{2L}{L+1}\right)^{\frac{1}{2}} \frac{1}{\Phi W} Q(b^2), \tag{95}$$

where

$$b^2 = \left(\sum_{r=1}^{L} R_{ir}\right)^2 < 1. \tag{96}$$

We have shown in Section IV that

$$\sum_{m=1}^{\infty} \frac{b^{2m}}{m^3} < b^2 U(b^2), \qquad b^2 < 1. \tag{97}$$

Therefore, the minimum baseband signal-to-interference ratio is bounded by

$$S/I > \frac{1}{0.54}\left(\frac{\pi}{3}\right)^{\frac{1}{2}}\left(\frac{L+1}{2L}\right)^{\frac{1}{2}} \frac{\Phi^3}{b^2 U(b^2)}, \qquad \Phi > (5)^{\frac{1}{2}}\text{ rad}, \quad b < 1. \tag{98}$$

From equation (87) we can also show that

$$S_I(0) > \frac{1}{2}\left(\sum_{r=1}^{L} R_{ir}^2\right) G_{1s}(0). \tag{99}$$

Equations (89) and (99) yield

$$S_I(0) > 0.23\left(\frac{3}{\pi}\right)^{\frac{1}{2}} \frac{1}{W\Phi}\left(\sum_{r=1}^{L} R_{ir}^2\right), \tag{100}$$

or

$$S/I < \frac{1}{0.46}\left(\frac{\pi}{3}\right)^{\frac{1}{2}} \frac{\Phi^3}{\sum_{r=1}^{L} R_{ir}^2}. \tag{101}$$

Hence we have

$$\frac{1}{0.46}\left(\frac{\pi}{3}\right)^{\frac{1}{2}} \frac{\Phi^3}{\sum_{r=1}^{L} R_{ir}^2} > S/I > \frac{1}{0.54}\left(\frac{\pi}{3}\right)^{\frac{1}{2}}\left(\frac{L+1}{2L}\right)^{\frac{1}{2}} \frac{\Phi^3}{b^2 U(b^2)},$$

$$\Phi > (5)^{\frac{1}{2}}\text{ rad}, \quad b = \sum_{k=1}^{L} R_{ik} < 1. \tag{102}$$

For any set of values of R_{ik}'s, $1 \leqq k \leqq L$, and for any Φ, bounds to the signal-to-interference ratio S/I can be calculated from equation (102), and a proper Φ can then be chosen to keep the baseband interference below any desired level.

Notice that the upper bound is a function of the total interference power, and the lower bound a function of the sum of the amplitudes of all the interfering carriers. In such cases, the distribution of R_{ik}'s generally determines the closeness of the two bounds. However, it may be observed that both these bounds are proportional to the cube of the modulation index Φ (for a high index system).

VI. RESULTS AND CONCLUSIONS

In this paper we consider the effect of interchannel interference on angle-modulated systems. We also derive an expression for the baseband interchannel interference when two (or more) waves angle modulated by gaussian noise interfere with each other. This formula can be used even when the baseband signal is passed through a linear network such as a pre-emphasis—de-emphasis network. We show that the calculation of the RF spectral density is essential to the evaluation of the baseband interchannel interference.

We then consider band-limited white gaussian noise modulation and show that, in the case of co-channel interference, maximum baseband interference occurs at the lowest baseband frequency present in the system. For moderately high modulation index, we show that we can derive upper and lower bounds to this minimum signal-to-interference ratio and that these bounds are proportional to the cube of the modulation index. It therefore follows that co-channel interference in PM systems can be reduced by expanding bandwidth, and that phase modulation can be used with advantage in combating interference. We also show that the first term in the power series expansion for the baseband interchannel interference gives most of the contribution if the carrier-to-interference ratio is greater than about 10 dB (the error is less than 5.36 per cent for a carrier-to-interference ratio greater than 10 dB).

In this paper we also give some results about the effects of adjacent channel interference on angle-modulated systems. We assume that all the incident signals at the receiver are in the passband of the PM receiver used in the system. This assumption is justified in the case of co-channel interference, but is not realistic in the case of adjacent channel interference. However, we feel that the results given in this paper for the adjacent channel interference may serve as a guide in

determining the deterioration in performance produced by adjacent channel interference.

VII. ACKNOWLEDGMENT

Some of the results presented here were obtained earlier by Clyde L. Ruthroff. We are grateful to him for his consent to publish those results in this paper.

REFERENCES

1. Black, H. S., *Modulation Theory*, Princeton, New Jersey: D. Van Nostrand, 1953.
2. Rowe, H. E., *Signals and Noise in Communication Systems*, Princeton, New Jersey: D. Van Nostrand, 1965, pp. 98–203.
3. Tillotson, L. C., "Use of Frequencies above 10 GHz for Common Carrier Applications," B.S.T.J., *48*, No. 7 (July–August 1969), pp. 1563–1576.
4. Ruthroff, C. L., and Tillotson, L. C., "Interference in Dense Radio Networks," B.S.T.J., *48*, No. 7 (July–August), pp. 1727–1743.
5. Medhurst, R. G., Hicks, E. M., and Grossett, W., "Distortion in Frequency Division Multiplex FM Systems Due to an Interfering Carrier," Proc. IEE, *105B*, No. 5 (May 1958), pp. 282–292.
6. Medhurst, R. G., "FM Interfering Carrier Distortion: General Formula," Proc. IEE, *109B*, No. 3 (March 1962), pp. 149–150 and 519–523.
7. Medhurst, R. G., and Roberts, J. H., "Expected Interference Levels Due to Interactions between Line-of-Sight Radio-Relay Systems and Broadband Satellite Systems," Proc. IEE, *111*, No. 3 (March 1964), pp. 519–523.
8. Hamer, R., "Radio-Frequency Interference in Multi-Channel Telephone FM Radio Systems," Proc. IEE, *108B*, No. 1 (January 1961), pp. 75–89.
9. Curtis, H. E., and Rice, S. O., unpublished work.
10. Borodich, S. Y., "Calculating the Permissible Magnitude of Radio Interference in Multi-Channel Radio Relay Systems," Electrosviaz, *1*, No. 1 (January 1962), pp. 13–24.
11. Hayashi, S., "On the Interference Characteristics of the Phase Modulation Receiver for the Multiplex Transmission," IEE (Japan), *35*, No. 11 (November 1952), pp. 522–528.
12. Curtis, H. E., "Interference between Satellite Communication Systems and Common Carrier Surface Systems," B.S.T.J., *41*, No. 3 (May 1962), pp. 921–943.
13. Bennett, W. R., Curtis, H. E., and Rice, S. O., "Interchannel Interference in FM and PM Systems under Noise Loading Conditions," B.S.T.J., *34*, No. 3 (May 1955), pp. 601–636.
14. Ruthroff, C. L., unpublished work.
15. Middleton, D., *Introduction to Statistical Communication Theory*, New York: McGraw-Hill, 1960, pp. 599–678.
16. Prabhu, V. K., and Rowe, H. E., "Spectral Density Bounds of a PM Wave," B.S.T.J., *48*, No. 3 (March 1969), pp. 789–831.
17. Abramowitz, M., and Stegun, I. A., *Handbook of Mathematical Functions*, Nat. Bureau of Standards, Washington, D. C., 1967, pp. 227–251.

Interference into Angle-Modulated Systems Carrying Multichannel Telephony Signals

BENJAMIN A. PONTANO, JORGE C. FUENZALIDA, AND NAND KISHORE M. CHITRE

Abstract—An investigation of interference between angle-modulated systems carrying multichannel telephony, which is directly applicable to existing satellite and terrestrial radio-relay systems, has been undertaken. This study includes a general solution for arbitrary narrow-band interference into an angle-modulated system with arbitrary modulation.

The algorithm has been implemented on a digital computer to provide calculations of interference between two FDM/FM systems. The analysis is valid for all modulation indices, and may be applied to any PM, FM, or preemphasized FM baseband. A comparison of calculated results and experimental measurements shows good agreement.

An extensive investigation, covering a wide range of modulation indices and basebands, has been conducted for the particular case of preemphasized FDM/FM signals. Wherever possible, a generalized presentation of interference results is given. Interleaving criteria are also presented for certain cases of interference involving two identical signals.

Paper approved by the Space Communications Committee of the IEEE Communications Society for publication without oral presentation. Manuscript received August 25, 1971; revised December 14, 1972. This work was sponsored by the International Telecommunications Satellite Consortium (INTELSAT).

The authors are with COMSAT Laboratories, Clarksburg, Md. 20734.

LIST OF SYMBOLS

A_1	Desired carrier amplitude.
A_2	Interfering carrier amplitude.
C/I	Ratio of desired to interfering unmodulated carrier power $(A_1/A_2)^2$.
f_{ch}	Midfrequency of baseband channel under consideration.
f_{m_1}	Maximum baseband frequency of desired signal.
f_{m_2}	Maximum baseband frequency of interfering signal.
f_s	Mean carrier separation between the desired and interfering carriers.
$H_p(f)$	Frequency response of preemphasis network.
m_1	rms modulation index of desired signal.
m_2	rms modulation index of interfering signal.
m_2'	$m_2 f_{m_2}/f_{m_1}$.
M	$(m_1^2 + m_2'^2)^{1/2}$.
P	Multichannel signal peak-to-rms voltage ratio.

Reprinted from *IEEE Trans. Comm.*, vol. COM-21, pp. 714-726, June 1973.

R	Ratio of interfering to desired unmodulated carrier voltage.
$R_{x_1}(\tau)$	Autocorrelation function of desired modulating baseband signal.
$R_{x_2}(\tau)$	Autocorrelation function of interfering modulating baseband signal.
$S_{\dot{x}}(f)$	Power spectral density after preemphasis of frequency-modulating baseband signal.
$S_{v_1}(f)$	Low-pass equivalent power spectrum of the desired RF signal (normalized to unity power).
$S_{V_2}(f)$	Low-pass equivalent power spectrum of interfering RF signal.
$S_{v_1^m}(f)$	Low-pass equivalent power spectrum of the mth power of the desired signal (normalized to unity power).
$S_{V_2^m}(f)$	Low-pass equivalent power spectrum of the mth power of the interfering RF signal.
$x_1(t)$	Phase-modulating signal of the desired carrier.
$x_2(t)$	Phase-modulating signal of the interfering carrier.
α	Normalized frequency (f/f_m).
Δf	Telephone channel bandwidth.
ϵ	Ratio of minimum to maximum baseband frequencies.
$\sigma_{x_1}(f)$	Normalized baseband spectrum of desired signal.
$\sigma_{x_2}(f)$	Normalized baseband spectrum of interfering signal.
μ	Arbitrary initial phase between desired and interfering carriers.

Introduction

PROBLEMS of mutual interference exist between communication satellites and/or terrestrial radio relays sharing the same frequency bands. A better utilization of the RF spectrum is possible if the mutual interference can be predicted accurately. In this paper we provide tools that are useful for calculating certain cases of interference arising between: 1) communication satellites; 2) radio relays; and 3) communication satellites and radio relays. Specifically, we address the problem of interference into FDM/FM systems carrying multichannel telephony.

The problem of mutual interference between angle-modulated systems has been considered by others [1]-[5], but each study has certain limitations, either in scope or mathematical rigor.

The most general treatment of the problem to date is given by Prabhu and Enloe [5]. In their analysis, the low-interference restriction (i.e., small interfering to desired carrier amplitude) common to the previous works is relaxed.

In this paper, the method of Prabhu and Enloe [5], which was originally developed for angle-modulated systems, is generalized to provide a means of considering an arbitrary interference entry, i.e., an interference entry that is not necessarily caused by an angle-modulated carrier. The assumption of Gaussian modulation for the desired signal has also been eliminated. In special cases, the results presented here reduce those given by [1]-[5].

Essential for computing the baseband interference is a knowledge of the RF power spectral densities of both the desired and interfering signals. Previous studies dealing with the generation of adequate spectrum models have resulted in an assortment of mathematical approximations, each applicable to a limited range of parameters. In this study we use a spectrum model recently developed by Fuenzalida *et al.* [6]. This model can be used for generating both PM and FM RF spectra for all realistic basebands and modulation indices, with the option of including preemphasis.

The computation of interference has been implemented on a digital computer for preemphasized FDM/FM and FDM/PM. Laboratory measurements show good agreement with the computation, even at large carrier separations.

An extensive investigation of interference, covering a wide range of operating parameters, was conducted. Wherever possible, the results are presented in a generalized format to eliminate redundancy and include a wide variety of cases. Interleaving criteria are presented for certain cases of interference involving two identical signals separated by a frequency somewhat larger than Carson's rule half-bandwidth. Results to evaluate the advantages of interleaving over cochannel operation are also given.

Arbitrary Interference into an Angle-Modulated System with Arbitrary Modulation

In this section we consider the problem of an arbitrary narrow-band signal interfering with an angle-modulated carrier with arbitrary modulation. The desired and interfering signals are generally characterized by

$$e_1(t) = \mathrm{Re}\,[Z_1(t)] = \mathrm{Re}\,\{A_1 \exp j\,[\omega_1 t + x_1(t) + \mu]\} \quad (1)$$
$$= \mathrm{Re}\,\{A_1\, v_1(t) \exp j\omega_1 t\}$$

$$e_2(t) = \mathrm{Re}\,[Z_2(t)] = \mathrm{Re}\,\{V_2(t) \exp\,[j\omega_2 t]\}. \quad (2)$$

At the receiver an ideal demodulator that exhibits perfect limiting is assumed. In addition, each signal is assumed to reach the demodulator undistorted by the transmission path or by the receiver filter. It should be emphasized that no assumptions regarding the statistics of $x_1(t)$ have been made and that the most general narrow-band characterization of $e_2(t)$ has been used. It is assumed, however, that $|Z_2(t)/V_1| < 1$.[1] Finally, statistical independence and wide-sense stationarity of $x_1(t)$ and μ, $e_1(t)$, and $e_2(t)$ have been assumed. The details of the analysis are given in the Appendix.

It has been shown that the power spectrum of the baseband interference is (see the Appendix)

$$I(f) = \sum_{m=1}^{\infty} \frac{1}{4m^2 A_1^{2m}} [T_m(f - mf_s) + T_m(-f - mf_s)] \quad (3)$$

where

$$T_m(f) = S_{V_2^m}(f) \otimes S_{v_1^m}(f) \quad (4)$$

$$S_{V_2^m}(f) = \mathfrak{F}\,[R_{V_2^m}(\tau)] \quad (5)$$

[1] $|Z_2(t)/V_1|$ need not be << 1.

$$S_{v_1^m}(f) = \mathcal{F}[R_{v_1^m}(\tau)] \tag{6}$$

and A_1 is the wanted carrier amplitude.

The solution to the problem of interference into an angle-modulated system, in its most general form therefore comprises two convolutional terms. Each term can be generated by convolving the power spectra of the mth power of the complex envelopes. For certain special cases that we will consider, the convolutional terms may be calculated from only the desired and interfering power spectra.

The Low Interference Condition

The resulting baseband interference power spectrum for the special case of a low interference entry, i.e., $|Z_2(t)/V_1| << 1$, may be approximated by the first term in (3):

$$I(f) = \frac{1}{4A_2^2}[S_{V_2} \otimes S_{v_1}(f - f_s) + S_{V_2} \otimes S_{v_1}(-f - f_s)]. \tag{7}$$

The baseband interference can be generated by simply convolving the desired and interfering power spectra. This result has been reported in [3]-[5] for the special case of an interfering angle-modulated carrier. Because of the generality of our analysis, however, we are now justified in using (6) to calculate the baseband interference resulting from all forms of narrow-band interference including: 1) amplitude-modulated signals; 2) video modulations (amplitude and phase); 3) band-limited thermal noise; 4) certain types of intermodulation; 5) filtered signals; and 6) multiple interference entries.

Angle-Modulated Interference

We now address the problem of an interfering signal having the form

$$e_2(t) = \text{Re}[Z_2(t)] = \text{Re}[A_2 \exp\{j\omega_2 t + x_2(t)]\}]$$
$$= \text{Re}\{A_2 v_2(t) \exp[j\omega_2 t]\} \tag{8}$$

and being angle modulated by $x_2(t)$. Both transmissions are assumed to carry FDM telephony. The baseband amplitude statistics of each carrier may then be expressed as a Gaussian distribution with zero mean [7]. The complex envelope of each signal has the form $A \exp jx(t)$; thus the mth power is given by $A^m \exp jmx(t)$. The autocorrelation function of the complex envelope raised to the mth power is thereby

$$R_{V^m}(\tau) = A^{2m} \exp\{-m^2[R_x(0) - R_x(\tau)]\} \tag{9}$$

and the power spectrum of $V^m(t)$ is then

$$S_{V^m}(f) = \mathcal{F}[R_{V^m}(\tau)]$$
$$= A^{2m}\{\exp[-(R_x(0) - R_x(\tau))]\}$$
$$\overset{m^2}{\otimes} \mathcal{F}\{\exp[-(R_x(0) - R_x(\tau))]\}$$
$$= m\text{th-order convolution of the power spectrum.} \tag{10}$$

Substituting from (10) into (3) results in

$$I(f) = \sum_{m=1}^{\infty} \frac{1}{4m^2}\left(\frac{A_2}{A_1}\right)^{2m} \left\{\left[\left(S_{v_1} \overset{m^2}{\otimes} S_{v_1}\right) \otimes \left(S_{v_2} \overset{m^2}{\otimes} S_{v_2}\right)\right](f - mf_s) + \left[\left(S_{v_1} \overset{m^2}{\otimes} S_{v_1}\right) \otimes \left(S_{v_2} \overset{m^2}{\otimes} S_{v_2}\right)\right]\left(-f - mf_s\right)\right\}, \tag{11}$$

which can be shown to agree with the results of Prabhu and Enloe [5]. It can now be seen from (10) that all terms in the expansion are calculated directly from the power spectra of the two carriers.

The desired and interfering normalized power spectra can be precisely determined for small and intermediate modulation indices by using the following expansions[2] (see the Appendix):

$$S_{v_1}(f) = \sum_{n=0}^{\infty} \exp[-R_{x_1}(0)] \frac{[R_{x_1}(0)]^n}{n!} \cdot \left[\sigma_{x_2}(f) \overset{n}{\otimes} \sigma_{x_1}(f)\right]. \tag{12}$$

$$S_{v_2}(f) = \sum_{n=0}^{\infty} \exp\left[-R_{x_2}(0)\right] \frac{[R_{x_2}(0)]^n}{n!} \cdot \left[\sigma_{x_2}(f) \overset{n}{\otimes} \sigma_{x_2}(f)\right]. \tag{13}$$

The factor $\exp[-R_x(0)]$ represents the residual carrier power (i.e., power not contained in the sidebands). For calculations of interference into preemphasized FM systems, (11) is modified by the factor $[f/Hp(f)]^2$.

The noise power ratio (NPR), or the ratio of the desired signal to the demodulated interference in a telephony channel centered at f_{ch}, may be written as

$$\text{NPR} = \frac{m_1^2 f_{m_1} \Delta f}{(1 - \epsilon)} \left[\int_{f_{ch}-(\Delta f/2)}^{f_{ch}+(\Delta f/2)} I(f)\, df\right]^{-1}. \tag{14}$$

When a restricted range of operating parameters, i.e., carrier separations, modulation indices, and basebands is considered, the general equations presented here reduce to provide analytical interference approximations in closed form. At low carrier separations, a Gaussian representation of the RF spectra provides accurate results for interference between either two high modulation index systems or between a high modulation index system and an unmodulated carrier. The Gaussian representation of the normalized RF power spectrum is given by [8].

$$S_v(f) = \mathcal{F}[R_v(\tau)] = \frac{1}{\sqrt{2\pi}\, mf_m} \exp\left(-\frac{f^2}{2(mf_m)^2}\right). \tag{15}$$

Substituting from (15) into (10), (11), and (14) for both the

[2] For large modulation indices, a different expansion about the Gaussian is used.

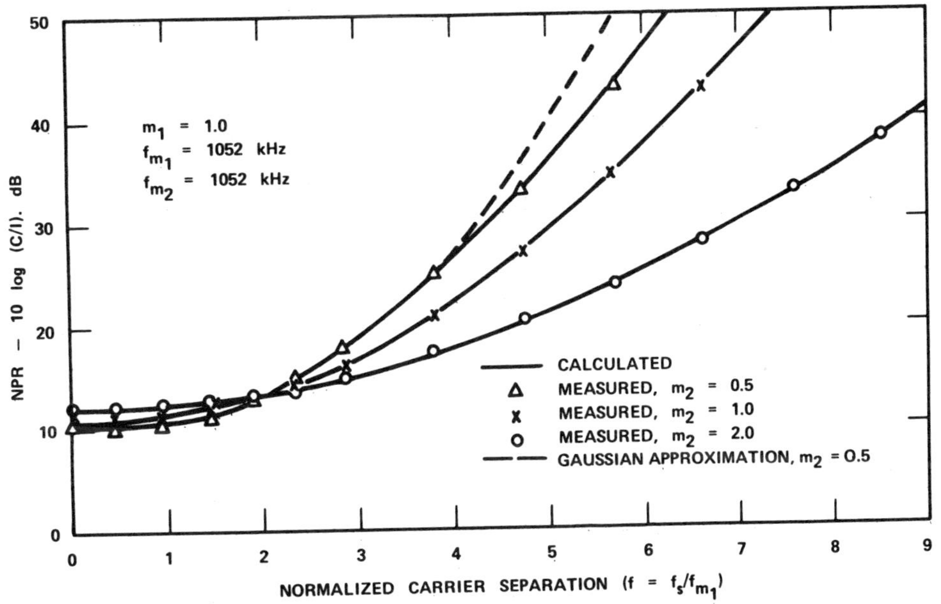

Fig. 1. Interference calculations and measurements for carriers having identical baseband sizes (m_1 = 1.0).

desired and interfering spectra yields

$$\mathrm{NPR} = \frac{2\sqrt{2\pi}\, f_{m_1}^2 \sqrt{m_1^2 + m_2'^2}}{(A_2/A_1)^2 (1-\epsilon) f_{ch}^2} \left\{ \exp\left[\frac{(f_{ch}+f_s)^2}{2f_{m_1}^2 (m_1^2 + m_2'^2)}\right] + \exp\left[\frac{(f_{ch}-f_s)^2}{2f_{m_1}^2 (m_1^2 + m_2')^2}\right]\right\}^{-1}. \quad (16)$$

which can be shown to agree with the results of Hamer [3].

Comparison of Theory with Laboratory Measurement

General

The algorithm for calculating interference between FDM/FM or FDM/PM carriers (with the option of CCIR preemphasis) has been implemented on a digital computer. An extensive investigation covering a wide range of modulation indices and basebands has been conducted for the particular case of preemphasized FDM/FM.

Extensive interference measurements for 600 and 960 channels can be found in [8]. Laboratory measurements of the NPR caused by an interfering FDM/FM carrier were made for a wide variety of cases in order to test our model. Our measurements were made through a system consisting of high-quality broadband modulators and demodulators. An ultra-stable frequency translator permitted frequency shifts of up to 10 MHz. The measurements were made at a carrier-to-interference ratio of 15 dB, thus satisfying the low interference condition. Results of these laboratory measurements are presented next and compared to calculations.

Preemphasized FDM/FM Measurements

Measurements of the NPR in the top baseband channel were made for a range of moderate to large modulation indices having basebands extending from 12 to 1052 kHz. The interference data for two carriers having identical baseband sizes are shown in Fig. 1 for m_1 = 1.0 and in Fig. 2 for m_1 = 0.5. The measured values are shown to be well within 0.5 dB of the calculated curves. A plot of the commonly employed Gaussian approximation formula [3] is also given to illustrate the inaccuracies caused by using this assumption. Fig. 3 shows interference caused by an unmodulated undesired carrier and modulation indices of moderate size for the desired carrier.

Low Modulation Indices

NPR measurements were made for two identical baseband signals, each having an rms modulation index of 0.1 and a baseband extending from 12 to 1052 kHz. From the previous section, it can be observed that a spike term, caused by the "beating" of the two residual carriers, appears in the baseband at a frequency equal to the mean carrier separation. This baseband component is, of course, most pronounced at low modulation indices when the residual carriers are quite significant. The worst channel in this case will occur at the mean carrier separation. As an example, the worst channel is compared with the top channel results in Fig. 4 for $m_1 = m_2 = 0.1$. As can be seen, for separations greater than fm_1, the top channel becomes the worst channel since the spike term falls outside of the desired baseband.

In the previous analysis, interference calculations for low modulation indices required actual experimental measurement of the desired and interfering spectra. In this work, it should be reemphasized that the spectra were calculated from the baseband parameters.

Measurements of interference, which are given by Hamer [3], are presented for the low modulation case. Our calculations using the same baseband parameters are shown to agree quite well with Hamer's measurements (Fig. 5).

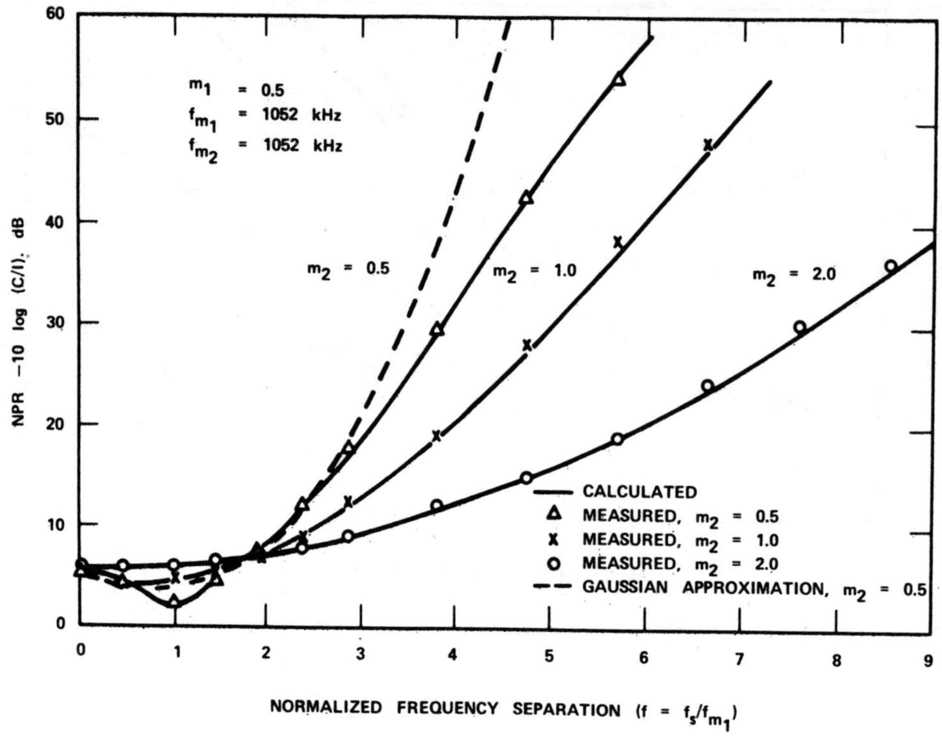

Fig. 2. Interference calculations and measurements for carriers having identical baseband sizes ($m_1 = 0.5$).

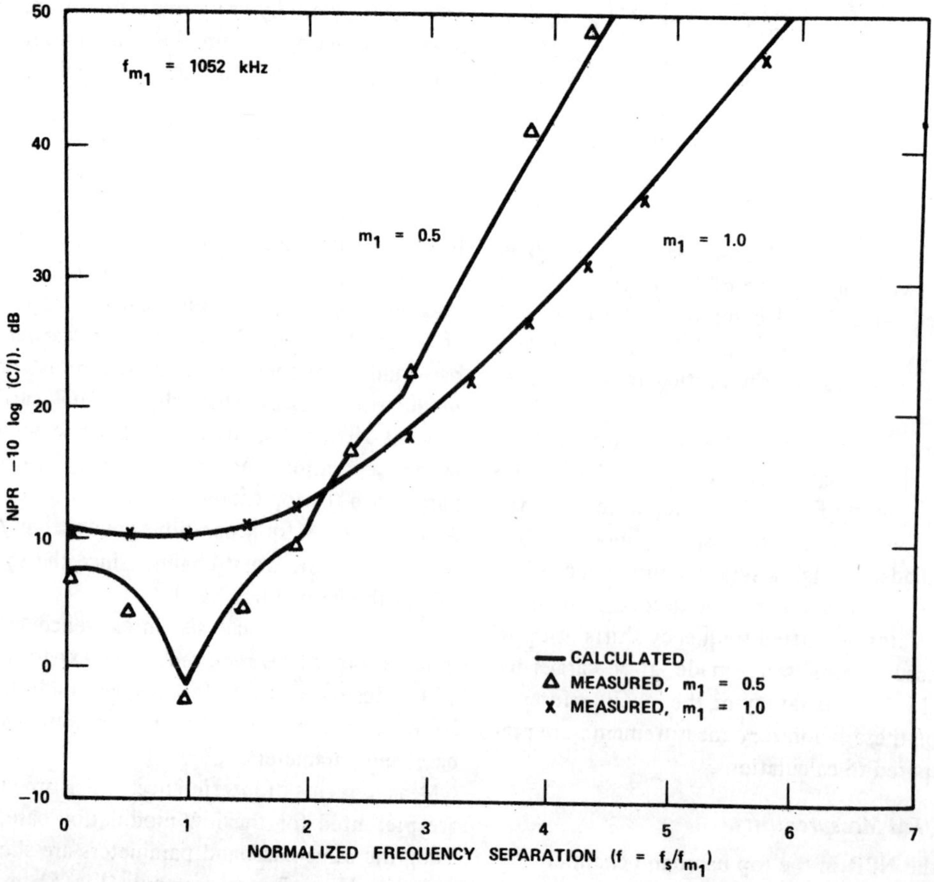

Fig. 3. Interference calculations and measurements for an unmodulated carrier interference.

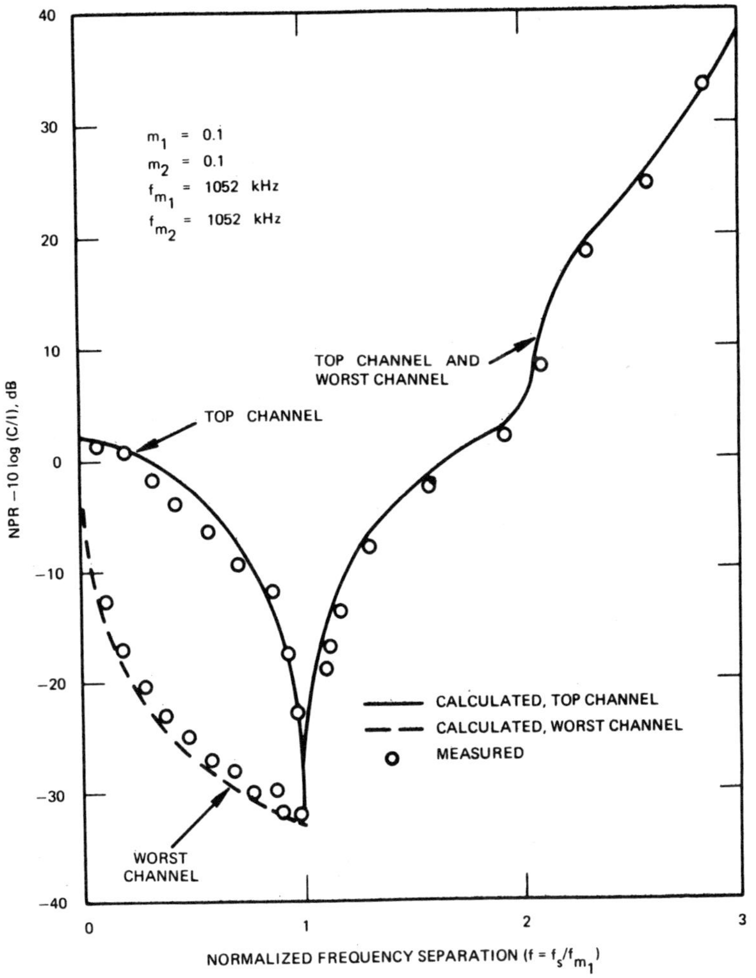

Fig. 4. Interference calculations and measurements for a low modulation index.

Large Interfering Basebands

Measurements were conducted to verify the results of an interference case in which the interfering signal has a much larger baseband than the desired signal. Initial calculations revealed a surprisingly large amount of interference, even at separations of 10 times the highest baseband frequency of the desired signal. Such results could not be predicted by using the Gaussian approximation, even for moderately high m_1. The results were verified experimentally for a case in which the interfering baseband was nearly 10 times the size of the desired carrier (see Fig. 6).

Generalized Interference Curves for Preemphasized FDM/FM

The baseband interference has been calculated over a wide range of baseband parameters for 10 evenly spaced baseband channels. An attempt was made to generalize the results and to present the results in a compact format. An example of a compact presentation (generalized format) is given in Fig. 7 for the top channel interference results. The resulting dependent variable = NPR − 20 log $(m_1 A_1/A_2)$ is plotted in terms of the "composite" modulation index M:

$$M = \left[m_1^2 + \left(m_2 \frac{f_{m_2}}{f_{m_1}} \right)^2 \right]^{1/2} \tag{17}$$

and is a function of the normalized carrier separation (f_s/f_{m_1}).

It is apparent from Fig. 7 that, even for a desired signal with a relatively large modulation index, it would not be possible to simply characterize both signals by a composite modulation index M without a further restriction on the relative baseband size of both signals.

Depending upon the relative size of the two basebands, three situations may arise: 1) both basebands may be identical; 2) the desired baseband may be larger; or 3) the interfering baseband may be larger. It has been observed (see Fig. 7) that cases *a* and *b* consistently give quite similar interference results; however, if the interfering baseband is larger and, furthermore, if the modulation index of the interfering signal is small, large variations in the resulting interference are obtained, depending upon the particular ratio of the baseband sizes considered. Thus for this case, M is insufficient to properly describe the interference.

Nevertheless, since it is desirable to make generalized interference results available, an attempt has been made to develop

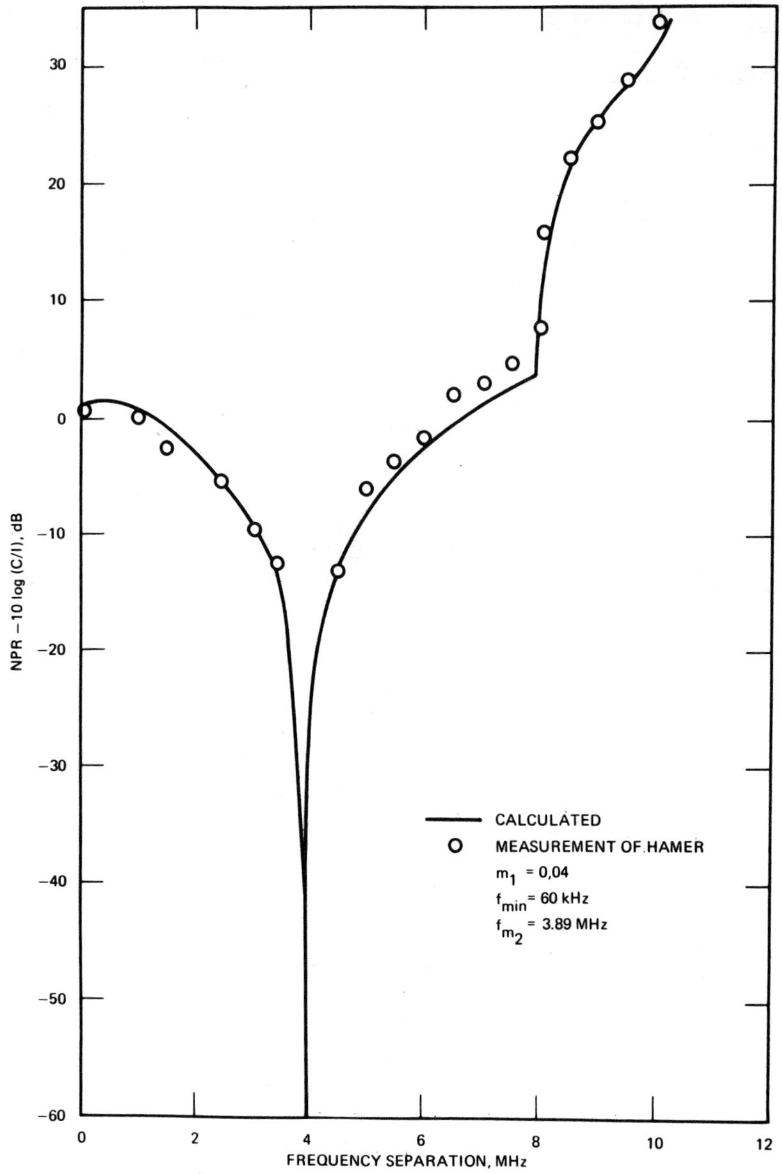

Fig. 5. Comparison of calculation with measurements by Hamer for identical baseband sizes.

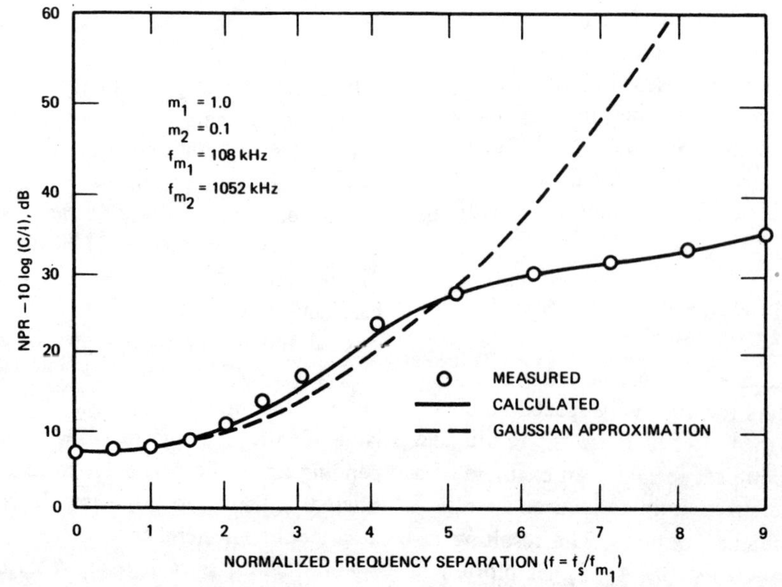

Fig. 6. Interference calculation and measurement for the case of a large interfering baseband.

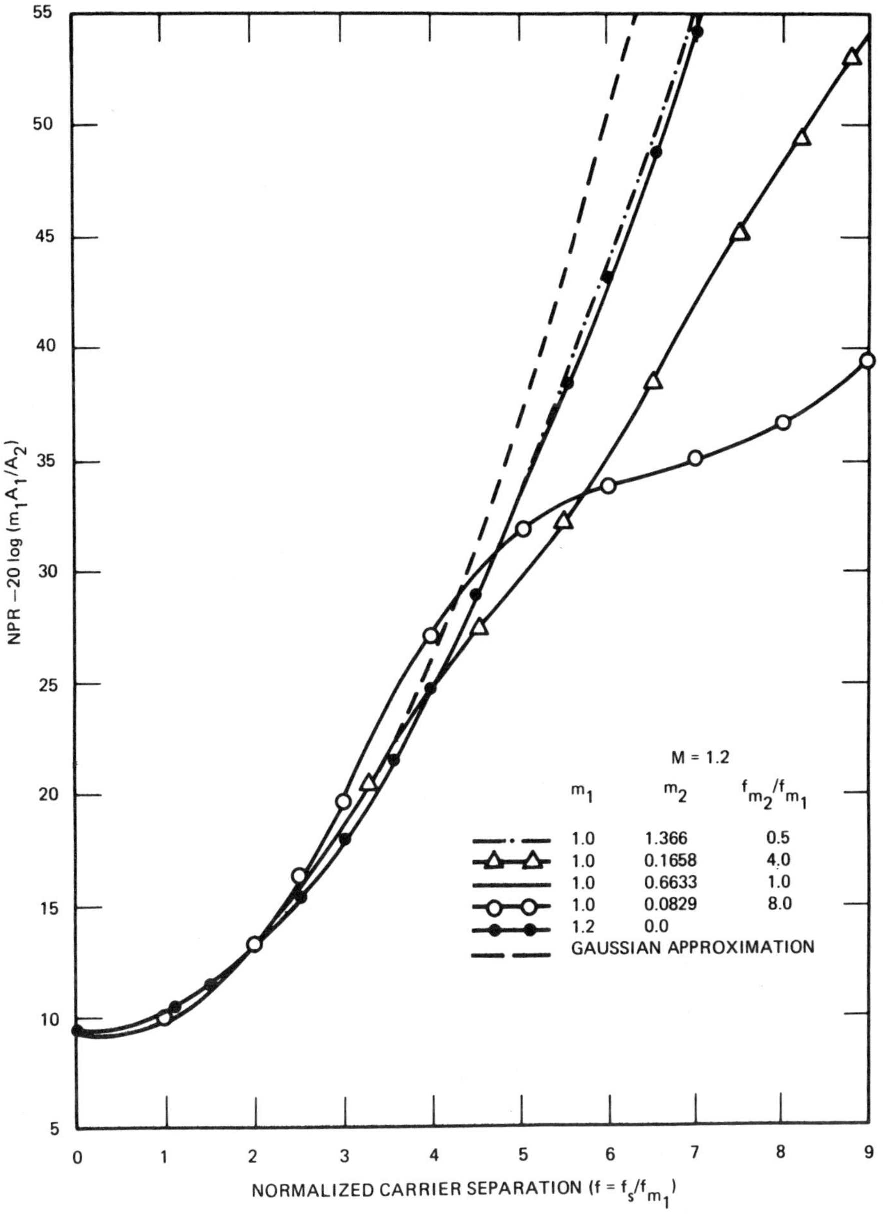

Fig. 7. Comparison of calculated interference for $M = 1.2$.

a presentation that will show interference calculations in a compact form for a range of parameters. From the preceding observations, it appears reasonable to perform interference calculations for a range of M and to consider identical signals. Such curves result in interference that coincides with the results obtained by the Gaussian approximation for small carrier separations and provide a better estimate of the resulting interference at large carrier separations.

For the values of M given in Fig. 8, such curves give precise values for the interference resulting from identical basebands and a broad range of modulation indices. The interference is computed at the worst channel, which usually happens to be the top channel for those cases for which such generalized curves can be given. The preemphasis improvement is included in the curves shown in Fig. 8.

If the interfering signal has the smaller baseband, Fig. 8 provides a good estimate of the resulting interference. Furthermore, it appears that in this case, the results given in this figure provide a good upper bound to the interference. If the interfering signal has the larger baseband, however, the results given in Fig. 8 are likely to show a smaller interference than is obtained from exact calculation.

For small carrier separations, the curves of Fig. 8 converge to yield the same results obtained from the Gaussian approximation. In this case, the resulting interference is independent of the baseband size of either signal and therefore the previous restrictions on the signal parameters can be removed.

Low Modulation Index Limitation on Generalizing the Interference Calculation

Generalizing the interference calculations by using the composite modulation index can lead to extremely inaccurate results at intermediate and low modulation indices. However, a generalized presentation of the interference is possible in cer-

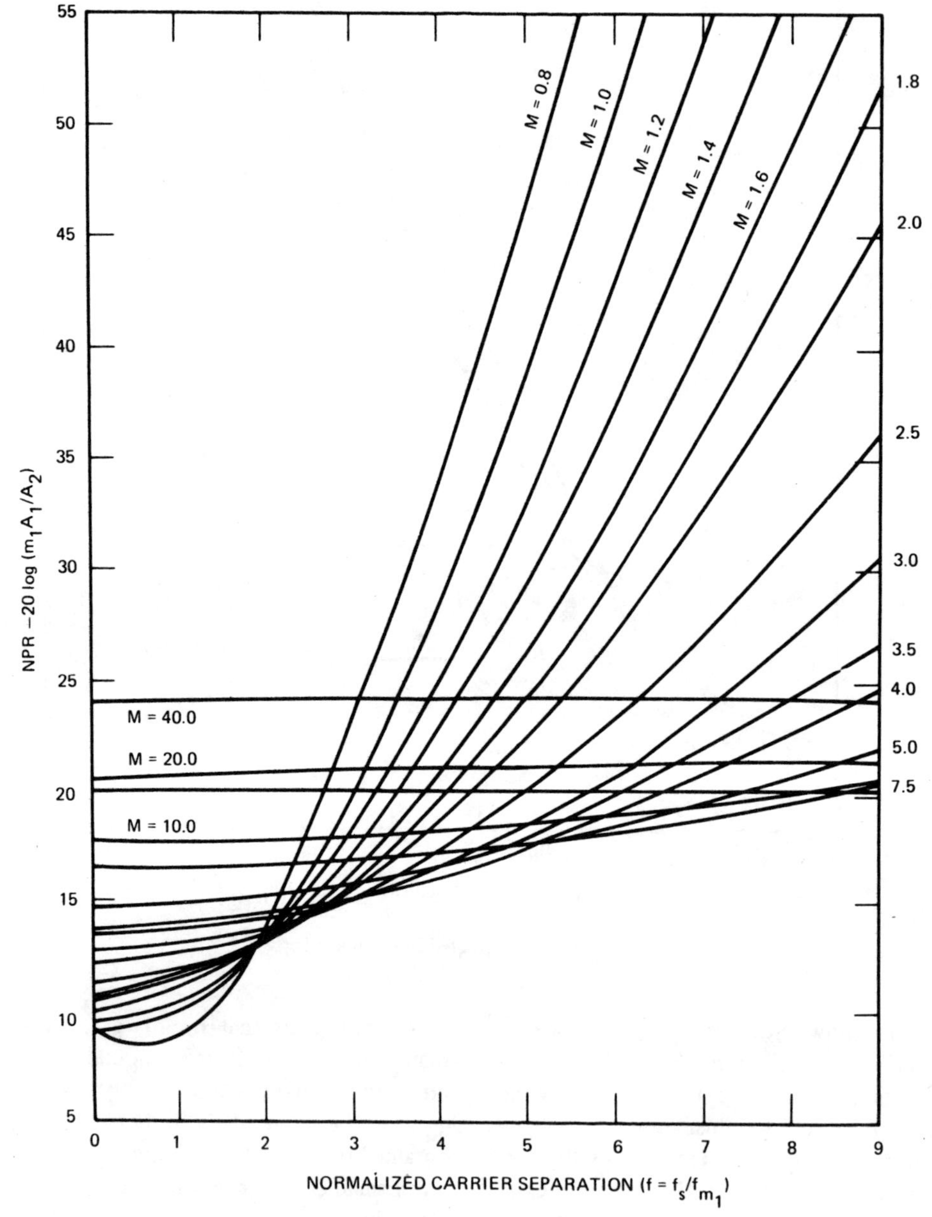

Fig. 8. Generalized curves for interference between two FDM/FM signals.

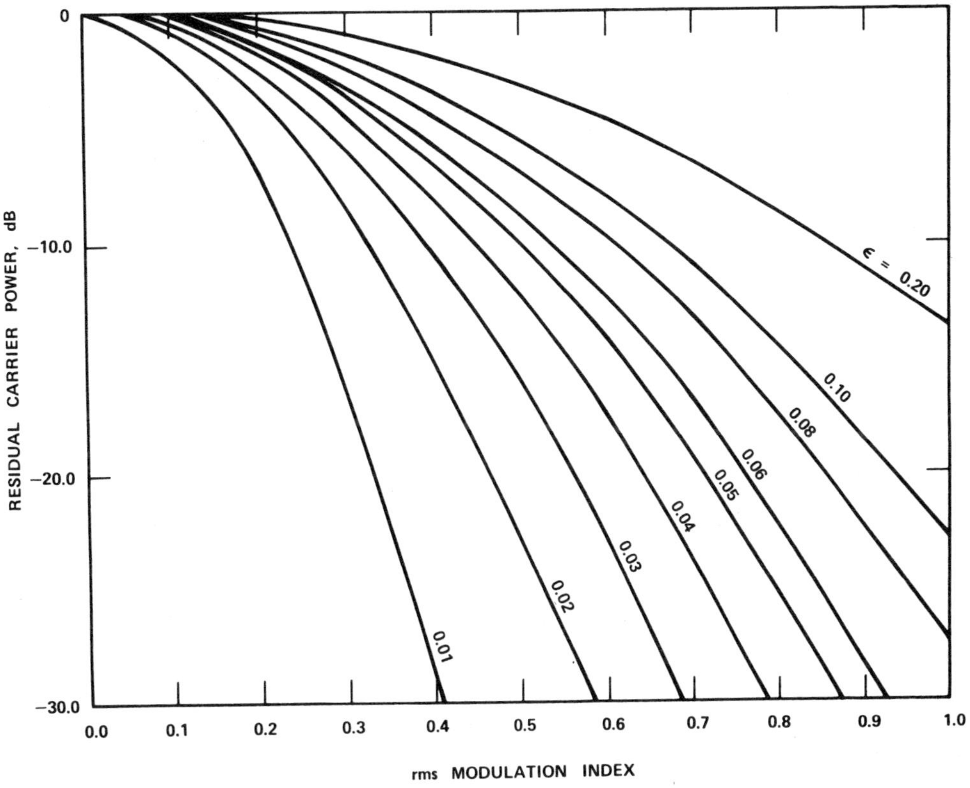

Fig. 9. Residual carrier power of the desired and interfering signals.

tain cases in which the primary source of interference in the worst baseband channel originates from the beating of the two residual carriers. For carrier separations less than 1 $f m_1$, an expression for the NPR in the worst channel (located at the mean carrier separation) can be obtained from (7), (12), and (14):

$$\mathrm{NPR} = \frac{2m_1^2 \, |H_p(f_s)|^2 \, \Delta f}{(A_2/A_1)^2 (f_s/f_{m_1})^2 f_{m_1}} \cdot \exp\left[-R_{x_1}(0) - R_{x_2}(0)\right]^{-1} \tag{18}$$

where

$$R_x(0) = m \frac{\int_\epsilon^1 H_p(\alpha)/\alpha^2 \, d\alpha}{\int_\epsilon^1 H_p(\alpha) \, d\alpha} \tag{19}$$

for preemphasized FM.

The residual carrier convolution term $\exp[-R_{x_1}(0) - R_{x_2}(0)]$ may be obtained directly from Fig. 9 by adding the residual carrier power of both the desired and interfering signals (in decibels). For other cases of low modulation index and for cases similar to the previous, but with carrier separations larger than f_{m_1}, the interference must be calculated separately for the specific parameters of both systems.

Interleaving Criteria

General

Formulas that may be helpful to the system designer in comparing the advantages of interleaving and cochannel operation are presented. In the development, two identical interfering signals have been considered at carrier separations that are somewhat larger than Carson's rule half-bandwidth.

Closed-Form Approximations Applicable to Certain Cases of Large m

The interference NPR present at the top baseband channel given by the Gaussian approximation is proportional to

$$\mathrm{NPR} \propto \operatorname{sech}\left[\frac{f_s/f_{m_1}}{m_1^2 + m_2'^2}\right] \exp\left[\frac{f_s^2 + f_{m_1}^2}{2f_{m_1}^2 (m_1^2 + m_2'^2)}\right]. \tag{20}$$

At cochannel, i.e., $f_s = 0$, the NPR is given by

$$\mathrm{NPR} \propto \exp\left[\frac{1}{2(m_1^2 + m_2'^2)}\right] \tag{21}$$

and the interleaving separation is given by

$$f_{s_I} = a(p m_1 + 1.0) f_{m_1} \tag{22}$$

where

- a additional margin over the Carson's rule half-bandwidth;
- p multichannel signal peak factor = 3.16.

The interleaving improvement factor T is defined as

$$T = 2 \frac{\mathrm{NPR}\,(f_s = f_{s_I})}{\mathrm{NPR}\,(f_s = 0.0)}, \tag{23}$$

which represents the improvement in the NPR resulting from operating with two interfering carriers separated by f_{s_I} instead of operating with a single cochannel interfering carrier. Thus,

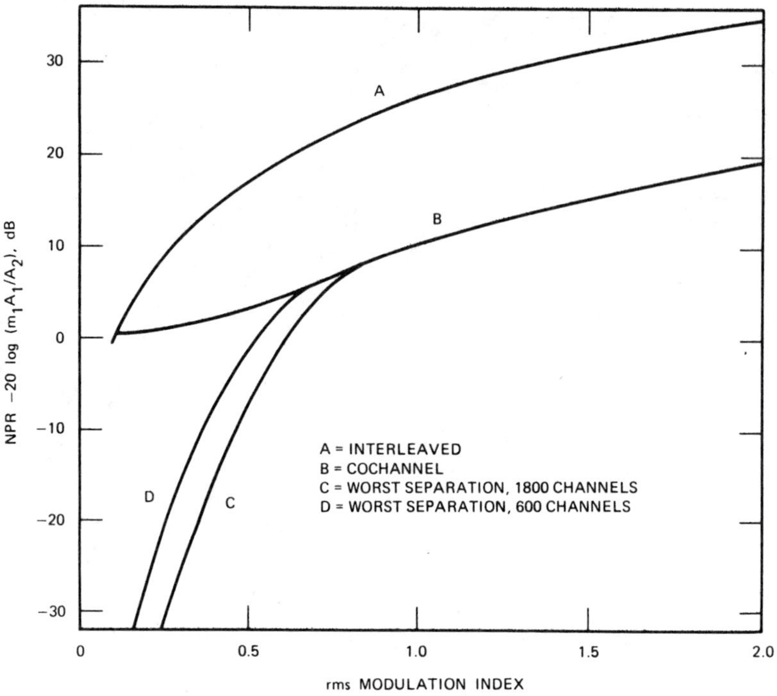

Fig. 10. Interference between two identical FDM/FM signals versus rms modulation index.

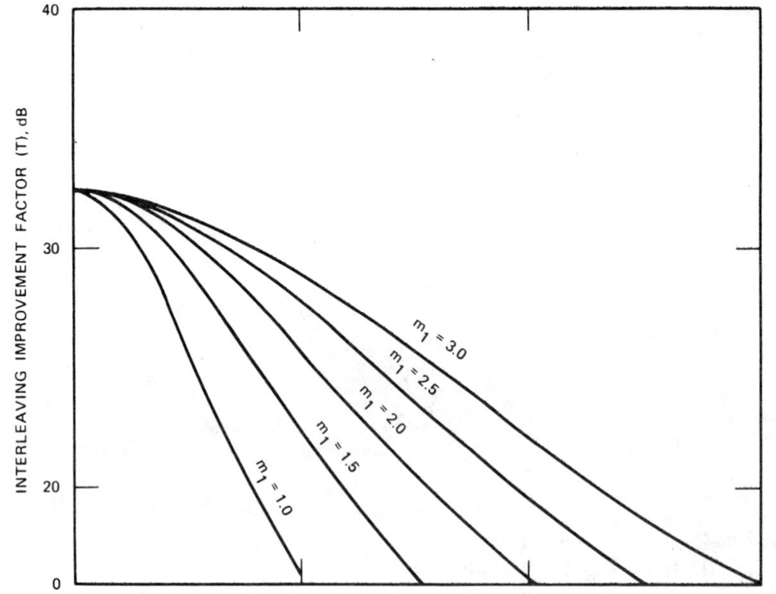

Fig. 11. Interleaving improvement as a function of loading on interleaved carriers.

the interleaving improvement factor T may be written as

$$T = \frac{1}{2}\,\text{sech}\left[\frac{a(pm_1 + 1.0)}{m_1^2 + m_2'^2}\right]\exp\left[\frac{a^2(pm_1 + 1.0)^2}{2(m_1^2 + m_2'^2)}\right]. \quad (24)$$

When cases for which $m_1 > m_2'$ and $m_1 <\simeq 3$ are considered,

$$T = \exp\frac{a^2(pm_1 + 1.0)^2/2.0 - a(pm_1 + 1.0)}{m_1^2 + m_2'^2}, \quad (25)$$

which may be applied with reasonable accuracy to large $M = m_1^2 + m_2'^2$.

Interleaving Calculations for a Homogeneous System

For a homogeneous system (i.e., $m_1 = m_2'$ and $f_{m_1} = f_{m_2}$) that has a multichannel signal peak factor of 10 dB and carrier separations 20 percent larger than Carson's rule half-bandwidth, the interleaving improvement factor becomes

$$T = 15.6 + 10 \log \exp\left(\frac{0.38}{m_1} - \frac{0.24}{m_1^2}\right). \quad (26)$$

For large m_1, the interleaving improvement can be seen to approach 15.6 dB.

In Fig. 10, the results of exact interference calculations are shown for three conditions: cochannel configuration; worst-case carrier separation; and interleaved operation. At large modulation indices, the cochannel case is also the worst interference case and the resulting interference can be reduced by approximately 15 dB by using an interleaved frequency plan. For progressively lower modulation indices, a lesser advantage results from interleaved carrier operation than from a cochannel operation. In both cases, the resulting interference is less than that corresponding to the worst case carrier separation.

Interleaving Calculations for Identical Interleaved Carriers with very Light Loading

For interfering carriers with very light loading, we may set $m_2' = 0.0$. From (25) we obtain the following expression for the improvement factor:

$$T = 31.3 + 10 \log \exp\left(\frac{0.75}{m_1} - \frac{0.28}{m_1^2}\right). \quad (27)$$

For large m_1, the improvement is seen to approach 31.3 dB.

Fig. 11 shows how the improvement varies as the loading is changed on the interfering carrier. The calculations are performed by using the Gaussian approximation; at the extremes (i.e., $m_2 = m_1$ and $m_2 = 0$) these calculations agree with the aforementioned results.

Summary and Conclusions

An investigation of interference between preemphasized FDM/FM signals has been undertaken. The theoretical development is very general in scope and includes cases of low and intermediate modulation index systems for which the spectra cannot be characterized by simple expressions. A generalized presentation of the results has been given whenever possible. Formulas that may be used to compare the advantages of interleaving and cochannel operation are presented along with plotted results for specific cases of interest. It is observed that the interleaving improvement decreases from about 32 to 15 dB as the loading on the interfering carriers increases from no loading to a loading equal to that of the desired carrier.

Appendix

The Analysis of Arbitrary Narrow-Band Interference into an Angle-Modulated Carrier

Let the desired angle-modulated signal $e_1(t)$ and an arbitrary narrow-band interfering signal $e_2(t)$ be given by

$$e_1(t) = \text{Re}\,[Z_1(t)] = \text{Re}\,[A_1 \exp\{j\,[\omega_1 t + x_1(t) + \mu]\}] = \text{Re}\,\{A_1 v_1(t)\,[\exp j\omega_1 t]\} \quad (\text{A-1})$$

$$e_2(t) = \text{Re}\,[Z_2(t)] = \text{Re}\,V_2(t)\exp\,[j\omega_2 t]\}, \quad (\text{A-2})$$

respectively. It is assumed that $e_1(t)$ and $e_2(t)$ are both wide-sense stationary and are generated from separate sources; thus they are statistically independent of each other. Furthermore, $x_1(t)$ and μ are assumed to be independent and μ is assumed to be uniformly distributed in $\{0 - 2\pi\}$. The excess phase angle (caused by the presence of the interference) at the output of an ideal demodulator is given by

$$\lambda(t) = \text{Im}\ln\left[1 + \frac{Z_2(t)}{Z_1(t)}\right]. \quad (\text{A-3})$$

For $|Z_2(t)/Z_1(t)| < 1$, $\lambda(t)$ can be expanded as

$$\lambda(t) = \text{Im}\sum_{m=1}^{\infty}\frac{(-1)^{m+1}}{m}\left(\frac{Z_2(t)}{Z_1(t)}\right)^m = \sum_{m=1}^{\infty}\lambda_m(t). \quad (\text{A-4})$$

The baseband power spectrum of the demodulated interference is obtained from the autocorrelation function of the total detected phase $\phi(t)$ where

$$\phi(t) = x_1(t) + \lambda(t) \quad (\text{A-5})$$

and the autocorrelation function is thereby

$$R_\phi(\tau) = \langle [x_1(t) + \lambda(t)]\,[x_1(t+\tau) + \lambda(t+\tau)]\rangle = R_{x_1}(\tau) + R_\lambda(\tau) \quad (\text{A-6})$$

since the cross terms vanish when averaged over μ.

The mth term of $\lambda(t)$ can be written as

$$\lambda_m(t) = \text{Im}\,\{V_2^m(t)\exp(jm\omega_2 t)\exp[jx_m(t)]\}\,K_m = \frac{K_m}{2j}\{V_2^m(t)\exp[jm\omega_2 t]\exp[jx_m(t)] - V_2^m(t)^*\exp(-jm\omega_2)\exp[-jx_m(t)]\} \quad (\text{A-7})$$

where

$$x_m(t) = -m\,[\omega_1 t + x_1(t) + \mu] \quad (\text{A-8})$$

and

$$K_m = \frac{-1^{m+1}}{mA_1^m}. \quad (\text{A-9})$$

The mth–nth term of the autocorrelation function of $\lambda(t)$ is then given by

$$R_\lambda^{mn}(\tau) = \langle \lambda_m(t)\, \lambda_n(t+\tau)\rangle$$
$$= -\frac{K_m K_n}{4}\, \{\langle V_2^m(t) \exp[jm\omega_2 t] \exp[jx_m(t)]$$
$$\cdot V_2^n(t+\tau) \exp[jn\omega_2(t+\tau)] \exp[jx_n(t+\tau)]\rangle$$
$$- \langle V_2^m(t) \exp(jm\omega_2 t) \exp[jx_m(t)]\; V_2^n(t+\tau)^*$$
$$\cdot \exp[-jn\omega_2(t+\tau) \exp[-jx_n(t+\tau)]\rangle$$
$$- \langle V_2^m(t)^* \exp(-jm\omega_2 t) \exp[-jx_m(t(]\; V_2^n(t+\tau)$$
$$\cdot \exp[jn\omega_2(t+\tau) \exp[jx_n(t+\tau)]\rangle + \langle V_2^m(t)^*$$
$$\cdot \exp[-jm\omega_2 t] \exp[-jx_m(t)]\; V_2^n(t+\tau)^*$$
$$\cdot \exp[-jn\omega_2(t+\tau)] \exp[-jx_n(t+\tau)]\rangle\}. \quad \text{(A-10)}$$

For $m = n$, $R^{mn}(\tau) = 0$ and for $m = n$, the first and last terms vanish because $\langle \exp jnu\rangle = 0$.

Therefore, (10) can be rewritten as follows:

$$R_\lambda^{mn}(\tau) = \Bigg[\frac{1}{4m^2 A_1^{2m}} R_{V_2^m}(\tau)\, R^*_{v_1^m}(\tau) \exp[jm(\omega_2 - \omega_1)\tau]$$
$$+ R^*_{V_2^m}(\tau)\, R_{v_1^m}(\tau) \exp[-jm(\omega_2 - \omega_1)\tau]\Bigg] \quad \text{(A-11)}$$

where

$R_{V_2^m}(\tau)$ autocorrelation function of $V_2^m(t)$,
$R^*_{v_1^m}(\tau)$ autocorrelation function of $v_1^m(t)$. (A-12)

The power spectrum of the baseband interference is then given by

$$I(f) = \sum_{m=1}^{\infty} \frac{1}{4m^2 A_1^{2m}} [T_m(f - mf_s) + T_m(-f - mf_s)] \quad \text{(A-13)}$$

where

$$T_m(f) = S_{V_2^m}(f) \otimes S_{v_1^m}(f)$$

and

$$S_{V_2^m}(f) = \mathfrak{F}[R_{V_2^m}(\tau)]$$
$$= \text{power spectral density of } V_2^m(t)$$
$$S_{v_1^m}(f) = \mathfrak{F}[R_{v_1^m}(\tau)]$$
$$= \text{power spectral density of } v_1^m(t). \quad \text{(A-14)}$$

Power Spectra of PM and FM Signals

To evaluate the baseband interference, expressions for both the desired and interfering spectra must be provided. The normalized low-pass equivalent power spectral density of a phase-modulated signal is given by [9]

$$S(f) = \int_{-\infty}^{\infty} \exp[-R_x(0) + R_x(\tau)] \exp[(-j2\pi f\tau)]\, d\tau \quad \text{(A-15)}$$

where

$R_x(0)$ average power in the phase-modulating baseband signal $x(t)$,

$R_x(\tau)$ correlation function of the phase-modulating baseband signal.

As in [6] and [10], this result can be expanded for small and intermediate modulation indices:

$$S(f) = \sum_{n=0}^{\infty} \exp[-R_x(0)] \frac{[R_x(0)]^n}{n!} \sigma_x(f) \overset{n}{\otimes} \sigma_x(f) \quad \text{(A-16)}$$

where

$$\sigma_x(f) \overset{0}{\otimes} \sigma_x(f) \equiv \delta(f) \quad \text{(A-17)}$$

and $\sigma_x(f)$ is the normalized power spectral density of the phase-modulating signal after preemphasis. The corresponding expansion for large modulation indices, given in [6], is more complicated.

For FM, the expressions for the PM spectrum are modified to provide the corresponding FM spectrum. Equation (A-16) can be used to calculate a frequency-modulated spectrum if the phase-modulating parameters are written in terms of the "frequency-modulation parameters," which are given in [6]. Thus for FM,

$$R_x(0) = m^2 \frac{\int_\epsilon^1 [H_p(\alpha)/\alpha^2]\, d\alpha}{\int_\epsilon^1 H_p(\alpha)\, d\alpha} \quad \text{(A-18)}$$

and

$$\sigma_{\dot{x}}(f) = \frac{S_{\dot{x}}(f)/f^2}{\int_{-\infty}^{\infty} S_{\dot{x}}(f)/f^2\, df} \quad \text{(A-19)}$$

where $S_{\dot{x}}(f)$ is the power spectral density after preemphasis of the frequency-modulating baseband signal.

References

[1] R. G. Medhurst, E. M. Hicks, and W. Grossett, "Distortion in frequency division multiplex FM system due to an interfering carrier," *Proc. Inst. Elec. Eng.* (London), vol. 105B, pp. 282–292, May 1958.

[2] R. G. Medhurst, "RF spectra and interfering carrier distortion in FM trunk radio systems with low modulation ratios," *IRE Trans. Commun. Syst.*, vol. CS-9, pp. 107–115, June 1961.

[3] R. Hamer, "Radio-frequency interference in multi-channel telephony FM radio systems," Inst. Elec. Eng., Paper 3326, Jan. 1971, pp. 75–89.

[4] R. G. Medhurst, "FM interfering carrier distortion general formula," *Proc. Inst. Elec. Eng.* (London), vol. 109B, pp. 149–150, Mar. 1962.

[5] V. K. Prabhu and L. H. Enloe, "Interchannel interference considerations in angle-modulated systems," *Bell Syst. Tech. J.*, pp. 2333–2357, Sept. 1969.

[6] J. C. Fuenzalida, O. Shimbo, and B. A. Pontano, "Spectral characteristics of angle-modulated signals," to be published.

[7] W. R. Bennett, H. E. Curtis, and S. O. Rice, "Interchannel interference in FM and PM systems under noise loading conditions," *Bell Syst. Tech. J.*, vol. 34, pp. 601–636, May 1955.

[8] B. R. Hallford and J. N. Ratti, "Microwave receiver interference characteristic measurements," *IEEE Trans. Commun. Technol.*, vol. COM-14, pp. 455–469, Aug. 1966.

[9] D. Middleton, *An Introduction to Statistical Communication Theory*. New York: McGraw-Hill, 1960.

[10] C. C. Ferris, "Spectral characteristics of FDM-FM signals," *IEEE Trans. Commun. Technol.*, vol. COM-16, pp. 233–238, Apr. 1968.

Interchannel Interference in FM and PM Systems Under Noise Loading Conditions

By W. R. BENNETT, H. E. CURTIS, and S. O. RICE

(Manuscript received January 5, 1955)

One of the principal sources of interchannel interference in multichannel FM and PM systems is the dependence of the attenuation and phase shift of the transmission path on frequency. Here we study the interference produced by a simple kind of such a dependence, namely that due to a single echo. Besides being important in themselves, the results are of interest because they may be used to estimate the interference in other, more complicated cases.

In this paper expressions are derived for the interchannel interference produced by echoes of relatively small amplitude. Several important special cases are studied in detail and curves that simplify the computation of the interference power are presented.

CONTENTS

INTRODUCTION

The systems we shall consider are of the FDM-FM and FDM-PM types; that is, systems in which the composite signal wave (the "base-band signal") from a group of carrier telephone channels in frequency division multiplex (FDM) is transmitted by frequency modulation (FM) or phase modulation (PM). Such methods are currently being used in the Bell System to send large groups of telephone channels by microwave radio. If the FM signal is accompanied by echoes, which may be due to reflections in the equipment or in the transmission medium, the wave of instantaneous frequency versus time is distorted in a nonlinear manner and interchannel interference occurs. Here we shall be concerned with this interference.

The distortion has been analyzed in a number of previous publications[1-4] for the case in which the base-band signal may be represented by one or more sine waves. However, when the number of telephone channels is not small, the sine wave representation becomes unwieldy because a large number of both low and high order modulation products must be considered. Here we avoid this difficulty, at the cost of somewhat more complex analysis, by using a band of random noise to represent the multiplex signal.

It has been found in practice that such a random noise signal of appropriate bandwidth and power adequately simulates a composite speech signal. For studies involving interchannel interference the energy corresponding to some particular telephone channel is removed. When such a wave is impressed on the frequency modulator and the resulting FM wave is transmitted, detected, and finally demodulated, the received output in the originally clear channel represents interchannel interference. Measurements of this type have been discussed by Albersheim and Schafer.[3]

One of the principal sources of interchannel interference is the variation of the attenuation and phase shift of the transmission path with frequency. An echo produces a simple form of such a variation. In fact, it is often possible to estimate the effect produced by a more complex type of variation by comparing it with a roughly equivalent echo.

It is the purpose of this paper to develop formulas for the interchannel interference produced by echoes of relatively small amplitude in FM and PM systems. General expressions are given in the form of definite integrals which may be evaluated by numerical integration. Approximations are obtained in a number of cases of importance and numerical tables and curves are furnished to facilitate applications to specific problems.

We wish to express our thanks to Miss Mary Corr and Miss Barbara Fischer who have performed the rather lengthy computations required for this investigation.

1. DEVELOPMENT OF GENERAL FORMULAS

Since the tolerable amount of interchannel interference in multiplex telephony is small, the practically important situations are characterized

Reprinted with permission from *Bell Syst. Tech. J.*, vol. 34, pp. 601-636, May 1955.

by echoes which are relatively small relative to the main transmitted wave. It is necessary to consider in detail only the case of one small echo. Effects of small multiple echoes may then be calculated by superposition of the effects of single echoes. Our distortion problem then reduces to the following: An original signal

$$E_0(t) = E \sin [pt + \varphi(t)] \tag{1.1}$$

is received along with an echo

$$E_1(t) = rE \sin [p(t - T) + \varphi(t - T)] \tag{1.2}$$

Here r represents the ratio of echo amplitude to the amplitude of the principal received component, p is the unmodulated carrier frequency, T is the delay difference of the paths and $\varphi(t)$ is the phase variation caused by the multichannel signal.

Calculation of the phase of the original signal plus its echo is equivalent to calculating the value of θ in the representation

$$\sin x + r \sin (x + y) = V \sin (x + \theta) \tag{1.3}$$

with x, y, r, V, and θ real numbers. V and θ are determined as functions of r and y by the two equations obtained from the sin x and cos x portions of (1.3). When $r \ll 1$, V approaches unity, and θ is proportional to r; direct expansion in powers of θ shows that to a first order approximation

$$\theta = r \sin y \tag{1.4}$$

In our case,

$$x = pt + \varphi(t) \tag{1.5}$$

$$y = \varphi(t - T) - \varphi(t) - pT \tag{1.6}$$

Hence the phase error produced by the echo is

$$\theta(t) = r \sin [v(t) - pT], \tag{1.7}$$

$$v(t) = \varphi(t - T) - \varphi(t). \tag{1.8}$$

Our problem is to study the power spectrum of $\theta(t)$ when $\varphi(t)$ is a random noise wave.

The treatment will be restricted to the practically important case in which the noise source is free from dc and discrete sinusoidal components. A sufficient mathematical condition, which we shall adopt to insure this, is that the power spectrum of the noise have limited total fluctuation. We shall assume also that the noise wave is a member of a Gaussian ensemble and hence that all the statistical properties are derivable from

either its power spectrum or autocorrelation function. Which of the two quantities we use at any stage of the computations is a matter of relative expediency. We shall adopt a uniform notation in that an ensemble of time functions $x(t)$ will be said to have a power spectrum $w_x(f)$ and an autocorrelation function $R_x(\tau)$, i.e.,

$$R_x(\tau) = \text{ave } [x(t)x(t + \tau)] = \int_0^\infty w_x(f) \cos 2\pi\tau f \, df \tag{1.9}$$

$$w_x(f) = 4 \int_0^\infty R_x(\tau) \cos 2\pi f\tau \, d\tau \tag{1.10}$$

The power spectrum is proportional to the mean square of the response of a filter of bandwidth df centered at f when $x(t)$ is applied as input.

In the echo distortion problem, the power spectrum $w_\varphi(f)$ is the quantity given and the power spectrum $w_\theta(f)$ is the one desired. The corresponding autocorrelation functions $R_\varphi(\tau)$ and $R_\theta(\tau)$ are calculable from the corresponding power spectra and vice versa. The power spectrum is the more convenient choice as a function to compute when we deal with the noise response of a linear system with known transfer admittance $Y(if)$, since the power spectrum of the response is merely $| Y(if) |^2$ times that of the input. For example, differentiation is equivalent to transmission through a transfer admittance $i2\pi f$ and hence multiplies the power spectrum by $4\pi^2 f^2$. The inverse process of integration divides the power spectrum by $4\pi^2 f^2$. When a highly nonlinear operation is performed on the noise, or when the operation is a linear one more simply described in the time domain, the autocorrelation function may be simpler to compute.

The first step in the solution of the problem is to evaluate the statistics of the noise wave ensemble,

$$\{v(t)\} = \{\varphi(t - T) - \varphi(t)\} \tag{1.8}$$

in terms of the statistics of $\varphi(t)$. This can readily be done in terms of power spectra by calculating the transfer admittance of a two-path transmission system such as the right-hand member of (1.8) defines, or in terms of autocorrelation functions by averaging the appropriate time functions. The latter procedure gives:

$$\begin{aligned} R_v(\tau) &= \text{ave } [v(t)\, v(t + \tau)] \\ &= \text{ave } \{[\varphi(t - T) - \varphi(t)][\varphi(t - T + \tau) - \varphi(t + \tau)]\} \\ &= \text{ave } [\varphi(t - T)\, \varphi(t - T + \tau)] + \text{ave } [\varphi(t)\, \varphi(t + \tau)] \\ &\qquad - \text{ave } [\varphi(t - T)\, \varphi(t - T + \tau + T)] \\ &\qquad - \text{ave } [\varphi(t)\, \varphi(t + \tau - T)] \\ &= 2R_\varphi(\tau) - R_\varphi(\tau + T) - R_\varphi(\tau - T) \end{aligned} \tag{1.11}$$

In the averaging process we have made use of the facts that the average of the sum is equal to the sum of the averages and that $\varphi(t)$ is a stationary ensemble, allowing us to replace $t - T$ by t when averaging. Since the two-path transmission system is linear and invariable, the $v(t)$ ensemble remains Gaussian. The corresponding power spectra relations are, from (1.10),

$$w_v(f) = 4\, w_\varphi(f) \sin^2 \pi f T \qquad (1.12)$$

Our next step called for by (1.7) requires the evaluation of the autocorrelation function of a sine function of a band of noise. The solution may be obtained* from (3.2 — 7) of Reference 5, which gives a general theorem for a Gaussian noise ensemble expressible in our notation as

$$\begin{aligned} \text{ave}\,\{\exp\,[iax(t) + ibx(t+\tau)]\} \\ = \exp\left[-\frac{a^2+b^2}{2} R_x(0) - abR_x(\tau)\right] \end{aligned} \qquad (1.13)$$

In this equation a and b are real constants. Multiplying both sides by the constant $\exp\,(i\beta)$, β real, and equating real parts gives the more directly applicable result:

$$\begin{aligned} \text{ave}\,\{\cos\,[ax(t) + bx(t+\tau) + \beta]\} \\ = \exp\left[-\frac{a^2+b^2}{2} R_x(0) - abR_x(\tau)\right] \cos\beta \end{aligned} \qquad (1.14)$$

Referring now to (1.7) we write

$$\begin{aligned} R_\theta(\tau) &= \text{ave}\,\{r \sin\,[v(t) - pT]\, r \sin\,[v(t+\tau) - pT]\} \\ &= \frac{r^2}{2}\,\text{ave}\,\{\cos\,[v(t) - v(t+\tau)]\} \\ &\qquad - \frac{r^2}{2}\,\text{ave}\,\{\cos\,[v(t) + v(t+\tau) - 2pT]\} \\ &= \frac{r^2}{2}\exp\,[-R_v(0) + R_v(\tau)] \\ &\qquad - \frac{r^2}{2}\exp\,[-R_v(0) - R_v(\tau)]\cos 2pT \\ &= \frac{r^2}{2} e^{-R_v(0)}[e^{R_v(\tau)} - e^{-R_v(\tau)}\cos 2pT] \end{aligned} \qquad (1.15)$$

* This application of the characteristic functions method of attack has been employed in similar situations by D. Middleton in Reference 6 and M. K. Zinn in unpublished memoranda.

Note that since the operation of taking the sine is nonlinear, the $\theta(t)$ ensemble is not Gaussian and neither the power spectrum nor the autocorrelation is sufficient to give a complete statistical description. Either will be sufficient for our purposes however.

The power spectrum of the distortion ensemble $\theta(t)$ is now determinable from the Fourier cosine transform of $4\,R_\theta(\tau)$ as indicated by (1.10). The remaining steps are simplified, however, if we perform two preliminary operations on $R_\theta(\tau)$ before the final Fourier transform is calculated.

We first observe that if $R_\theta(\tau)$ contains a component C which does not vary with τ, the ensemble $\theta(t)$ contains a dc component $\bar{\theta} = \sqrt{C}$. The presence of C complicates the integration and hence we subtract out such a term before taking the Fourier transform. The value of C is given by

$$\begin{aligned} C &= \lim_{\tau\to\infty} R_\theta(\tau) \\ &= \frac{r^2}{2} e^{-R_v(0)} \lim_{\tau\to\infty} [e^{R_v(\tau)} - e^{-R_v(\tau)}\cos 2pT] \\ &= \frac{r^2}{2} e^{-R_v(0)}(1 - \cos 2pT) \end{aligned} \qquad (1.16)$$

In calculating the limit we have made use of the fact that, for our assumed noise wave, $R_v(\tau)$ must approach zero as τ becomes infinite. Our autocorrelation function of interest is thereby reduced to

$$R_{\theta-\bar\theta}(\tau) = \frac{r^2}{2} e^{-R_v(0)}\,\{e^{R_v(\tau)} - 1 - [e^{-R_v(\tau)} - 1]\cos 2pT\} \qquad (1.17)$$

which is the autocorrelation function of the ensemble $\theta(t) - \bar{\theta}$.

Our second preliminary operation on $R_\theta(\tau)$ is suggested by our ultimate goal, namely the calculation of the interchannel interference spectrum $w_c(f)$ (i.e., $w_c(f)\,df$ is the average power received in an idle channel of width df when the other channels are carrying signals). We note that the spreading of the original spectrum into initially vacant frequency ranges occurs solely because of the nonlinear dependence of $R_{\theta-\bar\theta}(\tau)$ on $R_v(\tau)$. The disturbance received in an idle channel is therefore produced by this nonlinearity. The part of $R_{\theta-\bar\theta}(\tau)$ which varies linearly with $R_v(\tau)$ may be expected to represent the linear transmission, and the difference between $R_{\theta-\bar\theta}(\tau)$ and its linear portion to represent the nonlinear transmission. In other words, subtracting the linear portion from $R_{\theta-\bar\theta}(\tau)$ is equivalent to removing the linear transmission from any channel without disturbing the nonlinear contributions from the remaining channels. These considerations lead us to set the autocorrelation function $R_c(\tau)$ corresponding to $w_c(f)$ equal to $R_{\theta-\bar\theta}(\tau)$ minus its linear portion. The work of Appendix I shows that this equality is rigorously true.

In order to perform the subtraction it is convenient to write (1.17) in the form

$$R_{\theta-\bar\theta}(\tau) = F[R_v(\tau)] \tag{1.18}$$

where in our case for a general variable z,

$$F(z) = \frac{r^2}{2} e^{-R_v(0)}[e^z - 1 - (e^{-z} - 1) \cos 2pT] \tag{1.19}$$

The portion of $R_{\theta-\bar\theta}(\tau)$ which varies linearly with $R_v(\tau)$ is $F'(0)\ R_v(\tau)$ where

$$F'(z) = \frac{r^2}{2} e^{-R_v(0)}(e^z + e^{-z} \cos 2pT) \tag{1.20}$$

$$F'(0) = \frac{r^2}{2} e^{-R_v(0)}(1 + \cos 2pT) \tag{1.21}$$

According to the foregoing discussion and Appendix I, the autocorrelation function and the power spectrum of the interchannel interference are given by

$$\begin{aligned} R_c(\tau) &= R_{\theta-\bar\theta}(\tau) - F'(0)R_v(\tau) \\ &= \frac{r^2}{2} e^{-R_v(0)}\{e^{R_v(\tau)} - 1 - R_v(\tau) \\ &\qquad - [e^{-R_v(\tau)} - 1 + R_v(\tau)] \cos 2pT\} \end{aligned} \tag{1.22}$$

$$w_c(f) = 4 \int_0^\infty R_c(\tau) \cos 2\pi f\tau\, d\tau \tag{1.23}$$

The power spectrum of $\theta(t) - \bar\theta$ may be simply expressed in terms of $w_c(f)$ without further integration. We find by applying the Fourier transform relationship of (1.9) and (1.10) to (1.22):

$$\begin{aligned} w_{\theta-\bar\theta}(f) &= 4 \int_0^\infty [R_c(\tau) + F'(0)R_v(\tau)] \cos 2\pi f\tau\, d\tau \\ &= w_c(f) + F'(0)w_v(f) \\ &= w_c(f) + 4F'(0)w_\varphi(f) \sin^2 \pi fT \\ &= w_c(f) + 2r^2e^{-R_v(0)}(1 + \cos 2pT)w_\varphi(f) \sin^2 \pi fT \end{aligned} \tag{1.24}$$

where we have used (1.12) and (1.21). The power spectrum $w_\theta(f)$ of θ is obtained by adding to (1.24) a spire at $f = 0$ to represent the power in $\bar\theta$, the dc portion of θ.

The results obtained up to this point may be summarized as follows: Let $\varphi(t)$ be the phase variation produced by the impressed multichannel signal. The echo (1.2) produces a phase error $\theta(t)$ (1.7) in the received signal. $\theta(t)$ has a dc component $\bar\theta$ given by the square root of (1.16). $\theta(t) - \bar\theta$ is a function of time which fluctuates about zero and has the power spectrum $w_{\theta-\bar\theta}(f)$ given by (1.24). Now consider the problem of computing the interchannel interference. One procedure would be to consider the case in which all but one of the channels are loaded. Then the power spectrum $w_\varphi(f)$ of $\varphi(t)$ would have a narrow slot in it corresponding to the zero power in the unloaded channel. The values of $w_{\theta-\bar\theta}(f)$ computed from this $w_\varphi(f)$ for values of f within the slot would give the channel interference spectrum (for phase modulation). However, it turns out that as the slot width approaches zero, these values of $w_{\theta-\bar\theta}(f)$ approach those of $w_c(f)$ where $w_c(f)$ is computed from (1.23) on the assumption that all channels are loaded. In other words the $w_\varphi(f)$ used in the calculation of (1.23) has no slot. (Actually the same value of $w_c(f)$ is obtained whether $w_\varphi(f)$ has a slot or not. Of course, the slot must be vanishingly narrow).

Almost all of the preceding work pertains to the power spectrum of the phase error $\theta(t)$. For the sake of completeness we shall give the power spectrum of the complete phase angle, $Q(t) = \varphi(t) + \theta(t)$, of the output. In order to obtain all of the $0(r^2)$ term it is necessary to add another term to the approximation (1.7):

$$\theta(t) = r \sin [v(t) - pT] - \frac{r^2}{2} \sin [2v(t) - 2pT] + 0(r^3) \tag{1.25}$$

The autocorrelation function for $Q(t)$ may be shown to be

$$\begin{aligned} R_Q(\tau) &= R_\varphi(\tau) + R_\theta(\tau) + \text{ave}\ [\varphi(t)\ \theta(t + \tau) + \varphi(t + \tau)\ \theta(t)] \\ &= R_\varphi(\tau) + R_\theta(\tau) - rR_v(\tau)\ e^{-R_v(0)/2} \cos pT \\ &\qquad + r^2R_v(\tau)\ e^{-2R_v(0)} \cos 2pT + 0(r^3) \end{aligned} \tag{1.26}$$

The average has been evaluated by using

$$\begin{aligned} &\text{ave}\ \{\varphi(t \pm \tau) \sin [\varphi(t - T) - \varphi(t) - pT]\} \\ &\quad = [R_\varphi(T \pm \tau) - R_\varphi(\tau)] \cos pT \exp [R_\varphi(T) - R_\varphi(0)] \end{aligned} \tag{1.27}$$

and the corresponding result for 2φ and $2pT$ ($R_{2\varphi}(\tau) = 4R_\varphi(\tau)$). Equation (1.27) may be obtained from the result similar to (1.13) which contains the three variables $\varphi(t)$, $\varphi(t \pm \tau)$, $\varphi(t - T)$ instead of $x(t)$ and $x(t + \tau)$.

The power spectrum $w_Q(f)$ of $Q(t)$ may be obtained from (1.26) by replacing the autocorrelation functions in the coefficients by the corresponding power spectra. By using (1.24) we may obtain the following expression for the power spectrum of the fluctuating portion $Q(t) - \bar{Q}$ of $Q(t)$, $\bar{Q} = \bar{\theta}$ being the dc portion of $Q(t)$:

$$w_{Q-\bar{Q}}(f) = w_\varphi(f) + w_c(f) + w_v(f)[-r\, e^{-R_v(0)/2} \cos pT + r^2 e^{-R_v(0)} \cos^2 pT + r^2 e^{-2R_v(0)} \cos 2pT] \quad (1.28)$$

The expression (1.12) for $w_v(f)$ shows that it is proportional to $w_\varphi(f)$. Hence, assuming $w_\varphi(f)$ to have a narrow slot corresponding to an idle channel, it is seen that (1.28) is in agreement with the fact that only $w_c(f)$ contributes to the interchannel interference spectrum.

We remark that for echoes produced by reflections in wave guides the angle $2pT$ in (1.22) and (1.24) usually contains many multiples of 2π. In most cases it would in fact be reasonable to average the effect of the term $\cos 2pT$ by assuming the angle to be uniformly distributed throughout 2π radians. This gives the result zero for the term since plus and minus values are symmetrically distributed. We shall, however, carry the term along in our formulas since it is of importance when the delays are small, as in multipath radio propagation, and when our results are used to estimate interchannel interference in general.

2. APPLICATIONS TO FLAT NOISE SIGNAL

In general $w_\varphi(f)$ depends on the type of preemphasis used in the channel multiplex signal. Two representative conditions will be studied: (1) phase modulation (PM) in which channels at equal level are impressed on a phase modulator, or the more usual equivalent situation of equal level channels which are differentiated before being impressed on a frequency modulator, and (2) frequency modulation (FM) in which channels at equal level are impressed on a frequency modulator. The appropriate power spectra are:

$$\text{PM:} \quad w_\varphi(f) = P_0, \qquad f_a < f < f_b \quad (2.1)$$

$$\text{FM:} \quad w_\varphi(f) = P_0/4\pi^2 f^2, \qquad f_a < f < f_b \quad (2.2)$$

The latter form results because the phase is the time integral of the instantaneous frequency, which has a flat spectrum. In PM, P_0 is expressed in $(\text{radians})^2/\text{cps}$, but in FM, P_0 is expressed in $(\text{radians/sec.})^2/\text{cps}$. In

the case of most common practical interest, the band of frequencies between 0 and f_a is relatively narrow compared to the range $f_b - f_a$. We have accordingly assumed that f_a approaches zero. In the FM case we cannot set $f_a = 0$ immediately because the power spectrum would then become unbounded at the origin. It is found, however, that finite limits are approached for the actual quantities of interest which we compute.

When the spectrum $w_\varphi(f)$ of the signal has the form (2.1) or (2.2), expression (1.23) for the interchannel interference spectrum may be written as

$$w_c(f) = r^2(2\pi f_b)^{-1}(G - H \cos 2pT) \quad (2.3)$$

where $w_c(f)\, df$ is measured in $(\text{radians})^2$ and

$$G = 2e^{-R_v(0)} \int_0^\infty [e^{R_v(\tau)} - R_v(\tau) - 1] \cos au\, du \quad (2.4)$$

$$H = 2e^{-R_v(0)} \int_0^\infty [e^{-R_v(\tau)} + R_v(\tau) - 1] \cos au\, du \quad (2.5)$$

$$u = 2\pi f_b \tau \qquad U = 2\pi f_b T \qquad a = f/f_b \quad (2.6)$$

For PM:

$$R_v(\tau) = P_0 f_b \left[\frac{2 \sin u}{u} - \frac{\sin (u + U)}{u + U} - \frac{\sin (u - U)}{u - U}\right] \quad (2.7)$$

where $P_0 f_b$ is the mean square value, in $(\text{radians})^2$, of the PM signal $\varphi(t)$.

For FM:

$$R_v(\tau) = A[-2(1 - \cos U)\cos u - 2uSi(u) + (u + U)Si(u + U) + (u - U)Si(u - U)] \quad (2.8)$$

$$A = (P_0 f_b)/(2\pi f_b)^2$$

In (2.8) $P_0 f_b$ is the mean square value (measured in $(\text{radians per second})^2$) of the FM signal $\varphi'(t)$. Consequently, $A = (\sigma/f_b)^2$ where σ is the rms frequency deviation, in cycles per second, of the signal. Equations (2.7) and (2.8) follow from (2.1), (2.2) and (1.11).

When $R_v(0)$ is small compared to unity, the same is true of $R_v(\tau)$, and (2.4) and (2.5) lead to the approximations

$$G \approx H \approx \int_0^\infty R_v^2(\tau) \cos au\, du \quad (2.9)$$

The interchannel interference calculated from (2.9) is that produced by the "second order modulation products" and is studied, together with other approximations, in Sections 3 and 4.

The quantity of interest in practice is the ratio of average interference power in the idle channel to average signal power in an adjacent channel. For PM the average interference and signal powers (in a narrow channel centered on frequency f) are the channel bandwidth times $w_c(f)$ and $w_\varphi(f)$, respectively. When the number of channels is large, the power spectrum does not change appreciably in going from one channel to the next and we may write

$$\frac{P_I}{P_S} = \frac{w_c(f)}{w_\varphi(f)} = \frac{r^2}{2\pi P_0 f_b}(G - H \cos 2pT) \tag{2.10}$$

as the desired interference-to-signal power ratio.

For FM the average interference and signal powers in a narrow channel are $(2\pi f)^2 \times$ (channel bandwidth) times $w_c(f)$ and $w_\varphi(f)$, respectively. These powers are measured in (radians per second)^2. When we take the ratio P_I/P_S, the $(2\pi f)^2 \times$ (channel bandwidth) cancels out and we have from (2.2) and (2.3)

$$\frac{P_I}{P_S} = \frac{r^2 a^2}{2\pi A}(G - H \cos 2pT) \tag{2.11}$$

3. APPROXIMATIONS FOR G AND H — PHASE MODULATION

Table 3.1 contains various approximate expressions for the quantities G and H which enter expression (2.3) for the channel interference spectrum $w_c(f)$. The notation is explained in Section 2, and the results apply to the case in which $w_\varphi(f)$ has the flat power spectrum (2.1).

Case 1 gives the exact expressions developed in Section 2. Case 2 is the "second order modulation approximation", valid when $R_v(0) \ll 1$. Evaluation of the integral (2.9) for G with the help of

$$\int_{-\infty}^{\infty} \frac{\sin u \sin (u - U)}{u(u - U)} \cos au \, du = \begin{cases} \pi U^{-1} \cos\left(\frac{aU}{2}\right) \sin\left(U - \frac{aU}{2}\right), & 0 \leqq a \leqq 2 \\ 0, & 2 \leqq a \end{cases}$$

leads to the following expression for the quantity J appearing in Table 3.1:

TABLE 3.1 — PHASE MODULATION

Case No.	Restrictions on Parameters U and $P_0 f_b$	G and H	Notes
1	No restrictions, $w_\varphi(f)$ defined by (2.1)	G defined by (2.4) H defined by (2.5) $R_v(\tau)$ defined by (2.7)	$P_0 f_b = \text{ave}\,[\varphi^2(t)]$ $\frac{P_I}{P_s} = \frac{r^2}{2\pi P_0 f_b}(G - H \cos 2pT)$, $U = 2\pi f_b T$, $a = f/f_b$
2	$2P_0 f_b \left(1 - \frac{\sin U}{U}\right) \ll 1$	$G \approx H \approx 2\pi(P_0 f_b)^2 J$, J defined by (3.1)	"2nd Order Modulation" approx., J tabulated in Table 3.2, 10 log$_{10}$ J plotted in Fig. 5.1
3	$P_0 f_b U^2/3 \ll 1$, $U \ll 1$	$G \approx H \approx 2\pi(P_0 f_b)^2 \frac{U^4}{240} \cdot [12 - 30a + 20a^2 - a^5]$	Special case of Case 2
4	$2P_0 f_b \ll 1$, $U \gg 1$	$G \approx H \approx 2\pi(P_0 f_b)^2 \cdot \left(1 - \frac{a}{2}\right)\left(1 + \frac{\cos aU}{2}\right)$	Special case of Case 2 G is a rapidly oscillating function of a
5	$P_0 f_b U^2/3 \gg 1$, $U \ll 1$	$G \gg H$, $G \approx (10\pi/P_0 f_b U^2)^{1/2} \exp(-10a^2/4P_0 f_b U^2)$	When $U \ll 1$ Case 3 applies when $P_0 f_b$ is small Case 5 applies when $P_0 f_b$ is large
6	$U \gg 1$	$G \approx e^{-b_0}[I(b_0, a) + 2I(b_1, a) \cos aU]$, $H \approx e^{-b_0}[I(-b_0, a) + 2I(-b_1, a) \cos aU]$, $b_0 = 2P_0 f_b$, $b_1 = -P_0 f_b$ $I(b, a)$ studied in Appendix III	When $P_0 f_b \ll 1$, Case 6 reduces to Case 4. When $P_0 f_b \gg 1$, $G \gg H$. As a increases, G oscillates between G^+ and G^- given in Table 3.3. See Table 5.1

$$J = \left(1 - \frac{a}{2}\right)\left(1 + \frac{\cos aU}{2}\right) + \frac{\sin (2U - aU)}{4U} - \frac{2}{U} \cos\left(\frac{aU}{2}\right) \sin\left(U - \frac{aU}{2}\right) \tag{3.1}$$

Values of J are tabulated in Table 3.2 for various values of a and k where $U = k\pi/4$ (or $T = k/(8f_b)$). J is zero when a exceeds 2.

Cases 3 and 4 in Table 3.1 follow directly from Case 2. In order to

TABLE 3.2 — VALUES OF J

k	$a = 0$	0.25	0.50	0.75	1.00	1.25
0	0	0	0	0	0	0
1	0.018	0.009	0.003	0.001	0.002	0.004
2	0.227	0.115	0.041	0.011	0.023	0.057
3	0.794	0.434	0.161	0.049	0.098	0.231
4	1.500	0.903	0.352	0.123	0.250	0.524
5	1.924	1.284	0.527	0.224	0.458	0.816
6	1.924	1.385	0.585	0.332	0.659	0.936
7	1.712	1.231	0.505	0.427	0.773	0.800
8	1.500	0.994	0.375	0.506	0.750	0.494
9	1.335	0.800	0.325	0.579	0.602	0.228
10	1.245	0.658	0.425	0.657	0.405	0.180
11	1.307	0.544	0.643	0.725	0.262	0.358
12	1.500	0.472	0.883	0.752	0.250	0.601

obtain Case 5 we note that when $U \ll 1$, expression (2.7) gives

$$R_v(\tau) \approx -U^2 P_0 f_b \frac{d^2}{du^2} \frac{\sin u}{u}. \tag{3.2}$$

The corresponding integrals for G and H could, if required, be investigated by the methods used to study Lewin's integral in Appendix III. However here we consider only the case where $U^2 P_0 f_b$ is so large that (1) $\exp R_v(\tau)$ is the dominant term in the integral (2.4) for G, and (2) most of the contribution to the value of the integral comes from the region around $u = 0$. The results for Case 5 then follow from (2.4) and the fact that (3.2) becomes

$$R_v(\tau) \approx U^2 P_0 f_b \left(\frac{1}{3} - \frac{u^2}{10}\right)$$

When $U \gg 1$, expression (2.7) shows that $R_v(\tau)$ is small except when u is near 0 or near U:

$$\begin{aligned} R_v(\tau) &\approx 2\, P_0 f_b\, u^{-1} \sin u, && u \text{ near } 0 \\ R_v(\tau) &\approx -P_0 f_b (u - U)^{-1} \sin (u - U), && u \text{ near } U \end{aligned} \tag{3.3}$$

These approximations, expression (2.4) for G, and the definition (A3–1) of $I(b, a)$ give the results stated in Case 6. Just as in Case 4, G is a rapidly oscillating function of a when U is large. It oscillates between G^+ and G^- where

$$G^{\pm} = e^{-b_0}[I(b_0, a) \pm 2I(b_1, a)] \tag{3.4}$$

Values of G^+ and G^- are given in Table 3.3 for various values of $a = f/f_b$ and $P_0 f_b$ [in (radians)^2].

TABLE 3.3 — VALUES OF G^+ AND G^-. THE UPPER NUMBER OF AN ENTRY IS G^+ AND THE LOWER NUMBER IS G^-.

$P_0 f_b$ (radians)^2	$a = 0$	0.25	0.50	0.75	1.00	1.25
0.25	0.384 0.160	0.338 0.144	0.292 0.126	0.245 0.107	0.197 0.087	0.148 0.066
0.50	1.018 0.504	0.906 0.464	0.789 0.415	0.665 0.357	0.537 0.291	0.406 0.222
0.75	1.55 0.89	1.39 0.83	1.22 0.75	1.04 0.65	0.845 0.533	0.645 0.411
1.00	1.90 1.22	1.73 1.15	1.53 1.05	1.32 0.92	1.07 0.76	0.830 0.596
2.00	2.16 1.86	2.04 1.80	1.88 1.68	1.68 1.52	1.43 1.31	1.17 1.07
4.00	1.59 1.57	1.56 1.54	1.50 1.48	1.40 1.40	1.28 1.28	1.14 1.14

4. APPROXIMATIONS FOR G AND H — FREQUENCY MODULATION

The various cases which we shall consider for FM are roughly similar to those considered in Section 3 for PM, and are listed in Table 4.1. The power spectrum $w_\varphi(f)$ is assumed to be that given by (2.2). As pointed out in Section 2, the average FM signal power in a narrow channel of width Δf centered on frequency f is assumed to be $P_0 \Delta f$ (radians/second)^2 if $0 < f < f_b$ and zero if $f_b < f$. The average interchannel interference power is $(2\pi f)^2\, w_c(f) \Delta f$ (radians/second)^2. When $f > f_b$ this gives all of the power present in the frequency interval Δf.

Case 1, Table 4.1, gives the exact expressions for G and H, and Case 2 corresponds to the "second order modulation" approximation which holds when $R_v(0) \ll 1$, $R_v(0)$ being computed from (2.8). K is a function of $a = f/f_b$ and $U = 2\pi f_b T$ which may be obtained by writing the integral (2.9) as

$$G \approx \pi f_b \int_{-\infty}^{\infty} R_v^2(\tau) e^{2\pi i f \tau}\, d\tau = 4^{-1} \pi f_b \int_{-\infty}^{\infty} w_v(x) w_v(f - x)\, dx \tag{4.1}$$

where $R_v(\tau)$ and $w_v(f)$ are taken to be even functions. From the definitions (1.12) and (2.2) for $w_v(f)$ and $w_\varphi(f)$ we have

$$w_v(f) = \begin{cases} P_0(\pi f)^{-2} \sin^2 \pi f T, & |f| < f_b \\ 0, & |f| > f_b \end{cases} \tag{4.2}$$

TABLE 4.1 — FREQUENCY MODULATION

Case No.	Restrictions on Parameters U and A	G and H	Notes
1	No restriction, $w_\varphi(f)$ defined by (2.2)	G defined by (2.4) H defined by (2.5) $R_v(\tau)$ defined by (2.8)	$A = (\sigma/f_b)^2$, σ = rms $[d\varphi(t)/dt]/2\pi$, σ = rms frequency deviation of signal measured in cycles/sec. $P_I/P_S = [r^2a^2/(2\pi A)] \cdot [G - H \cos 2pT]$
2	$2A[USi(U) - 1 + \cos U] \ll 1$	$G \approx H \approx 2\pi A^2a^{-2}UK$ K defined by (4.5) and (4.6)	"2nd Order Modulation" approx. K tabulated in Table 4.2. See Fig. 5.2
3	$AU^2 \ll 1$, $U \ll 1$.	$G \approx H \approx \pi A^2U^4(2 - a)/4$, $0 \leqq a \leqq 2$	Special case of Case 2 $UK \approx a^2U^4(2 - a)/8$
4	$AU\pi \ll 1$, $U \gg 1$	$G \approx H \approx \begin{cases} 2\pi^2A^2Ua^{-2}, & 0 < a < 1 \\ \pi^2A^2U, & a = 1 \\ 0, & a > 1 \end{cases}$	Special case of Case 2
5	$U \ll 1$, Lewin's case	$G \approx e^{-b}I(b, a)$ $H \approx e^{-b}I(-b, a)$ $b = AU^2$	Case 5 agrees with Case 3 when $b \ll 1$. $G \gg H$ for $b \gg 1$. $I(b, a)$ defined and tabulated in Appendix III. See Fig. 5.4.
6	$4A \gg 1$, $U \gg 1$	$G \approx [\pi y_1/(A \sinh y_1)]^{1/2} \exp[-2A(\cosh y_1 - 1)]$ y_1 defined by $\frac{a}{2A} = \int_0^{y_1} \frac{\sinh v}{v} dv$	$G \gg H$ Value of G is independent of U. See Fig. 5.3.
7	$B \gg 1$, $U \approx 1$, $B = A[1 - U^{-1}\sin U]$	$G \approx [\pi/B]^{1/2} \exp[-a^2/(4B)]$	$G \gg H$

where the lower limit f_a of the frequency band is taken to be very close to zero. When $f > 2f_b$ the value of (4.1) is zero. When we take $0 < f < 2f_b$ the limits of integration in (4.1) are $x = f - f_b$ and $x = f_b$. Changing the variable of integration in (4.1) from x to $y = 2\pi xT$ converts (4.1) into

$$4\pi A^2U^3 \int_{\alpha-U}^{U} y^{-2}(\alpha - y)^{-2} \sin^2 \frac{y}{2} \sin^2\left(\frac{\alpha - y}{2}\right) dy \tag{4.3}$$

where $\alpha = 2\pi fT = aU$. By partial fractions

$$\alpha^2y^{-2}(\alpha - y)^{-2} = y^{-2} + 2\alpha^{-1}y^{-1} + (\alpha - y)^{-2} + 2\alpha^{-1}(\alpha - y)^{-1} \tag{4.4}$$

Considerations of symmetry show that the $\alpha - y$ terms on the right contribute the same amount to (4.3) as do the y terms. When the y^{-2} term is converted into y^{-1} by an integration by parts, (4.3) may be expressed in terms of $Si(x)$ and $Ci(x)$ functions. In this way it may be shown that the approximation (4.3) for G and H has the value

$$2\pi A^2U^3K/\alpha^2 = 2\pi A^2U K/a^2$$

where

$$\begin{aligned} K = {} & (-U^{-1} + \beta^{-1})(1 - \cos U)(1 - \cos \beta) \\ & + (Si\, U - Si\, \beta)(1 + \cos \alpha - 2\alpha^{-1} \sin \alpha) \\ & + (Si\, 2U - Si\, 2\beta)(-\cos \alpha + \alpha^{-1} \sin \alpha) \\ & + (Ci\, 2U - Ci\, 2|\beta| - Ci\, U + Ci\, |\beta|)(\sin \alpha + \alpha^{-1} \cos \alpha) \\ & + \alpha^{-1}(2 + \cos \alpha)[\log_e (U/|\beta|) - Ci\, U + Ci\, |\beta|] \end{aligned} \tag{4.5}$$

Here $\alpha = aU$ and $\beta = \alpha - U = (a - 1)U$. When $f = 2f_b$, a has the value 2 and K is zero, as it should be. When $f = f_b$, i.e., when $a = 1$ and $\alpha = U$,

$$\begin{aligned} K = {} & (1 + \cos U - 2U^{-1} \sin U)\, Si\, U \\ & + (-\cos U + U^{-1} \sin U)\, Si\, 2U \\ & + (Ci\, 2U - Ci\, U - \log_e 2)(\sin U + U^{-1} \cos U) \\ & + U^{-1}(2 + \cos U)(\log_e U + .577.. - Ci\, U) \end{aligned} \tag{4.6}$$

Values of K are tabulated in Table 4.2 for various frequencies and delays ($a = f/f_b$ and $k = 4U/\pi = 8f_bT$). Cases 3 and 4 in Table 4.1 show that when U is very small $K \approx a^2U^3(2 - a)/8$ and when U is very large $K \approx \pi$, $\pi/2$, or 0 according to whether $a < 1$, $a = 1$, or $a > 1$.

Case 3 in Table 4.1 is a special case of Case 2 which may be obtained

TABLE 4.2 — VALUES OF K

k	$a = 0$	0.25	0.50	0.75	1.00	1.25
0	0	0.0	0.0	0.0	0.0	0.0
1	0	0.006	0.022	0.041	0.058	0.068
2	0	0.048	0.164	0.305	0.422	0.473
3	0	0.141	0.490	0.890	1.20	1.26
4	0	0.286	0.961	1.74	2.19	1.45
5	0	0.453	1.57	2.64	3.05	2.62
6	0	0.673	2.16	3.36	3.41	2.45
7	0	0.897	2.69	3.68	3.13	1.73
8	0	1.14	3.13	3.65	2.40	0.883
9	0	1.39	3.45	3.34	1.62	0.370
10	0	1.66	3.68	2.95	1.20	0.378
11	0	1.93	3.79	2.64	1.32	0.741
12	0	2.20	3.79	2.51	1.89	1.10

by letting U become very small in (4.3). The value of P_I/P_S corresponding to Case 3 has been given by Albersheim and Schafer.[3] Case 4 may be obtained by letting U become large in (4.5) and (4.6).

When U is very small, expression (2.8) for $R_v(\tau)$ becomes

$$R_v(\tau) \approx AU^2u^{-1} \sin u \tag{4.7}$$

Assuming $AU^2 \ll 1$ and substituting (4.7) in the "second order modulation" approximation (2.9) for G and H gives us another derivation of Case 3 (see (A3–2)). However, if the rms frequency deviation of the signal is so large that AU^2 is not small, even though $U \ll 1$, we have Case 5, the case investigated by Lewin.[2] The formulas given in Table 4.1 are obtained when (4.7) is set in the integrals (2.4) and (2.5) for G and H, and the results compared with the definition (A3–1) of $I(b, a)$.

When U is very large, expression (2.8) for $R_v(\tau)$ becomes

$$R_v(\tau) \approx \begin{cases} A[\pi U - 2 \cos u - 2uSi(u)], & 0 \le u < U \\ 0, & U < u \end{cases} \tag{4.8}$$

Substituting (4.8) in (2.9) and integrating by parts twice leads to another derivation of Case 4. The expression for G given in Case 6 is obtained from (4.8) and (2.4) by the method outlined in Appendix II. $A\pi U$ is assumed to be so large that most of the contribution to the value of the integral (2.4) for G comes from the region around $u = 0$. It is also assumed that $R_v(\tau) + 1$ is negligible in comparison with exp $R_v(\tau)$ in this region. This leads to the approximation

$$G \approx 2 \int_0^\infty e^{2A[1-\cos u - uSi(u)]} \cos au \, du \tag{4.9}$$

which holds when $U \gg 1$ and $A\pi U \gg 1$.

The expression for G in Case 6 is merely the leading term in the asymptotic expansion arising from the saddle point at $u = iy_1$. When further terms in this expansion are obtained (using, for example, equation (10.4) of Reference 7) it is found that the expression for Case 6 should be multiplied by

$$1 + \frac{[4cs - 5y_1 + s^2(y_1^{-1} - 2y_1)]}{48As^3} + \cdots. \tag{4.10}$$

where c and s denote cosh y_1 and sinh y_1, respectively. The next term consists of $1/A^2$ times a function of y_1, and so on. When y_1 becomes small, as it does when A becomes large or a becomes small, (4.10) becomes

$$1 + \frac{1}{48A} + \frac{103}{13824A^2} + \cdots.$$

However, comparison of Tables 4.3 and 4.4 shows that the formula of Case 6 gives fairly reliable values of G when A is as small as 0.5 (U must be large, of course).

When $(1 - a)$ is small and $A \ll 1$, but U still large enough to make $A\pi U \gg 1$, (4.9) gives

$$G \approx Aa^{-2}\left[\pi + 2 \text{ arc tan } \frac{(1-a)}{A\pi}\right] + 0(A^2/a^2) \tag{4.11}$$

This may be obtained by letting A become small in

$$G \approx 4(A/a)^2 \int_0^\infty e^{2Ay} Si(u) F \, du \tag{4.12}$$

$$y = 1 - \cos u - uSi(u)$$

$$F = -2 \cos auSi(u) + Si[(1 - a)u] + Si[(1 + a)u]$$

which may be obtained from (4.9) by integrating by parts twice.

The formula for G given in Case 7, Table 4.1, is obtained when U is taken to be of order unity and A is assumed to be so large that only the exponential term in the integrand of (2.4) is of importance. Most of the contribution comes from around $u = 0$ where

$$R_v(\tau) = R_v(0) - u^2A(1 - U^{-1} \sin U) + \cdots.$$

When $U \gg 1$, and $A \gg 1$, Cases 6 and 7 both give

$$G \approx (\pi/A)^{1/2} \exp[-a^2/(4A)]$$

which leads to an expression for P_I/P_S similar to one given by Alber-

TABLE 4.3 — VALUES OF G AND H OBTAINED BY NUMERICAL INTEGRATION

	$a = 0$	0.25	0.50	0.75	1.00	1.25
G for $U = 3$						
$A = 0.125$	0.690	0.651	0.599	0.466	0.326	0.224
0.25	1.53	1.46	1.29	1.06	0.808	0.561
0.50	2.17	2.09	1.97	1.63	1.38	0.993
H for $U = 3$						
$A = 0.125$	0.424	0.393	0.335	0.260	0.182	0.111
0.25	0.574	0.528	0.447	0.339	0.233	0.136
0.50	0.283	0.255	0.211	0.156	0.104	0.039
G for $U = 6$						
$A = 0.125$	2.62	2.42	1.85	1.18	0.631	0.257
0.25	3.26	3.05	2.50	1.81	1.17	0.695
0.50	2.55	2.46	2.21	1.86	1.47	1.10
H for $U = 6$						
$A = 0.125$	0.802	0.705	0.465	0.227	0.067	0.0031
0.25	0.266	0.231	0.148	0.064	0.0126	0.0035
0.50	0.0092	0.0079	0.0048	0.0018	0.00017	0.00019
	$A = 0.0625$	0.125	0.25	0.50		
G for $a = 1$						
$U = 1.5$	0.0126	0.045	0.155	0.446		
3	0.111	0.326	0.808	1.38		
6	0.245	0.631	1.17	1.47		
12	0.300	0.657	1.16	1.47		

sheim and Schafer[3] for long delay. When $U \ll 1$ and $AU^2 \gg 1$, Cases 5 and 7 both give

$$G \approx (6\pi A^{-1} U^{-2})^{1/2} \exp [-3a^2 A^{-1} U^{-2}/2]$$

Before the approximations listed in Table 4.1 were developed, a number of values of G and H were obtained from (2.4) and (2.5) by numerical integration. These values, given in Table 4.3, are the best we have and may be used to check the various approximations.

As an example of the values given by our approximations we take the case $U = 6$. In Table 4.4, "G – Case 6" has been computed from the formula given in Case 6, Table 4.1 (which assumes $U \to \infty$). When these values are compared with the corresponding ones in Table 4.3, it is seen that the agreement is not good for $A = 0.125$. Better agreement is shown by "G — Improved Case 6" in which the values are computed from (2.4) and (4.8) by the method of Appendix II. It differs from "Case 6" in that $F(u)$ of (A2–1) is $\exp [R_v(\tau)] - R_v(\tau) - 1$ instead of merely $\exp [R_v(\tau)]$.

TABLE 4.4 — APPROXIMATE VALUES OF G FOR $U = 6$

	$a = 0$	0.25	0.50	0.75	1.00	1.25
G—Case 6						
$A = 0.125$	5.01	4.13	2.54	1.32	0.63	0.28
0.25	3.54	3.26	2.58	1.80	1.14	0.66
0.50	2.51	2.42	2.17	1.82	1.43	1.06
G—Improved Case 6						
$A = 0.125$	2.62	2.37	1.74	1.06	0.56	0.26
0.25	3.17	2.94	2.39	1.72	1.10	0.66
0.50	2.50	2.41	2.17	1.82	1.42	1.06

5. INTERCHANNEL INTERFERENCE POWER

The values of P_I/P_S, the ratio of the interchannel interference power to the signal power, may be computed from the formulas (2.10) and (2.11) when G and H are known. One would like to have curves giving P_I/P_S for representative combinations of echo delay, signal power, and channel position which are likely to occur in practice. However, the large number of such combinations coupled with the difficulty of computing G and H leads us to restrict ourselves mostly to curves for Cases 2 and 6 in Tables 3.1 and 4.1. In all cases the signal power P_S (per cps) is taken to be equal to the constant value P_0 (measured in (radians)^2/cps for PM and in (radians/sec)^2/cps for FM) over the signal band $(0, f_b)$, and is zero outside this band.

Case 2 is the "second order modulation" approximation which, roughly speaking, applies when the echo delay is very short or when the rms deviation of the phase angle (for PM) or of the frequency (for FM) is small. Case 6 applies when the echo delay is very long.

(a) *"Second order modulation" approximation for PM — Table 3.1, Case 2.* Since $G \approx H$, equation (2.10) may be written as

$$10 \log_{10} (P_I/P_S) \approx \rho + D_1 + D_2 + D_3$$

$$\rho = 10 \log_{10} r^2$$

$$D_1 = 10 \log_{10} (1 - \cos 2pT) \qquad (5.1)$$

$$D_2 = 10 \log_{10} (P_0 f_b)$$

$$D_3 = 10 \log_{10} J$$

where ρ is the reflection coefficient expressed in decibels, D_1 is a quantity which varies rapidly with T and whose representative value is zero. J is the function of a and U defined by equation (3.1). The quantities P_0,

f_b, a, U are defined by equation (2.1) and Case 1 of Table 3.1. $P_0 f_b$ is the average signal power in (radians)^2, $a = f/f_b$ gives the channel position and $U = 2\pi f_b T$ measures the echo delay.

The approximation (5.1) holds when $2P_0 f_b(1 - U^{-1} \sin U)$ is small in comparison with unity.

Fig. 5.1 shows D_3 plotted as a function of $f_b T$ for various values of a. The values of J which were used were taken from Table 3.2. Values of J for $U \ll 1$ may be obtained from the expression given for G in Case 3, Table 3.1. Case 4 gives another special case.

(b) *Large delay, $U \gg 1$ for PM — Table 3.1, Case 6.* When U is very large, G is a rapidly oscillating function of a. When in addition $P_0 f_b$ is small, Case 4 shows that J fluctuates between $(1 - a/2)/2$ and $3(1 - a/2)/2$. The corresponding fluctuations in D_3 are noticeable in Fig. 5.1 for the larger values of $f_b T$. If $2P_0 f_b$ is large compared to unity (and the delay is large), equation (5.1) no longer holds. In this case $G \gg H$ and we may write (2.10) as

$$10 \log_{10} (P_I/P_S) \approx \rho + D_4$$

$$\rho = 10 \log_{10} r^2$$

$$D_4 = 10 \log_{10} (2\pi P_0 f_b)^{-1} G \qquad (5.2)$$

$$G \approx e^{-b_0}[I(b_0, a) + 2I(b_1, a) \cos aU]$$

where b_0 and b_1 are defined in Case 6, Table 3.1. It is seen that as a increases from 0 to 1, D_4 oscillates rapidly between limits D_4^+ and D_4^- corresponding to G^+ and G^- which are defined by equation (3.4) and tabulated in Table 3.3. Table 5.1 gives values of D_4^+ and D_4^- computed from Table 3.3.

The entries corresponding to the values 0.25 and 0.50 for $P_0 f_b$ must be used with caution in equation (5.2) since they do not satisfy $2P_0 f_b \gg 1$ and H is not negligible in comparison with G.

(c) *"Second order modulation" approximation for FM — Table 4.1, Case 2.* By making use of the expressions for G and H given in Case 2 we may write equation (2.11) as

$$10 \log_{10}(P_I/P_S) \approx \rho + D_1 + D_2' + D_3'$$

$$D_2' = 10 \log_{10} A \qquad (5.3)$$

$$D_3' = 10 \log_{10} UK$$

where ρ and D_1 are defined by (5.1) and A by (2.8). K is the function of a and U defined by (4.5). A is proportional to the signal power: $A =$

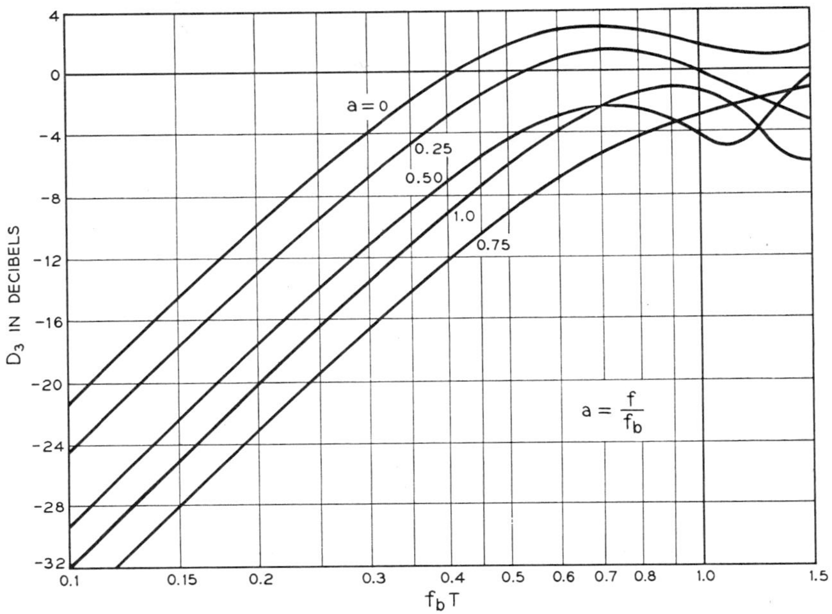

Fig. 5.1 — For "Second Order Modulation" the ratio P_I/P_S for PM depends upon D_3 as shown by equation (5.1).

TABLE 5.1 — VALUES OF D_4^+ AND D_4^-. THE UPPER NUMBER OF AN ENTRY IS D_4^+ AND THE LOWER ONE IS D_4^-

$P_0 f_b$ (radians)^2	$a = 0$	0.25	0.50	0.75	1.00—
0.25	−6.12 −9.92	−6.68 −10.37	−7.31 −10.96	−8.07 −11.67	−9.01 −12.56
0.50	−4.89 −7.94	−5.40 −8.31	−6.00 −8.79	−6.75 −9.45	−7.67 −10.34
0.75	−4.83 −7.24	−5.31 −7.56	−5.87 −8.01	−6.57 −8.63	−7.47 −9.46
1.00	−5.20 −7.12	−5.61 −7.38	−6.13 −7.77	−6.78 −8.33	−7.67 −9.17
2.00	−7.65 −8.30	−7.89 −8.44	−8.25 −8.74	−8.74 −9.17	−9.44 −9.82
4.00	−11.99 −12.05	−12.08 −12.13	−12.25 −12.31	−12.55 −12.55	−12.94 −12.94

$(\sigma/f_b)^2$ where σ is the rms frequency deviation of the signal in cycles per second. D_2' and D_2 play similar roles in (5.3) and (5.1).

The approximation (5.3) holds when $2A[USi(U) - 1 + \cos U]$ is small in comparison with unity.

The values of K given in Table 4.2 lead to the curves for D_3' shown in Fig. 5.2. When $U \ll 1$, Case 3 shows that $D_3' \approx 10 \log_{10} [a^2U^4(2 - a)/8]$, and Case 4 shows that when $U \gg 1$ (provided $a < 1$ and $A\pi U \ll 1$) $D_3' \approx 10 \log_{10} (\pi U) = 12.95 + 10 \log_{10} f_bT$.

(*d*) *Large delay, $U \gg 1$ for FM — Table 4.1, Case 6*. It has just been pointed out that when U becomes very large, P_I/P_S depends upon the delay only through the term $D_3' \approx 10 \log_{10} \pi U$ (neglecting the rapidly varying term D_1) if $AU\pi \ll 1$. If $AU\pi \gg 1$, P_I/P_S becomes independent of U as $U \to \infty$. This follows from the fact that the formulas of Case 6 allow us to write (2.11) as

$$10 \log_{10} (P_I/P_S) = \rho + D_4'$$
$$D_4' \approx 10 \log_{10} [Ga^2(2\pi A)^{-1}] \tag{5.4}$$

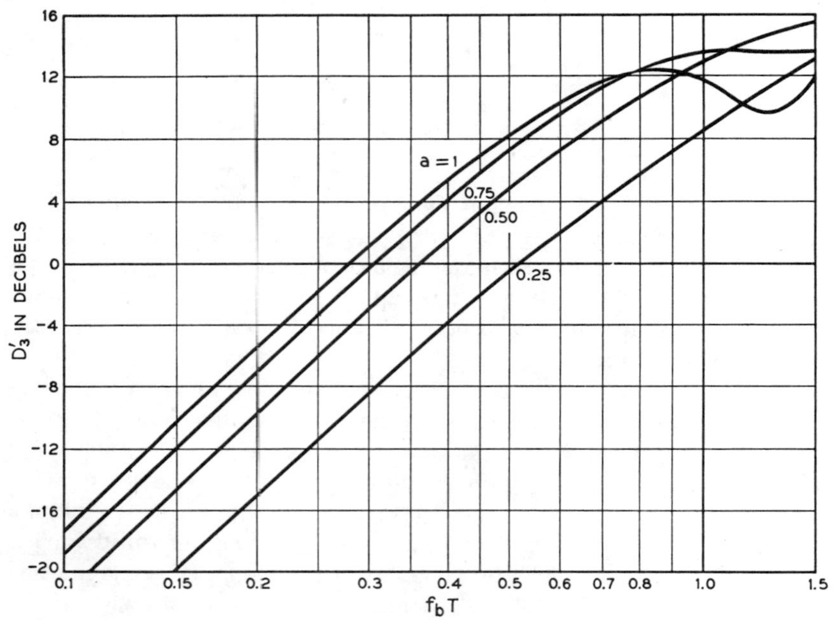

Fig. 5.2 — For "Second Order Modulation" the ratio P_I/P_S for FM depends on D_3' as shown by equation (5.3).

where ρ is defined by (5.1) and G depends only on a and A through

$$G \approx [\pi y_1/A \sinh y_1]^{1/2} \exp[-2A (\cosh y_1 - 1)]$$
$$\frac{a}{2A} = \int_0^{y_1} \frac{\sinh v}{v} dv \tag{5.5}$$

Fig. 5.3 shows D_4' plotted as a function of A for various values of a. It is assumed that $A\pi U \gg 1$.

(*e*) *Small delay and large rms frequency deviation for FM — Table 4.1, Cases 5 and 7*. It turns out that Case 5 (Lewin's, case, $U \ll 1$) and Case 7 ($A \gg 1$ and U of order unity) may be combined into a single case by taking the quantity b in the formulas of Case 5 to be $6A(1 - U^{-1} \sin U)$ instead of AU^2. When $U \ll 1$, Case 5 is obtained. When $A \gg 1$ the asymptotic expansion for $I(b, a)$ leads to Case 7 if U is 0(1).

In order to put this combined case in a form suited to calculation we write (2.11) as

$$\frac{P_I}{P_S} = \frac{r^2a^2G}{2\pi A}\left(1 - \frac{H}{G} \cos 2pT\right)$$
$$\approx \frac{r^2 6(1 - U^{-1} \sin U)a^2e^{-b}I(b, a)}{2\pi b}\left(1 - \frac{H}{G} \cos 2pT\right) \tag{5.6}$$
$$10 \log_{10} (P_I/P_S) \approx \rho + D_1' + D_5' + D_6'$$

where ρ is given by (5.1) and

$$D_1' = 10 \log_{10} (1 - (H/G) \cos 2pT)$$
$$D_5' = 10 \log_{10} (1 - U^{-1} \sin U)$$
$$D_6' = 10 \log_{10} 6a^2 e^{-b} I(b, a)/(2\pi b) \tag{5.7}$$
$$b = 6A(1 - U^{-1} \sin U)$$

Fig. 5.4 shows values of D_6', computed from the values of $I(b, a)$ given in Appendix III, plotted as a function of b for various values of a. The maximum value of 3 db for D_1' occurs when $AU^2 \ll 1$ and $\cos 2pT = -1$. When AU^2 is large D_1' is approximately zero.

Fig. 5.5 and 5.6 show, in a rough way, the regions in which the various approximations apply. For PM, the delay and the rms phase deviation (measured by $U = 2\pi f_bT$ and $(P_0f_b)^{1/2}$, respectively) are the parameters which determine the type of approximation to be used. The regions in the $[(P_0f_b),^{1/2} U]$ plane shown in Fig. 5.5 are marked with the numbers 2a, 3, 4, 5, 6b where the integer indicates the case number in Table 3.1 and the letters a and b refer to Cases (a) and (b) in this section. Fig. 5.6 is

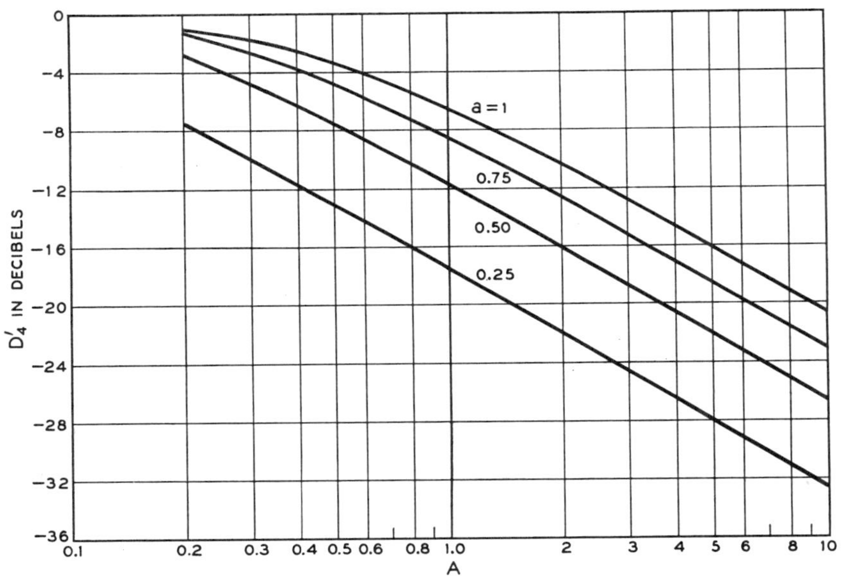

Fig. 5.3 — For long delayed echoes the ratio P_I/P_S for FM depends upon D_4' as shown by equation (5.4).

the corresponding figure for FM. The coordinates are U and $A^{1/2}$, where $A^{1/2}(=\sigma/f_b)$ measures the rms frequency deviation. The region numbers are 2c, 3, 4, 5, 6d, 7e where the integers indicate the case number in Table 4.1. It will be noted that there are regions where no approximation is available. However, an answer may always be obtained by numerical integration of equations (2.4) and (2.5) for G and H.

Fig. 5.7 shows values of P_I/P_S for the top channel ($a = 1$) where the interference is often at a maximum in an FM system. The coordinates $(A^{1/2}, f_bT)$ are essentially the same as those of Fig. 5.6. In order to simplify the plotting, the phase angle $2pT$ is assumed to be such that $\cos 2pT$ is zero so that the contours are given by

$$\text{Constant} = 10 \log_{10}\left(\frac{P_I}{r^2 P_S}\right) = 10 \log_{10}\left(\frac{G}{2\pi A}\right)$$

The contours have been obtained in part from the various approximations where applicable and in part from values obtained by numerical computation from the exact expression. While there are, of necessity, some areas of uncertainty in Fig. 5.7, it should be adequate for most engineering purposes. No corresponding curves have been computed for the case of phase modulation.

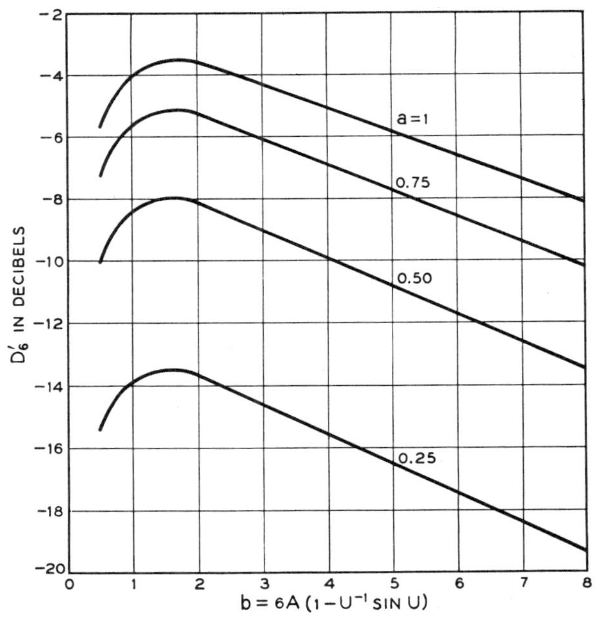

Fig. 5.4 — Under the condition of large rms frequency deviation and small delay the ratio P_I/P_S for FM depends upon D_6' as shown by equation (5.6).

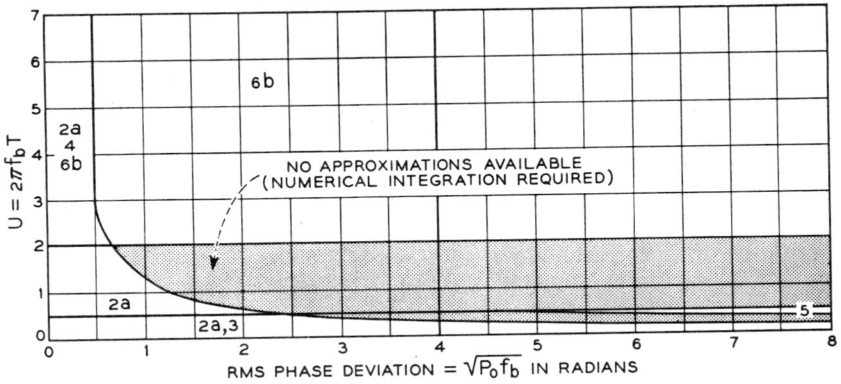

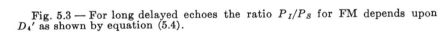

Fig. 5.5 — Regions of validity for the various approximations for PM. The integers refer to case numbers in Table 3.1 and the letters to cases discussed in Section 5. This figure and Fig. 5.6 are intended to give only an idea of the relative positions of the regions.

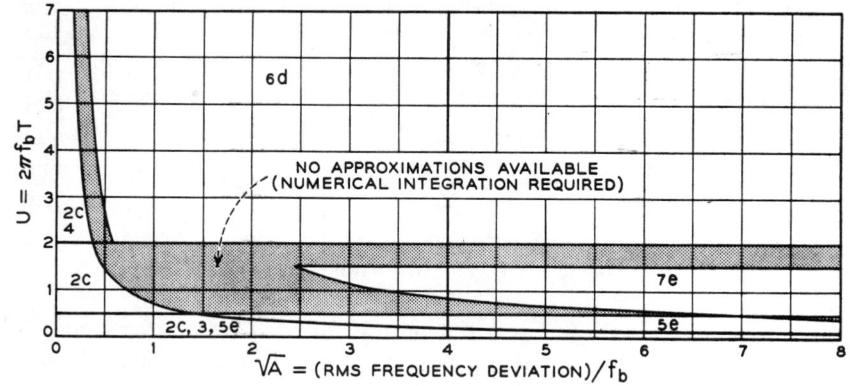

Fig. 5.6 — Regions of validity for the various approximations for FM. The integers refer to case numbers in Table 4.1 and the letters to cases discussed in Section 5.

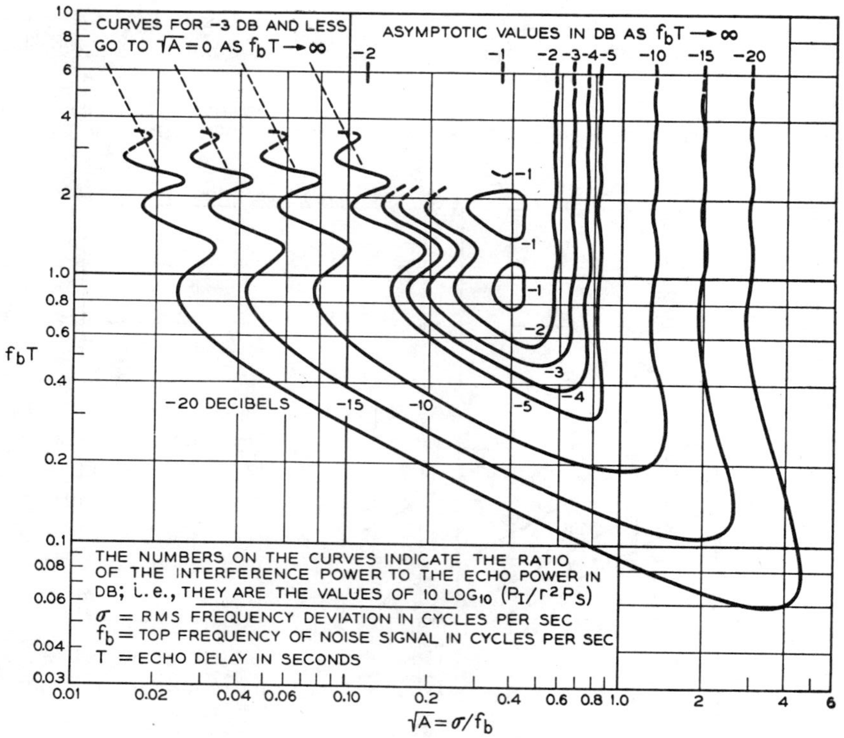

Fig. 5.7 — Contours of constant interference in the top channel of a multi-channel FM system.

It should be noticed that Fig. 5.7 is plotted for the case $a = f/f_b = 1$. If Fig. 5.7 were plotted for values of a slightly less than unity there would not be much change except in the upper left hand corner (A small and f_bT large), where the interference would tend to be 3 db stronger. This discontinuous behavior as a passes through unity is shown by Case 4, Table 4.1 where $U \gg 1$ and $A\pi U \ll 1$. When $U \gg 1$ and $A\pi U \gg 1$, as occurs when $f_bT \to \infty$ (with A held fixed), equation (4.11) gives

$$10 \log_{10} \frac{P_I}{r^2 P_S} \approx 10 \log_{10} \left[\frac{1}{2} + \frac{1}{\pi} \text{arc tan} \left(\frac{1-a}{A\pi}\right)\right]$$

when $A \ll 1$. This shows that at $f_bT = \infty$, the discontinuity arises for values of A near $(1 - a)/\pi$. When $A \gg 1$ and $U \gg 1$, we have

$$10 \log_{10} \frac{P_I}{r^2 P_S} \approx -15 \log_{10} A - 10 \log_{10} \frac{2\pi^{1/2}}{a^2} - \frac{4.34a^2}{4A}$$

which changes only slowly as $a \to 1$.

6. USE OF EQUIVALENT ECHO TO ESTIMATE INTERCHANNEL INTERFERENCE

When a steady sinusoid $\exp(i\omega t)$ is applied to a transmission medium of the sort we have under consideration, the output is $\exp(i\omega t - \alpha - i\beta)$ where α and β are the attenuation and phase shift, respectively. Distortionless transmission occurs when α is constant and β has a constant slope over the essential range of frequencies. Departures from these ideal conditions cause interchannel interference in multichannel FM and PM systems. Our evaluation of the interference caused by a small echo may equally well be regarded as an evaluation of interference for a particular kind of amplitude and phase distortion, namely that given by

$$\alpha \approx -r \cos \omega T, \qquad \beta \approx r \sin \omega T. \tag{6.1}$$

These expressions are obtained by writing $\exp(i\omega t) + r \exp[i\omega(t - T)]$ in the form $\exp(i\omega t - \alpha - i\beta)$ when $|r| \ll 1$.

The analysis given by E. D. Sunde in Section 1 of Reference 8 shows that a minimum phase system in which $\beta = r \sin \omega T$ also has $\alpha = -r \cos \omega T$ as in (6.1). This suggests a procedure for calculating interchannel interference from phase data alone when (1) the distortion is known to be of the minimum phase type and (2) the variation of phase with frequency can be approximated by a sine function. In such cases we can apply our echo analysis directly by identifying r as the amplitude of the phase oscillation and T as the reciprocal of the period.

In carrier multiplex systems the sinusoidal approximation need hold only in the region around the carrier frequency f_0 where $2\pi f_0 = p$ and p is the radian frequency appearing in equation (1.1). For FM we shall, in this section, arbitrarily take the region to extend from $f_0 - 4\sigma$ to $f_0 + 4\sigma$ where σ is the rms frequency deviation of the signal. For PM and the signal power spectrum given by equation (2.1) we may take the region to be $f_0 \pm 4f_b(P_0f_b/3)^{1/2}$.

A special case occurs when the nonlinear portion of β may be represented as $a_2(f - f_0)^2/2$ in the region of interest. We can think of $r \sin \omega T$ as going through several oscillations between $f = 0$ and $f = f_0$, and that a maximum, if $a_2 < 0$, (or a minimum, if $a_2 > 0$) of $r \sin \omega T$ occurs at $f = f_0$, i.e. at $\omega = p$. The band of interest is taken to be narrow enough to lie in the immediate vicinity of the maximum. This sort of curve fitting is permissible since equation (2.3) shows that the interchannel interference depends on the carrier frequency only through the term $\cos 2pT$. Furthermore, constant terms and terms linear in frequency in the expression for β do not affect the amount of interchannel interference.

At the maximum mentioned in the preceding paragraph $\omega = p$, $\sin pT = 1$ and $\cos 2pT = -1$. Near this maximum $r \sin \omega T$ is

$$r \cos 2\pi(f - f_0)T \approx r - r4\pi^2(f - f_0)^2T^2/2 \qquad (6.2)$$

In order that this approximation may hold over the region $f_0 \pm 4\sigma$ (for FM), we require

$$2\pi(4\sigma)T \leqq 1$$

We take T to be as large as possible, namely

$$T = 1/8\pi\sigma \qquad (6.3)$$

in order to make r as small as possible, since our work assumes $r \ll 1$. Comparison of $a_2(f - f_0)^2/2$ with (6.2) gives

$$r = -a_2(2\pi T)^{-2} = -16a_2\sigma^2 \qquad (6.4)$$

which must be small compared to unity if our results are to be used. This result holds for $a_2 < 0$. When $a_2 > 0$ expression (6.3) still holds for T but now $r = 16a_2\sigma^2$. Similar expressions hold for PM. These values of r and T may now be inserted in our formulas to determine the interchannel interference.

From the definitions of A and U it may be shown that (6.3) is equivalent to $AU^2 = 1/16$. Therefore the "second order modulation" approximation given by Case 2 in Table 4.1 may be used. Also $G - H \cos 2pT$ is

approximately equal to $2G$ because $\cos 2pT = -1$. It turns out that the second order modulation approximation may also be used in the PM case.

When a_2 is sufficiently small, considerations such as those above show that the ratio of the interference power for $\beta = a_2(f - f_0)^2/2$ (radians) to the received signal power at the frequency f is

$$\begin{aligned} P_I/P_S &= (a_2\sigma f_b/2)^2\, a^2(2 - a) \text{ for FM} \\ P_I/P_S &= (a_2 f_b^2/2)^2(P_0f_b)(12 - 30a + 20a^2 - a^5)/30 \text{ for PM} \end{aligned} \qquad (6.5)$$

Here $a = f/f_b$ where $(0, f_b)$ is the frequency band of the signal. The rms frequency deviation of the signal for FM is σ cps, and the rms phase deviation of the signal for PM is $(P_0f_b)^{1/2}$ radians. The first equation in (6.5) comes from a special case of Case 2, namely, Case 3 of Table 4.1, and requires the additional assumption $f_b/4\sigma \ll 1$ (corresponding to $U \ll 1$). The second equation requires a similar additional assumption.

Appendix I

Derivation of a General Theorem on the Interchannel Interference Spectrum

Let $w_a(f)$ be a finite power spectrum having limited total fluctuation for $0 \leqq f \leqq \infty$. Let the total power be finite so that the integral of $w_a(f)$ from $f = 0$ to $f = \infty$ converges absolutely. We define two auxiliary spectra

$$w_\epsilon(f) = \begin{cases} w_a(f), & |f - f_0| < \epsilon \\ 0, & |f - f_0| > \epsilon \end{cases}$$

$$w_b = w_a(f) - w_\epsilon(f) \qquad (A1\text{-}1)$$

and note that the autocorrelations corresponding to these spectra must satisfy

$$R_b(\tau) = R_a(\tau) - R_\epsilon(\tau) \qquad (A1\text{-}2)$$

We consider the problem of transmitting the ensemble having the spectrum $w_b(f)$ through a system in which the input and output autocorrelation functions, $\Psi_1(\tau)$ and $\Psi_2(\tau)$, respectively, are related by

$$\Psi_2(\tau) = F[\Psi_1(\tau)]$$

Here $F(z)$ and its derivatives $F'(z)$ and $F''(z)$ are assumed to be finite continuous functions which exist over the range of z of interest.

Let the output spectrum corresponding to the input spectrum $w_b(f)$ be $w_B(f)$. We shall show that as $\epsilon \to 0$ the value of $w_B(f)$ in the range

$|f - f_0| < \epsilon$ approaches

$$4 \int_0^\infty \{F[R_a(\tau)] - F'(0)R_a(\tau)\} \cos 2\pi f\tau \, d\tau \tag{A1-3}$$

When we multiply this expression by 2ϵ and set $f = f_0$, we obtain the power appearing at the output of the system in an unloaded channel of width 2ϵ centered on f_0. Here we are not interested in values of $w_B(f)$ outside the range $|f - f_0| < \epsilon$.

First we note several properties of autocorrelation functions. From

$$R(\tau) = \int_0^\infty w(f) \cos 2\pi f\tau \, df$$

it follows that $|R(\tau)| \leqq R(0)$. Also, if $w(f)$ has limited total fluctuation in the interval $(0, \infty)$, the Riemann-Lesbesgue lemmas* and the absolute convergence of the integral for $R(0)$ show that $R(\tau) = 0(1/\tau)$ as $\tau \to \infty$. Thus we may find positive numbers A, B, C such that for $0 < \tau$ and any ϵ less than some fixed value

$$|R_a(\tau)| < A/\tau$$

$$|R_\epsilon(\tau)| \leqq R_\epsilon(0) = \int_{f_0-\epsilon}^{f_0+\epsilon} w_a(f)\, df < B\epsilon \tag{A1-4}$$

$$|R_\epsilon(\tau)| < C/\tau$$

By the extended theorem of the mean the autocorrelation function corresponding to $w_B(f)$ is

$$\begin{aligned} R_B(\tau) &= F[R_b(\tau)] = F[R_a(\tau) - R_\epsilon(\tau)] \\ &= F[R_a(\tau)] - F'[R_a(\tau)]R_\epsilon(\tau) + r \end{aligned} \tag{A1-5}$$

$$|r| = 2^{-1}R_\epsilon^2(\tau)\, |F''[R_a(\tau) - \theta R_\epsilon(\tau)]| < R_\epsilon^2(\tau)D$$

where $0 \leqq \theta \leqq 1$ and D is a positive number such that $|F''(z)| < D$. Then

$$\begin{aligned} w_B(f) &= 4\int_0^\infty R_B(\tau) \cos 2\pi f\tau \, d\tau = I_1 - I_2 + I_3 \\ I_1 &= 4\int_0^\infty F[R_a(\tau)] \cos 2\pi f\tau \, d\tau \\ I_2 &= 4\int_0^\infty F'[R_a(\tau)]R_\epsilon(\tau) \cos 2\pi f\tau \, d\tau \\ I_3 &= 4\int_0^\infty r \cos 2\pi f\tau \, d\tau \end{aligned} \tag{A1-6}$$

* See, for example, Whittaker and Watson, Modern Analysis, 4th edition, p. 172.

Since $F'[R_a(\tau)] = F'(0) + s$, where by the mean value theorem

$$|s| = |R_a(\tau)\, F''[\theta R_a(\tau)]| < |R_a(\tau)|\, D$$

$$0 \leqq \theta \leqq 1$$

I_2 may be written as the sum of two integrals, the second of which has an absolute value not greater than

$$\begin{aligned} 4\int_0^\infty |R_a(\tau)DR_\epsilon(\tau)|\, d\tau &= 4D\int_0^T |R_a(\tau)R_\epsilon(\tau)|\, d\tau \\ &+ 4D\int_T^\infty |R_a(\tau)R_\epsilon(\tau)|\, d\tau < 4DR_a(0)B\epsilon T \\ &+ 4D\int_T^\infty AC\tau^{-2}\, d\tau = 4D[R_a(0)B\epsilon T + AC/T] \end{aligned}$$

where T is an arbitrary number and we have used the inequalities (A1-4). Choosing $T = \epsilon^{-1/2}$ shows that the last expression is $0(\epsilon^{1/2})$. Hence

$$\begin{aligned} I_2 &= 4F'(0)\int_0^\infty R_\epsilon(\tau) \cos 2\pi f\tau \, d\tau + 0(\epsilon^{1/2}) \\ &= 0(\epsilon^{1/2}) + F'(0)\begin{cases} w_a(f), & |f - f_0| < \epsilon \\ 0, & |f - f_0| > \epsilon \end{cases} \end{aligned}$$

Therefore in the range $|f - f_0| < \epsilon$, which comprises the only frequencies of interest in the channel interference spectrum,

$$I_2 = 4F'(0)\int_0^\infty R_a(\tau) \cos 2\pi f\tau \, d\tau + 0(\epsilon^{1/2})$$

By use of the inequalities for $|r|$ and $|R_\epsilon(\tau)|$ we see that

$$\begin{aligned} |I_3| < 4\int_0^\infty R_\epsilon^2(\tau)D\,d\tau &< 4D\int_0^T B^2\epsilon^2 d\tau + 4D\int_T^\infty C^2\tau^{-2} d\tau \\ &= 4D(B^2\epsilon^2 T + C^2/T) \end{aligned}$$

If we choose $T = 1/\epsilon$ this expression is $0(\epsilon)$.

When we collect our results and let ϵ become vanishingly small, we see that expression (A1-6) for $w_B(f)$ approaches (A1-3) for frequencies in the range $|f - f_0| < \epsilon$. Thus in the limit as $\epsilon \to 0$ we may use the autocorrelation function

$$F[R_a(\tau)] - F'(0)R_a(\tau)$$

to compute the interchannel interference spectrum. This is the result used in (1.22).

Appendix II

APPROXIMATE EVALUATION OF INTEGRALS OF A CERTAIN TYPE

The problem of evaluating the integral G defined by (2.4) is quite a difficult one. Here we shall outline a method which often may be used to obtain an idea of the order of the magnitude of such an integral.

Let $F(u)$ be an even analytic function of u such that the major contribution to the value of

$$I(a) = 2\int_0^\infty F(u)\cos au\,du = \int_{-\infty}^\infty F(u)e^{iau}\,du \qquad \text{(A2-1)}$$

comes from a saddle point on the positive imaginary u axis. Then the "method of steepest descents" suggests that an approximate value of $I(a)$ may be obtained by the following procedure.

1. Set $f(y) = F(iy)$ and plot $z = d[\log f(y)]/dy = f'(y)/f(y)$ as a function of y.
2. Draw the horizontal line $z = a$. Suppose its first intersection with the curve obtained in step 1 is at $y = y_1$, and let the slope of the curve, determined either graphically or by differentiation, be $(dz/dy)_{y_1}$ at y_1.
3. Then $I(a) \approx [2\pi/(dz/dy)_{y_1}]^{1/2} f(y_1)e^{-ay_1}$ (A2-2)

It should be noted that (A2-2) cannot be used indiscriminately. Thus, it does not work well for $F(u) = 1/(1 + u^4)$ because there is no saddle point on the imaginary u-axis. However, when it is applied to integrals of the type encountered in our study it appears to do fairly well, as Table 4.4 shows.

Appendix III

LEWIN'S INTEGRAL

Here we study the integral

$$I(b, a) = \int_{-\infty}^\infty [e^{bu^{-1}\sin u} - 1 - bu^{-1}\sin u]\cos au\,du \qquad \text{(A3-1)}$$

which occurs in several limiting cases in our work and which has been studied by Lewin[2] for $a = 0$ and $a = 1$.

When the exponential term is expressed as a power series in $(b \sin u)/u$ and the result integrated termwise, we obtain

$$I(b, a) = \sum_{n=2}^\infty A_n b^n/n! \qquad \text{(A3-2)}$$

$$A_n = \int_{-\infty}^\infty \left(\frac{\sin u}{u}\right)^n \cos au\,du$$

$$= \frac{2\pi}{2^n(n-1)!}\sum_{m=0} (-)^m C_m{}^n (n - 2m + a)^{n-1}$$

$$A_n \sim (6\pi/n)^{1/2}\exp[-3a^2/(2n)]$$

where $C_m{}^n = n!/m!(n-m)!$ and the last term in the summation for A_n is the last one for which $n - 2m + a$ is positive (assuming $a \neq$ integer) and for which $m \leqq n$. When $n \geqq 2$, A_n is a continuous function of a. Tables III A and III B were computed from (A3-2).

When b is small, the first term in (A3-2) gives, for $0 \leqq a \leqq 2$,

$$I(b, a) \approx b^2\pi(2 - a)/4 \qquad \text{(A3-3)}$$

and when b is a large positive number the contribution of the exponential term in the region around $u = 0$ gives

$$I(b, a) \approx (6\pi/b)^{1/2}\exp\left(b - \frac{3a^2}{2b}\right) \qquad \text{(A3-4)}$$

Lewin has given more careful approximations for the $a = 0$ and $a = 1$ cases.

When b is large and negative most of the contribution comes from around $u = \pm 3\pi/2$ where $\exp[(b \sin u)/u]$ attains its largest values.

It is found that

$$I(-\beta, a) = 10.76\,\beta^{-1/2}\cos(4.49a)e^{0.217\beta - 2.30a^2\beta^{-1}} + R \qquad \text{(A3-5)}$$

where $\beta = -b$ is a large positive number and R is a remainder term. The numbers in (A3-5) are related to the value $u_0 = 4.493\ldots$ where $(\sin u)/u$ has a minimum.

Computation shows that the value $\beta = 8$ is not large enough to make the leading term in (A3-5) a good approximation for $I(-\beta, a)$. In order to obtain a better approximation we write

$$I(-\beta, a) \approx 2\int_0^y \exp[-\beta u^{-1}\sin u]\cos au\,du$$
$$+ 2\int_0^y [-1 + \beta u^{-1}\sin u]\cos au\,du \qquad \text{(A3-6)}$$
$$+ \int_y^\infty (\beta u^{-1}\sin u)^2\cos au\,du$$

TABLE III A — $e^{-b}I(b, a)$ FOR $b > 0$

b	e^b	$a = 0$	0.25	0.50	0.75	1.00	1.25
0	1	0.0	0.0	0.0	0.0	0.0	0.0
0.5	1.649	0.272	0.241	0.209	0.176	0.142	0.107
1.0	2.718	0.761	0.685	0.602	0.511	0.414	0.314
2.0	7.389	1.560	1.440	1.291	1.117	0.919	0.713
3.0	20.08	1.913	1.801	1.645	1.448	1.215	0.968
4.0	54.60	1.974	1.888	1.751	1.566	1.341	1.098
5.0	148.4	1.905	1.844	1.731	1.571	1.372	1.153
6.0	403.4	1.794	1.751	1.660	1.525	1.356	1.166
7.0	1097.	1.680	1.649	1.575	1.463	1.320	1.157
8.0	2981.	1.576	1.552	1.492	1.398	1.277	1.138

TABLE III B — $I(b, a)$ FOR $b < 0$

b	$a = 0$	.25	.50	.75	1.0	1.25
0	0.0	0.0	0.0	0.0	0.0	0.0
−0.5	0.349	0.300	0.254	0.210	0.167	0.125
−1.0	1.25	1.06	0.885	0.723	0.576	0.432
−2.0	4.16	3.41	2.76	2.20	1.76	1.34
−3.0	8.03	6.37	4.97	3.88	3.14	2.46
−4.0	12.6	9.66	7.23	5.49	4.55	3.74
−5.0	17.8	13.2	9.40	6.89	5.93	5.19
−6.0	23.6	16.8	11.4	8.00	7.25	6.85
−7.0	30.0	20.7	13.1	8.71	8.48	8.78
−8.0	37.2	24.8	14.5	8.93	9.59	11.0

where y is such that the quantity within the brackets in (A3-1), with $b = -\beta$, is approximately $(\beta u^{-1} \sin u)^2/2$ when $u > y$. For rough work we may take $y = \beta$. The leading term in (A3-5) arises from the contribution of the region around $u = 3\pi/2$ to the value of the first integral in (A3-6). The contributions from the regions around $u = 7\pi/2, 11\pi/2, \ldots$ (if y is large enough) add to the value of R but they are generally small in comparison with the leading term in (A3-5).

Thus we are led to approximate R in (A3-5) by the sum of the second and third integrals (expressed in terms of integral sines and cosines) in (A3-6) with $y = \beta$. When R is replaced by this sum, expression (A3-5) gives values for $b = -8$ which agree fairly well with those in the table.

REFERENCES

1. S. T. Meyers, Nonlinearity in Frequency-Modulation Radio Systems Due to Multipath Propagation, Proc. I.R.E., **34**, pp. 256–265, 1946.
2. L. Lewin, Interference in Multi-Channel Circuits, Wireless Engineer, **27**, pp. 294–304, 1950.
3. W. J. Albersheim and J. P. Schafer, Echo Distortion in the FM Transmission of Frequency-Division Multiplex, Proc. I.R.E., **40**, pp. 316–328, 1952.

4. J. P. Vasseur, Calcul de la distortion d'une onde modulée in fréquence, Annales de Radioélectricité, **8**, pp. 20–35, 1953.
5. S. O. Rice, Mathematical Analysis of Random Noise, B.S.T.J., **23**, pp. 282–332, 1944 and **24**, pp. 46–156, 1945.
6. D. Middleton, The Distribution of Energy in Randomly Modulated Waves, Phil. Mag., **42**, pp. 689–707. July, 1951.
7. S. O. Rice, Diffraction of Plane Radio Waves by a Parabolic Cylinder, B.S.T.J., **33**, pp. 417–504, 1954.
8. E. D. Sunde, Theoretical Fundamentals of Pulse Transmission, B. S. T. J., **33**, pp. 721–788, and **33**, pp. 987–1010, May and July, 1954.

An Extended Analysis of RF Interference in FDM-FM Radio Relay Systems

G. J. GARRISON

REFERENCE: Garrison, G. J.: AN EXTENDED ANALYSIS OF RF INTERFERENCE IN FDM-FM RADIO RELAY SYSTEMS, Lenkurt Electric Company of Canada, Ltd., Burnaby, British Columbia. Rec'd 11/2/66; revised 5/15/67. Paper 19TP67-1269, presented at the 1967 IEEE International Conference on Communications, Minneapolis, Minn. IEEE TRANS. ON COMMUNICATION TECHNOLOGY, 15-5, October 1967, pp. 705–713.

ABSTRACT: In a conventional microwave receiver a number of *critical* frequencies exist such that, when an interfering RF signal is introduced in their vicinity, a spurious IF response results. The existing methods for computing intermodulation due to an interfering carrier are found to be only partially satisfactory, in that, they do not explicitly define the manner in which the receiver mixing mechanism modifies the RF interference with respect to the level relative to that of the wanted carrier, frequency displacement in the IF from the primary IF response, and modification of the modulation index of the interference modulation. This paper extends, in a general manner, the theories of RF interference to include interfering signals falling on or near all major critical frequencies of the receiver. It is shown that, with the relative conversion efficiencies of the mixer available from empirical data and an understanding of the manner in which the mixer will modify the frequency displacement and modulation index of the interfering signal, these interference situations can be analyzed within the scope of a previously defined *general interference equation.* As an illustrative example some interference problems in a 1200-channel system are analyzed and substantiated with measurement data.

KEYWORDS: Analysis, Communication Theory, Detection, Distortion, Frequency Modulation, Microwave Relay System, Multiplex, Radio Frequency Interference, Signal-to-Noise Ratio.

Nomenclature

ω_c	Wanted carrier frequency (rad/s)
ω_s	unwanted carrier frequency (rad/s)
ω_p	receiver local oscillator frequency (rad/s)
$\phi(t)$	phase modulation of wanted carrier (rad)
$\xi(t)$	phase modulation of unwanted carrier (rad)
ω_{IF}	center IF frequency of receiver (rad/s)
Vc, Vs	peak values of wanted and unwanted RF inputs to mixer (volts)
Wc, Ws	absolute power levels of wanted and unwanted RF inputs (dBm)
$f_n(V_p)$	functional representation of local oscillator signal applied to mixer
$K_1 \ldots K_4 \ldots$	coefficients of power series expansion of mixer voltage transfer function
C_{crit}	relative mixer conversion efficiency of interfering signal falling near receiver critical frequency
ω_D	departure of interference frequency from receiver critical frequency (rad/s)
ω_Δ	composite rms deviation due to wanted modulation (rad/s)
ρ	representative baseband frequency (rad/s)
ρ_o	minimum baseband modulating frequency (rad/s)
ρ_m	maximum baseband modulating frequency (rad/s)
$W(\omega)$	spectral power of wanted modulated carrier centered at frequency $\omega_{IF} \pm \omega$ in bandwidth w_b
$W_2(\omega)$	second order portion of $W(\omega)$
$[F_2(\omega)]^2$	function related to second order portion of the power spectrum $W_2(\omega)$
$f(\rho)$	expression defining power transfer function of CCIR pre-emphasis network as function of baseband frequency ρ
$g(\rho)$	approximation to the power transfer function of the CCIR pre-emphasis network
$[R_v(\tau)]_2$	autocorrelation of wanted modulated IF carrier (second order portion)
$R_\phi(\tau)$	autocorrelation of wanted carrier phase modulation
$s_\phi(p)$	power spectral density of wanted carrier phase modulation.

I. Introduction

THE ever-increasing demand for long distance telephone facilities has, in recent years, been met by a continuous extension of the channel capacity of frequency-division-multiplex (FDM)-FM radio relay systems. Already in operation are 960-, 1200-, and 1800-channel systems while 2700-channel systems are under consideration. These systems, concurrent with their capacity expansions, have been requested to meet, or exceed, the established international objectives[1] for such networks. For FDM-FM systems this increased channel capacity requires, in general, an increase in the significant RF bandwidth occupied by the modulation sidebands.[2],[3] This, as one would expect, leads to an increase in the vulnerability of the system to extraneous RF interference. Thus, one of the major considerations required in specifying equipment and system parameters economically to meet the current day performance objectives is an accurate estimate of the intermodulation effects of RF interference on the radio baseband.

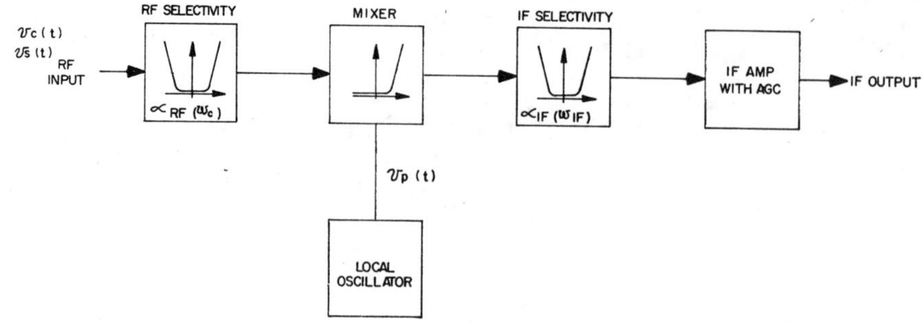

Fig. 1. Generalized receiving system.

The problem has previously been analyzed in detail[4]–[6] for a range of likely modulation characterisitics and finalized by Medhurst[7] with a general formula covering all characteristics of modulation on the wanted and unwanted carriers. In practical situations, however, these results find themselves limited to the cases where the RF interference falls in the vicinity of the receiver passband or the receiver image frequency since, in these instances only, is there a guarantee of a one-to-one correspondence between the RF characteristics of the wanted/unwanted carriers and their hetrodyned IF resultants. Although a recent treatment[8] empirically describes intermodulation effects at other than these primary response locations it is, by its nature, restricted to the modulation conditions employed in the measurement program and the specific type of receiver under test.

It is thus the objective of this paper to extend the analysis of RF interference to cover those cases where the interfering carrier falls on, or near, any of the receiver critical[1] frequencies and to show that, with a limited amount of empirical data defining the relative mixer conversion efficiency of the receiver to a spurious input, an accurate estimate of the resulting baseband intermodulation noise can be established for a given type of receiver.

II. Generalized Model of the Receiving System

Figure 1 represents a diagrammatic model of the generalized receiving system. The RF selectivity ($\propto_{RF}$) represents the effective attenuation to the interference from all sources including filtering, cross-polarization discrimination, propagation loss, etc., while $\propto_{IF}$ represents the selectivity of the IF circuitry. For an analysis of a specific interference problem it is important to distinguish between the two sources of filtering due to the nonlinear conversion effects of the mixer to the interference signal.

The principal critical frequencies of the receiving system are shown in Fig. 2. Here the terminology of Jacobsen[9] has been retained with the primary receiver response established as the origin and the receiver critical frequencies specified in terms of their harmonic and/or subharmonic relationship to the IF frequency. The

[1] Receiver critical frequency is defined as a frequency of the receiver that will yield a spurious frequency within the IF passband when the receiver is excited by LO, RX and spurious RF signals.

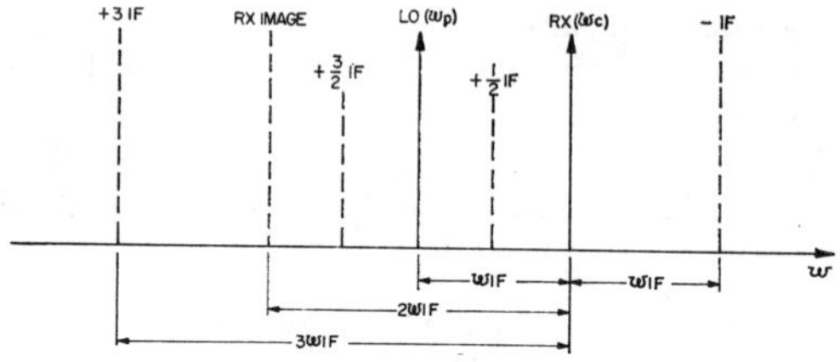

Fig. 2. Principal critical frequencies of receiver.

relative frequency nomenclature as shown is defined as positive for frequencies on the LO side of the receiver frequency. For inversion of the absolute relationship of the receiver and LO frequencies this polarity definition is still retained.

III. Analysis of Receiver Response

The nonlinear voltage transfer function of the mixer is usually represented[10], [11], [12] by a power series of the form

$$V_{OUT} = K_1 V_{IN} + K_2 V_{IN}^2 + K_3 V_{IN}^3 + K_4 V_{IN}^4 + \ldots \quad (1)$$

Experimental evidence obtained from a number of practical microwave receivers shows that, while the power series representation is adequate to define the mixer operation for low-level signals (wanted and unwanted carrier inputs), this description does not hold true for the relatively high level local oscillator signal. Figures 3 and 4 are typical of measured data obtained from conventional receivers. Here, the somewhat erratic variation of the spurious IF outputs simplies that prediction of output levels based only on assigned values of the power series coefficients in (1) will yield inaccurate results.

Since a direct analytical analysis based on (1) is found to be incorrect it is necessary to express the effect of the local oscillator signal in functional notation. The analysis will also be restricted, for simplicity, to a three frequency input which is of the form

$$V_{IN} = v_c(t) + v_s(t) + v_p(t) \quad (2)$$

where

$$v_c(t) = V_c \sin(\omega_c t + \phi(t)) \quad (3)$$

$$v_s(t) = V_s \sin(\omega_s t + \xi(t)) \quad (4)$$

$$v_p(t) = f_n(V_p) \sin \omega_p t. \quad (5)$$

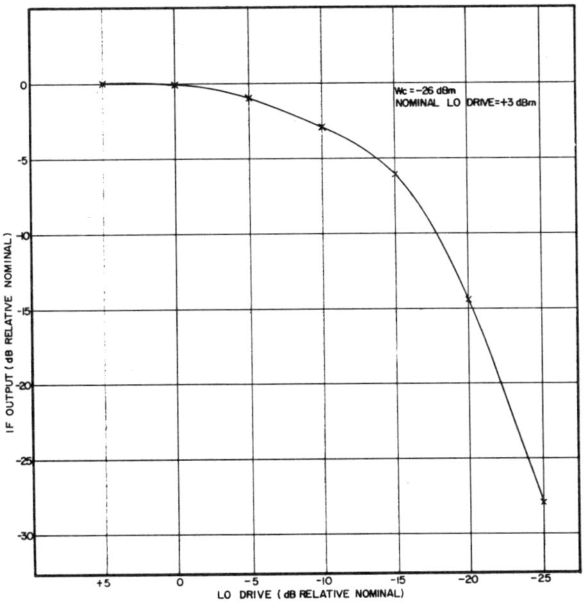

Fig. 3. Measured variation of mixer IF output (primary response) as function of applied LO signal.

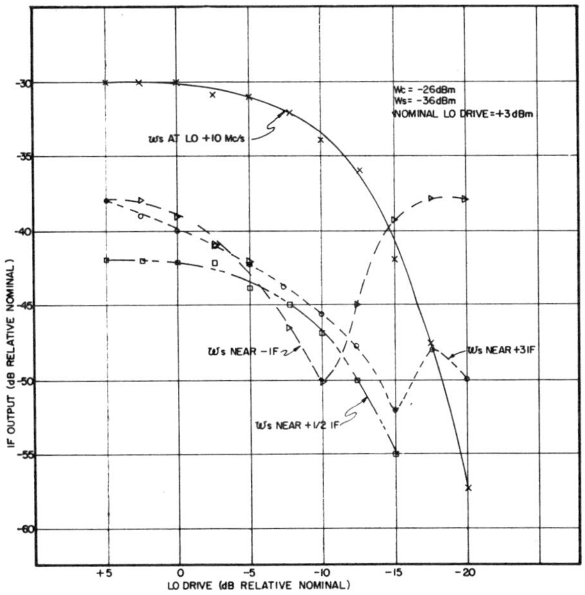

Fig. 4. Measured variation of mixer IF output (spurious response) as function of applied LO signal.

TABLE I

PRINCIPAL MIXER OUTPUTS FOR THREE-FREQUENCY INPUT (LO, RX AND INTERFERENCE)

Case	RF Mixing Product	IF Output
a	$\omega_c - \omega_p$	$K_2 f_2(V_p) V_c \cos\{(\omega_c - \omega_p)t + \phi(t)\}$
b	$\omega_p - \omega_s$	$K_2 f_2(V_p) V_s \cos\{(w_p - \omega_s)t - \xi(t)\}$
c	$\omega_c - \omega_s$	$K_2 V_c V_s \cos\{(\omega_c - \omega_s)t + \phi(t) - \xi(t)\}$
d	$2(\omega_p - \omega_s)$	$3/4 K_4 f_4(V_p) V_s^2 \cos\{2(\omega_p - \omega_s)t - 2\xi(t)\}$
e	$\omega_c + \omega_s - 2\,\omega_p$	$-3/2 K_4 f_4(V_p) V_c V_s \sin\{(\omega_c + \omega_s - 2\omega_p)t + \phi(t) + \xi(t)\}$

For each order of mixing $f_n(V_p)$ in (5) is described by a different function. Insertion of (2) through (5) into (1) will give the desired primary IF output

$$K_2 f_2(V_p) V_c \cos\{(\omega_c - \omega_p)t + \phi(t)\} \tag{6}$$

and, in addition, a number of spurious outputs, only a limited number of which could potentially fall near the IF frequency $w_{\rm IF}$. These are listed in Table I. Only the lowest order mixing product pertaining to a given frequency is shown as these tend to suffice to cover most practical interference problems. Although additional IF output frequencies exist for the same orders of mixing listed in Table I those listed exceed, for likely levels of V_c and V_s, all others by one or more orders of magnitude.

From an examination of Table I it is evident that the spurious outputs will fall within the IF passband only for specific values of the interference frequency w_s. It is thus convenient to specify the interference frequency in terms of its departure from a receiver critical frequency, i.e., for a given type of interference we will write

$$\omega_s = \omega_{\rm crit} \pm \omega_D \tag{7}$$

where w_D represents the departure of the interference frequency from the receiver critical frequency.

Substitution of (7) into the listings of Table I will yield the spurious IF outputs in terms of the receiver critical frequencies. For example, consider the interference falling in the vicinity of the $+1/2$ IF frequency. In this case

$$\omega_s = \omega_{+1/2\rm IF} \pm \omega_D \tag{8}$$

Substitution into Table I, case d, gives

$$[v_s(t)]_{\rm IF} = 3/4 K_4 f_4(V_p) V_s^2 \cos\{2(\omega_p - \omega_{+1/2\rm IF} \mp \omega_D)t - 2\xi(t)\} \tag{9}$$

Noting that

$$\omega_p = \omega_c + \omega_{\rm IF} \tag{10}$$

and

$$\omega_{+1/2\rm IF} = \omega_c + 1/2\omega_{\rm IF} \tag{11}$$

(9) becomes

$$[v_s(t)]_{\rm IF} = 3/4 K_4 f_4(V_p) V_s^2 \cos\{(\omega_{\rm IF} \mp 2\omega_D)t - 2\xi(t)\}. \tag{12}$$

In a similar manner interference inputs falling near other of the receiver critical frequencies can be identified. A full listing is given in Table II.

One final step is still necessary in order to establish the aforementioned in a form comparable to the general interference equations. This requires summing the desired

TABLE II

IDENTIFICATION OF SPURIOUS IF OUTPUTS IN TERMS OF RELATIVE LOCATION OF RF INTERFERENCE TO CRITICAL RECEIVER FREQUENCIES

Case	Spurious Location	IF Output
a	Wanted carrier	$K_2 f_2(V_p) V_c \cos(\omega_{IF} t - \phi(t))$
b	Near RX ($\omega_s = \omega_c \pm \omega_D$)	$K_2 f_2(V_p) V_s \cos\{(\omega_{IF} \mp \omega_D)t - \xi(t)\}$
c	Near RX image ($\omega_s = \omega_{IM} \pm \omega_D$)	$K_2 f_2(V_p) V_s \cos\{(\omega_{IF} \pm \omega_D)t + \xi(t)\}$
d	Near LO($\omega_s = \omega_p \pm \omega_D$)	*1)* $K_2 V_c V_s \cos\{(\omega_{IF} \pm \omega_D)t - \phi(t) + \xi(t)\}$ *2)* $3/2 K_4 f_4(V_p) V_c V_s \sin\{(\omega_{IF} \mp \omega_D)t - \phi(t) - \xi(t)\}$
e	Near +1/2 IF($\omega_s = \omega_{+1/2IF} \pm \omega_D$)	$3/4 K_4 f_4(V_p) V_s^2 \cos\{(\omega_{IF} \mp 2\omega_D)t - 2\xi(t)\}$
f	Near +3/2 IF($\omega_s = \omega_{+3/2IF} \pm \omega_D$)	$3/4 K_4 f_4(V_p) V_s^2 \cos\{(\omega_{IF} \pm 2\omega_D)t + 2\xi(t)\}$
g	Near −IF($\omega_s = \omega_{-IF} \pm \omega_D$)	$K_2 V_c V_s \cos\{(\omega_{IF} \mp \omega_D)t + \phi(t) - \xi(t)\}$
h	Near +3 IF($\omega_s = \omega_{+3IF} \pm \omega_D$)	$-3/2 K_4 f_4(V_p) V_c V_s \sin\{(\omega_{IF} \pm \omega_D)t + \phi(t) + \xi(t)\}$

TABLE III

GENERALIZED IF INTERFERENCE EQUATION FOR INTERFERENCE FALLING NEAR RECEIVER CRITICAL FREQUENCIES

Case	Spurious Location	Phase Modulated Sum of the Wanted and Unwanted Carriers
a	Near RX	$\cos\{\omega_{IF} t - \phi(t) + V_s/V_c \sin(\mp\omega_D t + \phi(t) - \xi(t))\}$
b	Near RX image	$\cos\{\omega_{IF} t - \phi(t) + V_s/V_c \sin(\pm\omega_D t + \phi(t) + \xi(t))\}$
c	Near LO	1) $\cos\{\omega_{IF} t - \phi(t) + C_{LO1}(\omega_D) V_s \sin(\pm\omega_D + \xi(t))\}$ 2) $\cos\{\omega_{IF} t - \phi(t) + C_{LO2}(\omega_D) V_s \sin(\mp\omega_D - \xi(t) + 3/2\pi)\}$
d	Near +1/2IF	$\cos\{\omega_{IF} t - \phi(t) + C_{+1/2IF} V_s^2/V_c \sin(\mp 2\omega_D t + \phi(t) - 2\xi(t))\}$
e	Near +3/2IF	$\cos\{\omega_{IF} t - \phi(t) + C_{+3/2IF} V_s^2/V_c \sin(\pm 2\omega_D t + \phi(t) + 2\xi(t))\}$
f	Near −IF	$\cos\{\omega_{IF} t - \phi(t) + C_{-IF} V_s \sin(\mp\omega_D t + 2\phi(t) - \xi(t))\}$
g	Near +3 IF	$\cos\{\omega_{IF} t - \phi(t) + C_{+3IF} V_s \sin(\pm\omega_D t + 2\phi(t) + \xi(t) + 1/2\pi)\}$

and spurious IF outputs as a modulation on the desired output and redefining the functional ratios of the local oscillator level dependence in a more convenient form.

Employing the +1/2 IF interference as an example,[13] for $V_c \gg V_s$, the composite IF output is given by

$$v_{IF}(t) = \cos\left\{\omega_{IF} t - \phi(t) + C_{+1/2IF} \frac{V_s^2}{V_c} \sin(\mp 2\omega_D t + \phi(t) - 2\xi(t))\right\} \quad (13)$$

where

$$C_{+1/2IF} \triangleq 3/4 \frac{K_4 f_4(V_p)}{K_2 f_2(V_p)} \quad (14)$$

is the relative mixer conversion efficiency of an interfering signal falling near the +1/2 IF frequency.

In (13) it must be noted that all AM has been neglected on the assumption of perfect limiting and that the value of $C_{+1/2IF}$ is fixed only for a given local oscillator level. Since in most practical receivers the local oscillator power is fixed quite closely about its nominal value this does not represent a serious restriction. For a given type of receiver $C_{+1/2IF}$ is, of course, required from empirical data. It is also convenient to evaluate C with Vc and Vs expressed relative to the *nominal* input level of the receiver.

Examination of (13) now allows us to identify the manner in which the RF interference has been modified after transmission through the mixer. For this example it is noted that

1) RF separation frequency (w_D) has been doubled

2) peak deviation of interference modulation has been doubled

3) intermodulation effect will vary dB/dB with respect to wanted carrier RF level and 2 dB/dB with respect to unwanted carrier RF level.

Using the same approach the effect of w_s located near other receiver critical frequencies can be determined. These results are summarized in Table III for all frequencies under consideration where the remaining relative mixer conversion efficiencies are defined as

$$C_{-IF} \equiv V_{LO_1} = \frac{1}{f_2(V_p)} \quad (15)$$

$$C_{+3/2IF} \equiv C_{+1/2IF} = 3/4 \frac{K_4 f_4(V_p)}{K_2 f_2(V_p)} \quad (16)$$

$$C_{+3IF} \equiv C_{LO_2} = 3/2 \frac{K_4 f_4(V_p)}{K_2 f_2(V_p)} \quad (17)$$

IV. COMPUTATIONAL EXAMPLE AND COMPARISON WITH MEASURED RESULTS

As an illustration of the application of the foregoing, a specific case will be analyzed and the results compared with measured values.

A 1200-channel system with the baseband simulated by the conventional band-limited white-noise test-signal will be considered.

The relevent transmission parameters are[14]

Baseband range	316 kHz–5564 kHz
RMS deviation per channel	140 kHz
Multichannel rms deviation	863 kHz

A. Computational Analysis

In order to compute the intermodulation effects of an interfering signal it is first necessary to determine the RF spectrum of the wanted and unwanted carriers. Since the mean-square phase deviation $(0.458\ \omega_\Delta^2/p_o p_m)$ is less than unity the 1200-channel spectrum falls into the *low modulation ratio* class and can be evaluated from Medhurst, [equation (3), modified to take approximate account of pre-emphasis and restricting the analysis to a second order expansion of $\phi(t)$].[6] With these changes the spectrum equation becomes

$$W(\omega) = \omega_b \exp\left(-0.458 \frac{\omega_\Delta^2}{p_o p_m}\right)\left\{\frac{1}{\omega_b}\Big/_{\omega=0} + \frac{1}{2}\frac{\omega_\Delta^2}{(p_m - p_o)}\frac{f(p)}{\omega^2}\Big/_{p_o \le \omega \le p_m} + \frac{1}{4}\frac{\omega_\Delta^4}{(p_m - p_o)^2} \times [F_2(\omega)]^2/_{0 \le \omega \le 2p_m}\right\}. \quad (18)$$

The pre-emphasized form of $[F_2(w)]^2$ has been evaluated (Appendix I) assuming the CCIR pre-emphasis network[15] can be approximated as an exponential of the form

$$g(p) = k_o \exp(\propto p/p_m) \quad (19)$$

where

$$k_o = 0.3$$

$$\propto = 2.1$$

Here the values of k_2 and $\propto$ are chosen to approximate simultaneously the CCIR network with respect to multichannel power, crossover frequency, and overall curve matching vs. frequency. The accuracy of the specified fit is within ±0.5 dB over the major portion of the baseband, decreasing to a maximum error of −0.8 dB at the lowest modulating frequency.

For the 1200-channel case computation of (18) is shown in Fig. 5 normalized to p_m/ω_b. The residual carrier spike is shown without this normalizing factor and its level is thus directly comparable with that of the unmodulated carrier.

In general, analysis of interference problems requires recourse to Medhurst [see equations (2) and (3)][7] and a semigraphical analysis is required for evaluation of the convolution type integrals involved. Here, however, we will restrict ourselves to interference from an extraneous 1200-channel system to effect some simplification and thus the use of another work by Medhurst [see equation (5)].[6]

Medhurst's equation (5)[6] may be rewritten to include the effect of pre-emphasis, for the distortion to signal ratio (D/S)

$$\text{D/S} = r^2 \frac{p_m p^2}{2\omega_b \omega_\Delta^2} \frac{1}{f(p)} \{W(\omega_D + p) + W(|\omega_D - p|)\} \quad (20)$$

where r^2 is the power ratio of the unmodulated unwanted/wanted carriers, in this case at IF. For the unmodulated

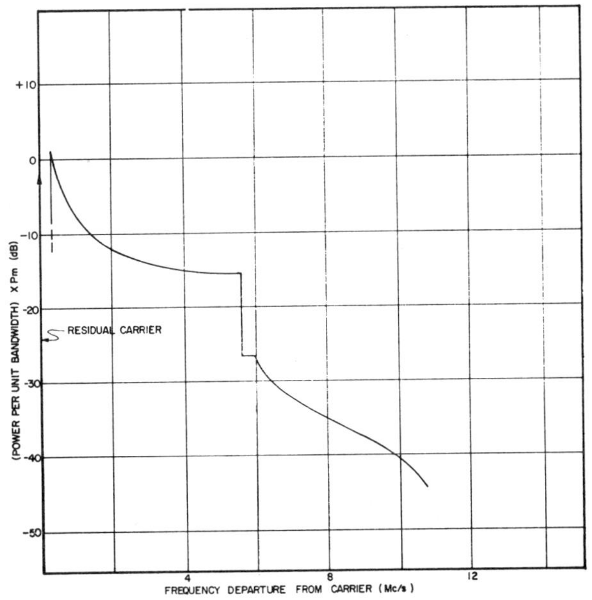

Fig. 5. Spectral density of frequency-modulated 1200-channel carrier (with pre-emphasis). Unity power in modulated carrier.

cases (18) and (20) can be applied directly. For the interference converted to the IF with either the same or *double* deviation, following Medhurst *et al.*[4] the modulated interference effects can be evaluated by replacing ω_Δ in W of (20) by $\sqrt{2}\ \omega_\Delta$ and $\sqrt{5}\ \omega_\Delta$, respectively.

For the three cases, computation of (18) and insertion in (20) yields the results of Fig. 6 where P represents the distortion to interference ratio under the condition of equal wanted and unwanted carrier levels, i.e., $P \equiv D/S|_{r^2 = 1}$.

Some care must be used when applying the *double deviation* curve of Fig. 6. As previously noted, an interference falling near the +1/2 IF is converted to the IF with both a doubling of its separation frequency (ω_D) and its modulation index. Inspection of Table III shows however, that for interference falling near the −IF and +3 IF the separation frequency remains unchanged but the spurious modulation is effectively doubled due to the $2\phi(t) \pm \xi(t)$ expression within the intermodulation term. The double deviation case of Fig. 6 as plotted must thus be considered as relating to carrier separations at IF.

B. Measurement Evaluation

Figure 7 illustrates in detail the test equipment configuration employed in a laboratory investigation of interference while Figs. 8 and 9 show measured values of the distortion/interference ratio for spurious inputs representative of Table III. The receiver under evaluation operated at 3990 MHz with a local oscillator frequency of 3920 MHz. The receiver, of the balanced type, employs 1N21F diodes in the mixer and is in commercial use for systems carrying from 960 to 1800 channel

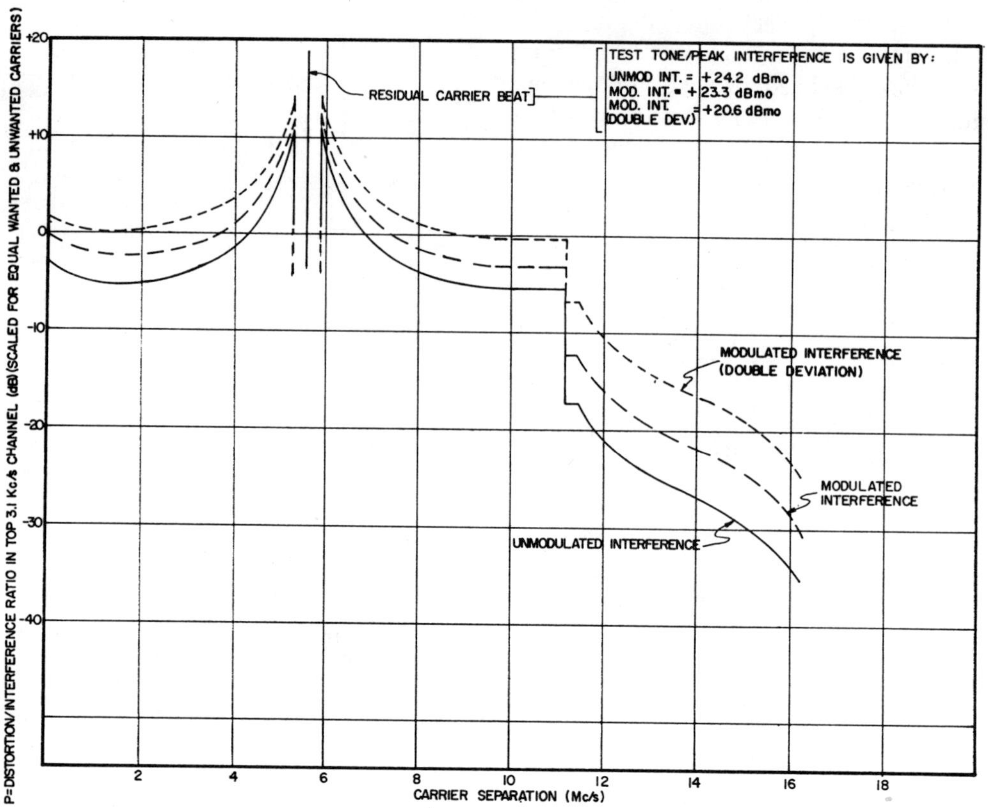

Fig. 6. Distortion due to interfering carrier in 1200-channel system (with pre-emphasis).

traffic loading. The nominal received signal for this receiver design has been set at −26 dBm.

The wanted and unwanted carriers were generated from test transmitters normally used in conjunction with the receiver. Modulation was applied to the transmitters via high quality 70-MHz FM modulators. A similar high-quality demodulator was employed for demodulation of the IF output of the interfered-with receiver.

Standard commercial test equipment was utilized for the application of the noise test signals and the calibration of signal levels and frequencies.

The RF interference frequency was varied by attaching a tunable LO to the interfering transmitter. For each set of measurements near a critical frequency care was taken to ensure adequate RF filtering on the interfering carrier outside of the frequency range of interest. For Figs. 8 and 9 the measurement data was taken with an RF wanted/unwanted carrier ratio (W_c/W_s) of 20 dB. The value of C_{RX} is, of course, zero (dB) while the values of $C_{+1/2}$IF and C_{+3IF} were found to be −20 dB and −30 dB, respectively. As shown, the agreement between theory and measurement is considered satisfactory and within the range of experimental accuracy expected.

Except for measurements where the level of the spurious IF response was allowed to approach the level of the desired primary response the variation of baseband intermodulation with change in V_c and V_s was found to be in accordance with that predicted in Table III. For exceptionally high levels of V_s (above the nominal receiver input level and/or approaching that of the LO) additional spurious outputs of the mixer become significant and must be considered.

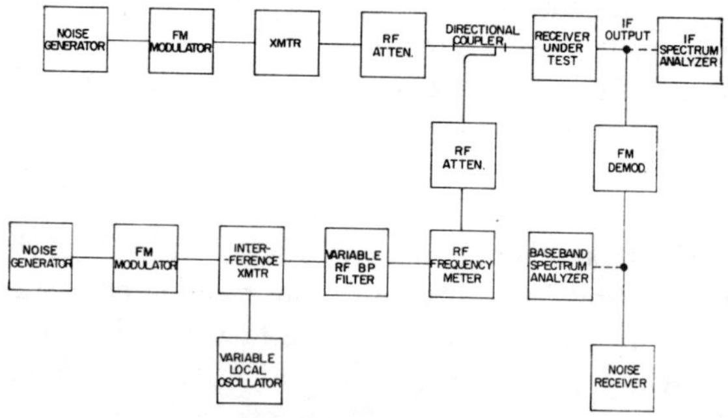

Fig. 7. Measuring equipment.

For the receiver under test the remaining values of C are given for general interest. These were

$$C_{IM} = 0$$

$$C_{LO1} = -23 \text{ dB } (\omega_D = \pm 5 \text{ MHz})$$

$$C_{LO2} = -24 \text{ dB } (\omega_D = \pm 5 \text{ MHz})$$

$$C_{+3/2IF} = -20 \text{ dB}$$

$$C_{-IF} = -29 \text{ dB}$$

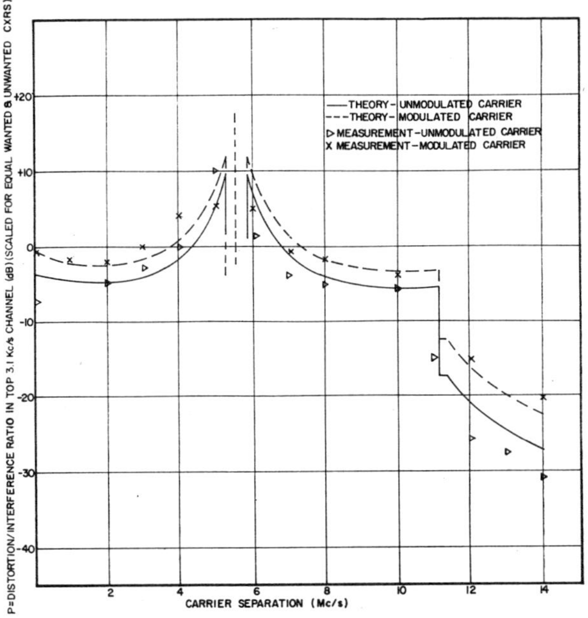

Fig. 8. Measured distortion due to interfering carrier in 1200-channel system (with pre-emphasis). Interference near receiver frequency.

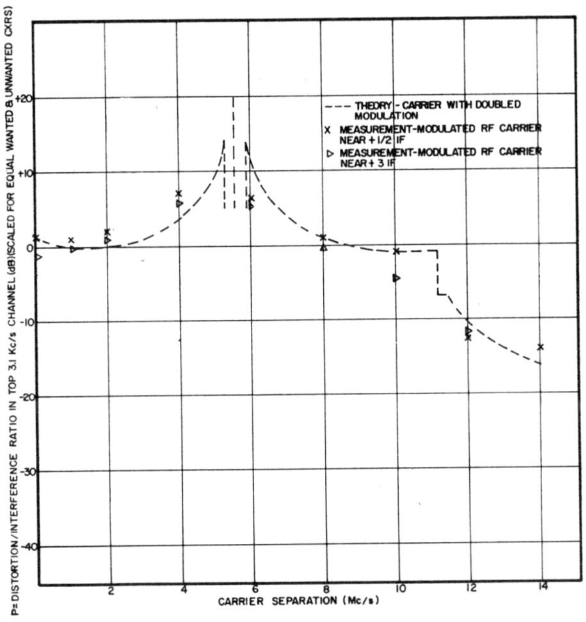

Fig. 9. Measured distortion due to interfering carrier in 1200-channel system (with pre-emphasis). Interference near + 1/2 IF and +3 IF frequencies.

Examination of Tables I and II indicates that, in addition to the equality of $C_{+1/2\text{IF}}$ and $C_{+3/2\text{IF}}$ a corresponding equality should exist between C_{LO1}, $C_{-\text{IF}}$ and C_{LO2}, $C_{+3\text{IF}}$. The experimental results did not confirm this with C_{LO1} and C_{LO2} being noted to be significantly frequency dependent as a function of w_D (see Fig. 10). This frequency dependence is thought to be related to the tuned circuitry associated with the mixer in the vicinity of the LO frequency although no attempt was made to explore this phenomena in detail. For this reason, however, all C values in Table III are specified independently and C_{LO1}, C_{LO2} are assigned a frequency dependence.

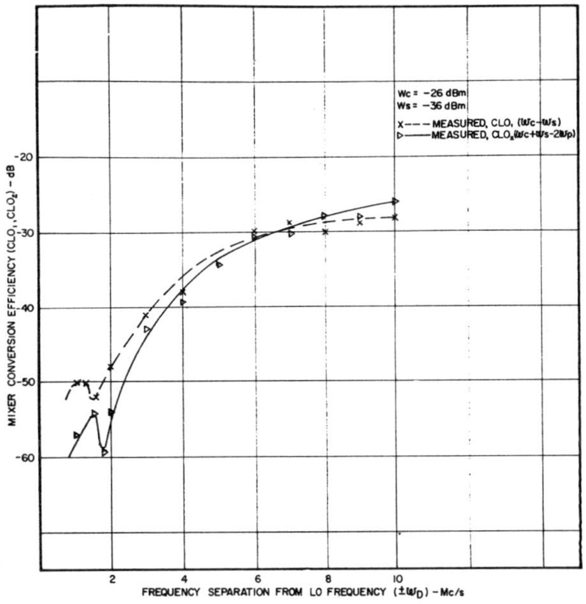

Fig. 10. Measured variation of LO mixer conversion efficiency as function of separation frequency (ω_D).

V. Conclusion

The theory of an interfering carrier falling in the vicinity of the receiver critical frequencies has been developed and substantiated by measurement. It has been shown that, with a knowledge of the relative mixer conversion efficiencies and an appreciation of the effects of the mixer conversion on the interfering carrier; a broad group of RF interference problems can be analyzed within the scope of the existing theories of interfering carrier distortion.

One other major interference problem of interest has been left undiscussed within the scope of this paper. It is the so-called direct adjacent channel interference problem[16] (DACI) involving closely spaced carriers. Although a theory[17] has been presented to account for this effect experimental evidence indicates that it is only partially complete. Here a full resort to empirical data still appears to be required and methods of specifying a trade off between acceptable interference rejection via filtering and potential degradation of the wanted channel (primarily limitations on equalization of the filter group delay) are somewhat of a *cut and try* technique. It is hoped that subsequent investigations will more fully clarify the mechanisms involved in DACI.

Appendix

Evaluation of $[F_2(w)]^2$ *(with emphasis)*

Following the treatments of Medhurst[3], [6] and Medhurst *et al.*,[4] it may be noted that with the modulated IF carrier having unity power and of the form

$$v_{\text{IF}}(t) = \sqrt{2}\cos(\omega_{\text{IF}}t + \phi(t)). \tag{21}$$

autocorrelation of $v_{\text{IF}}(t)$ (second order portion only) is given by

$$[R_v(\tau)]_2 = \frac{1}{2}\exp(-0.458\omega_\Delta^2/p_op_m)\cos\omega_{IF}\tau R_\phi^2(\tau). \quad (22)$$

The second order power spectrum in a bandwidth w_b at a frequency $w_{IF} \pm \omega$ is thus obtained from

$$W_2(\omega) = \frac{\omega_b}{2}\exp(-0.458\omega_\Delta^2/p_op_m)\int_{-\infty}^{\infty} \times \cos\omega_{IF}\tau R_\phi^2(\tau)e^{-j\omega\tau}d\tau \quad (23)$$

which can be transformed to

$$W_2(\omega) = \frac{\omega_b}{4}\exp(-0.458\,\omega_\Delta^2/p_op_m)S_\phi(p)*S_\phi(p) \quad (24)$$

where $*$ denotes the process of convolution and $S_\phi(p)$ is the power spectral density of the phase modulation $\phi(t)$ now considered to be centered about ω_{IF}.

In order to evaluate (24) we express $\phi(t)$ as

$$\phi(t) = \frac{\sqrt{2}\omega_\Delta}{\sqrt{p_m - p_o}}\sum_{p_o}^{p_m}\frac{\sqrt{f(p)}}{p}\sin(pt - \gamma_p) \quad (25)$$

here p increases in unit steps and γ_p is a random phase angle.

The power spectral density of the phase modulation is, from inspection of (25)

$$S_\phi(p) = \frac{\omega_\Delta^2}{p_m - p_o}\frac{f(p)}{p^2} \quad (26)$$

Hence, the one sided second order power spectrum becomes

$$W_2(\omega) = \exp(-0.458\omega_\Delta^2/p_op_m)\frac{\omega_b}{4}\frac{\omega_\Delta^4}{(p_m - p_o)^2} \times \left(\frac{f(p)}{p^2}\right)*\left(\frac{f(p)}{p^2}\right) \quad (27)$$

where $f(p)$ is approximated as

$$f(p) \approx g(p) = K_0\exp(\propto p/p_m), p_o \le p \le p_m$$

$$f(p) = 0, \text{ elsewhere.}$$

Equation (27) can be rewritten as

$$W_2(\omega) = \exp(-0.458\omega_\Delta^2/p_op_m)\frac{\omega_b}{4}\frac{\omega_\Delta^4}{(p_m - p_o)^2} \times \int_{\infty-}^{\infty} K_o^2\frac{g(p)}{p^2}\frac{g(\omega - p)}{(\omega - p)^2}dp \quad (28)$$

where the integral in (28) represents $[F_2(w)]^2$.

Performing the integration involved in (28) yields,

$$[F_2(\omega)]^2 = \frac{k_0^2}{2}\Big\{\exp(\propto\omega)\left[\frac{2(\omega - 2p_o)}{\omega^2p_o(\omega - p_o)} + \frac{4}{\omega^3}\ln\frac{\omega - p_o}{p_o}\right]_{2p_o \le \omega \le p_m + p_o} + \exp(\propto\omega)\left[\frac{2(2p_m - \omega)}{\omega^2p_m(\omega - p_m)} + \frac{4}{\omega^3}\ln\frac{p_m}{\omega - p_m}\right]_{p_m + p_o \le \omega \le 2p_m} + [\exp(-\propto\omega)A(\omega) + \exp(\propto\omega)B(\omega)]_{o \le \omega \le p_m - p_o}\Big\} \quad (29)$$

where

$$A(\omega) = \frac{1}{\omega^2}\left[\frac{\omega + 2p_o}{p_o(\omega + p_o)} - \frac{2p_m - \omega}{p_m(p_m - \omega)}\right] + \frac{p_m - p_o - \omega}{p_o(p_m - \omega)}\left[\frac{2\propto}{\omega} + 2\propto^2 + 4/3\propto^3\omega + 2/3\propto^4\omega^2\right] + \left[\ln\frac{p_op_m}{(\omega + p_o)(p_m - \omega)}\right]\left[\frac{2}{\omega^3} + \frac{2\propto}{\omega^2}\right] + \left[\ln\frac{p_m - \omega}{p_o}\right] \times [4/3\propto^3 + 4/3\propto^4\omega] + 2/3\propto^4[p_m - p_o - \omega] \quad (30)$$

and

$$B(\omega) = \frac{1}{\omega^2}\left[\frac{\omega + 2p_o}{p_o(\omega + p_o)} - \frac{2p_m - \omega}{p_m(p_m - \omega)}\right] + \frac{p_m - p_o - \omega}{p_m(\omega + p_o)}\left[-\frac{2\propto}{\omega} + 2\propto^2 - 4/3\propto^3\omega + 2/3\propto^4\omega^2\right] + \left[\ln\frac{p_op_m}{(\omega + p_o)(p_m - \omega)}\right]\left[\frac{2}{\omega^3} - \frac{2\propto}{\omega^2}\right] + \left[\ln\frac{p_m}{p_o + \omega}\right] \times [4/3\propto^3 - 4/3\propto^4\omega] + 2/3\propto^4[p_m - p_o - \omega]. \quad (31)$$

The third term in (29) defined for $0 \le w \le p_m - p_o$ requires evaluation of integrals of the form

$$\int_{\omega - p_m}^{-p_o}\frac{e^{\propto\omega}e^{-2\propto p}}{p^2(p - \omega)^2}dp \quad \text{and} \quad \int_{\omega + p_o}^{p_m}\frac{e^{-\propto\omega}e^{2\propto p}}{p^2(p - \omega)^2}dp.$$

Here the exponentials involving $e^{\pm 2\propto p}$ have been approximated as a truncated five-term power series in $2\propto p$.

Acknowledgment

The author is indebted to his colleagues, C. A. G. Delory, L. A. Fraser, and M. Mayer, who supplied the measurement data associated with this paper.

References

[1] Internat'l Radio Consultative Committee, Geneva, 1963, vol. 4, Recommendation 395, p. 63.
[2] R. Hamer, "Power spectrum of a carrier modulated in phase or frequency by white noise," *Electronic and Radio Engineer*, pp. 248–253, July 1957.
[3] R. G. Medhurst, "RF spectra of waves frequency modulated with white noise," *Proc. IEE*, vol. 107C, pp. 314–323, May 1960.
[4] R. G. Medhurst, E. M. Hicks, and W. Grossett, "Distortion in frequency-division-multiplex FM systems due to an interfering carrier," *Proc. IEE*, vol. 105B, pp. 282–292, May 1958.
[5] R. Hamer, "Radio frequency interference in multichannel telephony FM radio systems," *Proc. IEE*, vol. 108B, pp. 75–89, January 1961.
[6] R. G. Medhurst, "RF spectra and interfering carrier distortion

in FM trunk radio systems with low modulation ratios," *IRE Trans. Communication Systems*, vol. CS-9, pp. 107–115, June 1961.

[7] ——,"FM interfering carrier distortion: general formula," *Proc. IEE*, vol. 109B, pp. 149–150, March 1962.

[8] B. R. Hallford and J. N. Ratti, "Microwave receiver interference characteristic measurements," *IEEE Trans. Communication Technology*, vol. COM-14, pp. 455–469, August 1966.

[9] B. B. Jacobsen, "Frequency patterns for multiple-radio channel routes," *Proc. IEE*, vol. 107B, pp. 241–252, May 1960.

[10] James W. Steiner, "An analysis of radio frequency interference due to mixer intermodulation products," *IEEE Trans. Electromagnetic Compatibility*, vol. EMC-6, pp. 62–68, January 1964.

[11] L. M. Orloff, "Intermodulation analysis of crystal mixer," *Proc. IEEE*, vol. 52, pp. 173–179, February 1964.

[12] R. D. Trammell, Jr., "A method of determining mixer spurious response rejection," *IEEE Trans. Electromagnetic Compatibility*, vol. EMC-8, pp. 81–89, June 1966.

[13] S. Goldman, *Frequency Analysis, Modulation and Noise*. New York: McGraw Hill, 1948, p. 164.

[14] Internat'l Radio Consultative Committee, *11th Plenary Assembly Documents*, Oslo, 1966, Draft Recommendations for 1200/1260 channel radio-relay systems.

[15] Op. cit, reference 1, Recommendation 275, p. 78.

[16] H. E. Curtis, T. R. D. Collins, and B. C. Jamison, "Interstitial channels for doubling TD-2 radio system capacity," *Bell Sys. Tech. J.*, vol. 39, pp. 1505–1527, November 1960.

[17] C. L. Ruthroff, "A mechanism for direct adjacent channel interference," *Proc. IRE* (*Corres.*), vol. 49, pp. 1091–1092, June 1961.

Multipath Intermodulation Associated with Operation of FM–FDM Radio Relays in Heavily Built Areas

DECIO ONGARO

***Abstract*—Microwave FM radio link paths concerning heavily built areas may be found to suffer from high amounts of intermodulation noise caused by multiple reflections on buildings and reflecting obstacles lying alongside the direction of propagation. Little direct information may be gained on these effects using standard techniques for measuring transmission performance. The limitations of those techniques are discussed and some additional ones are proposed allowing a more direct insight into the phenomenon. The results obtained in measurements on a real hop suffering from marked echo effects of this type are reported and discussed.**

Paper approved by the Radio Communication Committee of the IEEE Communication Technology Group for publication after presentation at the 1970 International Conference on Communications, San Francisco, Calif., June 8–10. Manuscript received July 1, 1970; revised December 7, 1970.

The author is with the Radio Communications Division of Società Italiana Telecomunicazioni, Siemens, Milan, Italy.

I. Introduction

RADIO RELAY terminals are normally located in densely populated areas where the conveyed information must be utilized. It is widely known that RF interference is the main problem in such stations, especially when many routes converge to the same terminal. The RF channel allocation plans regulating RF spectrum utilization take this fact into full account. It is perhaps less known that additional troubles may arise due to

Reprinted from *IEEE Trans. Comm. Tech.*, vol. COM-19, pp. 293-301, June 1971.

propagation anomalies caused by the peculiarities of the environment wherein the electromagnetic energy is radiated (this is true in particular when entering into the center of European-built cities). These effects are normally negligible if the terminal antennas are installed well above the maximum height reached by the buildings lying near and along the direction of propagation. However, this may call for the use of radio towers hundreds of feet tall, a solution which in some cases may be too expensive or even impossible to implement.

If a simpler approach is followed of placing the antennas on top of a tall building, even if it is reasonably high, care should be taken to evaluate the effect of the chosen path on transmission performance of the link as soon as possible. In fact anomalous propagation effects in the form of echoes generated by single and multiple reflections might be found to affect transmission performance so much as to make further implementation of the link unadvisable without making some kind of change. It must be pointed out that first Fresnel zone clearance rules sometimes do not appear to be sufficient to warrant acceptable results in these cases. The reason for this is that large modern buildings in general contain sizable amounts of metal, to say nothing of steel towers, tin roofs, etc., all of which make these obstacles capable of reflecting any incident electromagnetic energy with a surprisingly high degree of efficiency. Hence even a small amount of energy available well off the main beam may undergo unfavorable sequences of reflections and eventually arrive at the receive antenna in the form of echoes. Since the delay of these echoes is very large, they may be capable of having a considerable effect on the intermodulation (IM) performance.

Estimation of these effects is probably impossible to carry out in advance on a quantitative basis without performing direct measurements in the final configuration (especially in regard to antenna used). It might be mentioned in support of this consideration that measurements carried out in some cases of "radar cross section" of obstacles whose effect had been identified on echo measurements yielded values which did not apparently justify the level of the echoes subsequently measured in more direct ways. However, difficulties also arise in this respect with measuring techniques, particularly with the conventional ones which generally do not provide the necessary information. For instance, it will be shown that the measurement of envelope delay distortion (EDD) is useless in most cases and that conventional noise loading techniques for measuring IM performance are capable of giving only indirect answers as to the presence of such effects, and only when carried out in connection with an antenna system and equipment configuration duplicating the final one as closely as possible.

Much more can be learned and better data obtained by using special techniques allowing a deeper and more direct insight into the phenomenon. In the following the limitations of the conventional measuring techniques will be pointed out, and the advantages of some additional ones will be discussed. The results obtained on an actual RF path exhibiting these effects to a marked degree will also be reported.

II. Echo Characterization and Limitation of Validity of Envelope Delay Measurement

Weak echoes cause a ripple form of EDD which may be approximated by the well-known formula

$$\tau_E = r\tau_0 \sin [\Delta\omega\tau_0 + \varphi] \qquad (1)$$

where

- τ_E EDD caused by echo
- τ_0 time delay of echo with respect to main signal
- r relative (voltage ratio) echo amplitude with respect to main signal amplitude
- $\Delta\omega$ angular frequency deviation of carrier with respect to its center frequency
- φ relative phase angle of main signal and echo when they recombine.

The period of the EDD ripple is $1/\tau_0$ or $c/2l$, where c is the velocity of light and l is the distance between two successive reflections. If l is of the order of some thousands of feet, τ_0 is of the order of several μs, and the ripple period consequently lies in the range of 100–200 kHz or even less. The latter fact makes this effect wholly transparent to a measure of EDD carried out with standard methods. Hence it is practically useless in this respect.

The reason for this lies in the fact that the actual EDD is subjected to a considerable amount of smoothing before being displayed. This smoothing is due to two causes. The first is that EDD is measured as phase variation of a fixed-frequency tone that frequency modulates an IF carrier swept along the band to be tested. The test-tone frequency ordinarily is in the range of several hundreds of kilohertz and cannot be too low due to noise-rejection problems. A typical value is 500 kHz. This yields a result which, at each IF frequency, is actually the average of the original EDD over a span of twice the test-tone frequency.

Moreover, the signal obtained by phase demodulating the test tone is subjected to an additional amount of post-detection filtering, again for noise rejection purposes. The latter effect can be reduced to a certain degree by lowering the sweep rate and/or by reducing the sweep width when it is desired to observe the small grain of the EDD curve. However, nothing can be done about the first cause.

It can be easily shown that in case of an EDD in the form of a sinusoidal ripple the sensitivity of the measuring method, i.e., the ratio between the displayed value of the EDD and its real value, exhibits a $(\sin x)/x$ shape with respect to the ripple rate (in ripple cycles per MHz of swept band). Fig. 1 shows the sensitivity plotted for a test-tone frequency of 250 kHz. It is apparent that in the latter case for ripple rates of about 10 cycles per MHz, as would be appropriate for double-reflection echoes with delays as assumed in the preceding, the sensitivity is at

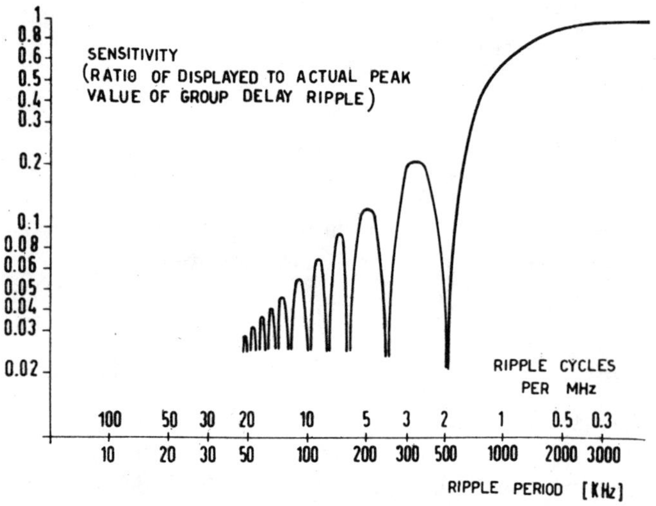

Fig. 1. Sensitivity in measure of group delay with 250-kHz test tone as function of ripple rate (ripple cycles per MHz) of group delay distortion.

best only 0.1, or 10 percent of the correct value. An estimate of IM noise level carried out on the basis of these observations would be in error by 1 to 100.

III. Effect of Echoes with Large Delay on IM Noise in FM Radio Links

The effect of echoes on IM noise in FM radio links has been derived, both theoretically and experimentally, in a great number of papers [1]–[3], [10]; universal charts are also available where IM values may be obtained, given echo delay and amplitude [3]. However, these data generally do not contain all the information needed when faced with the converse problem, i.e., when it is desired to distinguish in an IM measurement what part of the overall IM noise power should be ascribed to the presence of echoes and what part is due to equipment circuitry. Correlation between echo-generated and equipment-generated IM noise is the first information which ought to be known before proceeding further in this direction [7], [8]. When investigating the effects of some unknown cause of IM it is not sufficient to keep the IM noise generated within the measuring equipment reasonably lower than that which is presumably due to the unknown cause to warrant acquisition of adequate quantitative information on the latter. Correlation between the different IM causes might affect the final result in a very heavy measure. If, e.g., a case is considered where the measuring equipment contributes $\simeq$10 picowatts psophometrically weighted at a point of zero relative level ($pW0_p$) noise when the cause under investigation is removed, and where the latter's contribution would actually amount to 100 $pW0_p$, an overall measurement can be expected to yield any IM noise level between 46 and 174 $pW0_p$ depending on the degree of correlation existing between the IM causes. This amounts to an uncertainty range of almost 6 dB, in contrast to the limited effect of about 1 dB if straight power addition did apply.

It will be shown in the following that for transmission capacities in the order of 960 voice channels and with standard per channel frequency deviations, the IM noise generated in the upper modulation channels by a highly delayed echo has the following characteristics which may be of help in separating it from other kinds of noise.

1) It has no appreciable correlation with other kinds of equipment-generated IM noise.

2) IM (echo) noise level tends to be a constant value with respect to variations of signal deviation over a very wide range. This is contrary to the behavior of equipment-generated noise, which increases in level by at least 1 dB for every dB increase of signal deviation [7], [8].

It should be observed that the delay time τ_0 must meet the criterion $f_m\tau_0 \gg 1$ to be considered "long," where f_m is the maximum modulating signal frequency.

If reference is made to the phasor diagram of Fig. 2, it can be seen that the error angle produced by a single weak echo on the modulated RF carrier is given by

$$\epsilon = r \sin [\Phi(t) - \Phi(t - \tau_0) + \varphi] \qquad (2)$$

where r and φ have already been defined and $\Phi(t)$ is the phase modulating information signal which is assumed to be Gaussian [multichannel frequency division multiplex (FDM) case] in nature.

In Fig. 2 a block diagram is also shown, derived from (2), which graphically represents the formation process of the IM noise generated by the echo. The frequency modulating signal is passed through the linear transfer function $F(j\omega)$ and is then mixed with the constant angle φ into the nonlinear block $r \sin [\cdot]$. Since the signal entering into the latter block is Gaussian, the $\sin [\cdot]$ function can be developed in a sum of Hermite polynomials of increasing order [5]. Each polynomial of the $-n$th order gives rise exclusively to a phase intermodulation spectrum of the $-n$th order. This spectrum can then be converted into a frequency IM noise by weighting it with the subsequent differentiating block $j\omega$.

Different cases occur with respect to correlation to other kinds of IM noise and to behavior with regard to signal deviation, in dependence on the echo time delay, which in fact heavily affects the linear transfer function $F(j\omega)$. The absolute value of the latter is reported in Fig. 3 where it appears to exhibit the familiar $(\sin x)/x$ shape. If the modulating frequency f_m and echo time delay τ_0 are such that $f_m\tau_0 < 1$, only a portion of the first lobe of $F(j\omega)$ (where the latter has values near unity) affects the information signal. In this way an appreciable degree of correlation between $\dot{\Phi}(t)$ and $\Psi(t)$ is maintained [4]. This in turn results in some degree of correlation between IM spectra generated by the $r \sin [\cdot]$ block and spectra generated within the equipment by other nonlinear operations on the modulating signal $\dot{\Phi}(t)$ [6].

It must be pointed out that in the latter case the EDD curve would exhibit a ripple of less than one cycle for each portion of the swept band f_m MHz wide (4 MHz for the 960-channel case) and would be easily measurable with standard EDD measuring equipment. On the other

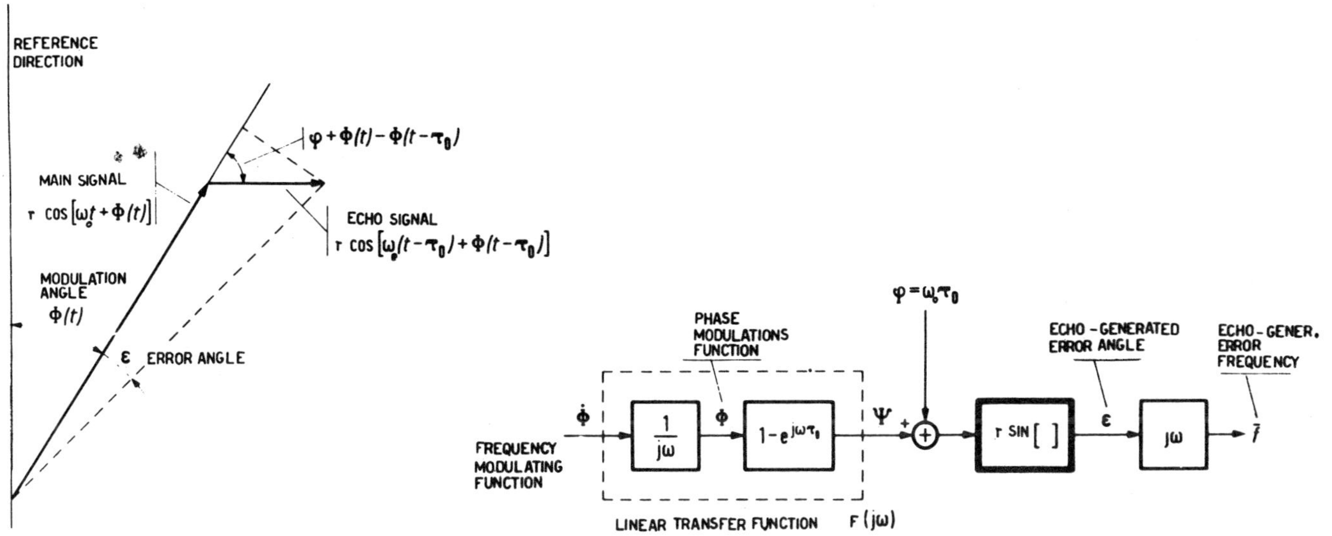

Fig. 2. Phasor and block diagram representing mechanism by which echo generates IM noise in baseband.

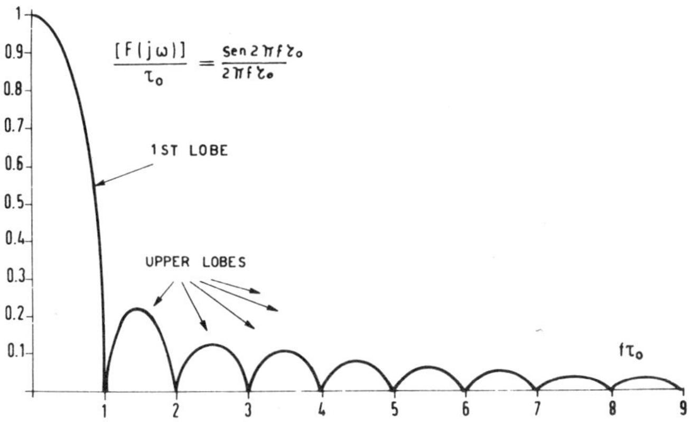

Fig. 3. Normalized absolute value of transfer function $F(j\omega)$ as function of normalized frequency $F\tau_0$.

hand, if τ_0 is such that $f_m\tau_0 \gg 1$ (e.g., 10 or more), many lobes of $F(j\omega)$ would be of interest, and the behavior of the corresponding IM noise would be very different. Since the higher lobes of $F(j\omega)$ have very low values (e.g., about 0.03 for the tenth lobe), cross correlation between $\dot{\Phi}(t)$ and $\Psi(t)$ is correspondingly lower, and the same also holds for the IM noise produced by the $r \sin [\cdot]$ block and the corresponding equipment-generated noise.

As for the second property mentioned in the foregoing, it may be observed that when τ_0 is large, signal $\Psi(t)$ can be decomposed into the sum of two terms Ψ_1 and Ψ_2; Ψ_1 is the part of $\Psi(t)$ processed by the first lobe of $F(j\omega)$, and Ψ_2 is the part of $\Psi(t)$ concerning all other lobes. It can be seen that Ψ_1 accounts for more than 90 percent of the overall power of $\Psi(t)$. It may be shown that the power contained in the first f MHz of $\Psi(t)$ is proportional to

$$y = \int_0^{2\pi f_m} |F(j\omega)|^2 \, d\omega = \int_0^{f_m} \left[\frac{\sin 2\pi f\tau_0}{2\pi f}\right]^2 df$$

$$= \text{Si}(4\pi f\tau_0) - \frac{\sin^2 2\pi f\tau_0}{2\pi f\tau_0}$$

which rapidly converges to its asymptotic value. This holds of course if for the sake of simplicity the assumption is made that the spectrum of $\Psi(t)$ is constant and extends all the way down to zero frequency.

It follows that if the error angle $\epsilon(\Psi)$ is expanded in a Taylor series, a sufficient approximation is provided by the first two terms so that the output of the nonlinear block $r \sin [\cdot]$ may be written as

$$\epsilon \simeq r \left\{\sin [\Psi_1 + \varphi] + \left[\frac{d}{d\Psi} \sin (\Psi + \varphi)\right]_{\Psi = \Psi_1} \cdot \Psi_2\right\}. \quad (3)$$

The first term gives rise to an IM spectrum which does not extend into the region of the upper channels, even if high-order IM terms are actually present since Ψ_1 is very narrow band with respect to the band occupied by the information signal. Only the second term substantially accounts for the IM noise in the upper channels. The multiplying factor

$$\alpha(t) = \left[\frac{d}{d\Psi} \sin (\Psi + \varphi)\right]_{\Psi = \Psi_1(t)}$$

is a random function of time, i.e., a random process whose properties depend both on those of the random process $\Psi_1(t)$ and on φ. If α_0 is the expected value of $\alpha(t)$, the latter can be written as $\alpha(t) = \alpha_0 + \alpha_1(t)$, where both α_0 and the variance of $\alpha_1(t)$ depend on the variance of $\Psi_1(t)$. It may be seen that when the latter increases, α_0 tends to zero for all values of φ, and the variance of $\alpha_1(t)$ tends to a constant value [5], [11].

This holds with a good approximation already when the variance of $\Psi_1(t)$ is larger than 1 rad. Under these conditions the effect of $\alpha(t) \simeq \alpha_1(t)$ is that of spreading the power of $\Psi_2(t)$ contained in each channel slot into the adjacent channel slots. Hence the IM power depends directly on the power of $\Psi_2(t)$, i.e., on signal deviation. It follows that the IM noise level is constant with respect to signal deviation.

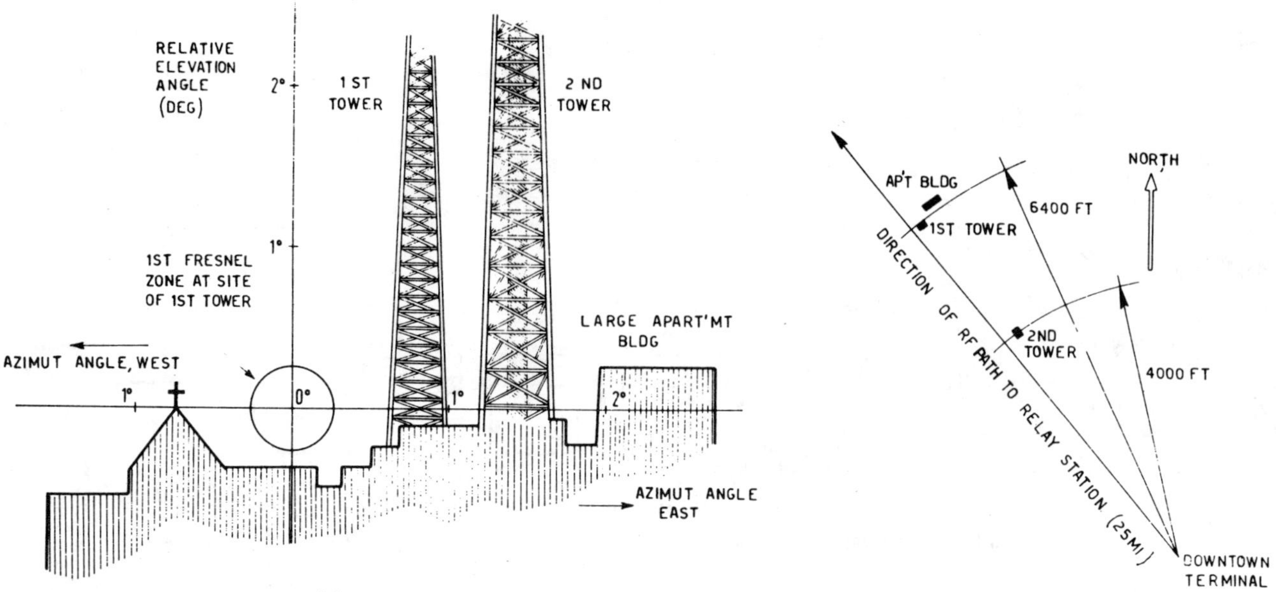

Fig. 4. (a) Skyline as seen from downtown terminal. (b) Layout of RF path in city area with position of main landmarks of (a).

IV. Experimental Results

Reference will be made in the following to a typical case of envelope delay and IM noise measurements on an RF path where long-delay echo effects were found. These effects were observed on the first hop of a new 4-GHz heavy route connecting two large cities of northern Italy. Planned capacity of each RF channel was 960 voice channels with a standard per channel deviation of 200 kHz rms.

Tests were carried out at different sites for the terminal station in order to determine the most convenient RF path to the first relay station of the route. The equipment used was the same in all cases, and the antennas were 10-ft diameter horn reflectors. The results obtained were consistently found to suffer from long-delay echo effects unless the terminal antennas were positioned more than 300 ft above ground. This was due to the particularly heavy building concentration in the section of interest.

A typical case occurred when the tests were carried out siting the antenna on an 11-story building already in use as a terminal for other radio routes long in operation. This solution would have been the most advisable on obvious economic grounds. In Fig. 4(a) the skyline is shown as viewed from the antenna's position. In Fig. 4(b) the radio path is plotted, and the main landmarks of Fig. 4(a) are indicated. It should be noted that there are several high buildings and steel towers in the general direction of the outgoing beam. However, there is reasonable clearance of the first Fresnel zone throughout.

Noise loading IM tests carried out at nominal capacity in both senses yielded IM noise powers of about 300 $pW0_p$. This is far in excess of any contribution which could be expected from the radio equipment and the antenna feeder runs. In the present case the feeder run was rectangular waveguide and was about 100 ft long. Feeder VSWR at input flange when checked yielded values within acceptable limits. Two separate antennas with associated feeder runs were employed at each terminal for transmission and reception. The signal to IM noise ratio when measured varying the frequency deviation of the signal according to a test-tone variation from 140 to 450 kHz rms, i.e., over a range of about 10 dB, turned out to remain constant over the whole range in both the middle and top measuring channels. These results are shown in Fig. 5(a) together with the corresponding IM values measured on a typical hop over the same route and with antenna system and radio equipment of the same characteristics. The latter IM noise levels may be seen to be lower by an order of magnitude at nominal deviation while increasing by approximately 2 dB for each dB variation of signal deviation. The IM noise levels in the said hop were found to be independent of the RF channel frequency and, of course, independent of the direction of propagation.

EDD measurements were also carried out with standard test equipment (test signal frequency was 250 kHz) on the same hop and yielded the results reported in Fig. 5(b). Curve A was obtained with maximum filtering after the phase detection circuit and looks quite acceptable. Curve B was obtained with minimum filtering, some degree of ripple distortion now being visible. The ripple period is about 2.5 MHz, but its peak value does not correlate with the amount of IM noise previously found. The ripple period of about 2.5 MHz may be associated (according to (1)) to an echo delayed by about 400 ns, corresponding to 400 ft in free space and about 240 ft of waveguide. This checks with the known length of the waveguides which was slightly more than 100 ft and sup-

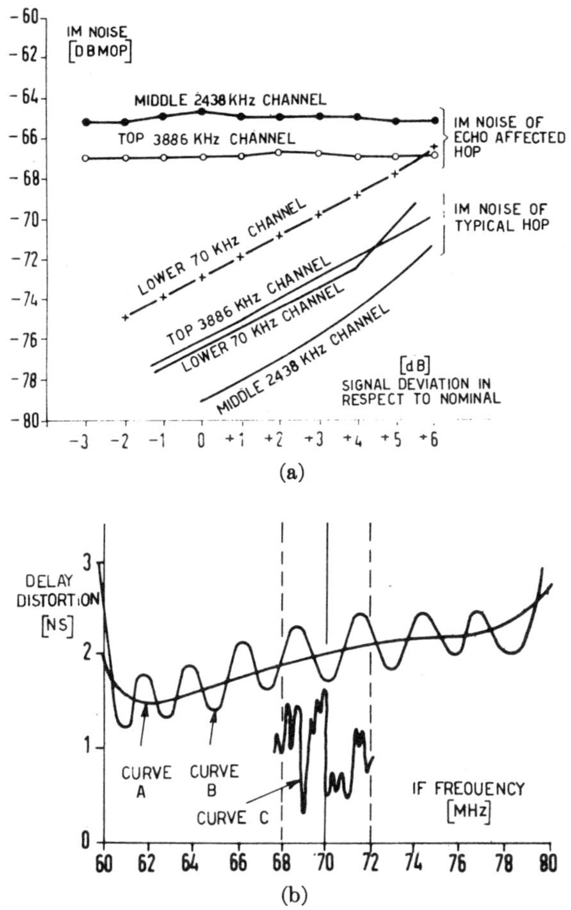

Fig. 5. (a) IM noise versus signal deviation for echo-affected hop and for normal hop with similar equipment. (b) Typical group delay distortion of RF channel on echo-affected hop as measured in different post-detection filtering conditions.

ports the belief that this echo was probably due to slight mismatchings at both feeder ends. No EDD ripple sensitivity problems as discussed in Section II and Fig. 1 exist in this case due to the ripple period of 2.5 MHz being much larger than the test-tone frequency of 250 kHz. Thus (1) allows us to obtain that, as the peak-to-peak EDD ripple was lower than 0.5 ns, the corresponding echo was attenuated by $r^{-1} = 2\tau_0/\tau_E = 400/0.25 = 1600$ or 64 dB which is in approximate agreement with the feeder VSWR measurements. According to the universal curves of [3, fig. 5.7] such an echo attenuation and delay correspond to a worst case IM noise contribution of about 4 $pW0_p$ for a 960-channel system with no de-emphasis. Even leaving out the favorable effect of the latter, the aforementioned figure may be seen to be lower by almost two orders of magnitude than the overall IM noise observed on the hop.

Curve C was obtained under the same conditions as curve B but with sweep width reduced from 20 to 4 MHz. A larger ripple distortion is now visible and additional components whose period is considerably smaller than in the previous case can be observed. Even in this case, however, the measured delay distortion is not such as to justify the IM noise found [3].

The conclusions which could be arrived at based on the preceding results were then that long-delay echo effects appeared to be present, but their contribution to the total measured IM noise was not apparent. It is interesting to note that similar results were obtained with the downtown terminal relocated at a distance of about one mile from the first one. This removed the two steel towers from a close proximity to the radio path. Different kinds of measurement techniques were then tried in order to gain a clearer and deeper insight into the phenomenon.

A. First Method: Measurement of Level Variations of Upper Harmonics Generated in System When Carrier is Modulated by Baseband Test Tone Varying in Frequency

A technique long out of use, but a variation of which proved to be of considerable help in the present case, is that of observing the harmonic distortion suffered by a sinusoidal baseband test signal as a function of its frequency [9]. The information obtained is not blurred as in the case of a wide-band test signal such as the white noise signal, and high resolution is obtainable by the use of a narrow-band selective voltmeter, frequently available at most FDM installations.

By going back to the block diagram of Fig. 2 which represents the mechanism by which an echo produces distortion on a modulating signal $\dot{\Phi}(t)$, the case may be examined where $\dot{\Phi}(t)$ is a sinusoidal test signal of the form

$$\dot{\Phi}(t) = \Delta f \sin 2\pi ft \qquad \text{Hz}.$$

In this case a signal is presented to the nonlinear block $r \sin[\cdot]$ of the form

$$\Psi = |F(j\omega)| \cdot \Delta f \sin 2\pi ft = A \sin 2\pi ft \tag{4}$$

where

$$A = |F(j\omega)| \cdot \Delta f.$$

The phase information is left out since it is irrelevant. The output of the nonlinear block is given by

$$\epsilon = r \sin\{A \sin[2\pi ft] + \varphi\} \tag{5}$$

which can be expanded into a series using Bessel functions as follows:

$$\epsilon = r\{J_0(A) \sin\varphi + 2\cos\varphi \cdot J_1(A) \sin 2\pi ft + 2\sin\varphi \cdot J_2(A) \cos 4\pi ft + 2\cos\varphi \cdot J_3(A) \sin 6\pi ft + \cdots\}. \tag{6}$$

Then by taking the time derivative of this expression, the output of the receiver discriminator is obtained. As might be expected, the echo-generated distortion is composed of a fundamental frequency and its upper harmonics. If the baseband test-tone frequency f is varied through a number of successive lobes of the transfer function $F(j\omega)$, the parameter A, i.e., the amplitude of the signal arriving to the input of the nonlinear block $r \sin[\cdot]$, goes through a succession of nulls corresponding to the nulls of $F(j\omega)$. As a consequence, the same happens for all distortion terms. As a matter of fact it is possible that additional nulls other than those due to $F(j\omega)$ occur if the frequency deviation of the test tone is chosen too large. However, this cannot occur if the maximum value of A is limited to 2 rad. This is always the case for the range of usable frequency deviations if the test-tone frequency is sufficiently high (e.g., in the MHz range).

Thus a measurement of the level of any of the preceding terms would show a characteristically alternating form whose period would give definite information on the time delay of the echo which causes the distortion, and whose amplitude would give information on the echo attenuation r. The effect of the $\cos \varphi$ and $\sin \varphi$ terms, respectively, can be canceled simply by a slight readjustment of the carrier center frequency in order to maximize the readings of the term of interest (it must be kept in mind that $\varphi = 2\pi f_c \tau_0$, where f_c is the carrier center frequency).

The second-harmonic term $2J_2(A) \sin 4\pi ft$ has proved to be the most suitable source of information. In fact the term at the fundamental frequency is unusable since it is swamped by the actual test signal, and the third-harmonic term is also unsuitable since its amplitude is generally strongly influenced by a comparatively high third-harmonic contribution generated within the measuring equipment. To use higher harmonics would require the use of a too wide baseband spectrum.

If the test-tone deviation is made as low as necessary to allow $J_2(A)$ to be approximated by the quadratic form $A^2/8$, the amplitude of the second harmonic at the output of the discriminator may be written, making reference to Fig. 3 and to the time derivative of (6):

$$\frac{rf}{2} A^2 \quad \text{or} \quad \frac{rf}{2} F^2(j\omega)\Delta f^2 = \frac{r\Delta f^2}{f}(1 - \cos 2\pi f\tau_0) \qquad \text{Hz} \tag{7}$$

i.e., the sum of a constant term and a sinusoidal term with period $1/\tau_0$ Hz. If more than one echo is present, the second-harmonic amplitudes generated by each of them simply add. In this case a Fourier analysis of the second-harmonic amplitude curve would yield information on the number of echoes involved and on their time delays and relative amplitudes.

In Fig. 6 the attenuation of the second harmonic with respect to the test-tone signal is shown as measured on the echo-affected hop already mentioned. The test-tone signal deviation was 400 kHz rms, and its frequency was varied from 2 to 3 MHz. The minimum value of the second-harmonic attenuation shown is about −53 dB at 2 MHz.

It is evident that the second-harmonic amplitude swings through a succession of maxima and minima, and besides that several sinusoidal components are present. This means that according to the previous discussion more than one echo is present. Echo attenuation r can be estimated by making use of (7). It must be kept in mind, however, that this is only an approximate procedure, as (7) holds only for the case of a single echo. According to this equa-

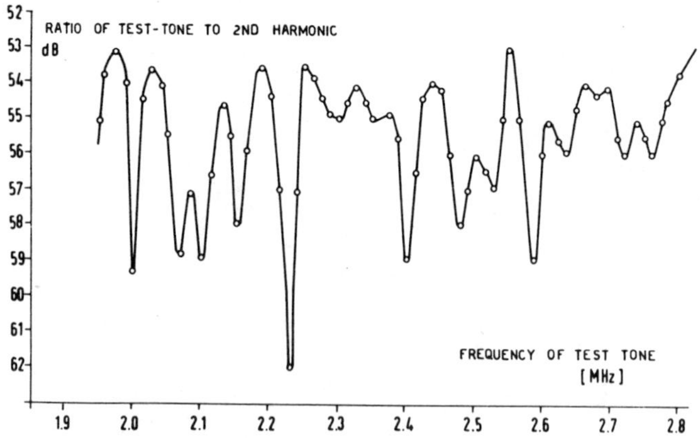

Fig. 6. Typical plot of second-harmonic amplitude versus test-tone frequency on echo-affected hop.

tion, the maximum relative value of second-harmonic amplitude is $r\Delta f/f$. As the measured value is −53 dB, by substituting the values of Δf ($\sqrt{2} \times 400$ kHz) and f (2 MHz), a value for r may be found of about −48 dB. This is in reasonable agreement with the measured IM noise of 300 pW0_p.

[In fact, reverting to use of the curves in [3], it may be seen that for highly delayed echoes the baseband signal-to-noise ratio equals the signal-to-echo ratio at RF. The curves refer to a case without emphasis. However, it may be shown that in this case the presence of emphasis makes no appreciable difference. The earlier discussion proving the property of the constancy of baseband signal-to-IM noise ratio as a function of signal deviation may serve also for this purpose. Hence the IM noise in dBm0_p corresponding to an r value of −48 dB may be obtained simply by subtracting the level of the single channel (−15 dBm0) and the effect of the psophometric weighting (3.6 dB). This yields an IM noise of −66.6 dBm0_p or about 220 pW0_p.]

The second-harmonic data measured were then used to obtain information on the echo pattern by the use of the Fourier transform as reported in the preceding. A fast Fourier transform computer program was used. A straightforward rectangular data window with no shaping was employed since the available data runs appeared to be of sufficient length to allow correct detection of the higher order second-harmonic ripples which were clearly present. The squared transform of the second-harmonic plot of Fig. 6 is shown in Fig. 7 (similar results were obtained for other second-harmonic data measured on different RF channels).

Viewing the results shown on Fig. 7, a number of peaks can be observed. The first and largest peak has a delay of about 2 μs, but the meaning of its presence is very doubtful since the data run is too short to give useful information on ripples of so low an order. Additional peaks may be seen with delays of 6, 10, 14, and 18 μs. It is apparent that even the closest peak would give rise to a group delay distortion unmeasurable with normal equipment.

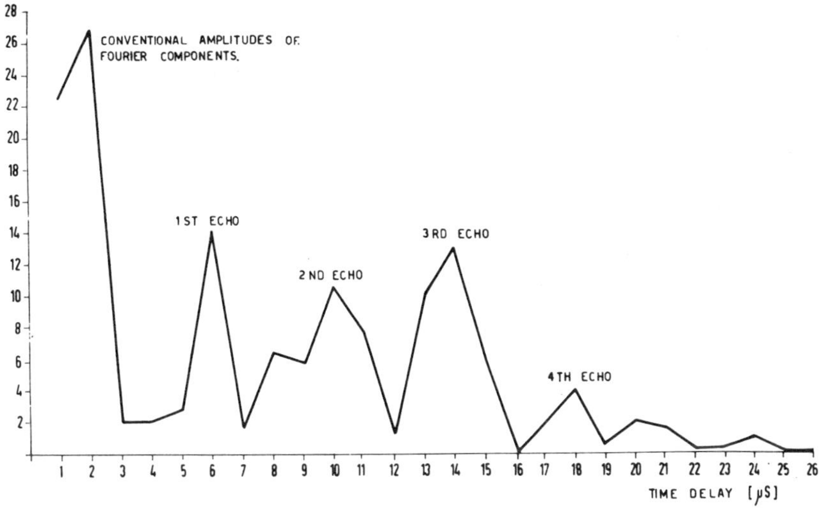

Fig. 7. Square of Fourier transform of second-harmonic plot amplitude derived from Fig. 6.

B. Second Method: Echo Determination with Pulse Techniques

The most direct way of determining the presence of echoes over an RF path is to transmit a sequence of RF pulses in one direction and observe the echoes trailing each pulse at the receive site. In this kind of measurement particular attention must be paid to the carrier leak occurring when the AM modulator used is in its "off" condition since echo pulses 40 dB down or more must be detected. If echo delays of the order of several microseconds must be observed, the pulsewidth may be chosen as high as about 1 μs. This allows for the use of narrow-band noise-rejection filters at the receiving end.

In the case reported here a reflex klystron, pulse-modulated at the repeller, was used to obtain the required high off-condition carrier rejection. The output signal was amplified by the transmitter TWT to a power level of several watts and then fed to the antenna. The received signal was converted to the 70-MHz IF frequency in a standard receiver. After amplification and IF filtering it was displayed (at 70 MHz) on a 100-MHz oscilloscope (which proved to be the most linear detecting device available).

Fig. 8 shows the block diagram of the measuring setup, together with an oscillogram of the received signal. Four main echoes are visible, together with a number of smaller ones. The four main echoes are shown to be delayed by 5, 14, 19, and 27 μs, respectively, which give rise to envelope delay ripple distortions that cannot be detected by standard group delay measuring equipment as discussed earlier.

It can also be seen that the echo delays determined by this direct method are in fair agreement with those derived from the harmonic distortion measurements previously reported. It is interesting to note that the 5-μs delay of the first echo corresponds to twice the distance between the two steel towers. However, the longer delayed echoes cannot be explained in a like manner. Moreover, echo attenuation was measured with a comparison method

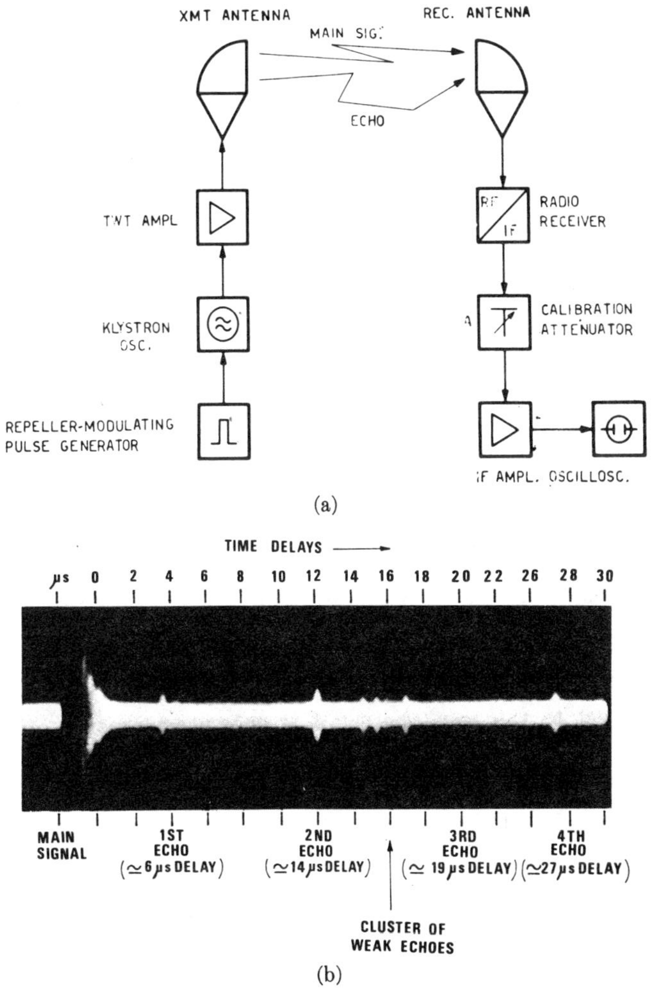

Fig. 8. (a) Layout of equipment for echo determination with pulse measurement. (b) Oscillogram of main signal and echoes with measurement layout as shown in (a). Time origin is at −2 μs with reference to upper time scale.

by making use of a calibrated attenuator A shown in Fig. 8. The values observed were slightly greater than 40 dB for each echo, which again is in good agreement with the results previously obtained with the other method.

Conclusions

The characteristics of multipath IM noise which is liable to degrade radio links located in a heavily built environment have been discussed. Experimental evidence from the case presented shows that echoes with a very large delay may develop, generating IM noise levels which degrade transmission performance in an unacceptable way. These effects are very easily overlooked during preliminary measurements on a prospective RF path. In fact they tend to be transparent to EDD measurements carried out with standard test equipment. Moreover, even white-noise IM measurements are capable at best of giving information only on the presence or absence of such effects, and that only if the test equipment has been equalized well enough to sufficiently reduce its own contribution to IM.

Two additional ways of assessing the presence of long-delay echo effects have been discussed. The first is the measurement of the second-harmonic distortion of a baseband test tone whose frequency is varied. The second is the technique of directly observing the echoes trailing the main signal at the receiver site, using pulse techniques. Experimental results have been presented which indicate that both methods are capable of giving information not only on the presence of echo effects but also on their characteristics such as number of echoes, delay, and amplitude. The pulse method is perhaps the more exact technique by which to obtain the desired data. On the other hand, it requires the use of nonstandard test equipment not normally available at an FDM site, while the first method requires only normally available signal-generating and measuring instruments.

Acknowledgment

A particular acknowledgment is due to P. Stangalino who supervised and carried out all measurements with success. The author also wishes to thank Prof. B. Peroni of the University of Rome for helpful advice and discussions on some points of this work.

References

[1] S. V. Borodich, "Statistical calculation of nonlinear interchannel interference caused by reflections from antenna feeders in multichannel radio relay systems," *Telecommun. Radio Eng.*, 1963, nos. 8, 9.
[2] G. D. Rose, "Laboratory simulation of multipath propagation effects on FM systems," *IEEE Trans. Commun. Technol.*, vol. COM-14, Aug. 1966, pp. 495–498.
[3] W. R. Bennett, H. E. Curtis, and S. O. Rice, "Interchannel interference in FM and PM systems under noise loading conditions," *Bell Syst. Tech. J.*, May 1955.
[4] A. Papoulis, *Probability, Random Variables and Stochastic Processes.* New York: McGraw-Hill, 1965.
[5] E. Kettel, "Die nichtlineare Verzerrung eines normalverteilten Rauschsignals," *Arch. Elek. Übertragung*, Apr. 1968.
[6] J. H. Laning and R. H. Battin, *Random Processes in Automatic Control.* New York: McGraw-Hill, 1966.
[7] T. G. Cross, "Power density spectrum of the sum of two correlated intermodulation noise contributions in FM systems," *Bell Syst. Tech. J.*, vol. 46, 1967, no. 10.
[8] G. J. Garrison, "Intermodulation distortion in frequency-division-multiplex FM systems—a tutorial summary," *IEEE Trans. Commun. Technol.*, vol. COM-16, Apr. 1968, pp. 289–303.
[9] L. Lewin, J. J. Muller, and R. Basard, "Phase distortions in feeders," *Wireless Eng.*, May 1950.
[10] A. Murphy, "Intermodulation distortion due to an echo in FM/FDM radio systems" (in English), *Alta Freq.*, Aug. 1969.
[11] E. D. Sunde, *Communication Systems Engineering Theory.* New York: Wiley, 1969.

Intermodulation Distortion in Frequency-Division-Multiplex FM Systems—A Tutorial Summary

G. J. GARRISON

REFERENCE: Garrison, G. J.: INTERMODULATION DISTORTION IN FREQUENCY-DIVISION-MULTIPLEX FM SYSTEMS—A TUTORIAL SUMMARY, Lenkurt Electric Company of Canada, Ltd., Burnaby, British Columbia. Rec'd 9/20/67; revised 1/15/68. Paper 68TP23-COM, approved by the IEEE Radio Communication Committee for publication without oral presentation. IEEE TRANS. ON COMMUNICATION TECHNOLOGY, 16-2, April 1968, pp. 289–303.

ABSTRACT: In the design of FDM-FM communication systems, a great deal of importance is attached to an appreciation of those transmission irregularities that yield second- and third-order intermodulation spectra. Such spectra are readily generated by amplitude, phase, and AM/PM conversion anomalies within the FM transmission path. In this paper the results of the available journal literature have been reworked for presentation in a generalized form appropriate for use by equipment designers and system analysts. Intermodulation noise magnitudes and spectral distributions are expressed in terms of the transmission network and system modulation parameters with the AM/PM conversion characteristic extended to that of a second-order power series in radian frequency.

To fully exploit the usefulness of the foregoing, the coherency between the intermodulation spectra of all distortion mechanisms is considered. The effects of this correlation is then exemplified for a tandemly interconnected set of distorters. It is found that all second- and third-order distortion mechanisms may be grouped under the classifications of: amplitude, delay, equivalent amplitude/delay, and quasi-equivalent amplitude/delay distortion. It is concluded that, as a result of the correlative structure of the distortion processes, an overall system measurement of intermodulation noise may exhibit an unexpected level of magnitude and/or have a spectral distribution significantly different from any of the inherent system distortions.

I. Introduction

For a number of years considerable attention has been devoted to the study of intermodulation distortion mechanisms associated with frequency-division-multiplex (FDM)-FM communication systems. In particular, a number of significant contributions have been presented pertaining to the consequences of amplitude and phase variations (direct transmission deviations) within the FM signal path. More recently, attention has been given to consideration of the combined effects of these transmission deviations coupled with AM/PM conversion (coupled transmission deviations) in an attempt to explain some of the apparent performance anomalies observed on ultrahigh capacity radio relay systems. Unfortunately, this information has been accumulated piecemeal in the technical literature. Additionally, it is often expressed in an advanced mathematical terminology, or, has been developed for a specific distorting effect and/or a specific set of transmission parameters. As a result, there exists a requirement for a generalized summary typifying the nature of these deleterious transmission irregularities. It is the principal purpose of this paper to fill this need for the important case of "low-order" direct and coupled transmission deviations.

Of particular interest in the design of FDM-FM communication systems is the effects of those transmission irregularities which yield predominately second- and third-order intermodulation spectra, as these low-order distortions tend to limit the performance of a well designed system. (We may note here that, in general, systems which exhibit significant amounts of higher order distortion at nominal traffic loading levels would be classed as totally unsatisfactory due to their tendency to exhibit an abrupt overload.) In terms of equipment requirements our interest with these low-order distortions is related to: the specification of individual subassembly transmission parameters, the allowed interconnection of such subassemblies, and finally the prediction of the performance of a system in which many similar subassemblies may be located in tandem. For any one transmission irregularity the principal facets of interest are thus the magnitude and spectral distribution of the resulting baseband intermodulation noise and the phase coherency of this noise with that from any other transmission variance. Within the scope of a tutorial summary we thus wish to:

1) rework and unify the results of the journal literature for presentation in general form,

2) extend the existing analyses to include a frequency dependent AM/PM converter representable by up to a second-order power series in radian frequency,

3) comment on the correlation of spurious phase distortions resulting from a tandem interconnection of transmission irregularities.

II. General Analysis

A. Direct Transmission Deviations

In order to establish a framework of continuity for subsequent sections let us recapitulate the nature of the modulation perturbations incident to an FM signal when it is transmitted through a linear transducer having nonuniform amplitude and/or phase as illustrated in Fig. 1. Because of the inherent nonlinearity of the FM modulation process, the output of the transducer is, in general, both amplitude modulated ($P(t)$) and phase modulated ($Q(t)$) for a transmission deviation of either amplitude or phase. These spurious modulations can be expressed in terms of the desired phase modulation $\phi(t)$ and the appropriate

Reprinted from *IEEE Trans. Comm. Tech.*, vol. COM-16, pp. 289-304, Apr. 1968.

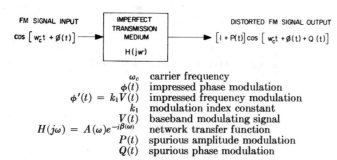

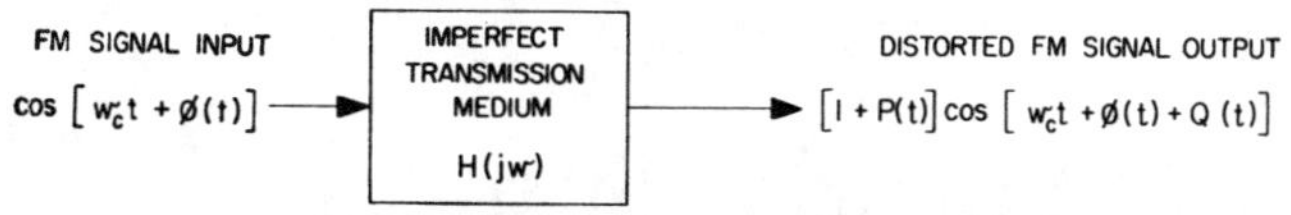

ω_c carrier frequency
$\phi(t)$ impressed phase modulation
$\phi'(t) = k_1 V(t)$ impressed frequency modulation
k_1 modulation index constant
$V(t)$ baseband modulating signal
$H(j\omega) = A(\omega)e^{-j\beta(\omega)}$ network transfer function
$P(t)$ spurious amplitude modulation
$Q(t)$ spurious phase modulation

Fig. 1. Model of an imperfect FM communications system (direct transmission deviations).

TABLE I

SPURIOUS AMPLITUDE AND PHASE MODULATIONS RESULTING FROM DIRECT TRANSMISSION DEVIATIONS

Transmission Deviation	Spurious Amplitude Modulation, $P(t)$	Spurious Phase Modulation, $Q(t)$
Linear gain g_1	$g_1\phi' - 1/2\, g_1^2\phi'^2 + 1/3\, g_1^3\phi'^3$	
Parabolic gain g_2	$g_2\phi'^2 + 1/2\, g_2^2\phi''^2$	$-g_2\phi'' + g_2^2\phi'^2\phi''$
Cubic gain g_3	$-g_3\phi''' + g_3\phi'^3$	$-3g_3\phi'\phi''$
Quartic gain g_4	$-4g_4\phi'\phi''' - 3g_4\phi''^2$	$g_4\phi'''' - 6g_4\phi'^2\phi''$
Parabolic phase b_2	$b_2\phi'' + 2b_2^2\phi'\phi''' + b_2^2\phi''^2$	$-1/2\, b_2^2\phi'''' + b_2\phi'^2$
Cubic phase b_3	$3b_3\phi'\phi''$	$-b_3\phi''' + b_3\phi'^3$
Quartic phase b_4	$-b_4\phi'''' + 6b_4\phi'^2\phi''$	$-4b_4\phi'\phi''' - 3b_4\phi''^2$
Interaction·		
g_1/g_2	$-g_1g_2\phi'^3$	$g_1g_2\phi'\phi'' - g_1^2g_2\phi'^2\phi''$
g_1/g_3	$g_1g_3\phi'\phi'''$	$3g_1g_3\phi'^2\phi''$
g_1/b_2	$2g_1b_2\phi'\phi'' - 2g_1^2b_2\phi'^2\phi''$	$-g_1b_2\phi'''$
g_1/b_3	$-g_1b_3\phi'''' + 3g_1b_3\phi'^2\phi''$	$-4g_1b_3\phi'\phi''' - 3g_1b_3\phi''^2$
g_2/b_2	$g_2b_2\phi'''' + 4g_2b_2\phi'^2\phi''$	$-4g_2b_2\phi'\phi''' - 3g_2b_2\phi''^2$

parameters of the network transfer function $H(j\omega)$. For a transfer function representable by up to fourth-order polynomials in amplitude and phase versus frequency, the corresponding spurious amplitude and phase modulations[1] have been tabulated by Cross[1] and are rewritten here for convenience (Table I). The parameters of the transmission medium apply, as discussed, to a transfer function of the form

$$H(j\omega) = A(\omega)e^{-j\beta(\omega)} \tag{1}$$

where

$$A(\omega) = 1 + g_1(\omega - \omega_c) + g_2(\omega - \omega_c)^2 + g_3(\omega - \omega_c)^3 + g_4(\omega - \omega_c)^4 \tag{2}$$

and

$$-\beta(\omega) = b_2(\omega - \omega_c)^2 + b_3(\omega - \omega_c)^3 + b_4(\omega - \omega_c)^4. \tag{3}$$

[1] The approximations involved in the derivation of Table I require that the transmission deviations be "small." This is normally the case encountered in practice.

B. Coupled Transmission Deviations

For a system in which a perfect limiter follows the imperfect linear transducer of Fig. 1, the spurious amplitude modulations $P(t)$ are not of concern. In practical situations this configuration is generally not possible and we are required to consider a system model such as that of Fig. 2 in which an AM/PM converter is located in tandem with the nonuniform transducer. The effect of the AM/PM converter is to translate the spurious AM modulation $P(t)$ to a spurious phase modulation $\hat{Q}(t)$. This additional phase modulation can again be expressed in terms of the desired modulation $\phi(t)$ and the network parameters of $H(j\omega)$, and additionally, in terms of the AM/PM conversion coefficient $\theta(\omega)$.

In most analyses both the level and frequency dependence of $\theta(\omega)$ are ignored and a constant magnitude value is assigned to this parameter. While it may be appropriate, for low levels of amplitude modulation, to ignore the level dependence of $\theta(\omega)$, we will find it desirous to assign a frequency dependence to this coefficient. This requirement occurs as a result of the nature of many equipment assemblies. Some devices, such as traveling-wave tube (TWT) amplifiers, are inherently broadband and can be assigned a constant AM/PM conversion coefficient. Other devices, such as intermediate frequency (IF) limiters and amplifiers with automatic gain control (AGC), may exhibit a noticeable frequency dependent AM/PM coefficient. Some sample measurement data taken on such units is illustrated in Fig. 3. Thus, for the system model of Fig. 2, we wish to consider an AM/PM coefficient given by

$$\theta(\omega) = \theta_0 + \theta_1(\omega - \omega_c) + \theta_2(\omega - \omega_c)^2. \tag{4}$$

Here $\theta(\omega)$ is the total AM/PM conversion coefficient given in radians per incremental amplitude change while θ_0, θ_1, and θ_2 are the respective constant, linear, and parabolic components of this parameter.

With the generalized hybrid signal

$$[1 + P(t)] \cos[\omega_c(t) + \phi(t) + Q(t)] \tag{5}$$

applied to the input of the nonlinear AM/PM device, the additional spurious phase modulation $\hat{Q}(t)$ has been evaluated in Appendix I and found to be

$$\hat{Q}(t) = \theta_0 P(t) + \theta_1 P(t)\phi'(t) + \theta_2 P(t)\phi'^2(t). \tag{6}$$

Substitution of specific types of $P(t)$ (Table I) into (6) thus yields the set of $\hat{Q}(t)$ phase distortion terms listed in Table II.

III. SPECTRAL ANALYSIS OF SPURIOUS PHASE MODULATIONS

In the preceding section a time representation for second- and third-order phase perturbations has been established. We now wish to develop appropriate definitions for the baseband intermodulation noise (magnitude and spectral distribution) for each spurious modulation. The analysis is basically that described by Liou[2]

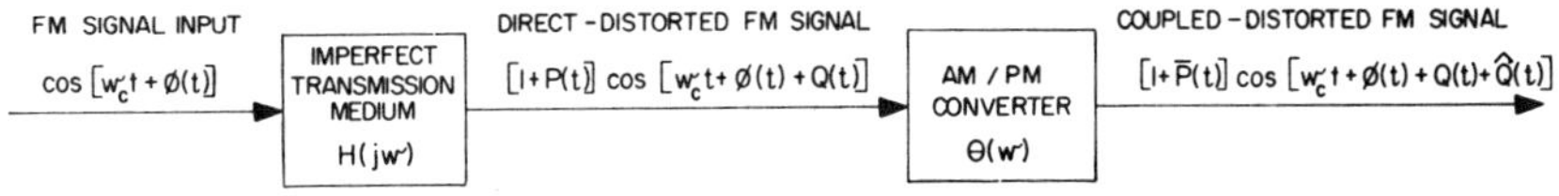

$\theta(\omega)$ frequency dependent AM/PM conversion coefficient
$\hat{Q}(t)$ spurious phase modulation resulting from AM/PM conversion
$\bar{P}(t)$ spurious amplitude modulation at output of AM/PM converter
$\bar{P}(t) \neq P(t)$, in general.

Fig. 2. Model of an imperfect FM communications system (coupled transmission deviations).

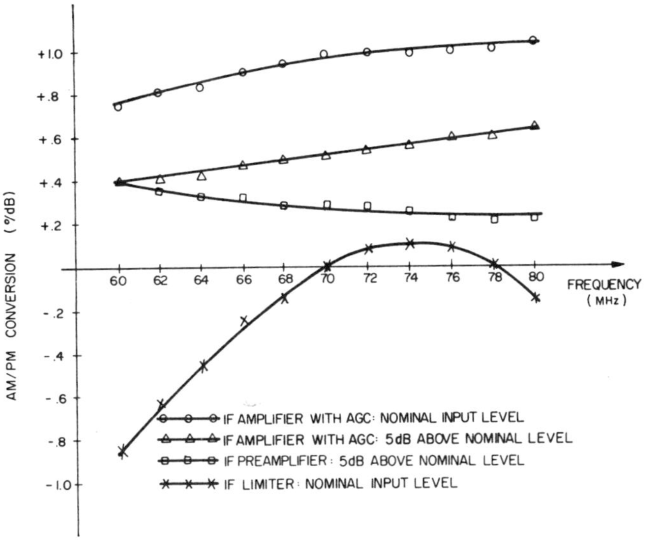

Fig. 3. Representative measurement data—AM/PM conversion of IF equipment assemblies.

TABLE II

Spurious Phase Modulations Resulting From Coupled Transmission Deviations

Transmission Deviation	Spurious Phase Modulation
Linear gain + AM/PM $(g_1 + \theta)$	$g_1\theta_0\phi' - 1/2\ g_1{}^2\theta_0\phi'^2 + 1/3\ g_1{}^3\theta_0\phi'^3$ $g_1\theta_1\phi'^2 - 1/2\ g_1{}^2\theta_1\phi'^3$ $g_1\phi_2\phi'^3$
Parabolic gain + AM/PM $(g_2 + \theta)$	$g_2\theta_0\phi'^2 + 1/2\ g_2{}^2\theta_0\phi''^2$ $g_2\theta_1\phi'^3 + 1/2\ g_2{}^2\theta_1\phi'\phi''^2$
Cubic gain + AM/PM $(g_3 + \theta)$	$-g_3\theta_0\phi''' + g_3\theta_0\phi'^3$ $-g_3\theta_1\phi'\phi'''$ $-g_3\theta_2\phi'^2\phi'''$
Quartic gain + AM/PM $(g_4 + \theta)$	$-g_4\theta_0(4\phi'\phi''' - 3\phi''^2)$ $-g_4\theta_1(4\phi'^2\phi''' - 3\phi'\phi''^2)$
Parabolic phase + AM/PM $(b_2 + \theta)$	$b_2\theta_0\phi'' + b_2{}^2\theta_0(2\phi'\phi''' + \phi''^2)$ $b_2\theta_1\phi'\phi'' + b_2{}^2\theta_1(2\phi'^2\phi''' + \phi'\phi''^2)$ $b_2\theta_2\phi'^2\phi''$
Cubic phase + AM/PM $(b_3 + \theta)$	$3b_3\theta_0\phi'\phi''$ $3b_3\theta_1\phi'^2\phi''$
Quartic phase + AM/PM $(b_4 + \theta)$	$-b_4\theta_0\phi'''' - 6b_4\theta_0\phi'^2\phi''$ $-b_4\theta_1\phi'\phi''''$
Interaction + AM/PM	$-b_4\theta_2\phi'^2\phi''''$
$g_1/g_2 + \theta$	$-g_1g_2\theta_0\phi'^3$
$g_1/g_3 + \theta$	$g_1g_3\theta_0\phi'\phi'''$ $g_1g_3\theta_1\phi'^2\phi'''$
$g_1/b_2 + \theta$	$2g_1b_2\theta_0\phi'\phi'' - 2g_1{}^2b_2\theta_0\phi'^2\phi''$ $2g_1b_2\theta_1\phi'^2\phi''$
$g_1/b_3 + \theta$	$-g_1b_3\theta_0\phi'''' + 3g_1b_3\theta_0\phi'^2\phi''$ $-g_1b_3\theta_1\phi'\phi''''$ $-g_1b_3\theta_2\phi''\phi''''$
$g_2/b_2 + \theta$	$g_2b_2\theta_0\phi'''' + 4g_2b_2\theta_0\phi'^2\phi''$ $g_2b_2\theta_1\phi'\phi''''$ $g_2b_2\theta_2\phi''\phi''''$

with an appropriate substitution of the CCIR parameters and a generalization allowing calculation for all system capacities. Since the analysis tends to be algebraically involved, only a sample analysis will be performed and the results for the remaining spurious modulations simply stated.

As an appropriate illustrative example let us consider a quartic gain transmission irregularity. Noting the spurious phase modulation (Table I) to be

$$Q(t) = -2g_4 \frac{d}{dt}[\phi'^3(t)] \tag{7}$$

the corresponding generalized baseband intermodulation spectral density $S_{Q'}(\omega)$ has been evaluated in Appendix II as

$$S_{Q'}(\omega) = 24\ g_4{}^2k_1{}^6\omega^4[S_V\ (\omega)]_3 \tag{8}$$

where $[S_V(\omega)]_3$ is a third-order convolution of the baseband modulation spectrum $S_V(\omega)$, i.e.,

$$[S_V(\omega)]_3 = S_V(\omega)*S_V(\omega)*S_V(\omega) \tag{9}$$

with $*$ denoting the process of convolution.

Convolutions such as (9) can readily be performed by hand for simple baseband modulation spectra or, for complicated spectra, the aid of a digital computer can be enlisted. We wish to evaluate (9) for the case of a band-limited Gaussian noise test signal, either white or colored by a pre-emphasis network in accordance with the relevant CCIR recommendations.[3] Thus

$$\begin{aligned} S_V(\omega) &= N_0, \quad 0 \leqslant \omega \leqslant \omega_m \\ &= 0, \quad \text{elsewhere} \end{aligned} \tag{10}$$

or

$$\begin{aligned} S_V(\omega) &= N_0K_0e^{\alpha\omega/\omega_m}, \quad 0 \leq \omega \leq \omega_m \\ &= 0, \qquad \text{elsewhere.} \end{aligned} \tag{11}$$

Here, N_0 is the baseband signal power per unit bandwidth without pre-emphasis and $K_0e^{\alpha\omega/\omega_m}$ is an appropriate shaping factor to approximate the CCIR pre-emphasis network, with ω/ω_m being the ratio of a representative baseband frequency to the top modulating frequency. A choice of $K_0 = 0.3$ and $\alpha = 2.1$ is found to provide a satisfactory fit to this network.

For the case without pre-emphasis inserting (10) into (9) and performing the necessary convolutions we find that $[S_V(\omega)]_3$ becomes

TABLE III

INTERMODULATION DISTORTION SUMMARY—DIRECT TRANSMISSION DEVIATIONS

Transmission Irregularity	Order of Distortion	NPR at Top Modulating Frequency (Without Pre-Emphasis)	Spectral Distribution	Distortion (pwp) in an 1800-Channel System. "Nominal" 10 MHz Variance: Amplitude, 0.1 dB Delay, 1.0 ns
Parabolic gain α_2, dB/MHz^2	Third	$\frac{1.72 \times 10^4}{\alpha_2^4 F^4 f_m^4}$	Fig. 4(g)	4.1×10^{-6}
Cubic gain α_3, dB/MHz^3	Second	$\frac{33.6}{\alpha_3^2 F^2 f_m^4}$	Fig. 4(a)	18.5
Quartic gain α_4, dB/MHz^4	Third	$\frac{6.32}{\alpha_4^2 F^4 f_m^4}$	Fig. 4(g)	1.12
Linear delay S, ns/MHz	Second	$\frac{10^6}{\pi^2 S^2 F^2 f_m^2}$	Fig. 4(c)	92.3
	Third	$\frac{7.5 \times 10^{11}}{\pi^4 S^4 F^4 f_m^4}$	Fig. 4(g)	9.4×10^{-5}
Parabolic delay P, ns/MHz^2	Third	$\frac{7.5 \times 10^5}{\pi^2 P^2 F^4 f_m^2}$	Fig. 4(i)	1.4
Cubic delay C, ns/MHz^3	Second	$\frac{1.19 \times 10^6}{\pi^2 C^2 F^2 f_m^6}$	Fig. 4(d)	35.5
Interaction α_1/α_2	Second	$\frac{1.43 \times 10^3}{\alpha_1^2 \alpha_2^2 F^2 f_m^4}$	Fig. 4(a)	4.2×10^{-3}
	Third	$\frac{1.30 \times 10^6}{\alpha_1^4 \alpha_2^2 F^4 f_m^4}$	Fig. 4(g)	5.6×10^{-10}
α_1/α_3	Third	$\frac{1.91 \times 10^3}{\alpha_1^2 \alpha_3^2 F^4 f_m^4}$	Fig. 4(g)	3.7×10^{-5}
α_2/S	Second	$\frac{1.61 \times 10^6}{\pi^2 \alpha_2^2 S^2 F^2 f_m^6}$	Fig. 4(d)	2.6×10^{-2}
α_1/P	Second	$\frac{5.07 \times 10^7}{\pi^2 \alpha_1^2 P^2 F^2 f_m^6}$	Fig. 4(d)	8.4×10^{-3}

TABLE IV

INTERMODULATION DISTORTION SUMMARY—COUPLED TRANSMISSION DEVIATIONS

Transmission Irregularity	Order of Distortion	NPR at Top Modulating Frequency (Without Pre-Emphasis)	Spectral Distribution	Distortion (pwp) in an 1800-Channel System. "Nominal" 10-MHz Variance: Amplitude, 0.1 dB Delay, 1.0 ns AM/PM, 1.0°/dB
Linear gain + constant AM/PM $\alpha_1 + k_{p_0}$	Second	$\frac{9.93 \times 10^5}{k_{p_0}^2 \alpha_1^4 F^2 f_m^2}$	Fig. 4(c)	9.5×10^{-6}
	Third	$\frac{5.62 \times 10^7}{k_{p_0}^2 \alpha_1^6 F^4 f_m^2}$	Fig. 4(i)	1.9×10^{-11}
Linear gain + linear AM/PM $\alpha_1 + k_{p_1}$	Second	$\frac{3.28 \times 10^3}{k_{p_1}^2 \alpha_1^2 F^2 f_m^2}$	Fig. 4(c)	0.29
	Third	$\frac{3.3 \times 10^5}{k_{p_1}^2 \alpha_1^4 F^4 f_m^2}$	Fig. 4(i)	3.2×10^{-7}
Linear gain + parabolic AM/PM $\alpha_1 + k_{p_2}$	Third	$\frac{1.09 \times 10^3}{k_{p_0}^2 \alpha_1^2 F^4 f_m^2}$	Fig. 4(i)	9.6×10^{-3}
Parabolic gain + constant AM/PM $\alpha_2 + k_{p_0}$	Second	$\frac{3.28 \times 10^3}{k_{p_0}^2 \alpha_2^2 F^2 f_m^2}$	Fig. 4(c)	0.29
Parabolic gain + linear AM/PM $\alpha_2 + k_{p_1}$	Third	$\frac{1.09 \times 10^3}{k_{p_1}^2 \alpha_2^2 F^4 f_m^2}$	Fig. 4(i)	9.6×10^{-3}

TABLE IV (*Cont'd*)

Cubic gain + constant AM/PM $\alpha_3 + k_{p_0}$	Third	$\frac{1.09 \times 10^3}{k_{p_0}^2 \alpha_3^2 F^4 f_m^2}$	Fig. 4(i)	9.6×10^{-3}
Cubic gain + linear AM/PM $\alpha_3 + k_{p_1}$	Second	$\frac{2.81 \times 10^4}{k_{p_1}^2 \alpha_3^2 F^2 f_m^6}$	Fig. 4(e)	1.5×10^{-2}
Cubic gain + parabolic AM/PM $\alpha_3 + k_{p_2}$	Third	$\frac{7.76 \times 10^3}{k_{p_2}^2 \alpha_3^2 F^4 f_m^6}$	Fig. 4(j)	6.1×10^{-4}
Quartic gain + constant AM/PM $\alpha_4 + k_{p_0}$	Second	$\frac{9.75 \times 10^2}{k_{p_0}^2 \alpha_4^2 F^2 f_m^6}$	Fig. 4(d)	0.45
Quartic gain + linear AM/PM $\alpha_4 + k_{p_1}$	Third	$\frac{5.56 \times 10^2}{k_{p_1}^2 \alpha_4^2 F^4 f_m^6}$	Fig. 4(k)	8.6×10^{-3}
Linear delay + constant AM/PM $S + k_{p_0}$	Second	$\frac{6.18 \times 10^{13}}{\pi^4 k_{p_0}^2 S^4 F^2 f_m^6}$	Fig. 4(f)	6.7×10^{-4}
Linear delay + linear AM/PM $S + k_{p_1}$	Second	$\frac{1.73 \times 10^8}{\pi^2 k_{p_1}^2 S^2 F^2 f_m^4}$	Fig. 4(a)	0.36
	Third	$\frac{2.82 \times 10^{13}}{\pi^4 k_{p_1}^2 S^4 F^4 f_m^6}$	Fig. 4(l)	1.6×10^{-5}
Linear delay + parabolic AM/PM $S + k_{p_2}$	Third	$\frac{1.3 \times 10^8}{\pi^2 k_{p_2}^2 S^2 F^4 f_m^4}$	Fig. 4(g)	5.4×10^{-3}
Parabolic delay + constant AM/PM $P + k_{p_0}$	Second	$\frac{4.33 \times 10^7}{\pi^2 k_{p_0}^2 P^2 F^2 f_m^4}$	Fig. 4(a)	1.45
Parabolic delay + linear AM/PM $P + k_{p_1}$	Third	$\frac{3.24 \times 10^7}{\pi^2 k_{p_1}^2 P^2 F^4 f_m^4}$	Fig. 4(g)	2.15×10^{-2}
Cubic delay + constant AM/PM $C + k_{p_0}$	Third	$\frac{1.44 \times 10^7}{\pi^2 k_{p_0}^2 C^2 F^4 f_m^4}$	Fig. 4(g)	4.9×10^{-2}
Cubic delay + linear AM/PM $C + k_{p_1}$	Second	$\frac{1.2 \times 10^9}{\pi^2 k_{p_1}^2 C^2 F^2 f_m^8}$	Fig. 4(b)	2.3×10^{-2}
Cubic delay + parabolic AM/PM $C + k_{p_2}$	Third	$\frac{6.06 \times 10^8}{\pi^2 k_{p_2}^2 C^2 F^4 f_m^8}$	Fig. 4(h)	5.2×10^{-4}
Interaction+ AM/PM $\alpha_1/P + k_{p_0}$	Third	$\frac{2.45 \times 10^9}{\pi^2 k_{p_0}^2 \alpha_1^2 P^2 F^4 f_m^4}$	Fig. 4(g)	2.85×10^{-6}
$\alpha_1/P + k_{p_1}$	Second	$\frac{5.12 \times 10^{10}}{\pi^2 k_{p_1}^2 \alpha_1^2 P^2 F^2 f_m^8}$	Fig. 4(b)	5.5×10^{-6}
$\alpha_1/P + k_p$	Third	$\frac{1.71 \times 10^8}{\pi^2 k_{p_2}^2 \alpha_1^2 P^2 F^4 f_m^8}$	Fig. 4(h)	1.85×10^{-5}
$\alpha_2/S + k_{p_0}$	Third	$\frac{6.14 \times 10^8}{\pi^2 k_{p_0}^2 \alpha_2^2 S^2 F^4 f_m^4}$	Fig. 4(g)	1.15×10^{-5}
$\alpha_2/S + k_{p_1}$	Second	$\frac{2.28 \times 10^{10}}{\pi^2 k_{p_1}^2 \alpha_2^2 S^2 F^2 f_m^8}$	Fig. 4(b)	1.25×10^{-5}
$\alpha_2/S + k_{p_2}$	Third	$\frac{2.28 \times 10^{10}}{\pi^2 k_{p_2}^2 \alpha_2^2 S^2 F^4 f_m^8}$	Fig. 4(h)	1.4×10^{-7}
$\alpha_1/S + k_p$	Second	$\frac{3.28 \times 10^9}{\pi^2 k_{p_0}^2 \alpha_1^2 S^2 F^2 f_m^4}$	Fig. 4(a)	1.9×10^{-4}
	Third	$\frac{1.86 \times 10^{11}}{\pi^2 k_{p_0}^2 \alpha_1^4 S^2 F^4 f_m^4}$	Fig. 4(g)	3.6×10^{-10}
$\alpha_1/S + k_{p_1}$	Third	$\frac{2.46 \times 10^9}{\pi^2 k_{p_1}^2 \alpha_1^2 S^2 F^4 f_m^4}$	Fig. 4(g)	2.9×10^{-6}
$\alpha_1/\alpha_2 + k_{p_0}$	Third	$\frac{8.26 \times 10^4}{k_{p_0}^2 \alpha_1^2 \alpha_2^2 F^4 f_m^2}$	Fig. 4(i)	1.25×10^{-6}
$\alpha_1/\alpha_3 + k_{p_0}$	Second	$\frac{2.13 \times 10^6}{k_{p_0}^2 \alpha_1^2 \alpha_3^2 F^2 f_m^6}$	Fig. 4(e)	2.0×10^{-6}
$\alpha_1/\alpha_3 + k_{p_1}$	Third	$\frac{6.18 \times 10^5}{k_{p_1}^2 \alpha_1^2 \alpha_3^2 F^4 f_m^6}$	Fig. 4(h)	7.8×10^{-8}

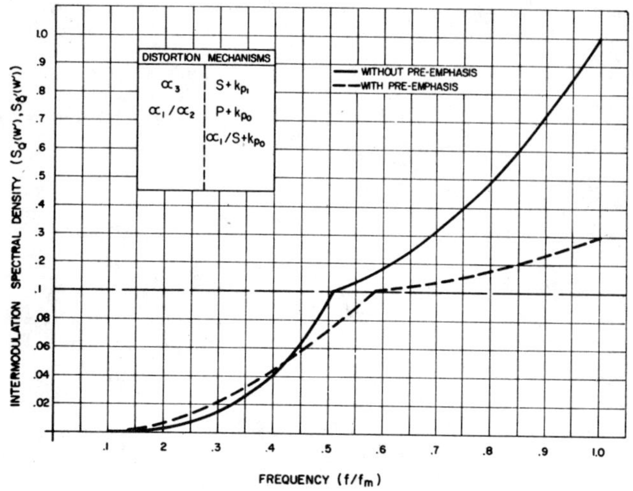

Fig. 4(a).

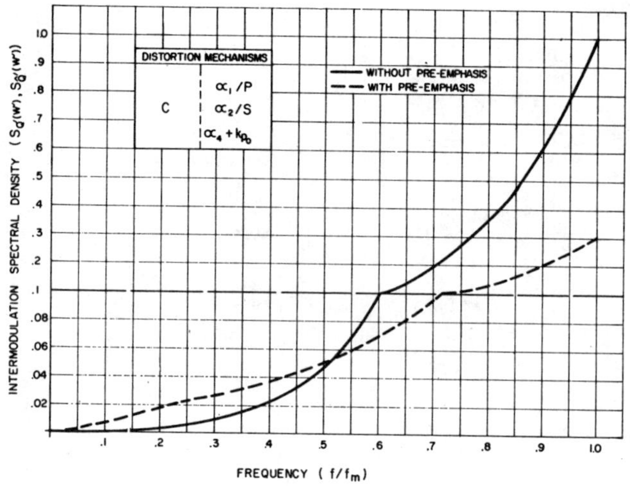

Fig. 4(d).

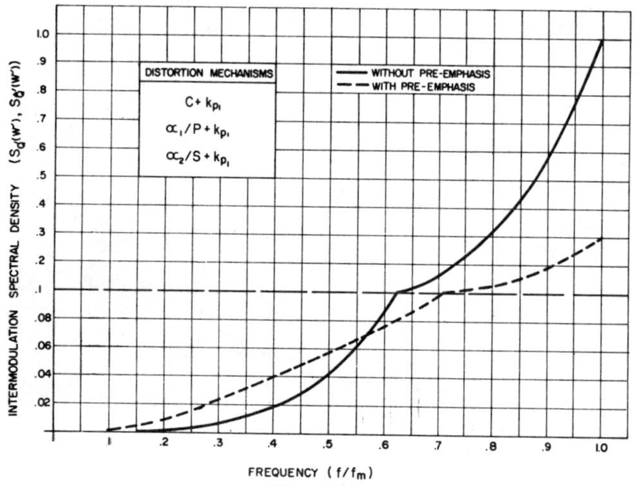

Fig. 4(b).

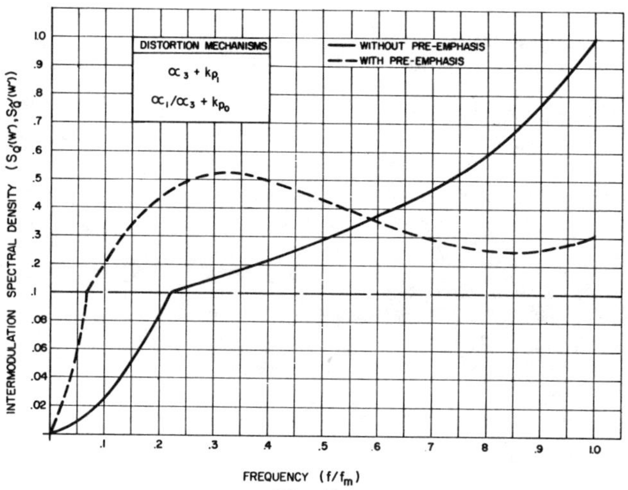

Fig. 4(e).

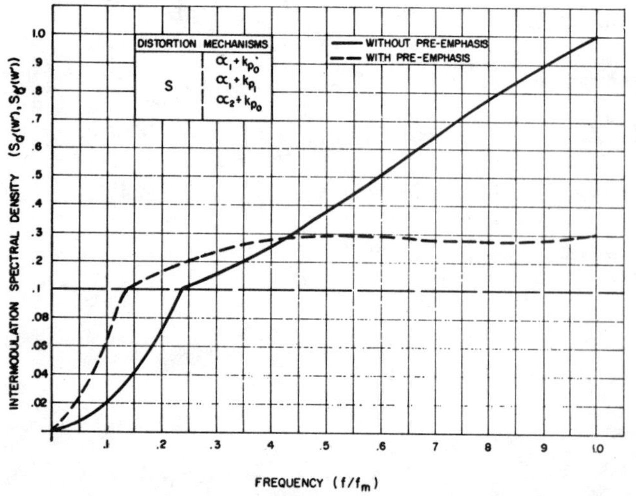

Fig. 4(c).

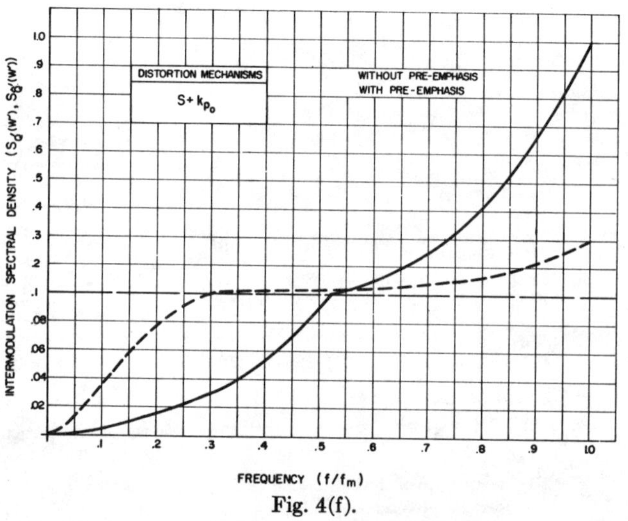

Fig. 4(f).

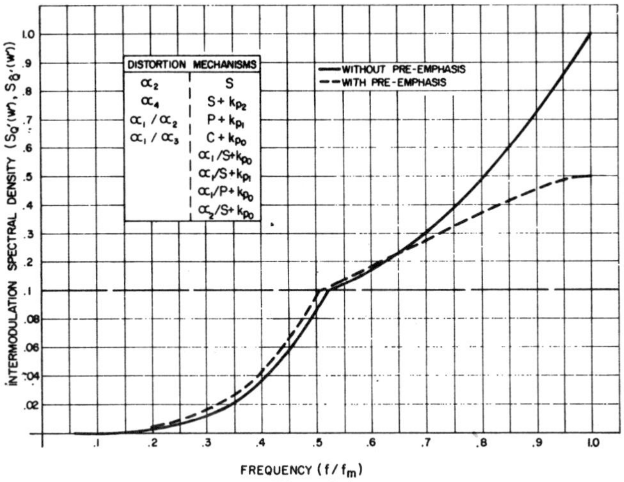

Fig. 4(g).

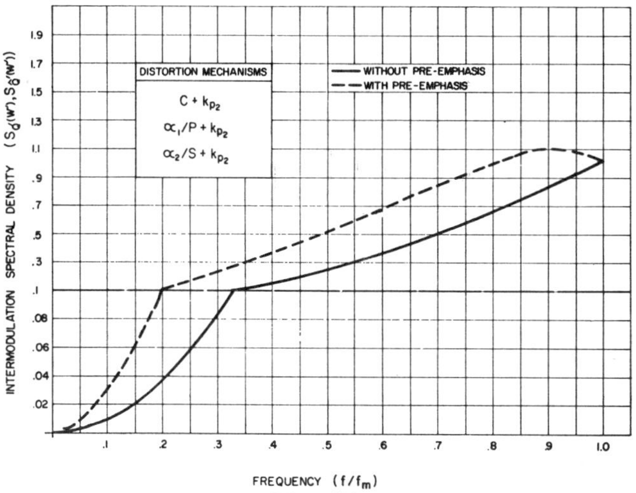

Fig. 4(h).

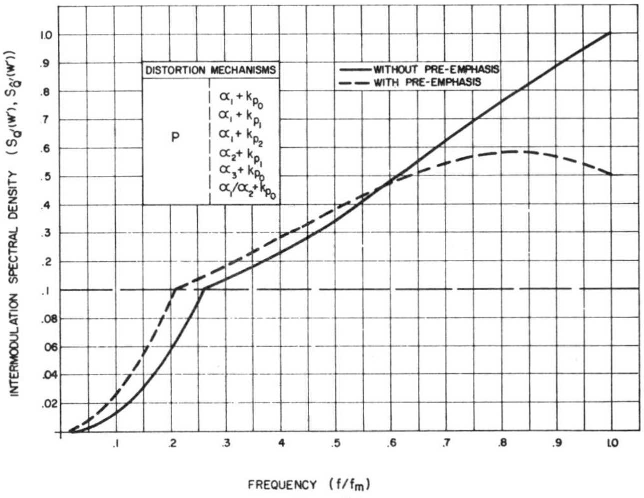

Fig. 4(i).

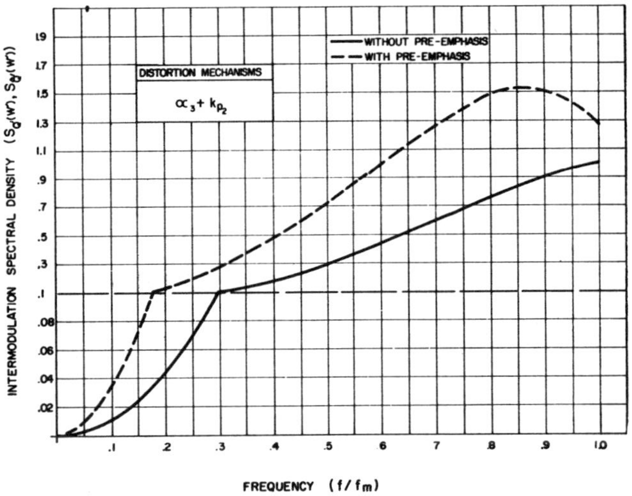

Fig. 4(j).

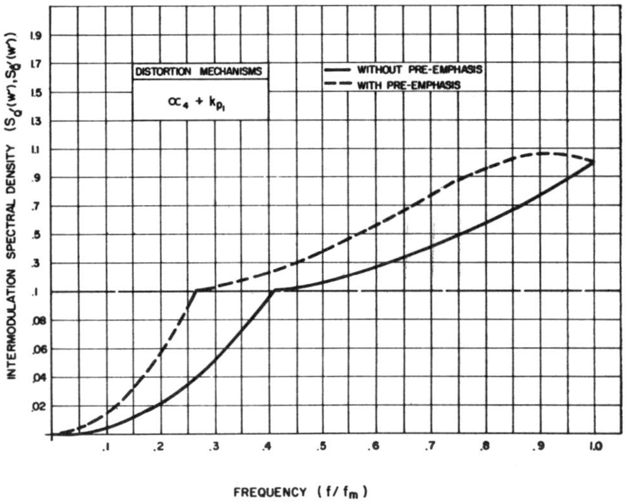

Fig. 4(k).

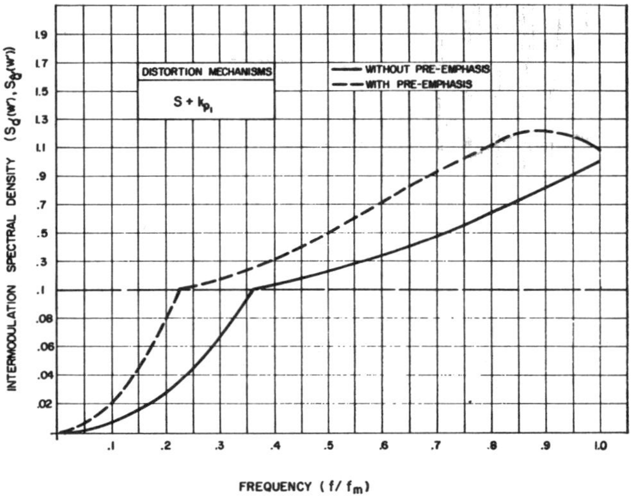

Fig. 4(l).

Fig. 4. Spectral distributions of intermodulation noise due to direct and coupled transmission deviations. (a) Second-order amplitude and equivalent amplitude distortion. (b) Second-order quasi-equivalent amplitude distortion. (c)–(d) Second-order delay and equivalent delay distortion. (e)–(f) Second-order quasi-equivalent delay distortion. (g) Third-order amplitude and equivalent amplitude distortion. (h) Third-order quasi-equivalent amplitude distortion. (i) Third-order delay and equivalent delay distortion. (j)–(l) Third-order quasi-equivalent delay distortion.

$$[S_V(\omega)]_3 = N_0^3/4(3\omega_m^2 - \omega^2)\big|_{0 \leq \omega \leq \omega_m}. \quad (12)$$

Hence (8) may be written

$$S_{Q'}(\omega) = 6g_4^2 k_1^6 N_0^3 \Phi_3(\omega) \quad (13)$$

where

$$\Phi_3(\omega) = \omega^4(3\omega_m^2 - \omega^2)_{0 \leq \omega \leq \omega_m} \quad (14)$$

and represents the frequency dependent portion of $S_{Q'}(\omega)$.

The noise power ratio (NPR) at the top modulating frequency is, by definition,

$$\text{NPR} = \frac{k_1^2 N_0}{S_{Q'}(\omega_m)} = \frac{1}{12\ g_4^2 k_1^4 N_0^2 \omega_m^6}. \quad (15)$$

Noting that $k_1\sqrt{N_0\omega_m} = \Omega$ where Ω is the multichannel rms frequency deviation we can rewrite (15) as

$$\text{NPR} = \frac{1}{12 g_4^2 \Omega^4 \omega_m^4}. \quad (16)$$

Finally, expressing all frequency nomenclature in MHz ($\omega_m \leftrightarrow f_m$, $\Omega \leftrightarrow F$) and, for $g_4 \ll 1$, equating $g_4 \leftrightarrow \alpha_4$(dB/MHz^4) we may rewrite (16) in a form convenient for numerical calculation as

$$\text{NPR} = \frac{6.32}{\alpha_4^2 F^4 f_m^4}. \quad (17)$$

For the case with pre-emphasis substitution of (11) into (9) and subsequent substitution into (8) will yield an appropriate $\Phi_3(\omega)$, (9) being evaluated by means of a digital computer program. Since we wish to compare intermodulation spectra at a baseband point of flat spectral density, it is necessary to reshape the emphasized spectra by the appropriate de-emphasis network, i.e.,

$$S_{Q'}(\omega)\big|_{\text{flat}} = S_{Q'}(\omega)\big|_{\text{emp}} \times \frac{1}{K_0} e^{-\alpha\omega/\omega_m}. \quad (18)$$

We will find it convenient, for calculation purposes, to normalize both $S_{Q'}(\omega)$ and ω to the top modulating frequency without pre-emphasis, i.e.,

$$S_{Q'}(\omega_m)\big|_{\omega_m = 1} = 1$$

and

$$\omega_{\text{norm}} = \omega/\omega_m.$$

With this normalization the spectral distributions with and without pre-emphasis may be plotted as illustrated in Fig. 4(g), the normalization allowing simple adaptation to a variety of system parameters.

We may, in a similar manner, evaluate appropriate NPRs and spectral distributions for all transmission irregularities considered in Tables I and II. These results are summarized in Tables III and IV and Fig. 4. Tables III and IV specify the NPR at the top modulating frequency (without pre-emphasis) in terms of "conventional" definitions of the transmission irregularities. These conventional definitions require appropriate substitutions[2] for a restatement of the transmission irregularities as follows:

$$g_1, g_2, g_3, g_4 \leftrightarrow \alpha_1(\text{dB/MHz}), \alpha_2(\text{dB/MHz}^2), \alpha_3(\text{dB/MHz}^3), \alpha_4(\text{dB/MHz}^4)$$

$$b_2, b_3, b_4 \leftrightarrow S(\text{ns/MHz}), P(\text{ns/MHz}^2), C(\text{ns/MHz}^3)$$

$$\theta_0, \theta_1, \theta_2 \leftrightarrow k_{p_0}(°/\text{dB}), k_{p_1}(°/\text{dB/MHz}), k_{p_2}(°/\text{dB/MHz}^2).$$

A numerical calculation for an 1800-channel system is also provided in Tables III and IV as an aid to estimation of the significance[3] of any one distortion term. This calculation has been performed in terms of "nominal" transmission deviations of 0.1-dB amplitude, 1.0-ns group delay, and 1.0°/dB AM/PM, all at 10 MHz from the transmission passband center. For system capacities other than 1800 channels, the relative significance of individual distortion terms may be modified due to the weighting of f_m and F in the intermodulation equations.

Fig. 4 provides the corresponding intermodulation spectral distributions to the listings of Tables III and IV. For each case the distributions are given with and without pre-emphasis and normalized as previously discussed.

IV. Correlation of Distortion Effects

From observation of Tables III and IV we note that the magnitude of many of the distortion terms can be expected to be negligible for representative values of the transmission irregularities and most established system capacities. We should not, however, ignore any of these terms without first considering the correlation that may exist between a tandem interconnection of differing distorting mechanisms, or, the correlation that may occur from a tandem interconnection of similar distorting irregularities. Specifically we wish to examine whether or not the combined intermodulation effects of a specified set of transmission irregularities are:

1) power additive (uncorrelated or phase quadrature),
2) voltage additive (in phase or antiphase),
3) voltage additive and cumulative.

[2] More specifically, these substitutions are:

$$\text{amplitude}\quad g_n \cong \frac{\alpha_n}{8.7\,(2\pi \times 10^6)^n}, \quad n \geq 1, g_n \ll 1$$

$$\text{delay}\quad b_n = \frac{(\text{delay})_{n-1}}{n(2\pi \times 10^6)^{n-1}}, \quad n \geq 2$$

$$\text{AM/PM}\quad \theta_n \cong \frac{0.152\,K_{p_n}}{(2\pi \times 10^6)^n}, \quad n \geq 0.$$

[3] Although only the intermodulation aspects of the transmission irregularities are being considered within the scope of this paper, the "first-order" distortion effects, as they pertain to modification of the system frequency response, are also of interest to the transmission analyst. In general, the relative significance of a transmission deviation as a "first-order distorter" differs from its significance as an "intermodulation distorter." Analysis of these first-order effects may be achieved by noting those terms in Tables I and II that will yield a fundamental output ($S_V(\omega)$ or its derivatives) and subsequently comparing the amplitude and phase spectrum of these distortion outputs with that of the desired fundamental output.

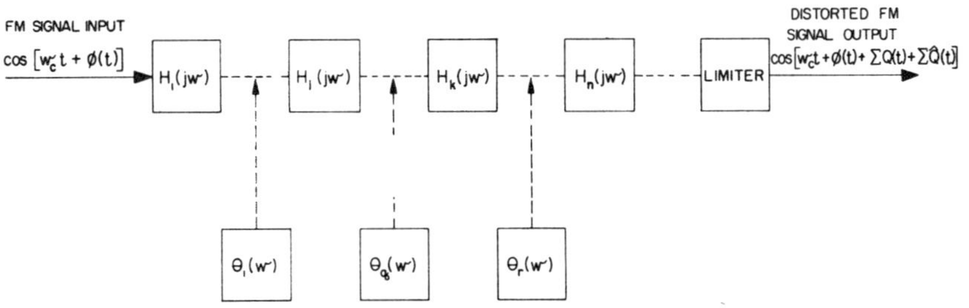

Fig. 5. Model of an imperfect FM communications system (tandem interconnection of direct and coupled transmission deviations).

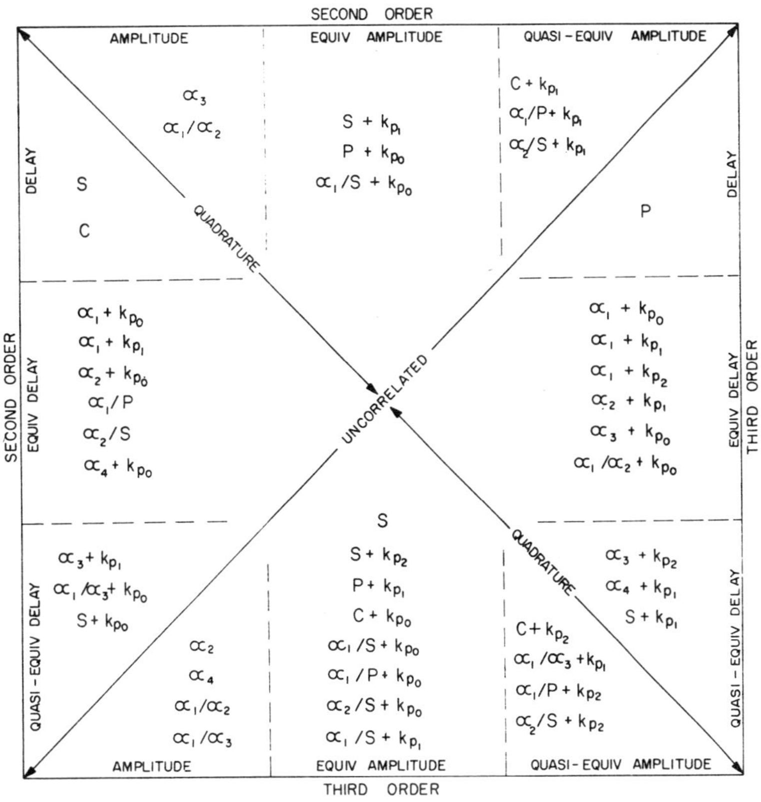

Fig. 6. Intermodulation correlation chart—direct and coupled transmission deviations.

To begin, let us extend our systems model to consider the configuration of Fig. 5. As illustrated, we have a tandem interconnection of direct transmission deviations irregularly interspaced with AM/PM nonlinearities and an ultimate connection to a station limiter. Such a configuration is representative of one hop of a radio relay system. Firstly we wish to consider the correlation between any two of the spurious phase modulations $Q(t)$, $\hat{Q}(t)$ listed in Tables III and IV. To extract this correlation information we examine the cross-correlation functions $R_{xy}(\tau)$, $R_{yx}(\tau)$ of any two phase perturbations $x(t)$ and $y(t)$ as described in Appendix III. From this analysis we may conclude that:

1) If

$$R_{xy}(\tau) = R_{yx}(\tau) = 0$$

the spectrums of $x(t)$ and $y(t)$ are uncorrelated.

2) If

$$\mathfrak{F}[R_{xy}(\tau)] = \mathfrak{F}[R_{yx}(\tau)] > 0$$

the spectrums of $x(t)$ and $y(t)$ are fully correlated, in phase and hence voltage additive.

3) If

$$\mathfrak{F}[R_{xy}(\tau)] = \mathfrak{F}[R_{yx}(\tau)] < 0$$

the spectrums of $x(t)$ and $y(t)$ are fully correlated, antiphase, and hence voltage cancelling.

4) If

$$R_{xy}(\tau) = -R_{yx}(\tau)$$

the spectrums of $x(t)$ and $y(t)$ are in phase quadrature and hence power additive.

We may now proceed to evaluate the correlation between all pairs of spurious distortions considered in Tables III and IV. This procedure involves a straightforward manipulation of random variables[4] but is, however, algebraically unrewarding; hence the results of the analysis will simply be tabulated as shown in Fig. 6. For Fig. 6 we note that:

1) All second-order distortions are uncorrelated with all third-order distortions.

2) Those terms grouped as correlated sets may be either in phase or antiphase with one another, this relationship being determined, in general, by the relative polarities of the transmission deviation coefficients and possibly the polarity of the FM deviation coefficient k_1.[4]

3) For each direct transmission deviation (amplitude or phase) a number of additional distortions exist that exhibit a full intermodulation equivalence. These may appropriately be described[5] as "equivalent amplitude" and "equivalent delay" distortion as shown by the groupings of the correlation table. As well, other terms exist that have a full phase correlation with the direct deviations although their intermodulation spectra and behavior with system transmission parameters differs from that of the direct deviations considered here. These we define to be "quasi-equivalent" amplitude and delay distortion in view of their phase equivalence with the direct deviations.

Having established the correlation relationships between individual distortion terms, let us now comment on the effects of a tandem interconnection of like distortions. For any one of the direct transmission deviations $Q(t)$ perfect limiting should not affect the phase coherency, this being a voltage sum or difference (as determined by the polarity criteria discussed above). With some transmission deviations, such as linear group delay S, the distortion sign tends to be random (positive or negative) at each distortion location. The total system distortion may thus range from zero to a full voltage addition of all irregularities. For other direct deviations, such as parabolic delay P and quartic gain α_4, most equipment tendencies are such as to ensure a similarity of sign at each distortion location. Consequently, a voltage additive condition is to be expected for such effects.

The situation with coupled transmission deviations is somewhat more involved. Here, distortion in excess of voltage addition is possible. Consider, for example, a cubic gain deviation coupled with a constant AM/PM distortion $(g_{3_A} + \theta_{0_A})$. The spurious phase distortion resulting from a first coupled pair will be $\theta_{0_A} g_{3_A} k_1^3 V^3(t)$ in accordance with the analysis of Section II. After transmission through a second amplitude deviation g_{3_B}, the total amplitude irregularly becomes $g_{3_A} + g_{3_B}$ (assumes that $P(t) = \bar{P}(t)$).[5] The total coupled phase distortion at the output of a second AM/PM converter θ_{0_B} is thus

$$\hat{Q}(t) = k_1^3 V^3(t)[\theta_{0_A} g_{3_A} + \theta_{0_B}(g_{3_A} + g_{3_B})]$$

and a cumulative effect of the first amplitude irregularity g_{3_A} is evident.[6] We may, of course, extend the analysis to include additional distortion pairs.

With the inclusion of a limiter between coupled distortions, the cumulative effects described above may be avoided. For many radio equipment designs, full limiting is applied only once at each radio repeater. Thus, although this cumulative effect is avoided between successive radio repeaters, it may still occur for transmission irregularities existing within each individual repeater. A practical illustration of the concepts described above is given in Appendix IV where the α_3 and $P + k_{p_0}$ deviations have been analyzed for a representative radio system.

V. Conclusion

The effects of direct and coupled transmission deviations occurring within an FM system have been described and a generalized tabulation of the resulting intermodulation noise presented for the use of equipment designers and system analysts. The phase coherency between different distortion mechanisms has been considered and it has been found that a number of equivalent amplitude and delay distortions exist. As a result we may conclude that a system optimization of amplitude (between limiters) and group delay (between successive radio stations or end terminals) may not yield a minimization or intermodulation noise due to the transparency[7] of some of the equivalent distortions to conventional testing techniques.

It has also been determined that a number of quasi-equivalent distortion terms exist that share a phase coherency with the direct transmission deviations but exhibit noticeably different spectral distributions. It is thus possible for the "summed" spectral distribution of a number of such distortions to exhibit a marked difference in shape from any of the participating distortions. Such a characteristic has been observed from measurements taken on actual radio systems where, for example, it is often found that the intermodulation contribution of the middle measurement slot is substantially worse than that of the top measurement slot. Infrequently, some system measurement data has shown a different "controlling order of distortion" between the middle and top measurement slots, a result that is also explainable by consideration of

[4] The frequency shifting techniques (IF-RF-IF) employed in most radio equipment designs can also invert the polarity of the phase distortion. A "sum or difference" assignment for a correlated set of distorters will require, in general, a knowledge of the location of each distorter (IF or RF) and the shifting arrangement employed (high or low LO) in addition to any other polarity criteria involved.

[5] Most circuit devices exhibit some degree of compression that, for a cascaded set of distorters, can significantly alter the magnitude of cumulative distortion expected. A specific example of interest is that of delay equalization at a location subsequent to amplitude compression. Although an equalization of the phase distortion may be achieved, the amplitude modulation introduced by the correction device will overequalize the system. This modulation, if allowed to pass without limiting to a subsequent AM/PM distorter, will represent an additional intermodulation source that must be considered in the transmission analysis.

[6] This distortion effect could also be cancelling depending on the relative polarities of θ_{0A} and θ_{0B}.

[7] Most group delay measuring equipments utilize a low deviation search frequency of the order of 200 kHz that frequency-modulates a slowly swept IF carrier. Such a test signal can be equated to a combined high-low index FM signal. An analysis of the system response to this signal indicates that the equivalent delay distortions should be interpreted as group delay by the measurement equipment since they will add a spurious quadrature component to the search frequency. Confirmation of this analysis is being attempted as part of a laboratory measurement program related to the transmission properties of some of the equivalent delay distortions. The equivalent amplitude distortions, however, add only an in-phase spurious component to the search frequency and thus should display an effective transparency to the group delay measurement technique.

the correlation and spectral distributions of a given set of distortion terms.

Additionally we note that the phase coherency between distortions may yield a significant magnitude of total system distortion even though the magnitude of individual transmission irregularities is quite small. We further note that the measurement of such small individual irregularities, whether directly or indirectly via noise loading tests, requires an instrumentation capability that tends to be outside the range of that generally available. In view of this difficulty, an initial equipment design criteria taking cognizance of the correlation possibilities described would appear well advised.

An optimum design criterion would appear to be that of "correction immediately after generation" although this may not always be possible or economically desirable. An alternate approach is that of "mop up" equalization where advantage of the phase coherency of system distortions is taken and an inverse distorter is introduced for correction. Such correction may, however, be only partially effective due to possible differences in the spectral densities of some of the system distortions.

Appendix I

AM/PM Distortion Analysis

Consider a system model such as illustrated in Fig. 2 with an AM/PM nonlinearity θ (radians per incremental amplitude change). A hybrid AM + PM signal as given by (5) is applied to the system input and we wish to determine the PM output.

By definition of the AM/PM process, an AM modulation $P(t)$ incident to the input signal will produce a distortion phase modulation $\hat{Q}(t)$ on the output signal given by

$$\hat{Q}(t) = \theta P(t). \tag{19}$$

θ is, however, frequency dependent as defined by (4), and hence we must establish an appropriate transformation of $\theta(\omega)$ to $\theta(t)$. Since, in general, $\phi(t) \gg Q(t)$, we may approximate the instantaneous phase of the input signal to be

$$\xi(t) \cong \omega_c t + \phi(t). \tag{20}$$

The instantaneous input frequency $\mu(t)$ is thus

$$\mu(t) = \xi'(t) = \omega_c + \phi'(t). \tag{21}$$

The input signal thus "sees" an instantaneous value of θ corresponding to the instantaneous input frequency $\mu(t)$. Equivalently, substituting (21) into (4)

$$\theta(t) = \theta(\phi'(t)) = \theta_0 + \theta_1\phi'(t) + \theta_2\phi'^2(t). \tag{22}$$

Thus, from (19), the spurious AM/PM modulation may be written

$$\check{Q}(t) = \theta_0 P(t) + \theta_1 P(t)\phi'(t) + \theta_2 P(t)\phi'^2(t). \tag{23}$$

Appendix II

Intermodulation Spectrum of a Quartic Gain Transmission Deviation

From Table I, the spurious phase modulation due to a quartic gain transmission irregularity is

$$Q(t) = -2g_4 \frac{d}{dt} [\phi'^3(t)]. \tag{24}$$

With the substitutuion $\phi'(t) = k_1V(t)$, the corresponding FM distortion may be written as

$$Q'(t) = -2g_4k_1^3 \frac{d^2}{dt^2} [V^3(t)]. \tag{25}$$

Autocorrelation of (25) is given by

$$R_{Q'}(\tau) = 4g_4^2k_1^6R_{V^{3''}}(\tau) \tag{26}$$

where $R_{Q'}(\tau)$ and $R_{V^{3''}}(\tau)$ are the respective autocorrelation functions of $Q'(t)$ and $V^{3''}(t)$.

From Downing[6] and Laning and Battin[4] it may be derived that

$$R_{V^{3''}}(\tau) = R_{V^3}''''(\tau) \tag{27}$$

and

$$R_{V^3}(\tau) = 6R_V^3(\tau) - 9R_V(0)R_V(\tau). \tag{28}$$

The intermodulation portion of (26) may thus be rewritten as

$$R_{Q'}(\tau) = 24\, g_4^2k_1^6R_{V^3}''''(\tau). \tag{29}$$

Now the power spectral density of the FM distortion $S_{Q'}(\omega)$ is obtained by taking the Fourier transform of (29), i.e.,

$$\mathfrak{F}[R_{Q'}(\tau)] = S_{Q'}(\omega) = 24\, g_4^2k_1^6\mathfrak{F}[R_{V^3}''''(\tau)]. \tag{30}$$

However,

$$\mathfrak{F}[R_{V^3}''''(\tau)] = (j\omega)^4[S_V(\omega)]_3 \tag{31}$$

where $[S_V(\omega)]_3$ is a third-order convolution of the baseband modulation spectrum $S_V(\omega)$. Finally, by substitution of (31) into (30), we may write

$$S_{Q'}(\omega) = 24\, g_4^2k_1^6\omega^4[S_V(\omega)]_3. \tag{32}$$

Thus (32) establishes as desired a generalized intermodulation spectral noise equation in terms of frequency domain convolutions of the modulating signal $S_V(\omega)$.

Appendix III

Cross-Correlation Analysis of Spurious Modulations

Consider a system configuration such as Fig. 7 whose summed output $z(t)$ is given by

$$z(t) = x(t) + y(t). \tag{33}$$

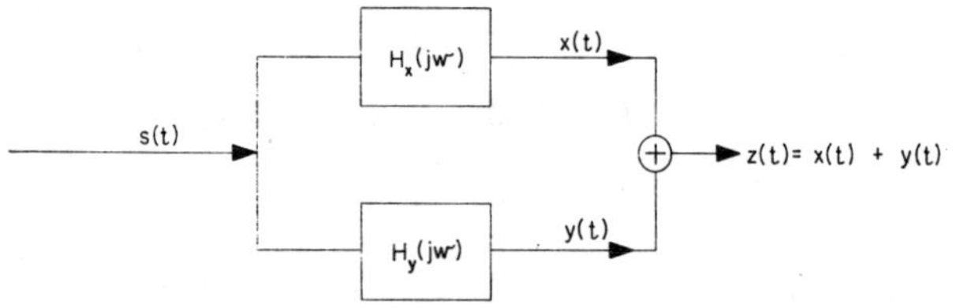

Fig. 7. Block diagram configuration for a cross-correlation analysis.

The autocorrelation of $z(t)$ may be written as

$$R_z(\tau) = R_x(\tau) + R_y(\tau) + R_{xy}(\tau) + R_{yx}(\tau) \quad (34)$$

where $R_{xy}(\tau)$ and $R_{yx}(\tau)$ represent the cross-correlation functions of $x(t)$ and $y(t)$. Now the power spectral density of $z(t)$ is obtained by taking the Fourier transform of (34), i.e.,

$$S_z(\omega) = S_x(\omega) + S_y(\omega) + S_{xy}(\omega) + S_{yx}(\omega) \quad (35)$$

where

$$S_{xy}(\omega) = \mathfrak{F}[R_{xy}(\tau)] \quad (36)$$

and

$$S_{yx}(\omega) = \mathfrak{F}[R_{yx}(\tau)]. \quad (37)$$

We now note, from consideration of (33) and (35) that the magnitude of the summed intermodulation spectra $S_z(\omega)$ is dependent on the phase coherency between the two signals $x(t)$ and $y(t)$. This coherency is defined by the cross spectral densities $S_{xy}(\omega)$ and $S_{yx}(\omega)$. Alternatively, we may interpret this coherency by consideration of the cross-correlation functions $R_{xy}(\tau)$ and $R_{yx}(\tau)$ due to the transform equivalence of these expressions[8] with the cross spectral densities given by (36) and (37). Let us now consider some specific situations of interest.

A. Second- and Third-Order Distortion Relationships

Assume $H_x(j\omega)$ and $H_y(j\omega)$ to be such that they introduce second- and third-order distortions, respectively. We note, from the analysis of Section II, that $x(t)$ and $y(t)$ are of the form

$$x(t) = S_1^{\alpha}(t)S_2^{\beta}(t), \quad (\alpha + \beta = 2) \quad (38)$$

and

$$y(t) = S_3^{\delta}(t)S_4^{\sigma}(t)S_5^{\eta}(t), \quad (\delta + \sigma + \eta = 3). \quad (39)$$

Here $S_1(t) \ldots S_5(t)$ are the input function $S(t)$ and/or its derivatives and $\alpha, \beta \ldots \eta$ are integer exponents.

The autocorrelation function $R_{xy}(\tau)$ may be written as

$$R_{xy}(\tau) = E[x(t)y(t + \tau)] = E[S_1^{\alpha}(t)S_2^{\beta}(t)S_3^{\delta}(t + \tau) \times S_4^{\sigma}(t + \tau)S_5^{\eta}(t + \tau)] \quad (40)$$

[8] The power density spectrums $S_z(\omega)$, $S_x(\omega)$, and $S_y(\omega)$ are required to be real, even, and non-negative functions whereas, in general, the crosspower density spectrums $S_{xy}(\omega)$ and $S_{yx}(\omega)$ do not necessarily exhibit these characteristics. As a result, the polarity of $S_{xy}(\omega)$ or $S_{yx}(\omega)$ is not always inferred simply by observation of $R_{xy}(\tau)$ or $R_{yx}(\tau)$ and an actual evaluation of $S_{xy}(\omega)$ and $S_{yx}(\omega)$ may be required.

but

$$\alpha + \beta + \delta + \sigma + \eta = 5 = \text{odd}.$$

Thus it can be shown that[7]

$$R_{xy}(\tau) = 0.$$

Similarly

$$R_{yx}(\tau) = 0.$$

Therefore

$$S_z(\omega) = S_x(\omega) + S_y(\omega)$$

and we conclude that $x(t)$ and $y(t)$ are uncorrelated.

B. In-Phase, Antiphase, and Phase Quadrature Relationships

Assume $H_x(j\omega)$ and $H_y(j\omega)$ to be such that one of the stated phase relationships between $x(t)$ and $y(t)$ results, i.e., we may write

$$H_x(j\omega) = X(\omega)e^{-j\beta(\omega)} \quad (41)$$

and

$$H_y(j\omega) = Y(\omega)e^{j\beta(\omega)}, \quad \text{(in-phase)} \quad (42)$$

$$H_y(j\omega) = Y(\omega)e^{j(\beta(\omega)+\pi)}, \quad \text{(antiphase)} \quad (43)$$

$$H_y(j\omega) = Y(\omega)e^{j(\beta(\omega)+\pi/2)}, \quad \text{(phase quadrature)}. \quad (44)$$

We may express the output cross-correlation functions in terms of the network transfer functions and the input spectral density $S_s(\omega)$ by noting that

$$R_{xy}(\tau) = \frac{1}{2\pi}\int_{-\infty}^{\infty} S_s(\omega)H_x(j\omega)H_y^*(j\omega)e^{j\omega\tau}d\omega \quad (45)$$

and

$$R_{yx}(\tau) = \frac{1}{2\pi}\int_{-\infty}^{\infty} S_s(\omega)H_x^*(j\omega)H_y(j\omega)e^{j\omega\tau}d\omega. \quad (46)$$

By successive substitutions of (41) and (42)–(44) into (45) and (46) we may thus establish the nature of the cross-correlation functions as follows:

1) In-phase

$$\mathfrak{F}[R_{xy}(\tau)] = \mathfrak{F}[R_{yx}(\tau)]; > 0.$$

Hence

$$S_z(\omega) = S_x(\omega) + S_y(\omega) + 2\, S_{xy}(\omega).$$

2) Antiphase

$$\mathfrak{F}[R_{xy}(\tau)] = \mathfrak{F}[R_{yx}(\tau)]; < 0.$$

Hence

$$S_z(\omega) = S_x(\omega) + S_y(\omega) - 2S_{xy}(\omega).$$

3) Phase quadrature

$$R_{xy}(\tau) = -R_{yx}(\tau).$$

Hence

$$S_z(\omega) = S_x(\omega) + S_y(\omega).$$

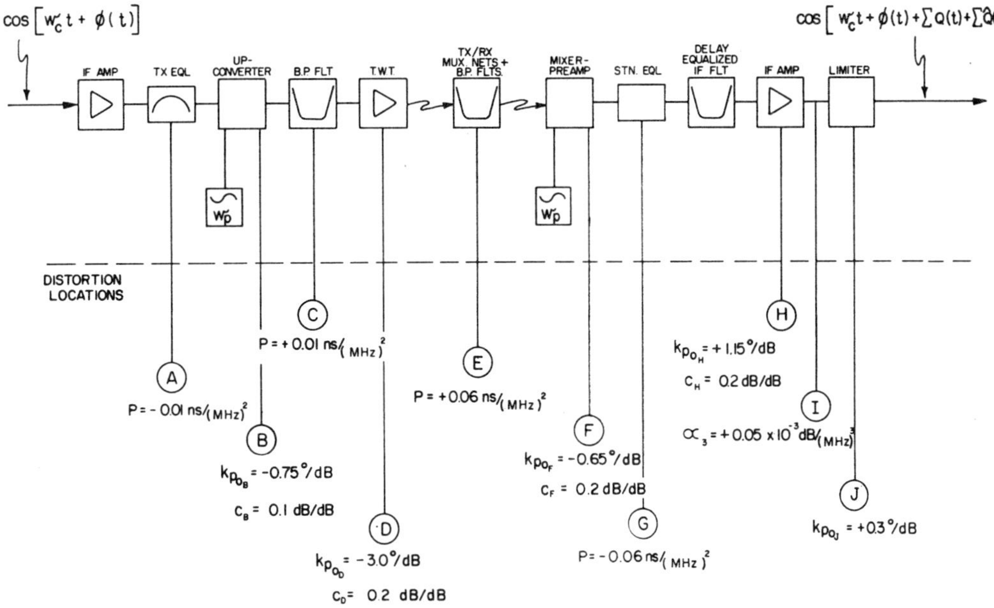

Fig. 8. Representative model of a heterodyne radio repeater.

Appendix IV

Illustrative Intermodulation Analysis

Consider an equipment configuration such as Fig. 8 that is representative of a conventional heterodyne radio repeater. Our analysis will be restricted to a discussion of the second-order distortions resulting from a cubic gain deviation α_3 and that from parabolic delay coupled with a frequency independent AM/PM converter $P + k_{p0}$. Although a number of additional distortion mechanisms of interest may exist within such a repeater assembly, the two chosen have been found to be particularly troublesome in many radio installations and are sufficient to demonstrate the application of the concepts developed within the text of this paper.

As illustrated, the α_3 deviation may be assigned a value consistent with the overall repeater measurement whereas the P and k_{p0} deviations must be broken down to relate to actual physical locations. An additional parameter required to complete the analysis is the degree of dynamic compression c (usually expressed in dB/dB) associated with the subassemblies making up the repeater. A knowledge of this parameter is essential in order to accurately estimate[9] the cumulative nature of the equivalent amplitude distortions. Typical values for these transmission deviations have been chosen for the analysis that are considered representative of the magnitudes one might experience in an actual equipment design.

Firstly, we note the major sources of AM/PM conversion to be the transmit upconverter (B), the TWT (D), the mixer preamp (F), the main IF amplifier (H) and the limiter (J). (See Fig. 8.) The major sources of parabolic delay distortion are the upconverter sideband filter (B) preceding the TWT, and the RF multiplexing network plus TX/RX BP filters (E). Delay equalization is provided at (A) with an IF equalizer intended to partially offset the effects of the upconverter filter, and at (G), where the station equalizer is assumed to provide an exact inverse equalization to the accumulated parabolic delay of the repeater. One final prerequisite to the analysis is the assignment of local oscillator frequencies, these being assumed high (above the RF signal frequency) for the purpose of this discussion, and thus introducing a polarity inversion (IF to RF) of the signal and distortion modulations.

We may now proceed to calculate the FM distortions as they would appear at the output of the repeater. The designations (+) and (−) are employed to define the relative phase nature of the distortions, the (+) designation arbitrarily being assigned to an $|\alpha_5| > 0$ or to a $|k_{p0}P| > 0$. Polarity reversals within the repeater will, as discussed previously, invert the relative phase of an RF distortion between its point of origin and that resulting at the repeater IF output. For each equivalent amplitude distortion contribution the intermodulation noise (pwp) has been evaluated by equating the total amplitude modulation to a corresponding magnitude of parabolic delay and subsequently utilizing the appropriate design equations developed in Section III. These computations are summarized in Table V.

The total distortion for the radio repeater is a voltage sum of the individual contributors and may be computed to be

$$\text{Total Noise}_{1\ \text{hop}} = 5.75 \text{ pwp}.$$

For a 10-hop system consisting of identical repeaters the

[9] Within the scope of this analysis it will be assumed that the only alteration to the amplitude modulation $P(t)$ incident to the FM signal is compression by the parameter c, i.e., the hybrid input signal $[1 + P(t)] \cos [\omega_c t + \phi(t)]$ is altered to the form $[1 + (1 - c) P(t)] \cos [\omega_c t + \phi(t) + \theta P(t)]$ after transmission through a nonlinear device characterized by a compression coefficient c and an AM/PM coefficient θ. For small amounts of compression the values of c(dB scale) can be equated with those on a voltage scale.

TABLE V

DISTORTION TABULATION—HYPOTHETICAL RADIO REPEATER

Distortion Mechanism	AM/PM Source kp_0, °/dB	Accumulated AM at Input of AM/PM Converter $\Sigma P(t)$	Equivalent Parabolic Delay Corresponding to AM Total P, ns/MHz²	Intermodulation Noise at Top Modulating Frequency in an 1800-Channel System (With Pre-Emphasis)
α_3				1.4 pwp (+)
$P + k_{p_0}$	TX upconverter $kp_{0B} = +0.75$	$P_A(t)$	−0.01	0.245 pwp (−)
	TWT $kp_{0D} = -3.0$	$(1 - c_B)P_A(t) + P_C(t) = 0.9P_A(t) + P_C(t)$	+0.0011	0.05 pwp (+)
	Mixer-preamp $kp_{0F} = -0.65$	$(1 - c_D)(0.9\,P_A(t) + P_C(t) + P_E(t)$ $\cong P_E(t)$	+0.06	6.6 pwp (+)
	IF amp $kp_{0H} = +1.15$	$(1 - c_F)P_E(t) + P_G(t) = 0.8\,P_E(t) + P_G(t)$	−0.012	0.83 pwp (−)
	IF limiter $kp_{0J} = +0.3$	$(1 - c_H)(0.8\,P_E(t) + P_G(t)) = 0.64\,P_E(t) + 0.8\,P_G(t)$	−0.0096	0.032 pwp (−)

TABLE VI

PERTURBATION SUMMARY—HYPOTHETICAL RADIO SYSTEM

System Configuration	Intermodulation Noise at Top Modulating Frequency in an 1800-Channel System (With Pre-Emphasis): 1 Hop Total	10 Hop Total	dB Difference From Table V Configuration
As per Fig. 8	5.75 pwp	575 pwp	...
As per Fig. 8 except for a random (±) variation of α_3 that cancels over 10 hops	0.04/5.75 pwp	150 pwp	−5.85 dB
As per Fig. 8 except for an inversion of the LOs and a subsequent inversion of the TWT and mixer preamp distortion polarities	10 pwp	1000 pwp	+2.4 dB
As per Fig. 8 except for a reduction of the IF amp AM/PM to +0.2 °/dB	9.8 pwp	980 pwp	+2.3 dB
As per Fig. 8 except for a lower RX signal input resulting in a $kp_{0F} = -0.5°$/dB and a $c_F = 0.1$ dB/dB	4.6 pwp	460 pwp	−0.97 dB

distortion would become

$$\text{Total Noise}_{10\text{ hops}} = (5.75) \times (10)^2 = 575 \text{ pwp}.$$

It is now of interest to mildly perturb the system characteristics and observe the variations in intermodulation noise that can result. Some sample perturbations are given in Table VI. Examination of Table VI indicates that a surprising degree of variation in the system intermodulation noise may occur with what one might normally conclude to be inconsequential changes in the system transmission characteristics.

An actual system would not be expected to exhibit the same degree of uniformity as assumed in the calculations, this being a variation that would alter the tabulated intermodulation totals. Minor rearrangements of the equipment configuration, such as relocation of the station equalizer (G) to follow the main IF amplifier (H), could also significantly modify the results of the analysis. Possible solutions to the situation described are:

1) reduction of the AM/PM conversion factor of those equipment assemblies that noticeably affect the system intermodulation noise,

2) RF delay equalization of the microwave filters,

3) provision of an α_3 "mop up" equalizer at the end terminal.

Solutions 1) and 2) would appear to be the most desirable in that they automatically eliminate other potential intermodulation effects.

ACKNOWLEDGMENT

The author wishes to extend his thanks to his colleague L. A. Fraser who undertook the preparation of the digital computer programs essential to the preparation of this paper and who also assisted with the evolution of suitable instrumentation for the measurement of AM/PM conversion.

REFERENCES

[1] T. G. Cross, "Intermodulation in FM systems due to transmission deviations and AM/PM conversion," *Bell Sys. Tech. J.*, vol. 45, pp. 1749–1773, December 1966.

[2] M. L. Liou, "Noise in an FM system due to an imperfect linear transducer," *Bell Sys. Tech. J.*, vol. 45, pp. 1537–1561, November 1966.

[3] Internat'l Radio Consultive Committee, *Doc. of 11th Plenary Assembly* (Oslo, 1966), vol. 4, Recommendation 275-1.

[4] J. H. Laning and R. H. Battin, *Random Processes in Automatic Control.* New York: Wiley, 1961, pp. 82–85.

[5] S. Morimoto and R. Yamamoto, "Distortion due to equivalent amplitude characteristics resulting from delay distortions and AM-PM conversion," *Electronics and Communications* (Japan), pp. 54–63, September 1963.

[6] J. J. Downing, *Modulation Systems and Noise.* Englewood Cliffs, N. J.: Prentice-Hall, 1964, p. 48.

[7] J. M. Wozencroft and I. M. Jacobs, *Principles of Communication Engineering.* New York: Wiley, 1965, pp. 205–206.

Time-domain analysis of intermodulation effects caused by nonlinear amplifiers

J. C. FUENZALIDA, O. SHIMBO, AND W. L. COOK

Abstract

Nonlinear amplifiers are widely used in microwave radio communications. If the input signal to such a device exhibits a time-varying envelope, then the output signal is distorted. This situation arises in communications satellites when a number of angle-modulated carriers share a common nonlinear amplifier so that the output contains intermodulation products in addition to the carriers.

This paper uses a time-domain analysis to express the baseband distortion caused by intermodulation in angle-modulated carriers. The combined effects of amplitude nonlinearity and AM-PM conversion are considered, and the particular case of an input consisting of angle-modulated carriers and a band of Gaussian noise is treated in detail.

The physical interpretation of the analytical results and the computational method used to implement these results are described. The computational method is the basis for a flexible, user-oriented computer program which provides considerable generality while minimizing computational requirements. The computer program is used to investigate several problems of practical interest for which analytical results have been previously unobtainable.

This paper is based upon work performed at COMSAT Laboratories under the sponsorship of the International Telecommunications Satellite Organization (INTELSAT). Views expressed in this paper are not necessarily those of INTELSAT.

Introduction

When a nonlinear amplifier is used simultaneously by a number of carriers, intermodulation products are generated which cause distortion in the signals. Many studies dealing with calculation of the levels of intermodulation products at the output of the amplifier have been reported [1]–[6]; the most complete to date is the general analysis by Shimbo [5]. Chitre and Fuenzalida [7] have shown that the baseband distortion caused by such intermodulation products cannot always be calculated from the power spectral density of the signal at the output. The analysis of Reference 5 has therefore been extended to permit the evaluation of baseband distortion in the time-domain representation.

This paper presents a unified theory which combines the effects of amplitude nonlinearity and AM/PM conversion for the memoryless case. The analysis is performed initially for an input consisting of arbitrary bandpass signals. The general expressions are then applied to obtain both the RF and the angle-demodulated power spectra for an input consisting of a band of Gaussian noise in addition to multiple angle-modulated carriers.

The steps in the analytical derivation may be summarized as follows. First, the nonlinearity is modeled. Then, an analysis is performed to obtain a time-domain representation of the output of the nonlinear device, as well as general expressions for the power spectrum of the angle-demodulated output. It is shown that the expression for the output power spectrum reported in Reference 5 can be derived from these results.

The theoretical development is followed by a description of a computer program, based on these results, which has been found to be a useful tool in the determination of intermodulation spectra and baseband distortion in multicarrier systems. The design objectives of the program are discussed, with emphasis on the important computational techniques. Results of the calculations are then given for several interesting problems for which a rigorous solution was not previously available.

Modeling and Analysis

The essential features of the model used in the analysis and the important results generated by the analysis are summarized in the following subsections. Detailed mathematical derivations of the time-domain representation of the output of a nonlinear device and the resulting baseband and RF power spectra are included in Appendix A.

Reprinted with permission from *COMSAT Tech. Rev.*, vol. 3, pp. 89–141, Spring 1973.

Mathematical model

The nonlinear device is characterized in terms of the input and output envelopes, following the mathematical model of Reference 5. Two nonlinear functions are necessary to completely describe the properties of such a nonlinearity: the nonlinear amplitude and phase functions. For an input consisting of a single unmodulated carrier, the output of the nonlinear amplifier contains the fundamental frequency as well as higher harmonics. The laboratory measurements necessary to characterize the nonlinearity involve only the fundamental component. Therefore, the envelope characterization is ideally suited for the purposes of this analysis and physically more meaningful than instantaneous voltage characterizations.

Assume that the general representation of the input bandpass signal is

$$e_i(t) = Re\,\{V(t)\exp(j\omega_o t)\} = Re\,\{\rho(t)\exp[j\omega_o t + j\Theta(t)]\} \quad (1)$$

where

$V(t) = \rho(t)\exp[j\Theta(t)]$

$\rho(t) =$ amplitude of $V(t)$

$\Theta(t) =$ angle of $V(t)$

$\omega_o =$ midband angular frequency.

The fundamental signal output in a single-carrier test can be represented by

$$e_o(t) = Re\,\{g(A)\exp[j\omega_o t + jf(A)]\} \quad (2)$$

for the cosine wave input $Re\,\{A\exp[j\omega_o t]\}$, where $g(A)$ and $f(A)$ are the output amplitude and phase functions, respectively. It is assumed that $g(A)$ and $f(A)$ are independent of frequency.* Therefore, the bandpass output caused by an arbitrary bandpass input signal [as given in equation (1)] can be represented by

$$e_o(t) = Re[g\{\rho(t)\}\exp\{j\omega_o t + jf[\rho(t)] + j\Theta(t)\}] \quad . \quad (3)$$

*This case is sometimes referred to as "memoryless."

The output can also be written in terms of a complex envelope gain function:

$$e_o(t) = Re\,\{G[\rho(t)]\,V(t)\exp[j\omega_o t]\}$$

where

$$G[\rho(t)] = \frac{g[\rho(t)]}{\rho(t)}\exp\{jf[\rho(t)]\} \quad (4)$$

In this paper a bandpass envelope characterization of the nonlinear device is used since the functions $g(\rho)$ and $f(\rho)$ can be directly measured in the laboratory. The instantaneous voltage transfer function (including all harmonics) cannot be uniquely defined on the basis of envelope measurements. However, this transfer function is related to the equivalent bandpass envelope transfer function by a Chebychev transformation [8]. In the absence of AM/PM conversion, Chebychev transformation of the real transfer function $g(\rho)$ is required; otherwise the bandpass envelope transfer function is complex. The real and imaginary components can be transformed separately, resulting in two instantaneous voltage transfer functions in quadrature, as reported in Reference 6.

Time-domain output of the nonlinear device

INPUT CONSISTING OF NARROWBAND BANDPASS SIGNALS

Consider the case in which the input to the nonlinear device is formed by the sum of m narrowband bandpass signals (as defined in Reference 9):

$$e_i(t) = Re\sum_{l=1}^{m} A_l(t)\exp[j\omega_o t + j\theta_l(t)] = Re\,\{\rho(t)\exp[j\omega_o t + j\Theta(t)]\} \quad (5)$$

If the bandpass output is represented by

$$e_o(t) = Re\,[g\{\rho(t)\}\exp\{j\omega_o t + j\Theta(t) + jf[\rho(t)]\}] \quad (6)$$

it is shown in Appendix A [equation (A10)] that $e_o(t)$ can also be expressed as

$$e_o(t) = Re\left\{\exp[j\omega_o t]\sum_{\substack{k_1,k_2,\ldots,k_m=-\infty\\(k_1+k_2+\ldots+k_m=1)}}^{\infty}\exp\left[j\sum_{l=1}^{m}k_l\theta_l(t)\right]\cdot M(k_1,k_2,\ldots,k_m)\right\} \quad (7)$$

The condition $\sum_{l=1}^{m} k_l = 1$ is a consequence of the bandpass representation of the output. The complex amplitude $M(k_1, k_2, \ldots, k_m)$, which can be expressed as [equation (A12)]

$$M(k_1, k_2, \ldots, k_m) = \int_0^\infty \gamma \left\{ \prod_{l=1}^{m} J_{k_l}[A_l(t)\,\gamma] \right\} d\gamma \int_0^\infty \rho g(\rho) \exp[jf(\rho)]\, J_1(\gamma\rho)\, d\rho \quad (8)$$

can be calculated for a number of series expansions of $g(\rho) \exp[jf(\rho)]$, as described in Reference 5. The Bessel function expansion

$$g(\rho) \exp[jf(\rho)] = \sum_{s=1}^{L} b_s J_1(\alpha s \rho) \quad (9)$$

where b_s are complex coefficients, leads to a particularly simple result.

Berman and Podraczky [3] used a Fourier series expansion with real coefficients of the instantaneous voltage transfer function. The corresponding single-carrier bandpass output is given by a Bessel function series. More recently Kaye et al. have introduced complex coefficients to account for the effects of AM/PM conversion. Then, as shown in Appendix A, introducing equation (9) into equation (8) yields

$$M(k_1, k_2, \ldots, k_m) = \sum_{s=1}^{L} b_s \prod_{l=1}^{m} J_{k_l}[\alpha s A_l(t)] \quad . \quad (10)$$

In the particular case of an input consisting solely of angle-modulated carriers, the bandpass output of the nonlinearity is also formed solely by angle-modulated signals. Each component (carriers as well as intermodulation products) is described by a set of integers $k_1, k_2, \ldots, k_m$. The amplitude (which is constant) is $M(k_1, k_2, \ldots, k_m)$, and its angle modulation is given by $\sum_{l=1}^{m} k_l \theta_l(t)$.

In the general case of an input consisting of narrowband bandpass signals, $M(k_1, k_2, \ldots, k_m)$ is a function of time. Consequently, the set of integers k_l is no longer sufficient to describe the intermodulation products. In the next subsection the case of an input consisting of angle-modulated signals and a band of noise is analyzed, and it is shown that an additional parameter is required to account for the effects of the amplitude variations of the noise signals.

INPUT CONSISTING OF ANGLE-MODULATED CARRIERS AND NOISE

Consider the case of $m - 1$ angle-modulated carriers plus noise. The input to the nonlinear device is

$$e_i(t) = Re \left\{ \sum_{l=1}^{m-1} A_l \exp[j\omega_o t + j\theta_l(t)] + [N_c(t) + jN_s(t)] \cdot \exp[j\omega_o t + j\omega_m t] \right\}$$

$$= Re \left\{ \sum_{l=1}^{m-1} A_l \exp[j\omega_o t + j\theta_l(t)] + A_m(t) \exp[j\omega_o t + j\theta_m(t)] \right\} \quad (11)$$

where

$$\theta_l(t) = \omega_l t + \phi_l(t) + \lambda_l, \qquad l = 1, 2, \ldots, m-1$$

$$\theta_m(t) = \omega_m t + \tan^{-1} \frac{N_s(t)}{N_c(t)}$$

$$A_l = \text{constant}, \qquad l = 1, 2, \ldots, m-1$$

$$A_m(t) = \sqrt{N_c^2(t) + N_s^2(t)} \quad . \quad (12)$$

The output, $e_o(t)$, is therefore given by

$$e_o(t) = Re \left\{ \exp[j\omega_o t] \sum_{\substack{k_1, k_2, \ldots, k_m = -\infty \\ (k_1 + k_2 + \ldots + k_m = 1)}}^{\infty} \exp\left[j \sum_{l=1}^{m-1} k_l \theta_l(t) \right] \cdot M(k_1, k_2, \ldots, k_m) \exp[jk_m \theta_m(t)] \right\} \quad . \quad (13)$$

In the absence of a noise signal at the input of the device, the output, $e_o(t)$, consists of the angle-modulated carriers and intermodulation products which also have the properties of angle-modulated carriers. With the introduction of noise at the input, the output may be divided into two categories:

a. the original output components with modified complex amplitudes, and

b. additional intermodulation components caused by the introduction of noise.

These two classes of output are represented by $e_S(t)$ and $e_N(t)$, respectively:

$$e_o(t) = e_S(t) + e_N(t) \quad . \tag{14}$$

To obtain $e_S(t)$, it is necessary to take the expected value of $e_o(t)$ on $N_c(t)$ and $N_s(t)$. For the particular case of Gaussian noise whose rms power is $R(0)$, this yields

$$e_S(t) = Re \left\{ \exp[j\omega_o t] \sum_{\substack{k_1,k_2,\ldots,k_{m-1}=-\infty \\ (k_1+k_2+\ldots+k_{m-1}=1)}}^{\infty} \exp\left[j \sum_{l=1}^{m-1} k_l \theta_l(t) \right] \cdot M_S(k_1, k_2, \ldots, k_{m-1}) \right\} \tag{15}$$

where, in genera',

$$M_S(k_1, k_2, \ldots, k_{m-1}) = \int_0^\infty \gamma \prod_{l=1}^{m-1} J_{k_l}(\gamma A_l) \exp\left[-\frac{\gamma^2}{2} R(0) \right] d\gamma \cdot \int_0^\infty \rho g(\rho) \exp[jf(\rho)] J_1(\gamma\rho)\, d\rho \quad . \tag{16}$$

For the Bessel function characterization, the carrier and intermodulation product levels are given by

$$M_S(k_1, k_2, \ldots, k_{m-1}) = \sum_{s=1}^{L} b_s \exp\left[-\frac{\alpha^2 s^2}{2} R(0) \right] \prod_{l=1}^{m-1} J_{k_l}(\alpha s A_l) \tag{17}$$

where $R(\tau)$ is the autocorrelation function of $N_c(t)$ or $N_s(t)$. As suggested in equation (17), the effects of noise in this case may be introduced merely by modifying each complex coefficient, b_s, with the factor

$$\exp\left[-\frac{\alpha^2 s^2}{2} R(0) \right] \quad .$$

An equivalent expression for $e_N(t)$ in which all intermodulation products can be identified was not derived here. Nevertheless, in the following expressions for the *RF* and baseband power spectra caused by intermodulation, the individual products can be identified. Identification of these products is necessary to incorporate the effects of noise in the computational algorithm.

RF intermodulation power spectrum

The general expression for the *RF* intermodulation power spectrum is given in Reference 5. For the particular case of a Bessel function expansion of the nonlinearity, the autocorrelation function of $e_o(t)$ is

$$\text{avg}\,[e_o(t)\, e_o(t+\tau)] = \frac{1}{2} Re \sum_{k_1,k_2,\ldots,k_{m-1}=-\infty}^{\infty} \exp\left(j\omega_o\tau + j \sum_{l=1}^{m} k_l \omega_l \tau \right) \cdot \text{avg}\left[\exp \sum_{l=1}^{m-1} jk_l\psi_l \right] H(k_1, k_2, \ldots, k_{m-1}) \tag{18}$$

where

$$\psi_l = \phi_l(t) - \phi_l(t+\tau) \quad ,$$

and

$$H(k_1, k_2, \ldots, k_{m-1}) = \sum_{s=1}^{L} \sum_{p=1}^{L} b_s b_p^* \left[\prod_{l=1}^{m-1} J_{k_l}(\alpha s A_l) \right] \left[\prod_{l=1}^{m-1} J_{k_l}(\alpha p A_l) \right] \cdot \exp\left[-\frac{\alpha^2(s^2+p^2)}{2} R(0) \right] I_{k_m}[R(\tau)\, \alpha^2 sp] \tag{19}$$

where $I_n(x)$ is the modified Bessel function of the first kind and nth order. In equation (18), the average on the left-hand side is taken on t, $\phi_l(t)$, $\phi_l(t+\tau)$, $N_c(t)$, $N_s(t)$, $N_c(t+\tau)$, and $N_s(t+\tau)$. The average on the right-hand side is taken on $\phi_l(t)$ and $\phi_l(t+\tau)$. In all subsequent equations, it will be assumed that $k_m = 1 - \sum_{l=1}^{m-1} k_l$, since this selection ensures that $\sum_{l=1}^{m} k_l = 1$, as required.

A series expansion of $I_n(x)$ leads to

$$I_{k_m}[R(\tau)\, \alpha^2 sp] = \sum_{q=0}^{\infty} \frac{\left[\frac{1}{2} R(\tau)\, \alpha^2 sp \right]^{2q+|k_m|}}{q!(|k_m|+q)!} \quad . \tag{20}$$

The term $H(k_1, k_2, \ldots, k_{m-1})$ therefore includes the noise contributions to the spectrum through $R(\tau)$ in the argument of the modified Bessel function. Rewriting makes it possible to obtain

$$H(k_1, k_2, \ldots, k_{m-1}) = \sum_{q=0}^{\infty} |N(k_1, k_2, \ldots, k_{m-1}; q)|^2 \cdot [\rho_o(\tau)]^{2q+|k_m|} \tag{21}$$

where
$$\rho_o(\tau) = \frac{R(\tau)}{R(0)} \qquad (22)$$

$$N(k_1, k_2, \ldots, k_{m-1}; q) = \sum_{s=1}^{L} b_s \exp\left[-\frac{\alpha^2 s^2}{2} R(0)\right] \cdot \prod_{l=1}^{m-1} J_{k_l}(\alpha s A_l)\, T(q, |k_m|, s) \qquad (23)$$

and
$$T(q, |k_m|, s) = \left\{ \frac{\left[\frac{1}{2} R(0)\, \alpha^2 s^2\right]^{2q+|k_m|}}{q!(|k_m|+q)!} \right\}^{1/2} . \qquad (24)$$

When equation (21) is introduced into equation (18), the output autocorrelation function becomes

$$\text{avg}\,[e_o(t)\, e_o(t+\tau)] = \frac{1}{2} Re \sum_{k_1,k_2,\ldots,k_{m-1}=-\infty}^{\infty} \exp\left(j\omega_o\tau + j\sum_{l=1}^{m} k_l\omega_l\tau\right) \cdot \text{avg}\left\{\exp\left[\sum_{l=1}^{m-1} jk_l\psi_l\right]\right\} \sum_{q=0}^{\infty} [\rho_o(\tau)]^{2q+|k_m|} \cdot |N(k_1, k_2, \ldots, k_{m-1}; q)|^2 \; . \qquad (25)$$

In the absence of noise, this autocorrelation function reduces to

$$\text{avg}\,[e_o(t)\, e_o(t+\tau)] = \frac{1}{2} Re \sum_{\substack{k_1,k_2,\ldots,k_{m-1}=-\infty \\ (k_1+k_2+\ldots+k_{m-1}=1)}}^{\infty} \exp\left(j\omega_o\tau + j\sum_{l=1}^{m-1} k_l\omega_l\tau\right) \cdot \text{avg}\left\{\exp\left[\sum_{l=1}^{m-1} jk_l\psi_l\right]\right\} |M(k_1, k_2, \ldots, k_{m-1})|^2 \qquad (26)$$

since
$$N(k_1, k_2, \ldots, k_{m-1}; 0) = M(k_1, k_2, \ldots, k_{m-1}) \qquad (27)$$

when
$$k_m = 1 - \sum_{l=1}^{m-1} k_l = 0.$$

The power spectral density of the output is given by the Fourier transform of equation (18). After a term-by-term transformation is performed, the following expression for the output power spectrum is obtained:

$$S(f) = \frac{1}{2} Re \sum_{k_1,k_2,\ldots,k_{m-1}=-\infty}^{\infty} \sum_{q=0}^{\infty} |N(k_1, k_2, \ldots, k_{m-1}); q|^2 \cdot \Omega\left(k_1, k_2, \ldots, k_{m-1}; q; f - f_o - \sum_{l=1}^{m} k_l f_l\right) \qquad (28)$$

where
$$\Omega(k_1, k_2, \ldots, k_{m-1}; q; f) = \left(\sigma_{\Sigma k_l\phi_l} \otimes \sigma_n \overset{2q+|k_m|}{\otimes} \sigma_n\right)(f)$$

and $\sigma_n(f)$ = power spectral density of $N_c(t)$ or $N_s(t)$ normalized to unit power

$\sigma_{\Sigma k_l\phi_l}$ = low-pass equivalent power spectral density, normalized to unit power, of a carrier angle modulated by $\sum_{l=1}^{m-1} k_l\phi_l(t)$.

The convolution operator $\overset{i}{\otimes}$ is defined as follows:

$$\sigma_n(f) \overset{0}{\otimes} \sigma_n(f) = \delta(f)$$
$$\sigma_n(f) \overset{1}{\otimes} \sigma_n(f) = \sigma_n(f)$$

and $\sigma_n(f) \overset{i}{\otimes} \sigma_n(f)$ represents $i - 1$ convolutions of $\sigma_n(f)$.

It has been shown that the output power spectrum is given by the summation of components indentified by the integers k_l and q. Analogous to the noise-free case, each component can be regarded as an intermodulaiont product. The total power of each product is $\frac{1}{2}|N(k_1, k_2, \ldots, k_{m-1}; q)|^2$, the corresponding normalized spectrum is $\Omega(k_1, k_2, \ldots, k_{m-1}; q; f)$, and its center frequency is $f_o + \sum_{l=1}^{m} k_l f_l$.

The normalized spectrum of a carrier which is angle modulated by a series of baseband functions can be calculated by convolving the normalized spectra of carriers modulated by each of the baseband functions; i.e.,

$$\sigma_{\Sigma k_l\phi_l} = \sigma_{k_1\phi_1} \otimes \sigma_{k_2\phi_2} \otimes \cdots \otimes \sigma_{k_{m-1}\phi_{m-1}} \; . \qquad (29)$$

Baseband intermodulation power spectrum

In the case of an interfering signal which is statistically independent of the angle-modulated desired carrier, the RF spectra of both the interfering signal and the desired carrier determine the spectrum of the baseband signal generated by the interference. Most intermodulation products fall into this category, with the exception of intermodulation products that contain the desired carrier [7]. Of particular interest are those products that are modulation sidebands of the desired carrier.

The mathematical derivation of the baseband power spectrum generated by the intermodulation process is included in Appendix A. In this section, only the most significant results will be discussed.

Again consider the case of $m - 1$ angle-modulated carriers and a band of noise, as shown in equation (11). The signal at the input to the angle demodulator can be represented by three terms:

a. the angle-modulated carrier (undistorted),
b. the modulation sidebands caused by the noise, and
c. the other intermodulation components.

Without loss of generality, consider the demodulation of the first carrier. The three corresponding terms are shown in the following equation:

$$\begin{aligned} e_{o1}(t) = {} & Re\,\{\exp[j\omega_o + j\theta_1(t)]\,M_o\} \\ & + Re\,\{\exp[j\omega_o t + j\theta_1(t)]\,[M(1, 0, \ldots, 0; t) - M_o]\} \\ & + Re\left\{\exp[j\omega_o t] \sum_{\substack{k_1, k_2, \ldots, k_m = -\infty \\ (k_1 + k_2 + \ldots + k_m = 1)}}^{\infty} \exp\left[j \sum_{l=1}^{m-1} k_l\theta_l(t)\right] \right. \\ & \left. \cdot\, M(k_1, k_2, \ldots, k_m; t) \exp[jk_m\theta_m(t)]\right\} \\ = {} & Re\,\{\exp[j\omega_o t + j\theta_1(t)]\,M_o[1 + R(t) + jI(t)]\} \end{aligned} \quad (30)$$

where, in the case of a Bessel function expansion,

$$M_o = \sum_{s=1}^{L} b_s \exp\left[-\frac{\alpha^2 s^2}{2} R(0)\right] J_1(\alpha s A_1) \prod_{l=2}^{m-1} J_0(\alpha s A_l)$$

and

$$\begin{aligned} M(k_1, k_2, \ldots, k_m; t) = {} & \sum_{s=1}^{L} b_s J_{k_m}\left[\alpha s \sqrt{N_c^2(t) + N_s^2(t)}\right] \\ & \cdot \prod_{l=1}^{m-1} J_{k_l}(\alpha s A_l) \quad . \end{aligned} \quad (31)$$

For small intermodulation levels relative to the demodulated carrier, the detected angle can be approximated by*

$$\phi_1(t) + \tan^{-1}\left[\frac{I(t)}{1 + R(t)}\right] \approx \phi_1(t) + I(t) \quad . \quad (32)$$

To evaluate the power spectral density of $I(t)$, its autocorrelation function is first determined. The two types of components for which the cross-correlation terms do not vanish are identified as $M(2 - k_1, -k_2, -k_3, \ldots, -k_m)$ and $M(k_1, k_2, \ldots, k_m)$. Particular components satisfying this condition are the $A + B - C$ and $A + C - B$ intermodulation products, which are modulation sidebands of carrier A.

The autocorrelation function can be written as

$$\begin{aligned} \text{avg}\,[I(t)\, I(t + \tau)] = {} & R_1(\tau) + \sum_{k_1, k_2, \ldots, k_{m-1} = -\infty}^{\infty}{}'' \\ & \cdot\, R(k_1, k_2, \ldots, k_{m-1}; \tau) \end{aligned} \quad (33)$$

where the components for which the cross-correlation terms do not vanish have been combined in the summation $\sum''$ [see equation (A40)], and $R_1(\tau)$ represents the effects of noise components falling on the demodulated carrier, i.e., intermodulation products of the form $N + A - N$ or $2N + A - 2N$. The other terms of particular interest are those for which $k_1 = 1$, since these terms are modulation sidebands of the demodulated carrier.

After a detailed derivation, which can be found in Appendix A, $R_1(\tau)$ and $R(1, k_2, \ldots, k_{m-1}; \tau)$ are expressed as

$$R_1(\tau) = \sum_{q=1}^{\infty} \left\{Im\left[\frac{1}{M_o} N(1, 0, \ldots, 0; q)\right]\right\}^2 [\rho_o(\tau)]^{2q} \quad (34)$$

and

$$\begin{aligned} & R(1, k_2, \ldots, k_{m-1}; \tau) \\ & = 2 \sum_{q=0}^{\infty} \left\{Im\left[\frac{1}{M_o} N(1, k_2, \ldots, k_{m-1}; q)\right]\right\}^2 [\rho_o(\tau)]^{2q + |k_m|} \\ & \quad \cdot \exp\left[j \sum_{l=2}^{m} k_l \omega_l \tau\right] \text{avg}\left\{\exp\left[j \sum_{l=2}^{m-1} k_l \psi_l\right]\right\} \end{aligned} \quad (35)$$

respectively. For all other components, the phase between the carrier and the intermodulation products can be ignored, and the correlation is given by

*This approximation is good in all practical cases since the distortion objectives for intermodulation are low.

$$R(k_1, k_2, \ldots, k_{m-1}; \tau)$$
$$= \sum_{q=0}^{\infty} \frac{|N(k_1, k_2, \ldots, k_{m-1}; q)|^2}{|M_o|^2} [\rho_o(\tau)]^{2q+|k_m|} \exp\left[\sum_{l=2}^{m} k_l \omega_l \tau\right]$$
$$\cdot \text{ avg}\left\{\exp\left[j \sum_{l=2}^{m-1} k_l \psi_l\right] \exp[j(k_1 - 1)\psi_1]\right\}$$
$$\cdot \exp[j(k_1 - 1)\omega_1 \tau] \quad . \tag{36}$$

The corresponding baseband power spectrum after angle demodulation is

$$S_\phi(f) = \sum_{q=1}^{\infty} \left\{Im\left[\frac{N(1, 0, \ldots, 0; q)}{M_o}\right]\right\}^2 \Omega(0, 0, \ldots, 0; q; F)$$
$$+ \sum_{k_2, \ldots, k_{m-1} = -\infty}^{\infty} \sum_{q=0}^{\infty} 2\left\{Im\left[\frac{N(1, k_2, \ldots, k_{m-1}; q)}{M_o}\right]\right\}^2$$
$$\cdot \; \Omega(0, k_2, \ldots, k_{m-1}; q; F)$$
$$+ \sum_{k_1, k_2, \ldots, k_{m-1} = -\infty}^{\infty} \sum_{q=0}^{\infty} \frac{|N(k_1 \neq 1, k_2, \ldots, k_{m-1}; q|^2}{|M_o|^2}$$
$$\cdot \; \Omega(k_1 - 1, k_2, \ldots, k_{m-1}; q; F) \tag{37}$$

where
$$F = f - (k_1 - 1) f_1 - \sum_{l=2}^{m} k_l f_l \quad . \tag{38}$$

The corresponding baseband power spectrum after frequency demodulation is

$$S_f(f) = f^2 S_\phi(f) \quad . \tag{39}$$

The power spectrum of the demodulated output is formed by components which can be easily related to those appearing in the RF power spectrum [equation (28)]. However, in the power spectrum of the angle-modulated output, three distinct categories of terms can be identified:

a. *Noise intermodulation components centered on the demodulated carrier A*: $N(1, 0, \ldots, 0; q)$. These components are of the type which includes $N + A - N$ and $2N + A - 2N$ products. The spectrum shape Ω is determined by convolutions of the input noise spectrum. Only the component in phase quadrature appears in the demodulator.

b. *Modulation sidebands of the demodulated carrier A*: $N(1, k_2, \ldots, k_{m-1}; q)$. These components are of the type which includes $A + B - C$ and $A + C - B$ products. The spectrum shape Ω is determined by convolutions of the spectra of the other signals generating the product. Only the component in phase quadrature appears in the demodulator.

c. *All other components*: $N(k_1 \neq 1, k_2, \ldots, k_{m-1}; q)$. These components are of the type which includes $B + C - D$ and $2B - C$ products. Such products may also contain the desired carrier, in which case the index corresponding to the demodulated carrier in the spectrum shape Ω is $k_1 - 1$. Both the in-phase and the in-quadrature components of the product appear in the demodulator.

Computational methods

In previous sections, the pertinent aspects of the theory have been discussed. Attention is now directed to the manner in which these techniques have been implemented in a versatile, user-oriented computer program for intermodulation analysis. The design goals, as well as the means by which these goals have been achieved, will be discussed.

Program overview

The Intermodulation Analyzer Program has been designed with the following objectives in mind:

a. *Arbitary nonlinearity*. The user may enter any arbitrary set of transfer curves to determine the real and imaginary coefficients of the Bessel function expansion.

b. *Choice of input signals*. The program accepts three distinct carrier types (FM telephony, FM television, and PCM/PSK) in addition to bands of equal carriers or thermal noise.

c. *Analysis options*. Both 3rd- and 5th-order intermodulation products, as well as the resulting intermodulation spectra and baseband distortion for all carrier types may be calculated.

d. *Output options*. Intermodulation spectra and intermodulation product frequencies may be generated in printed, plotted, or punched form. A listing of intermodulation products which includes all such products, only those in a specified range, or only those exceeding a specified threshold value may be generated.

e. *Ease of use*. A large number of user convenience features, including a mnemonic-controlled, free-field input format and the capability for rerun with modified parameters, are provided.

Figure 1 is an idealized flow chart of the program. After the input data deck is processed and the nonlinear coefficients are determined, the intermodulation products are selected and processed one at a time. The only

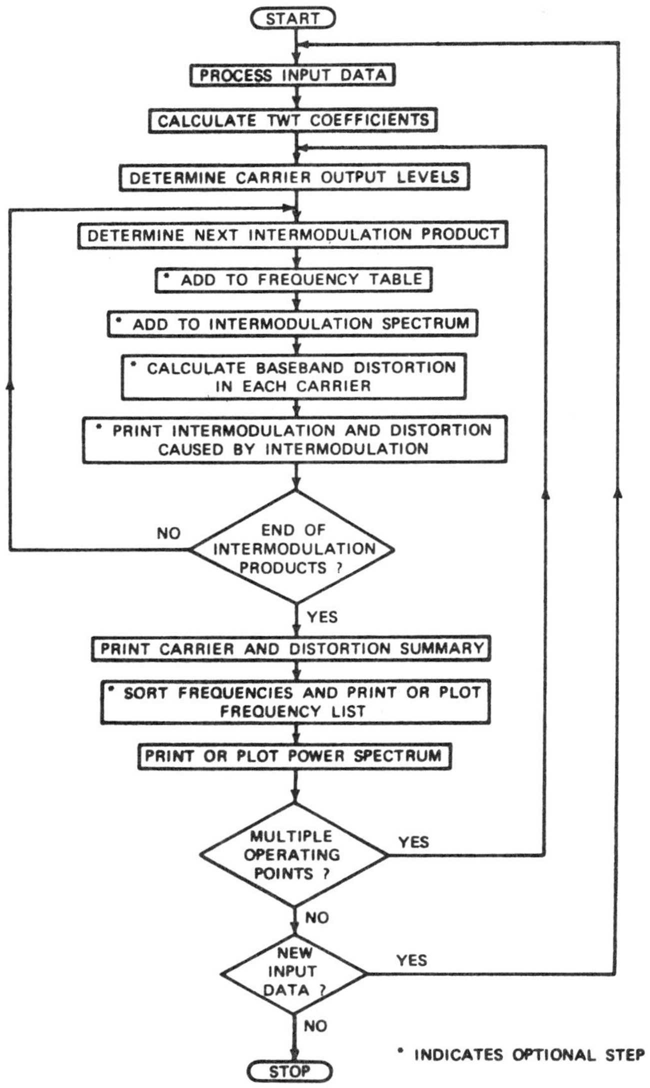

Figure 1. *Program Flow Chart*

information retained from this process is a list of distinct product frequencies, a table of power spectrum values, and the total baseband distortion in each carrier. If data pertaining to particular intermodulation products are desired, these data are printed as each intermodulation product is generated.

Following the processing of the intermodulation products, an output processor is called to print a summary of carriers and baseband distortion, and to print, plot, or punch the power spectrum and intermodulation product frequency data. Provisions are included in the program for considering multiple operating points, or for modifying any parameters in the problem in multiple executions.

Characterization of the nonlinear device

The characteristics of the nonlinear device are derived from the single-carrier transfer curves relating the output power, P_o, and phase, ψ_o, to the input power, P_i. Assume that the complex output envelope may be approximated by the Bessel function expansion shown in equation (9). Then the real and imaginary components of equation (9) are

$$g(\rho) \cos f(\rho) = \sum_{s=1}^{L} b_{sr} J_1(\alpha s \rho) \tag{40a}$$

$$g(\rho) \sin f(\rho) = \sum_{s=1}^{L} b_{si} J_1(\alpha s \rho) \tag{40b}$$

respectively. It has been found that 10 terms are sufficient for typical nonlinear characteristics. A procedure has been included in the program for determining appropriate values of b_{sr} and b_{si} from a set of measured values for P_i, P_o, and ψ_o. The normalized envelope levels corresponding to the measured input and output power levels may be found relative to the levels at saturation:

$$\bar{\rho} = \sqrt{\frac{2P_i}{P_{is}}} \tag{41a}$$

$$\bar{g} = \sqrt{\frac{2P_o}{P_{os}}} \tag{41b}$$

$$\bar{f} = \psi_o, \text{ in radians} \tag{41c}$$

where P_{is} and P_{os} are the power levels at saturation.

Error functions E_1 and E_2, which represent the squared difference between the envelope quantities in equation (41) and the expansion in equation (40), summed over all N measured points on the transfer curves, are defined as follows:

$$E_1(b_{sr}) = \sum_{l=1}^{N} \left[g_l \cos(\bar{f_l}) - \sum_{s=1}^{10} b_{sr} J_1(\alpha s \bar{\rho}_l) \right]^2 \tag{42a}$$

$$E_2(b_{si}) = \sum_{l=1}^{N} \left[g_l \sin(\bar{f_l}) - \sum_{s=1}^{10} b_{si} J_1(\alpha s \bar{\rho}_l) \right]^2 \quad . \tag{42b}$$

The limit of the output phase angle as the input power approaches zero is given by

$$\lim_{\rho \to 0} \psi_o = \frac{\sum_{s=1}^{10} s b_{si}}{\sum_{s=1}^{10} s b_{sr}} \tag{43}$$

To ensure that this limit is zero, it has been found necessary to impose the following constraint:

$$\sum_{s=1}^{10} s b_{si} \equiv 0 \quad . \tag{44}$$

The Fletcher-Powell optimization scheme [10] is used to determine values of the coefficients b_{sr} and b_{si} which minimize the error functions $E_1(b_{sr})$ and $E_2(b_{si})$, subject to the constraint given by equation (44). As an example of the application of this technique, a typical traveling wave tube (TWT) amplifier on the INTELSAT IV satellite will be considered. The measured transfer curves for this device are shown in Figure 2, where the values presented to the program are indicated by dots. The resulting real and imaginary coefficients generated by the program for $\alpha = 0.6$ are shown in Table 1. The contributions of selected terms to the total real and imaginary components of the envelope are indicated in Figures 3 and 4.

The difference between the measured data and the calculated values based on these coefficients is smaller than the measurement error. For low input levels, the calculated real output component is characterized by a linear behavior, whereas the imaginary component follows the third power of the input level. Furthermore, the calculated output angle is

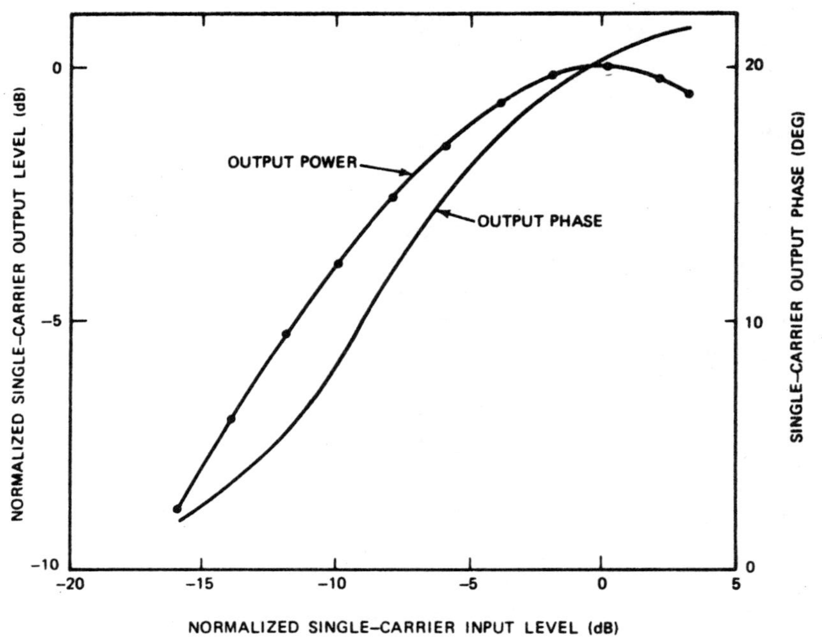

Figure 2. *Measured Nonlinear Transfer Curves*

TABLE 1. CALCULATED TRANSFER CURVE COEFFICIENTS ($\alpha = 0.6$)

Index, s	1	2	3	4	5
b_{sr}	3.089	−0.09465	−0.2075	1.399	−0.1674
b_{si}	1.045	−1.034	1.992	−0.900	−0.6464
Index, s	6	7	8	9	10
b_{sr}	−0.4258	0.3040	0.4548	−0.5160	0.2435
b_{si}	0.6189	1.017	−2.342	1.837	−0.6750

linear with input power, indicating a convergence to zero as the input level approaches zero.

Since the Bessel function expansions for the real and imaginary components are obtained independently, a different α might have been used

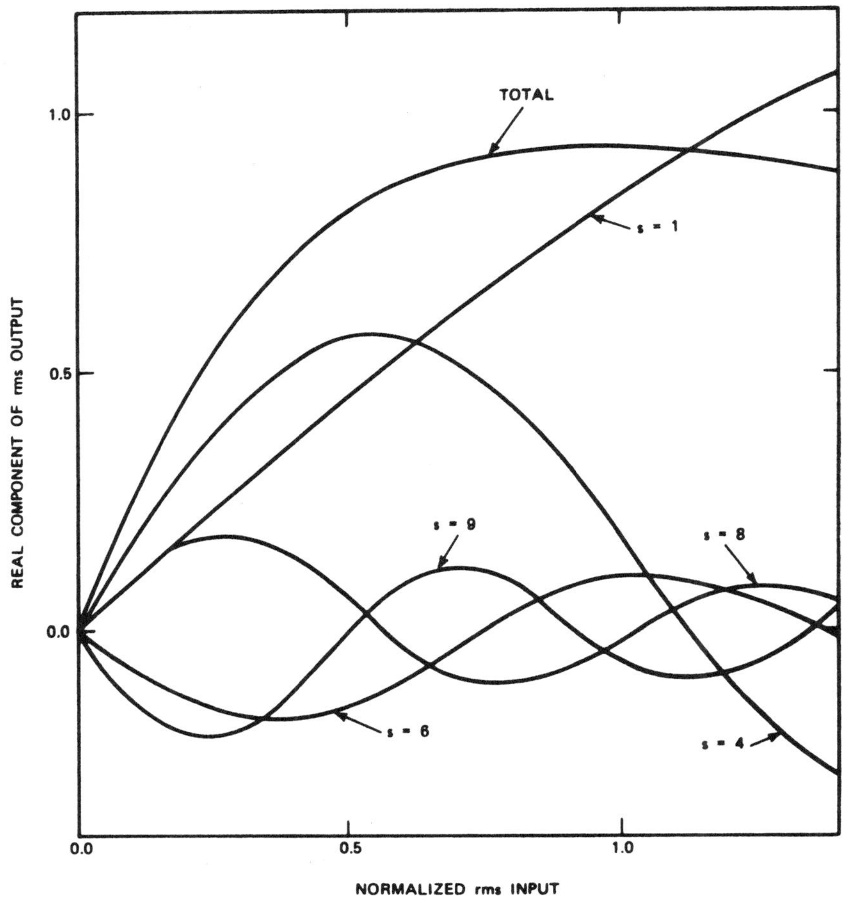

Figure 3. *Bessel Function Expansion of the Real Component of the Single-Carrier Transfer Function*

in each expansion. Furthermore, an optimum α could be found for each component by using the same optimization scheme. For the typical nonlinear transfer curves encountered, this has not been found to be necessary, since an adequate fit may be obtained for almost any reasonable value of α.

The coefficients b_s might also be determined from measurements of intermodulation product output power and phase as functions of input

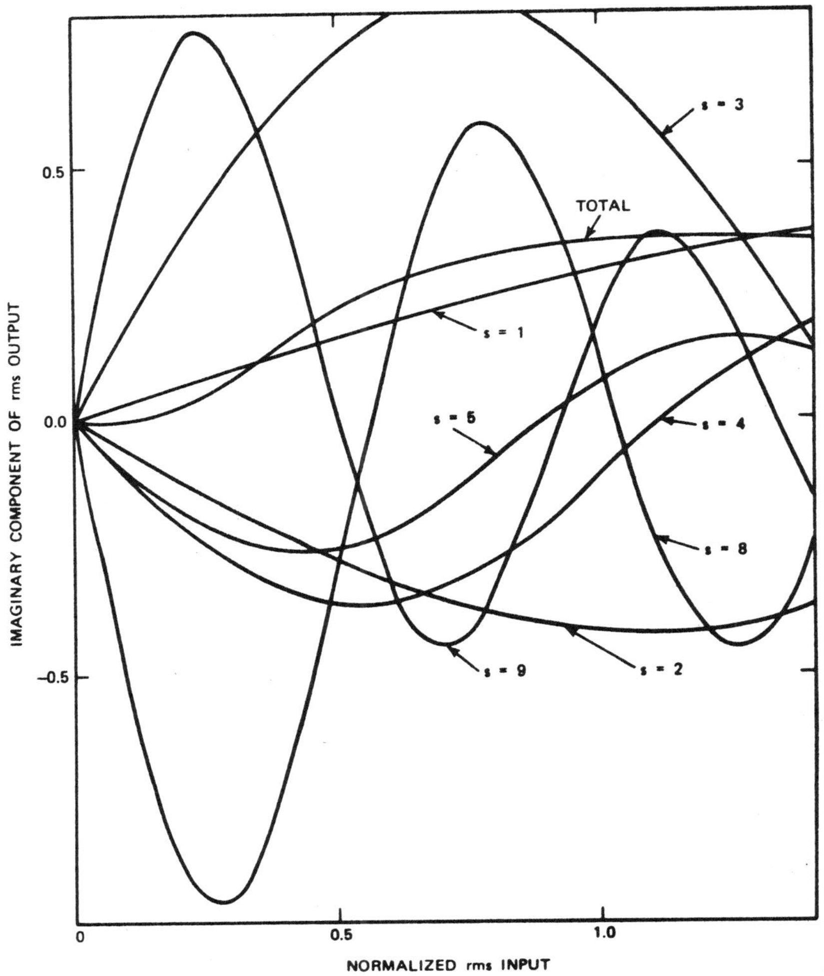

Figure 4. *Bessel Function Expansion of the Imaginary Component of the Single-Carrier Transfer Function*

level. This approach, although mathematically attractive, has practical disadvantages.

Characterization of input signals

The input signals are defined by a series of parameters specifying the input power level, the center frequency, and other information necessary to calculate the RF spectrum and baseband distortion caused by intermodulation. The power spectral density of all input signals is assumed to be either Gaussian, rectangular (i.e., uniform over a finite range), or a convolution of both. This restriction has been imposed to eliminate the need for time-consuming numerical convolution techniques; all convolutions in the program are based on closed-form solutions.

The spectral shape assumed for each signal type is summarized in the following:

a. *FM telephony carriers.* All fully loaded telephony carriers are represented by a Gaussian spectrum whose rms deviation is specified by the user. For partially loaded carriers, the rms deviation of the signal is reduced accordingly. If the addition of a triangular spectrum-spreading waveform is desired, the resulting spectrum is the convolution of the spectrum of the partially loaded carrier and a uniform power density spectrum. This uniform spectrum represents the RF spectrum of the carrier modulated by the spreading waveform.

b. *FM television carriers.* No attempt was made to model the TV spectrum in the presence of a baseband signal, since each baseband signal generates a different RF spectrum. Instead, the case of a black picture, in which the RF spectrum is uniform and its peak-to-peak frequency deviation is determined by the spreading waveform, was considered. Unlike the multichannel telephony case, the spreading waveform power was not changed during transmission.

c. *Digital PCM/PSK carriers.* The RF spectrum (normalized to unit power) of a PCM/PSK carrier is given by

$$S_D(f) = T\left[\frac{\sin(\pi f T)}{(\pi f T)}\right]^2 \tag{45}$$

where $1/T$ is the baud rate and f is the frequency measured from the carrier frequency. However, because of the self-imposed restriction on Gaussian and rectangular spectra, an alternate representation had to be derived. It can be shown that the convolution of a Gaussian function with a rectangle of appropriate width closely approximates the function $S_D(f)$.

If an rms deviation equal to $0.246/T$ and a uniform spectrum of width $0.826/T$ are chosen, then the mean squared error between $S_D(f)$ and the convolved function, summed over a representative number of points on the main lobe of $S_D(f)$, is minimized. The effect of filtering the PCM/PSK carrier and the resulting time-varying envelope has not been considered.

d. *Noise or carrier band.* The noise or carrier band spectrum is assumed to be rectangular.

Calculation of intermodulation product levels, spectra, and baseband distortion intermodulation products

Intermodulation products are generated by the program one at a time, according to generic form. Table 2 shows all possible 3rd- and 5th-order intermodulation products, including those arising in the presence of noise.

The center frequency of each intermodulation product is given by

$$F(k_1, k_2, \ldots, k_m; q) = \sum_{k=1}^{m} k_l f_l \tag{46}$$

where f_l is the center frequency of the lth input signal. The indices k_i can take integer values between -2 and $+3$. Furthermore, the condition for an in-band center frequency imposes the constraint that $\sum_{l=1}^{m} k_l \equiv 1$.

Note that the intermodulation products which include the noise term are characterized by $k_m \neq 0$ and/or $q \neq 0$. The index q affects the order of the intermodulation product, the spectrum, and the level, but does not alter the center frequency. Analogous to the noise-free case, the order of an intermodulation product may be defined as

$$\text{order} = \sum_{l=1}^{m} |k_l| + 2q \quad . \tag{47}$$

The selection of intermodulation products in this manner may be extended to odd orders higher than the fifth order, although the computation time will become prohibitive for a reasonable number of carriers. However, since the intermodulation product level decreases rapidly with increasing order, the inclusion of only 3rd-order products has been found to be adequate for most analyses.

INTERMODULATION PRODUCT LEVELS

The complex amplitude of carrier and intermodulation products, $M_S(k_1, k_2, \ldots, k_{m-1})$, is determined from equation (17). The corresponding complex amplitude of intermodulation products which include noise, $N(k_1, k_2, \ldots, k_{m-1}; q)$, is found from equation (23).

TABLE 2. SUMMARY OF INTERMODULATION PRODUCT TYPES

Generic Name*	Order	k_A	k_B	k_C	k_D	k_E	k_m	$2q$
$A + B - C$	3	1	1	−1	0	0	0	0
$2A - B$	3	2	−1	0	0	0	0	0
$A + B + C - D - E$	5	1	1	1	−1	−1	0	0
$2A + B - C - D$	5	2	1	−1	−1	0	0	0
$3A - B - C$	5	3	−1	−1	0	0	0	0
$A + B + C - 2D$	5	1	1	1	−2	0	0	0
$3A - 2B$	5	3	−2	0	0	0	0	0
$A + N - B$	3	1	−1	0	0	0	1	0
$A + B - N$	3	1	1	0	0	0	−1	0
$2A - N$	3	2	0	0	0	0	−1	0
$2N - A$	3	−1	0	0	0	0	2	0
$N + A - N$	3	1	0	0	0	0	0	2
$N + N - N$	3	0	0	0	0	0	1	2
$A + B + N - C - D$	5	1	1	−1	−1	0	1	0
$A + B + C - D - N$	5	1	1	1	−1	0	−1	0
$2A + N - B - C$	5	2	−1	−1	0	0	1	0
$2A + B - C - N$	5	2	1	−1	0	0	−1	0
$2N + A - B - C$	5	1	−1	−1	0	0	2	0
$3N - A - B$	5	−1	−1	0	0	0	3	0
$3A - B - N$	5	3	−1	0	0	0	−1	0
$A + B + N - 2C$	5	1	1	−2	0	0	1	0
$A + B + C - 2N$	5	1	1	1	0	0	−2	0
$3N - 2A$	5	−2	0	0	0	0	3	0
$3A - 2N$	5	3	0	0	0	0	−2	0
$2N + A - 2N$	5	1	0	0	0	0	0	4
$2N + N - 2N$	5	0	0	0	0	0	1	4
$N + A + B - C - N$	5	1	1	−1	0	0	0	2
$N + 2A - B - N$	5	2	−1	0	0	0	0	2
$N + A + N - B - N$	5	1	−1	0	0	0	1	2
$N + A + B - N - N$	5	1	1	0	0	0	−1	2
$N + 2A - N - N$	5	2	0	0	0	0	−1	2
$N + 2N - A - N$	5	−1	0	0	0	0	2	2

*A, B, C, D, and E are angle-modulated carriers; N is a noise or carrier band.

In the calculation of the intermodulation product levels, it is found that a small number of factors recur in various combinations. These factors, which are calculated only once at the beginning of the execution, are $J_{k_l}(\alpha s A_l)$ and $T(0, |k_m|, s)$ for k_l and k_m between 0 and 3.

SORTING OF INTERMODULATION PRODUCT FREQUENCIES

Recall that the intermodulation products are calculated one at a time, on the basis of generic form; hence, sorting is required if an ordered listing of distinct intermodulation frequencies and the corresponding total power is desired. Because of the large number of intermodulation products generated in most problems of practical interest, improper choice of a sorting algorithm can drastically affect the program running time.

Throughout the execution of the program, an array of intermodulation product frequencies and the total level at each frequency are maintained in core. As each new intermodulation product frequency and level are calculated, it must be determined whether an intermodulation product at that frequency has been previously calculated. If so, the levels are added; if not, the new intermodulation product is added to the array and will appear in the sorted list generated by the output processor.

A technique which has been found to be ideally suited to this purpose is a binary distributive sort [11]. Figure 5 is an example of a tree structure demonstrating the sorting algorithm. Each node in the structure corresponds to a distinct intermodulation product frequency, which, in turn, points to at most two other frequencies in the list, one higher in value and one lower in value. These points take the form of integer arrays UP and DWN.

As each new frequency is generated, it must be compared with no more than $\log_2 n$ frequencies in the list, where n is the total length of the list,

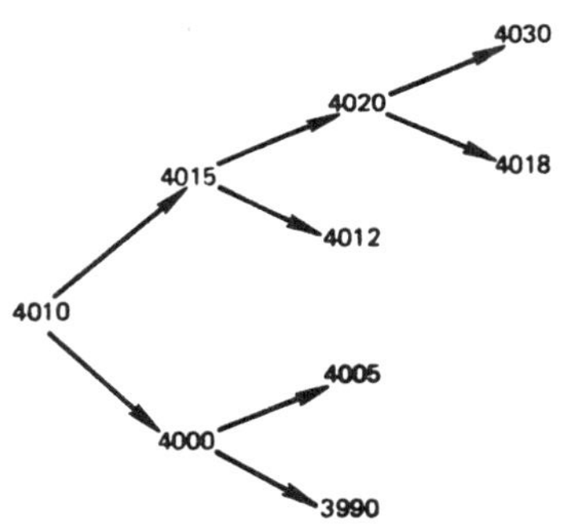

LOC	FREQ	ARRAY	
		UP	DWN
1	4010	2	3
2	4015	6	8
3	4000	4	5
4	4005	—	—
5	3990	—	—
6	4020	9	7
7	4018	—	—
8	4012	—	—
9	4030	—	—

Figure 5. *Tree Structure for Sorting Algorithm*

to determine whether an identical frequency has been previously entered. If not, the new frequency is entered in the list and an appropriate pointer is set in the UP or DWN array. After all frequencies have been entered in the table, the arrays UP and DWN contain sufficient information to generate a sorted list of frequencies without referencing the actual frequency values.

CONVOLUTION TECHNIQUE

Convolutions of spectra are necessary to calculate both RF spectra and baseband distortion. Hence, the process of convolution is an important aspect of the program design. The convolution algorithm is restricted to accept only Gaussian or rectangular spectra. This restriction was established to avoid both the excessive computation times associated with numerical convolution and the complexity involved in accommodating a wider range of spectrum types.

A further limitation of the algorithm is that an exact closed-form solution is obtained for only three or fewer rectangles. If additional rectangles are present, the narrowest ones are approximated by Gaussian functions with equivalent rms deviations. Since convolving two or more Gaussian functions results in another Gaussian function, the convolution algorithm inputs are limited to a single Gaussian function and up to three rectangular functions.

BASEBAND DISTORTION

The distortion measures of interest depend on the modulation and baseband signals used. Three cases have been considered:

a. frequency modulation with multichannel telephony signals,
b. frequency modulation with various TV signals, and
c. PCM/PSK.

Although an exact analysis is available for the first two cases, for PCM/PSK, approximations are used which correspond to treating the intermodulation effect as white Gaussian additive noise.

Telephony Carriers. The *NPR* (noise power ratio) is defined as the ratio of signal power to noise power in a specified frequency band. The *NPR* caused by intermodulation is calculated by using the formula

$$NPR = \frac{P(f_r) \cdot f_{rms}^2}{(1-\epsilon) f_r^2 S_f} \tag{48}$$

where S_f is the frequency demodulated power spectral density, f_{rms} is the *rms* frequency deviation, ϵ is the ratio of minimum to maximum baseband frequencies, f_r is the frequency for which the *NPR* is calculated, and $P(f_r)$ is the pre-emphasis weighting factor.

TV Carriers. In TV transmission the signal-to-noise ratio, S/N, is defined as the peak-to-peak luminance power divided by the noise power (either weighted or unweighted). In the baseband, the peak-to-peak luminance voltage is 0.714 volt. The peak test-tone deviation, f_d, referenced to a 1-volt peak-to-peak luminance power, L_{p-p}, in Hz^2, is given by

$$L_{p-p} = (2 \times 0.714 f_d)^2 \tag{49}$$

and the resulting S/N is

$$S/N = \frac{(2 \times 0.714 f_d)^2}{\int_{f_1}^{f_m} S_f(f)\, df} . \tag{50}$$

The effects of the de-emphasis network and noise weighting are introduced into the S/N equation as follows:

$$(S/N)_D = \frac{(2 \times 0.714 f_d)^2}{\int_{f_1}^{f_m} S_f(f)\, P_D(f)\, df} \tag{51}$$

$$(S/N)_{D,W} = \frac{(2 \times 0.714 f_d)^2}{\int_{f_1}^{f_m} S_f(f)\, P_D(f)\, P_W(f)\, df} \tag{52}$$

where $P_D(f)$ and $P_W(f)$ are respectively the network responses of the de-emphasis and noise weighting networks.

Digital Carriers. The baseband impairment in a carrier with digital traffic is measured in terms of the bit-error probability, P_{be}. To estimate this parameter, it has been necessary to make two assumptions:

a. That P_{be} can be calculated from the following formula, which is valid for Gaussian noise and 4-phase PSK:

$$P_{be} = \frac{1}{2} erfc(\sqrt{\gamma}) \tag{53}$$

where γ is the S/N at the filter output at the sampling instant. For the carrier-to-IF noise power, the total power must be divided between the two quadrature components; hence,

$$\gamma = \frac{C}{2N}$$

where C/N is the IF carrier-to-noise power.

b. That the noise N is strictly the IF noise, and that no special consideration is made to incorporate time-domain results since no rigorous mathematical solutions are available.

Sample calculations

This section presents some results obtained from the computer program. Two separate examples have been selected, each illustrating a different aspect of the analysis. The first example deals exclusively with RF calculations. A number of RF spectra are included to demonstrate the validity of approximating a band of carriers by an equivalent noise input.

In a second example, an input consisting of three carriers is assumed, and the resulting intermodulation spectrum and demodulated output are determined. This case is especially significant since it involves the generation of intermodulation products which are modulation sidebands of the input carriers.

RF spectrum calculations

Three cases have been considered here. All three are characterized by an input signal made up of an angle-modulated carrier (FM with a Gaussian baseband signal) and a second signal which can take one of three forms, each having a uniform RF spectrum:

case 1: 10 small, equally spaced carriers at a uniform level;
case 2: a band of Gaussian noise of constant level; and
case 3: a single carrier, frequency modulated by a triangular waveform. A large modulation index is assumed.

It is assumed that both signals are at the same level and that they saturate the amplifier. The FM carrier is located at 4,000 MHz and the other signal in the range between 4,015 MHz and 4,025 MHz. The resulting intermodulation power spectra are calculated for all three cases.

CASE 1

The unmodulated intermodulation products resulting from the single carrier and the band of 10 small carriers are shown in Figure 6. If both 3rd- and 5th-order products are calculated, the plot shown in Figure 7 is obtained. The effect of the 5th-order products, which may be observed by comparing both figures, is to fill in the valleys between 3rd-order intermodulation products and to change slightly the level at frequencies where 3rd-order products fall.

The corresponding power spectra are shown in Figure 8 for the case in which all 10 carriers are modulated with 500-MHz rms deviation. Note that, if the level of the input is decreased, the 5th-order products

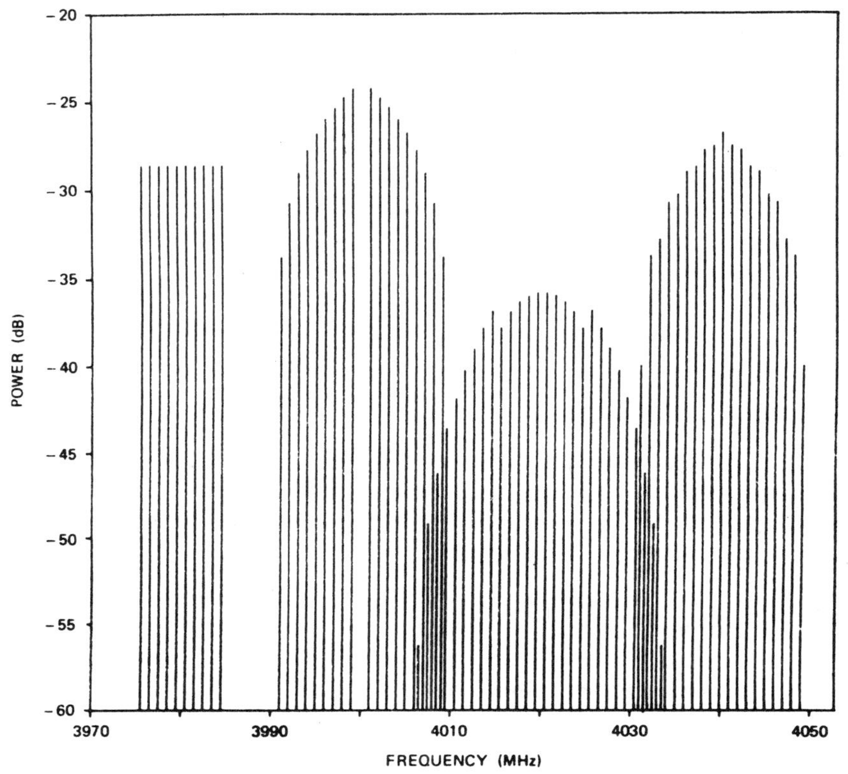

Figure 6. *3rd-Order RF Intermodulation Products Generated by 10 Equal Carriers and a Single Carrier*

decrease very rapidly and can be ignored in most cases. Also note that there are many intermodulation products clustered about the single carrier at 4,000 MHz. (The spectrum is triangular in a linear scale.) There is no product falling exactly on the single carrier, which accounts for the dip in the spectrum at 4,000 MHz.

CASE 2

The input is characterized by a single carrier at 4,000 MHz and a band of noise (or carriers) centered at 4,020 MHz. The intermodulation power spectrum plots generated by the program are shown in Figure 9. Note that the spectrum is strikingly similar to that of case 1, except that the dip at 4,000 MHz does not occur, since calculating the power spectrum of a band of noise is equivalent to taking the limit as the interval between carriers approaches zero.

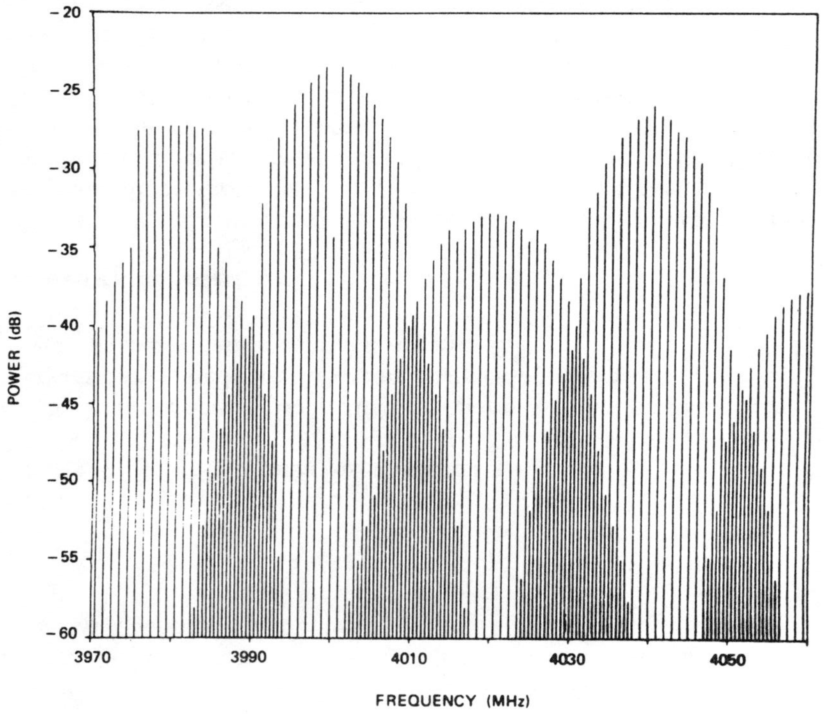

Figure 7. *3rd- and 5th-Order RF Intermodulation Products Generated by 10 Equal Carriers and a Single Carrier*

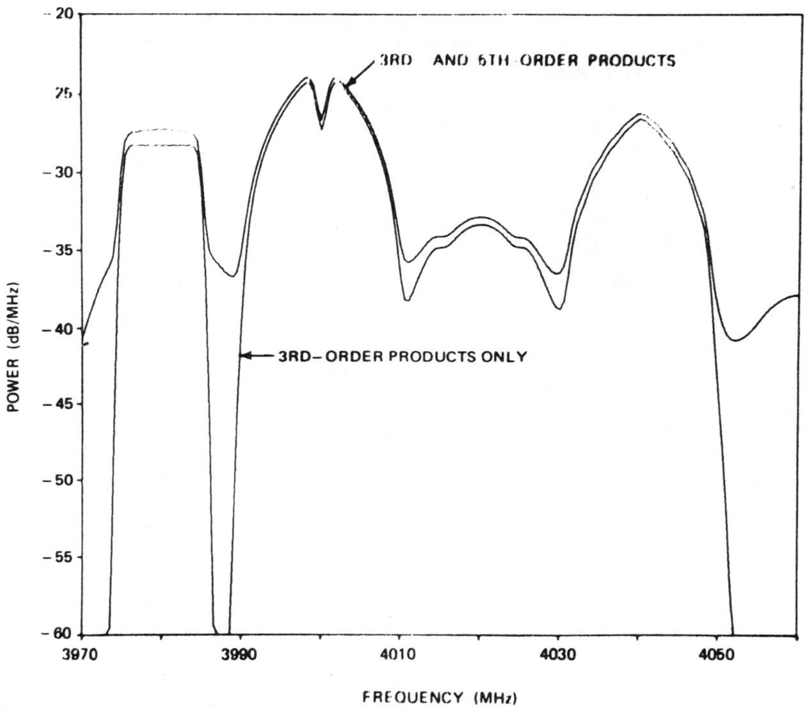

Figure 8. *RF Intermodulation Power Spectral Density Generated by 10 Equal Carriers and a Single Carrier*

CASE 3

The same single carrier is now assumed to share an amplifier with another carrier which is frequency modulated with a triangular waveform so that the input spectrum is identical to that of case 2. The resulting intermodulation spectrum is shown in Figure 10. There is a significant difference between this spectrum and the previous spectra because, although the input spectra are identical in all cases, in this last case, both input signals have a constant envelope and hence generate considerably fewer intermodulation products.

3-Carrier Input

A 3-carrier input to a nonlinear device is an interesting configuration, since there are intermodulation products which are modulation sidebands

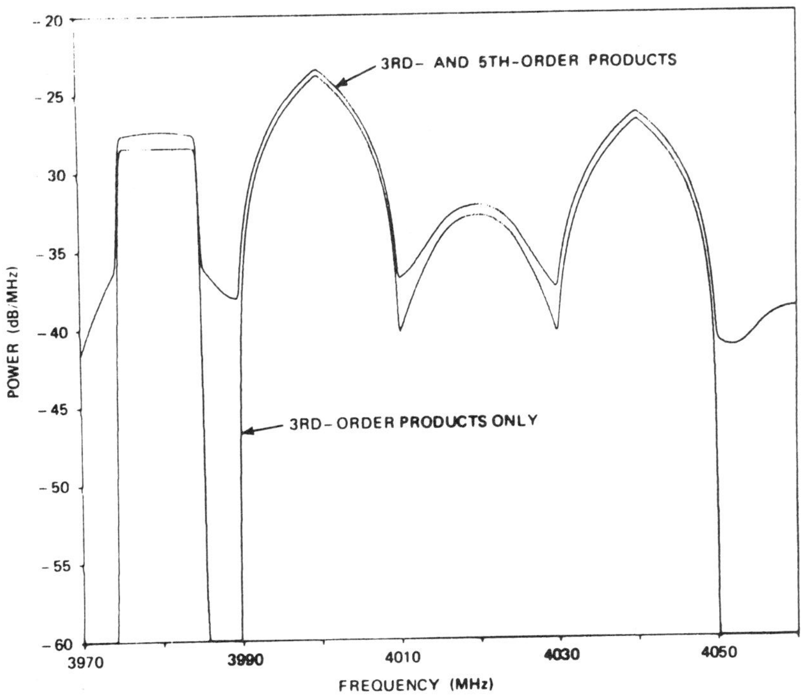

Figure 9. *RF Intermodulation Power Spectral Density Generated by a Carrier Band and a Single Carrier*

and consequently cannot be treated as an independent interference entry. The three carriers considered are listed in Table 3. Changes in the relative carrier level, Δ, were found to alter dramatically the character of the carrier-to-intermodulation curves and the resulting demodulated output of reference carrier A.

TABLE 3. INPUT CARRIER PARAMETERS

Carrier	Frequency (MHz)	Level (dB)	Modulation
A	4,020.0	0.0	TV
B	4,005.0	Δ	Telephony
C	4,007.5	Δ	Telephony

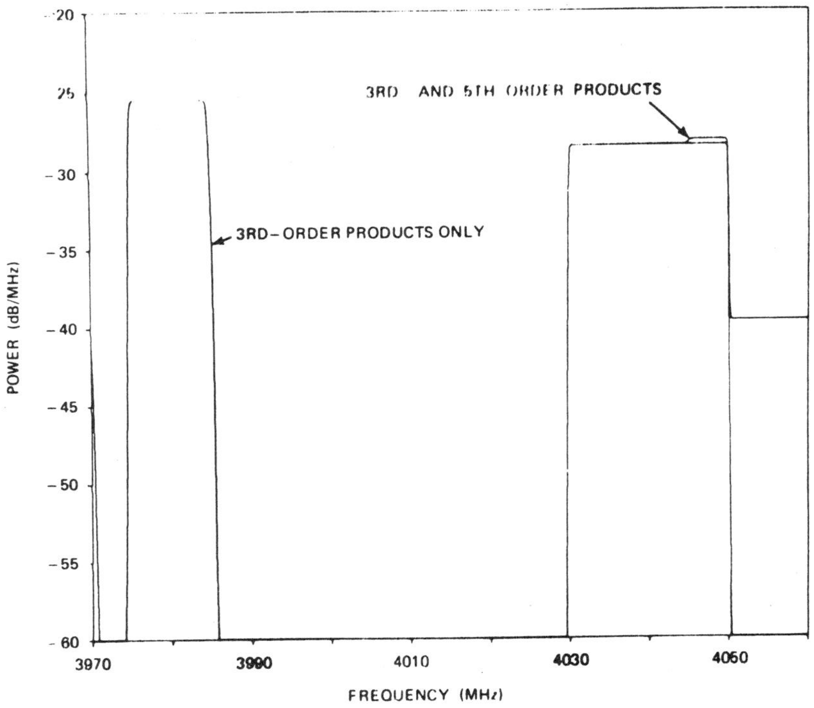

Figure 10. *RF Intermodulation Power Spectral Density Generated by a Carrier with a Rectangular Spectrum and a Single Carrier*

RF INTERMODULATION CALCULATIONS FOR $\Delta = -12$

Initially the levels of carriers B and C are assumed to be 12 dB below that of carrier A ($\Delta = -12$ dB), which is the typical level for program channels sharing a satellite transponder with a TV carrier. The RF intermodulation power spectrum in the vicinity of carrier A is shown in Figure 11 for input backoffs varying from 0 to -14 dB. Third-order $A + B - C$ and $A + C - B$ products occur at 4,017.5 MHz and 4,022.5 MHz, respectively. Fifth-order $A + 2B - 2C$ and $A + 2C - 2B$ products occur at 4,015 MHz and 4,025 MHz, respectively.

It appears that the 3rd-order intermodulation products decrease monotonically as the tube is backed off, while in the regions where 5th-order intermodulation products dominate, this is no longer the case. This is

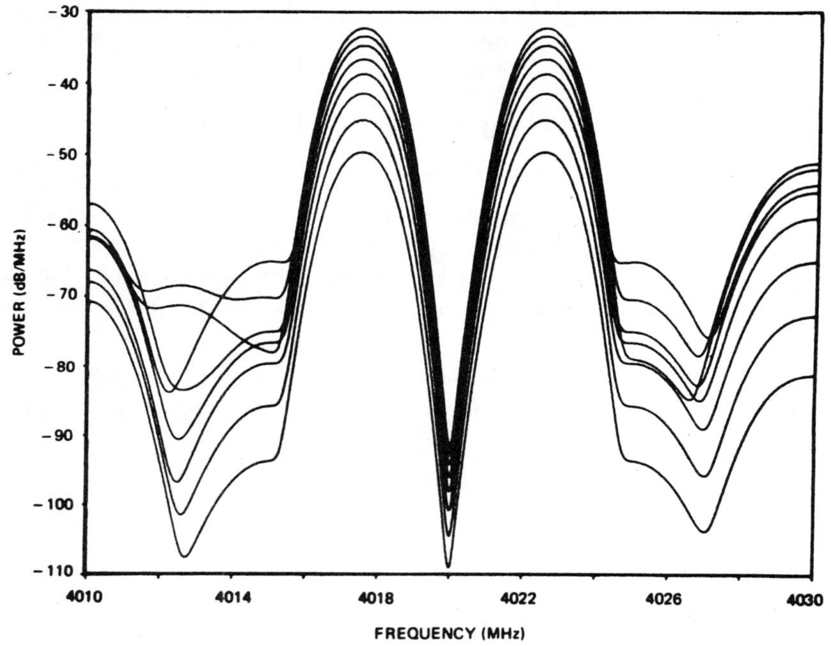

Figure 11. *RF Intermodulation Power Spectral Density Generated by a TV Carrier (spreading only) and Two Program Channels*

confirmed in Figure 12, where the levels of the A + B − C and A + 2B − 2C intermodulation products are plotted as functions of input backoff. It is apparent that, at some points, the levels of 5th-order products may actually increase as the input level is decreased. This effect has also been reported in Reference 12.

BASEBAND INTERMODULATION CALCULATIONS FOR $\Delta = -12$

The major distortion of carrier A is caused by the 3rd-order A + B − C and A + C − B intermodulation products located symmetrically about A at 4,017.5 MHz and 4,022.5 MHz, respectively. Although the RF power of these products decreases with increasing backoff, the baseband impairment (measured as signal-to-weighted noise) does not behave in the same fashion. Figure 13 shows the calculated S/N ratio, as well as the ratios between the carrier and both the in-phase and in-quadrature components of the intermodulation products. It is apparent that the signal-to-noise ratio is considerably higher than the estimate based on the RF

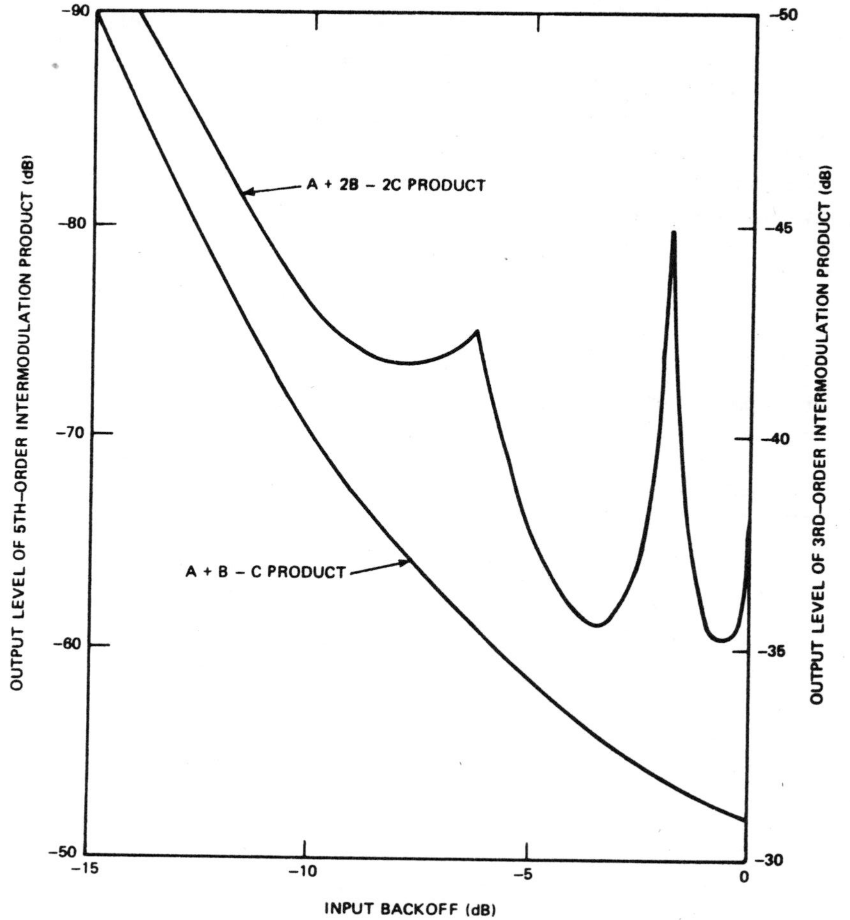

Figure 12. *Intermodulation Product Level as a Function of Input Backoff*

intermodulation spectral density. This phenomenon was observed experimentally during the INTELSAT IV simulation and is reported in Reference 7.

EFFECT OF THE RELATIVE CARRIER LEVELS, Δ

The angle between the A + B − C intermodulation product and the demodulated carrier is shown in Figure 14. Initially the angle is positive, but it becomes negative with increasing backoffs. For $\Delta = -12$ dB, the

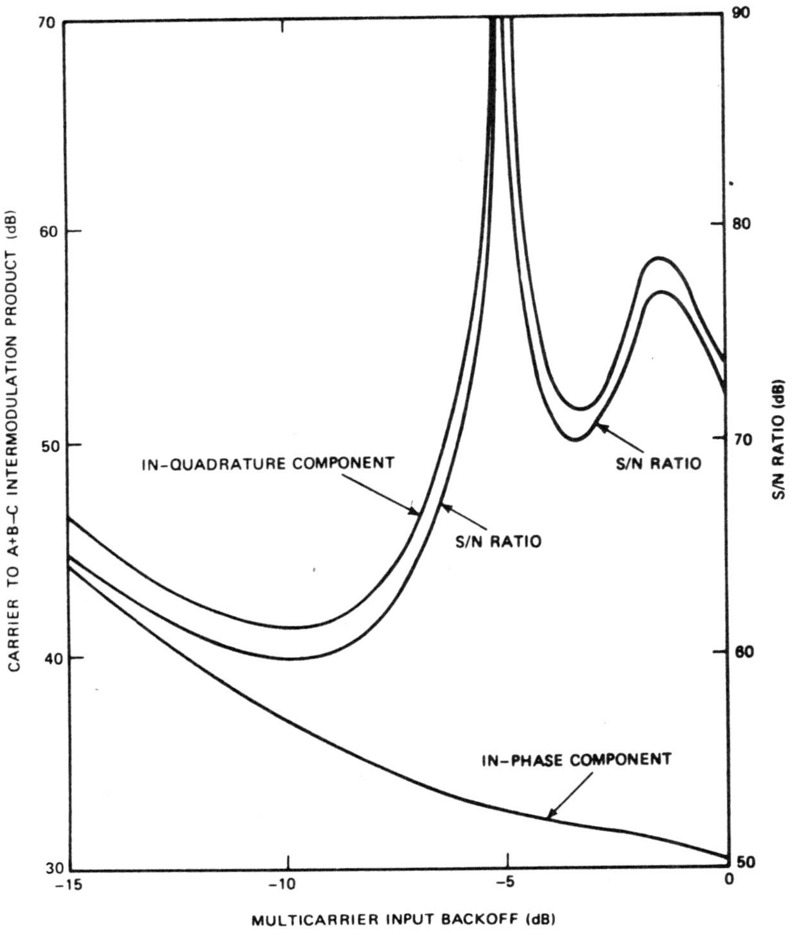

Figure 13. *Carrier-to-Intermodulation Ratio and S/N vs Multicarrier Input Backoff*

angle is zero at the backoff level for which S/N becomes infinite. Also shown in Figure 14 are the angles obtained for other relative levels of the carriers.

Figures 15 and 16 show the corresponding ratios between the carrier and the two intermodulation product components. The improvement in the ratio of carrier-to-in-quadrature components with increasing input level becomes pronounced as the level of the reference carrier is increased.

This explains why the measurement reported in Reference 4 for $\Delta = 20$ dB does not show this effect.

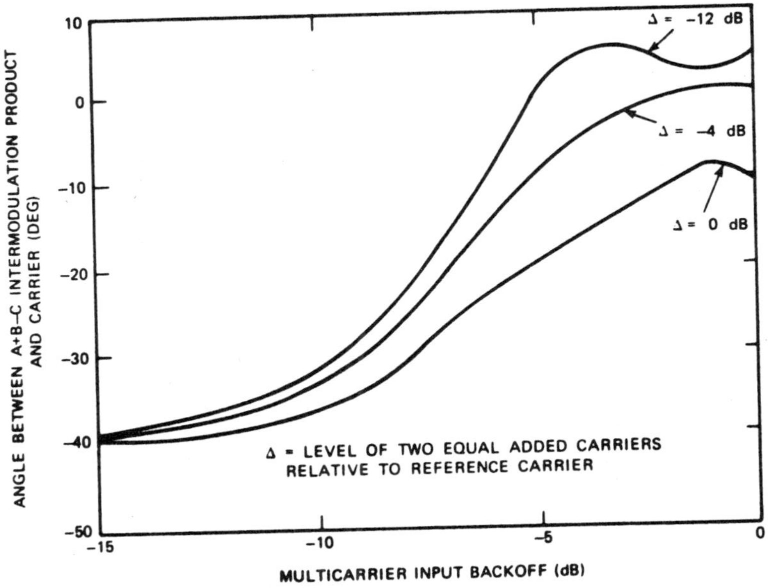

Figure 14. *Relative Phase Angle Between Intermodulation Product and Carrier vs Multicarrier Input Backoff*

Conclusions

A unified approach to the calculation of intermodulation effects in a "memoryless" nonlinear amplifier has been presented. Both amplitude nonlinearity and AM/PM conversion effects have been considered. An expression has been derived for the output of the device when the input consists of a number of bandpass band-limited signals. The particular case of an input formed by angle-modulated carriers and a band of Gaussian noise has been treated in detail.

The power spectral density of the baseband signal caused by intermodulation has been derived by using a time-domain approach. This technique permits an exact calculation of baseband distortion since the statistical dependence between the intermodulation noise and carriers is elucidated.

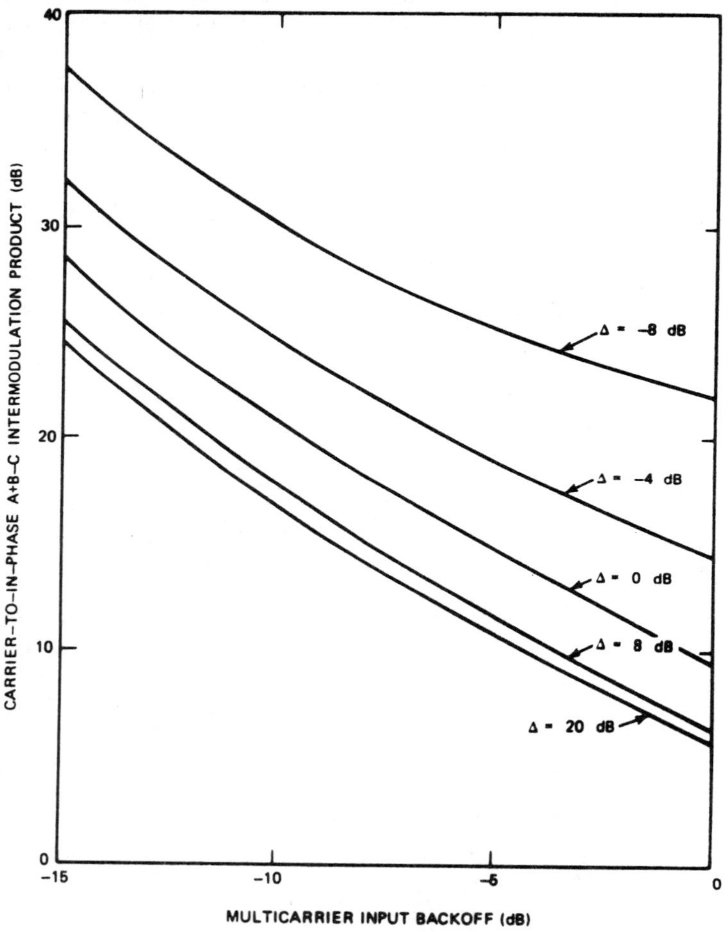

Figure 15. *Carrier-to-In-Phase Intermodulation Ratio vs Multicarrier Input Backoff*

A flexible computer program has been developed and exercised on a wide range of intermodulation problems. Comparison of calculations with reported measurements indicates that the memoryless assumption is valid for helix-type TWTs such as those presently used in communications satellites.

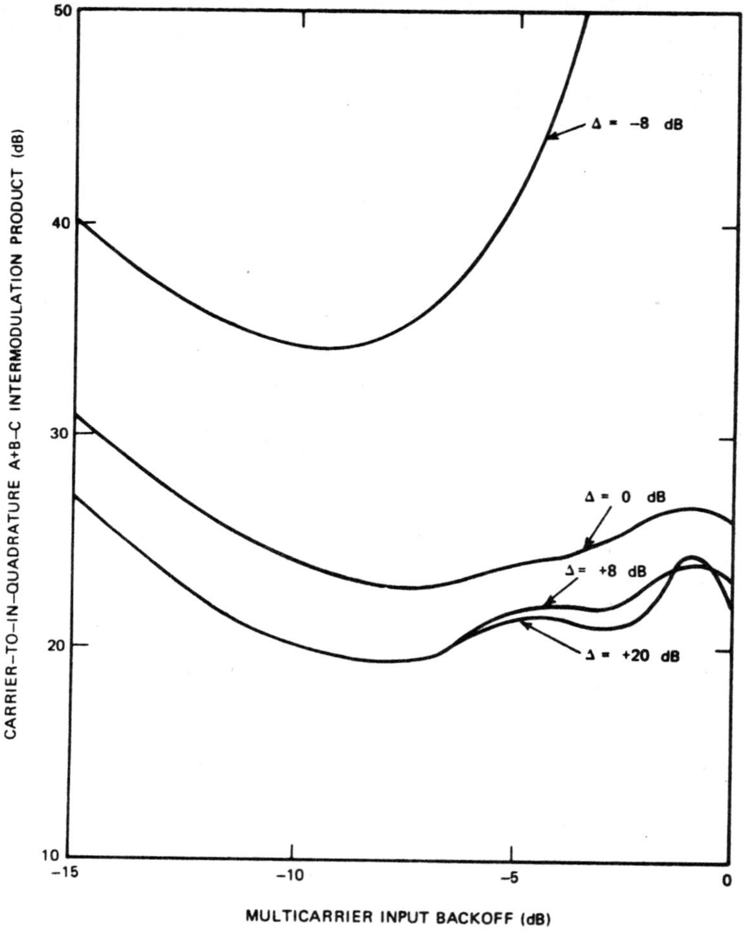

Figure 16. *Carrier-to-In-Quadrature Intermodulation Ratio vs Multicarrier Input Backoff*

For want of a more rigorous model, the program has also been used to evaluate the performance of high-power cavity TWTs using a representative nonlinear transfer function. Comparison with measurements has indicated that such nonlinearities can be accurately described as memoryless over a narrow bandwidth.

References

[1] E. D. Sunde, "Intermodulation Distortion in Multicarrier FM Systems," IEEE International Convention, New York, March 1965, *Convention Record*, Vol. 13, Part 2, pp. 130–146.

[2] R. J. Westcott, "Investigation of Multiple FM/FDM Carriers Through a Satellite TWT Operating Near to Saturation," *Proceedings of the IEE*, Vol. 114, No. 6, June 1967, pp. 726–750.

[3] A. L. Berman and E. Podraczky, "Experimental Determination of Intermodulation Distortion Produced in a Wide-Band Communication Repeater," IEEE International Convention, New York, March 1967, *Convention Record*, Vol. 15, Part 2, pp. 69–88.

[4] A. L. Berman and C. E. Mahle, "Nonlinear Phase Shift in Traveling Wave Tubes as Applied to Multiple-Access Communications Satellites," *IEEE Transactions on Communications Technology*, COM-18, No. 1, February 1970, pp. 37–48.

[5] O. Shimbo, "Effects of Intermodulation, AM-PM Conversion, and Additive Noise in Multi-Carrier TWT Systems," *Proceedings of the IEEE*, Vol. 59, No. 2, February 1971, pp. 230–238.

[6] A. R. Kaye, D. A. George, and M. J. Eric, "Analysis and Compensation of Bandpass Nonlinearities for Communications," *IEEE Transactions on Communications*, COM-20, No. 5, October 1972, pp. 965–972.

[7] N. K. M. Chitre and J. C. Fuenzalida, "Baseband Distortion Caused by Intermodulation in Multicarrier FM Systems," *COMSAT Technical Review*, Vol. 2, No. 1, Spring 1972, pp. 147–172.

[8] N. M. Blachman, "Detectors, Bandpass Nonlinearities and Their Optimization: Inversion of the Chebyshev Transform," *IEEE Transactions on Information Theory*, IT-17, No. 4, July 1971, pp. 398–404.

[9] S. Stein and J. J. Jones, *Modern Communication Principles*, New York: McGraw-Hill, 1967, p. 75.

[10] R. Fletcher and M. J. D. Powell, "A Rapidly Convergent Descent Method for Minimization," *The Computer Journal*, Vol. 6, 1963–1964, pp. 163–168.

[11] T. N. Hibbard, "Some Combinatorial Properties of Certain Trees with Application to Searching and Sorting," *Journal of the Association for Computing Machinery*, January 1962, pp. 13–28.

[12] O. Shimbo, "Nonlinear Distortion of Frequency Division Multiplexed Signals," *Journal of the Institute of Electrical Communications Engineers of Japan*, February 1961 (in Japanese).

Appendix A. Mathematical Derivations

Time-domain analysis

The input signal is represented by

$$e_i(t) = Re\left\{\sum_{l=1}^{m} A_l(t)\exp\left[j\omega_o t + j\theta_l(t)\right]\right\} \qquad (A1)$$

where $A_l(t)$ and $\theta_l(t)$ are arbitrary baseband time functions. Equation (A1) can be rewritten as

$$e_i(t) = Re\,\{\hat{e}_i(t)\} \qquad (A2a)$$

$$\hat{e}_i(t) = \rho(t)\exp\left[j\omega_o t + j\Theta(t)\right] \qquad (A2b)$$

where

$$\rho = \sqrt{x^2 + y^2} \qquad (A3a)$$

$$\Theta = \tan^{-1}\frac{y}{x} \qquad (A3b)$$

$$x = \sum_{l=1}^{m} A_l(t)\cos\theta_l(t) \qquad (A3c)$$

$$y = \sum_{l=1}^{m} A_l(t)\sin\theta_l(t) \quad . \qquad (A3d)$$

Cartesian coordinates will be introduced during the following steps in the derivation, since a double Fourier transformation is required.

The fundamental component of the output is represented by

$$e_o(t) = Re\left\{g\left(\sqrt{x^2+y^2}\right)\exp\left[j\omega_o t + j\tan^{-1}\frac{y}{x} + jf\left(\sqrt{x^2+y^2}\right)\right]\right\}$$

$$= Re\left\{\frac{g\left(\sqrt{x^2+y^2}\right)}{\sqrt{x^2+y^2}}\exp\left[jf\left(\sqrt{x^2+y^2}\right)\right](x+jy)\exp(j\omega_o t)\right\}$$

$$= Re\left\{\frac{g\left(\sqrt{x^2+y^2}\right)}{\sqrt{x^2+y^2}}\exp\left[jf\left(\sqrt{x^2+y^2}\right)\right]\hat{e}_i(t)\right\} \quad . \qquad (A4)$$

For the sake of convenience, the following double Fourier transformation is defined as

$$L(u,v) = \int_{-\infty}^{\infty} \int_{-\infty}^{\infty} \frac{g(\sqrt{x^2 + y^2})}{\sqrt{x^2 + y^2}} \exp [jf(\sqrt{x^2 + y^2})]$$

$$\cdot (x + jy) \exp (-jux - jvy) \, dx \, dy \qquad \text{(A5)}$$

or, alternatively,

$$\frac{g(\sqrt{x^2 + y^2})}{\sqrt{x^2 + y^2}} \exp [jf(\sqrt{x^2 + y^2})] (x + jy)$$

$$= \frac{1}{4\pi^2} \int_{-\infty}^{\infty} \int_{-\infty}^{\infty} L(u,v) \exp (jux + jvy) \, du \, dv \quad . \qquad \text{(A6)}$$

Substituting the function $L(u, v)$ into equation (A4) makes it possible to represent the output by

$$e_o(t) = \frac{1}{4\pi^2} Re \left\{ \exp [j\omega_o t] \int_{-\infty}^{\infty} \int_{-\infty}^{\infty} L(u,v) \right.$$

$$\left. \cdot \exp [jux + jvy] \, du \, dv \right\} \quad . \qquad \text{(A7)}$$

Substituting the expressions for x and y from equation (A3) into equation (A7) results in

$$e_o(t) = \frac{1}{4\pi^2} Re \left\{ \exp [j\omega_o t] \int_{-\infty}^{\infty} \int_{-\infty}^{\infty} L(u,v) \prod_{l=1}^{m} \exp \left[jA_l \sqrt{u^2 + v^2} \right. \right.$$

$$\left. \left. \cdot \sin \left\{ \theta_l(t) + \tan^{-1} \frac{u}{v} \right\} \right] du \, dv \right\} \quad . \qquad \text{(A8)}$$

Expanding the exponential as a Bessel function series, i.e.,

$$\exp [jz \sin \theta] = \sum_{k=-\infty}^{\infty} J_k(z) \exp [jk\theta] \qquad \text{(A9)}$$

makes it possible to represent equation (A8) by

$$e_o(t) = Re \left\{ \exp [j\omega_o t] \sum_{\substack{k_1, k_2, \ldots, k_m = -\infty \\ (k_1 + k_2 + \ldots + k_m = 1)}}^{\infty} \exp \left[j \sum_{l=1}^{m} k_l \theta_l(t) \right] \right.$$

$$\left. \cdot M(k_1, k_2, \ldots, k_m) \right\} \qquad \text{(A10)}$$

where

$$M(k_1, k_2, \ldots, k_m) = \frac{1}{4\pi^2} \int_{-\infty}^{\infty} \int_{-\infty}^{\infty} L(u,v) \left[\prod_{l=1}^{m} J_{k_l}(A_l \sqrt{u^2 + v^2}) \right]$$

$$\cdot \exp \left[j \sum_{l=1}^{m} k_l \tan^{-1} \frac{u}{v} \right] du \, dv$$

$$= \int_{-\infty}^{\infty} \int_{-\infty}^{\infty} \int_{-\infty}^{\infty} \int_{-\infty}^{\infty} \frac{g(\sqrt{x^2 + y^2})}{\sqrt{x^2 + y^2}}$$

$$\cdot \exp [jf(\sqrt{x^2 + y^2})] (x + jy)$$

$$\cdot \left[\prod_{l=1}^{m} J_{k_l} (A_l \sqrt{u^2 + v^2}) \right]$$

$$\cdot \exp \left[j \sum_{l=1}^{m} k_l \tan^{-1} \frac{u}{v} \right]$$

$$\cdot \exp (-jux - jvy) \, du \, dv \, dx \, dy \quad . \qquad \text{(A11)}$$

Using the polar coordinate transformations,

$$x = \rho \cos \theta \qquad u = \gamma \sin \eta$$

$$y = \rho \sin \theta \qquad v = \gamma \cos \eta$$

and performing the integrations on θ and η makes it possible to simplify equation (A11):

$$M(k_1, k_2, \ldots, k_m) = \int_0^{\infty} \gamma \left[\prod_{l=1}^{m} J_{k_l} (\gamma A_l) \right] d\gamma$$

$$\cdot \int_0^{\infty} \rho g(\rho) \exp [jf(\rho)] J_1(\gamma\rho) \, d\rho \quad . \qquad \text{(A12)}$$

In the summation in equation (A10), it is assumed that $M(k_1, k_2, \ldots, k_m) \equiv 0$ if $k_1 + k_2 + \ldots + k_m \neq 1$ and that M is given by equation (A12) if $k_1 + k_2 + \ldots + k_m = 1$. Thus,

$$e_o(t) = Re \left\{ \exp [j\omega_o t] \sum_{\substack{k_1, k_2, \ldots, k_m = -\infty \\ (k_1 + k_2 + \ldots + k_m = 1)}}^{\infty} \exp \left[j \sum_{l=1}^{m} k_l \theta_l(t) \right] \right.$$

$$\left. \cdot M(k_1, k_2, \ldots, k_m) \right\} \qquad \text{(A13)}$$

where $k_l \, (l = 1, 2, \ldots, m)$ can take negative, positive, and zero values under the constraint that

$$\sum_{l=1}^{m} k_l = 1 \quad . \tag{A14}$$

If

$$\theta_l(t) = \omega_l t + \phi_l(t) + \lambda_l, \qquad l = 1, 2, \ldots, m-1$$

$$\theta_m(t) = \omega_m t + \tan^{-1}\left[\frac{N_s(t)}{N_c(t)}\right] = \omega_m t + \phi_m(t) \tag{A15}$$

the input, $e_i(t)$, represents the $m-1$ carriers, modulated in amplitude and phase, plus narrowband noise. In the output, $e_o(t)$, the component for which $k_p = 1$ and all of the other $k_l = 0$ represents the pure output signal corresponding to the pth pure signal of the input, $e_i(t)$. The pure output signals and intermodulation products which are not disturbed by the noise (N_c or N_s) may be obtained by taking the average of equation (A13) on N_c and N_s.

In the following, it is assumed that the noise angle, $\phi_m(t)$, is uniformly distributed on $[0, 2\pi]$ and is independent of the noise amplitude,

$$\xi = \sqrt{N_c^2(t) + N_s^2(t)} \quad . \tag{A16}$$

Furthermore,

$$\exp\left[jk_m \tan^{-1}\frac{N_s(t)}{N_c(t)}\right] = \left[\frac{N_c(t) + jN_s(t)}{\sqrt{N_c^2(t) + N_s^2(t)}}\right]^{k_m} \quad . \tag{A17}$$

Thus, the average of the factors including $N_c(t)$ and $N_s(t)$ is

$$C(\gamma) = \underset{N_c N_s}{E}\left\{J_{km}\left[\gamma\sqrt{N_c^2(t) + N_s^2(t)}\,\right]\exp\left[jk_m \tan^{-1}\frac{N_s(t)}{N_c(t)}\right]\right\}$$

$$= 0 \ , \qquad k_m \neq 0$$

$$= \int_0^\infty J_0(\xi\gamma)\, p(\xi)\, d\xi \ , \qquad k_m = 0 \tag{A18}$$

where $p(\xi)$ is the probability density function of the noise amplitude, ξ.

Now, when the average of equation (A18) is represented by

$$C(\gamma) = \int_0^\infty J_0(\xi\gamma)\, p(\xi)\, d\xi \tag{A19}$$

and the noise is Gaussian, $p(\xi)$ is Rayleigh distributed; hence,

$$C(\gamma) = \exp\left[-\frac{\gamma^2}{2} R(0)\right] \tag{A20}$$

where $R(0)$ represents the RF noise power or the average of N_c^2 or N_s^2. Thus, the pure signals and intermodulation products are given by

$$e_S(t) = Re\left\{\exp[j\omega_o t] \sum_{\substack{k_1,k_2,\ldots,k_{m-1}=-\infty \\ (k_1+k_2+\ldots+k_{m-1}=1)}}^{\infty} \exp\left[j\sum_{l=1}^{m-1} k_l\theta_l(t)\right] \cdot M_S(k_1, k_2, \ldots, k_{m-1})\right\} \tag{A21}$$

where

$$M_S(k_1, k_2, \ldots, k_{m-1}) = \int_0^\infty \gamma\left[\prod_{l=1}^{m-1} J_{kl}(\gamma A_l)\right] C(\gamma)\, d\gamma \int_0^\infty \rho g(\rho) \cdot \exp[jf(\rho)]\, J_1(\gamma\rho)\, d\rho \tag{A22}$$

and the components at the output disturbed by noise, which represent new noise at the output, are given by

$$e_N(t) = e_o(t) - e_S(t) \quad . \tag{A23}$$

Reference A1 describes a method for evaluating $M(k_1, k_2, \ldots, k_m)$ when the nonlinear characteristic $g(\rho)\exp[jf(\rho)]$ is approximated by a power or Fourier series. The method used here is that suggested in Reference A2.

First, $g(\rho)\exp[jf(\rho)]$ is approximated by

$$g(\rho)\exp[jf(\rho)] = \sum_{s=1}^{L} b_s J_1(\alpha s\rho) \quad . \tag{A24}$$

Substituting this expression into equation (A12) and using the formula

$$\int_0^\infty G(\gamma)\, d\gamma \int_0^\infty \rho J_1(B\rho)\, J_1(\gamma\rho)\, d\rho = \frac{G(B)}{B} \tag{A25}$$

results in

$$M(k_1, k_2, \ldots, k_m) = \sum_{s=1}^{L} b_s\left[\prod_{l=1}^{m} J_{kl}(\alpha s A_l)\right] \quad . \tag{A26}$$

In a similar fashion, M_S of equation (A22) is found to be

$$M_S(k_1, k_2, \ldots, k_{m-1}) = \sum_{s=1}^{L} b_s C(\alpha s)\left[\prod_{l=1}^{m-1} J_{kl}(\alpha s A_l)\right] \quad . \tag{A27}$$

If there is no noise and the input consists of $m-1$ modulated carriers,

$$M = M_S = \sum_{s=1}^{L} b_s \left[\prod_{l=1}^{m-1} J_{k_l}(\alpha s A_l) \right] \quad . \tag{A28}$$

Similarly, for pure signals and intermodulation products, if noise is present, the nonlinear characteristic $g(\rho) \exp [jf(\rho)]$ becomes

$$g(\rho) \exp [jf(\rho)] = \sum_{s=1}^{L} b_s C(\alpha s) J_1(\alpha s \rho) \quad . \tag{A29}$$

Demodulated output

In equation (A1), it has been assumed that all of the amplitudes $A_l (l = 1, 2, \ldots, m-1)$ are constants and that the last component $(l = m)$ represents thermal noise. The problem is to demodulate an angle $\theta_p(t)$ at the output. It can be assumed that $p = 1$ without loss of generality. The output can then be represented by the following three categories:

a. the main carrier to be demodulated;

b. the intermodulation products and noise falling in the receiver filter of this main carrier; and

c. the other carriers, intermodulation products, and noise falling away from the main carrier, which can be filtered out.

Only categories a and b are important in demodulating the main carrier. The components of b and c are represented as follows:

$$\begin{aligned} e_{o1}(t) &= Re \{\exp [j\omega_o t + j\theta_1(t)] M_o\} \\ &+ Re \{\exp [j\omega_o t + j\theta_1(t)] [M(1, 0, \ldots, 0; t) - M_o]\} \\ &+ Re \left\{ \exp [j\omega_o t] \sum_{\substack{k_1, k_2, \ldots, k_m = -\infty \\ (k_1 + k_2 + \ldots + k_m = 1)}}^{\infty}{}' \exp \left[j \sum_{l=1}^{m-1} k_l \theta_l(t) \right] \right. \\ &\left. \cdot M(k_1, k_2, \ldots, k_m; t) \exp [jk_m \theta_m(t)] \right\} \end{aligned} \tag{A30}$$

where

$$\begin{aligned} M_o &= \int_0^\infty \gamma \left[\prod_{l=2}^{m-1} J_o(\gamma A_l) \right] J_1(\gamma A_1) C(\gamma) \, d\gamma \\ &\cdot \int_0^\infty \rho g(\rho) \exp [jg(\rho)] J_1(\gamma \rho) \, d\rho \end{aligned} \tag{A31}$$

$$\begin{aligned} M(1, 0, \ldots, 0; t) &= \int_0^\infty \gamma \left[\prod_{l=2}^{m-1} J_0(\gamma A_l) \right] J_1(\gamma A_1) \\ &\cdot J_0 \left[\gamma \sqrt{N_c^2(t) + N_s^2(t)} \right] d\gamma \\ &\cdot \int_0^\infty \rho g(\rho) \exp [jf(\rho)] J_1(\gamma \rho) \, d\rho \end{aligned} \tag{A32}$$

$$\begin{aligned} &M(k_1, k_2, \ldots, k_m) \exp [jk_m \theta_m(t)] \\ &= \int_0^\infty \gamma \left[\prod_{l=1}^{m-1} J_{k_l}(\gamma A_l) \right] J_{k_m} \left[\gamma \sqrt{N_c^2(t) + N_s^2(t)} \right] d\gamma \exp [jk_m \omega_m t] \\ &\cdot \int_0^\infty \rho g(\rho) \exp [jf(\rho)] J_1(\gamma \rho) \, d\rho \left[\frac{N_c(t) + jN_s(t)}{\sqrt{N_c^2(t) + N_s^2(t)}} \right]^{k_m} \quad . \end{aligned} \tag{A33}$$

The prime following the summation indicates that categories a and c are excluded. Equation (A30) is represented as follows:

$$e_{o1}(t) = Re \exp [j\omega_o t + j\theta_1(t)] M_o [1 + R(t) + jI(t)] \quad . \tag{A34}$$

The demodulated output can now be represented by

$$\phi_1(t) + \tan^{-1} \left[\frac{I(t)}{1 + R(t)} \right] \quad . \tag{A35}$$

Since, in the normal situation, $R(t)$ and $I(t)$ are small, equation (A35) can be approximated as

$$\phi_1(t) + \tan^{-1} \left[\frac{I(t)}{1 + R(t)} \right] \approx \phi_1(t) + I(t) \quad . \tag{A36}$$

The demodulated intermodulation products and noise are given by

$$\begin{aligned} I(t) &= Im \{M_o^{-1} [M(1, 0, \ldots, 0; t) - M_o]\} \\ &+ Im \left\{ M_o^{-1} \sum_{\substack{k_1, k_2, \ldots, k_m = -\infty \\ (k_1 + k_2 + \ldots + k_m = 1)}}^{\infty}{}' M(k_1, k_2, \ldots, k_m; t) \right. \\ &\cdot \exp \left[jk_m \left\{ \omega_m t + \tan^{-1} \frac{N_s(t)}{N_c(t)} \right\} \right] \\ &\left. \cdot \exp \left[j(k_1 - 1) \theta_1(t) + j \sum_{l=2}^{m-1} k_l \theta_l(t) \right] \right\} \end{aligned} \tag{A37}$$

where, for the particular case of the Bessel function expansion,

$$M_o = \sum_{s=1}^{L} b_s \exp\left[-\frac{\alpha^2 s^2}{2} R(0)\right] J_1(\alpha s A_1) \prod_{l=2}^{m-1} J_0(\alpha s A_l) \tag{A38}$$

and

$$M(k_1, k_2, \ldots, k_m; t) = \sum_{s=1}^{L} b_s J_{k_m}\left[\alpha s \sqrt{N_c^2(t) + N_s^2(t)}\right] \prod_{l=1}^{m-1} J_{k_l}(\alpha s A_l) \quad . \tag{A39}$$

The autocorrelation function of $I(t)$, i.e., avg $\{I(t) \cdot I(t + \tau)\}$, may now be obtained in the following manner. In the average of $I(t) \cdot I(t + \tau)$, the cross-terms are not all zero; i.e., the cross-correlation functions of the two components $(k_1, k_2, \ldots, k_m)$ and $(2 - k_1, - k_2, - k_3, \ldots, - k_m)$ are not zero. (All other terms in the cross-correlation function are zero.) Then $I(t)$ of equation (A37) can be expressed as

$$\begin{aligned} I(t) = Im \Bigg\{ & M_o^{-1}[M(1, 0, \ldots, 0; t) - M_o] \\ & + M_o^{-1} \sum_{\substack{k_1,k_2,\ldots,k_m=-\infty \\ (k_1+k_2+\ldots+k_m=1)}}^{\infty}{}'' \exp\left[j(k_1 - 1)\,\theta_1(t) + j \sum_{l=2}^{m-1} k_l \theta_l(t)\right] M(k_1, k_2, \ldots, k_m; t) \\ & \cdot \exp\left[jk_m\left\{\omega_m t + \tan^{-1}\frac{N_s(t)}{N_c(t)}\right\}\right] \\ & + M(2 - k_1, - k_2, \ldots, - k_m; t) \\ & \cdot \exp\left[-j(k_1 - 1)\,\theta_1(t) - j\sum_{l=2}^{m-1} k_l \theta_l(t)\right] \\ & \cdot \exp\left[-jk_m\left\{\omega_m t + \tan^{-1}\frac{N_s(t)}{N_c(t)}\right\}\right]\Bigg\} \end{aligned} \tag{A40}$$

where Σ' is rewritten as Σ'' by combining the components $(k_1, k_2, \ldots, k_m)$ and $(2 - k_1, - k_2, \ldots, -k_m)$.

The relationship between arbitrary complex numbers a and b,

$$[Im\, a]\,[Im\, b] = \frac{1}{2} Re\,\{ab^* - ab\} \tag{A41}$$

and

$$\begin{aligned} \text{avg} \Bigg\{ & J_{k_m}\left[\alpha s\sqrt{N_c^2(t) + N_s^2(t)}\right] J_{k_m}\left[\alpha p\sqrt{N_c^2(t + \tau) + N_s^2(t + \tau)}\right] \\ & \cdot \exp\left[\pm k_m j \tan^{-1}\frac{N_s(t)}{N_c(t)}\right] \exp\left[\mp k_m j \tan^{-1}\frac{N_s(t + \tau)}{N_c(t + \tau)}\right]\Bigg\} \\ = & \exp\left[-\frac{\alpha^2(s^2 + p^2)}{2} R(0)\right] I_{k_m}[\alpha^2 sp R(\tau)] \end{aligned} \tag{A42}$$

yield

$$\text{avg}\,\{I(t)\, I(t + \tau)\} = R_1(\tau) + \sum_{k_1,k_2,\ldots,k_{m-1}=-\infty}^{\infty}{}'' R(k_1, k_2, \ldots, k_{m-1}; \tau) \tag{A43}$$

where

$$\begin{aligned} R_1(\tau) = & \frac{1}{2} Re \Bigg[\sum_{s=1}^{L}\sum_{p=1}^{L} \frac{b_s}{M_o}\left(\frac{b_p}{M_o}\right)^* J_1(\alpha s A_1)\, J_1(\alpha p A_1) \\ & \cdot \exp\left\{-\frac{\alpha^2(s^2 + p^2)}{2} R(0)\right\} \prod_{l=2}^{m-1} J_0(\alpha s A_l) \prod_{l=2}^{m-1} J_0(\alpha p A_l) \\ & \cdot \{I_0[\alpha^2 sp R(\tau)] - 1\} - \sum_{s=1}^{L}\sum_{p=1}^{L} \frac{b_s}{M_o}\frac{b_p}{M_o} J_1(\alpha s A_1)\, J_1(\alpha p A_1) \\ & \cdot \exp\left\{-\frac{\alpha^2(s^2 + p^2)}{2} R(0)\right\} \prod_{l=2}^{m-1} J_0(\alpha s A_l) \prod_{l=2}^{m-1} J_0(\alpha p A_l) \\ & \cdot \{I_0[\alpha^2 sp R(\tau)] - 1\}\Bigg] \\ = & \sum_{q=1}^{\infty}\left\{Im\left[\frac{N(1, 0, \ldots, 0; q)}{M_o}\right]\right\}^2 [\rho_0(\tau)]^{2q} \quad . \end{aligned} \tag{A44}$$

In equation (A44),

$$N(k_1, k_2, \ldots, k_{m-1}; q) = \sum_{s=1}^{L} b_s \exp\left[-\frac{\alpha^2 s^2}{2} R(0)\right] \prod_{l=1}^{m-1} J_{k_l}(\alpha s A_l) \cdot \left\{\frac{\left[\frac{1}{2} R(0)\, \alpha^2 s^2\right]^{2q+|k_m|}}{q!(|k_m| + q)!}\right\}^{1/2}$$

$$\rho_0(\tau) = \frac{R(\tau)}{R(0)}$$

$$R(k_1, k_2, \ldots, k_{m-1}; \tau)$$

$$= \frac{1}{2} Re\, |M_o|^{-2} \sum_{s=1}^{L} \sum_{p=1}^{L} b_s b_p^* \exp\left[-\frac{\alpha^2(s^2+p^2)}{2} R(0)\right] J_{k_1}(\alpha s A_1)$$

$$\cdot J_{k_1}(\alpha p A_1) \prod_{l=2}^{m-1} J_{k_l}(\alpha s A_l) \prod_{l=2}^{m-1} J_{k_l}(\alpha p A_l)\, I_{k_m}[\alpha^2 sp R(\tau)]$$

$$\cdot \text{avg}\left\{\exp\left[j(k_1 - 1)(\omega_1\tau + \psi_1) + \sum_{l=2}^{m-1} jk_l\psi_l + \sum_{l=2}^{m} j\omega_l\tau\right]\right\}$$

$$+ \frac{1}{2} Re\, |M_o|^{-2} \sum_{s=1}^{L} \sum_{p=1}^{L} b_s b_p^* \exp\left[-\frac{\alpha^2(s^2+p^2)}{2} R(0)\right] J_{2-k_1}(\alpha s A_1)$$

$$\cdot J_{2-k_1}(\alpha p A_1) \prod_{l=2}^{m-1} J_{-k_l}(\alpha s A_l) \prod_{l=2}^{m-1} J_{-k_l}(\alpha s A_l)\, I_{-k_m}[\alpha^2 sp R(\tau)]$$

$$\cdot \text{avg}\left\{\exp\left[-j(k_1 - 1)(\omega_1\tau + \psi_1) - \sum_{l=2}^{m-1} jk_l\psi_l - \sum_{l=2}^{m} j\omega_l\tau\right]\right\}$$

$$- \frac{(-1)^{k_m}}{2} Re\left\{M_o^{-2} \sum_{s=1}^{L} \sum_{p=1}^{L} b_s b_p \exp\left[-\frac{\alpha^2(s^2+p^2)}{2} R(0)\right]\right.$$

$$\left.\cdot J_{k_1}(\alpha s A_1)\, J_{2-k_1}(\alpha p A_1)\right\}$$

$$\cdot \prod_{l=2}^{m-1} J_{k_l}(\alpha s A_l) \prod_{l=2}^{m-1} J_{-k_l}(\alpha p A_l)\, I_{k_m}[\alpha^2 sp R(\tau)]$$

$$\cdot \text{avg}\left\{\exp\left[j(k_1 - 1)(\omega_1\tau + \psi_1) + \sum_{l=2}^{m-1} jk_l\psi_l + \sum_{l=2}^{m} j\omega_l\tau\right]\right\}$$

$$- \frac{(-1)^{k_m}}{2} Re\left\{M_o^{-2} \sum_{s=1}^{L} \sum_{p=1}^{L} b_s b_p \exp\left[-\frac{\alpha^2(s^2+p^2)}{2} R(0)\right]\right.$$

$$\left.\cdot J_{2-k_1}(\alpha s A_1)\, J_{k_1}(\alpha p A_1)\right\}$$

$$\cdot \prod_{l=2}^{m-1} J_{-k_l}(\alpha s A_l) \prod_{l=2}^{m-1} J_{k_l}(\alpha p A_l)\, I_{k_m}[\alpha^2 sp R(\tau)]$$

$$\cdot \text{avg}\left\{\exp\left[-j(k_1 - 1)(\omega_1\tau + \psi_1) - \sum_{l=2}^{m-1} jk_l\psi_l - \sum_{l=2}^{m} j\omega_l\tau\right]\right\}$$

$$= \frac{1}{2} Re \sum_{q=0}^{\infty}\left\{\left|\frac{N(k_1, k_2, \ldots, k_{m-1}; q)}{M_o}\right|^2\right.$$

$$\cdot \Delta(k_1 - 1, k_2, \ldots, k_{m-1}; q; \tau)$$

$$+ \left|\frac{N(k_1 - 2, k_2, \ldots, k_{m-1}; q)}{M_o}\right|^2$$

$$\cdot \Delta^*(k_1 - 1, k_2, \ldots, k_{m-1}; q; \tau)$$

$$+ \frac{N(k_1, k_2, \ldots, k_{m-1}; q)\, N(k_1 - 2, k_2, \ldots, k_{m-1}; q)}{M_o^2}$$

$$\cdot \Delta(k_1 - 1, k_2, \ldots, k_{m-1}; q; \tau)$$

$$+ \frac{N(k_1 - 2, k_2, \ldots, k_{m-1}; q)\, N(k_1, k_2, \ldots, k_{m-1}; q)}{M_o^2}$$

$$\left.\cdot \Delta^*(k_1 - 1, k_2, \ldots, k_{m-1}; q; \tau)\right\} \qquad \text{(A45)}$$

where

$$\psi_l = \phi_l(t) - \phi_l(t + \tau) \qquad \text{(A46)}$$

and

$$\Delta(k_1, k_2, \ldots, k_{m-1}; q; \tau)$$

$$= \text{avg}\left\{\exp\left[j \sum_{l=1}^{m-1} k_l(\omega_l\tau + \psi_l) + jk_m\omega_m\tau\right]\right\} [\rho_o]^{2q+|k_m|} \quad . \qquad \text{(A47)}$$

If Δ is real, equation (A45) becomes

$$R(k_1, k_2, \ldots, k_{m-1}; \tau)$$

$$= \frac{1}{2} \sum_{q=0}^{\infty} \frac{|N(k_1, k_2, \ldots, k_{m-1}; q) + N^*(k_1 - 2, k_2, \ldots, k_{m-1}; q)|^2}{|M_o|^2}$$

$$\cdot \Delta(k_1 - 1, k_2, \ldots, k_{m-1}; q; \tau) \quad . \qquad \text{(A48)}$$

For the special case of $k_1 = 1$, equation (A45) can be reduced to

$$R(k_1, k_2, \ldots, k_{m-1}; \tau) = 2 \sum_{q=0}^{\infty}\left\{\left[Im \frac{N(1, k_2, \ldots, k_{m-1}; q)}{M_o}\right]^2\right\}$$

$$\cdot Re\, \Delta(0, k_2, \ldots, k_{m-1}; q; \tau) \qquad \text{(A49)}$$

where

$$N(1, k_2, \ldots, k_{m-1}; q) = -N(-1, k_2, \ldots, k_{m-1}; q) \quad . \qquad \text{(A50)}$$

Note that, in the preceding derivations, the power series expansion for $I_{k_m}(z)$, namely,

$$I_{k_m}(z) = \sum_{q=0}^{\infty} \frac{(1/2)^{|k_m|+2q}}{q!(|k_m| + q)!} z^{|k_m|+2q} \qquad \text{(A51)}$$

was used. Note also that, in the case of $k_1 = 1$, if there is no AM/PM conversion, equation (A49) becomes zero. However, in all other cases ($k_1 \neq 1$), even if there is no AM/PM conversion, equation (A49) does not become zero.

The order of the intermodulation products is defined as

$$|k_1| + |k_2| + \ldots + |k_{m-1}| + |k_m| + 2q \quad . \qquad \text{(A52)}$$

In the expressions for $R(k_1, k_2, \ldots, k_{m-1}; \tau)$, the magnitude and spread of the power spectrum are represented separately; i.e., the Fourier transform of Δ gives the power spectrum spread whose total power is unity, and the other coefficient gives the magnitude.

Power spectrum analysis

In this section, it is assumed that the input consists of $m - 1$ angle-modulated carriers and a Gaussian noise signal. As in Reference A1, the autocorrelation function of the output is given by*

$$\text{avg}\,[e_o(t)\, e_o(t+\tau)] = \frac{1}{2} Re \sum_{k_1,k_2,\ldots,k_{m-1}=-\infty}^{\infty} \exp\left[j\omega_o\tau + j\sum_{l=1}^{m} \omega_l \tau \right] \text{avg}\left\{ \exp\left[j \sum_{l=1}^{m-1} k_l \psi_l \right] \right\}$$
$$\cdot \int_0^\infty t \exp\left[-\frac{t^2}{2} \right] \cdot |Q(k_1, k_2, \ldots, k_m; t)|^2\, dt \qquad \text{(A53)}$$

where the averages are taken over the values

$$\phi_1(t), \phi_2(t), \ldots, \phi_{m-1}(t) \quad .$$

Then, rewriting equation (13) in [A1] yields

$$Q(k_1, k_2, \ldots, k_{m-1}; t)$$
$$= \int_0^\infty \gamma \prod_{l=1}^{m-1} J_{k_l}(\gamma A_l)\, J_{k_m}[\sqrt{R(\tau)}\,\gamma t] \exp\left\{ -[R(0) - R(\tau)] \frac{\gamma^2}{2} \right\} d\gamma$$
$$\cdot \int_0^\infty \rho g(\rho) \exp[jf(\rho)]\, J_1(\gamma\rho)\, d\rho \qquad \text{(A54)}$$

*The result obtained in this section can also be obtained directly from equations (A12), (A13), and (A26) if it is assumed that all of the cross-terms in avg $\{e_i(t)\, e_i(t+\tau)\}$ are zero.

where

$$k_m = 1 - \sum_{l=1}^{m-1} k_l \, .$$

The method used to evaluate $Q(k_1, k_2, \ldots, k_m; t)$ is described in Reference A1. The case in which the nonlinearity characteristics are approximated by a sum of Bessel functions, as in equation (A24), is analyzed here, since this case is not analyzed in Reference A1. Expansion of equation (A24) and use of equation (A25) yield

$$|Q(k_1, k_2, \ldots, k_{m-1}; t)|^2 = \left| \sum_{s=1}^{L} b_s \left[\prod_{l=1}^{m-1} J_{k_l}(\alpha s A_l) \right] \cdot \exp\left\{ -\frac{\alpha^2 s^2}{2} [R(0) - R(\tau)] \right\} \cdot J_{k_m}[\sqrt{R(\tau)}\,\alpha s t] \right|^2 \quad . \qquad \text{(A55)}$$

Therefore, the integral with respect to t in equation (A53) is given by

$$\sum_{s=1}^{L} \sum_{p=1}^{L} b_s b_p^* \left[\prod_{l=1}^{m-1} J_{k_l}(\alpha s A_l) \right] \left[\prod_{l=1}^{m-1} J_{k_l}(\alpha p A_l) \right]$$
$$\cdot \exp\left\{ -\frac{\alpha^2(s^2+p^2)}{2} [R(0) - R(\tau)] \right\} \int_0^\infty \exp\left[-\frac{t^2}{2} \right] t$$
$$\cdot J_{k_m}[\sqrt{R(\tau)}\,\alpha p t]\, J_{k_m}[\sqrt{R(\tau)}\,\alpha s t]\, dt \quad . \qquad \text{(A56)}$$

The formula given in Reference A3 [equation (1), p. 395] makes it possible to evaluate the integral and reduce equation (A56):

$$H(k_1, k_2, \ldots, k_{m-1}) = \sum_{s=1}^{L} \sum_{p=1}^{L} b_s b_p^* \left[\prod_{l=1}^{m-1} J_{k_l}(\alpha s A_l) \right]$$
$$\cdot \left[\prod_{l=1}^{m-1} J_{k_l}(\alpha p A_l) \right] \exp\left[-\frac{\alpha^2(s^2+p^2)}{2} R(0) \right]$$
$$\cdot I_{k_m}[R(\tau)\, \alpha^2 s p] \quad . \qquad \text{(A57)}$$

Expanding the Bessel function as a power series

$$H(k_1, k_2, \ldots, k_{m-1}) = \sum_{q=0}^{\infty} |N(k_1, k_2, \ldots, k_{m-1}; q)|^2 \cdot [\rho_o(\tau)]^{2q+|k_m|} \qquad \text{(A58)}$$

yields the autocorrelation function of the output:

$$\begin{aligned}
&\operatorname{avg}\left[e_o(t)\, e_o(t+\tau)\right] \\
&= \frac{1}{2} Re \sum_{k_1,k_2,\ldots,k_{m-1}=-\infty}^{\infty} \exp\left[j\omega_o\tau + j \sum_{l=1}^{m-1} k_l\omega_l\tau \right] \\
&\quad \cdot \operatorname{avg}\left\{ \exp\left[j \sum_{l=1}^{m-1} k_l\psi_l \right] \right\} H(k_1, k_2, \ldots, k_{m-1}) \\
&= \frac{1}{2} Re \sum_{k_1,k_2,\ldots,k_{m-1}=-\infty}^{\infty} \sum_{q=0}^{\infty} |N(k_1, k_2, \ldots, k_{m-1}; q)|^2 \\
&\quad \cdot \Delta(k_1, k_2, \ldots, k_{m-1}; q; \tau) \quad . \qquad \text{(A59)}
\end{aligned}$$

Acknowledgment

The authors wish to acknowledge the helpful suggestions of numerous colleagues concerning the assumptions inherent in the mathematical model and the validity of the results. In particular, the insight and encouragement offered by Dr. N. K. M. Chitre and the contributions of Dr. R. Fang have been invaluable.

References

[A1] O. Shimbo, "Effects of Intermodulation, AM-PM Conversion, and Additive Noise in Multicarrier TWT Systems," *Proceedings of the IEEE*, Vol. 59, No. 2, February 1971, pp. 230–238.

[A2] A. R. Kaye, D. A. George, and N. J. Eric, "Analysis and Compensation of Bandpass Nonlinearities for Communications," *IEEE Transactions on Communications*, COM-20, No. 5, October 1972, pp. 965–972.

[A3] G. N. Watson, *A Treatise on the Theory of Bessel Functions*, 2nd Edition, Cambridge, U.K.: University Press, 1966.

Expected interference levels due to interactions between line-of-sight radio relay systems and broadband satellite systems

R. G. Medhurst, B.Sc. and J. H. Roberts, B.Sc.

Synopsis

The paper is concerned with intermodulation distortion effects when interference occurs between satellite communication systems and trunk radio line-of-sight systems. Curves are presented showing distortion levels (which are proportional to the ratio of unwanted to wanted carrier powers) for various expected modulation conditions and carrier separations. These curves form the basis for recommendations in a companion paper regarding allowable radiated power levels in both the satellite and the line-of-sight systems. It appears that a particularly dangerous situation exists when the satellite system is lightly loaded. Permanent artificial dispersal of the satellite carrier seems necessary, and this is considered in some detail.

List of symbols

r^2 = r.f. power of the interfering carrier relative to that of the interfered-with carrier.

ω_d r.m.s. frequency deviation per channel of the interfered-with system (usually 140kc/s for a system carrying 1800 speech channels, 200kc/s for numbers of channels between 300 and 960), rad/s

$f(p)$ = ratio of baseband power with and without pre-emphasis in the baseband of the interfered-with system

k_1, k_2 = residual carrier powers of the interfering and interfered-with carrier systems, relative to the respective total r.f. powers

ω_D = carrier separation, rad/s

p = baseband frequency in interfered-with system at which the distortion is being evaluated, rad/s

p_s = bandwidth of channel over which distortion is measured, rad/s

$\delta(\omega_D - p)$ = a Dirac delta function at $p = \omega_D$

$P_1(\omega)$ = continuous portion of the power spectrum of the wanted carrier

$P_2(\omega)$ = continuous portion of the power spectrum of the interfering carrier

ω = departure from the respective carrier frequencies

ω_B = semibandwidth of the receiver of the wanted carrier

1 Introduction

The proposed use of earth satellites as repeaters in a worldwide broadband microwave system involves many problems concerning the preferred technical characteristics of such systems. Because of the difficulty of accommodating their signals in an already overcrowded microwave spectrum, some of the more urgent questions relate to the frequencies and permissible power levels for transmission to and from the satellite repeaters. One current proposal which has considerable support is that the 6 and 4Gc/s bands, at present allocated to line-of-sight communication systems, should be used on a shared basis by the satellite systems, the higher frequency band being employed for transmission from ground to satellite, and the lower from satellite to ground. This proposal involves the possibility of interference between the two systems, and it has to be decided whether interference levels can be kept acceptably low without involving excessive limitations on the allowable transmission powers.

Four principal interference modes between the satellite system and any one link of an existing line-of-sight radio relay system have to be considered. These are

(*a*) interference into a line-of-sight receiver operating in the 4Gc/s band from the transmitter situated on the satellite
(*b*) interference into the receiver situated on the satellite from a line-of-sight transmitter operating in the 6Gc/s band
(*c*) interference into the satellite-system ground station receiver from a line-of-sight transmitter operating in the 4Gc/s band
(*d*) interference into a line-of-sight receiver operating in the 6Gc/s band from the satellite-system ground station transmitter

Of these four interference modes, (*c*) and (*d*) involve a siting problem. In principle they can be held to an acceptable level by situating the satellite-system ground-station transmitter and ground-station receiver sufficiently far away from existing terrestrial line-of-sight stations. In evaluating required separations the possibility of interference via the tropospheric scatter mechanism, as well as by line-of-sight propagation, has to be kept in mind. The problem is analysed in detail in Reference 2.

At first sight it might seem, in view of the low receiver powers involved, that interference modes (*a*) and (*b*) would be of negligible importance. It turns out, however, that these are likely to be the interference mechanisms which will determine maximum allowable powers in both the satellite and line-of-sight systems. The reasons for this are firstly that it will be impossible to ensure that satellites cannot, on occasions, cross the main beam of line-of-sight transmitters or receivers, and secondly that many interactions, all causing additive interference, may be possible between satellite systems and a single line-of-sight system. This second consideration involves statistical considerations which are examined in a companion paper[1]. The present paper is concerned with the evaluation of distortion levels experienced by systems employing frequency modulation, attention being focused on a single satellite/line-of-sight interaction, this

Paper 4212E, first received 24th October 1962 and in revised form 18th January 1963. It was presented at the International Conference on Satellite Communication, 22nd November 1962
Mr. Medhurst and Mr. Roberts are with the General Electric Co. Ltd., Hirst Research Centre, Wembley, England

Reprinted with permission from *Proc. Inst. Elec. Eng.*, vol. 111, pp. 519-523, Mar. 1964.

information forming the basis of the statistical evaluation in Reference 1. It is assumed that the satellite system will employ broadband f.m.; this being the current proposal that seems most likely of acceptance in the immediate future. Since it will emerge that the most dangerous interaction occurs when the satellite system is very lightly loaded (e.g. during a substantially all-white television transmission), it is probable that relatively minor modifications will be needed to take account of other methods of modulation which could be used in the satellite system, but further study is needed.

2 General distortion formula

The basic theory of an f.m. interfering carrier is covered in References 3 and 4, while Reference 5 gives a general formula directly applicable to the present situation. According to expression 3 of Reference 5, the distortion power appearing in a narrow band in the baseband of the interfered-with f.m. system is given by

$$\frac{r^2 p^2 p_s}{2}\frac{1}{\omega_d^2\,\mathrm{f}(p)}P_D(p)+\frac{r^2}{2}\frac{k_1k_2}{\mathrm{f}(\omega_D)}\left(\frac{\omega_D}{\omega_d}\right)^2\delta(\omega_D-p) \quad \text{milliwatts} \quad . \quad . \quad (1)$$

where

$$P_D(p)=\int_{-\omega_B}^{\omega_B}[P_1(|p_x+p|)+P_1(|p_x-p|)]P_2(|\omega_D-p_x|)dp_x$$
$$+k_1P_2(|\omega_D-p|)+k_1P_2(\omega_D+p)$$
$$+k_2P_1(|\omega_D-p|)+k_2P_1(\omega_D+p) \quad . \quad . \quad . \quad (2)$$

Expression 1 shows that the baseband distortion spectrum is of two types: firstly an interference having a continuous spectral distribution over the baseband channels and secondly a single-tone interference (given by the δ function term) which appears at the baseband channel position corresponding to the frequency separation of the two carriers.

3 Modulation conditions

3.1 Satellite system

The satellite system is assumed to be capable of transmitting 1200 speech channels, wide-deviation f.m. being employed. An r.f. bandwidth of 50 Mc/s is envisaged. If the r.m.s. frequency deviation ω_Δ is sufficiently large relative to the maximum baseband frequency p_m, the r.f. spectrum will approximate to a Gaussian shape. The necessary condition[7] is that $\omega_\Delta/p_m > 1$. Then, taking unit power in the modulated wave, the power spectrum (power per unit bandwidth) is given by the expression

$$\frac{1}{\omega_\Delta\sqrt{(2\pi)}}\exp\left(-\frac{\omega^2}{2\omega_\Delta^2}\right) \quad . \quad . \quad . \quad . \quad . \quad . \quad . \quad (3)$$

where ω is the departure from carrier.

By integrating this expression over the ± 25 Mc/s around carrier which is allotted to the satellite signal, a value of 7·5 Mc/s is found for ω_Δ in order that this band should contain about 99·9% of the transmitted power. There being no definite CCIR recommendation for the baseband range for a 1200-speech-channel system, an interpolation was made amongst the recommendations for 300, 600, 960 and 1800 channel systems[6] to produce a figure of 5·2 Mc/s for p_m. The ratio ω_Δ/p_m is then 1·44, confirming the adequacy of the Gaussian approximation.

Application of pre-emphasis, as recommended by the CCIR, will not change the spectrum shape, since the pre-emphasis is such as not to alter the r.m.s. deviation, and, as seen from expression 3, in wide-deviation f.m. the form of the spectrum is independent of the baseband power distribution.

3.2 Line-of-sight system

Systems of 960 and 1800 channels are considered, since the larger-capacity systems will be the most vulnerable to the type of interference with which we are concerned. Table 1 shows the relevant parameters. The expression for the mean-square phase deviation, given in the last column, assumes that the CCIR recommended pre-emphasis is used.

The mean-square phase deviation is a critical parameter in deciding the spectrum shape[7,9]. Since both values of this parameter in Table 1 are a good deal less than unity, the r.f. spectra are in the 'low-deviation' class; i.e. they consist of a residual carrier surrounded by sharply defined sideband spectra which are derived from successive low integral powers of the phase modulation.

With unit power in the modulated wave, the low-deviation type of spectrum in units of power per unit bandwidth is given, in the notation of Table 1, by

$$W(\omega)=\exp\left(-0{\cdot}458\frac{\omega_\Delta^2}{p_0p_m}\right)\left\{\delta(\omega)+\left[\frac{1}{2}\frac{\omega_\Delta^2\mathrm{f}(\omega)}{(p_m-p_0)\omega^2}\right]_{p_0<\omega<p_m}\right.$$
$$\left.+\left[\frac{1}{4}\frac{\omega_\Delta^4\mathrm{F}_2(\omega)}{(p_m-p_0)^2}\right]_{0<\omega<2p_m}\ldots\right\} \quad . \quad . \quad . \quad (4)$$

where $\mathrm{f}(\omega)$ is the ratio of baseband power at frequency ω with and without pre-emphasis, ω is departure from carrier frequency and $W(+\omega) = W(-\omega)$. The expressions in square brackets have their analytical values over the ranges shown

Table 1

MODULATION PARAMETERS OF HIGH-CAPACITY LINE-OF-SIGHT SYSTEMS

Number of channels	Baseband range $p_0/2\pi$	$p_m/2\pi$	R.M.S. frequency deviation per channel*	Equivalent noiseband r.m.s. frequency deviation $\left(\frac{\omega_\Delta}{2\pi}\right)$	Mean-square phase deviation $\left(0{\cdot}458\frac{\omega_\Delta^2}{p_0p_m}\right)$
960	0·316 Mc/s	4·188 Mc/s	200 kc/s	1·102 Mc/s	0·420
1800	0·316 Mc/s	8·204 Mc/s	140 kc/s	1·056 Mc/s	0·197

* For a 1 mW 800 c/s tone at a point of zero reference level

and are zero elsewhere, while $F_2(\omega)$ is derived from the square of the phase modulation. $F_2(\omega)$ is given explicitly in Reference 9 for the case of no pre-emphasis: in the present work, the form of F_2 has been modified to take approximate account of the pre-emphasis.

4 Intermodulation distortion levels

Figs. 1 and 2 show the intermodulation distortion levels arising as a result of substituting the spectra given in the previous Section into expression 1, the resulting convolution type of integrals appearing in expression 2 for the various cases considered having been evaluated by graphical means.

distortion levels shown in the Figures. The effect of weighting on the distributed distortion levels will be to reduce them by 2·5dB. The effect on the single-tone distortion levels would depend on the position of the tone in the speech channel, and could vary from a reduction of 6dB at the edge of the channel to an increase of 1dB near the centre. Carrier drift and jitter, and variation of Doppler shift, will modify the picture to a certain extent. However, as will appear in the next Section, it is proposed that permanent dispersal should be included in the satellite signal, and this will ensure that what appears in the Figures as single-tone distortion is spread out over a number of channels, so that the full psophometric weighting will apply.

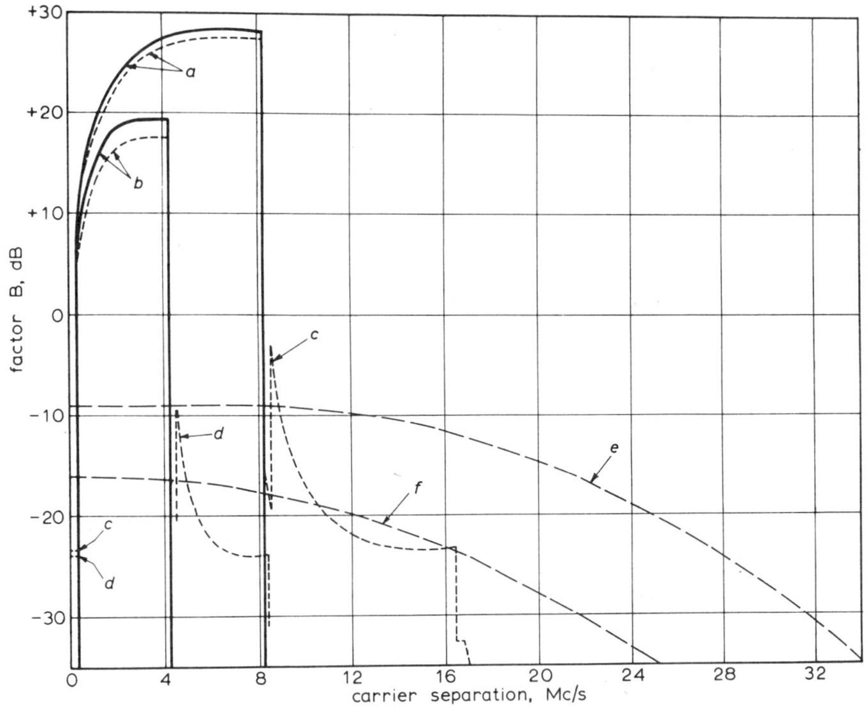

Fig. 1

Interference in line-of-sight systems

Satellite, when loaded, carrying 1200 speech channels.
Satellite deviation per channel—1·2Mc/s
a 1800 channels line-of-sight (distortion in channel at carrier separation frequency; single-tone distortion)
b 960 channels line-of-sight (distortion in channel at carrier separation frequency; single-tone distortion)
c 1800 channels line-of-sight (distortion in top channel; distributed distortion)
d 960 channels line-of-sight (distortion in top channel; distributed distortion]
e 1800 channels line-of-sight (distortion in top channel; distributed distortion)
f 960 channels line-of-sight (distortion in top channel; distributed distortion)
———— Satellite and line-of-sight systems both lightly loaded
– – – – – Satellite lightly loaded, line-of-sight fully loaded
—— —— Satellite heavily loaded, line-of-sight lightly loaded
Heavy loading on line-of-sight changes values plotted by less than 1dB
For the 1800-channel system, baseband range is 316–8204kc/s
For the 960-channel system, baseband range is 316–4188kc/s
Satellite system's receiver bandwidth taken as ±25Mc/s
Line-of-sight system's receiver bandwidth taken as ± twice the maximum modulating frequency
B factors shown are not psophometrically weighted
Use of a low-frequency dispersal waveform will reduce the single-tone distortion levels
Pre-emphasis assumed on both line-of-sight and satellite systems

The quantity given in these Figures as factor *B* is the ratio (expressed in decibels) of the intermodulation noise power in the specified 3·1kc/s channel divided by the ratio of unwanted to wanted carrier powers. The intermodulation noise is to be regarded as expressed in decibels relative to 1mW at a point of zero reference level (dbm0), and the carrier ratio as in decibels.

Psophometric weighting has not been applied to the

Consideration of expressions 1 and 2 shows that the distortions, whose levels are plotted in the Figures, may contain, besides unintelligible distortion, a certain amount of intelligible crosstalk. This could be of considerable concern, since specifications on intelligible crosstalk are considerably more stringent than those on noise. However, the incorporation of dispersal (see Section 5) into the satellite signal will ensure that intelligible crosstalk cannot arise.

5 Distortion under conditions of light loading

It is seen in Figs. 1 and 2 that the highest interference levels are associated with single-tone distortion. For this type of distortion to exist, the r.f. spectra of both the line-of-sight and the satellite systems must possess a large residual carrier component. This will be the case all the time for 960- and 1800-channel line-of-sight systems, since these carry low-deviation f.m. The satellite signal will include a large residual carrier only if it happens to be lightly loaded. Such a state of affairs will exist in telephony when only a small number of channels are active, or in television when the picture is carrying little detail, e.g. when it is in a nearly all-white condition. Though it may be hoped that such occasions will be rare and of short duration, they are of great importance.[1] The reason for this is involved in the nature of existing international specifications relating to interference in line-of-sight systems, the specifications being intended to safeguard the system even against infrequent disturbances of short duration.

The high interference levels shown in Figs. 1 and 2 for the single-tone distortion condition would be substantially reduced if light loading of the satellite signal were made impossible. This would be achieved if some form of dispersal[10] were to be permanently incorporated in the satellite signal. This could be expected to take the form of a subaudio waveform, say of the order of 25–50c/s repetition frequency, added to the modulation of the satellite carrier. The deviation due to this dispersal signal will have to be restricted, so as to avoid disturbance of the wanted telephony or television baseband owing to nonlinear distortion effects.

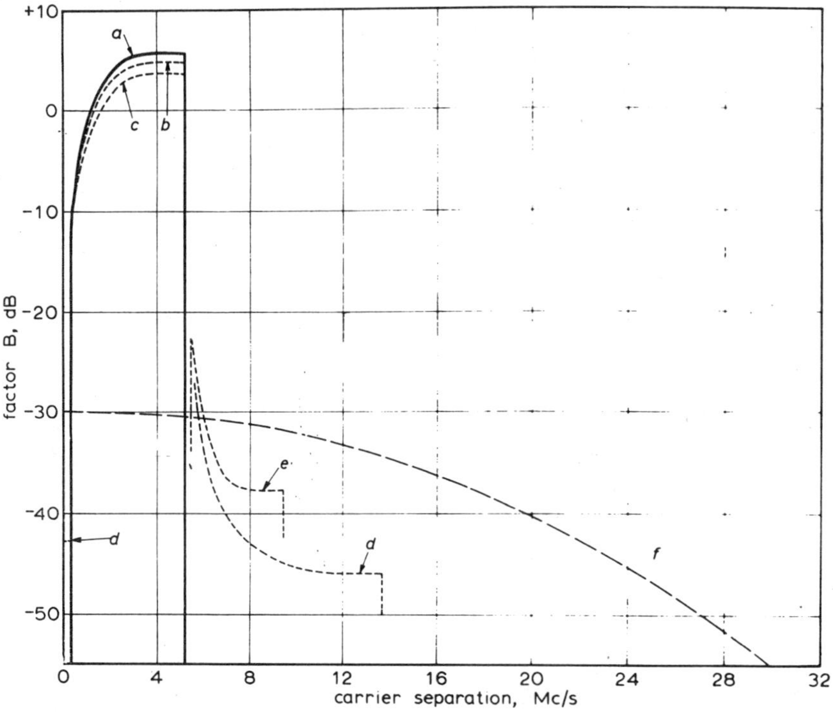

Fig. 2

Interference in satellite systems

Satellite, when loaded, carrying 1200 speech channels.
Satellite deviation per channel—1·2Mc/s

a 1800 or 960 channel line-of-sight (distortion in channel at carrier separation frequency; single-tone distortion)
b 1800 channel line-of-sight (distortion in channel at carrier separation frequency; single-tone distortion)
c 960 channel line-of-sight (distortion in channel at carrier separation frequency; single-tone distortion)
d 1800 channel line-of-sight (distortion in top channel; distributed distortion)
e 960 channel line-of-sight (distortion in top channel; distributed distortion)
f 1800 or 960 channel line-of-sight (distortion in top channel; distributed distortion)

——— Satellite and line-of-sight systems both lightly loaded
- - - - - Satellite lightly loaded, line-of-sight fully loaded
— — Satellite fully loaded, line-of-sight lightly loaded

Heavy loading on line-of-sight systems changes values plotted by less than 1dB
For the 1800-channel system, baseband range is 316–8204kc/s
For the 960-channel system, baseband range is 316–4188kc/s
Satellite system's receiver bandwidth taken as ±25Mc/s
Line-of-sight system's receiver bandwidth taken as ± twice the maximum modulating frequency
B factors shown are not psophometrically weighted
Use of a low-frequency dispersal waveform will reduce the single-tone distortion levels
Pre-emphasis assumed on both line-of-sight and satellite systems

The simplest dispersal waveform would be a single tone. It turns out, however, that this is a most unsuitable waveform for the purpose. This may be shown by a very simple analysis. Since, to achieve any useful effect, it is necessary to employ a frequency deviation considerably greater than the 25–50c/s modulating frequency, the modification of the satellite signal by the dispersal tone can be estimated using a quasistationary approach (this has been verified in specific cases by a direct summation of sidebands). The quasistationary argument is that the r.f. energy, in a particular frequency band, is proportional to the fraction of time (on a long-term basis) that the carrier spends in this band. Now, let the instantaneous frequency when the dispersal tone is present be

$$\omega_c + \omega_d \cos \omega_a t \quad . \quad . \quad . \quad . \quad . \quad . \quad . \quad . \quad . \quad . \quad (5)$$

where ω_c and ω_a are, respectively, carrier and modulation frequencies, and ω_d is the peak frequency deviation. Consideration of the form of expression 5 shows that the proportion of the time spent by the instantaneous frequency in a particular narrow band is considerably greater when this band is situated at the edges of the deviation range rather than at its centre. Consequently, the dispersed carrier energy is sharply peaked at the two ends of the r.f. band, so that the efficiency of the energy spreading process will be substantially less than might have been hoped.

It is easy to show that, with single-tone dispersal, the energy in the worst channel, relative to the undispersed carrier energy, is

$$\frac{1}{\pi}\cos^{-1}\frac{\omega_d - \omega_b}{\omega_d} \text{ (where } \omega_b \text{ is the channel bandwidth)}$$

which is approximately

$$\frac{1}{\pi}\sqrt{\frac{2\omega_b}{\omega_d}} \quad . \quad . \quad . \quad . \quad . \quad . \quad . \quad . \quad . \quad . \quad . \quad . \quad . \quad (6)$$

This expression gives the reduction due to dispersal in the single-tone distortion levels shown in Figs. 1 and 2. It is to be noticed that the reduction in the audio-channel interference improves only rather slowly with increase in dispersal-tone deviation, being proportional, in terms of power, only to the square root of the frequency deviation. The left-hand curve of Fig. 3 gives numerical values for the dispersal improvement factor calculated from expression 6, when ω_b is taken as 3·1 kc/s.

Dispersal would be most satisfactorily achieved by a triangular waveform, since this would produce a uniform spread of energy between the deviation limits. In this case, the energy reduction in a channel of width ω_b is given by

$$\omega_b/2\omega_p \quad . \quad . \quad . \quad . \quad . \quad . \quad . \quad . \quad . \quad . \quad . \quad . \quad . \quad (7)$$

where ω_p is the peak frequency deviation. The improvement is now proportional to the frequency deviation, rather than to its square root, as in the case of single-tone dispersal. Expression 7 is plotted in Fig. 3, and it is seen that a 20 dB improvement factor, or higher, is readily achievable.

The triangular waveform, however, has the disadvantage that substantial amounts of power reside in quite high harmonics of its fundamental repetition frequency, and this would lead to an unacceptable situation when the modulation is by a television signal. The best compromise will probably be a fundamental tone with suitable amounts of the first few odd harmonics, properly phased. The centre curve of Fig. 3 shows the improvement due to a fundamental plus third harmonic, the harmonic level being chosen so as to equalise the peaks of r.f. energy that occur at the carrier position and at the deviation limits. With this arrangement, and reasonable tolerances on the harmonic amplitude and phase, an improvement factor between 17 and 18 dB is attainable with not too large a peak deviation. Higher improvement factors would seem to require either a substantial increase in the peak deviation or the exploitation of more favourable waveforms. Such waveforms could be obtained by the addition of higher harmonics, or by the use of tones not harmonically related. The latter alternative is attractive from the telephony viewpoint but might produce undesirable effects on a television signal.

A few existing line-of-sight systems already use dispersal.[11] It is possible for dispersal arrangements existing simultaneously in two interfering systems to cancel themselves out.

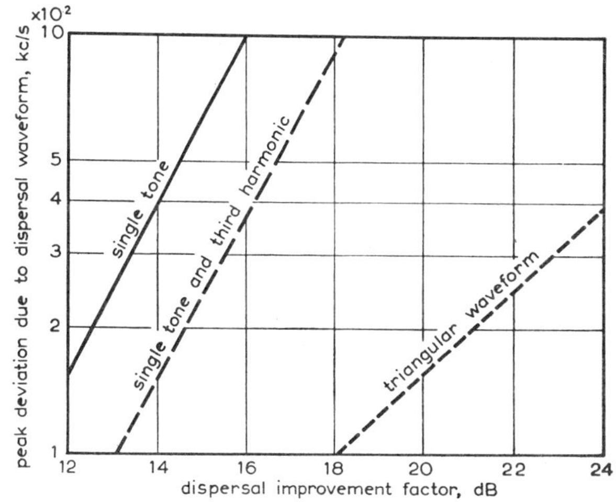

Fig. 3

Interference reduction due to dispersal of the satellite carrier under light-loading conditions

——— Single tone
— — — Single tone and third harmonic
- - - - - - Triangular waveform

This would happen, for example, if dispersal on each system happened to be by means of tones having equal amplitudes and frequencies, and in the same phase. Thus, the dispersal waveform for the satellite system must be chosen with due regard for possible dispersal arrangements on interacting line-of-sight systems.

References

1 CHAMBERLAIN, J. K., and MEDHURST, R. G.: 'Mutual interference between communication satellites and terrestrial radio relay systems', see next page

2 CURTIS, H. E.: 'Interference between satellite communication systems and common carrier surface systems', *Bell Syst. tech. J.*, 1962, **41**, p. 921

3 MEDHURST, R. G., HICKS, E. M., and GROSSET, W.: 'Distortion in frequency-division-multiplex f.m. systems due to an interfering carrier', *Proc. IEE*, 1958, **105** B, p. 282

4 HAMER, R.: 'Radio-frequency interference in multichannel telephony f.m. radio systems', *ibid.*, 1961, **108** B, p. 75

5 MEDHURST, R. G.: 'F.M. interfering carrier distortion: general formula', *ibid.*, 1962, **109** B, p. 149

6 CCIR Documents of IXth Plenary Assembly, Vol. 1, Recommendations (International Telecommunication Union. Geneva, 1959)

7 MEDHURST, R. G.: 'R.F. bandwidth of frequency-division-multiplex systems using frequency modulation', *Proc. Inst. Radio Engrs*, 1956, **44**, p. 189

8 HAMER, R., and ACTON, R. A.: 'Power spectrum of a carrier modulated in phase or frequency by white noise', *Electronic and Radio Engr*, 1957, **34**, p. 246

9 MEDHURST, R. G.: 'R.F. spectra and interfering carrier distortion in f.m. trunk radio systems with low modulation ratios', *IRE Trans*, 1961, **CS-9**, p. 107

10 CURTIS, H. E.: 'Frequency modulated wave transmissions', US Patent 2352254, June 1944

11 JACOBSEN, B. B.: 'Frequency patterns for multiple-radio-channel routes', *Proc. IEE*, 1960, **107** B, p. 241

Interference between Satellite Communication Systems and Common Carrier Surface Systems

By HAROLD E. CURTIS

(Manuscript received December 6, 1961)

Various published papers have discussed in quite general terms the problem of interference between satellite systems and ground systems. These studies have been largely qualitative, rather than quantitative, in nature. The magnitude of the interference between a satellite system and ground system, however, depends greatly on the frequency plans involved, the character and degree of modulation used, and the parameters of the equipment. Bell Telephone Laboratories has under construction experimental satellite equipment designed to operate in the heavily used 4- and 6-kmc common carrier bands, and the present paper is directed to the potential interference between this satellite equipment and ground point-to-point systems.

Interference involving a satellite station and the TD-2 and TH systems is analyzed specifically, and it is shown that the separation between systems must be of the order of 100 to 120 miles or more when the antenna of the common carrier transmitter or receiver is pointed directly at the satellite ground station. If the antenna is beamed 90 degrees or more from the satellite site, the minimum distance may be of the order of 10 miles even when line-of-sight propagation exists between locations. This assumes the use of the Bell System's horn-reflector antenna on the terrestrial system. With a parabolic dish antenna the latter distance must be increased to about 40 miles and adequate blocking must exist in the interference path. These distances provide adequate freedom from mutual interference for both telephone and television modulating signals.

I. INTRODUCTION

Satellite systems will of necessity use ground transmitter powers of several thousand watts. Present microwave systems in the United States operating in the common carrier band utilize transmitters of the general order of one watt output power and in this sense are consequently only about one-thousandth as interfering as a satellite ground transmitter. Furthermore, the inherent noise per cycle of bandwidth at the input of the satellite ground receiver will be about 20 db less than that of the present-day commerical common carrier receiver, thus making it correspondingly more sensitive to interference from other systems.

Several general studies of interference[1,2] have used criteria of interference intended to encompass all varieties of ground radio relay systems, but inevitably the decision as to whether interference between particular sites is tolerable or intolerable must be made on the basis of the specific radio systems involved and the frequency bands in which they operate.

The F.C.C., as a result of its studies, has recommended for consideration a number of bands between 3700 and 8400 mc, including the two common carrier bands 3700 to 4200 mc for the spacecrafts and 5925–6425 mc for the earth stations.[3]

The experimental satellite equipment presently under construction by Bell Telephone Laboratories will operate in the top 100 megacycles of the 4- and 6-kmc common carrier frequency bands mentioned above and thus, potentially, interference may occur between the satellite system and the many ground commerical systems operating in these bands.

The Western Union Company has a radio relay system operating in the 4-kmc band, and a transcontinental system in the 6-kmc band is under construction. There are also a relatively large number of short haul common carrier systems in the 6-kmc band. However, the most extensive user of each of these two bands in the United States is undoubtedly the Bell System which had in operation at the beginning of 1961 approximately 300,000 one-way broadband channel miles of microwave systems. A large fraction of this service utilizes the TD-2 system[4] operating in the 3700- to 4200-mc band, and, at present, a small fraction utilizes the recently developed TH system[5] operating in the 6-kmc band. For this reason the interference study described herein is directed quite specifically to the TD-2 and TH systems, but the basic philosophy is readily applicable to other microwave systems.

Such a complicated network of microwave routes as presently exists in the United States and Canada with its numerous sources of inter- and intra-system radio interferences has necessitated most careful attention to this problem in order that the interference at baseband at the end of a long circuit would be in reasonable balance with other sources of system impairment, such as intermodulation between elements of the system load and the noise arising in the converters. This problem has been discussed at some length in an earlier paper,[6] and the philosophy

Reprinted with permission from *Bell Syst. Tech. J.*, vol. 41, pp. 921-943, May 1962.

developed therein will be applied in the present paper to interference involving satellite systems.

The experimental equipment under construction by Bell Laboratories will transmit from ground to spacecraft in the 6-kmc frequency band, and in the reverse direction in the 4-kmc band, so that the only appreciable interferences between ground stations are those from satellite ground transmitters into TH receivers and from TD-2 transmitters into satellite ground receivers. For completeness, consideration is also given to the two complementary interferences that would exist if the frequencies were interchanged. Possible interference from the spacecraft into ground receivers is discussed briefly.

Contours of permissible minimum separation between satellite ground station and TH and TD-2 microwave stations are presented in this paper. It should be emphasized at this point that the results are based on values of parameters for the three systems that are pertinent at the present time. Changes that may be made in the future such as increases in transmitter power, improvements in receiver noise figure or change in frequency plan would, of course, alter the conclusions reached herein.

While the contours are based on propagation under "average" terrain conditions, it is believed that they should be of considerable value in the early phases of site selection. However, in any particular case, if the profile of the path so indicates, the power of the interference should be calculated and compared with the objectives given later in Table III.

II. OBJECTIVES

Microwave systems with which we are concerned may handle television or multichannel data and telephone signals, and in the latter case the signal load may range from busy-hour full load to light early-morning load. Interference objectives must be sufficiently stringent to protect the systems under all normal conditions; moreover, interference powers should be sufficiently less than the total receiver input noise so as not to impair significantly the fading margin of the interfered-with system.

Basically the amount of RF interference that can be tolerated depends on the interference that it produces at baseband frequencies. The spectrum of the interference at baseband frequencies resulting from RF interference between two FM or PM waves is made up of beats between each frequency component of the spectrum of one RF wave and each frequency component of the other. The frequency of any baseband component is that of the frequency difference between the two RF com-

ponents that produced it; and finally the power of the baseband component depends on the powers of the RF components.

Therefore the baseband interference spectrum can, for convenience, be thought of as the result of (a) a tone resulting from the beat between the two carrier frequencies, (b) the sidebands of one wave beating against the carrier of the second and vice-versa, and (c) the sidebands of one beating against the sidebands of the other. The beat tone ideally may appear as sinusoidal interference in a video signal, or as a tone in some particular telephone channel. Actually, because of the very low frequency noise normally present on FM transmitters using klystron deviators, the tone is more like a "burble" spread over a number of channels, the particular channels affected at any time depending on the difference frequency between the carriers at that instant. The second and third classes of interference appear normally as unintelligible crosstalk.

The relative magnitudes of carrier and sidebands at any time depend on the degree and type of modulation applied to the radio transmitter. Consequently the RF interference objectives must be sufficiently restrictive that the baseband interference is adequately low for all conditions of modulation. The procedure here will be to develop objectives on the basis of full load telephone considerations and then to make certain that they are adequate for all other loads and signals, whether telephone or television. In general, the interference objective set by telephone considerations is sufficiently controlling so that it is satisfactory for other types of signals.

All long-haul microwave systems are subject to a number of sources of transmission impairments. For example, a 4000-mile TD-2 system may have approximately 140 sources of thermal-type noise due to the converters, an equal number of sources of cross modulation due to repeater phase and amplitude nonlinearity, 280 sources of waveguide echoes, 280 sources of intersystem co-channel interference, together with a number of somewhat less important contributions.

Good engineering practice indicates that for telephone service, the rms sum of these impairments should, during busy hours, be not over 38 dba0, i.e., 38 dba* at a point of zero-db transmission level. This is equivalent to −43 dbm in a 4-kc band, and the signal-to-noise ratio is 27 db where the signal in each telephone channel is random noise equal in power to an rms talker, one quarter active, or −15.8 dbm in a 4-kc band, using values obtained from the Holbrook and Dixon paper.[7]

* The unit dba identifies a particular weighting characteristic for which 82 dba is equivalent to one milliwatt of thermal noise in a 3-kc band.

The division of this total permissible impairment into all the various individual sources cannot be done by any set rule. The total number of significant sources of noise impairment in a 4000-mile system, as enumerated above, totals very roughly one thousand. Therefore if each were given an equal share of the total, the individual allotment would be 8 dba0. Since interference and cross modulation are generally more annoying than thermal-type noise, it is normally a desirable goal that the baseband noise due to the converters in a repeater should be slightly greater than inter-modulation, which in turn should exceed the baseband interference from RF crosstalk. Then during a fade of the desired carrier, the converter noise and interference in any telephone channel will rise together and the interference will not predominate over the noise.

The distance between any satellite system site and a potentially involved interfering or interfered-with TD-2 and TH station may vary from a few miles to more than 100 miles. Thus, propagation between them may range from line-of-sight, in which free-space propagation normally exists for a very large fraction of the time over a full year, to tropospheric scatter propagation, in which the long-term distribution of path loss is normal in db and in which the chance of the received signal being, let us say, 32 db or more above the median value is about 0.01 per cent.

For this reason two interference objectives are proposed. The first applies to line-of-sight interference paths in which the propagation is very close to that of free-space nearly 100 per cent of the time. Up-fades greater than about 5 db occur less than about 0.3 per cent of a year's time, and down-fades of the interfering carrier simply decrease the baseband interference. This is referred to herein as the "100 per cent" objective, and it is intended that it be met for line-of-sight paths during free-space transmission.

For interference signals which are constantly fading both up and down, such as on tropospheric scatter paths, a second objective is proposed, which should be exceeded only 0.01 per cent of the time. This objective in terms of baseband noise may obviously be higher than the "100 per cent" objective, and it appears reasonable to let the 0.01 per cent objective be 15 db more lenient than the "100 per cent" objective. The "100 per cent" objective is so chosen that the unintelligible crosstalk type of interference in the worst telephone channel in the full-load telephone case may be expected to be 9 dba0 during nonfading periods. It may be 24 dba0 or greater 0.01 per cent of the time when the interfering signal path is well beyond line-of-sight. This may be compared with the contribution of about 10 dba0 per repeater due to noise arising in

the converters of a 4000-mile system during nonfading conditions. Interference may also manifest itself as tones in certain telephone channels or in television transmission. The magnitudes of these effects will be discussed in a subsequent section.

III. FREQUENCY PLAN

Interference between two FM or PM systems depends upon such parameters as frequency deviation, top baseband frequency, and upon the frequency separation between the carrier frequencies of the systems involved.

The CCIR recommends for the 6-kmc common carrier band a plan based on a spacing of 29.65 mc starting at 5945.20 mc, and this is identical with the plan used in the United States by the Bell System for the TH system and also by Western Union for its 6-kmc system. Thus eight satellite assignments, each about 50-mc wide, can be obtained in the same band with a minimum of mutual interference by placing the satellite carriers midway between the common carrier assignments as shown on Fig. 1.

Coordination in the 4-kmc common carrier range is less satisfactory. The TD-2 system, when a route is fully developed, will have a channel

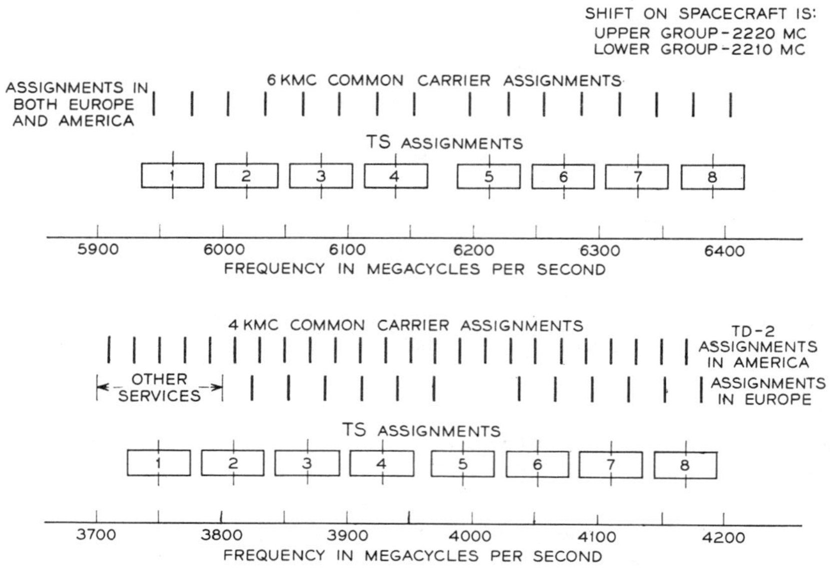

Fig. 1 — Frequency plan.

every 20 mc from 3710 to 4170 mc, while in Europe a spacing of exactly 29 mc is used.

A frequency shift from the 6-kmc to the 4-kmc band must obviously be made on the spacecraft, the optimum value of the shift for each of the channels depending to a considerable extent on problems outside the scope of this paper.

However, one possibility might be to use a shift of 2220 mc for the upper four channels, and 2210 mc for the lower four channels. The satellite assignments in the 4-kmc band would then be as shown on Fig. 1, and this plan is assumed in the present study. It will be noted that the satellite carrier frequencies would, in all cases, be very close to certain of those used in the TD-2 system. The effect on interference of moderate departures in frequency from the plan shown on Fig. 1 is discussed below.

IV. INTERFERENCE BETWEEN TWO FM OR PM WAVES

Signal and interference are customarily specified in units of watts per cycle of bandwidth. Since there is a linear relationship between signal power per cycle and carrier deviation in mean square radians per cycle of bandwidth, it is convenient in this paper to express both signal and interference in the latter units.

The method used herein to compute quantitatively the baseband interference due to the presence of a weak interfering FM or PM wave can be demonstrated by the limiting case when neither interfered-with nor interfering carrier is modulated. Let the peak amplitudes of the two carriers be E and kE, respectively, where k is small relative to unity, and let the frequency difference between them be f. The interfering carrier phase modulates the stronger carrier by $k \cos 2\pi ft$ radians, and the baseband interference power is proportional to $k^2/2$ mean square radians.

However, the present problem is most closely approximated by two carriers separated an appropriate amount in frequency and modulated with random noise so as to simulate a number of telephone channels arranged in frequency division multiplex. This problem has been treated for the case of pure frequency modulation by several writers.[8,9]

The Appendix gives an expression for the interference in integral form that is valid when the interfering carrier is substantially weaker than the interfered-with carrier. This integral is difficult to deal with numerically. However, following the argument used in the Appendix a practical, but basically exact, method of evaluating the interference has been developed, subject to the same premise of weak interference. This method

consists simply of normalizing the level of the unmodulated power or voltage of the stronger carrier to a reference value of 0 db, and expressing the spectrum of each of the two waves in db below this reference value. The two spectra thus described are convoluted (adding the db values). The interference thus computed in a very narrow band, such as one cycle, is varying with time, and in this case to obtain the mean square value, or power, the result must be decreased by 3 db.

This procedure, then, gives the distributed interference spectrum in mean square radians per cycle of bandwidth, together with a sine wave component at the difference frequency, whose power is expressed in mean square radians. The noise signal simulating a typical talker can be expressed in the same units as the distributed baseband interference, and thus the baseband signal-to-interference ratio can be obtained. Since the noise signal, for the loading constants assumed herein, is 65 dba0, the interference can be expressed in dba0 by subtracting the signal-to-interference ratio in db from 65. Baseband signals are usually pre-emphasized so that the higher frequencies phase modulate the carrier and the lower ones frequency modulate it. In this study, pure phase modulation is assumed for simplicity of analysis for all baseband frequencies.

Nominal values of the parameters determining the spectra for the three systems considered here are given in Table I.

For both the TD-2 and the TH carriers the frequency deviation is sufficiently low that, for purposes of this study, sidebands of order greater than unity are sufficiently small that they may be neglected.

The spectral power per cycle of bandwidth for first order sidebands

TABLE I

Item	Symbol	System		
		TD-2	TH	TSX-1*
Top baseband frequency	f_b	2 mc**	10 mc	2 mc**
RMS frequency deviation due to noise load	f_d	0.71 mc	0.71 mc	5 mc

* This is the designation for the Telstar experimental satellite equipment being constructed at Bell Telephone Laboratories. The ground station will be located in Andover, Maine. The spacecraft will have active repeater equipment for one broadband channel. Transmission to the spacecraft will be at a frequency of about 6390 mc. Transmission from spacecraft to ground will be at about 4170 mc. The parameters given in Table I above for the TSX-1 equipment represent values that appear to be reasonable at the time of preparation of the paper.

** The top frequency of 2 mc is appropriate to 480 telephone channels arranged in frequency division multiplex.

only, relative to the unmodulated carrier, for a low index of modulation is[10]

$$P_i = \frac{\exp\left[-3f_d^2/f_b^2\right]}{2f_b}\frac{3f_d^2}{f_b^2} \quad \text{for} \quad |f| < f_b \tag{1}$$
$$\text{and zero for} \quad |f| > f_b$$

where f is any frequency relative to that of the unmodulated carrier.

In the case of the TSX-1 system, the deviation with noise load is sufficiently great that the wide deviation approximation may be used. Assume the power of the carrier when unmodulated is unity. When the phase deviation, as defined in (4) below, is substantially greater than unity, the power per cycle of the PM wave at a frequency $\pm f$ from that of the carrier frequency is very closely

$$P = \frac{1}{\sqrt{2\pi}f_d}\exp\left(-f^2/2f_d^2\right). \tag{2}$$

There is also a carrier spike present whose power relative to that of the unmodulated carrier of unity power is

$$P(\text{spike}) = \exp\left(-3f_d^2/f_b^2\right). \tag{3}$$

Fig. 2 shows the spectra for the three systems with noise loading as computed using the parameters and formulae given above.

The mean square phase deviation, D, of the carrier is related to the above defined constants in the following way:

$$D = 3f_d^2/f_b^2 \text{ mean square radians.} \tag{4}$$

The applied signal power, S, per cycle of bandwidth is

$$S = D/f_b = 3f_d^2/f_b^3 \text{ mean square radians per cycle of bandwidth.} \tag{5}$$

The signal power in db is given by 10 log S and is tabulated in Table II for the three systems of interest, using the constants from Table I. The symbol dbR/cbw will be used to denote mean square radians per cycle of bandwidth expressed in decibels.

The application of the method of determining the interference in a specific case will illustrate the procedure. Let us consider, as an example,

TABLE II

Signal Power	TD-2	TH	TSX-1	Units
10 log S	−67.3	−88.2	−50.3	dbR/cbw

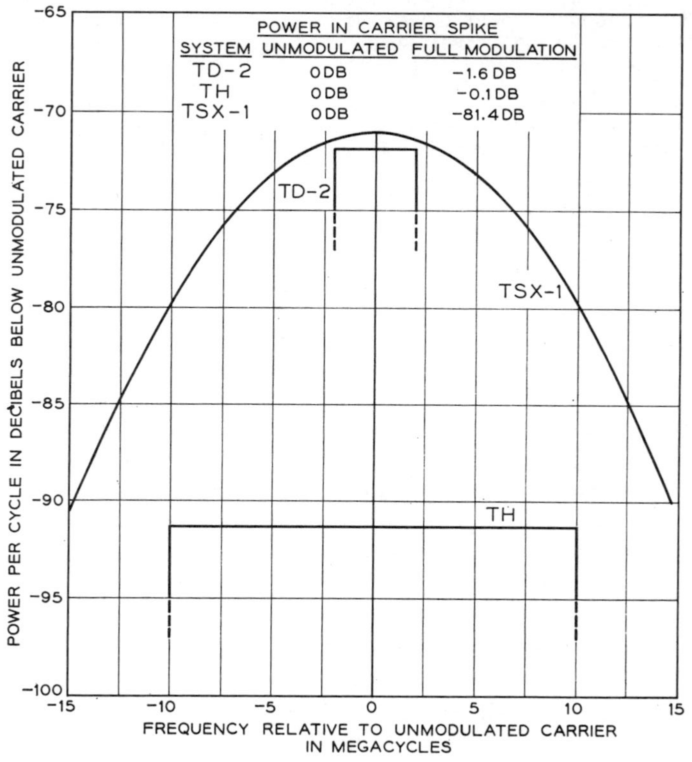

Fig. 2 — Spectra of phase-modulated waves.

interference into the satellite system from the TD-2 system and compute, specifically, the dba value of interference corresponding to a particular ratio of unmodulated TD-2 carrier power to unmodulated satellite carrier power at the input to the TSX-1 receiver, such as −35 db. The unmodulated carrier frequencies in this case will be nearly co-channel.

Fig. 2 shows that in the case of the satellite system modulated with random noise, the sideband power per cycle near the carrier will be 71 db below the unmodulated satellite carrier power. The power of the TD-2 carrier in turn was assumed to be 35 db below the carrier power of the interfered-with TSX-1 carrier.

The TD-2 carrier spike then beats with each component of the TSX-1 spectrum to produce baseband interference as described above. Consider specifically the beat between the unmodulated TD-2 carrier and the TSX-1 spectral power one kilocycle removed therefrom. The resulting interference that will fall in a bandwidth of one cycle at a baseband frequency of one kilocycle can be obtained, as described above, by adding the db values of the appropriate points on the two spectra and subtracting 3 db from the result. This gives an interference value of $-71 + (-35) -3$, or -109 dbR/cbw.

From Table II we find that the TSX-1 signal power is -50.3 dbR/cbw. The signal-to-noise ratio in a narrow band then is 55.7 db, which is equivalent to approximately +9 dba0. This establishes the position of the linear relationship shown on Fig. 3, for interference from the TD-2 system into the TSX-1 system.

If the speech load on the satellite system is very low, then the continuous portion of baseband interference of the satellite system due to the TD-2 system can be obtained by convoluting the TSX-1 carrier spike and the TD-2 spectrum. By following the above procedure it can be shown that the baseband interference is 0.9 db less than when the TSX-1 system was fully modulated.

Interference from the satellite system into the TH system behaves quite differently. With the frequency plan shown on Fig. 1, the unmodulated carriers are approximately 15 mc apart, and only when the satellite system is substantially fully loaded will there be appreciable interference into the TH system.

Assume a ratio of TH carrier to satellite carrier power of 68 db. The maximum interference into TH falls at 10 mc and is the result of a beat between the TH carrier and satellite sidebands 10 mc removed, i.e., 5

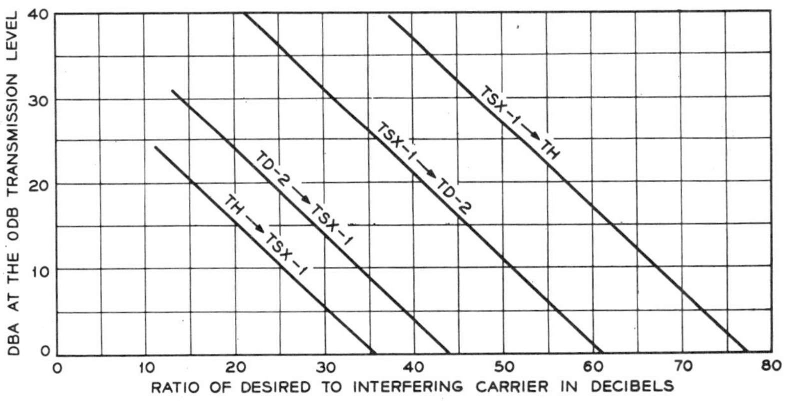

Fig. 3 — Expected message channel interference, all-message case.

mc from the satellite carrier itself. Fig. 2 shows the spectral power is 73 db below the TSX-1 unmodulated carrier or 141 db below the TH carrier. Therefore the interference on the TH carrier corresponds to phase modulation of -144 dbR/cbw. The TH signal power (see Table II) is -88.2 dbR/cbw. The signal-to-interference ratio is 56 db and the interference is +9 dba0.

Interference into the TSX-1 system from the TH system increases rapidly with baseband frequency. In order not to jeopardize the potential use of the satellite system above 2 mc, interference is computed at 5 mc but is referred to the signal as specified in Table I, thus giving a conservative value for the minimum allowable separation.

Finally, interference from the TSX-1 system into the TD-2 system is a maximum at the bottom baseband frequencies since the two systems are nearly co-channel. Computations are made in a manner similar to that for interference in the reverse direction. Fig. 3 shows the relationships between baseband interference to the ratio of desired carrier to interfering carrier for these four cases. These relationships are linear, and are valid for carrier ratios greater than about 10 db.

In the frequency plan shown on Fig. 1, the satellite channels are uniformly spaced between TH assignments. If a plan were used with a spacing different from 15 mc, the magnitude of the interference would, of course, be affected. Fig. 4 shows the computed baseband interference spectrum as a function of carrier spacing, and it will be noted that the increase due to reducing the separation to 10 mc is only about 2 db at the top baseband frequency.

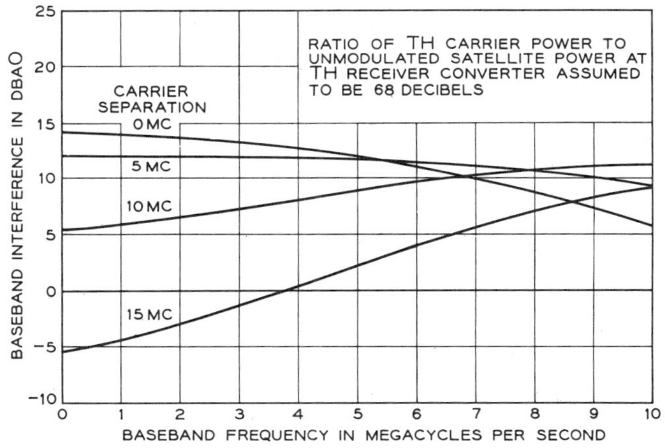

Fig. 4 — Interference — satellite transmitter into TH system.

TABLE III

System		Normal Received Carrier	Required Interference Ratio	Interfering Carrier Objectives at receiver-converter	
Interfering	Interfered-with			100%	0.01%
TH	TSX-1	−98 dbm	27 db	−125 dbm	−110 dbm
TSX-1	TH	−27	68	−95	−80
TD-2	TSX-1	−98	35	−133	−118
TSX-1	TD-2	−38	52	−90	−75

V. PERMISSIBLE INTERFERENCE

On the basis of the normal received carrier power for the TD-2 and TH systems and the minimum received satellite carrier, permissible interference carrier powers at the input to the receiver-converter of the interfered-with system can be obtained (Table III). The values tabulated in the "required interference ratio" column in Table III are from Fig. 3, and correspond to a baseband interference value of 9 dba0.

The above objectives have been based on busy-hour telephone interference. During such periods the power in the carrier of the satellite system is negligible, but under light loads it may be strong enough to make the beat note between it and the carrier of the ground common carrier system a serious source of interference. Ideally this interference would be a pure tone, but actually it will be spread over a group of telephone channels, its location depending on the frequency difference between the two carriers at any moment

An order of magnitude estimate of this interference can readily be made as follows for the case of interference from the TD-2 system into the satellite system when the latter is very lightly loaded:

a. Permissible carrier to interference ratio from Table III = +35 db
b. Resulting tone interference = −38 dbR
c. Signal (noise load) from Table II = −50.3 dbR/cbw
d. Signal in a 3-kc band = −15.5 dbR
e. Signal-to-interference ratio = +22.5 db
f. If the interference is assumed to have the character of noise and spread over one channel only, it will read = +42.5 dba0.

This, it will be noted, is 4.5 db above the total 4000-mile objective. Actually it will probably be spread over a number of telephone channels,

depending upon the steadiness of the carriers, but still it is a potentially important interference source. The interference from the satellite system into the TD-2 system computed in the same manner will have the same value.

This source of interference can be decreased substantially by applying a very low frequency baseband signal to the transmitter to keep the carrier in motion even when the total speech load is low.[11]

The above objectives have been based on telephone channel interference considerations. It appears possible to produce interference into the satellite system when the interfering carrier is weakly modulated, thus having a strong carrier spike, and when at the same time the interfered-with carrier is handling a television picture with a large gray area, thus also having a large component of energy in a fairly narrow band.

For example, assume the television signal content in the TSX-1 channel is such that there is a concentration of energy one mc removed from an assumed interfering TD-2 carrier. A sinusoidal baseband tone will result whose peak-to-peak amplitude relative to the peak-to-peak amplitude of the desired baseband signal for full deviation is given by the ratio of the carrier powers involved in db, plus the FM improvement. The assumed carrier ratio from Table III is 35 db; the FM improvement for a TSX-1 deviation of 20 mc is 26 db. Therefore, the signal-to-tone ratio is 61 db as shown later in Table IV. Bearing in mind that line-of-sight signals during fading conditions may "up-fade" as much as 5 db, this TV signal-to-tone ratio would then be reduced to about 56 db. This is of the same order of magnitude as the tolerable tone interference ratio in a television picture.

The various impairments that may be expected using the interference ratios given in Table III are summarized in Table IV. Since the ground satellite station in the TSX-1 experiment transmits in the 6-kmc band

TABLE IV

System		Telephone Interference		TV Sp-p/Tp-p	Loss of Fade Margin
Interfering	Interfered-with	Full Load	No Load		
TH	TSX-1	+9 dba0	nil	53 db	0.3 db
TSX-1	TH	9 dba0	nil	80	1.6
TD-2	TSX-1	9 dba0	42.5 dba0*	61	0.1
TSX-1	TD-2	9 dba0	42.5 dba0*	64	2.2

* This interference will appear in a few telephone channels only, and there is also considerable uncertainty in the values given.

and receives in the 4-kmc band, it is evident that TD-2 interference into TSX-1 is controlling from the standpoint of interference to television.

The loss of fading margin is computed on the basis of the increase in peak noise voltage at the converter when the interfering carrier is present. A noise peak factor of 12 db is assumed for the noise at the converter. The interference values used in computing loss of fading margin are the "100 per cent" values, so-called, on Table III since the use of the "0.01 per cent" values seemed unduly conservative.

An examination of Table IV appears to indicate that relaxing any of the RF interference objectives will result in undesirable impairment increases in one or more of the categories listed.

VI. REQUIRED PHYSICAL SEPARATION

The separation between satellite station and common carrier station should be such that the received signal does not exceed the values stated on Table III. This separation depends, of course, upon the transmitter power, path loss and antenna discrimination involved.

In this study the satellite transmitter power is assumed to be 2 kw. The transmitter power in the TH system is 5 watts. Because of atmospheric effects it is expected that the minimum useful elevation of the satellite antenna above the horizontal will be limited to about 7.5°. If so, the effective gain of the satellite antenna in the horizontal direction may be expected to be 0 db above isotropic or less. The gain of the conventional horn-reflector antenna used with the TD-2 and TH microwave systems at a given angle off-beam is conveniently expressed as the on-beam gain (about 40 db above isotropic at 4 kmc) reduced by the relative directivity pattern of the antenna, a typical example of which is shown later as Fig. 6. Thus, the net response of the horn-reflector in the backward direction is +40 −70, or 30 db below isotropic.

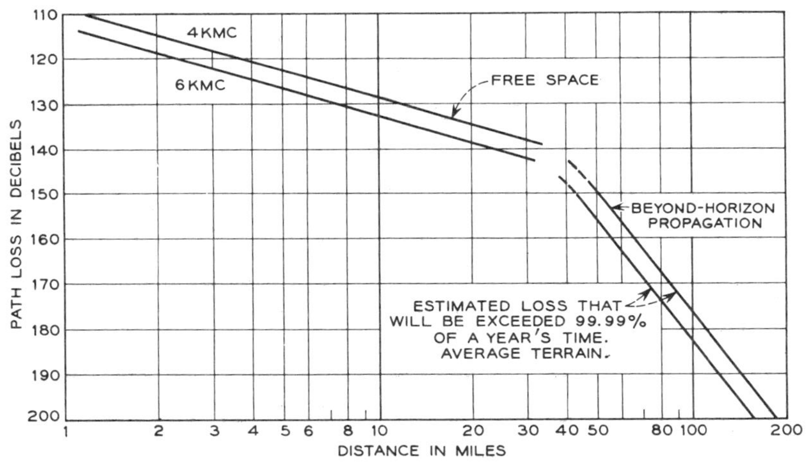

Fig. 5 — Path loss — line-of-sight and beyond-horizon.

Typical path-loss curves between isotropic antennas as a function of distance are shown on Fig. 5. For distances up to about 30 miles, free-space loss values are plotted; from about 40 miles to 140 miles, scatter propagation over average terrain is assumed. The scatter loss curve is that which will be exceeded 99.99 per cent of a year's time as estimated by K. Bullington of Bell Telephone Laboratories.[12]

The interfering carrier power in dbm is given by

$$P_R = P_T + A_T - L + A_R \qquad (6)$$

where P_T = effective transmitter power

A_T = transmitter antenna gain in the direction of the interfered-with receiver

L = path loss (from Fig. 5)

A_R = receiving antenna gain in direction of the interfering transmitter.

As an example, interference from the TD-2 system into the TSX-1 system will be considered. Transmitter power and on-beam antenna gain for the TD-2 system are approximately +27 dbm and 40 db, respectively. The required separation when the antenna of the interfering transmitter is pointed at the satellite site will be such that propagation will presumably be by scatter and hence the 0.01 per cent objective of −118 dbm given in Table III will apply. The total required path loss will be then 185 db, corresponding to 123 miles.

If the satellite receiving station is off the beam of the TD-2 transmitting antenna, the energy radiated thereto is decreased by the antenna discrimination, and the required separation is therefore decreased. Fig. 6 shows the measured directivity pattern of an individual horn-reflector antenna relative to its forward gain, and it will be noted that the backward gain is 70 db below the on-beam gain. Tentatively applying the line-of-sight objective of −133 dbm from Table III, and substituting the appropriate values into (6), we have

$$-133 = +27 + (40-70) - L + 0$$

from which L = 130 db. Reference to Fig. 5 shows that this loss corre-

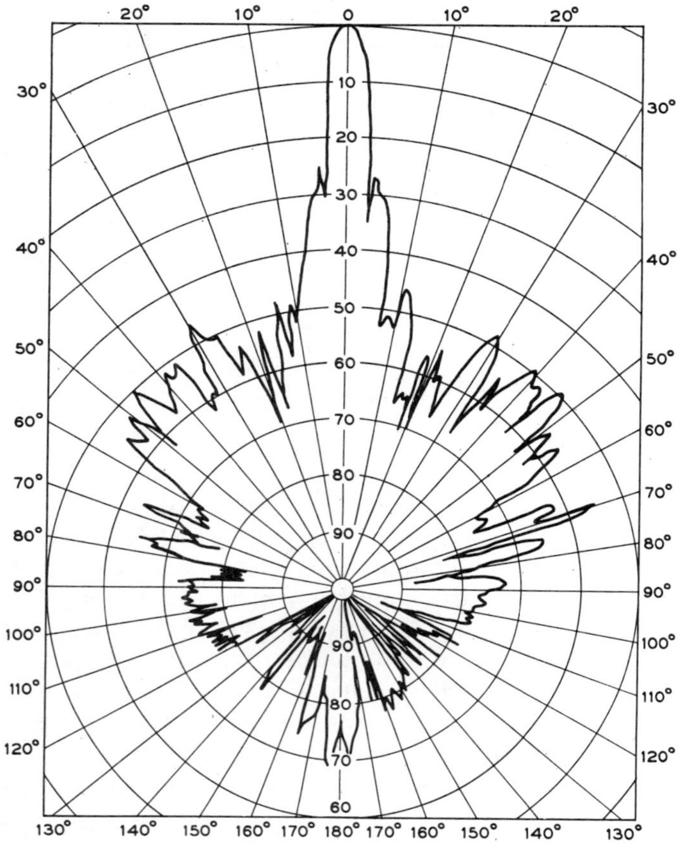

Fig. 6 — Measured pattern in db of horn-reflector antenna at 3740 mc.

sponds to a separation of about 11 miles. Thus, this distance is adequate when the horn-reflector antenna is beamed directly away from the satellite site, even with line-of-sight propagation.

The measured antenna pattern shows a large number of nulls, the position and depth of which will vary from antenna to antenna. Therefore, in developing contours of minimum permissible separation, it has seemed a reasonable and conservative approach to use not the measured patterns but instead an envelope drawn through the maximum values of the pattern as in Fig. 8 of Ref. 6. An additional element of conservatism lies in the fact that the TSX-1 antennas will use circularly polar-

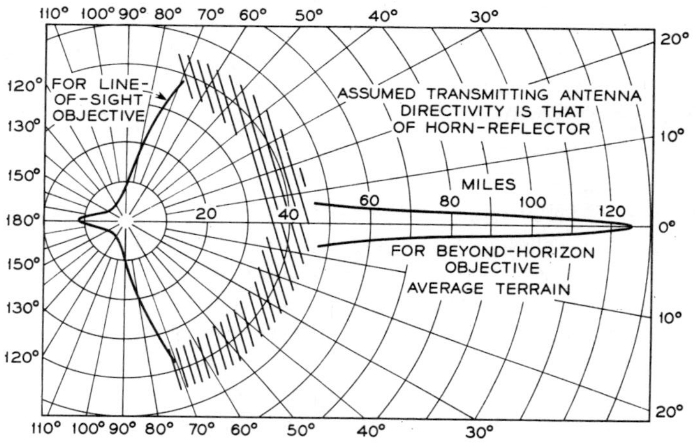

Fig. 7 — Contours of minimum permissible separation between TD-2 transmitter and satellite ground receiver.

ized waves, whereas plane polarized waves are used by the ground systems.

Fig. 7 shows a contour of minimum permissible separation between a TD-2 transmitter and a satellite receiver as a function of the angle between the bearing of the TD-2 antenna and the bearing of the satellite site, based on a smoothed envelope of the antenna discrimination pattern of Fig. 6. For angles greater than 90°, the separation can be as low as about 5 miles. For angles less than about 5° it must be 60 miles or greater. Distances in the order of about 40 miles are cross-hatched because of the uncertainties in propagation. Specific cases that fall in this range should be examined individually to ascertain whether line-of-sight or beyond-horizon objectives should be applied.

Figs. 8, 9 and 10 show similar contours of minimum permissible separation for the three other interference combinations.

It should be emphasized that these values of separation may vary substantially, depending on the local terrain. If the path is obviously line-of-sight, the "100 per cent" objective shown on Table III should be used and met on a free-space propagation basis. In the case of stations only somewhat beyond line-of-sight, it must be kept in mind that ducting may occur occasionally and the signal may become strong enough to give interference much stronger than expected under true scatter conditions. In this situation the chance of exceeding the 0.01 per cent objective given on Table III must be estimated for the specific case involved, due allowance being made, of course, for the directivity of the antennas; if the estimated probability is 0.01 per cent or less, the locations may be considered safe.

The close permissible spacing indicated when the horn-reflector antenna is oriented 90° or more from the direction of the satellite station is, of course, due to its very low backward response. An 8-foot parabolic

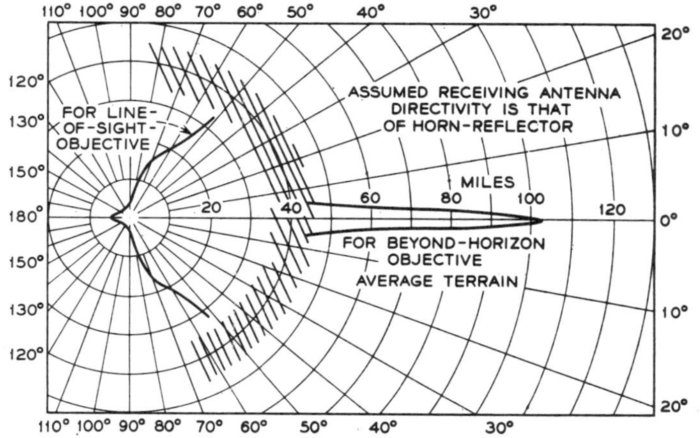

Fig. 8 — Contours of minimum permissible separation between satellite ground transmitter and TD-2 receiver.

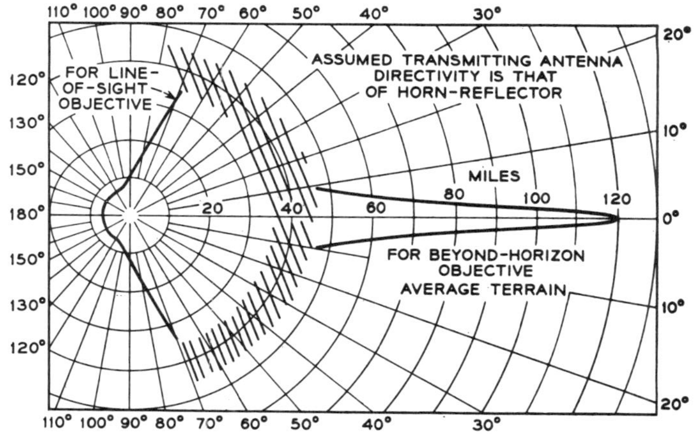

Fig. 9 — Contours of minimum permissible separation between TH transmitter and satellite ground receiver.

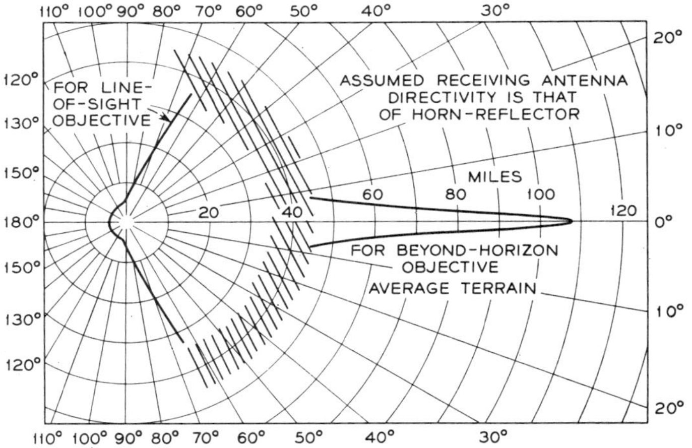

Fig. 10 — Contours of minimum permissible separation between satellite ground transmitter and TH receiver.

antenna may be expected to provide some 20 db to 30 db less discrimination in this general direction. The application of the line-of-sight objective would then result in a minimum distance so great that it would be highly improbable that line-of-sight transmission could take place. The application of the "0.01 per cent" objective indicates, however, a minimum separation of 30 to 40 miles.

In this case, it would be necessary to make certain that propagation is by scatter. The process of estimating propagation over relatively short non-line-of-sight paths involves a detailed study of the exact profile and knowledge of climatic conditions. This subject is outside the scope of the present paper.

While these contours indicate that separations of the order of 5 to 10 miles are possible under the conditions assumed, it should be emphasized that these conclusions are valid only in the absence of significant reflection paths between the interfering and interfered-with sites. Thus, reflecting objects such as houses or trees in the foreground of the horn-reflector antenna may degrade its directivity pattern in the backward direction so that the values shown on Fig. 6 are not attained. Normally it may, however, be expected that such reflections can be made adequately low by careful examination of the terrain illuminated by the ground system antenna.

Transient reflections may be produced by objects in the sky such as airplanes and birds. Also, reflections from rain clouds and precipitation represent a possible way in which interference might reach the satellite ground receiver. However, calculations of the probable magnitude of the effects of such reflections indicate that with station separations of the order of 10 miles or more, interference of this kind from TD-2 into TSX-1 will not exceed the objective of −118 dbm for any appreciable percentage of time.

VII. INTERFERENCE FROM SPACECRAFT TO GROUND RECEIVERS

Another possibility is, of course, interference from the spacecraft transmitter into ground microwave systems. For example, assume a power of one watt and an isotropic antenna on the spacecraft, and on-beam gain of 40 db for the ground antenna. The interference at the receiver converter would have a value of −90 dbm with the spacecraft at a distance of about 370 statute miles, therefore just meeting the "100 per cent" objective for interference from the satellite system into the TD-2 system under these assumed conditions. On this same basis, the maximum power on the spacecraft could be increased to 100 watts if the distance from the spacecraft to the earth were increased ten-fold.

In the case of the horn-reflector antenna, the beamwidth is sufficiently narrow that any particular spacecraft would be in the beam of a given antenna for only brief and infrequent intervals.

APPENDIX

Using an approach similar to that developed in Ref. 13, Messrs. S. O. Rice and L. H. Enloe of Bell Telephone Laboratories have shown independently that the distributed portion, W(f), of the interference spectrum is given by

$$W(f) = k^2 \int_0^\infty \exp[-R_w(0)](\exp[R_w(\tau)] - 1) \cdot [\cos(2\pi f - \omega_0)\tau + \cos(2\pi f + \omega_0)\tau]\, d\tau \tag{7}$$

where the amplitude of the stronger carrier is unity and

k = ratio of the weaker to the stronger carrier; this must be small compared to unity

f = any baseband frequency

$\omega_0/2\pi$ = frequency difference between the two carriers

$R_w(\tau)$ = sum of the autocorrelation functions of the two applied noise signals.

If the noise signals applied to the carriers have powers w_1 and w_2 mean square radians per cycle of bandwidth for the weaker and stronger

carriers, respectively, and f_1 and f_2 are the top baseband frequencies of the noise signals applied to the same two carriers, then

$$R_w(\tau) = R_{w_1}(\tau) + R_{w_2}(\tau) \tag{8}$$

$$= \frac{w_1 \sin 2\pi f_1\tau}{2\pi\tau} + \frac{w_2 \sin 2\pi f_2\tau}{2\pi\tau}. \tag{9}$$

Also, let

$w_2 f_2 = D_2$ = mean square phase deviation of the stronger carrier

and

$w_1 f_1 = D_1$ = mean square phase deviation of the weaker carrier;

then

$$R_w(0) = D_1 + D_2. \tag{10}$$

In addition to the distributed interference, there is a sinusoidal component of magnitude

$$k\, e^{-\frac{1}{2}R_w(0)} \sin \omega_0 t \text{ radians.} \tag{11}$$

The expression $W(f)$ as given by (7) is not readily evaluated except over a somewhat limited range of parameters. However, the quantity $\exp[R_w(\tau)]$ in (7) is the product of the two autocorrelation functions $\exp[R_{w_1}(\tau)]$ and $\exp[R_{w_2}(\tau)]$. It therefore follows from the convolution theorem for Fourier integrals that the power spectrum of the interference can be obtained by convoluting the power spectra of the two phase-modulated waves. Thus, baseband interference spectrum in mean square radians is given by twice the value obtained by convoluting the spectral power of the two waves, both in a resistance of one ohm. This provides the method* used in this paper for computing interference.

* While the paper was in page proof it was brought to the writer's attention that the convolution method of computing interference between two PM waves had previously been described in a paper (in Japanese) entitled "On the Interference Characteristics of the Phase Modulation Receiver for the Multiplex Transmission," Shinji Hayashi, The Journal of the Institute of Electrical Communication Engineers of Japan, **35,** pp. 522–528, November 1952.

REFERENCES

1. Fine, H., and Dixon, J. T., Suggested Criteria for Limiting Satellite Power in Channels Shared with Ground Systems, Office of the Chief Engineer, Technical Research Division, FCC, TRR Report No. 5.2.1, June 1, 1961.
2. Downing, J. J., Interference Considerations for Communications Satellites, Seventh National Communication Symposium Record, 1961, pp. 302–309.
3. Final Preliminary Views of the United States of America, as Submitted by the Federal Communications Commission, Frequency Allocations for Space

Radiocommunication, September 7, 1961. Hearings on Space Communications and Allocation of Radio Spectrum before the Communications Subcommittee on the Committee on Commerce, United States Senate, August 1, 23 and 24, 1961 — United States Government Printing Office, Washington, D. C.

4. Roetkin, A. A., Smith, K. D., and Friis, R. W., The TD-2 Microwave Radio Relay System, B.S.T.J., **30,** October 1951, p. 1041.
5. Mc Davitt, M. B., 6000-Megacycle-per-Second Radio Relay System for Broadband Long-Haul Service in the Bell System, AIEE Trans., **76,** Part 1, January 1958, p. 715.
6. Curtis, H. E., Radio Frequency Interference Considerations in the TD-2 Radio Relay System, B.S.T.J., **39,** March, 1960, p. 369.
7. Holbrook, B. D., and Dixon, J. T., Load Rating Theory for Multi-Channel Amplifiers, B.S.T.J., **18,** October 1939, p. 624.
8. Medhurst, R. G., Hicks, Mrs. E. M., and Grossett, W. G., Distortion in Frequency-Division-Multiplex FM Systems Due to an Interfering Carrier. Proc. I.E.E., **105,** Part B., No. 21, May 1958.
9. Hamer, R., Radio-Frequency Interference in Multi-Channel Telephony FM Radio Systems, Proc. I.E.E., **108,** Part B, No. 37, January 1961.
10. Stewart, J. L., The Power Spectrum of a Carrier Frequency Modulated by Gaussian Noise, Proc. I.R.E. **42,** October 1954, pp. 1539–1542.
11. U. S. Patent 2,352,254, Frequency Modulated Wave Transmission, June 27, 1944, H. E. Curtis.
12. Response of the A. T. & T. Co. before the Federal Communications Commission, Docket No. 13522, February 27, 1961.
13. Bennett, W. R., Curtis, H. E., and Rice, S. O., Interchannel Interference in FM and PM Systems Under Noise Load Conditions, B.S.T.J., **34,** May 1955, pp. 601–636.

A Survey of Interference Problems and Applications to Geostationary Satellite Networks

MICHEL C. JERUCHIM

Abstract–The fundamental limitation in the utilization of the geostationary orbit is the mutual interference generated by satellite networks. A review of the theory of operation of communication systems in an interference environment is provided, with particular attention to those systems of importance in space communications, namely multichannel FM telephony, frequency-modulated television, and coherent PSK digital signals. Examples are then given to show the limiting effects of interference on orbit utilization.

I. Introduction

FOR SOME TIME it has been recognized that the geostationary orbit is a natural resource that must be managed efficiently. Much has been written [1]–[16] about the technical, operational, economic, and regulatory aspects of orbit utilization. The paper by Withers [16] provides an up-to-date overview of the problem. While many factors are relevant to orbit utilization, in a sense the most fundamental is interference, since it is the unavoidable presence of interference that creates a problem in the first place.

The inevitability of interference is as physically fundamental as that of noise within a system. In the case of interference, however, the ultimate limitation arises because it is not possible to constrain an antenna beam to a prescribed volume. Hence, geostationary satellite networks operate in an environment in which interference can be considered to be as much a part of the design constraints as noise, or other parameter limitations. Thus, the basic problem to be considered here is the determination of network performance in an interference environment. In general terms it is of interest to evaluate an expression of the form

$$Q = f(S_w, D_w; S_I, D_I) \tag{1}$$

where

- Q wanted signal quality, e.g., signal-to-noise (S/N) ratio and error probability;
- S the set of parameters specifying the modulation characteristics, e.g., signal type, modulation index, and baseband bandwidth; subscripts w and I refer to wanted and interfering signals, respectively;
- D the set of network link parameters, e.g., *e.i.r.p.*, frequency, and antenna size; subscripts w and I refer to wanted and interfering networks, respectively.

The symbolic equation (1) will be defined in more specific terms subsequently. There are two aspects in the evaluation of (1). The first is the calculation of the interference levels themselves. This requires a relatively straightforward calculation involving geometrical parameters, power levels and antenna gains. The second aspect, which is the main concern of this paper, is the determination of signal quality Q, given specific interference levels.

The evaluation of interference effects depends upon the nature of both the wanted signal and the interference, including the modulating signals, the modulation methods, the frequency assignments, the definition of Q, and the operations (filtering, amplification, etc.) that these signals may be subjected to.

Manuscript received May 17, 1976; revised September 14, 1976.
The author is with the General Electric Company, Space Division, Valley Forge Space Technology Center, Philadelphia, PA 19101.

Reprinted from *Proc. IEEE*, vol. 65, pp. 317-331, Mar. 1977.

A virtually limitless number of combinations of these factors may arise in practice. Although certain sets of these combinations are amenable to general formulation, numerical evaluation of specific cases often requires specific treatment.

Thus while it is desirable to provide a context which is as general as possible, this paper will concentrate on a few cases of particular importance to satellite communications: FDM/FM multichannel telephony, FM television, and coherent phase-shift-keyed (CPSK) signals. The application of the theory to the orbit utilization problem will be illustrated with examples which will indicate the limiting role played by interference.

II. Interference into Analog Wanted Signals

In a formal sense the computation of the effect of interference consists of writing an expression for the output of the demodulator and separating this into a "signal" part and an "interference" part.[1] Assuming that this can be done, perhaps the most complex part of the problem arises in the characterization of the effect of interference on the final receptor. That is, the description of interference effects in the same terms in which the signal quality Q is measured, is of ultimate interest. Although this is rather difficult to do in general, this has been done in one case of particular importance, namely multichannel telephony systems.

Earlier studies of interference [17] arose principally in the broadcasting context, in which the effects of interference[2] were estimated by approximating the latter by sinusoids. The effect of interference on the capture phenomenon of FM receivers was also studied [18], [19] by observing the behavior of the instantaneous frequency.

With the evolution of microwave radio-relay systems other types of interference problems arose, namely that between FM multichannel telephony (also referred to as FDM/FM telephony) signals, between FM television signals, or between telephony and TV signals. The first of these three cases has received a great deal of attention [20]-[27]. With the advent of communication satellites, which also transmitted FM telephony and television (although with parameters different from those used by radio relays), the scope of application of these problems was enlarged to encompass interference between radio-relay systems and satellite systems, and between satellite systems themselves. This generated renewed interest in the problem, as evidenced by a number of studies [28]-[43].

Because multichannel speech can be represented by a Gaussian signal, almost all of the studies cited have concerned interference between FDM/FM telephony signals. Conversely, the difficulty of finding objective measures of the effect of interference on television signals has prevented the development of a systematic theory and forced reliance on empirical data embodied in "protection ratios." Consequently, the cases of wanted FM multichannel telephony signals and FM television signals will be treated separately.

Another class of transmissions that has recently become important in satellite communications is single-channel-per-carrier (SCPC) voice transmission. The modulating signal for SCPC signals can be either analog or digital. Very little work has been done on the susceptibility of these signals to interference. Although the problem can be formally approached by using the methods discussed herein, the parameters of SCPC signals relative to large-capacity systems can induce special problems. The only work of which the author is aware [44] concerns digital SCPC and is summarized in this paper.

A. Angle-Modulated Wanted Signal

Although this paper will concentrate on wanted FDM/FM telephony and FM television signals, it is useful to begin with no assumptions and thus establish a perspective from which specific assumptions can be appreciated in relationship to the general context. Thus following the treatment by Pontano *et al.* [42], consider a wanted angle-modulated signal

$$w(t) = A \cos [\omega_1 t + \phi(t)] \tag{2}$$

and an interfering signal

$$\begin{aligned} i(t) &= R(t) \cos [\omega_2 t + \psi(t) + \mu] \\ &= \text{Re}\, \{u(t) \exp (j\omega_2 t + \mu)\} \end{aligned} \tag{3}$$

where $u(t) = R(t) \exp [j\psi(t)]$, and the angle μ is assumed to be uniformly distributed on $(0, 2\pi]$, reflecting the general assumption in this paper that the signals involved are independently generated.[3] Note that a single interferer is not necessarily assumed. Any bandpass signal, which can be the sum of several signals, can be represented by (3). In addition these signals may be filtered, e.g., by the front-end of the wanted receiver, for the form (3) still applies. In the latter case, however, it may not be simple to find an explicit expression for $R(t)$ and $\psi(t)$ in terms of the original modulating functions.

The sum signal can be manipulated to yield

$$s(t) = \text{Re}\, \{Aa(t) \exp \{j[\omega_1 t + \phi(t) + \lambda(t)]\}\} \tag{4}$$

where

$$a(t) \exp [j\lambda(t)] = 1 + z(t) \exp \{j[2\pi f_d t - \phi(t) + \mu]\} \tag{5}$$

$z(t) = u(t)/A$ and $f_d = f_2 - f_1$. An ideal phase detector[4] operating on (4) provides the baseband output

$$e_0(t) = \phi(t) + \lambda(t)$$

and hence the effect of interference is embodied in the excess angle $\lambda(t)$ which from (5) can be expressed as

$$\lambda(t) = \text{Im} \ln \{1 + z(t) \exp \{j[2\pi f_d t - \phi(t) + \mu]\}\} \tag{6}$$

where Im indicates imaginary part. There is a standard series representation for $\ln (1 + \gamma)$ valid for $|\gamma| < 1$. Applying this expansion to (6) yields

$$\lambda(t) = \sum_{m=1}^{\infty} \frac{(-1)^{m+1}}{mA^m} R^m(t) \sin [m(2\pi f_d t - \phi(t) + \psi(t) + \mu)] \tag{7}$$

where $|z(t)|$ must be less than 1, i.e., the peak interference en-

[1] Strictly speaking one should include thermal noise to the extent that it may affect the interference term. Studies of interference usually omit noise, and in fact it is normally appropriate to treat the noise additively as a separate contribution. Of course, for linear systems no approximation is involved.

[2] This could include interference from adjacent transmitters or multipath (ghosts), which is a form of correlated interference.

[3] The paper by Bennett *et al.* [21] treats the case in which $i(t)$ is an echo of $w(t)$ so that $\mu = 2\pi f_1 T$ where T is the delay. As $T \to \infty$ the echo should become independent of the signal, and hence the analysis in [21] should become equivalent to that presented here.

[4] Although real receivers have frequency selectivity, imperfect limiters, and other departures from ideal, an analysis taking account of these impairments does not materially add insight to the problem. The effect of some of these impairments is discussed in [22].

velope must be less than the desired signal (constant) envelope. This condition is almost always met in practice although it is occasionally possible for $|z(t)| > 1$. This case has been analyzed [37], [40] and it has been shown that the excursions of $|z(t)|$ produce impulsive or "click" noise.

In a formal sense, equation (7) contains all the possible information on the effect of interference. The evaluation of that effect, however, is another matter. For example, if $\phi(t)$ is a television signal, how does $\lambda(t)$ relate to picture quality? For some purposes the power spectral density (PSD) of $\lambda(t)$ is sufficient for this quantification. It was implied above that $\phi(t)$ and $z(t)$ are independent processes, both independent of μ. If we further stipulate that these two processes are wide-sense stationary and now define

$$E[\bar{u}(t)u(t+\tau)]^m = R_{mi}(\tau) \tag{8a}$$

$$E[\exp\{jm[\phi(t) - \phi(t+\tau)]\}] = R_{mw}(\tau) \tag{8b}$$

$$S_m(f) = \int_{-\infty}^{\infty} R_{mi}(\tau)R_{mw}(\tau)\exp(-j2\pi f\tau)\,d\tau \tag{8c}$$

one can show that the output PSD due to interference can be written as

$$S_\lambda(f) = \sum_{m=1}^{\infty} c_m\{S_m(f - f_d) + S_m(-f - f_d)\} \tag{9}$$

where $c_m = [4m^2A^m]^{-1}$, the overbar represents complex conjugate, and E is the expectation operator. From the definitions (8) it is also easy to show that

$$S_m(f) = S_{mi}(f) * S_{mw}(f) \tag{10}$$

where $*$ represents convolution and $S_{mw}(f)$, which is the Fourier transform of (8b), represents the PSD of the mth power of the complex envelope of $w(t)$ or, equivalently, it represents the low-pass equivalent spectrum of the "multiplied" signal $w_m(t) = A\cos\{m[\omega_1 t + \phi(t)]\}$. Similarly, $S_{mi}(f)$, which is the Fourier transform of the autocorrelation function (8a), represents the PSD of the mth power of the complex envelope of the interference.

When the interference is small, the first term of (9) is sufficient to characterize it and the output PSD is simply

$$S_\lambda(f) = \frac{1}{4A^2}\{S_i(f - f_d) * S_w(f) + S_i(-f - f_d) * S_w(-f)\} \tag{11}$$

which is the well-known formulation leading to a convolution of spectra. The general solution (9) is thus the sum of the convolution of successively higher order spectra. Although this is a formal "solution," the computation of these spectra for arbitrary signals is generally a difficult problem.[5] The situation is somewhat simplified when the small interference condition holds for then only the "first-order" spectra [see (11)] are required. The computation of spectra can be regarded as a separate discipline and hence will not be considered here (see, e.g., [45]–[50]). Fortunately, for the practically important case of multichannel FM telephony, there are systematic procedures for computing the desired spectra.

[5] Recently, Koh and Shimbo [43] have outlined a method for obtaining the convolution without explicitly computing the PSD of the wanted signal, when the interfering PSD is given in tabular form (e.g., from measurements) or can be assumed to have an otherwise specified behavior.

1) Interference Between FDM/FM Telephony Signals: Most of the studies published on the performance of (analog) signals in an interference environment have concerned FDM/FM telephony signals [20]–[43]. The most extensive treatments, which are basically generalizations of the preceeding work, are found in [35], [42]. As noted previously, the basic fact that makes it possible to obtain a solution is that for most purposes a multichannel FDM speech signal can be represented by a white Gaussian noise process. Since it is assumed that both signals are of the same type, the wanted signal is

$$w(t) = A\cos[\omega_1 t + \phi(t)] \tag{12}$$

and the interference is

$$i(t) = rA\cos[\omega_2 t + \psi(t) + \mu], \qquad r < 1 \tag{13}$$

where $\phi(t)$ and $\psi(t)$ are independent Gaussian processes. They are not necessarily white, since the formulation can just as easily account for pre-emphasis.

When the aggregate interference is small it is sufficient to deal with the single interferer (13) since the effect of a number of independent interferences is the linear superposition of their individual effects, as can be seen from (11). When $\phi(t)$ is a Gaussian random process, it can be shown [21] that (8b) reduces to

$$R_{mw}(\tau) = \exp\{-m^2[R_\phi(0) - R_\phi(\tau)]\} \tag{14}$$

hence the corresponding PSD is

$$S_{mw}(f) = \int_{-\infty}^{\infty} \exp\{-m^2[R_\phi(0) - R_\phi(\tau)]\}\exp(-j2\pi f\tau)\,d\tau \tag{15}$$

where $R_\phi(0)$ is the mean-square value (average power) of $\phi(t)$, $R_\phi(\tau)$ is its autocorrelation function. If we write

$$S_w(f) = \int_{-\infty}^{\infty} \exp[R_\phi(0) - R_\phi(\tau)]\exp(-j2\pi f\tau)\,d\tau \tag{16}$$

and notice that $R_{mw}(\tau)$ can be considered as the product of m^2 autocorrelation functions, it is clear that (15) can be expressed as

$$S_{mw}(f) = S_w(f) \overset{N}{*} S_w(f)$$

where $N = m^2$ and $\overset{N}{*}$ denotes N convolutions. Thus, the mth-order spectrum $S_{mw}(f)$ can be calculated from the first-order spectrum $S_w(f)$. The numerical evaluation of (11) thus reduces to the evaluation of $S_w(f)$ and $S_i(f)$, the first-order PSD of the interference.

In general, closed-form expressions cannot be obtained for $S_w(f)$. Therefore, results must ultimately be expressed in graphical form (see, e.g., [22], [30], and [42]). However, some special cases of interest, i.e., those in which both carriers are "wide band" (high modulation index), one of the carriers is wideband and the other unmodulated, or both carriers have identical parameters are amenable to analytical formulation.

a) Noise power ratio (NPR): It is convenient to first express the preceeding results in terms of the appropriate quality factor Q. When the baseband is modeled as a noise process, the common unit of quality is the noise power ratio (NPR), which is the ratio of the wanted signal power (modeled by noise) in a particular telephone channel to that produced by the interference (or any other undesired effects). For FM signals with pre-emphasis it can easily be shown [5] that

$$\text{NPR}\,(f_c) = \frac{(2\pi)^2 M_1^2 fm_1 b}{r^2(1-\epsilon_1)} \cdot \left\{ \int_{f_c-(b/2)}^{f_c+(b/2)} (2\pi f)^2 G_d(f) S_\lambda(f)\, df \right\}^{-1} \tag{17}$$

where

- f_c center frequency of channel under consideration
- b telephone channel bandwidth (3.1 kHz)
- fm_1 top baseband frequency of wanted signal
- M_1 rms modulation index of wanted multichannel baseband
- ϵ_1 ratio of lowest to highest frequency of multichannel baseband
- $G_d(f)$ power transfer function of de-emphasis network.

In (17) the interference-to-carrier ratio r^2 has been factored out so that $S_\lambda(f)$ there corresponds to a normalized spectrum. The factor $(2\pi f)^2$ is due to the fact that we are dealing with an FM receiver, whereas $S_\lambda(f)$ has previously been taken as the output of a phase receiver. Usually $S_\lambda(f)$ can be considered constant over $f_c \pm b/2$. Hence (17) can be simplified to yield

$$\text{NPR}\,(f) = \frac{M_1^2 fm_1}{(1-\epsilon_1) r^2 f^2 S_\lambda(f) G_d(f)} \tag{18}$$

where the subscript on f_c has been omitted.

Normally the particular channel of interest is the one for which NPR(f) is minimum. The location of that channel generally varies with the particular system assumptions made.

b) High-index FM: A case particularly amenable to analysis occurs when the rms modulation index of both signals is reasonably high (e.g., $\gtrsim 1$). In this case, the spectrum $S_w(f)$ is given approximately by

$$S_w(f) = \frac{\exp\,[-f^2/2\Delta f_1^2]}{\sqrt{2\pi}\Delta f_1}$$

A similar expression holds for the interference spectrum. The convolution is straightforward, yielding, after some manipulation, the formula

$$\text{NPR}\,(f) = \frac{2\sqrt{2\pi} M_1^2 M G_p(f)}{(1-\epsilon_1) r^2 u^2 \{\exp\,[-(u+v)^2/2M^2] + \exp\,[-(u-v)^2/2M^2]\}} \tag{19}$$

where

- $G_p(f) = G_d^{-1}(f)$
- $u = f/f_{m_1}$
- $v = f_d/f_{m_1}$
- $M^2 = M_1^2 + M_2'^2$
- $M_2' = M_2 f_{m_2}/f_{m_1}$
- f_{m_2} top baseband frequency of interfering signal
- M_2 rms modulation index of interfering multichannel baseband $\Delta f_2/f_{m_2}$.

Thus, for a given channel, the NPR due to interference reduces in this case to a three parameter problem involving M, M_1, and v.

Equation (19) lends itself to presentation in generalized form if NPR $(r/M_1)^2$ is plotted as a function of v with M as a parameter. For the case considered the top channel is either the worst one or close to it so that $u \simeq 1$. Hence universal family of curves such as that shown in Fig. 1, (taken from Pontano *et al.* [42]) can be plotted. It should be noted that a family of curves derived from (19) would have the same *appearance* as that in Fig. 1, but not the same values. Since (19) is based on the high-index assumption it will evidently not produce the correct value for small modulation indices. However, equation (19) and Fig. 1 coincide in the range of M_1, M_2 that satisfies the high-index assumption. The more general interpretation of (19) is discussed subsequently.

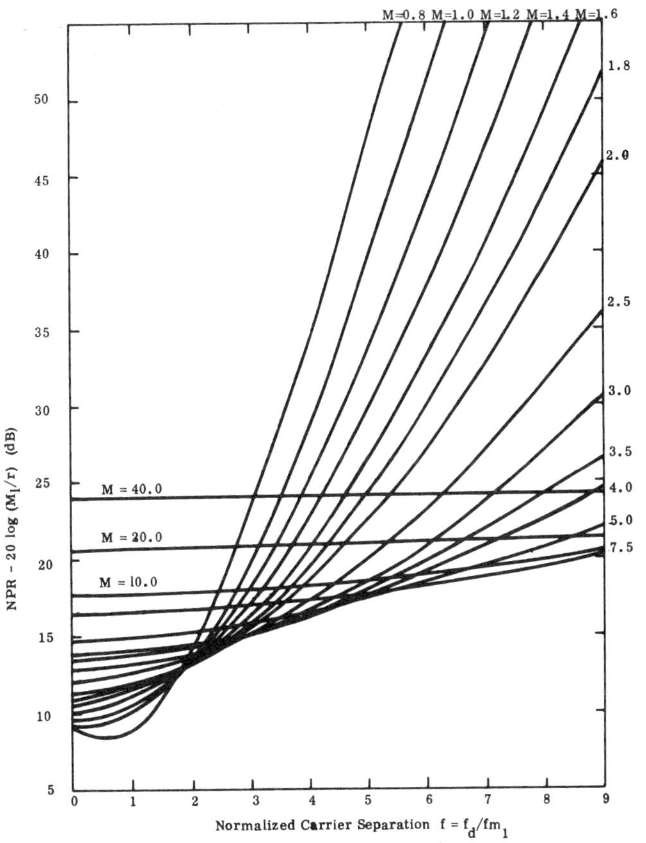

Fig. 1. Generalized curves for interference between FDM/FM telephony signals.

In the case of one high-index carrier and an interfering unmodulated carrier one obtains a form identical to (19) with $M_2 = 0$.

c) Identical cochannel carriers: In many practical cases the wanted and interfering carriers have identical or very similar parameters. For high-index cochannel carriers the NPR is given by (19) with $v = 0$, $M_1 = M_2 = M_2'$. For low modulation indices this expression no longer applies. However, it is still possible to obtain a single expression. Pontano *et al.* [42] have presented a curve of NPR as a function of rms modulation index for identical cochannel carriers (their Fig. 10). A simple expression fits this curve rather well for the top channel

$$\text{NPR} \simeq \frac{1 + 9.5 M_0^3}{r^2} \tag{20}$$

where M_0 is the common multichannel rms modulation index. This expression includes the effect of CCIR pre-emphasis.

d) Generalized interference curves: For arbitrary wanted and interfering signal parameters it is usually necessary to solve explicitly for the NPR using the process described previously. However, for a restricted but still useful range of parameters, generalized graphical results can be obtained. These results, originally presented in [42], appear herein as Fig. 1. This figure gives exact results for the worst-channel NPR when the wanted and interfering basebands are identical, but can also be applied to a larger set of cases under conditions described in [42].

It should be noted that computations are usually carried out under the assumption that the wanted and interfering carriers are fully loaded (which does not necessarily represent a worst case). This assumption underlies Fig. 1. At less than full load it is customary in satellite communication to employ energy dispersal, which is a technique whereby a waveform, with amplitude which varies inversely with the loading, is artificially added to the baseband. The idea is to maintain relatively constant deviation so as to avoid harmful spectral peaks. The standard dispersal waveform is a low-frequency (below baseband) triangular waveform. Thus at less than full load the carrier modulation is the sum of the dispersal waveform and the multichannel load. The corresponding RF spectrum is the convolution of the spectra produced by these two signals. With this modification, the general theory still applies.

e) Comparison with thermal noise: The effect of thermal noise, because of its importance as well as its intuitive value borne of long experience, provides a natural yardstick for assessing the effect of other disturbances. Therefore, a brief comparison of thermal noise and interference is useful. It can be easily shown [5] that in an FM receiver the worst channel NPR due to thermal noise (with CCIR pre-emphasis) is

$$\mathrm{NPR}_t = 2.5(C/N)M_0^2(W/f_{m_1}) \tag{21}$$

where C/N is the carrier-to-noise power ratio, W is the RF bandwidth, and M_0, f_{m_1} are as defined previously. Using Carson's rule for W one can show that

$$\mathrm{NPR}_t = 5(C/N)M_0^2(\alpha M_0 + 1) \tag{22}$$

where α is the peak-to-rms baseband ratio.

A useful reference case for interference is the identical cochannel case characterized by (20). It can be seen that, for equal C/N and C/I ratios, the same kind of dependence on M_0 obtains. For $\alpha = \sqrt{10}$, which is typical, comparison of (20) and (22) reveals that the susceptibility to interference is about 3 dB worse than that to thermal noise for $M_0 \gtrsim 0.75$, but as M_0 becomes very small thermal noise becomes more harmful than interference. Other interference conditions, of course, can alter the comparison. For instance, interleaved carrier operation [42] can improve the interference NPR by as much as 15 dB.

2) Interference into FM Television Signals: Unlike the telephony situation, there are no analytic theories to predict and explain the effect of interference into television signals. Although (7) can be used to describe the output of an FM receiver with interference, the question is how to relate it to a useful measure of performance. The problem lies with the difficulty of modeling the psychophysical process of vision. This is an extremely complex subject [51], [52], even when restricted to the relatively narrow context of quality assessment.

There are two basic methods of quality assessment, implicit evaluation by subjective testing and explicit (or objective) evaluation by meter or calculation [51], [53].

For establishing the quality of a transmission, the objective method is preferable. However, there does not seem to be an accepted procedure for establishing the quality of a picture in the presence of interference in terms that would be most useful in the present context, e.g., in units that are subjectively equivalent to the S/N ratio in the presence of thermal noise only. Ideally, such a procedure would set up a standardized calculation, e.g., the power spectrum in (9), appropriately weighted, using perhaps a "representative" spectral characteristic for the video process such as that discussed in [54]. However, the effects of interference have usually been evaluated by using the subjective method, i.e., the determination of carrier-to-interference ratios (protection ratios), necessary to produce a given subjective reaction for a given set of parameters.[6] Results of specific measurements are presented in the following subsection.

Before the protection ratios are discussed it may be instructive to look briefly at the information that is available "objectively" in the error signal itself, equation (7). Assuming the interference is small the error signal is given by

$$\lambda(t) = \frac{R(t)}{A} \sin\left[2\pi f_d t - \phi(t) + \psi(t) + \mu\right]. \tag{23}$$

If the interfering signal is another FM TV signal, cochannel with the desired carrier, with relative amplitude r, and further, if the modulation indices (and μ) are small, then

$$\lambda(t) \simeq r\left[\psi(t) - \phi(t) + \mu\right]. \tag{24}$$

Thus to a first approximation, the wanted signal amplitude is now reduced by the factor $(1 - r)$ and the interfering signal appears in a reduced but undistorted form at the receiver output.[7] This would be expected to give rise to the worst situation since the interference is "intelligible"; this is confirmed by the protection ratios listed in Table 1.

When (24) is approximately true, the baseband signal-to-interference (S/I) ratio can be expressed as

$$\frac{S}{I} = \left(\frac{1-r}{r}\right)^2 \frac{P_\phi}{P_\psi} \tag{25}$$

where $P_\phi = E[\phi^2(t)]$, $P_\psi = E[\psi^2(t)]$. Since P_ϕ, P_ψ are proportional to the square of the modulation index S/I would be expected to vary accordingly. In fact, Table I shows that the protection ratio is proportional to β^2, the wanted modulation index squared. A similar dependence on the interfering modulation index would not be expected (except where $\sin x = x$) since the peak limit $\sin x \leqslant 1$ would become significant. Nevertheless, some decrease in S/I, or increase in protection ratio, might be expected; this is also seen to be the case in Table I.

If $f_d \neq 0$, the baseband interference is spectrally shifted to f_d. This will tend to reduce the annoying effect of the interference for two reasons: the picture information is "chopped" by the subcarrier $\sin 2\pi f_d t$, which should reduce its intelligibility, and if f_d is sufficiently large some of the energy in $\lambda(t)$ should be filtered out. When f_d is larger than the wanted baseband bandwidth one would expect, for given wanted β^2, that the required protection ratio would increase with increasing modulation index of the interference. This follows from the fact that the angle-modulated interference term (23) would have a wider spectrum and hence extend further (to lower frequencies) into the passband of the receiver. These observations are also basically confirmed by the data in Table I.

[6] An objective approach is outlined in [55a] but this appears as yet in a rudimentary stage.

[7] The visibility of the interfering picture will depend on the standard used relative to the wanted signal, and on the relative synchronization in time.

TABLE I

MEASURED PROTECTION RATIOS FOR FM TV

Modulation Index		Carrier Separation (MHz)	Protection Ratio	
Wanted Signal	Unwanted Signal		μ (dB)[a]	σ (dB)
3	3	0	19.9 (19.95)	3.1
3	1	0	19.6 (19.95)	1.5
1	1	0	28.7 (29.5)	2.3
1	3	0	29.2 (29.5)	2.5
3	1	10	14.7 (14.75)	3.0
1	3	10	25.5 (25.25)	3.7
3	3	20	11.9 (12.07)	1.6
3	1	20	7.2 (7.22)	1.6
1	1	20	9.8 (9.5)	2.3
1	3	20	16.5 (16.2)	1.7
3	1	3.24	20.6 (18.53)	3.0

[a]The numbers in parentheses are obtained from equation (26).

It is not suggested that a quantitative theory of assessment can be developed from considerations similar to the foregoing. However these considerations may be useful, in providing qualitative explanations of the trends in measured protection ratios.

3) Protection Ratios: FM Television Interference: It might be presumed that protection ratios (PR) represent definitive statements concerning tolerable interference levels, stemming as they do from direct subjective evaluation. Unfortunately, this is not the case. First, the PR is a function of the permissible degree of impairment. Various scales have been devised [53], but the picture quality must ultimately be defined and each individual's inner interpretation of this definition will be different. The reaction to picture impairment also depends on such variables as the viewing distance, thermal noise level (which can mask other impairments), ambient illumination, scene material and system standard [53]. Thus, even though subjective testing may be the ultimate measure of quality, there is by no means unanimity on the protection ratios needed to achieve that quality. Nevertheless, it is useful to illustrate the possibilities with some specific measurements, which are shown in Table I [55a].[8] The measurements were fit to normal distributions whose means and standard deviations are shown in the table; the mean is taken as the PR. The latter correspond to "just perceptible" impairment, or better.

Even with subjective measurements available, as in Table I, there is still a need for "objective" interpretation or manipulation of these results to account for situations not specifically covered. This will always be the case since an experiment cannot be performed for every conceivable measurement of interest. The first extension of these results is for deviations and frequency offsets not covered in the table. This can be most expeditiously done by fitting an expression to the tabulated numbers, taking into account as much as possible apparent trends such as those discussed previously. The result of this process is the following equation [5][9]

$$\rho(M_w, M_i, f_d) = 29.5 - 20 \log M_w - f_d M_w^{-0.85} - 0.475(\mu^{-2.5}) \cdot (f_d^{0.645\mu}) \log \mu \quad (26)$$

where

ρ protection ratio (dB)
M_w wanted signal peak modulation index
M_i interfering signal peak modulation index
f_d difference in carrier frequency of wanted and interfering signals
$\mu = M_w/M_i$.

The values of ρ computed from (26) are shown in parentheses in Table I below the corresponding measured values; measured and computed values are seen to agree quite well.

Another situation which requires extrapolation from the measured data is that of multiple interfering signals. Here the concept of sensitivity factor [5] is useful. This is defined as

$$10 \log S(M_w, M_i, f_d) = \rho(M_w, M_w, 0) - \rho(M_w, M_i, f_d) \quad (27)$$

and represents the difference in PR between the case in which interfering signal has the same modulation index and carrier frequency as the wanted one, and that in which it has arbitrary parameters M_i, f_d. Thus it is assumed that an identical co-channel interfering signal reduced in power by $S(M_w, M_i, f_d)$ has the same subjective effect as the actual one. It is further conjectured that when the impairment is sufficiently small the aggregate interference power has the same effect as a single interfering signal of the same power. If C/X represents the total carrier-to-interference ratio (C/I) and $(C/X)_j$ the C/I for the jth interfering signal, it is postulated that

$$\left(\frac{C}{X}\right)^{-1} = \sum_{j=1}^{N} \left(\frac{C}{X}\right)_j^{-1} S^{-1}(M_w, M_j, f_j) \quad (28)$$

where M_j, f_j, are the modulation index and difference frequency of the jth interfering signal. To determine whether a PR requirement is met, (C/X) is then compared to $\rho(M_w, M_w, 0)$. Thus, this procedure provides at least an approximate way of determining if system requirements are satisfied.

Interference from FDM/FM telephony signals can be handled in completely analogous fashion. An empirical equation has been derived for the protection ratio [5] and one can also apply the idea of sensitivity factor to multiple FDM/FM interferers as well as to a mixture of telephony and TV interferers.

In summary, interference into FM television signals is treated empirically with perhaps some analytical extrapolation to account for cases that have not been measured. There is no theory which expresses the quality Q as a continuous function of the interference parameters. Instead the quality is assumed to be satisfactory if the interference level is below a prescribed figure. Thus, in practice, it is merely necessary to determine the interference level, which is a straightforward calculation.

[8] Specific conditions for the tests are given in [55a], which reports on other measurements as well; additional test results are discussed in [55b].

[9] A dependence on thermal SNR has also been proposed [55b].

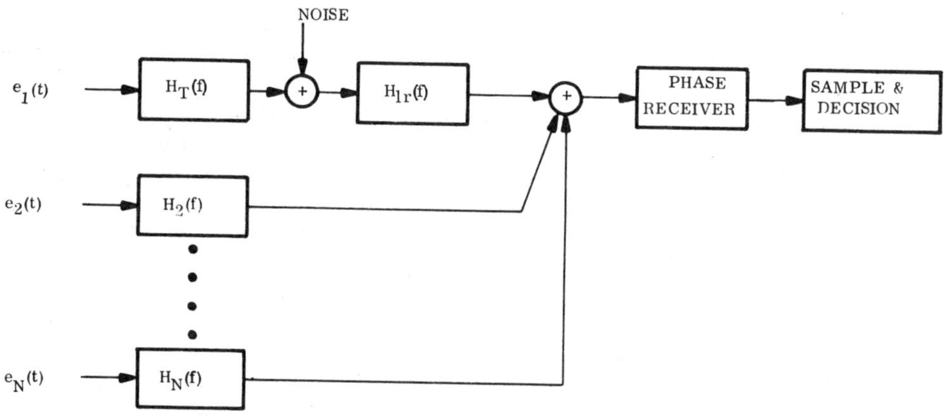

Fig. 2. Simplified block diagram of general (linear) interference problem.

III. Interference into Digital Signals

During the last few years, interest in the problem of digital signal performance[10] in an interference environment [44], [56]-[90] has increased significantly. Because of the inherent nonlinear nature of digital systems, there is no formal solution to this problem analogous to equation (9) for FDM/FM telephony signals. The performance of the digital system depends specifically and inextricably upon every detail of its design and environment, interference being only one aspect.

Thus effect of interference cannot be explicitly defined as can be done (in most instances) for analog systems. For an existing system design which produces a certain degradation D_1 (at a given BER) without interference and a degradation D_2 with interference, it is fair to say that $D_2 - D_1$ is degradation that can be attributed to the presence of interference with the given set of conditions. However, a problem arises when, as in this paper, it is necessary to conceptually isolate this effect. This problem, which is basically unresolvable, is of more than academic interest because in preliminary system planning it is desirable to make allocations for various sources of degradation. If many sources of impairment are included it becomes more difficult to extract one effect from another. In this case it is perhaps possible to show a generalized approach, but numerical results are necessarily *ad hoc*. The most difficult part of the problem, in fact, is the numerical evaluation itself. Thus, in the following it will be attempted to formulate the problem in somewhat general terms, indicate the variety of numerical approaches, including "exact" and bounding techniques, show an illustrative set of curves, and finally apply the theory to an orbit utilization problem. The discussion will consider only coherent phase-shift keyed (CPSK) systems as they currently represent the most practical implementations for communications satellites.

A. General Formulation, CPSK

Fig. 2 shows a general (linear) configuration.[11] The desired signal is given by

$$\nu_1(t) = \sum_{k=-\infty}^{\infty} p(t - kT) \cos [\omega_c t + a_k \theta(t - kT)] \quad (29)$$

where

- $p(t)$ possible amplitude shaping function,
- $\theta(t)$ possible phase pulse shaping function,
- T symbol interval,
- a_k kth phase symbol.

For an M-ary system, a_k is usually chosen in the set

$$a_k \in \left\{ \frac{2\pi n}{M} \right\}, \qquad n = 1, 2, \cdots, M \quad (30)$$

The ith interfering signal can be represented as

$$\nu_i(t) = R_i(t) \cos [\omega_i t + \psi_i(t) + \mu_i], \qquad i = 2, 3, \cdots, N \quad (31)$$

where

- $R_i(t)$ envelope of ith interfering signal,
- f_i ith interfering carrier frequency[12] $= \omega_i/2\pi$,
- $\psi_i(t)$ phase of the ith interfering signal,
- μ_i phase angle of ith interfering signal assumed to be independent of one another and of other functions, and uniformly distributed on $(0, 2\pi)$.

When the interfering signals are digital, they take on the form

$$\nu_i(t) = \sum_{k_i=-\infty}^{\infty} r_i p_i(t - k_i T_i - \tau_i) \cos [\omega_i t + a_{ik_i} \psi_i(t - k_i T_i - \tau_i) + \mu_i] \quad (32)$$

where T_i is the symbol duration, τ_i the relative time origin or symbol slip, r_i is the relative interference level, and a_{ik_i} is in a set similar, but not necessarily identical, to that in (30). Each signal is passed through a filter with low pass equivalent transfer function $H_i(f)$ and corresponding impulse response $h_i(t)$. Since all signals must pass through the same receiver, characterized by $H_{1r}(f)$ [or $h_{1r}(t)$], that transfer function is common to all $H_i(f)$. The receiver is assumed to be an ideal phase receiver[13] that, once per symbol, samples the instantaneous phase β and decides that $a_k = (2\pi n/M)$ was sent if

$$(2n - 1)\pi/M \leqslant \beta \leqslant (2n + 1)\pi/M. \quad (33)$$

Defining the complex envelopes

$$e_1(t) = \sum_{k=-\infty}^{\infty} p(t - kT) \exp [j a_k \theta(t - kT)] \quad (34)$$

[10] Performance in this context is defined to be the bit error probability P_e, or synonymously, bit-error rate (BER).

[11] This is as general a case as has been analyzed. The typical link in satellite communication is nonlinear, due primarily to TWT amplifiers. Interference in the nonlinear channel is an important problem that remains to be completely solved.

[12] An interfering signal is usually referred to as cochannel if $f_i = f_c$, and adjacent-channel if $f_i \neq f_c$.

[13] Receiver structure will be addressed subsequently.

$$e_i(t) = R_i(t) \exp\{j[\omega_i' t + \psi_i(t) + \mu_i]\}, \qquad i = 2, \cdots, N \tag{35}$$

where $\omega_i' = \omega_i - \omega_c$, the complex envelope at the input to the phase detector is

$$e(t) = n_c(t) + jn_s(t) + \sum_{i=1}^{N} e_i(t) * h_i(t) \tag{36}$$

where $n_c(t)$ and $n_s(t)$ are the in-phase and quadrature components of the noise at the output of the receiver filter. Suppose, without loss of generality, that the zeroth symbol a_0 for which the sampling occurs at $t = t_0$, is to be detected. The decision variable is

$$\beta_0 \triangleq \beta(t_0) = \tan^{-1} \frac{e_s(t_0)}{e_c(t_0)} \tag{37}$$

where $e_c(t)$ and $e_s(t)$ are the real and imaginary parts, respectively, of $e(t)$:

$$e_c(t_0) = s_{c0} + n_{c0} + x_{c0} + y_{c0} \tag{38}$$

$$e_s(t_0) = s_{s0} + n_{s0} + x_{s0} + y_{s0} \tag{39}$$

and the subscript 0 implies t_0, s is the "signal" component, x is the intersymbol interference component and y is the interference component. These functions are obtained from (34)–(36) as follows.

Let

$$C_k(t) \triangleq p(t - kT) \cos [a_k\theta(t - kT)]$$

$$S_k(t) \triangleq p(t - kT) \sin [a_k\theta(t - kT)]$$

$$A_i(t) \triangleq R_i(t) \cos [\omega_i' t + \psi_i(t) + \mu_i]$$

$$B_i(t) \triangleq R_i(t) \sin [\omega_i' t + \psi_i(t) + \mu_i]$$

$$h_i(t) = h_{ic}(t) + jh_{is}(t).$$

Then

$$s_c(t) = C_0(t) * h_{1c}(t) - S_0(t) * h_{1s}(t) \tag{40a}$$

$$s_s(t) = C_0(t) * h_{1s}(t) + S_0(t) * h_{1c}(t) \tag{40b}$$

$$x_c(t) = \sum_{k \neq 0} C_k(t) * h_{ic}(t) - S_k(t) * h_{1s}(t) \tag{40c}$$

$$x_s(t) = \sum_{k \neq 0} C_k(t) * h_{1c}(t) + S_k(t) * h_{1s}(t) \tag{40d}$$

$$y_c(t) = \sum_{i=2}^{N} A_i(t) * h_{1s}(t) + B_i(t) * h_{ic}(t) \tag{40e}$$

$$y_s(t) = \sum_{i=2}^{N} A_i(t) * h_{is}(t) + B_i(t) * h_{ic}(t). \tag{40f}$$

It can be seen that, if the interfering signals are digital, as in (32), equations (40e) and (40f) consist of $(N - 1)$ summations, each of which is identical in form to that in (40c) or (40d). Thus, the external inteference problem becomes formally identical to the intersymbol interference (ISI) problem, which is why many of the techniques recently developed for the latter problem have been used for the former.

To minimize the cumbersome notation, the following will deal primarily with binary wanted systems. More importantly, the character error probability of an M-ary system is bounded to within a factor of 2 by that of the binary system. The nature of the problem is the same, in any event. For a binary system the decision is incorrect if $e_c(t_0)$ has the wrong sign. Thus the probability of error is

$$P_2 = \tfrac{1}{2} \Pr [e_c(t_0) < 0/a_0 = 0] + \tfrac{1}{2} \Pr [e_c(t_0) > 0/a_0 = \pi].$$

In particular, when $n_c(t)$ is a Gaussian process with variance σ^2,

$$P_2 = \frac{1}{4} E\left\{\text{erfc}\left(\frac{s_1 + x_{c0} + y_{c0}}{\sqrt{2}\sigma}\right) + \text{erfc}\left(\frac{-s_2 - x_{c0} - y_{c0}}{\sqrt{2}\sigma}\right)\right\} \tag{41}$$

where

s_1	s_{c0} given $a_0 = 0$
s_2	s_{c0} given $a_0 = \pi$
erfc	complementary error function evaluated for given values of x_{c0}, y_{c0}
E	expectation operator applied to x_{c0}, y_{c0}

Thus the computational problem reduces to a conditional expectation of erfc. The different methods of attacking this problem constitute the essential difference among the various existing approaches.

The development leading to (41) was situation-specific, that is, based on the configuration of Fig. 2. It is possible to approach the problem abstractly [76] by determining the probability $\Pr(z \in D)$ that the n-dimensional real random vector z lies in a region D of n-dimensional space, where $z = S + N + \Sigma a_k$, and S, N, a_k are each n-dimensional. To reduce this formulation to the present case, let $n = 2$, $S \to (s_{c0}, s_{s0})$, $N \to (n_c, n_s)$ and $\Sigma a_k = (x_{c0} + y_{c0}, x_{s0} + y_{s0})$. However, since the formulation in [76] is not situation-specific it is applicable to a larger class of problems than that represented by Fig. 2.

1) Numerical Methods: The basic problem is the evaluation of a term of the form

$$I = E\left\{\text{erfc}\left(\frac{s + u}{\sqrt{2}\sigma}\right)\right\} \tag{42a}$$

or equivalently,

$$I = \int_{-\infty}^{\infty} \text{erfc}\left(\frac{s + u}{\sqrt{2}\sigma}\right) f(u)\, du \tag{42b}$$

where u denotes the random variables on which erfc is conditioned and $f(\cdot)$ is its probability density function (pdf). Still another equivalent form, using the characteristic function $\phi(v) = E[\exp(juv)]$, is

$$I = \frac{1}{2\pi} \int_{-\infty}^{0} \int_{-\infty}^{\infty} \exp[-(1/2)\sigma^2 v^2] \exp(-jv(x - s)). \tag{42c}$$

Except for minor variations there are basically four distinct approaches, using one of the forms of (42) as a point of departure, which have been used to evaluate P_2. Those approaches are briefly elaborated as follows:

a) Series method I: This approach, starting from (42a) is based on an expansion of erfc $(\cdot)$ and then taking the term-by-term expectation. The most complete discussions of this approach appear in [58] and [59], which consider the case of cochannel angle-modulated interference only (no filtering). Goldman [78] has also formulated the problem in similar terms, but within a more general framework.

b) Series method II: Another series approach [76] starts from the representation (48c) and consists of expanding the characteristic function, or the characteristic function times an exponential factor. Extensive discussions of the convergence properties of this method are given in [70], [72], and [76]. One technique used in this approach to speed convergence is to assign a fraction (Δ) of the interference power to the noise power and to treat a reduced interference source. However, no constructive way to find Δ is available so that trial and error method is necessary.

c) Gaussian quadrature method: This approach approximates the integral (42b) as

$$I \approx \sum_{j=1}^{L} w_j \operatorname{erfc}\left(\frac{s+u_j}{\sqrt{2}\sigma}\right)$$

where the set of pairs (w_j, u_j) is called a quadrature rule and can be derived from the first $2L + 1$ moments of u. One of the few comparisons of different approaches is given in [71], which indicates an apparent advantage for the quadrature method over the series method, at least for the examples chosen. This reference treats ISI only. When there is both ISI and adjacent channel interference, [74] and [89] treat virtually the same case, the former using the quadrature method and the latter the series method. The numerical results are very consistent, but no information is available to assess the relative computational cost.

d) Direct averaging method: This approach [11] is based on a per-letter evaluation of (42b). This requires an explicit representation of the pdf $f(u)$. In general this is practically impossible. In special cases, however, $f(u)$ is available, in particular when the interfering signals are all angle modulated and there is no filtering. For the case of a single interfering signal, Rosenbaum [57] has also used this approach in somewhat different terms, by directly integrating the pdf of the phase.

When u has certain properties, which are usually assured in practice, all of the methods converge. However, the pertinent expressions cannot be evaluated exactly. Truncations of one sort or another are unavoidable and for each method the consequent errors can be shown to be bounded and vanishing in the limit. However, simple and accurate expressions for the rate and manner of convergence seem practically impossible to obtain. Unfortunately no ranking of these methods is available since no general critical comparison seems to exist. It should also be noted that the "general" formulation has certain shortcomings. Many practical effects remain to be incorporated. Of chief importance perhaps is the influence of interference on carrier and timing recovery which appear [85], [90] to be seriously affected. Relatively little work has been done on the behavior of such circuits [66], [79], [88] in an interference environment, and certainly the effect of interference-dependent phase errors has not been analytically studied.[14] Another important omission is the effect of nonlinearities, although in some cases it may be possible to account for them separately [76]. The representation of the signal as in (33) is not typical of many modulators in which the waveform is sequence dependent. Practical impairments, such as modulator phase imbalance, phasor amplitude imbalance, static phase error, oscillator instability and data asymmetry are particularly important in high-data-rate systems and remain to be accounted for in a "general" analysis. It may well be that the most fruitful approach for such an "analysis" is computer simulation, and in fact there are simulations [86] that take account of most of the factors including interference.

[14] A notable exception is the extensive analysis of differentially coherent PSK systems [57], [59], [60], [64]. These analyses, however, have been for an idealized channel.

B. Bounding Approaches

Because of the substantial numerical complexity of the "exact" formulation just described, it is natural to seek bounds which are perhaps less accurate but easier to compute. Another reason for developing bounds is that in many cases the interference may not be known in such details as implied in the previous section, and a bound that uses only a few measurable interference parameters is useful. Two of these bounds are briefly described as follows.

1) Chernoff Bound: The point of departure for the Chernoff bound is the inequality $\text{Pe} \leqslant e^{g(\lambda)}$, where

$$g(\lambda) = \ln E(e^{\lambda V}) \tag{43}$$

λ is any positive integer, and V is the decision variable.

The tightness of the bound cannot be determined without studying specific cases. Equation (43) has been applied only to the unfiltered angle-modulated interference case [61], [68], [73]. For this case it is straightforward to show [68] that

$$g(\lambda) = \lambda(\lambda r^2/2 - 1) + \sum_{i=1}^{N} \ln I_0(r_i \lambda) \tag{44}$$

where I_0 is the zeroth-order modified Bessel function, $\frac{1}{2}r_i^2$ is the ith C/I ratio, r^2 is the noise variance, and N is the number of interferers. In the preceding equation, the desired envelope has been normalized to unity.

The bound is tightest for the value $\hat{\lambda}$ which minimizes (44). When N is small $\hat{\lambda}$ might perhaps be found numerically or graphically. In general this is tedious and prompts a search for alternate means. If inequalities are used for $I_0(\cdot)$ it can be shown that

$$g(\lambda) \leqslant \lambda \left\{ \frac{\lambda}{2}\left(r^2 + \frac{1}{2}\sigma_I^2\right) - (1 - v_I) \right\} \tag{45}$$

where I is any set of integers drawn from the set $(0, 1, \cdots, N)$ and

$$\sigma_I^2 = \sum_{i \in I} r_i^2, \quad v_I = \sum_{i \notin I} r_i.$$

Equation (45) involves a partition of the interference into a portion that effectively increases the noise power, and another that directly reduces (coherently cancels) the signal amplitude. The numerical problem is now reduced to finding the set I that minimizes (45). This is discussed in detail in [68] and [73] where several interesting results are given concerning the partition of I.

The bound is not very tight for large interference, but improves as the C/I ratio increases. Perhaps the greatest virtue of this approach is that it gives some insight into the behavior of the interference with respect to thermal noise, e.g., when the former can be treated in the same way as the latter. Incidentally, this approach also demonstrates [equation (45) with $v_I = 0$] that treating all of the interference as thermal noise gives a worst case for the conditions assumed.

2) Maximizing Distribution Bound: This bound [77], [82], [84] is based on the assumption the only information available about the interference envelope is its peak value V_i and its variance σ_i^2. The idea of the bound is then to find the distribution, among all distributions with parameters V_i and σ_i^2, that maximizes the probability of error. The approach is best elucidated by considering an undistorted binary CPSK wanted signal and a general interference $r(t) \cos [\omega_0 t + \psi(t)]$, where $r(t)$ and $\psi(t)$ are independent and the latter is uniformly distributed on $(0, 2\pi)$. The main result from [77] is that for all distributions $F(r_0)$ with the constraints $r_0 \leqslant V_i$ and $E(r_0^2) \leqslant \sigma_i^2$, the maximizing distribution is

$$F_m(r_0) = (1 - p)u(r_0) + pu(r_0 - V_i) \tag{46}$$

where $p = \sigma_i^2/V_i^2$ and $u(\cdot)$ is the unit step function. From the general development one has for this case

$$P_2 = \frac{1}{2} E\left\{\text{erfc}\left(\frac{s_0 + r_0 \cos \psi_0}{\sqrt{2}\sigma}\right)\right\}. \tag{47}$$

Expanding (47) in a power series and taking the expectation based on the distribution (46), one gets

$$P_2 \leqslant \frac{1}{2} \text{erfc}(s) + p \frac{e^{-s^2}}{\sqrt{\pi}} \sum_{n=1}^{\infty} H_{2n-1}(s) \frac{(V_i/\sqrt{2}\sigma)^n}{2^{2n}(n!)^2} \tag{48}$$

where $H_n(\cdot)$ is the Hermite polynomial of order n, and $s = s_0/\sqrt{2}\sigma$. Some restrictions on the maximum value of V_i are described in [77].

Equation (48) is not computationally trivial. However, it is a simplification in that it reduces all interference situations to a single expression. Moreover families of curves can be developed with the peak factor, $-10 \log p$, and the C/I ratio, $20 \log s_0/\sigma_i$, as parameters. These calculations need to be done only once if a sufficiently broad range of parameters is used, and values not specifically computed can be obtained by interpolating.

The accuracy of the bound is quite good for most purposes when the peak factor is relatively small and deteriorates as the peak factor increases. The method described in [77] has been expanded in [82], to permit tighter bounds to be obtained, and also to incorporate intersymbol interference. Recently another approach has been described [84] which starts from the same bounding theory, but which is apparently computationally simpler.

C. Receiver Structures

The studies outlined in the preceding sections have all assumed a phase receiver in which the decision variable is the instantaneous sampled phase. Although such a receiver is optimum in a noise-only environment, this is no longer the case when interference is present. It is therefore of interest to evaluate the relative performance of different receiver structures in the presence of interference, and to determine if possible the optimum receiver. This problem has so far received relatively little attention.

Wilmut and Campbell [65] have obtained the form of the optimum receiver for a binary PSK signal in the presence of an interfering sinusoid, and of a suboptimum receiver (called a "correlation" receiver), which is a correlation receiver whose reference reflects knowledge of the interference. It is shown that this receiver is superior to the standard correlation receiver (one which is matched to the desired signal only), but not greatly so when the interference is at the relatively low level expected in most cases.

In fact, Goldman has shown [87] that the standard correlation receiver is indeed optimum for a binary PSK signal when the interference consists of any number of independent cochannel angle-modulated interfering signals whose collective peak value is less than the desired signal envelope. It is further stated in [87] that, when the interference frequencies are unknown and the peak condition is satisfied, the standard correlation receiver is minimax in that it minimizes the worst performance over all receivers designed based on the assumption of a specific set of interference frequencies.

The results mentioned above were derived for an ideal transmission channel. For a real channel no structure has yet been advanced for an optimum receiver. It would be interesting to see whether such a structure can be obtained for a wide class of signals, systems, and interference, and whether the implied receiver complexity is warranted by the improvement.

D. Numerical Results

Unfortunately, unlike multichannel FM telephony (see Fig. 1), there cannot be a compact set of curves to describe the effect of interference to digital transmissions. A computer program is necessary to study the effect of interference in the particular cases of interest. However, many results (curves) have been published which lend a good deal of insight into digital system performance in the presence of interference. One representative set of results is presented in Fig. 3. This figure is basically [76, Fig. 4(b)] with some additional curves superimposed. In this figure the nonadditivity of degradations is evident. The curves also suggest that the degradation caused by interference is greater in an already degraded system, e.g., one with intersymbol interference, than in an ideal one. Some further general conclusions that can be drawn from this figure, and others, are as follows:

a) For a given total C/I ratio the degradation increases with the number of interfering signals.

b) When the C/I of the order of (or smaller than) the C/N ratio, the effect of interference is not as severe as an equal amount of thermal noise power. When the C/I is large compared to the C/N, it is a reasonable approximation to treat the interference power as an equal amount of noise power. The two observations suggest that in a multiple interference environment, small interferences can be added to the thermal noise, only larger ones must be treated as actual interference.

c) The vulnerability to interference increases substantially as the number of phases M increases.

d) The error probability of M-phase systems can be bounded by twice that of the binary P_e evaluated at values of C/I and C/N both increased by $-20 \log \sin (\pi/M)$.

e) In practice the error probability is substantially degraded by the effect of interference on carrier and timing recovery circuits and by channel nonlinearities. Further, when nonlinearities are present, treating the interference as additional thermal noise does not necessarily represent the worst case.

f) The situation discussed so far applies to "steady" interference. There may be cases in which transient effects are important. A case in point arises when a television carrier modulated only by a spreading signal (low-frequency triangular waveform) interferes with a digital SCPC signal. The latter is narrow band so that it effectively sees the full TV carrier a small fraction of the time. The signal performance in this case is not the same as that which would be predicted by using an interference power equal to the average power intercepted by the SCPC receiver.

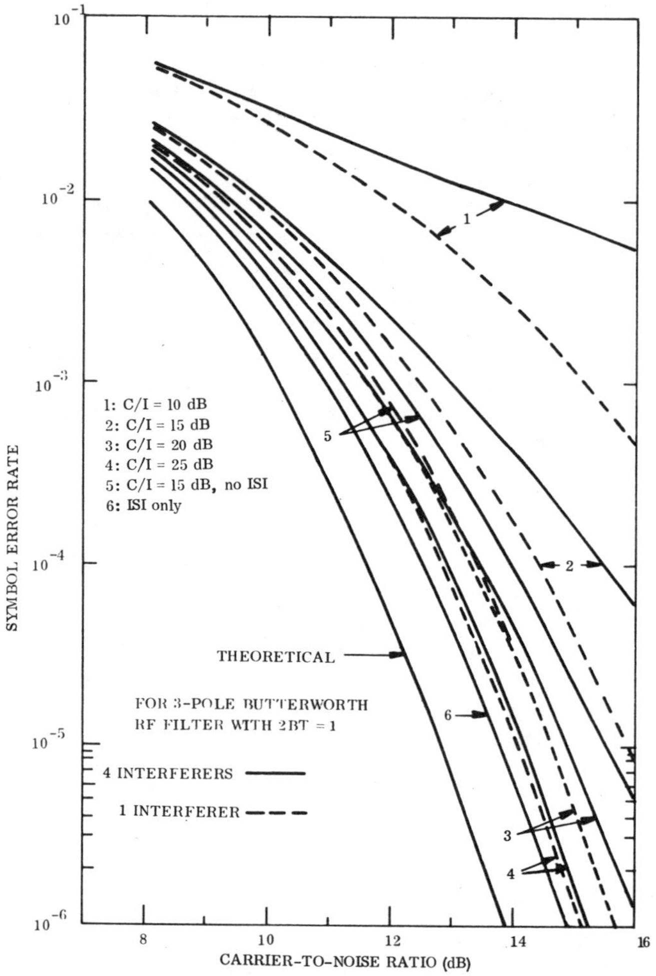

Fig. 3. Typical bit error rate curves for interference into 4-phase PSK system.

IV. Applications to Geostationary Orbit Utilization

The theory outlined in the preceeding sections has applications to a wide range of problems, including system design, filter specification, and transmission planning. This section will present some simple examples that indicate the fundamentally limiting role played by interference in the utilization of the geostationary orbit.

A. Orbit Utilization for Homogeneous FDM/FM Telephony Systems

In actuality the geostationary orbit is, and will be, occupied by satellites possessing a wide range of parameters, fulfilling a variety of functions. A general analysis of the behavior of any measure of orbit utilization for an arbitrary configuration of satellite networks is not feasible. Rather, computer programs must be used to examine specific cases and in fact such programs have been developed [5], [13]. If certain assumptions are made to reduce the dimensionality of the problem, then analysis can be applied. One particular set of assumptions that renders the problem amenable to formulation has been referred to as the homogeneous model [5]–[7]. The main assumptions are as follows:

a) a full ring of identical, equally spaced satellites;

b) all satellites having single-beam, earth-coverage antennas with identical polarizations;

c) clear weather conditions.

In addition, for the present example, a further stipulation is that all satellites transmit single-carrier per transponder FDM/FM telephony, with identical carrier frequencies, numbers of channels, and modulation indices. This model, while necessarily somewhat artificial, nevertheless provides a reasonable visualization of the tradeoffs involved.

1) Carrier-to-Interference Ratios: From assumptions a)–c) it follows that the C/I ratio for a given pair of wanted and interfering signals is approximately

$$\left(\frac{C}{I}\right)_j = \frac{G_{0d}}{G(\Delta\theta_j)}\left(\frac{1}{1+r_j}\right)$$

where G_{0d} is the on-axis gain of the earth station antenna on the down-link, $G(\Delta\theta_j)$ is the off-axis gain corresponding to the angular separation $\Delta\theta_j$ of the jth interfering signal, and r_j is the ratio of down-link to up-link C/I. It is convenient to use a sidelobe gain function of the form $G(\Delta\theta) = A/(\Delta\theta)^k$, which is used, for example, by the CCIR with $A = 10^{3.2}$ and $k = 2.5$. With the relationship $G_{0d} = \eta(\pi D/\lambda)^2$ where η is the antenna efficiency and D/λ is its diameter-to-wavelength ratio, the total C/I can be expressed as

$$(C/I)^{-1} = \sum (C/I)_j^{-1} = \frac{2A(1+r_0)}{\eta(\pi D/\lambda)^2(\Delta\theta)^k}\sum_{j=1}^{N} j^{-k}$$

where $2N$ is the number of interfering signals summetrically disposed about the wanted satellite at angular distances $\pm j\Delta\theta$, and because of the initial assumptions, $r_j = r_0$, for all j. For future reference the summation indicated is denoted $S(k)$.

2) Interference Noise: For present purposes it is convenient to define the signal quality in terms of baseband noise N which is related to an equivalent measure, the test-tone-to-noise (S/N) ratio, through

$$N = \frac{10^9}{S/N}$$

where N is measured in picowatts (at a point of zero relative level). Further for the jth interfering signal,

$$\left(\frac{S}{N}\right)_j = \left(\frac{C}{I}\right)_j R_j$$

where R_j, the receiver "transfer characteristic," depends on the particulars of wanted and interfering signals. Since all transmissions are identical; $R_j \triangleq R_0$, all j, where

$$R_0 \approx 76(1 + 9.5M_0^3), \qquad n \geqslant 240 \text{ channels}$$
$$\approx 3n^{0.6}(1 + 9.5M_0^3), \qquad 12 \leqslant n < 240,$$

which includes a 4-dB factor due to pre-emphasis and 2.5 dB of psophometric weighting.

3) Thermal Noise: Although interference noise is of interest here, thermal noise must also be briefly considered because the tradeoff between thermal and interference noise, for a given overall objective, affects the orbit utilization. It can easily be shown [5] that the test-tone-to-thermal noise ratio is given by

$$\left(\frac{S}{N}\right)_t = \left(\frac{C}{N_d}\right)\left(\frac{1}{1+r}\right)R_t$$

where $(C/N)_d$ is the down-link C/N ratio, r is the ratio of down-link to up-link C/N ratio and the thermal receiver trans-

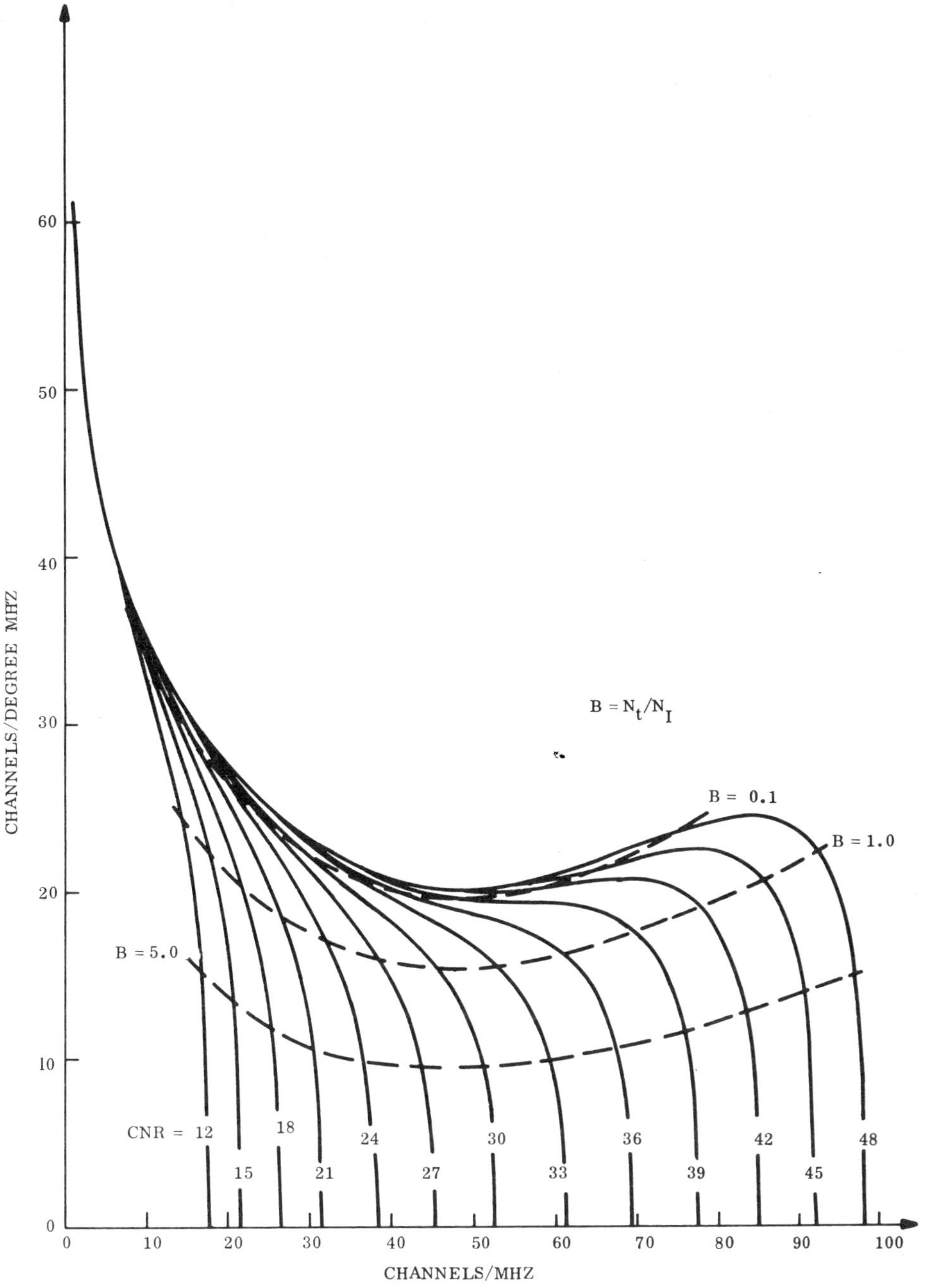

Fig. 4. Orbit/spectrum utilization measure for FDM/FM telephony signals.

fer characteristic (including pre-emphasis and weighting) is given by

$$R_t = 380(\alpha M_0^3 + M_0^2), \qquad n \geqslant 240 \text{ channels}$$
$$= 15n^{0.6}(\alpha M_0^3 + M_0^2), \qquad 12 \leqslant n < 240 \text{ channels}$$

where α^2 is the peak-to-average baseband power ratio.

4) Measure of Orbit Utilization: It is usually valid to find the signal quality by adding the separate contributions. Thus, combining the previous expressions, the total baseband noise N_T takes the form

$$N_T = 10^9 \left\{ \frac{1 + r}{(C/N)_d R_t} + \frac{2A(1 + r_0)S(k)}{\eta(\pi D/\lambda)^2(\Delta\theta)^k R_0} \right\}.$$

Normally N_T is fixed as a performance objective and there is a tradeoff among the other parameters within that constraint. To conduct a meaningful tradeoff bandwidth occupancy must also be considered because the radio spectrum is itself a resource and communications services inevitably exist within an allocation constraint. Thus it is useful to define a measure of orbit (and spectrum) utilization, $\ddot{\eta}$, as channels/degree/MHz. It is instructive to plot this quantity versus $\dot{\eta}$, the number of channels/MHz, which is a measure of individual satellite utilization. From the Carson's rule bandwidth $W = 8400n(\alpha M_0 + 1)$, one has

$$\dot{\eta} = \frac{119}{\alpha M_0 + 1}$$

and it follows that

$$\ddot{\eta} = \frac{\dot{\eta}}{\Delta\theta} = \frac{119}{(\alpha M_0 + 1)} \left\{ \left(\frac{N_T}{10^9} - \frac{1+r}{(C/N)_d R_t} \right) \cdot \frac{R_0 \eta (\pi D/\lambda)^2}{2A(1+r_0)S(k)} \right\}^{-1/k}. \quad (49)$$

For the particular case $r = r_0 = 0$ (nonzero values can be interpretated as modified C/N or D/λ), using the CCIR antenna pattern, with $\eta = 0.5, D/\lambda = 300, \alpha = \sqrt{10}$ and $N_T = 7500$ pW, equation (49) is plotted in Fig. 4. The region of operation on this figure is constrainted by the permissible interference noise, which is deducible from the loci of $B = N_t/N_I$ shown. It can be seen that, while a reasonable amount of interference noise must be allowed to reasonably utilize the resource, there is a diminishing return in that respect as the interference noise is allowed to increase.

B. Satellite Spacing for Homogeneous CPSK Systems

Another homogeneous model will now be considered with assumptions a-c as before. In this case however all satellites transmit CPSK carriers with identical parameters. In addition an ideal phase receiver that does not distort the input waveform is assumed. Thus, the degradation is purely due to interference, which yields a clearer picture of its limiting effect than if other impairments were present. The desired signal is

$$s(t) = \sqrt{2S} \cos [\omega_c t + \theta]$$

where θ is one of M phases; the received interference is

$$I(t) = \sum_{j=1}^{k} \sqrt{2I_j} \cos [\omega_j t + \theta_j + \mu_j] \triangleq \sum \sqrt{2I_j} \cos \lambda_j$$

where μ_j is uniformly distributed on $(0, 2\pi)$ and all θ_j, μ_j are mutually independent. Then, one can show that

$$P_{eM} = E\{\text{erfc}\,(p \sin \pi/M + p\eta)\} \quad (50)$$

where P_{eM} is the symbol error probability, exact for $M = 2$ and upper bounded by at most a factor of 2 for $M > 2$, p^2 is the C/N ratio, and $\eta = \Sigma R_j \cos \lambda_j$, with $R_j = \sqrt{I_j/S}$. For the downlink interference from the jth satellite east or west of the desired one, one gets

$$R_j \approx \frac{18}{|j\Delta\theta|^{1.25}(D/\lambda)}. \quad (51)$$

If (51) is substituted into (50) P_{eM} can be explicitly related to the satellite spacing, $\Delta\theta$. For convenience Fig. 5 is a reproduction of one such curve of error probability for binary PSK with $\Delta\theta$ as a parameter. Other curves for $M > 2$ can be found in [5] and [11]. Although antenna sidelobes which are better than those in the CCIR pattern can be obtained, other idealizing assumptions have been made (e.g., phase coherence) so that Fig. 5 can be considered to represent the minimum spacings possible for a given BER and C/N.

The character of these curves depends on compliance to the constraint $\Sigma R_j = [18S(k)(\lambda/D)/\Delta\theta^{k/2}] \leqslant \sin(\pi/M)$, for otherwise the error probability has an interference-induced floor. As discussed earlier, contributions from different impairments cannot actually be separated. However, for assigning an interference noise "budget," Fig. 5 suggests that a reasonable procedure is to permit a given increase in C/N over the interference-free situation, which in general would also reflect other impairments.

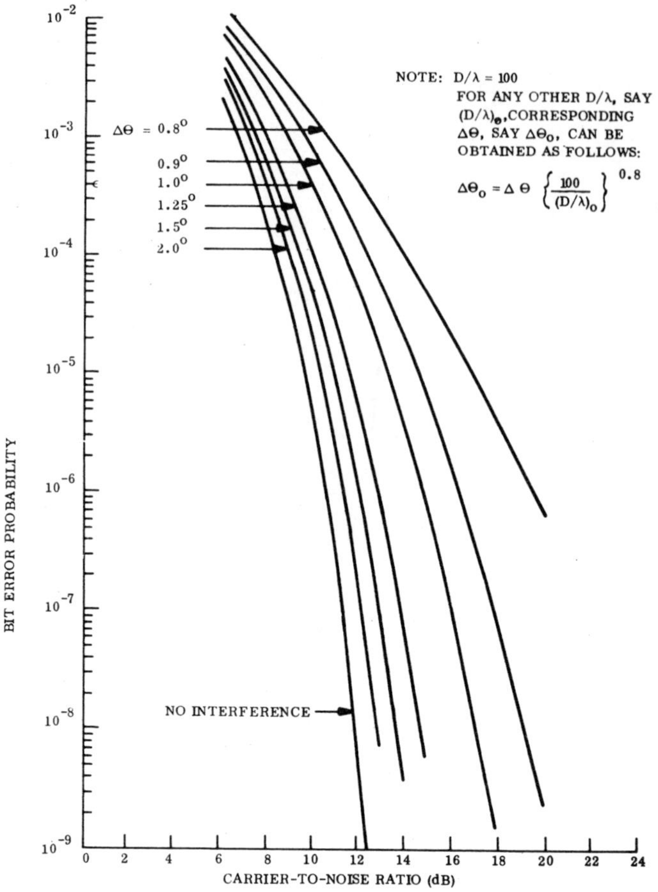

Fig. 5. Influence of satellite spacing on bit error rate of binary PSK signals.

V. Conclusions

Geostationary satellite networks, by their nature, operate in an interference environment. This mutual interference governs the angular spacing that must be maintained between satellites and thus ultimately controls the utilization of the geostationary orbit. International agreements currently specify the maximum permissible interference. In order to make the fullest use of this allowance, and hence maximize orbit utilization, proper methods must be used to relate the parameters of wanted and interfering signals to the appropriate measure of degradation induced by the latter. A fairly extensive body of theory concerning interference already exists and has been outlined as it applies to three classes of signals. However, many problems of importance still need to be solved. Some of these have been indicated earlier. As new types of signals gain importance their interaction with existing ones will need to be studied. A case in point mentioned earlier in SCPC, whose effect on and from other signals, e.g., FM-TV, requires analysis.

It might be presumed that maximum utilization of the orbit will be achieved for signal types that are most impervious to interference. The case is not so simple for it is not easy to make a comparison for which all things are otherwise equal. Bandwidth occupancy, for example, should be a parameter in the comparison. But in practice bandwidth is determined by the filtering which itself affects signal quality. And equal quality must surely be assumed in a meaningful comparison. This aspect becomes even more "fuzzy" when comparing analog and digital transmission of the same signal because quantization noise must be included in the final expression for quality. Thus, while digital transmission is generally admitted to be robust to interference, it has in fact been shown [5] that, depending upon the specifics of the case, it is sometimes inferior to FM and sometimes superior insofar as orbit utilization (measured in channels/degree/MHz) is concerned. It is difficult to generalize in this context. Rather one must usually study specific cases using the appropriate theoretical tools, much in the manner of the examples considered earlier.

References

[1] W. E. Bradley, "Communications strategy of geostationary orbit," *Astronaut. Aeronaut.*, vol. 6, no. 4, pp. 35–41, Apr. 1968.

[2] J. L. Hult *et al.*, "The technology potentials for satellite spacing and frequency sharing," RAND Corp, Santa Monica, CA, Memo. RM-5785-NASA, Oct. 1968.

[3] J. K. S. Jowett and A. K. Jefferis, "Ultimate communications capacity of the geostationary-satellite," *Proc. Inst. Elec. Eng.*, vol. 116, pp. 1304–1310, August 1969.

[4] J. C. Fuenzalida, "A comparative study of the utilization of the geostationary orbit," in *Proc. INTELSAT/IEE Conf. Digital Satellite Communication*, pp. 213–225, Nov. 1969.

[5] M. C. Jeruchim *et al.*, "Orbit/spectrum utilization study," General Electric, Valley Forge, Pa., vol. 1, doc. 69SD4270, 15 May 1969; vol. 2, doc. 69SD4348, 12 September 1969; vol. 3, doc. 70SD4246, 30 June 1970; vol. 4, doc. 70SD4293, 31 December 1970 (available through NTIS).

[6] D. M. Jansky and M. C. Jeruchim, "A technical basis for communication satellites to share the geostationary orbit," in *Proc. AIAA 3rd Communications Satellite Systems Conf.* (Los Angeles, CA), Paper 70-441, Apr. 1970.

[7] ——, "Technical factors and criteria affecting geostationary orbit utilization," in *Communications Satellite Systems for the 70's: Systems*, N. E. Feldman and C. N. Kelly, Ed. Cambridge, MA: M.I.T. Press, 1971.

[8] J. L. Hult and E. E. Reinhart, "Satellite spacing and frequency sharing for communication and broadcast services," *Proc. IEEE*, vol. 59, pp. 118–128, Feb. 1971.

[9] M. C. Jeruchim, "Orbit utilization," presented at Electronics and Aerospace Systems Conv., Washington, DC, Oct. 1971.

[10] D. A. Kane and M. C. Jeruchim, "Orbital and frequency sharing between broadcasting-satellite service and the fixed-satellite service," presented at Int. Conf. on Communications, Philadelphia, PA, June 19–21, 1972.

[11] M. C. Jeruchim and F. E. Lilley, "Spacing limitations of geostationary satellites using multilevel PSK signals," *IEEE Trans. Commun.*, vol. COM-20, Oct. 1972.

[12] M. Matsushita and J. Majima, "Efficient utilization of orbit/frequency for satellite broadcasting," *IEEE Trans. Aerosp. Electron. Syst.*, vol. AES-9, pp. 2–10, Jan. 1973.

[13] E. E. Reinhart, "Orbit-spectrum sharing between the fixed-satellite and broadcasting services with applications to 12 GHz domestic system," RAND Corp., Santa Monica, CA, R-1463-NASA, May 1974.

[14] International Radio Consultative Committee, *Documents of the XIIIth Plenary Assembly*, vol. IV, Report 453-1, Geneva, Switzerland, 1974.

[15] P. H. Sawitz, "Spectrum-orbit utilization: An overview," presented at National Telecommunications Conf., New Orleans, LA, pp. 43-1, 43-7. Dec. 1–3, 1975

[16] D. J. Withers, "Effective utilization of the geostationary orbit for satellite communications," *Proc. IEEE*, this issue, pp. 308–317.

[17] S. Goldman, *Frequency Analysis, Modulation, and Noise.* New York: McGraw-Hill Book Co., 1948.

[18] J. Granlund, "Interference in frequency-modulation reception," M.I.T. Research Lab. Electronics, Cambridge, MA, Tech. Rep., 42, Jan. 20, 1949.

[19] E. J. Baghdady, *Lectures on Communication System Theory.* New York: McGraw-Hill, 1961, pp. 483–508.

[20] S. Hayashi, "On the interference characteristics of the phase modulation receiver for multiplex transmission," *Proc. IECE (Jap.)*, vol. 35, pp. 522–528, Nov. 1952.

[21] W. R. Bennett, H. E. Curtis, and S. O. Rice, "Interchannel interference in FM and PM systems under noise loading conditions," *Bell Syst. Tech. J.*, vol. 34, pp. 601–636, May 1955.

[22] R. G. Medhurst, E. M. Hicks, and W. Grossett, "Distortion in frequency-division-multiplex FM systems due to an interfering carrier," *Proc. Inst. Elec. Eng.*, vol. 105B, pp. 282–292, May 1958.

[23] H. E. Curtis, "Radio frequency interference considerations in the TD-2 radio relay system," *Bell Syst. Tech. J.*, vol. 39, pp. 369–387, Mar. 1960.

[24] R. Hamer, "Radio-frequency interference in multi-channel telephony FM radio systems," *Proc. Inst. Elec. Eng.*, vol. 108B, pp. 75–89, January 1961.

[25] R. G. Medhurst, "RF Spectra and interfering carrier distortion in FM trunk radio systems with low modulation ratios," *IRE Trans. Commun. Syst.*, vol. CS-9, pp. 107–115, June 1961.

[26] S. Y. Boroditch, "Calculating the permissible magnitude of radio interference in multi-channel radio relay systems," *Electrosviaz*, vol. 1, no. 1, pp. 13–24, Jan. 1962.

[27] R. G. Medhurst, "FM interfering carrier distortion: general formula," *Proc. Inst. Elec. Eng.*, vol. 109B, pp. 149–150, Mar. 1962.

[28] H. E. Curtis, "Interference between satellite communication systems and common carrier surface systems," *Bell Syst. Tech. J.*, vol. 41, pp. 921–943, May 1962.

[29] R. G. Medhurst and J. H. Roberts, "Expected interference levels due to interactions between line-of-sight radio relay systems and broadband satellite systems," *Proc. Inst. Elec. Eng.*, vol. 111, no. 3, pp. 519–523, Mar. 1964.

[30] P. B. Johns, "Graphical method for the determination of interference transfer factors between interfering frequency-modulated multichannel telephony systems," *Electron. Lett.*, vol. 2, no. 3, pp. 84–86, Mar. 1966.

[31] P. B. Johns, "Interference between terrestrial line-of-sight radio-relay systems and communication-satellite systems," *Electron. Lett.*, vol. 2, no. 5, pp. 177–178, May 1966.

[32] R. J. Westcott, "Investigation of multiple FM/FDM carriers through a satellite TWT operating near to saturation," *Proc. Inst. Elec. Eng.*, vol. 114, no. 6, pp. 726–740, June 1967.

[33] H. W. Evans, "Technical background, AT&T domestic satellite proposal," presented at AIAA 2nd Communications Satellite Systems Conf., San Francisco, CA, Apr. 8–10, 1968, AIAA Paper 68-411.

[34] E. E. Reinhart, "Radio relay system performance in an interference environment," RAND Corp., Santa Monica, CA, Memo. RM-5786-NASA, October 1968.

[35] V. K. Prabhu and L. H. Enloe, "Interchannel interference considerations in angle-modulated systems," *Bell Syst. Tech. J.*, vol. 48, pp. 2333–2358, Sept. 1969.

[36] M. C. Jeruchim, "Interference in angle-modulated systems with predetection filtering," *IEEE Trans. Commun. Technol.*, vol. COM-19, pp. 723–726, Oct. 1971.

[37] M. Wachs, "Analysis of adjacent channel interference in a multi-carrier FM communication system," *COMSAT Tech. Rev.*, vol. 1, no. 1, pp. 139–170, Fall 1971.

[38] N. K. M. Chitre and J. C. Fuenzalida, "Baseband distortion caused by intermodulation in multicarrier FM systems," *COMSAT Tech. Rev.*, vol. 2, no. 1, pp. 147–172, Spring 1972.

[39] R. K. Khatri and J. E. Wilkees, "Convolution noise and distortion in FDM/FM systems," presented at Int. Conf. Communications, Philadelphia, PA, June 19–21, 1972.

[40] T. Furuya, A. Fujii, and K. Tezuka, "Impulse noise in FM receivers in the presence of adjacent channel interference and thermal noise," presented at Int. Conf. Communications, Philadelphia, PA, June 19–21, 1972.

[41] M. Horstein, "Satellite adjacent-channel interference due to multicarrier transponder operation," presented at Int. Conf. Communications, Philadelphia, PA, June 19–21, 1972.

[42] B. A. Pontano, J. C. Fuenzalida, and N. K. M. Chitre, "Interference into angle-modulated systems carrying multichannel telephony signals," *IEEE Trans. Commun.*, vol. COM-21, pp. 714–727, June 1973.

[43] E. K. Koh and O. Shimbo "Computation of interference into angle-modulated systems carrying multichannel telephone signals," *IEEE Trans. Commun.*, vol. COM-24, pp. 259–263, Feb. 1976.

[44] M. R. Wachs and D. Kurjan, "Interference into the SPADE system by a cochannel TV signal," *COMSAT Tech. Rev.*, vol. 6, no. 2, Fall 1976.

[45] D. Middleton, *An Introduction to Statistical Communication Theory.* New York: McGraw-Hill, 1960.

[46] N. Abramson, "Bandwidth and spectra of phase and frequency-modulated waves," *IEEE Trans. Commun. Syst.*, vol. CS-11, pp. 407–414, Dec. 1963.

[47] H. E. Rowe, *Signals and Noise in Communication Systems.* New York: Van Nostrand, 1965.

[48] C. C. Ferris, "Spectral characteristics of FDM-FM signals," *IEEE Trans. Commun. Technol.*, vol. COM-16, pp. 233–238, Apr. 1968.

[49] L. Lundquist, "Digital PM spectra by transform techniques," *Bell Syst. Tech. J.*, vol. 48, no. 2, pp. 397–411, Feb. 1969.

[50] H. E. Rowe and V. K. Prabhu, "Power spectrum of a digital, frequency-modulation signal," *Bell Syst. Tech. J.*, vol. 54, no. 6, pp. 1095–1125, July–Aug. 1975.

[51] G. Biernson, "A feedback-control model of human vision," *Proc. IEEE*, vol. 54, pp. 858–872, June 1968.

[52] Z. L. Budrikis, "Visual fidelity criterion and modeling," *Proc.*

IEEE, vol. 60, pp. 771–779, July 1972.
[53] D. E. Pearson, "Methods for scaling television picture quality: A survey," presented at Symp. Picture Bandwidth Compression, M.I.T., Cambridge, MA, Apr. 2–4, 1969.
[54] L. E. Franks, "A model for the random video process," *Bell Syst. Tech. J.*, vol. 45, pp. 609–630, Apr. 1966.
[55] a) International Radio Consultative Committee, *Documents of the XIIIth Plenary Assembly*, Report 449-1, Geneva, Switzerland, 1974.
b) ——, Report 634.
[56] M. A. Koerner, "Effect of interference on a binary communication channel using known signals," Jet Propulsion Laboratory, Pasadena, CA, Tech. Rep. 32-1281, Dec. 1, 1968.
[57] A. S. Rosenbaum, "PSK error performance Gaussian noise and interference," *Bell Syst. Tech. J.*, vol. 48, no. 2, pp. 413–442, Feb. 1969.
[58] V. K. Prabhu, "Error rate considerations for coherent phase-shift keyed systems with co-channel interference," *Bell Syst. Tech. J.*, vol. 48, no. 3, pp. 743–767, Mar. 1969.
[59] A. S. Rosenbaum, "Binary PSK error probabilities with multiple cochannel interferences," *IEEE Trans. Commun. Technol.*, vol. COM-18, pp. 241–253, June 1970.
[60] ——, "Error performance of multiphase DPSK with noise and interference," *IEEE Trans. Commun. Technol.*, vol. COM-18, pp. 821–824, Dec. 1970.
[61] V. K. Prabhu, "Error-probability upper bound for coherently detected PSK signals with cochannel interference," *Electron. Lett.*, vol. 5, no. 16, pp. 383–386, Aug. 1969.
[62] S. A. Cohen, "Interference effects of pseudo-random frequency-hopping signals," *IEEE Trans. Aeronaut. Electron. Syst.*, vol. AES-7, pp. 279–287, Mar. 1971.
[63] H. A. Bodner, "Error rate considerations of single sideband amplitude modulation systems with cochannel interference," presented at Int. Conf. Communications, June 14–16, 1971, Montreal, Canada.
[64] J. Goldman, "Multiple error-performance of PSK systems with cochannel interference and noise," *IEEE Trans. Commun. Technol.*, vol. COM-19, pp. 420–430, Aug. 1971.
[65] M. J. Wilmut and L. L. Campbell, "Signal detection in the presence of cochannel interference and noise," *IEEE Trans. Commun.*, vol. COM-20, pp. 1153–1158, Dec. 1972.
[66] R. E. Ziemer, "Perturbation analysis of the effect of CW interference in Costas loops," in *Proc. Nat. Telecommunications Conf.* (Houston, TX), pp. 20G-1–20G-6, Dec. 1972.
[67] "SSB interference earmarked as most damaging to digital communications systems," *Commun. Design*, Oct. 1972.
[68] J. M. Aein, "On the effects of undesired signal interference to a coherent digital carrier," Inst. for Defense Analyses, Paper P-812, Feb. 1972.
[69] C. Colavito and M. Sant'Agostino, "Multiple co-channel and interchannel interference effects in binary and quatermary PSK radio system," presented at Int. Conf. Communications, Philadelphia, PA, June 19–21, 1972.
[70] O. Shimbo, R. J. Fang, and M. Celebiler, "Performance of M-ary PSK systems in Gaussian noise and intersymbol interference," *IEEE Trans. Information Theory*, vol. IT-19, pp. 44–58, Jan. 1973.
[71] S. Benedetto, G. Devincentiis, and A. Luvison, "Error probability in the presence of intersymbol interference and additive noise for multilevel digital signals," *IEEE Trans. Commun.*, vol. COM-21, pp. 181–190, Mar. 1973.
[72] O. Shimbo and R. Fang, "Effects of cochannel interference and Gaussian noise in *M*-ary PSK systems," *COMSAT Tech. Rev.*, vol. 3, no. 1, pp. 183–207, Spring 1973.
[73] J. M. Aein and R. D. Turner, "Effect of cochannel interference on CPSK carriers," *IEEE Trans. Commun.*, vol. COM-21, pp. 783–790, July 1973.
[74] S. Benedetto, E. Biglieri, and V. Castellani, "Combined effects of intersymbol, interchannel, and co-channel interference in *M*-ary CPSK systems," *IEEE Trans. Commun.*, vol. COM-21, pp. 997–1008, Sept. 1973.
[75] C. Colavito and M. Sant'Agostino, "Binary and quaternary PSK radio systems in a multiple-interference environment," *IEEE Trans. Commun.*, vol. COM-21, pp. 1056–1067, Sept. 1973.
[76] R. Fang and O. Shimbo, "Unified analysis of a class of digital systems in additive noise and interference," *IEEE Trans. Commun.*, vol. COM-21, pp. 1075–1091, Oct. 1973.
[77] A. S. Rosenbaum and F. E. Glave, "An error-probability upper bound for coherent phase-shift keying with peak limited interference," *IEEE Trans. Commun.*, vol. COM-22, pp. 6–16, Jan. 1974.
[78] J. Goldman, "Statistical properties of a sum of sinusoids and Gaussian noise and its generalization to higher dimensions," *Bell Syst. Tech. J.*, vol. 53, no. 4, pp. 557–580, Apr. 1974.
[79] A. Blanchard, "Interference in phase-locked loops," *IEEE Trans. Aerosp. Electron. Syst.*, vol. AES-10, pp. 686–697 (corrections in vol. AES-11, pp. 285–286), Sept. 1974.
[80] L. Wang, "Error probability of a binary noncoherent FSK system in the presence of two CW tone interferers," *IEEE Trans. Commun.*, vol. COM-22, pp. 1948–1949, Dec. 1974.
[81] P. J. McLane, "Error rate lower bounds for digital communication with multiple interference," *IEEE Trans. Commun.*, vol. COM-23, pp. 539–543, May 1975.
[82] F. E. Glave and A. S. Rosenbaum, "An upper bound analysis, for coherent phase-shift keying with cochannel, adjacent-channel, and intersymbol interference," *IEEE Trans. Commun.*, vol. COM-23, pp. 586–597, June 1975.
[83] M. J. Massaro, "Error performance of *M*-ary noncoherent FSK in the presence of CW tone interference," *IEEE Trans. Commun.*, vol. COM-23, pp. 1367–1369, Nov. 1975.
[84] J. Krishnamurthy, "Bounds on probability of error due to cochannel interference," *IEEE Trans. Aerosp. Electron. Syst.*, vol. AES-11, pp. 1373–1377, Nov. 1975.
[85] M. R. Wachs and D. E. Weinreich, "A laboratory study of the effects of CW interference on digital transmission over non-linear satellite channels," in *Proc. INTELSAT 3rd Int. Conf. on Digital Communications* (Kyoto, Japan), pp. 65–72, Nov. 11–13, 1975.
[86] M. C. Jeruchim, "Digital computer simulation of satellite quadrature communication systems," presented at Nat. Telecommunications Conf., New Orleans, LA, December 1975.
[87] J. Goldman, "Detection in the presence of spherically symmetric random vectors," *IEEE Trans. Inform. Theory*, vol. IT-22, pp. 52–59, Jan. 1976.
[88] I. E. Kliger and C. F. Olenberger, "Phase-lock loop jump phenomenon in the presence of two signals," *IEEE Trans. Aerosp. Electron. Syst.*, vol. AES-12, pp. 55–64, Jan. 1976.
[89] V. K. Prabhu, "Bandwidth occurancy in PSK systems," *IEEE Trans. Commun.*, vol. COM-24, pp. 456–462, Apr. 1976.
[90] D. E. Weinreich and M. R. Wachs, "A laboratory simulation of multiple unmodulated interference sources on digital satellite channels," presented at Int. Conf. Communications, Philadelphia, PA, June 1976.

Part II
Digital-Signal Interference

THIS PART deals with interference into digitally modulated signals, i.e., signals generated by frequency or phase modulation of a carrier by an information signal which can be represented by a stream of pulses that may or may not have been obtained through a special processing of the primary information signal (digitization, etc.). A basic requirement for this type of transmission is the exact (or with some acceptable error) reproduction of these pulses at the receiver. Interference contributes to the increase of error. The purpose of the discussion on interference of this part is to illustrate the relationship between this error and interference for these types of signals.

In the first paper by C. R. Cahn, the performance of digital phase modulation systems in Gaussian noise is analyzed. The required S/N as a function of a number of phases and the desired error rate is determined. It is shown that multiphase modulation provides an efficient trade-off between bandwidth and S/N in comparison with multilevel amplitude modulation. The study is extended by A. S. Rosenbaum in the second paper to include an interferer which consists of a single constant amplitude, angle modulated sinusoid which lies within the bandwidth of the detector. Both coherent and differential detection schemes are examined. It is concluded that differential detection suffers more degradation from lack of phase synchronization, resulting in an increase of the error probability with interference. However, the effect is considerably less than if the interference were replaced by Gaussian noise. The case which includes both cochannel and adjacent channel interference is given in the paper by V. K. Prabhu. It is shown that the error rate is minimum when all the interference power is concentrated in a single interferer and that the error rate attains its maximum when the total interference power is equally distributed among all interferers. In the fourth paper, by A. S. Rosenbaum, comprehensive results for multiple interference for CPSK are obtained. The main result is the determination of the exact error probability for binary PSK with multiple interferers. A more general mathematical procedure is used by J. Goldman in the fifth paper to derive the multiple error performance of PSK systems disturbed by Gaussian noise and interference. The interference can be of a general nature, and the only requirement is that it be circularly symmetric. An application of these results to spacing limitations of geostationary satellites using multilevel coherent PSK signals is presented by M. C. Jeruchim and F. E. Lilley. A computer simulation analysis of a digital radio relay system during fading periods is given in the seventh paper, by C. Colavito and M. Sant'Agostino. The possibility of attaining a required fading margin when many cochannel and interchannel interferers are present is analyzed in this paper for both long-haul and short-haul systems.

A unified analysis of a particular class of digital systems is given by R. Fang and O. Shimbo in the eighth paper. This class of systems includes *M*-ary amplitude shift keying (ASK), *M*-ary phase shift keying (PSK), and *M*-ary amplitude and phase keying (APK) for both coherent and binary differential PSK. The noise is not necessarily Gaussian, and the interference can be intersymbol, cochannel, adjacent-channel, and intermodulation products. This approach essentially expands the characteristic function of the interferences into a power series so that the desired error probability can be evaluated as the sum of terms representing perturbation around the error probability due to additive noise alone. Examples are given to illustrate how the unified analysis can be applied to evaluate the error probabilities of various digital systems. In the paper by O. Shimbo and R. Fang computational results for *M*-ary PSK systems with Gaussian noise and cochannel interference are given. The combined effects of intersymbol, interchannel, and cochannel interference in *M*-ary CPSK systems are examined in the tenth paper, by S. Benedetto *et al.* This work shows that only impulse responses of the overall channels are needed to derive exact expressions for the error probability of binary CPSK systems.

Upper bounds on the error probability for CPSK with peak-limited interference are analyzed by A. S. Rosenbaum and F. E. Glave. Those bounds are particularly useful in analyzing the effect of cochannel and adjacent channel interference into coherent phase-shift keying (CPSK) systems. These bounds are often tight enough to be employed as system design tools in addition to being simple to apply. In the twelfth paper, by F. E. Glave and A. S. Rosenbaum, an extension of these results is made to include multiple mixed interference such as Gaussian noise, intersymbol interference, adjacent channel interference, and cochannel interference. Finally, the effect of TV/FM interference into SCPC with PCM-PSK modulation is presented in the thirteenth paper of this part by D. Kurjan and M. Wachs through experimental measurements.

Digital microwave transmission on line-of-sight paths occasionally experience severe fading associated with multiple transmission paths of differing propagation delays. Envelope delay distortion causes serious intersymbol interference. The value of the delay distortion required to cause excessive errors in a digital system is given in the paper by W. C. Jakes, Jr., and is compared with measured results. System performance degradation can also be caused in digital systems, as we saw in the previous part for analog systems, by nonlinearity effects of the transmission path, filter distortion and crosstalk. In [1] the effects of satellite transponder nonlinearity on the performance of CPSK systems are considered. The BER is computed in the form of an infinite series as function of uplink and downlink carrier to noise ratio. In [2] the detection of BPSK signals through a hard limiting repeater is studied. Asymptotic expressions for the BER are given for extreme values of C/N, and it is shown that the BER is smaller than that of a linear system. The combined effects of filter distortion and the associated intersymbol interference on CPSK signals is studied in [3]. The system performance degradation is defined as the additional amount of energy per bit-to-noise power spectral density ratio E_b/N_o needed to maintain a given average bit error rate in a specified filtering condition over that required for the ideal undistorted case. A measure of performance based on crosstalk between adjacent digital channels is analyzed in [4]. An approach for reducing degradation by using mismatched windows in a standard correlation detector is presented.

Oscillator phase and frequency instabilities (noise) of both random and deterministic nature have become of great concern to many scientists working in the field of communications. Since in many communication systems the phase or frequency of the transmitted signals carries the information, any inaccuracy or uncertainty on these parameters can degrade system performance. A good account of the various approaches that have been used to characterize phase and frequency noise mathematically as a result of natural mechanisms has already been documented [5].

Demodulation of PSK signals requires the generation of a local carrier phase reference for coherent detection. The effect of reference carrier noise on communication system performance is the subject covered by the fifteenth, sixteenth, and seventeenth papers of this part. It is very important to thoroughly understand the underlying mechanism that is responsible for the generation of the noise in order to determine the limitation of the models used for the analysis of communciation systems. One method for generating a reference signal requires the transmission of an auxiliary carrier which is tracked at the receiver by means of a phase-locked loop with the output of this loop used as a reference signal for performing a coherent detection. Another method, referred to as self-synchronization, uses the reference signal derived from the modulated data signal by means of a squaring loop. The probability distribution for the phase error is dependent on the process that generates this error. A special type of reference phase noise distribution, namely the Tikhonov type, has been used by Prabhu for both BPSK and QPSK systems when the transmit and receive filters are four-pole Butterworth. The bit error rate is then calculated for various values of phase noise variance. A similar approach is applied in the sixteenth paper, by Lindsey, to cases when the phase noise is produced by means of a narrow band filter. To take advantage of the modulation process, an offset QPSK scheme is usually used to increase the immunity of the system to reference phase noise. A significant feature of the detection properties of offset QPSK is that it allows the same detection efficiency to be achieved with a smaller value of signal to phase reference noise ratio. This result is shown, in the seventeenth paper, by S. A. Rhodes.

The problem of maintaining the performance of communication systems in the presence of interfering signals originating from nearby transmissions competing for the same range of the RF spectrum has gained considerable importance. Code division multiple access communication techniques are presently being considered as attractive alternatives to conventional time and frequency division multiplexing. Code division multiplexing is commonly implemented as a spread spectrum multiple access communication system in which each user is assigned a unique pseudorandom noise. Each data bit for a particular user is encoded into many code bits, resulting in a substantial increase in the signal bandwidth. In addition to providing multiple access capability, the large bandwidth is expected to reduce the effects of interference. Two papers on this subject are included here. In the paper by S. A. Musa and W. Wasylkiwskyj, the interference into a spread spectrum system is a random signal with fading envelope governed by a Rayleigh distribution and random phase, whereas in the nineteenth paper, by C. S. Gardner and J. A. Orr, the interferers are distributed in accordance with a specified probability density function. In both cases the system error rate is calculated for various spread spectrum systems. It is shown that the system performance is a function of the interfering signal correlation function and the particular coding and modulation scheme used.

References

[1] P. Hetrakul and D. P. Taylor, "The effects of transponder nonlinearity on binary CPSK signal transmission," *IEEE Trans. Commun.,* May 1976.

[2] P. C. Jain and N. M. Blackman, "Detection of a PSK signal transmitted through a hard-limited channel," *IEEE Trans.* Infor. Theory, vol. IT-19, Nov. 1973.

[3] J. J. Jones, "Filter distortion and intersymbol interference effects on PSK signals," *IEEE Trans. Commun. Technol.,* vol. COM-19, Apr. 1971.

[4] I. Kolet, "A look at crosstalk in quadrature carrier modulation systems," *IEEE Trans. Commun.,* vol. COM-25, Sept. 1977.

[5] D. Rutman, "Characterization of phase and frequency instabilities in precision frequency sources: Fifteen years of progress," *Proc. IEEE,* vol. 66, Sept. 1978.

Performance of Digital Phase-Modulation Communication Systems*

CHARLES R. CAHN†

Summary—**This paper analyzes the performance of digital phase modulation systems in Gaussian noise and determines required signal-to-noise ratio as a function of the number of discrete phases and the desired error rate, under conditions of no fading. Both coherent detection with a locally-derived reference carrier and phase comparison detection are considered. The calculations show that multiphase modulation provides an efficient trade of bandwidth for signal-to-noise ratio in comparison with multilevel amplitude modulation. It is also found that phase comparison detection introduces about a 3-db degradation over coherent detection except with binary modulation, for which the degradation is less than 1 db for error rates not exceeding about 0.001.**

Introduction

THE need for efficient utilization of various media for transmission of digital data has stimulated the investigation of advanced coding and modulation techniques. Because highly stable oscillators are available for practical application, it is now possible to establish a phase reference in a receiver to detect digital phase-modulated signals, and a number of communication systems employing such signals have been developed.[1,2]

This paper determines the performance of such systems for the important case of signals corrupted by Gaussian noise and considers both coherent and phase-comparison detection schemes. A steady received signal is assumed in the analysis in accordance with the common procedure of adding fading allowances to the transmitter power calculated for median propagation and noise conditions. If an error rate averaged over a fading cycle is desired, a further graphical or analytical integration is necessary.[3]

Optimum Detection

The signals of the type under consideration consist of phase-modulation pulses of specified width, transmitted at a known repetition rate. The signal is sampled in the receiver at the pulse peaks. Each sample has the form

$$s(t) = \sqrt{2S} \cos (\omega_0 t + \theta) \qquad (1)$$

where S is the received signal power, ω_0 is the angular center frequency, and θ may have any value in the discrete set $2\pi k/m$, $0 \leq k \leq m - 1$. The detection problem is to determine θ when the signal is received under conditions of no fading and with additive Gaussian noise of the form

$$n(t) = x(t) \cos \omega_0 t + y(t) \sin \omega_0 t. \qquad (2)$$

In (2), x and y are low-frequency random variables, each with zero mean and power $\overline{n^2} = N$. The bandwidth is assumed sufficient to resolve the individual pulses of the signal so that no overlap occurs at the sampled peaks.

Maximum likelihood detection is presumed for the theoretical analysis. With equal *a priori* probabilities for the possible phases, this type of detection corresponds to selection of that phase having the maximum *a posteriori* probability according to the particular processing scheme utilized in the receiver.

Coherent Detection

The basic digital phase-modulation system provides a coherent phase reference in the receiver to facilitate signal processing. The set of m possible transmitted signals may be described by a set of m equally-spaced phasors in the complex plane, as shown in Fig. 1 for $m = 8$. Noise is added in transmission and is indicated in Fig. 1 by the two quadrature components added to the zero-phase signal. These distort the signal, both in amplitude and in phase. The dotted lines placed symmetrically between the signal phasors indicate the optimum decision thresholds. The possible signal phase having maximum *a posteriori* probability is indicated by the sector in which the actual phase of signal-plus-noise falls.

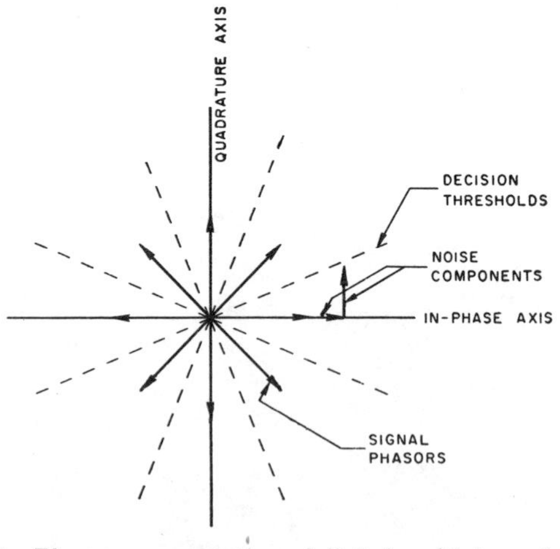

Fig. 1—Phasor representation of digital multiphase signals.

* Manuscript received by the PGCS, December 12, 1958.
† Ramo-Wooldridge, a division of Thompson Ramo Wooldridge Inc., Los Angeles, Calif.
[1] M. L. Doelz, E. T. Heald, and D. L. Martin, "Binary data transmission techniques for linear systems," Proc. IRE, vol. 45, pp. 657–661; May, 1957.
[2] F. A. Losee, "A digital data transmission system using phase modulation and correlation detection," *Proc. Natl. Conf. on Aeronautical Electronics*, pp. 88–93; May, 1958.
[3] G. F. Montgomery, "A comparison of amplitude and angle modulation for narrow-band communication of binary-coded messages in fluctuation noise," Proc. IRE, vol. 42, pp. 447–454; February, 1954.

Reprinted from *IRE Trans. Comm. Syst.*, vol. CS-7, pp. 3-6, May 1959.

It is seen that the probability of an error due to noise is the probability that the phase of signal-plus-noise will be distorted outside the sector $-\pi/m < \theta < \pi/m$, where zero is the undistorted or true phase. This probability can be calculated by integration of the probability density of the phase of a steady signal plus Gaussian noise. The probability density of phase can be evaluated from the joint probability density of envelope and phase[4] by integrating over-all values of envelope from 0 to ∞ and is found to be

$$p(\theta) = \frac{1}{2\pi} e^{-S/N}[1 + \sqrt{4\pi S/N} \cos \theta e^{(S/N)\cos^2\theta} \cdot \Phi(\sqrt{2S/N} \cos \theta)] \quad (3)$$

where $\Phi(x)$ is the probability integral defined by

$$\Phi(x) = \frac{1}{\sqrt{2\pi}} \int_{-\infty}^{x} e^{-x^2/2}\, dx. \quad (4)$$

The probability density of phase is plotted in Fig. 2 for several values of signal-to-noise ratio, S/N.

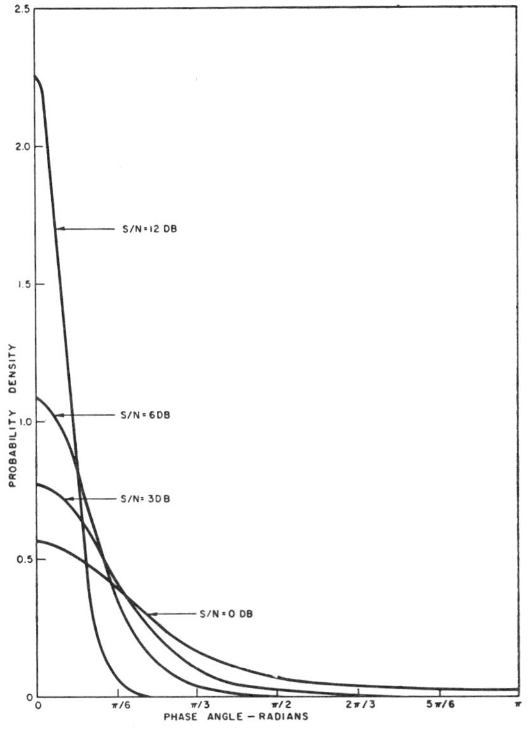

Fig. 2—Probability density of phase for sine wave plus Gaussian noise.

The probability of error P_e as a function of the signal-to-noise ratio and the number of phase positions can be obtained by numerical or graphical integration of (3), according to

$$P_e = 1 - \int_{-\pi/m}^{\pi/m} p(\theta)\, d\theta. \quad (5)$$

From probability distributions previously presented in the literature,[5] a set of working curves has been drawn in Fig. 3. These curves indicate the required signal-to-noise ratio for a specified error rate and number of phase positions.

The information conveyed per pulse is $\log_2 m$ when multiphase modulation is used. Hence, for a fixed information rate, the signal bandwidth is reduced by this factor. A convenient normalization is the signal-to-noise ratio measured in the equivalent bandwidth, *i.e.*, the bandwidth necessary with binary phase modulation at the specified information rate. Assuming flat Gaussian noise over the frequency band of interest, the noise power is directly proportional to bandwidth, and the resulting curves are given in Fig. 4. The ordinate of these curves is directly proportional to required signal power and allows direct comparison of power requirements for communication systems with different numbers of phase positions. Note that biphase and quadriphase modulations are essentially equivalent, while triphase modulation requires slightly less signal power than biphase modulation.

Phase-Comparison Detection

In certain cases of practical importance, obtaining and maintaining a coherent phase reference in the receiver is not feasible. In such cases, it may be possible to utilize phase comparison of successive samples for the detection process. Thus, the information is conveyed by the phase transitions between pulses rather than as the absolute phases of the pulses. It is apparent that this type of detection process is not as effective as coherent detection and will lead to a higher error rate for any specified signal-to-noise ratio.

For reasonably large positive values of x, the probability integral may be approximated by an asymptotic expansion such as

$$\Phi(x) \cong 1 - \frac{e^{-x^2/2}}{\sqrt{2\pi}\, x}. \quad (6)$$

If this expression is utilized in the probability density function (3) for the phase, the result is

$$p(\theta) \cong \frac{\cos \theta}{\sqrt{\pi N/S}} e^{-(\sin^2\theta)/N/S} \cong \frac{1}{\sqrt{\pi N/S}} e^{-(\theta^2)/N/S}, \quad |\theta| < \pi/2$$

$$\cong 0, \quad |\theta| > \pi/2 \quad (7)$$

with a smooth transition between the two expressions occurring at $\theta = \pm \pi/2$. That the probability density of the phase is approximately Gaussian for small angular deviations also may be seen from a formula previously derived by Rice.[6]

[4] S. O. Rice, "Mathematical analysis of random noise," in "Noise and Stochastic Processes," Nelson Wax, ed., Dover Publications, Inc., New York, N. Y., p. 238; 1954.

[5] K. A. Norton, E. L. Shultz, and H. Yarbrough, "The probability distribution of the phase of the resultant vector sum of a constant vector plus a rayleigh distributed vector," *J. Appl. Phys.*, vol. 23, pp. 137–141; January, 1952.

[6] S. O. Rice, "Statistical properties of a sine wave plus random noise," *Bell Sys. Tech. J.*, vol. 27, pp. 109–157; January, 1948.

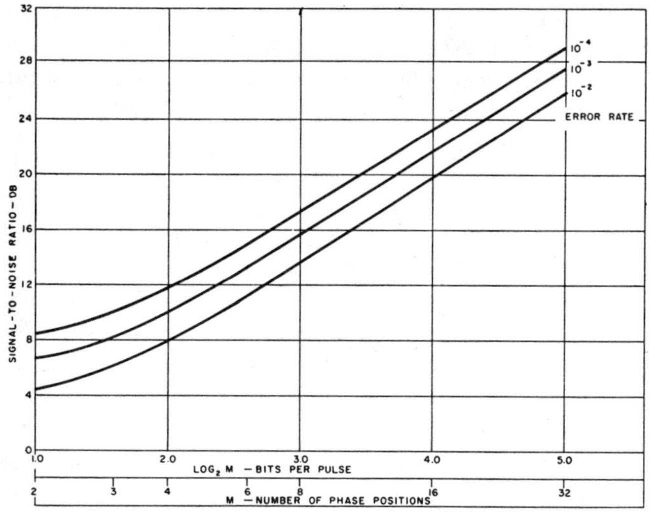

Fig. 3—Signal-to-noise ratio requirement for digital phase modulation.

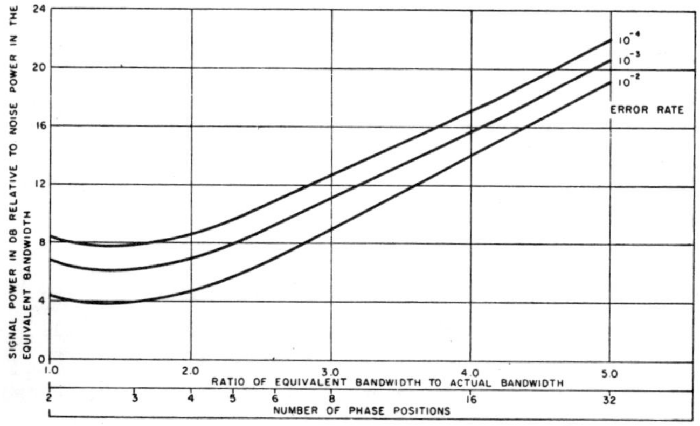

Fig. 4—Relative power requirement with digital phase modulation.

For phase comparison of successive samples with uncorrelated noise voltages, the resulting angle is a random variable defined by the difference of the phase angles of the two samples. For reasonably high signal-to-noise ratios and for small angular deviations, the difference phase has approximately a Gaussian distribution with twice the variance of the distribution for the phase of a single sample. Therefore, the degradation introduced by phase-comparison detection is essentially 3 db over coherent detection for a specified (low) error rate. This result applies, however, only with multiphase modulation for which the angular deviations of interest will be small.

For binary or phase-reversal modulation, the degradation at high signal-to-noise ratios is found to be less than 3 db. One method of calculating the error rate for a specified signal-to-noise ratio is to obtain the probability density function for the difference phase by graphical convolution of the density function for the phase of a single pulse.[7] The probability that the difference phase deviates by more than 90° from its proper value can then be obtained by graphical integration.

An alternate technique which gives a closed-form solution is now presented. Suppose that the phase of the noise-distorted first sample has the value θ with respect to its undistorted position. The conditional probability of error is then the probability that the phasor representing the noise-distorted second sample will cross the decision surface indicated in Fig. 5. Only the component of noise on the second sample parallel to the distorted position of the first sample can cause an error. The conditional probability of this is

$$p(\text{error} \mid \theta) = \frac{1}{\sqrt{2\pi N}} \int_{\sqrt{2S}\cos\theta}^{\infty} e^{-x^2/2N}\, dx$$

$$= 1 - \Phi(\sqrt{2S/N}\cos\theta). \quad (8)$$

The total probability of error is the integral of the joint density function, $p(\theta)\, p(\text{error} \mid \theta)$, over all possible values of θ, or

$$P_e = \int_{-\pi}^{\pi} p(\theta)[1 - \Phi(\sqrt{2S/N}\cos\theta)]\, d\theta. \quad (9)$$

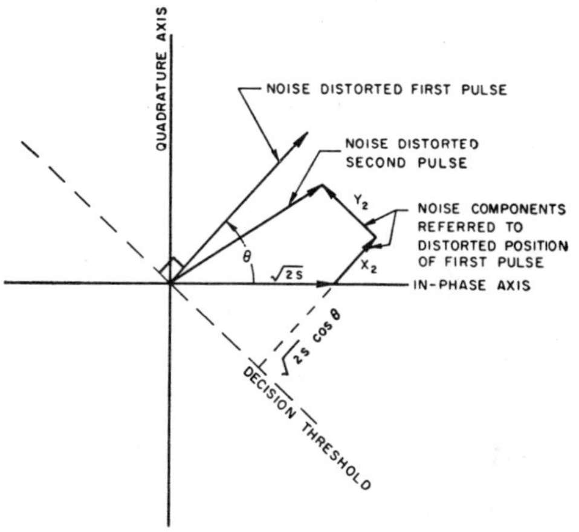

Fig. 5—Effects of noise on phase-comparison detection.

If the expression for $p(\theta)$ given in (3) is substituted into (9) and all functions expanded as (absolutely convergent) power series in $\sqrt{2S/N}\cos\theta$, it will be seen that all terms except a single constant term contain $\cos\theta$ raised to an odd power and will integrate to zero over a full cycle. The remaining constant term yields the probability of error,

$$P_e = \tfrac{1}{2}e^{-S/N}. \quad (10)$$

The calculated probability of error for binary transmission with phase-comparison detection is presented in Fig. 6 along with the corresponding curve for coherent detection. It is seen that the degradation is indeed less than 3 db for high values of S/N.

In previous analyses[3,8] of binary transmission systems, a formula identical in form to (10) is derived for the error rate with frequency-shift modulation under nonfading

[7] J. L. Doob, "Stochastic Processes," John Wiley and Sons, Inc., New York, N. Y., p. 78; 1953.

[8] G. L. Turin, "Error probabilities for binary symmetric ideal reception through nonselective slow fading and noise," Proc. IRE, vol. 46, pp. 1603–1619; September, 1958.

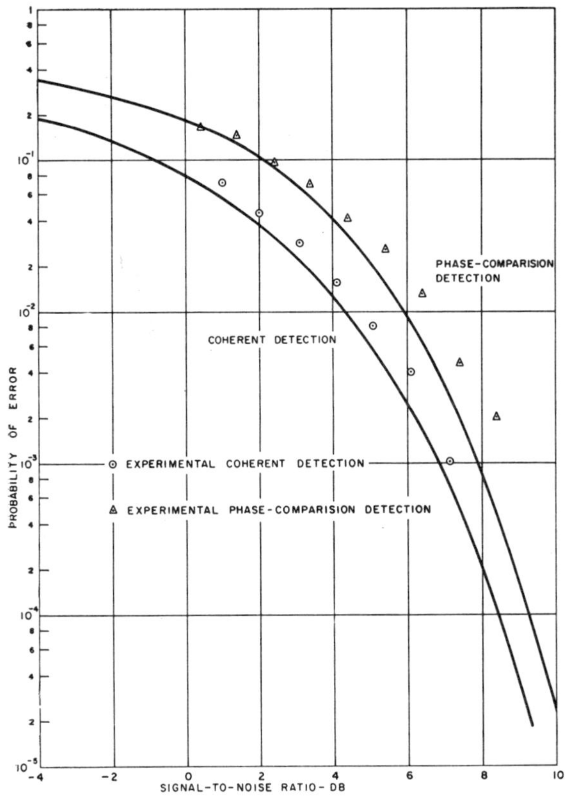

Fig. 6—Error rates with binary phase modulation.

conditions. This apparently is a coincidence, although not completely so, since phase transitions can be considered as the result of positive or negative shifts in the carrier frequency. It is important to note, however, that frequency-shift keying utilizes twice the bandwidth necessary with phase-reversal modulation and, therefore, requires 3 db more power for any specified error rate.

To verify the theoretical calculations for binary phase-modulated systems, experimental determinations of error rate vs signal-to-noise ratio were made with both coherent and phase-comparison detection. The results are shown along with the theoretical curves in Fig. 6. The agreement between theory and experiment regarding the relative performance of the two systems is excellent. The experimental curves show a small degradation with respect to the theoretical curves as may usually be expected due to non-ideal performance of practical phase detectors.

Significance of Error Rate

The error rate given in (5) and in Fig. 6 is actually a character error rate. That is, the size of the alphabet is m (number of phases), and P_e is the probability that the transmitted character will be received incorrectly. In many cases, however, the information to be transmitted is in binary form and must be coded for an m-phase system. For example, three bits could be combined to give any of eight phases. The character error rate and the bit error rate are not usually equal under such circumstances.

As a simple illustration, suppose that the three bits select any of the eight phases by a Gray code, so that adjacent phase positions correspond to bit combinations differing in only one place. The large majority of transmission errors will cause the phase to fall in a sector adjacent to that of the correct signal, and only one bit will be in error. The average probability of error in the binary information is then about 1/3 of the value from (5).

Similar analyses can be made for other coding techniques that translate binary data into a set of m characters for transmission over an m-phase system.

Conclusions

Multiphase digital modulation proves to be an efficient technique for trading bandwidth for signal-to-noise ratio to reduce spectrum congestion. That is, as the number of phases is increased beyond two, the signal power required goes up slowly at first, particularly in comparison with the behavior for multilevel amplitude modulation.[9] In fact, no additional power is required until the number of phases exceeds four, when coherent detection is employed.

For high signal-to-noise ratios (low error rates), phase-comparison detection yields about a 3-db degradation over coherent detection, except in the binary case. In that special case, the degradation approaches zero with high signal-to-noise ratios and is less than about 1 db for error rates below 0.001. Therefore, multiphase modulation is not as efficient, relative to binary modulation, as was the case with coherent detection.

Although not considered in this paper, combinations of digital amplitude and phase modulation are also possible. When a large-size alphabet is contemplated, such combinations may make some efficient use of power than either amplitude or phase modulation used alone.

Acknowledgment

The author wishes to thank J. Taber of Space Technology Laboratories, who suggested the approach used to calculate the probability of error for binary phase-comparison detection. Thanks are due also to B. Dell and H. Lee of Ramo-Wooldridge and to A. Gold of Space Technology Laboratories, who derived and plotted the probability density functions of phase and obtained the experimental data presented in Fig. 6.

[9] B. M. Oliver, J. R. Pierce, and C. E. Shannon, "The philosophy of pcm," Proc. IRE, vol. 36, pp. 1324–1331; November, 1948.

PSK Error Performance with Gaussian Noise and Interference

By ARNOLD S. ROSENBAUM

(Manuscript received September 13, 1968)

A single, constant amplitude, in-band, additive interference is included in the analysis of detecting phase shift keyed signals in gaussian noise. For coherent detection we give a method applicable to any M-phase system, and evaluate the symbol error probability for M = 2, 3, and 4. For differential detection we treat the important cases M = 2, 4, 8, and 16, offering comprehensive numerical results for each.

The analysis in each case is based on a single sinusoid with random phase adding to the noisy phase shift keyed signal. The results are then interpreted to include an angle modulation impressed on the continuous wave interferer. The receiver consists of an ideal phase discriminator with a perfect slicer. The channel is also assumed ideal in that intersymbol interference is not considered.

I. INTRODUCTION

Phase shift keying (psk) is becoming more popular as a modulation scheme for transmitting digital information. Lately much analysis has been done for both coherent and differentially coherent detection. Unfortunately the analyses done to date have generally considered only two signal degradations: channel anomalies (such as distortion, gain and delay variations, and so on) and thermally generated noise modeled by a gaussian random process. This article considers the effects of a spurious signal, or interference, falling in the band of the desired signal, as well as gaussian noise. It is understood that both the noise and the interference additively corrupt the desired signal; these are the only perturbing factors.

For coherent detection, the phase probability density function for the received composite of signal, noise, and interference is found. From this, the theoretical error probability may be evaluated for any M-phase system. We give comprehensive numerical results for the important cases, $M = 2, 3, 4$.

For differential detection we present analysis and results for $M = 2$, 4, 8, and 16. In the binary case a simple closed form solution was found which yields both a good approximation and exact bounds to the actual error probability. The solutions to the multilevel $M > 2$ differential detection problem, which are exact, required machine computation of a double integral; complete numerical results are given.

Finally, we draw general comparisons between coherent and differential detection error performance as affected by interference.

II. SIGNALS, NOISE, AND INTERFERENCE

A phase shift keyed signal has the form (ignoring any amplitude function)

$$s(t) = \cos [2\pi f_s t + \phi_s(t)] \tag{1}$$

where we choose to normalize the peak signal amplitude to unity. The digital modulation is carried in the angle of s by $\phi_s(t)$, which assumes discrete values from a set of M equally spaced points in $[0, 2\pi]$ at the sample times T seconds apart. Thus the Nth message or baud is modulated by

$$\phi_s(NT) = \frac{2\pi k}{M}, \qquad k = 0, 1, 2, \cdots, M - 1 \tag{2}$$

where each of the M values of k is equally probable.

For a coherent receiver an M-ary symbol is transmitted in one baud by the value of k. For a differential detection receiver the information is transmitted by the changes in k (or carrier phase) between adjacent bauds.

The noise is presumed to originate thermally and is therefore modeled in the usual fashion by a stationary zero mean gaussian random process with uniform spectral density. At the output of a symmetrical bandpass filter the noise voltage may be written as[1]

$$n(t) = u(t) \cos (2\pi f_s t) - v(t) \sin (2\pi f_s t) \tag{3}$$

where u and v are low-pass, stationary, independent, zero mean gaussian random processes with ensemble averages

$$\langle n^2 \rangle_{av} = \langle u^2 \rangle_{av} = \langle v^2 \rangle_{av} = \sigma^2, \tag{4}$$

equal to the noise power.

Reprinted with permission from *Bell Syst. Tech. J.*, vol. 48, pp. 413-442, Feb. 1969.

In the differential detection analysis we make the further restriction that the noise process autocorrelation vanish at the baud interval, thus

$$R_{nn}(T) = 0. \tag{5}$$

This assures the four gaussian random variables

$$u(t_0),\ u(t_0 + T),\ v(t_0),\ v(t_0 + T)$$

to be uncorrelated and hence independent.[2]

Interference shall consist of a constant amplitude, possibly angle modulated, sinusoid which lies within the bandwidth of the detector. It is assumed to originate independently of the signal, and so it is natural that its phase relationship to the signal is random with all angles equally probable. Therefore let

$$i(t) = b \cos [2\pi f_i t + \phi_i(t) + \Gamma] \tag{6}$$

which has a peak value of b, and is angle modulated by ϕ_i. The arbitrary phase angle Γ, independent of ϕ_i, is a random variable whose probability density function is $(2\pi)^{-1}$ when reduced modulo 2π.

For coherent detection, where the interference is observed only once per symbol, the random phase variable Γ vitiates the modulation ϕ_i because the sum $(\phi_i + \Gamma)$ is distributed exactly as if it were uniform. This is discussed in Section III.

III. COHERENT DETECTION

An ideal phase discriminator is assumed which compares the received wave (composed of signal, noise, and interference) with the unmodulated signal carrier (the reference) and produces instantly the signed phase difference between the two inputs.

The detector examines the discriminator output and announces an estimate of the transmitted symbol. The detector operates with no timing error and with zero width decision thresholds. Using maximum likelihood detection based on equal *a priori* symbol probabilities, the thresholds are at π/M, $(3\pi)/M$, . . . , $[(2M - 1)\pi]/M$. In a phasor diagram these thresholds correspond to $(2\pi)/M$ angular sections centered about the M signal positions.

The approach used to find Pe, the probability of a symbol error, is to find the probability density function of the phase of the received composite $(s + n + i)$, and then integrate the density over the error regions.

We first notice that the phase angle of the interference relative to the signal is, from equations (1) and (6),

$$\Phi(t) \equiv 2\pi(f_i - f_s)t + \phi_i(t) - \phi_s(t) + \Gamma \tag{7}$$

where Γ is independent of the other terms on the right-hand side. Since Γ is uniformly distributed modulo 2π, the relative interference phase process $\Phi(t)$ is also uniformly distributed modulo 2π. This is a general result for the modulo addition of several variables, one of which is uniform.[3]

Figure 1 is a phasor diagram of the receiver input components, signal, interference, and noise, at a sample time t_0. The phase reference is $2\pi f_s t + \phi_s$ so that the signal lies along the reference (vertical) axis. The orthogonal noise phasors are assumed to be at angles 0 and $\pi/2$, relative to the signal. We seek the probability density function of the resultant angle A, and begin by considering the two dimensional joint probability density function of the cartesian coordinates of the resultant phasor. Conditioned on Φ, it is clearly jointly gaussian with means

$$\langle x \rangle_{av} = b \sin \Phi, \qquad \langle y \rangle_{av} = 1 + b \cos \Phi \tag{8}$$

so that

$$f_{XY}(x, y \mid \phi) = \frac{1}{2\pi\sigma^2} \exp \left\{ -\frac{1}{2\sigma^2} [(x - b \sin \phi)^2 + (y - 1 - b \cos \phi)^2] \right\}. \tag{9}$$

Eliminating the Φ dependency gives

$$f_{XY}(x, y) = \frac{\exp \left\{ -\frac{1}{2\sigma^2} [x^2 + (y - 1)^2 + b^2] \right\}}{(2\pi\sigma)^2} \cdot \int_0^{2\pi} \exp \left\{ \frac{b}{\sigma^2} [x^2 + (y - 1)^2]^{\frac{1}{2}} \cos (\phi + \eta) \right\} d\phi \tag{10}$$

where $\eta = \tan^{-1}[(y - 1)/x]$ is not a function of ϕ.

This integrates directly to

$$f_{XY}(x, y) = \frac{1}{2\pi\sigma^2} \exp \left\{ -\frac{1}{2\sigma^2} [x^2 + (y - 1)^2 + b^2] \right\} \cdot I_0 \left\{ \frac{b}{\sigma^2} [x^2 + (y - 1)^2]^{\frac{1}{2}} \right\} \tag{11}$$

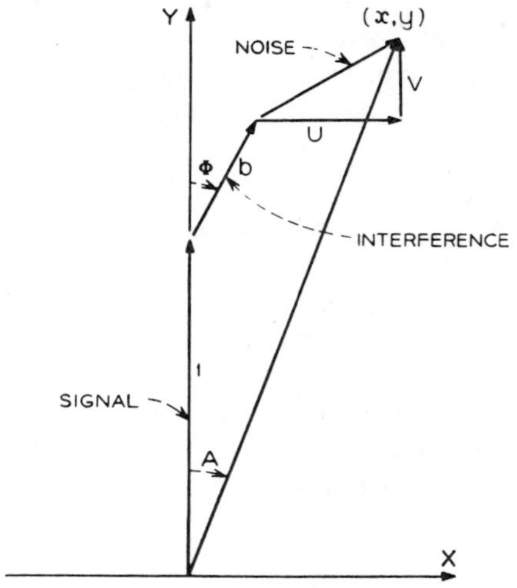

Fig. 1 — Phasor diagram of the signal, noise, and interference components at a sample time t_0.

where I_0 is the zero order modified Bessel function of the first kind.

We now convert equation (18) to polar coordinates through the usual transformation

$$x = r \sin \alpha \quad \text{and} \quad y = r \cos \alpha \tag{12}$$

which has the Jacobian r. Then the polar coordinate two-dimensional density is integrated over all radius values to yield the desired probability density function of the angle.

$$f_A(\alpha) = \frac{1}{2\pi\sigma^2} \int_0^\infty \exp \left\{ -\frac{1}{2\sigma^2} [r^2 + b^2 + 1 - 2r \cos \alpha] \right\} \cdot I_0 \left[\frac{b}{\sigma^2} (r^2 + 1 - 2r \cos \alpha)^{\frac{1}{2}} \right] r \, dr. \tag{13}$$

The above integration has been done numerically to generate exact $f_A(\alpha)$ curves for several values of

$-20 \log_{10} [2^{1/2}\sigma]$ = carrier to noise ratio in dB (CNR),

$-20 \log_{10} b$ = carrier to interference ratio in dB (CIR).

It is clear from equation (13) that $f_A(\alpha)$ has at least the symmetries of $\cos \alpha$.

Figure 2 offers typical families of f_A probability density function

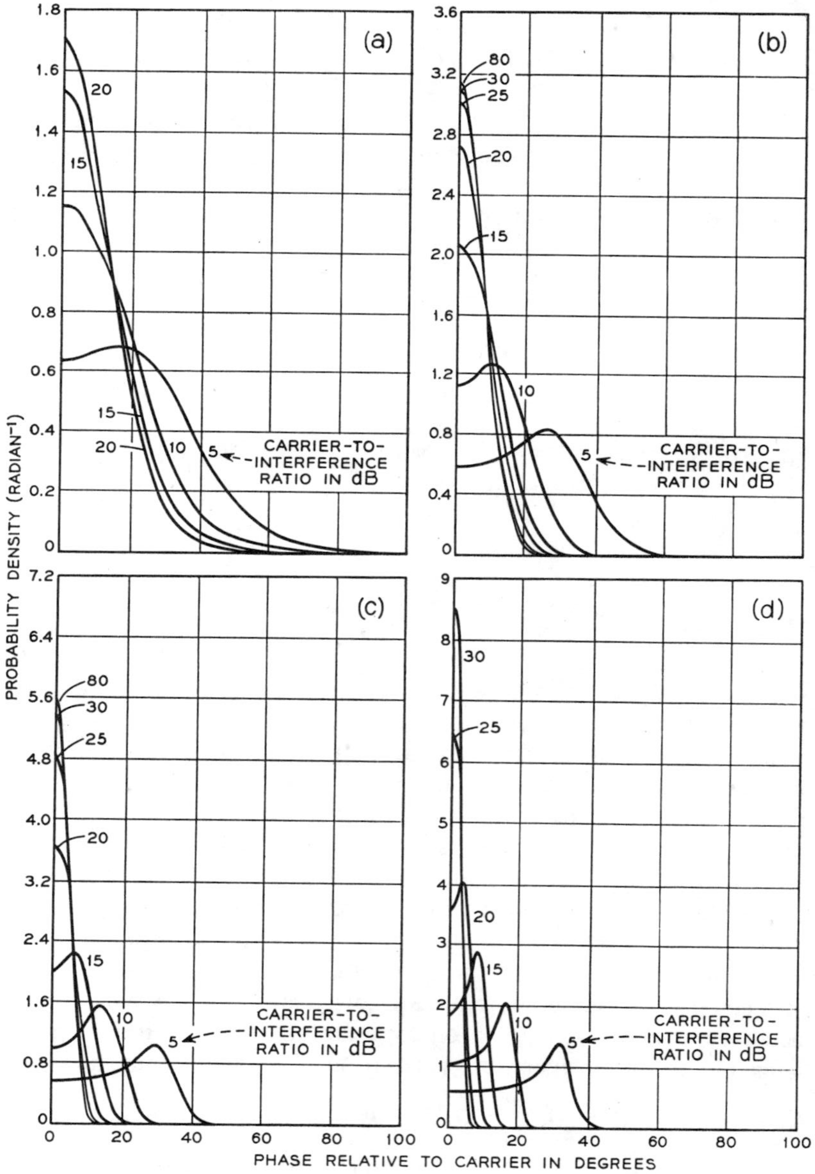

Fig. 2 — Probability density function of the phase of $s + n + i$ for various CIR values. (a) CNR = 10 dB, (b) CNR = 15 dB, (c) CNR = 20 dB, (d) CNR = 25 dB.

curves for representative CNR values. As the interference amplitude increases it is seen to control or to affect the shape of the curve to a greater degree. At high interference levels a saddle-like shape appears with peaks at roughly $\tan^{-1}b$, as one would expect in the absence of noise.

Since equation (13) is the probability density function of the angle of the complete receiver input, and since the probability of a symbol error is the probability that A lies outside the region $[-\pi/M, \pi/M]$ at t_0, we have

$$\text{Pe} = \int_{-\pi}^{-\pi/M} f_A(\alpha)\, d\alpha + \int_{\pi/M}^{\pi} f_A(\alpha)\, d\alpha \tag{14}$$

which by symmetry is

$$\text{Pe} = 2 \int_{\pi/M}^{\pi} f_A(\alpha)\, d\alpha. \tag{15}$$

Again, integral (15) was done numerically (a simple summation of the f_A data) for practical combinations of CNR, CIR, and for $M = 2, 3$, and 4. The results appear in Figs. 3–5.*

The above method, which employs two (rather simple) machine integrations, yields exact results but at the expense of not generating useful expressions for Pe. Therefore we now indicate one approach which yields Pe for $M = 2$, and bounds Pe for $M > 2$, as a convergent series. We begin by considering binary reception.

Referring back to Fig. 1, an error is made if $|\alpha| > 90°$ or, equivalently, if the resultant resides in the lower half plane $y < 0$. Then for fixed Φ,

$$\text{Pe} \mid \Phi = (2\pi\sigma^2)^{-\frac{1}{2}} \int_{-\infty}^{0} \exp\left[-\frac{1}{2\sigma^2}(y - 1 - b\cos\phi)^2\right] dy$$

$$= \tfrac{1}{2}\,\text{erfc}\left(\frac{1 + b\cos\phi}{(2)^{\frac{1}{2}}\sigma}\right). \tag{16}$$

Averaging over the uniformly weighted Φ gives

$$\text{Pe} = \frac{1}{\pi}\int_0^{\pi} \tfrac{1}{2}\,\text{erfc}\left(\frac{1 + b\cos\phi}{(2)^{\frac{1}{2}}\sigma}\right) d\phi. \tag{17}$$

This integral, which is virtually the cumulative distribution function of a sine wave of amplitude b plus gaussian noise of variance σ^2

* The abscissa values are true carrier-to-noise power ratios, and are not adjusted to reconcile bandwidth to bit-rate differences. One may do this by subtracting 2 dB (3 dB) from the abscissa values for $M = 3(4)$.

(evaluated at -1), has been examined by Rice[4] and others. It can be evaluated by expanding the integrand in a Taylor series about $(2\sigma^2)^{-\frac{1}{2}}$ and then integrating term by term. If the interference is small, $b \ll 1$, only the first several terms need be retained for reasonable accuracy.

The Pe values obtained for Binary may be used to bound the symbol Pe for $M > 2$. The decision thresholds are at $\pm\pi/M$ for the M-ary receiver. The error region consists of the union of two half planes formed by the extended detector thresholds. The probability that the resultant

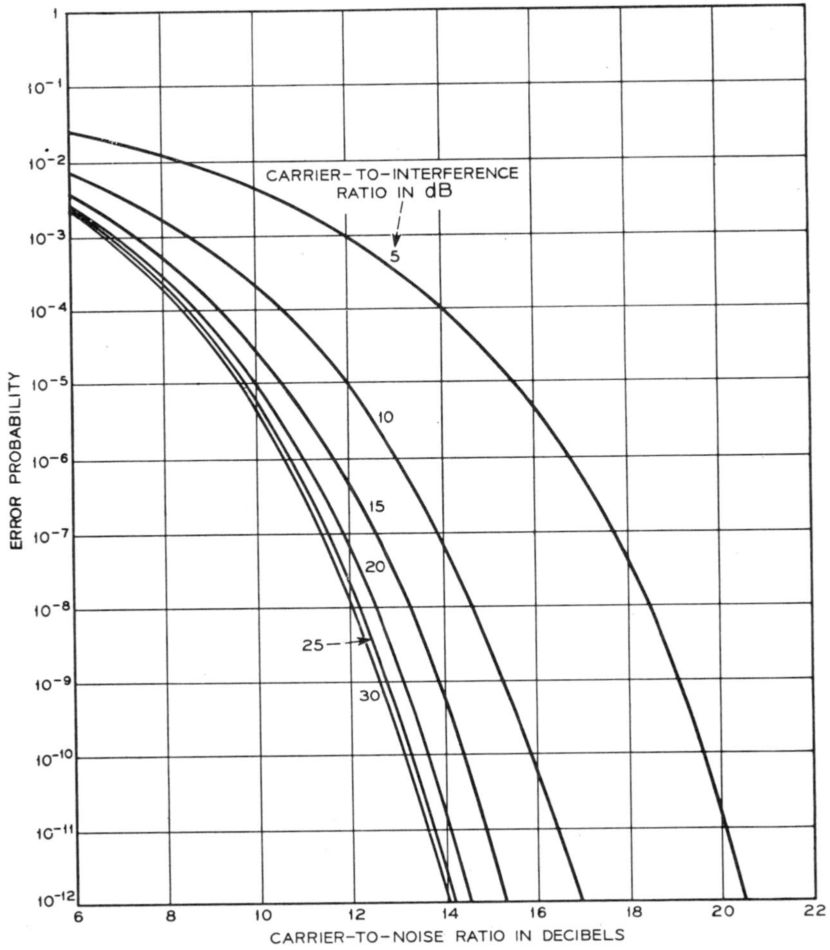

Fig. 3 — Binary ($M = 2$) Pe versus CNR. Coherent detection.

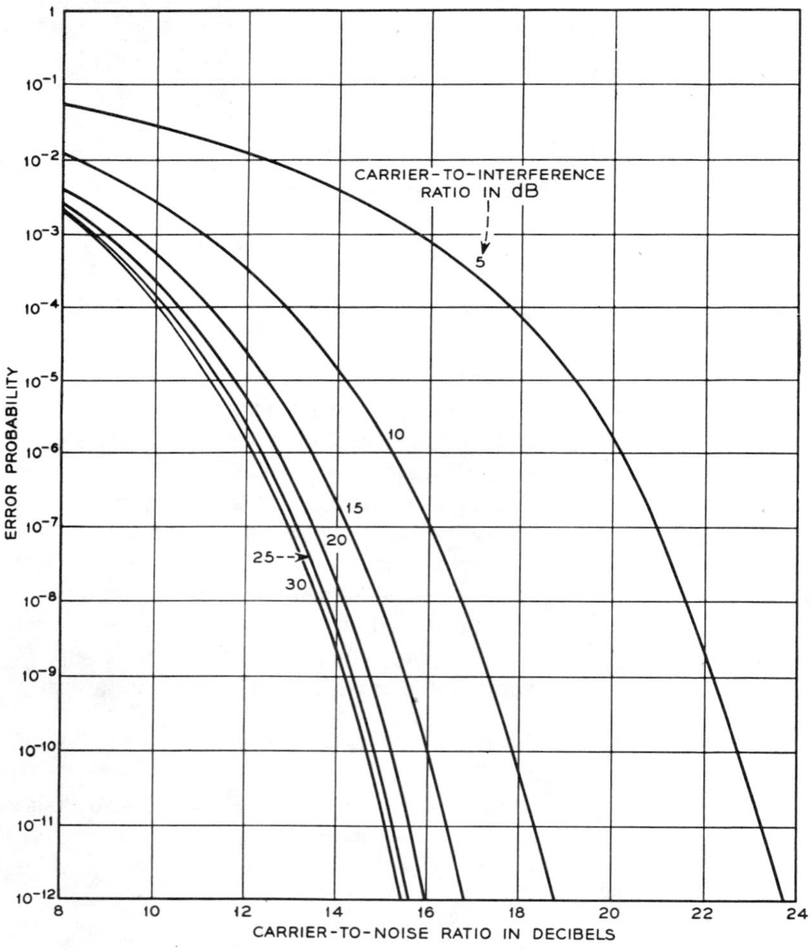

Fig. 4 — Ternary ($M = 3$) Pe versus CNR. Coherent detection.

phasor terminates in the error region, Pe, is thus bounded by the probability of terminating in either half plane. As M increases, the bound becomes a good approximation* because the size (hence the relative probability) of the doubly counted intersection decreases rapidly with M.[5] By symmetry, the probability of terminating in either half plane is twice that of one half plane.

* The approximation improves with CNR also.

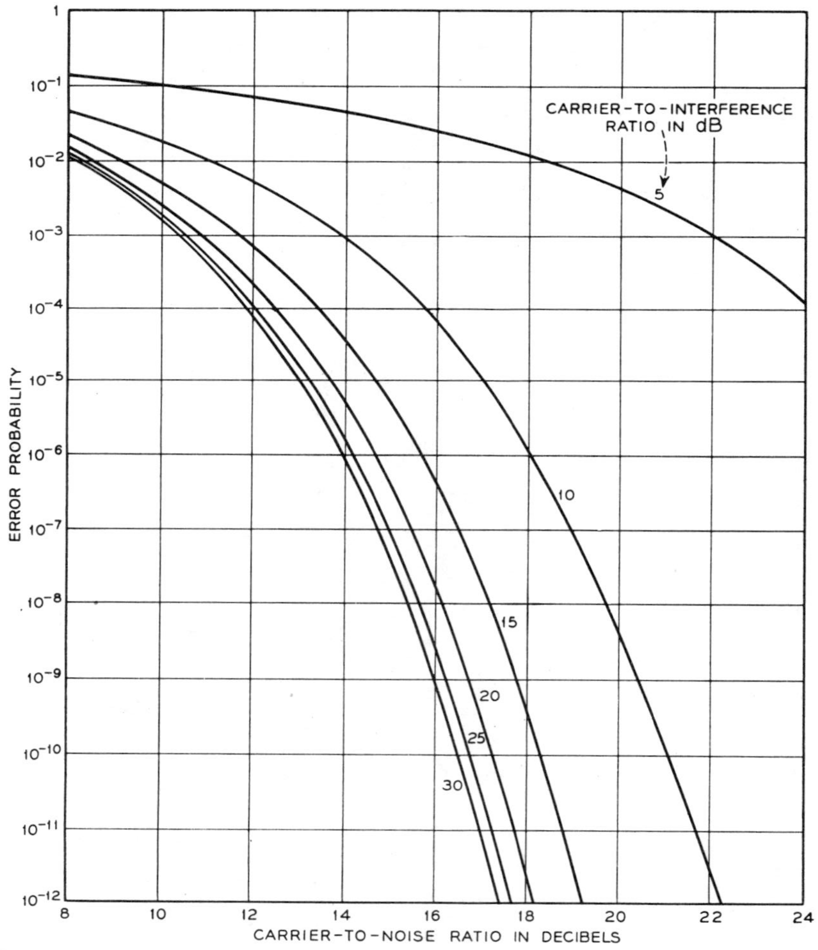

Fig. 5 — Quaternary ($M = 4$) Pe versus CNR. Coherent detection.

The probability of one half plane is related to the binary Pe very simply. The distance to the boundary from (0, 1) is $\sin(\pi/M)$. If now the interference phasor and noise phasor were scaled by the same factor, we see that the probability of the half plane is just the binary Pe with interference amplitude $b \sin(\pi/M)$ and noise variance $[\sigma \sin(\pi/M)]^2$.*

* The author is grateful to S. O. Rice for suggesting this notion.

Therefore we have the interesting relationship

$$\text{Pe}\left(M > 2, \sigma \sin \frac{\pi}{M}, b \sin \frac{\pi}{M}\right) \leqq 2\ \text{Pe}\ (M = 2, \sigma, b) \qquad (18)$$

which means that twice a given binary Pe value is an upper bound (or approximation) to an M-ary Pe where the binary CNR and CIR values are each increased by $-20 \log \sin (\pi/M)$ dB.

IV. DIFFERENTIAL DETECTION—BINARY

This type of detection has become widely considered lately because it eliminates the requirement of phase synchronism between the transmitter and receiver. The price one pays for nonsynchronous detection, however, is poorer performance.

The analysis for differential detection is complicated by the fact that the phase reference, being the previous signal, is subject to the same corruptions as the present signal being phase detected. (For the noise-only case, an exact solution in closed form is available for binary, and good approximations exist for $M \geqq 4$.) We begin with an analysis for binary differential psk (d-psk), considering again a single CW interference in addition to noise, which yields in closed form both bounds on Pe and a good approximation.

We are concerned with the reception of two successive bauds, where the data is encoded in the phase change of the signal. Arbitrarily, let no phase change represent a "0" and a π phase change represent a "1." For convenience, we will refer to the signal during a baud interval as a "pulse" of carrier at a certain phase angle, although in a pure phase modulated (PM) system the signal would not consist of carrier pulses.

From previous assumptions, the noise corrupting each signal pulse acts independently; the interference at two adjacent sample times t_0 and $t_0 + T$ does not. This dependency may be summarized by an angle

$$\theta \equiv 2\pi(f_i - f_s)T$$

which is the relative phase slip of the interference from one sample instant to the next. Assume, for the present, that the interference modulation is absent.

We will use the same equally spaced detection thresholds as in the coherent psk analysis. Ignoring interference, the probability of error in d-psk is not data dependent because of obvious symmetries.

However, the addition of interference destroys that symmetry, and causes Pe to be strongly data dependent for a given θ. Fortunately, however, the error probability for only one symbol (that is, "0" or "1") needs to be found because the probability of error for the other symbol(s) is derived directly from it. An over-all probability of error is then found by averaging the individual symbol error probabilities with equal weighting.

Consider the transmission of a "0" whereby two carrier pulses of the same phase are sent. A "double exposure" phasor diagram, Fig. 6, pictures the signal, noise, and interference components at the two successive sample instants t_0 and $t_0 + T$. The interference at t_0 assumes an angle ϕ relative to the signal, where ϕ is random and uniformly distributed in $[0, 2\pi]$. At time $t_0 + T$ the interference has progressed to an angle $\phi + \theta$.

The noise phasor amplitudes are the random variables

$$U_d \equiv u(t_0), \quad V_d \equiv v(t_0), \quad U \equiv u(t_0 + T), \quad \text{and} \quad V \equiv v(t_0 + T) \qquad (19)$$

which we recall are independent, equal variance, zero mean gaussians.

The two resultant phasors, Z and Z_d, are the actual phase discriminator inputs; the output being their phase difference, δ. Since a

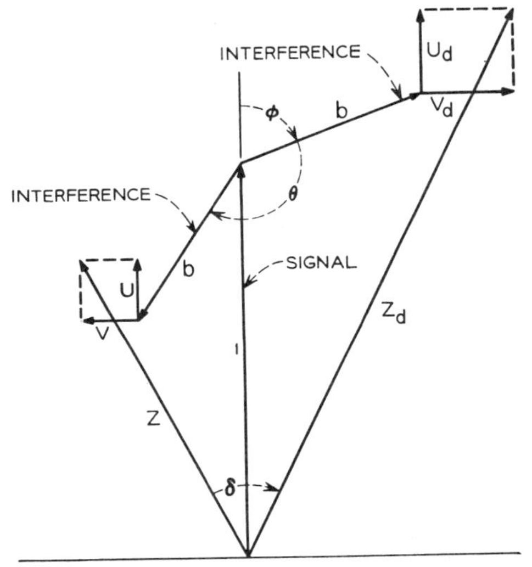

Fig. 6 — Signal, interference, and noise phasors at successive sample times t_0 and $t_0 + T$ for a transmitted "0."

"0" is transmitted, the probability of error is

$$\text{Pe} \mid \text{``0''} = \Pr\left\{ | \delta | > \frac{\pi}{2} \right\}, \qquad -\pi < \delta \leqq \pi \tag{20}$$

which is equivalent to $\Pr\{\cos \delta < 0\}$.

Also, because

$$\cos \delta = \frac{\text{Re}\,[ZZ_d^*]}{|Z|\,|Z_d|} \tag{21}$$

and the denominator is nonnegative, we have alternatively

$$\text{Pe} \mid \text{``0''} = \Pr\{\text{Re}\,[ZZ_d^*] < 0\}. \tag{22}$$

We now employ a technique used by Stein which leads to a closed form solution to equation (22).[6] Based on the simple identity

$$\text{Re}\,[ZZ_d^*] = \left| \frac{Z + Z_d}{2} \right|^2 - \left| \frac{Z - Z_d}{2} \right|^2 \tag{23}$$

we see that

$$\text{Pe} \mid \text{``0''} = \Pr\{|Z + Z_d| < |Z - Z_d|\} \tag{24}$$

For economy of notation we let

$$\begin{aligned} \Sigma &\equiv Z + Z_d = U + U_d + i(V + V_d) + C(\Sigma) \\ \Delta &\equiv Z - Z_d = U - U_d + i(V - V_d) + C(\Delta) \end{aligned} \tag{25}$$

where $C(\ )$ denotes the nonrandom components,* that is,

$$\begin{aligned} C(\Sigma) &= b[\sin \phi + \sin (\phi + \theta)] \\ &\qquad + i\{2 + b[\cos \phi + \cos (\phi + \theta)]\} \end{aligned} \tag{26}$$

$$C(\Delta) = b[\sin (\phi + \theta) - \sin \phi] + ib[\cos (\phi + \theta) - \cos \phi].$$

Thus Σ and Δ are complex (two dimensional) random variables whose jointly gaussian orthogonal components have equal variance $2\sigma^2$. We are concerned with the magnitudes, $|\Sigma|$ and $|\Delta|$, often referred to as a Ricean random variable.

The probability density function (of $|\Sigma|$, for example) is the well known result for the envelope of sine wave plus noise[4]

$$f_{|\Sigma|}(r) = \frac{r}{2\sigma^2} \exp\left[-\frac{r^2 + |C(\Sigma)|^2}{4\sigma^2} \right] I_0\left[\frac{r\,|C(\Sigma)|}{2\sigma^2} \right]. \tag{27}$$

Furthermore, $|\Sigma|$ and $|\Delta|$ are independent by the following argument. $U + U_d$ is independent of $V - V_d$. Consider $U + U_d$ and $U - U_d$. They are uncorrelated, hence independent, by virtue of their equal variance.

$$\langle [(U + U_d)(U - U_d)] \rangle_{\text{av}} = [U^2 - U_d^2] = 0 \tag{28}$$

Therefore all four random components of Σ and Δ are independent, from which the independence of $|\Sigma|$ and $|\Delta|$ follows.

The advantage of this approach is that equation (24), the probability that the amplitude of one complex gaussian (that is, a Ricean) exceeds another, is expressible in terms of the function[7]*

$$Q(A, B) \equiv \int_B^\infty \tau \exp\left(-\frac{A^2 + \tau^2}{2} \right) I_0(A\tau)\, d\tau. \tag{29}$$

One formulation is[8]

$$\begin{aligned} \Pr\{|\Sigma| < |\Delta|\} = \frac{1}{2}\Bigg\{ 1 &- Q\left[\frac{|C(\Sigma)|}{2\sigma}, \frac{|C(\Delta)|}{2\sigma} \right] \\ &+ Q\left[\frac{|C(\Delta)|}{2\sigma}, \frac{|C(\Sigma)|}{2\sigma} \right] \Bigg\}. \end{aligned} \tag{30}$$

Evaluating the magnitudes of the means from equation (26),

$$\begin{aligned} |C(\Sigma)| &= 2\left[1 + 2b \cos \frac{\theta}{2} \cos\left(\phi + \frac{\theta}{2} \right) + b^2 \cos^2 \frac{\theta}{2} \right]^{\frac{1}{2}} \\ |C(\Delta)| &= 2b \sin \frac{\theta}{2}. \end{aligned} \tag{31}$$

We have thus far found Pe|"0" exactly, conditioned on the initial interference angle ϕ. The desired result is obtained by averaging equation (30) over ϕ. However, inserting equation (31) into equation (30) leaves an expression which offers little promise for analytically performing the integration. As an alternative, the integral is both approximated and bounded in what follows, thereby avoiding a machine integration.

The integration parameter ϕ appears in $|C(\Sigma)|$ but is not a factor of $|C(\Delta)|$. Then with the identification

$$\begin{aligned} B &\equiv |C(\Sigma)|/2\sigma \\ A &\equiv |C(\Delta)|/2\sigma \end{aligned} \tag{32}$$

* θ is constant and we have conditioned the solution on ϕ.

* Q functions are tabulated, but not adequately in the argument ranges needed for these problems. Their usefulness lies in having good approximations which lead to easy machine calculations. See Ref. 8.

we are led to consider the behavior of the integrand

$$F(A, B) \equiv \tfrac{1}{2}[1 - Q(B, A) + Q(A, B)] \tag{33}$$

as a function of B only (fix A). Familiarity with $F(A, B)$ suggests that it is roughly exponential in $-B^2/2$.

To affirm this supposition we first show that

$$\frac{\partial F(A, B)}{\partial B} = -BK(A, B)F(A, B) \tag{34}$$

where $K(A, B)$ is bounded near 1:

$$1 \geqq K(A, B) \geqq 1 - \epsilon. \tag{35}$$

This approach is motivated by the recognition that if $F(A, B)$ could be approximated by an exponential in B^2, the integral of $F(A, B)$ over ϕ would be simple.

The derivation of equations (34) and (35) consists of elementary manipulations of the series and integral representations for $Q(A, B)$ which are given in Ref. 8. Henceforth, a prime designates partial differentiation with respect to B. For brevity, let

$$E \equiv \exp\left[-\frac{A^2 + B^2}{2}\right]. \tag{36}$$

Then using the series representations

$$Q(A, B) = E \sum_{m=0}^{\infty} \left(\frac{A}{B}\right)^m I_m(AB) \tag{37}$$

$$1 - Q(B, A) = E \sum_{m=1}^{\infty} \left(\frac{A}{B}\right)^m I_m(AB) \tag{38}$$

we see that

$$2F(A, B) = 2Q(A, B) - EI_0(A, B). \tag{39}$$

Then

$$2F'(A, B) = 2Q'(A, B) - EI_0'(AB) - E'I_0(AB) \tag{40}$$

$$= 2Q'(A, B) - EAI_1(AB) + BEI_0(AB). \tag{41}$$

Referring to the integral representation for the Q function we inspect

$$Q'(A, B) = -BEI_0(AB). \tag{42}$$

Therefore, from equations (42) and (41) we have

$$2F'(A, B) = -BE\left[I_0(AB) + \frac{A}{B} I_1(AB)\right]. \tag{43}$$

But from equations (37) and (38),

$$1 - Q(B, A) + Q(A, B) = E\left[I_0(AB) + 2 \sum_{m=1}^{\infty} \left(\frac{A}{B}\right)^m I_m(AB)\right]. \tag{44}$$

So that equation (34) is true for

$$K(A, B) = \frac{I_0(AB) + \left(\frac{A}{B}\right)I_1(AB)}{I_0(AB) + 2\left(\frac{A}{B}\right)I_1(AB) + 2 \sum_{m=2}^{\infty} \left(\frac{A}{B}\right)^m I_m(AB)}. \tag{45}$$

Since A, B, and the $I_m(x)$ are nonnegative the unity upper bound in equation (35) is obvious. For the lower bound, invert equation (45) and notice that

$$K(A, B)^{-1} \leqq 1 + 2 \sum_{m=1}^{\infty} \left(\frac{A}{B}\right)^m \frac{I_m(AB)}{I_0(AB)} \tag{46}$$

$$\leqq 1 + 2 \sum_{m=1}^{\infty} \left(\frac{A}{B}\right)^m \tag{47}$$

since the $I_m(x)$ decrease with the order m.

Referring to the definitions of equations (32) and (31) it can be shown that $A < B$ whenever $b < 1/(2)^{1/2}$, so that the summation converges for reasonable interference levels (CIR > 3 dB). Then

$$K(A, B)^{-1} \leqq 1 + \frac{2A}{B - A} \tag{48}$$

so that

$$\epsilon \leqq \frac{2A}{B + A}. \tag{49}$$

In most cases $B \gg A$ so that $K(A, B)$ is near 1.

We now solve the linear, first order differential equation (34) to obtain

$$F(A, B) = \mathrm{Pe}_0 \exp\left[\int_{B_0}^{B} - \tau K(A, \tau)\, d\tau\right] \tag{50}$$

with initial data $F(A, B_0) = \mathrm{Pe}_0$. The above expression for $F(A, B)$

is exact. We now approximate the exp argument by approximating $K(A, B)$ with a constant K_0. Then we are able to carry out the integration in equation (50) to obtain

$$F(A, B) \approx \text{Pe}_0 \exp\left[-\frac{K_0}{2}(B^2 - B_0^2)\right]. \tag{51}$$

We know the function at B_0, and we extrapolate to the function using an approximation of its derivative. Now the ϕ integration will range over some section of the exponentially varying $F(A, B)$ in equation (51) above. Because it is exponential, the significant contribution to the integral occurs over a relatively small range of B where $F(A, B)$ is near its maximum. This suggests that the initial data be specified at a maximum so that the approximation function is best in the important range of integration. Therefore we let

$$B_0 = B_{\min} \tag{52}$$

so that

$$\text{Pe}_0 = F(A, B_0) \geqq F(A, B), \qquad 0 \leqq \phi \leqq 2\pi. \tag{53}$$

We notice from equation (31) that

$$B_{\min} = \frac{1}{\sigma}\left(1 - b\cos\frac{\theta}{2}\right) \tag{54}$$

and so

$$B^2 - B_0^2 = \frac{2b}{\sigma^2}\cos\frac{\theta}{2}\left[1 + \cos\left(\phi + \frac{\theta}{2}\right)\right]. \tag{55}$$

In addition, we slope-match at B_0, approximating $K(A, B)$ by $K_0 = K(A, B_0)$. From equation (45)

$$K(A, B_0) \approx \frac{1 + \left(\frac{A}{B_0}\right)\frac{I_1(AB_0)}{I_0(AB_0)}}{1 + 2\left(\frac{A}{B_0}\right)\frac{I_1(AB_0)}{I_0(AB_0)}}. \tag{56}$$

Inserting equation (55) into equation (51) and integrating,

$$\frac{1}{2\pi}\int_0^{2\pi} F(A, B)\, d\phi = F(A, B_0)\exp\left[-\frac{bK_0}{\sigma^2}\cos\frac{\theta}{2}\right] \cdot \int_0^{2\pi}\frac{d\phi}{2\pi}\exp\left[-\frac{bK_0}{\sigma^2}\cos\left(\phi + \frac{\theta}{2}\right)\right] \tag{57}$$

gives the desired result

$$\text{Pe} \mid \text{“0”} = \tfrac{1}{2}[1 - Q(B_0, A) + Q(A, B_0)] \cdot \exp\left[-\frac{bK_0}{\sigma^2}\cos\frac{\theta}{2}\right] I_0\left[\frac{bK_0}{\sigma^2}\cos\frac{\theta}{2}\right]. \tag{58}$$

Exact bounds are now easily obtained by bounding $K(A, B)$. With B_0 still chosen to be $B_{\min}$, clearly

$$\int_{B_0}^{B} -\tau K(A, \tau)\, d\tau \leqq \int_{B_0}^{B} -\tau K_{\min}\, d\tau = -\tfrac{1}{2}(B^2 - B_0^2)K_{\min} \tag{59}$$

where $K_{\min}$ is the minimum value of $K(A, B)$ in $[B_0, B]$. Similarly,

$$\int_{B_0}^{B} -\tau K(A, \tau)\, d\tau \geqq -\tfrac{1}{2}(B^2 - B_0^2)K_{\max}. \tag{60}$$

Therefore

$$F(A, B) \gtrless \text{Pe}_0 \exp\left[-\tfrac{1}{2}(B^2 - B_0^2)K_{\substack{\max\\ \min}}\right] \tag{61}$$

so that

$$\text{Pe} \mid \text{“0”}, \theta \lesseqgtr \text{Pe}_0 \exp\left[-K_{\substack{\min\\ \max}}\frac{b}{\sigma^2}\cos\frac{\theta}{2}\right] I_0\left[K_{\substack{\min\\ \max}}\frac{b}{\sigma^2}\cos\frac{\theta}{2}\right]. \tag{62}$$

The ratio of the bounds in equation (62) may be bounded in order to ascertain their closeness. We omit the cumbersome derivation, but state the result below.

$$\frac{\text{Upper Bound}}{\text{Lower Bound}} \leqq \left[\frac{\frac{1}{b} - \cos\frac{\theta}{2} + \sin\frac{\theta}{2}}{\frac{1}{b} - \cos\frac{\theta}{2} - \sin\frac{\theta}{2}}\right]^{0.65} \tag{63}$$

For CIR $\geqq$ 10 (15) dB, the ratio is less than 1.6 (1.25).

The symbol “1” is transmitted as two pulses of carrier 180° apart. Then correct reception results if $90° < |\delta| < 180°$. If an analysis quite similar to the preceding were carried out for this case, the resulting expressions would be identical with those above except that θ is replaced by $\pi - \theta$. It follows that

$$\text{Pe} \mid \text{“1”}, \theta = \text{Pe} \mid \text{“0”}, \pi - \theta. \tag{64}$$

We will elaborate on the relationship of Pe for the different data symbols in Section V.

Notice that Pe|"0" is symmetric in θ about 0 and π. The symmetry follows from the averaging of ϕ, and the insignificance of which of the two pulses occurs first as far as the detector is concerned.

Since the overall error probability

$$\text{Pe}\,(\theta) = \tfrac{1}{2}[\text{Pe} \mid \text{“1”}, \theta + \text{Pe} \mid \text{“0”}, \theta] \tag{65}$$

we use equation (64) to write

$$\text{Pe}\,(\theta) = \tfrac{1}{2}[\text{Pe} \mid \text{“0”}, \theta + \text{Pe} \mid \text{“0”}, \pi - \theta] \tag{66}$$

which is easily shown to be evenly symmetric about 0, π and $\pi/2$. We therefore need examine Pe only in the range $0 \leqq \theta \leqq 90°$.

Pe(θ) does vary considerably as seen in Fig. 7. Here the maximum and minimum values of Pe(θ), which happen to occur at $\pi/2$ and 0 respectively, are given for interesting combinations of interference and noise levels. To further illustrate the effects of interference, we present curves of decibel degradation versus θ in Fig. 8. Degradation is defined as the dB reduction in carrier-to-noise ratio which is allowed to maintain the same Pe after removing the interference.

Finally, consider an angle modulation impressed on the interference. This situation may be viewed simply as a time varying θ. Then one may average the Pe(θ) results given here over the variations of θ. If this is undesirable, the curves of Fig. 7 are certainly bounds on Pe averaged over the θ variation.

V. DIFFERENTIAL DETECTION—QUATERNARY

We now examine the effect of a single interference on a differentially detected quaternary (4-phase) signal. We will refer to the four symbols as "0", "1", "2", and "3", where the associated baud to baud phase shifts are 0, $\pi/2$, π, and $-\pi/2$ respectively. As before, the phase discriminator examines the two composite phasors, Z and Z_d, and reports their angle difference δ. The ordering of the bauds is important, since we must distinguish between $\delta = \pi/2$ and $\delta = -\pi/2$. The "0" symbol possesses the same symmetry as in the binary case; we therefore base the analysis on Pe|"0". Then in an analogous fashion we relate Pe|"0" to the probability of error for the other symbols.

In the binary case the receiver tested for the sign of Re $[ZZ_d^*]$. This test was transformable to a test between the amplitudes of two Ricean random variables, one which enjoys a closed form solution. Unfortunately, in the present case the test, which is for a "0"

$$\text{Pe} \mid \text{“0”} = \text{Pr}\left\{\cos\delta = \frac{\text{Re}\,[ZZ_d^*]}{|Z|\,|Z_d|} \leqq \frac{(2)^{\frac{1}{2}}}{2}\right\}$$

is not known to be transformable to one which has a closed solution. On the other hand, we offer a very straightforward analysis which is exact and amenable to machine computation.

Figure 9 is a "double exposure" of the signal, noise, and interference components for a "0". The two carrier pulses are, of course, coincident and lie along the reference axis. We recall that the angle of

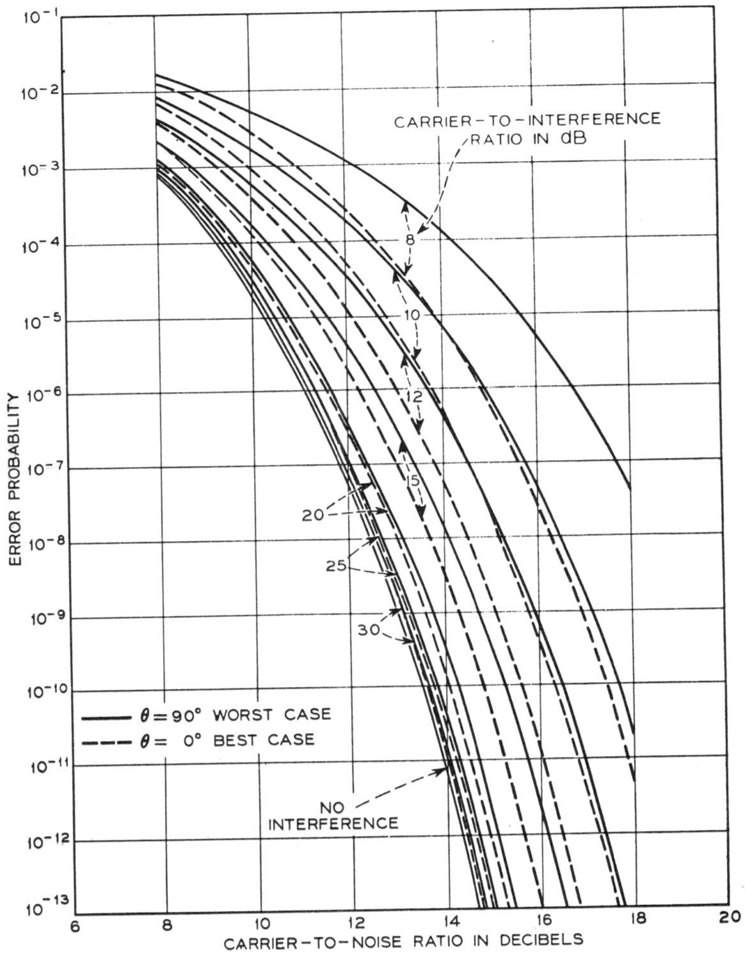

Fig. 7 — Binary Pe versus CNR. Differential detection.

the interference at t_0 is ϕ, and at $t_0 + T$ is $\phi + \theta$. We will again average over ϕ, leaving θ as a parameter.

At time t_0 the resultant of signal and interference, shown by a dashed line, has length W and angle ξ which by inspection are

$$W = (1 + b^2 + 2b \cos \phi)^{\frac{1}{2}} \tag{67}$$

$$\xi = \tan^{-1} \left[\frac{b \sin \phi}{1 + b \cos \phi} \right]. \tag{68}$$

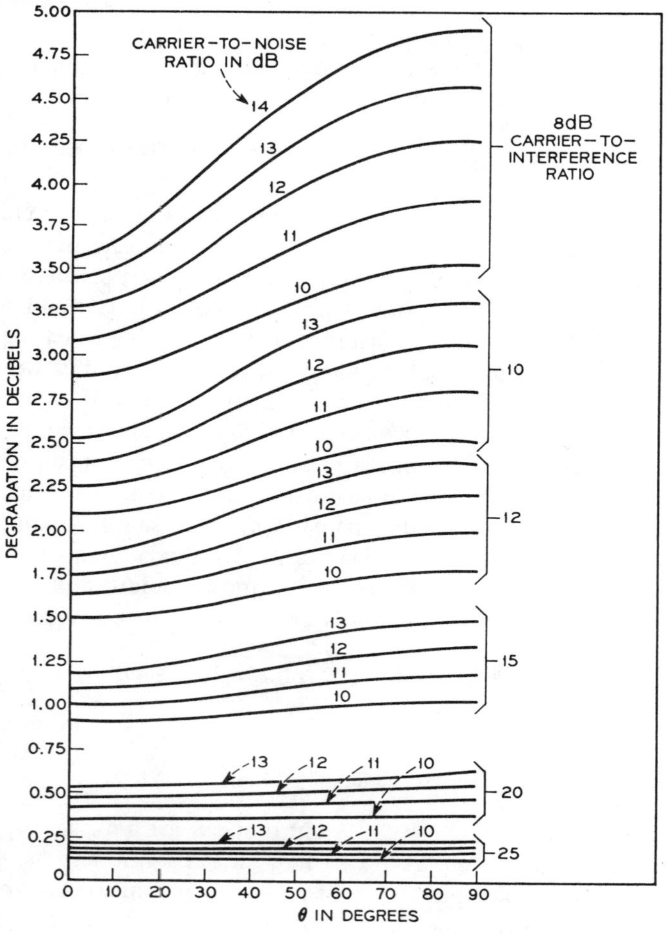

Fig. 8 — Degradation in CNR caused by interference versus θ for binary differential detection.

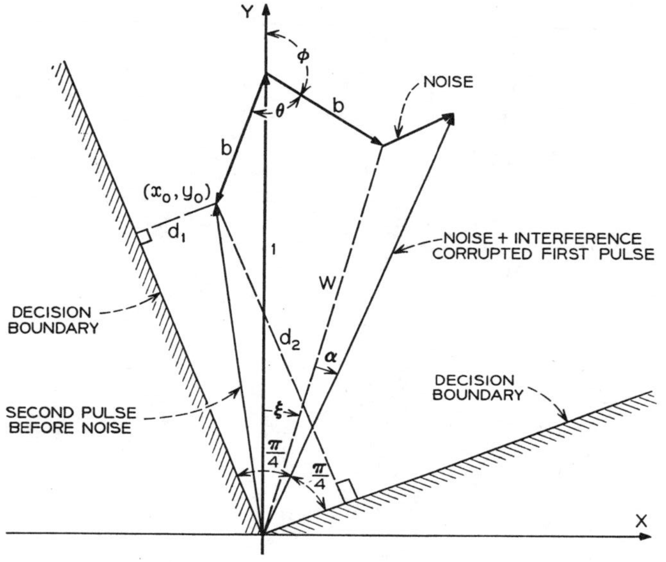

Fig. 9 — Phasor diagram for quaternary differential detection.

To the resultant of signal and interference adds a Rayleigh amplitude, uniform angle noise phasor. The resultant W is perturbed both in amplitude and angle by the noise. The resulting amplitude is unimportant, but the angle, $\xi + \alpha$, establishes the reference for detecting the second pulse. The probability density function of α is well known to be

$$f_\alpha(\alpha) = \frac{1}{2\pi} \exp(-\Psi) \{1 + \Psi^{\frac{1}{2}} \pi \cos \alpha \cdot \exp(\Psi \cos^2 \alpha)[1 + \operatorname{erf}(\Psi^{\frac{1}{2}} \cos \alpha)]\} \tag{69}$$

where

$$\operatorname{erf}(x) \equiv \frac{2}{\pi^{\frac{1}{2}}} \int_0^x \exp(-u^2)\, du \tag{70}$$

is the usual error function integral and

$$\Psi = \frac{W^2}{2\sigma^2} \tag{71}$$

is the power ratio (signal plus interference)/noise.

At $t_0 + T$ the second pulse is examined. It is disturbed by interference to the point (x_0, y_0),

$$x_0 = b \sin(\phi + \theta)$$
$$y_0 = 1 + b \cos(\phi + \theta). \tag{72}$$

The decoding region is a quarter plane bisected by the direction of the first pulse as shown by the orthogonal decision boundaries. The probability of correct reception is simply the probability that a random noise phasor originating at (x_0, y_0) will terminate inside this quadrant. Using the independent orthogonal phasor representation for noise, and choosing the components to lie alongside the perpendicular distances d_1 and d_2 from (x_0, y_0) to the boundaries, we write directly

$$1 - \text{Pe} \mid \text{``0''}, \phi, \alpha = \frac{1}{2}\left[1 + \text{erf}\left(\frac{d_1}{2^{\frac{1}{2}}\sigma}\right)\right]\frac{1}{2}\left[1 + \text{erf}\left(\frac{d_2}{2^{\frac{1}{2}}\sigma}\right)\right]. \tag{73}$$

Now using erf + erfc = 1 we have

$$\text{Pe} \mid \text{``0''}, \phi, \alpha = \tfrac{1}{2}\,\text{erfc}\left(\frac{d_1}{2^{\frac{1}{2}}\sigma}\right) + \tfrac{1}{2}\,\text{erfc}\left(\frac{d_2}{2^{\frac{1}{2}}\sigma}\right) - \tfrac{1}{4}\,\text{erfc}\left(\frac{d_1}{2^{\frac{1}{2}}\sigma}\right)\text{erfc}\left(\frac{d_2}{2^{\frac{1}{2}}\sigma}\right). \tag{74}$$

The distances d_1 and d_2 may be verified to be

$$d_1 = -y_0 \sin\left(\alpha + \xi - \frac{\pi}{4}\right) + x_0 \cos\left(\alpha + \xi - \frac{\pi}{4}\right)$$
$$d_2 = y_0 \sin\left(\alpha + \xi + \frac{\pi}{4}\right) - x_0 \cos\left(\alpha + \xi + \frac{\pi}{4}\right) \tag{75}$$

such that they take the positive sign if (x_0, y_0) lies on the correct reception side of the respective boundary.

Eliminating the ϕ and α dependency results in a finite limits double integral

$$\text{Pe} \mid \text{``0''} = \int_0^{2\pi} \frac{d\phi}{2\pi} \int_{-\pi}^{\pi} f_\alpha(\alpha)\, \text{Pe} \mid \text{``0''}, \phi, \alpha\, d\alpha \tag{76}$$

which was machine evaluated.

The relationship between Pe | "0" and the other symbols is easily demonstrated graphically. Figure 10 is a phasor diagram illustrating a typical noise and interference corrupted first pulse, and the four

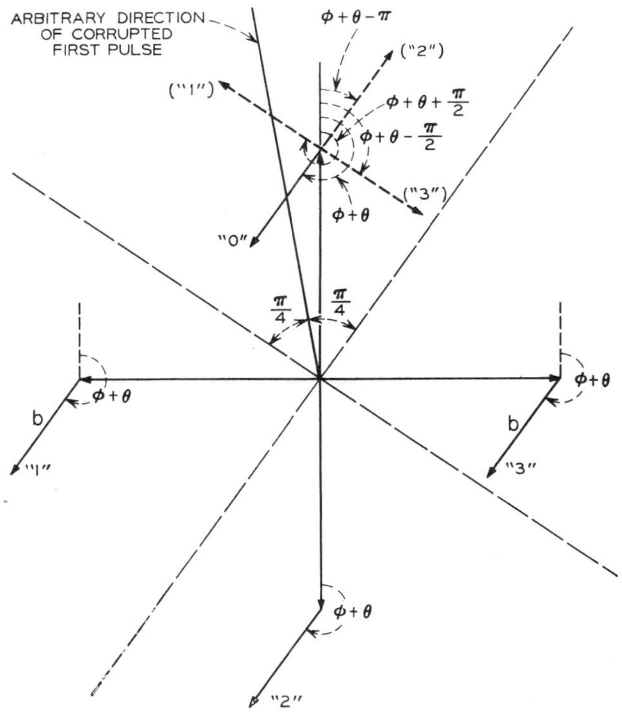

Fig. 10 — The relationship of Pe|θ for the individual symbols.

possible positions of the pre-interference second pulse. Decoding quadrants determined by the angle of the first pulse are shown by dashed lines. Adding to each of the possible second pulse carrier phasors, which correspond to the four data symbols, is an interference phasor having some angle $\phi + \theta$.

If a "0" had been sent, the probability of error conditioned on the present geometry (that is, the values of ϕ, α, and θ) is the probability that a noise phasor originating at the tip of the solid interference phasor (labeled "0") terminates outside the "0" quadrant. This probability is a function of σ and the distances from the tip of the interference phasor to the boundaries.

Now assume that a "3" had been sent instead. The associated second carrier pulse is shifted clockwise by $\pi/2$, but the interference phasor at $t_0 + T$ is still at $\phi + \theta$ relative to the diagram reference. Returning to the "0" quadrant, consider an interference phasor having the angle $\phi + \theta - \pi/2$, shown by the dashed line labelled "3". It is apparent (and trivial to show) that the distances of the dashed "3" phasor to the "0" quadrant boundaries are identical with the distances of the actual "3" interference phasor to the "3" quadrant boundaries. We conclude, therefore, that

$$\text{Pe} \mid \text{"3"}, \phi, \alpha, \theta = \text{Pe} \mid \text{"0"}, \phi, \alpha, \theta - \frac{\pi}{2}. \qquad (77)$$

Now integrating both sides of the equality over all ϕ, α yields the desired relationship

$$\text{Pe} \mid \text{"3"}, \theta = \text{Pe} \mid \text{"0"}, \theta - \frac{\pi}{2}. \qquad (78)$$

Similarly we have

$$\text{Pe} \mid \text{"2"}, \theta = \text{Pe} \mid \text{"0"}, \theta - \pi \qquad (79)$$

$$\text{Pe} \mid \text{"1"}, \theta = \text{Pe} \mid \text{"0"}, \theta + \frac{\pi}{2} \qquad (80)$$

so that the average symbol error probability becomes

$$\text{Pe}(\theta) = \frac{1}{4}\left[\text{Pe} \mid \text{"0"}, \theta + \text{Pe} \mid \text{"0"}, \theta - \frac{\pi}{2} + \text{Pe} \mid \text{"0"}, \theta + \frac{\pi}{2} + \text{Pe} \mid \text{"0"}, \theta - \pi\right] \qquad (81)$$

which is solely in terms of Pe | "0".

Notice that equation (79) is exactly the result obtained for a "1" in binary. This is not surprising, since a "2" constitutes a 180° phase shift of the second pulse. In fact, the arguments relating to Fig. 10 may be generalized for an M-phase d-psk signal with a "J" symbol phase shift of $(2\pi J)/M$, to be

$$\text{Pe} \mid \text{"}J\text{"}, \theta = \text{Pe} \mid \text{"0"}, \theta + \frac{2\pi J}{M}. \qquad (82)$$

Averaging over the symbols in equation (81) produces a Pe(θ) which is evenly symmetric about 0, $\pi/4$, $\pi/2$, . . . ; points half as far apart as in binary. Also, while Pe|"0" still varies over a considerable range, when four symbols are averaged instead of two the range of Pe(θ) is significantly decreased. This is evidenced by the numerical results plotted in Fig. 11. Again, the solid and dashed lines represent

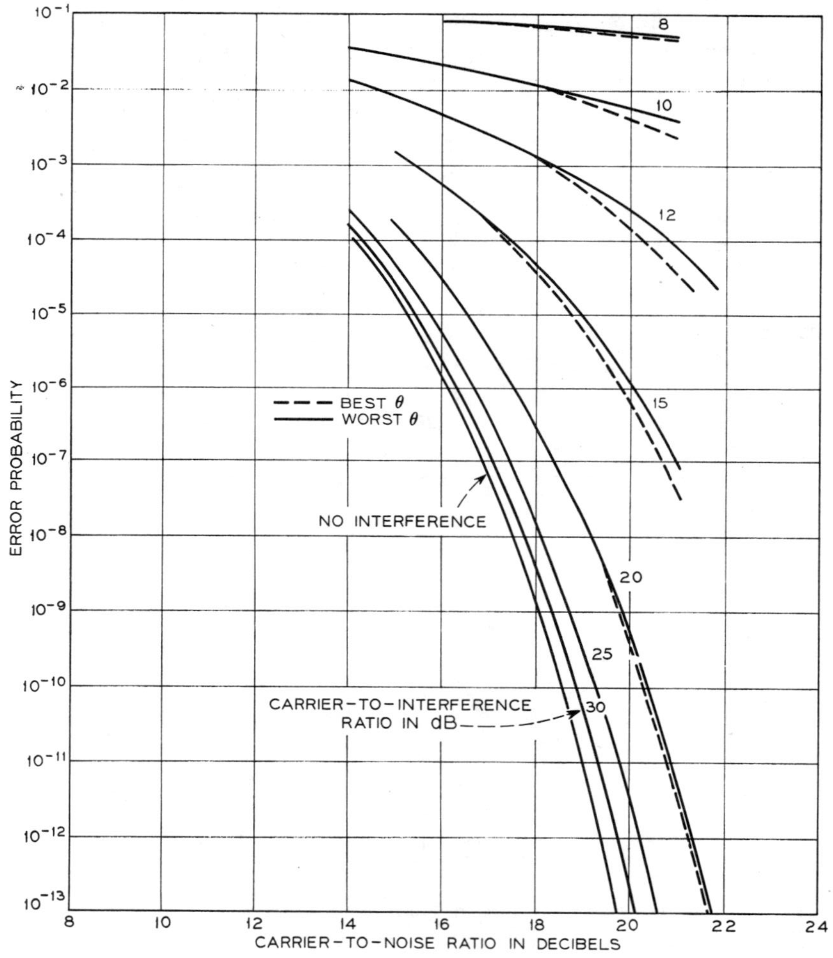

Fig. 11 — Quaternary symbol Pe versus CNR. Differential detection.

maximum and minimum values of the symbol Pe. The values of θ which correspond to the maximum and minimum are not the same for all noise and interference level combinations. However, Pe(θ) is generally lowest near 0, $\pi/2$, π . . . and highest near $\pi/4$, $3\pi/4$, . . . , and so on.

Since Pe(θ) fluctuates less severely for quaternary, it is more meaningful to average over θ. This was done, and the average used as a base for computing the degradation curves of Fig. 12.

VI. DIFFERENTIAL DETECTION — M > 4

For large M systems, for example $M \geqq 8$, we compute the average error probability

$$\text{Pe} = \frac{M}{\pi} \int_0^{\pi/M} \text{Pe}(\theta)\, d\theta \tag{83}$$

assuming now that θ is a uniformly distributed random variable. Because of the rapidly diminishing θ dependency noted in Section V, the average over θ is a useful measure of error performance. The choice of a uniform distribution for θ allows an approach which relies on the previously obtained $f_A(\alpha)$ data, rather than finding $\text{Pe}(\theta)$ and then integrating.

Let the interference phase angles at t_0 and $t_0 + T$ be random variables Φ and $\Phi + \Theta$, respectively. We first note that the sum $\Phi + \Theta$ modulo 2π is uniformly distributed since Φ (or Θ) is uniform. Hence both interference angles, Φ and $\Phi + \Theta$, are uniformly distributed. Furthermore, since for any $\Phi = \phi$ the sum $\phi + \Theta$ (modulo 2π) is uniform, we conclude that Φ and $\Phi + \Theta$ are independent. The adjacent interference angles then are independent; and consequently the phase angle, A, of $s + n + i$ is independent from sample to sample.

We therefore obtain the probability density function of the differ-

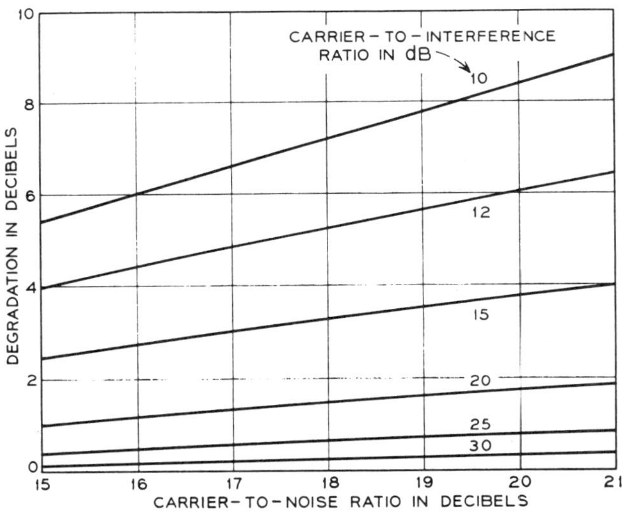

Fig. 12 — Degradation in CNR caused by interference for quaternary differential detection.

ence angle $\delta = A(t_0 + T) - A(t_0)$ as the modulo 2π convolution of $f_A(\alpha)$ with itself. Then the integral of the probability density function of δ over $|\delta| > (\pi/M)$ yields Pe. Numerical results were obtained in this fashion for $M = 4$, 8, and 16. The $M = 4$ data was in excellent agreement with Fig. 11. Pe for $M = 8$ and $M = 16$ is displayed in Fig. 13. When $b > \sin(\pi/2M)$ the interference alone can exceed the thresholds and cause errors. This is seen as a Pe floor for low CIR

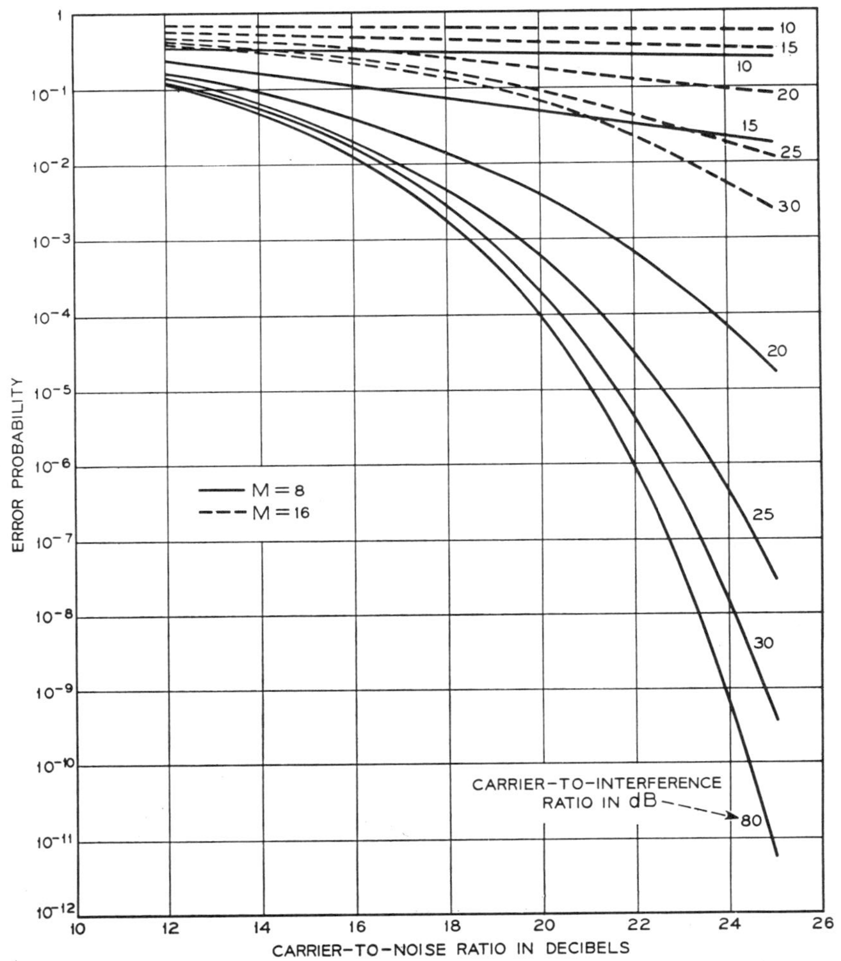

Fig. 13 — Symbol Pe versus CNR for 8 and 16 phase differential detection.

values; increasing CNR does not cause Pe to tend toward zero.

Finally, we remark that the averaging of Pe over the M symbols was implicitly done in the θ averaging. We see from equation (82) that all symbols have equal error probability when θ is uniformly distributed.

VII. SUMMARY AND CONCLUSIONS

We have evaluated the symbol error probabilities for both coherent and differential detection of low M psk signals in the presence of interference. The effect of the interference is readily observed in the curves of Figs. 3–5, 7, 11, and 13. Although the increase in Pe resulting from interference is large, it is considerably less than if the interference were replaced by gaussian noise of the same power, especially at low Pe levels.

Comparing Figs. 3, 4, and 5, we see that vulnerability increases with M. That is, at a given CNR, the Pe is raised least for $M = 2$ and greatest for $M = 4$ by the addition of interference. For example, without interference the error performance of $M = 2$ versus $M = 4$ differs by 3 dB. When a -10 dB interferer is added, it differs by approximately 5 dB, indicating a 2 dB CNR penalty for equal *Pe* values.

Drawing comparisons between coherent and differential detection reveals that differential detection clearly suffers more degradation. Binary differential, however, performs about as well with interference at optimum θ values as does binary coherent. This is in contrast with the performance disparity between the two $M = 4$ systems. With a -10 dB interference, differential detection suffers a degradation ranging from $5\frac{1}{2}$ to 8 dB; coherent detection is degraded only 4 to $4\frac{1}{2}$ dB for the same CNR range.

We use degradation rather than the raw Pe versus CNR curves to make the above comparisons because of the inherent difference in performance between differential and coherent psk for noise alone. That is, the degradation comparisons automatically reconcile any disparities in the noise-only performances of the various systems.

REFERENCES

1. Davenport, W. B. and Root, W. L., *Random Signals and Noise,* New York: McGraw-Hill, 1958, pp. 158–159.
2. Papoulis, A., *Probability, Random Variables and Stochastic Processes,* New York: McGraw-Hill, 1965, pp. 182–183.
3. Scire, F. J., "A Probability Density Function Theorem for the Modulo y Values of the Sum of Two Statistically Independent Processes," Proc. IEEE, *56,* No. 2 (February 1968), pp. 204–205.

4. Rice, S. O., "Statistical Properties of a Sine Wave Plus Random Noise," B.S.T.J., *27,* No. 1 (January 1948), pp. 109–114.
5. Arthurs, E. and Dym, H., "On the Optimum Detection of Digital Signals in the Presence of White Gaussian Noise—a Geometric Interpretation and a Study of Three Basic Data Transmission Systems," IRE Trans. on Communication Systems, *CS-10,* No. 4 (December 1962), pp. 336–372.
6. Stein, S., "Unified Analysis of Certain Coherent and Noncoherent Binary Communications Systems," IEEE Trans. on Information Theory, *IT-10,* No. 1 (January 1964), pp. 43–51.
7. Marcum, J. I., "Tables of the Q Function," Rand Corporation Memorandum RM-399 (January 1950).
8. Schwartz, M., Bennett, W. R., and Stein, S., *Communication Systems and Techniques,* New York: McGraw-Hill, 1966, Appendix A.

Error Rate Considerations for Coherent Phase-Shift Keyed Systems with Co-Channel Interference

By V. K. PRABHU

(Manuscript received July 23, 1968)

In this paper we present a theoretical analysis of the performance of an m-phase coherent phase-shift keyed system in the presence of random gaussian noise and interference. An explicit expression is given for the probability of error of the phase angle of the received signal; we show that this probability of error can be expressed as a converging power series. We show that the coefficients of this series are expressible in terms of well-known and well-tabulated functions, and we give methods of evaluating the character error rates of the systems. We also show that this error rate is minimum when all the interference power is concentrated in a single interferer, and that it attains its maximum $[P_m]_{max}$ *when the total interference power is equally distributed amongst the K interferers. The limiting case when K goes to infinity is considered. The cases of K = 1, and m = 2, 4, 8, and 16 are treated in some detail, and the results are given graphically. The usefulness of the results presented in this paper is that the designer can have at his disposal very simple expressions with which to evaluate the performance of any given Coherent Phase-Shift Keyed system when the received signal is corrupted by both interference and random gaussian noise.*

I. INTRODUCTION

The performance of coherent phase-shift keyed (CPSK) systems has been investigated by many authors;[1-5] in the transmission of information the CPSK system has been shown to be one of the most efficient techniques for trading bandwidth for signal-to-noise ratio. However, the type of noise considered by these authors is almost always limited to be random gaussian noise although most authors admit that interference other than normal noise must be considered in the design of any modulation scheme for digital transmission.

Consider the following situation. In the frequency bands above 10 GHz where the signal attenuation resulting from rain storms could be very severe, close spacings of the repeaters are almost always mandatory for reliable communication from point to point and for all periods of time.[6] In such cases the problem of interference may be much more important than the problem of noise in the optimum detection of the desired signal; hence it is very desirable to evaluate the performance of a CPSK system with co-channel and adjacent channel interference so that, for the selection of an optimum transmission scheme, comparative advantages of CPSK over other broadband modulation techniques (like FM) in combating interference can be determined.

We consider in this paper the performance of a CPSK system when the received signal is corrupted by both interference and random gaussian noise.* We first discuss binary (2-phase) and quaternary (4-phase) CPSK systems and show that exact expressions can be obtained for their probability of error P_m. These expressions are in the form of infinite power series which are shown to converge for all values of signal-to-noise ratio and for all signal-to-interference ratios above a certain level determined by the system. For $m = 2$ and 4, these error rates are calculated and the results are given in graphical form.

For $m = 3$ and for $m > 4$ we show that exact expressions for P_m are very complicated functions of signal-to-noise ratio, and signal-to-interference ratios; in this paper we only indicate how these expressions can be obtained. However, we do obtain expressions for upper and lower bounds to P_m and show that the difference between these two bounds is a monotonically decreasing function of signal-to-noise ratio, signal-to-interference ratios, and the number m of phases used in the system. For $m \geqq 4$, signal-to-noise ratio $\rho^2 \geqq 5$ dB,† and for signal-to-interference ratio $1/L^2 \geqq 20$ dB, we show that this difference is less than 5 percent, and that the upper bound can be used as a good approximation to P_m. For $m = 8$ and 16, we calculate these upper bounds and we present the results graphically.

For a given amount of interference power, we show that the character error rate is minimum when all the power is concentrated in a single interferer. If the total number of interferers is K we also show that the error rate P_m reaches its maximum $[P_m]_{max}$ when the interference power is equally distributed among all the interferers. It

* The word "noise" indicates random gaussian noise corrupting the desired received signal.

† We use the notation $b = a$ dB if $10 \log_{10} b = a$.

Reprinted with permission from *Bell Syst. Tech. J.*, vol. 48, pp. 743-767, Mar. 1969.

follows that $[P_m]_{\max}$ is a monotonically increasing function of K and attains its maximum when K goes to infinity. We show that the case of K going to infinity can be treated in a simple manner.

For the computation of error rates P_m (or upper bounds to P_m, $m > 2$) it is necessary to calculate the central moments μ_{2n}'s of a certain random variable η defined in terms of the K interfering carriers. For large values of K the conventional method of evaluating μ_{2n}'s can be rather tedious; we give some simple methods of evaluating these moments.

In conclusion, this paper determines the performance of m-phase CPSK systems for the important case of signals corrupted by random gaussian noise and interference. The cases of $m = 2, 4, 8,$ and 16 are treated in some detail.

II. PHASE ANGLE DISTRIBUTION IN CPSK SYSTEMS

Let us consider an m-phase CPSK system. We assume that there is a steady received signal* which is corrupted by random gaussian noise and interference. The gaussian noise is assumed to have zero mean and variance σ^2. The signals under consideration consist of phase-modulation pulses of specified width transmitted at a known repetition rate; we assume that there are K interferers, each interferer having the same form as the signal.

If we assume that each signal transmitted has a duration T, the received signal waveform in the absence of noise during the Nth interval can be represented as

$$s_N(t) = (2S)^{\frac{1}{2}} \cos (\omega_0 t + \theta),\ NT \leqq t \leqq (N+1)T, \tag{1}$$

where S is the received signal power, ω_0 is the angular frequency of the signal, and θ will have some value in the discrete set $2\pi k/m$, $0 \leqq k \leqq m - 1$, corresponding to the Nth message. All m messages are assumed to be equally likely. In the absence of noise and interference, the set of m possible received signals is described by a set of m equally-spaced vectors in the complex plane as shown in Fig. 1. The noise and interference corrupting the signal distort the signal both in amplitude and in phase; a zero-phase signal (corresponding to $k = 0$), as disturbed by noise and interference, is also shown in Fig. 1.

If we now assume that power in the jth interferer is I_j, the jth interferer as received during the Nth interval can be represented as*

$$i_{jN}(t) = (2I_j)^{\frac{1}{2}} \cos \{\omega_j t + \theta_j + \mu_j\},\ NT \leqq t \leqq (N+1)T \tag{2}$$

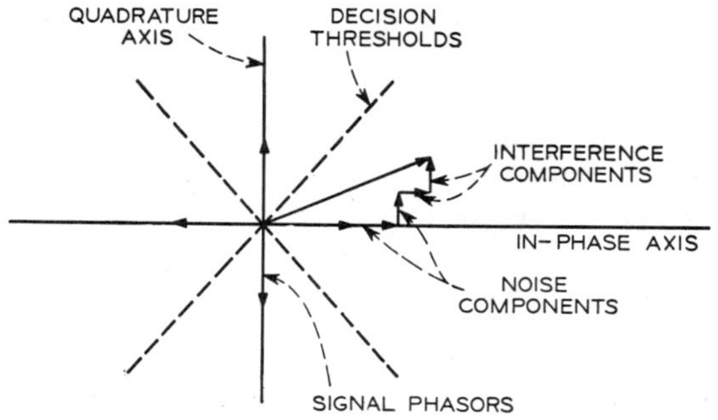

Fig. 1 — Phasor representation of CPSK signals for $m = 4$.

where ω_j is the angular frequency of the jth interferer, θ_j is some value in the discrete set $(2\pi/m)\,k$, $0 \leqq k \leqq m - 1$, and the probability density $\pi_{\mu_j}(\mu_j)$ of μ_j is given by

$$\pi_{\mu_j}(\mu_j) = \begin{cases} \dfrac{1}{2\pi}, & 0 \leqq \mu_j < 2\pi \\ 0, & \text{otherwise.} \end{cases} \tag{3}$$

Since the K interferers are assumed to originate from K different sources, it is reasonable to assume that all μ_j's are statistically independent of each other and are also independent of gaussian noise $n(t)$.

The total received signal during the Nth interval can then be written as

$$r_N(t) = (2S)^{\frac{1}{2}} \cos (\omega_0 t + \theta) + \sum_{j=1}^{K} (2I_j)^{\frac{1}{2}} \cos (\omega_j t + \theta_j + \mu_j) + n(t),$$
$$NT \leqq t \leqq (N+1)T \tag{4}$$

where $n(t)$ has zero mean and variance σ^2.

Assuming that the receiver used in the system detects only the phase angle Φ of $r_N(t)$ and does not respond to its amplitude variations,† we can write[8]

---

* In this paper we do not consider the effects of fading on the error rates of CPSK systems. The effects of fading can usually be accounted for by a further integration of error rates obtained in this paper.[7]

* We assume that all i_{jN}'s, $1 \leqq j \leqq K$, are in the passband of the CPSK receiver used in the system.

† This can be achieved in practice by using an ideal limiter at the front end of the receiver. If $A(t)e^{j\varphi(t)}$ is the input to an ideal limiter, its output is given by $A_0 e^{j\varphi(t)}$ where A_0 is a constant.

$$\Phi = \tan^{-1} \frac{\hat{r}_N(t)}{r_N(t)} - \omega_0 t \tag{5}$$

where $\hat{r}_N(t)$ is the Hilbert transform of $r_N(t)$ and is given by

$$\hat{r}_N(t) = \frac{1}{\pi} \int_{-\infty}^{\infty} \frac{r_N(\tau)}{t - \tau} \, d\tau. \tag{6}$$

Let us write

$$n(t) = I_c \cos (\omega_0 t + \theta) - I_s \sin (\omega_0 t + \theta). \tag{7}$$

We can show[9] that I_c and I_s are two independent gaussian random variables each distributed with mean zero and variance σ^2.* From (4)–(7), we can now show that

$$\Phi = \theta + \tan^{-1} \frac{I_s + \sum_{i=1}^{K} (2I_i)^{\frac{1}{2}} \sin [(\omega_i - \omega_0)t + \theta_i - \theta + \mu_i]}{(2S)^{\frac{1}{2}} + I_c + \sum_{i=1}^{K} (2I_i)^{\frac{1}{2}} \cos [(\omega_i - \omega_0)t + \theta_i - \theta + \mu_i]}. \tag{8}$$

Let us now write

$$\rho = \frac{S^{\frac{1}{2}}}{\sigma}, \tag{9}$$

$$\frac{I_s}{(2S)^{\frac{1}{2}}} = v, \tag{10}$$

$$\frac{I_c}{(2S)^{\frac{1}{2}}} = u, \tag{11}$$

$$\delta = \sum_{i=1}^{K} R_i \sin \lambda_i , \tag{12}$$

$$\eta = \sum_{i=1}^{K} R_i \cos \lambda_i , \tag{13}$$

where

$$R_i = \left(\frac{I_i}{S}\right)^{\frac{1}{2}}, \tag{14}$$

and

$$\lambda_i = (\omega_i - \omega_0)t + \theta_i - \theta + \mu_i . \tag{15}$$

* It is assumed that the spectrum of gaussian noise is symmetrical around the frequency $\omega = \omega_0$.

Let us also denote the set $\{\lambda_1, \lambda_2, \cdots, \lambda_i, \cdots, \lambda_K\}$ of random variables λ_i's by $\underline{\lambda}$.

We can now write eq. (8) as

$$\Phi = \theta + \tan^{-1} \frac{v + \delta}{1 + u + \eta} \tag{16}$$

where δ and η are functions of $\underline{\lambda}$.

If K is a finite number, we can show[10] that the probability density $p_\eta(\eta)$ can be represented as*

$$p_\eta(\eta) = \frac{1}{2\pi} \int_{-\infty}^{\infty} e^{i\eta t} \prod_{i=1}^{K} J_0(tR_i) \, dt, \tag{17}$$

where $J_0(x)$ is the Bessel function of the first kind and of order zero. For $K = 1$, we can show that[10]

$$p_\eta(\eta) = \begin{cases} \frac{1}{\pi} \frac{1}{(R_1^2 - \eta^2)^{\frac{1}{2}}}, & |\eta| \leqq R_1 \\ 0, & \text{otherwise.} \end{cases} \tag{18}$$

For $K = 2$, $p_\eta(\eta)$ can be expressed in terms of elliptic functions, and for $K > 2$, no closed form expressions can be obtained for $p_\eta(\eta)$. In Ref. 10 $p_\eta(\eta)$ has been expressed as a converging sum and has been evaluated for $K = 10$. It is easy to show that

$$p_\eta(\eta) = 0, \quad \text{for} \quad |\eta| > \sum_{i=1}^{K} R_i , \tag{19}$$

and

$$\int_{-\infty}^{\infty} p_\eta(\eta) e^{-i\eta t} \, d\eta = \prod_{i=1}^{K} J_0(tR_i). \tag{20}$$

III. CPSK RECEIVER

An ideal CPSK receiver is shown in Fig. 2. The ideal limiter removes all the amplitude variations of the received signal before it reaches the ideal phase detector of the system. We shall assume maximum likelihood detection for our analysis of the receiver. Let us assume that the receiver shown in Fig. 2 has zero-width decision thresholds as shown in Fig. 1.

* We can also write similar expressions for $p_\delta(\delta)$.

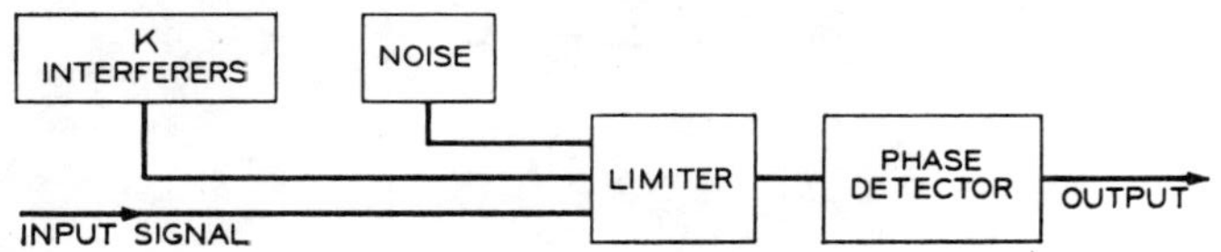

Fig. 2 — CPSK receiver.

3.1 *Error Rates for Binary CPSK Systems*

For a binary CPSK system the set of two possible received signals in the absence of noise and interference is shown in Fig. 3. The noise and interference corrupting the desired signal distort the signal both in amplitude and in phase; a zero-phase signal (corresponding to $k = 0$) as disturbed by noise and interference is also shown in Fig. 3.

When the message $k = 0$ is sent, and when the phase angle Φ of the received signal lies in the second and third quadrants of the complex plane shown in Fig. 3, an error is made in detecting the received signal. For a given ρ^2, and for an arbitrary set of λ_i's let us assume that the origin of the gaussian noise vector is at point G in Fig. 3. When the terminus or tip of the gaussian noise vector lies in the left half of the complex plane (the shaded portion of Fig. 3) an error is made by the receiver. Since I_c and I_s are two independent gaussian random variables and since they are distributed independently of λ_i's, the probability $P_2(\underline{\lambda})$ that the terminus of the gaussian noise vector lies in the

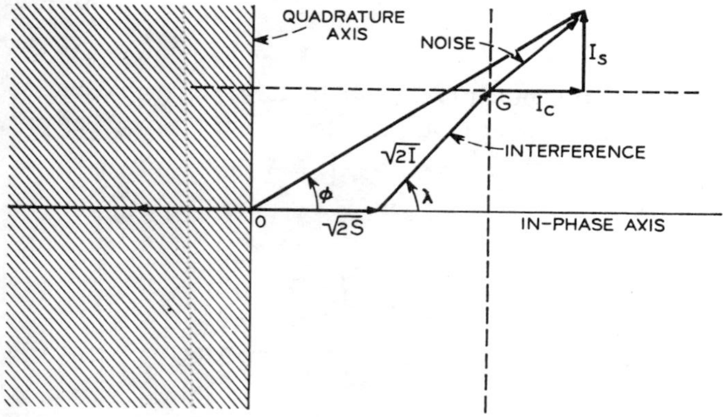

Fig. 3 — Phasor representation of CPSK signals for $m = 2$. I_c and I_s are the in-phase and quadrature components of gaussian noise corrupting the desired received signal.

left half of the complex plane is given by*

$$P_2(\underline{\lambda}) = \Pr[-\infty < I_s < \infty] \cdot \Pr\left[-\infty < I_c < -\left((2S)^{\frac{1}{2}} + \sum_{j=1}^{K} (2I_j)^{\frac{1}{2}} \cos\lambda_j\right)\right] = \frac{1}{(2\pi)^{\frac{1}{2}}\sigma} \int_{-\infty}^{-\left\{(2S)^{\frac{1}{2}} + \sum_{j=1}^{K}(2I_j)^{\frac{1}{2}}\cos\lambda_j\right\}} \exp(-t^2/2\sigma^2)\, dt. \quad (21)$$

We can show from Equation (21) that

$$P_2(\underline{\lambda}) = \tfrac{1}{2}\,\mathrm{erfc}\,[\rho + \rho\eta], \quad (22)$$

where

$$\mathrm{erf}\,(x) = \frac{2}{\pi^{\frac{1}{2}}} \int_0^x \exp(-u^2)\, du \quad (23)$$

and

$$\mathrm{erfc}\,(x) = 1 - \mathrm{erf}\,(x). \quad (24)$$

The character error rate P_2 for a binary CPSK system is, therefore, given by

$$P_2 = E[P_2(\underline{\lambda})], \quad (25)$$

where $E[P_2(\underline{\lambda})]$ represents the mathematical expectation of the random function $P_2(\underline{\lambda})$.

From Equations (22) and (25) we have

$$P_2 = \tfrac{1}{2}E[\mathrm{erfc}\,\{\rho + \rho\eta\}]. \quad (26)$$

We now note that we can write[11, 12]

$$\mathrm{erfc}\,[x + z] = \mathrm{erfc}\,[x] + \frac{2}{\pi^{\frac{1}{2}}} \exp(-x^2) \sum_{\ell=1}^{\infty} (-1)^{\ell} H_{\ell-1}(x) \frac{z^{\ell}}{\ell!}, \quad (27)$$

where $H_n(x)$ represents the Hermite polynomial of order n. The series converges for all values of $x + z$ such that

$$x + z \geqq 0. \quad (28)$$

From Equations (26) and (27) we have

$$P_2 = \tfrac{1}{2}\,\mathrm{erfc}\,(\rho) + \frac{1}{\pi^{\frac{1}{2}}} \exp(-\rho^2) \sum_{\ell=1}^{\infty} (-1)^{\ell} H_{\ell-1}(\rho) \frac{\rho^{\ell}}{\ell!} E(\eta^{\ell}). \quad (29)$$

* The notation $\Pr[a < x < b]$ denotes the probability that the random variable x satisfies the inequality $a < x < b$. It may also be noted that $P_2(\underline{\lambda})$ is a conditional probability conditioned on $\underline{\lambda}$.

Let us denote by μ_n the nth central moment of η.[8] It can then be shown that[10]

$$\mu_{2\ell+1} = 0, \qquad \ell = 0, 1, 2, \cdots . \tag{30}$$

We, therefore, have

$$P_2 = \tfrac{1}{2} \operatorname{erfc}(\rho) + \frac{1}{\pi^{\frac{1}{2}}} \exp(-\rho^2) \sum_{\ell=1}^{\infty} H_{2\ell-1}(\rho) \frac{\rho^{2\ell}}{(2\ell)!} \mu_{2\ell} . \tag{31}$$

The series given in Equation (31) converges for all values of ρ and R_j's such that

$$\rho + \rho\eta \geqq 0 \qquad \text{for all } \lambda. \tag{32}$$

From Equations (13) and (32) we can show that the series converges when

$$\Omega \leqq 1, \tag{33}$$

where

$$\Omega = \sum_{j=1}^{K} R_j . \tag{34}$$

Equation (34) states that the sum of the normalized amplitudes of all the interfering carriers may not exceed the normalized amplitude of the desired signal. This is not a very stringent requirement and it is almost always satisfied when low error rates are desired.

Since we also know that*

$$\left\{\frac{1}{K} \sum_{j=1}^{K} R_j\right\} \geqq \prod_{j=1}^{K} R_j^{1/K}, \tag{35}$$

when Equation (33) is satisfied, we have

$$\prod_{j=1}^{K} \left(\frac{I_j}{S}\right) \leqq \left(\frac{1}{K}\right)^{2K}. \tag{36}$$

The expression S/I_j denotes the signal-to-interference ratio of the jth interfering carrier.

When there is only one interfering carrier we can show that,

$$\mu_{2\ell} = R^{2\ell} \frac{(2\ell)!}{2^{2\ell}\{\ell!\}^2}, \tag{37}$$

* Equation (35) states that the arithmetic mean of a set of real variables is always greater than or equal to its geometric mean.

and equation (31) can be written as

$$P_2 = \tfrac{1}{2} \operatorname{erfc}(\rho) + \frac{1}{\pi^{\frac{1}{2}}} \exp(-\rho^2) \sum_{\ell=1}^{\infty} H_{2\ell-1}(\rho) \frac{\left[\frac{\rho R}{2}\right]^{2\ell}}{[\ell!]^2}. \tag{38}$$

The series in equation (38) converges for all signal-to-interference ratios such that

$$I/S \leqq 1. \tag{39}$$

The values of P_2 have been calculated from equation (38) and the results are given in graphical form in Fig. 4.*

Notice that we need to calculate only the even order moments μ_{2n}'s of the random variable η in determining P_2 from equation (31). Some methods of calculating these moments are given in Appendix A.

3.2 *Error Rates for Quaternary CPSK Systems*

Let us now consider a 4-phase CPSK system. For this system the set of four possible signal phasors and the four optimum decision thresholds are shown in Fig. 5. A signal phasor (corresponding to $k = 1$) as disturbed by noise and interference is also shown in Fig. 5.

For a given set of λ_j's let us assume that the gaussian noise is represented by a vector from the point G. If the message $k = 1$ is transmitted, an error is made if the received phase angle lies in areas marked 1, 2, and 3. The phase angle of the received signal will lie in areas marked 1, 2, and 3 if the terminus of the gaussian noise vector lies in this area of the plane.[14]

We notice that

$$GA = (2S)^{\frac{1}{2}} \sin \frac{\pi}{4} + \sum_{j=1}^{K} (2I_j)^{\frac{1}{2}} \sin\left(\frac{\pi}{4} + \lambda_j\right), \tag{40}$$

and

$$GB = (2S)^{\frac{1}{2}} \sin \frac{\pi}{4} + \sum_{j=1}^{K} (2I_j)^{\frac{1}{2}} \cos\left(\frac{\pi}{4} + \lambda_j\right). \tag{41}$$

Let us denote by $\Pi_{k1,k2,\cdots,kn}(\lambda)$ the probability that the terminus of the gaussian noise vector lies in area

$$\bigcup_{i=1}^{n} k_i .\dagger$$

* The results obtained in Fig. 4 indicate that the error rates obtained in Refs. 13, 14, and 15 agree well with those obtained in this paper.

† The notation $\cup_{i=1}^{n} k_i$ denotes the union of all elements of the set $\{k_1, k_2, \cdots, k_n\}$.

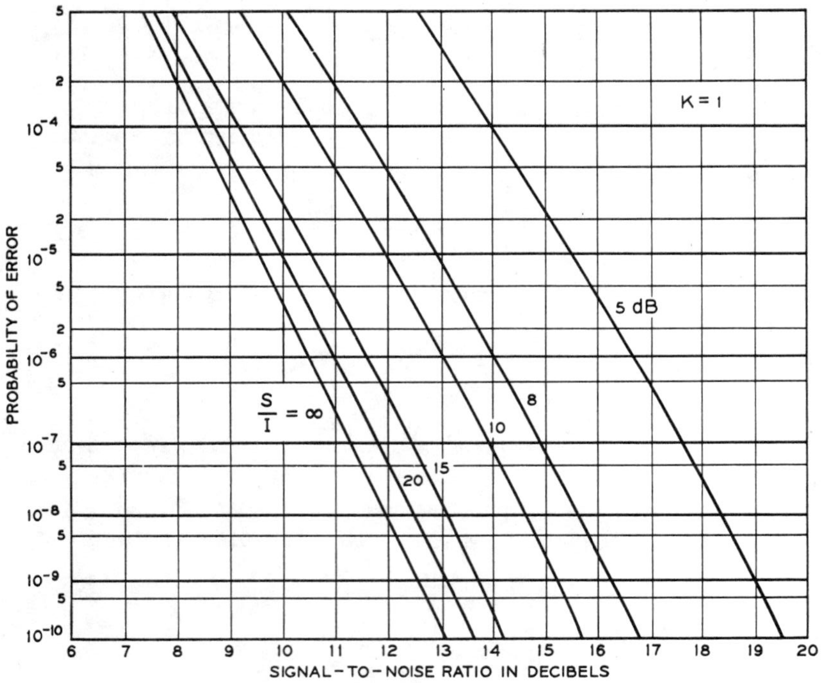

Fig. 4 — Error rates for a 2-phase CPSK system with one interferer.

We can show from Fig. 5 that

$$\Pi_{1,2}(\lambda) = \tfrac{1}{2}\,\text{erfc}\left[\rho \sin\frac{\pi}{4} + \rho \sum_{j=1}^{K} R_j \cos\left(\frac{\pi}{4} + \lambda_j\right)\right], \qquad (42)$$

$$\Pi_{2,3}(\lambda) = \tfrac{1}{2}\,\text{erfc}\left[\rho \sin\frac{\pi}{4} + \rho \sum_{j=1}^{K} R_j \sin\left(\frac{\pi}{4} + \lambda_j\right)\right], \qquad (43)$$

and

$$\Pi_2(\lambda) = \tfrac{1}{4}\,\text{erfc}\left[\rho \sin\frac{\pi}{4} + \rho \sum_{j=1}^{K} R_j \cos\left(\frac{\pi}{4} + \lambda_j\right)\right] \cdot \text{erfc}\left[\rho \sin\frac{\pi}{4} + \rho \sum_{j=1}^{K} R_j \sin\left(\frac{\pi}{4} + \lambda_j\right)\right]. \qquad (44)$$

The probability $P_4(\lambda)$ of an error due to noise is, therefore, given by

$$P_4(\lambda) = \Pi_{1,2}(\lambda) + \Pi_{2,3}(\lambda) - \Pi_2(\lambda). \qquad (45)$$

The probability of an error due to noise and interference is therefore

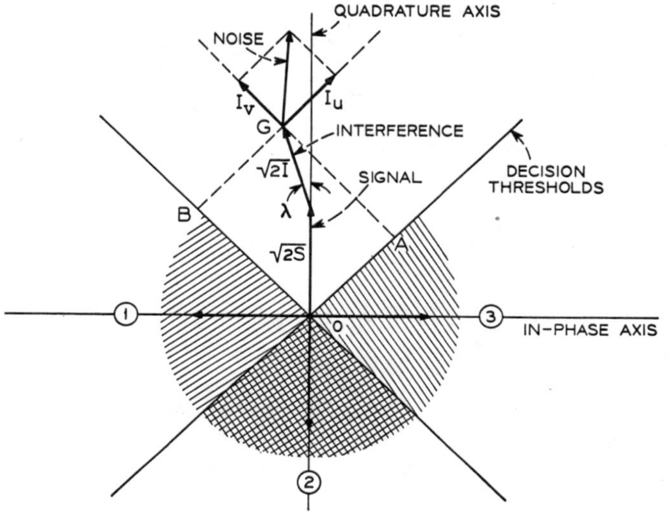

Fig. 5 — Phasor representation of CPSK signals for $m = 4$. I_u and I_v are two orthogonal components of gaussian noise.

given by

$$P_4 = E[P_4(\lambda)]. \qquad (46)$$

From equations (27), and (42) through (46) we can show that

$$P_4 = \text{erfc}\left[\rho \sin\frac{\pi}{4}\right] - \tfrac{1}{4}\,\text{erfc}^2\left[\rho \sin\frac{\pi}{4}\right] + \frac{1}{(\pi)^{\frac{1}{2}}} \exp\left(-\rho^2 \sin^2\frac{\pi}{4}\right)\left\{2 - \text{erfc}\left[\rho \sin\frac{\pi}{4}\right]\right\} \cdot \sum_{\ell=1}^{\infty} H_{2\ell-1}\left(\rho \sin\frac{\pi}{4}\right) \frac{\rho^{2\ell}}{(2\ell)!}\,\mu_{2\ell} - \frac{1}{\pi} \exp\left(-2\rho^2 \sin^2\frac{\pi}{4}\right) \cdot \sum_{\ell=1}^{\infty} \sum_{j=1}^{\infty} \frac{H_{2\ell-1}\left(\rho \sin\frac{\pi}{4}\right) H_{2j-1}\left(\rho \sin\frac{\pi}{4}\right)}{(2\ell)!\,(2j)!}\,\rho^{2(\ell+j)}\mu^*_{2\ell,2j} \qquad (47)$$

where $\mu^*_{2\ell,2j}$'s are given by

$$\mu^*_{2\ell,2j} = \frac{1}{(2\pi)^K} \int_0^{2\pi} d\theta_1 \int_0^{2\pi} d\theta_2 \cdots \int_0^{2\pi} d\theta_K \left\{\sum_{j=1}^{K} R_j \cos\theta_j\right\}^{2\ell} \cdot \left\{\sum_{\ell=1}^{K} R_\ell \sin\theta_\ell\right\}^{2j}. \qquad (48)$$

For a given set of R_i's, $\mu^*_{2\ell,2j}$'s may be evaluated from equation (48). For $K = 1$, we can show that

$$\mu^*_{2\ell,2s} = R^{2(\ell+s)} \frac{(2\ell)!\,(2s)!}{2^{2(\ell+s)}\ell!\,s!\,(\ell+s)!}. \tag{49}$$

For $K = 1$, we have calculated P_4 from equation (47) and the results are presented in Fig. 6.

We can again show that the series given in equation (47) converges for all values of ρ and R_j's such that

$$\Omega \leqq \sin\frac{\pi}{4} = \frac{1}{\sqrt{2}}. \tag{50}$$

For $K = 1$, equation (50) becomes

$$S/I \geqq 2. \tag{51}$$

Equation (50) is usually satisfied by systems encountered in practice.

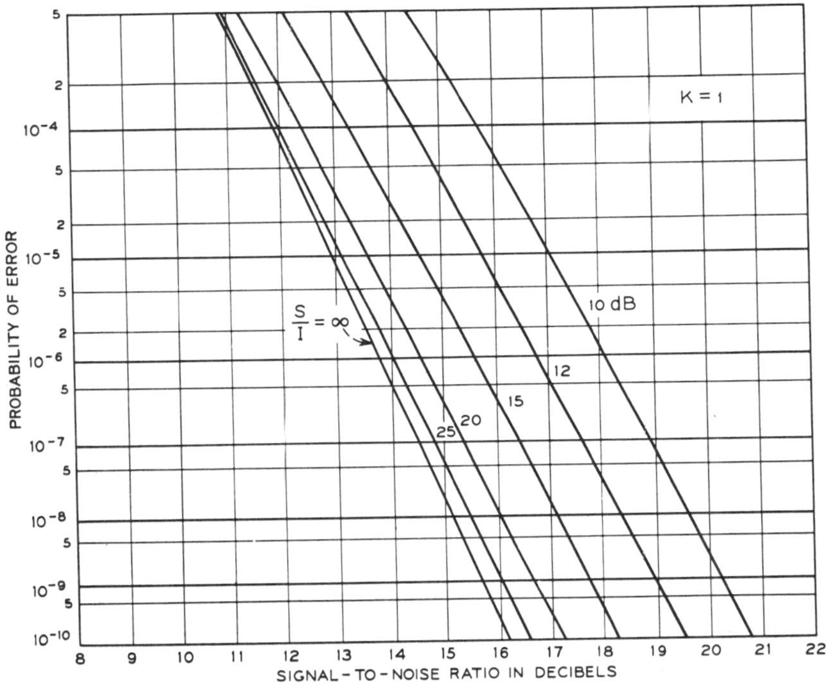

Fig. 6 — Error rates for a 4-phase CPSK system with one interferer.

3.3 *Error Rates for Multilevel CPSK Systems*

In this section we shall investigate the performance of a multilevel ($m \geqq 3$) CPSK Systems and indicate a method in which an exact expression can be obtained for the probability of error of the system. This exact expression is a very complicated function of signal-to-noise ratio and R_i's; we do not obtain this expression in this paper. However, we obtain upper and lower bounds to P_m and show that the difference between these two bounds is a monotonically decreasing function of ρ, m, and signal-to-interference ratios. For $K = 1$, $m \geqq 4$, $\rho^2 \geqq 5$ dB, and $S/I \geqq 20$ dB, we show that this difference is less than 5 percent of the lower bound, and hence the upper bound is a good approximation to P_m when low error rates are desired.

A signal phasor corresponding to $k = 0$ as disturbed by noise and interference is shown in Fig. 7. For a given set of λ_i's let us again assume that random gaussian noise is represented by a vector from the point G shown in Fig. 7. If the message $k = 0$ is transmitted, an error is made if the terminus of the noise vector lies in areas marked 1, 2, and 3. We can show that *

$$\Pi_{1,2}(\lambda) = \tfrac{1}{2}\,\mathrm{erfc}\left[\rho\sin\frac{\pi}{m} + \rho\sum_{i=1}^{K} R_i \sin\left(\frac{\pi}{m} - \lambda_i\right)\right] \tag{52}$$

and

$$\Pi_{2,3}(\lambda) = \tfrac{1}{2}\,\mathrm{erfc}\left[\rho\sin\frac{\pi}{m} + \rho\sum_{i=1}^{K} R_i \sin\left(\frac{\pi}{m} + \lambda_i\right)\right]. \tag{53}$$

The probability of error due to noise is, therefore, given by

$$P_m(\lambda) = \Pi_{1,2}(\lambda) + \Pi_{2,3}(\lambda) - \Pi_2(\lambda). \tag{54}$$

By looking at Fig. 7 we can see that no simple expression can be obtained for $\Pi_2(\lambda)$ (except when $m = 4$). $\Pi_2(\lambda)$ denotes the probability that the terminus of the gaussian noise vector lies in area 2; we shall now obtain upper and lower bounds to $\Pi_2(\lambda)$. Assume that

$$\rho\sin\frac{\pi}{m} - \rho\sum_{i=1}^{K} R_i \geqq 0 \tag{55}$$

---

* Note that

$$GA = (2S)^{\frac{1}{2}}\sin\frac{\pi}{m} + \sum_{i=1}^{K}(2I_i)^{\frac{1}{2}}\sin\left(\frac{\pi}{m} - \lambda_i\right)$$

and

$$GB = (2S)^{\frac{1}{2}}\sin\frac{\pi}{m} + \sum_{i=1}^{K}(2I_i)^{\frac{1}{2}}\sin\left(\frac{\pi}{m} + \lambda_i\right).$$

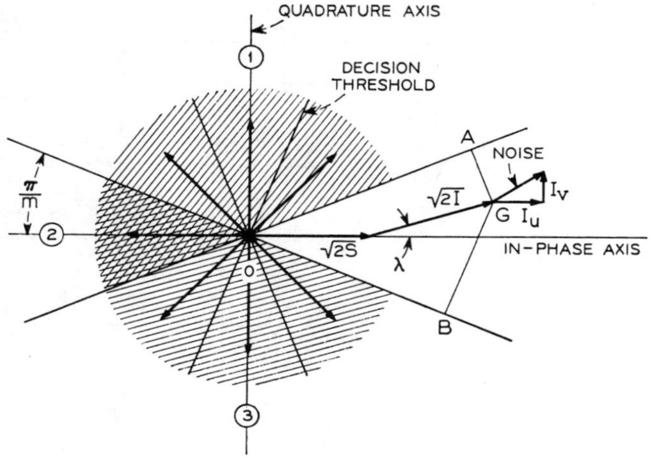

Fig. 7 — Phasor representation of CPSK signals for $m = 8$. I_u and I_v are the in-phase and quadrature components of gaussian noise.

so that $\Pi_{1,2}(\lambda)$ and $\Pi_{2,3}(\lambda)$ are nonnegative for all values of λ. If equation (55) is satisfied, it is easy to see (see Fig. 7) that

$$\Pi_2(\lambda) \geqq 0 \qquad \text{for all } \lambda, \tag{56}$$

and $\Pi_2(\lambda)$ reaches its maximum when*

$$\eta = -\sum_{j=1}^{K} R_j = -\Omega. \tag{57}$$

For this value of η it can be shown (see Fig. 8) that

$$\Pi_2(\lambda) = \frac{1}{\pi\sigma^2}\int_{-\infty}^{-y_0} \exp(-y^2/2\sigma^2)\, dy \int_0^{-(y+y_0)\tan \pi/m} \exp(-x^2/2\sigma^2)\, dx$$

or

$$\Pi_2(\lambda) = \frac{1}{\pi\sigma^2}\int_{y_0}^{\infty} \exp(-y^2/2\sigma^2)\, dy \int_0^{(y-y_0)\tan \pi/m} \exp(-x^2/2\sigma^2)\, dx \tag{58}$$

where

$$y_0 = (2S)^{\frac{1}{2}}[1 - \Omega]. \tag{59}$$

Since we always have

$$0 \leqq \exp(-x^2/2\sigma^2) \leqq 1 \text{ for all real } x, \tag{60}$$

* We can show that $\eta = -\Omega$ when all λ_j's are odd multiples of π.

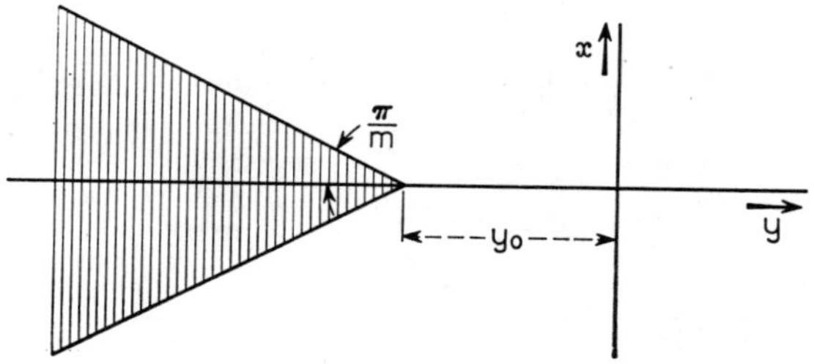

Fig. 8 — Computation of lower bound to P_m.

we have

$$\Pi_2(\lambda) \leqq \frac{\tan \pi/m}{\pi\sigma^2}\int_{y_0}^{\infty} (y - y_0)\exp(-y^2/2\sigma^2)\, dy. \tag{61}$$

Equation (61) can be simplified to*

$$\Pi_2(\lambda) \leqq Q_{m0} = \frac{\tan \pi/m}{\pi}\exp[-\rho^2(1-\Omega)^2]$$

$$\cdot [1 - (\pi)^{\frac{1}{2}}\rho(1-\Omega)\exp[\rho^2(1-\Omega)^2]\operatorname{erfc}\{\rho(1-\Omega)\}]. \tag{62}$$

From equations (54), (56), and (62) we have

$$\Pi_{1,2}(\lambda) + \Pi_{1,3}(\lambda) - Q_{m0} \leqq P_m(\lambda) \leqq \Pi_{1,2}(\lambda) + \Pi_{1,3}(\lambda). \tag{63}$$

Since

$$P_m = E[P_m(\lambda)], \tag{64}$$

we can show from equations (52), (53), (55), and (63) that

$$Q_m - Q_{m0} \leqq P_m \leqq Q_m \tag{65}$$

where

$$Q_m = \operatorname{erfc}\left(\rho \sin \frac{\pi}{m}\right)$$

$$+ \frac{2}{(\pi)^{\frac{1}{2}}}\exp\left(-\rho^2 \sin^2 \frac{\pi}{m}\right)\sum_{\ell=1}^{\infty} \frac{H_{2\ell-1}\left(\rho \sin \frac{\pi}{m}\right)}{(2\ell)!}\rho^{2\ell}\mu_{2\ell}. \tag{66}$$

* For large values of ρ and small values of Ω, Q_{m0} is approximately equal to

$$\frac{\tan \pi/m}{2\pi}\frac{\exp[-\rho^2(1-\Omega)^2]}{\rho^2(1-\Omega)^2}.$$

The series given in equation (66) converges if equation (55) is satisfied, or if

$$\Omega \leqq \sin \frac{\pi}{m}. \tag{67}$$

When low error rates are desired, equation (67) must be satisfied.

Equation (65) gives an upper and a lower bound to P_m; as can be seen from equation (62) the difference Q_{m0} between these two bounds is a monotonically decreasing function of ρ, m, and signal-to-interference ratios. From equation (65) we have

$$-\frac{Q_{m0}}{Q_m - Q_{m0}} \leqq \frac{P_m - Q_m}{P_m} \leqq 0. \tag{68}$$

For $K = 1$, $R_1 = \frac{1}{10}$, and for $m = 4$, 8, and 16, we have plotted in Fig. 9 $Q_{m0}/(Q_m - Q_{m0})$ as a function of ρ^2. From Fig. 9 we see that $Q_{m0}/(Q_m - Q_{m0})$ is less than 5 percent for $\rho^2 \geqq 5$ dB and for $m \geqq 4$. We can, therefore, use Q_m as a good approximation to P_m for high values of signal-to-noise ratio ($\rho^2 \geqq 5$ dB) and for high values of signal-to-interference ratio ($1/R_1 \geqq 10$ dB).

In these cases we then have

$$P_m \approx \operatorname{erfc}\left(\rho \sin \frac{\pi}{m}\right)$$

$$+ \frac{2}{(\pi)^{\frac{1}{2}}} \exp\left(-\rho^2 \sin^2 \frac{\pi}{m}\right) \sum_{k=1}^{\infty} \frac{H_{2k-1}\left(\rho \sin \frac{\pi}{m}\right)}{(2k)!} \rho^{2k} \mu_{2k} . \tag{69}$$

For $K = 1$, and for $m = 8$ and 16, the values of P_m obtained from equation (69) are given in Figs. 10 and 11. The error made in this approximation can be estimated from equation (68).

IV. ERROR RATE AS A FUNCTION OF NUMBER OF INTERFERERS

Let us now investigate how P_m varies as a function of K for a total given interference power. Let us assume that the total interference power is some number SL^2 where

$$\sum_{i=1}^{K} I_i = SL^2. \tag{70}$$

This power SL^2 can be distributed among the K interferers in a variety of ways; every one of these distributions will in general lead to a different character error rate of the system. Let us find out those

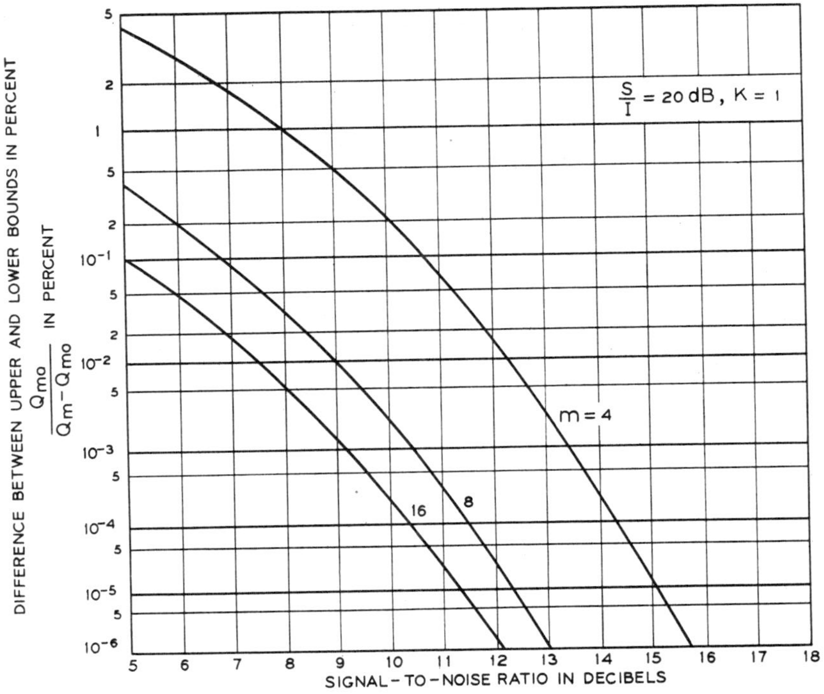

Fig. 9 — $Q_{m0}/(Q_m - Q_{m0})$ as a function of ρ.

distributions of power (if they exist) which make this character error rate a maximum or a minimum.

4.1 *Error Rates for K Interferers*

Let us first consider the case when $\rho \gg 1$ and $\Omega \ll 1$. In this case the series corresponding to P_m (or Q_m) converges very rapidly; let us say that the first N terms of the series are sufficient to evaluate P_m to the desired degree of accuracy.

For all ℓ and z, we have

$$H_{2\ell-1}(z) = 2zH_{2\ell-2}(z) - 2\,(2\ell - 2)H_{2\ell-3}(z). \tag{71}$$

From equation (71) we can show that

$$H_{2j-1}\left(\rho \sin \frac{\pi}{m}\right) \geqq 0, \qquad 1 \leqq j \leqq N, \tag{72}$$

if

$$\rho \geqq \frac{2N - \frac{3}{2}}{\sin \frac{\pi}{m}}, \qquad N > 1. \tag{73}$$

If Equations (72) and (73) are satisfied notice from Equations (31) and (69) that P_m's are monotonically increasing functions of $\mu_{2\ell}$'s, $\ell \geqq 1$. For a given μ_2, it can be shown from Equation (13) that $\mu_{2\ell}$'s $\ell \geqq 2$, reach their minimum when Ω is minimum and they reach their maximum when Ω is maximum.

We can then say that P_m's (or Q_m's) attain their minimum when Ω is minimum and that they are at their maximum when Ω is maximum.

From Figs. 3, 5, and 7 this seems to be true for all values of ρ and Ω which satisfy Equation (55).

Let us now find out when Ω is minimum for a given value of signal-

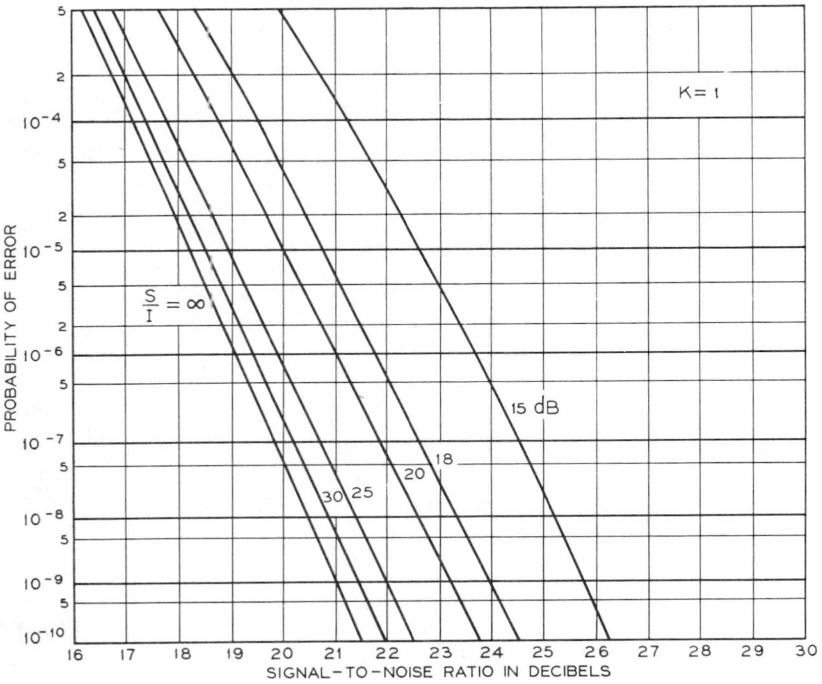

Fig. 10 — Error rates for an 8-phase CPSK system with one interferer.

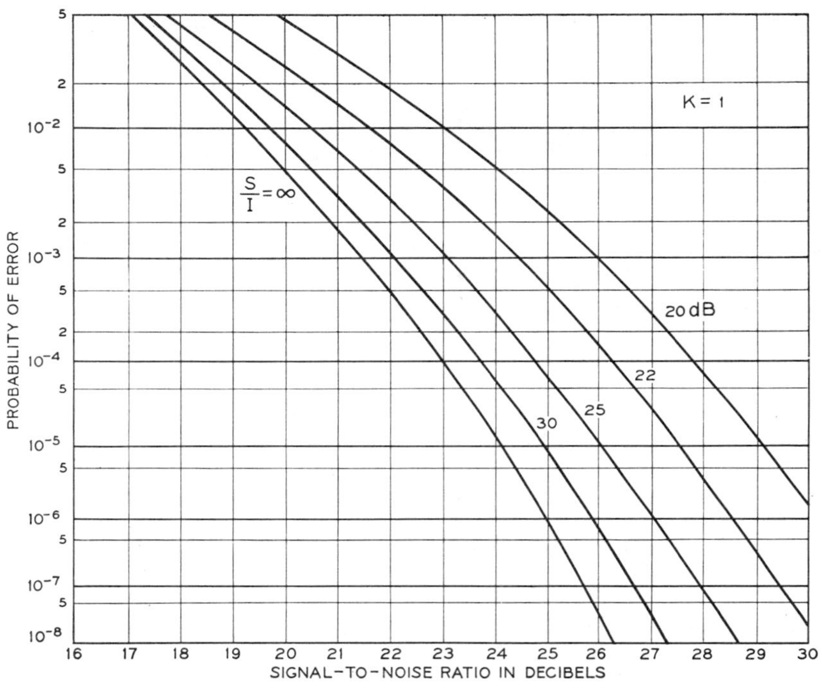

Fig. 11 — Error rates for a 16-phase CPSK system with one interferer.

to-interference ratio. The signal-to-interference ratio $1/L^2$ is given by

$$L^2 = \sum_{j=1}^{K} \frac{I_j}{S}. \tag{74}$$

Clearly Ω is minimum when

$$I_j = SL^2, \qquad 1 \leqq j \leqq K, \tag{75}$$

and

$$I_\ell = 0, \qquad 1 \leqq \ell \leqq K, \ell \neq j. \tag{76}$$

We can then say that the character error rate P_m is minimum when the total interference power is concentrated in a single interferer.

Now from equations (14), (34) and (74), Ω is a maximum when

$$\frac{\partial}{\partial I_j}\left[\sum_{j=1}^{K}\left(\frac{I_j}{S}\right)^{\frac{1}{2}} - \epsilon \sum_{j=1}^{K}\left(\frac{I_j}{S}\right)\right] = 0, \qquad 1 \leqq j \leqq K. \tag{77}$$

ϵ is a constant and is the Lagrange multiplier used in finding the extremum of Ω.

Solving equation (77) we observe that Ω is a maximum or that P_m is a maximum when

$$I_j = \frac{SL^2}{K}, \qquad 1 \leqq j \leqq K \tag{78}$$

or that the total interference power is equally distributed among the K interferers.

Let us now assume that K is a variable number. It is clear from equation (78) that $[P_m]_{\max}$ is a monotonically increasing function of K.

4.2 *Error Rates for a Large Number of Interferers**

Let us now consider the limiting case when K goes to infinity and

$$\sum_{j=1}^{K} I_j = SL^2. \tag{79}$$

We can show[10] that the probability distribution function of†

$$y(t) = \sum_{j=1}^{K} (2I_j)^{\frac{1}{2}} \cos \{\omega_j t + \theta_j + \mu_j\} \tag{80}$$

as K goes to infinity approaches that of gaussian noise with mean zero and variance SL^2 under certain conditions.

In this case we have from equation (4)

$$r_N(t) = (2S)^{\frac{1}{2}} \cos (\omega_0 t + \theta) + y(t) + n(t). \tag{81}$$

Since $y(t)$ and $n(t)$ are independent gaussian random variables their sum

$$b(t) = y(t) + n(t) \tag{82}$$

is also a random gaussian variable with mean zero and variance $SL^2 + \sigma^2$.

From equations (81) and (82) we can write

$$r_N(t) = (2S)^{\frac{1}{2}} \cos (\omega_0 t + \theta) + b(t) \tag{83}$$

where $b(t)$ is a gaussian random variable.

* The results of this section are applicable for any signal-to-noise ratio and any signal-to-interference ratio.

† Ruthroff has shown that for $K \geqq 50$ the distribution of $y(t)$ can be considered to be gaussian in practice for the computation of distortion in PM systems.[16]

The case where $r_N(t)$ can be described by equation (83) has been considered in detail in Ref. 17;* we can easily determine the deterioration in performance produced by interference from the results presented in that paper. For example, suppose that $m = 4$, $S/\sigma^2 = 16$ dB, and $L^2 = -16$ dB. Clearly

$$\frac{\sigma^2}{S} + L^2 = -13 \text{ dB} \tag{84}$$

and P_4 from Ref. 17 is given by

$$P_4 = 7.9 \times 10^{-6}. \tag{85}$$

For the calculation of the effect of interference in CPSK systems, we note that we have not shown the validity of the gaussian approximation of $y(t)$ for $K \gg 1$. However, this assumption seems to be justified for large signal-to-noise ratios and small interference-to-signal ratios.[15]

In conclusion, this section gives methods of evaluating character error rates of CPSK systems for all values of m and for all values of K. It shows that the error rate P_m is minimum when all the interference power is concentrated in a single interferer and that it attains its maximum value $[P_m]_{\max}$ when the interference power is equally distributed amongst all the interferers. We further show that $[P_m]_{\max}$ is a monotonically increasing function of the number K of interferers. We also show that the case, K going to infinity, can be treated and that the deterioration in performance produced by interference can be determined.

V. CONCLUSIONS

A method to evaluate the character error rates of CPSK systems has been presented in this paper. The received signal is assumed to be corrupted by both interference and random gaussian noise. When the number of interferers is very large it can be shown that the interference and random gaussian noise can be combined together to give rise to an equivalent noise source having gaussian properties. The variance of this random variable is the sum of variance of random gaussian noise and total interference power. In this case the analysis of the CPSK system can be done by methods presented in Ref. 17.

When K is a finite number and when $m = 2$ or 4, exact expressions

* The results presented in this paper for $S/I = \infty$ are also sufficient to determine P_m for a large number of interferers.

are given for the probability of error P_m. When $m \geqq 3$, upper and lower bounds to P_m are derived. We show that the difference between these two bounds is a monotonically decreasing function of signal-to-noise ratio ρ^2, signal-to-interference ratio $1/L^2$, and the number m of phases used in the system. For $K = 1$, $m \geqq 4$, $\rho^2 \geqq 5$ dB, and $1/R_1 \geqq 10$ dB we show that this difference is less than 5 per cent, and that the upper bound can be used as a good approximation to P_m.

We then show that for any m-phase CPSK system the character error rates can be expressed in terms of the central moments of a certain random variable η and that they can be calculated to any desired degree of accuracy by using a set of tables or by using a digital computer.

For a total given interference power we show that the character error rate P_m attains its minimum when all the power is concentrated in a single interferer, and that it reaches its maximum $[P_m]_{\max}$ when the power is equally distributed among all the K interferers. It is also shown that $[P_m]_{\max}$ is a monotonically increasing function of K.

The cases of $K = 1$, $m = 2$, 4, 8, and 16, have been treated in some detail and the results are given in graphical form. The required signal-to-noise ratio for any value of signal-to-interference ratio can be determined from these figures.

The usefulness of the presented results is that they provide the designer with some relatively simple expressions with which to evaluate the performance of any given CPSK system with interference and random gaussian noise. The only quantities he must have at his disposal are the central moments of a certain random variable η defined in terms of the K interfering carriers.

VI. ACKNOWLEDGMENT

I would like to express my very sincere thanks to Clyde L. Ruthroff for his help and advice in the preparation of this paper, as well as, for suggesting the approach taken in this paper.

APPENDIX

Evaluation of Central Moments of η

In the computation of character error rates for CPSK systems it is necessary to calculate the even order moments of the random variable η; we shall give in this section two alternate methods to evaluate these moments.

By definition μ_{2n} is given by

$$\mu_{2n} = \frac{1}{(2\pi)^K} \int_0^{2\pi} d\theta_1 \int_0^{2\pi} d\theta_2 \cdots \int_0^{2\pi} d\theta_K \left[\sum_{j=1}^{K} R_j \cos \theta_j \right]^{2n}. \tag{86}$$

By the multinomial theorem

$$\left[\sum_{j=1}^{K} R_j \cos \theta_j \right]^{2n} = \sum \frac{(2n)!}{\prod_{j=1}^{K} n_j!} \prod_{j=1}^{K} (R_j)^{n_j} \cos^{n_j} \theta_j \tag{87}$$

where n_j's are positive integers such that

$$\sum_{j=1}^{K} n_j = 2n. \tag{88}$$

Since θ_j's are statistically independent of each other, and since $\mu_{2\ell+1} = 0$ for all ℓ, we have from Equations (86) and (87)*

$$\mu_{2n} = \sum \frac{(2n)!}{\prod_{\ell=1}^{K} n_\ell!} \prod_{\ell=1}^{K} (R_\ell)^{n_\ell} \frac{(n_\ell)!}{2^{n_\ell}\left[\left(\frac{n_\ell}{2}\right)!\right]^2}, \tag{89}$$

where n_ℓ's are a set of even positive integers satisfying Equation (88).

Even though equation (89) gives an exact expression to evaluate μ_{2n}'s, it can be rather tedious to evaluate μ_{2n}'s from equation (89) for large values of n and K. We shall therefore give an alternate method to evaluate the central moments of the random variable η.

It can be shown that the probability density function $p_\eta(\eta)$ of the random variable η can be expressed as[10]

$$p_\eta(\eta) = \frac{1}{2\Omega} \left[1 + 2 \sum_{s=1}^{\infty} \cos \frac{s\pi\eta}{\Omega} \prod_{j=1}^{K} J_0\left(\frac{s\pi R_j}{\Omega}\right) \right]. \tag{90}$$

The $2n$th moment of η can be represented as

$$\mu_{2n} = \int_{-\Omega}^{\Omega} z^{2n} p_\eta(z)\, dz. \tag{91}$$

From equations (90) and (91) we can show that

$$\mu_{2n} = \Omega^{2n}\Bigg(\frac{1}{2n+1} + 2 \sum_{\ell=1}^{\infty} (-1)^{\ell+1} \cdot \left\{ \left[\prod_{j=1}^{K} J_0\left(\frac{\ell\pi R_j}{\Omega}\right) \right] \sum_{k=1}^{n} (-1)^k \frac{(2n)!}{[2n-2k+1]!\,(\ell\pi)^{2k}} \right\}\Bigg). \tag{92}$$

* For $K = 1$, equation (89) reduces to equation (37).

It can be seen that the infinite series appearing in equation (92) converges rapidly for all values of R_j's; we need take only a finite number of terms from equation (92) to estimate μ_{2n}'s. It is, therefore, easier to evaluate μ_{2n}'s from equation (92) than from equation (89) when there are a large number of interferers, and we have to take a large number of terms in estimating P_m.

REFERENCES

1. Lawton, J. G., "Comparison of Binary Data Transmission Systems," Proc. Nat'l. Convention on Military Electronics, Washington, D. C., August 1958, pp. 54–61.
2. Helstrom, C. W., "The Resolution of Signals in White, Gaussian Noise," Proc. IEEE, *43*, No. 9 (September 1955), pp. 1111–1118.
3. Cahn, C. R., "Performance of Digital Phase-Modulation Systems," IEEE Trans. on Communication Systems, *CS-7*, No. 1 (May 1959), pp. 3–6.
4. Cahn, C. R., "Comparison of Coherent and Phase-Comparison Detection of a Four-Phase Signal," Proc. IEEE, *47*, No. 9 (September 1959), pp. 1662.
5. Arthurs, E. and Dym, H., "On the Optimum Detection of Digital Signals in the Presence of White Gaussian Noise—A Geometric Interpretation and a Study of Three Basic Data Transmission Systems," IEEE Trans. on Communication Systems, *10*, No. 4 (December 1962), pp. 336–372.
6. Tillotson, L. C. and Ruthroff, C. L., "The Next Generation of Short Haul Radio Systems," unpublished work.
7. Montgomery, G. F., "A Comparison of Amplitude and Angle Modulation for Narrow-Band Communication of Binary-Coded Messages in Fluctuation Noise," Proc. IEEE, *42*, No. 2 (February 1954), pp. 447–454.
8. Rowe, H. E., *Signals and Noise in Communication Systems,* Princeton, N. J.: D. Van Nostrand Co., Inc., 1965, pp. 13–16.
9. Rice, S. O., "Mathematical Analysis of Random Noise," B.S.T.J., *23*, No. 3 (July 1944), pp. 282–332.
10. Bennett, W. R., "Distribution of the Sum of Randomly Phased Components," Quart. Appl. Math., *5*, No. 1 (April 1948), pp. 385–393.
11. Morse, P. M. and Feshbach, H., *Methods of Theoretical Physics,* New York: McGraw-Hill Book Co., Inc., 1953, pp. 786–787.
12. Magnus, W. and Oberhettinger, F., *Formulas and Theorems for the Functions of Mathematical Physics,* New York: Chelsea Publishing Co., 1943, pp. 80–82.
13. Rosenbaum, A. S., "Error Performance of Coherently Detected PSK Signals in the Presence of Gaussian Noise and Co-channel Interference," unpublished work.
14. Pagones, M. J., "Error Probability Upper Bound of a Coherently Detected PSK Signal Corrupted by Interference and Gaussian Noise," unpublished work.
15. Koerner, M. A., "Effect of Interference on a Binary Communication Channel Using Known Signals," Technical Rep. 32-1281, Jet Propulsion Lab., California Inst. Technology, Pasadena, Calif., December 1968.
16. Ruthroff, C. L., "Computation of FM Distortion in Linear Networks for Bandlimited Periodic Signals," B.S.T.J., *47*, No. 6 (July–August 1968), pp. 1043–1063.
17. Prabhu, V. K., "Error Rate Considerations for Digital Phase-Modulation Systems," to be published in IEEE Trans. on Communication Technology, *17*, No. 1 (February 1969).

Binary PSK Error Probabilities with Multiple Cochannel Interferences

ARNOLD S. ROSENBAUM

Abstract—**Exact error probabilities are given for a binary phase-shift-keyed (PSK) signal which is transmitted over a nondistorting channel, but additively corrupted by Gaussian noise and cochannel interference. Both coherent detection and quasi-coherent (noisy reference) detection are considered. The analysis admits a wide class of interferences, but numerical results are limited to interferences modeled by a sum of randomly phased sinusoïds. This includes most extraneous analog frequency-modulation (FM), phase-modulation (PM), or digital PSK signals, or combinations of the same which are cochannel with the desired PSK signal.**

I. Introduction

FOR a phase-shift-keyed (PSK) digital radio transmission system, perhaps in the microwave band, two main sources of performance degradation (i.e., errors) are the thermal noise present in the channel or generated in the receiver and unwanted signals emanating from the same or nearby stations. While it is recognized that other factors contribute to the overall system performance, notably fading and receiver anomalies (including intersymbol interference), attention will be fixed on noise and cochannel interference.

Paper 69TP465-COM, approved by the Radio Communication Committee of the IEEE Communication Technology Group for publication after presentation at the 1969 IEEE International Conference on Communications, Boulder, Colo., June 9–11. Manuscript received September 12, 1969.

The author is with the Bell Telephone Laboratories, Inc., Holmdel, N. J. 07733.

Previous contributions have dealt with this problem. The single interference case for both coherent [1]–[4] (CPSK) and differentially coherent [2] (DPSK) reception of binary and multiphase signals is well documented. An elaboration of the single interference CPSK approach handles multiple interferences [3], but no calculations were heretofore available. This paper presents comprehensive results for multiple interferences for CPSK, but its main result is the exact error probability for binary DPSK with multiple interferences.

Thus a fairly complete picture of binary PSK error performance is offered here for the important situation where the effects of "front end" receiver noise and cochannel interferences predominate in causing bit errors. In the Conclusion, comparisons are drawn between CPSK and DPSK with respect to interference immunity. Most importantly, the old resort of treating cochannel interference as additional Gaussian noise of equal power is dismissed as being quite pessimistic in many cases, even for a moderate number of interferences.

II. Signals, Noise, and Interference

A binary PSK signal $s(t)$ with baud rate T^{-1} is given by

$$s(t) = \sum_{k=-\infty}^{\infty} a_k g(t - kT) \sin \omega_0 t \qquad (1)$$

Reprinted from *IEEE Trans. Comm. Tech.*, vol. COM-18, pp. 241-253, June 1970.

where $g(t)$ is a low-pass pulse-like function which establishes the bandwidth of the signal. The a_k are ± 1 according to the binary information source, with appropriate (absolute or differential) encoding.

The transmission channel is assumed to be transparent, and so the signal arrives undistorted at the receiver. There, Gaussian noise $n(t)$ and interference $i(t)$ are added. The noise, which was originally zero mean, white, and stationary, can be expressed at the output of the symmetrical receiver filter as

$$n(t) = x(t) \cos \omega_0 t - y(t) \sin \omega_0 t \tag{2}$$

where now $x(t)$ and $y(t)$ are jointly normal, independent, and low-pass random processes.

Let the bandpass interference at the output of the receiver filter be written as

$$i(t) = R(t) \sin [\omega_0 t + \Psi(t)] \tag{3}$$

such that it has an envelope R and phase Ψ with respect to the carrier. This analysis requires that $\Psi(t)$ and $R(t)$ are independent, and that the probability density function (pdf) of Ψ is a constant $(2\pi)^{-1}$. Thus the time samples of Ψ are uniformly distributed over $(0,2\pi)$, and the position of $i(t)$ in a $\cos \omega_0 t$, $\sin \omega_0 t$ phasor diagram[1] is random with a circularly symmetric two-dimensional pdf,

$$f_{R,\Psi}(r,\psi) = (2\pi)^{-1} f(r). \tag{4}$$

The analysis for conventional DPSK requires two further restrictions: that adjacent (by T seconds) samples of the noise are independent, and that adjacent samples of the interference are independent. Note that the independence of the noise samples is obtained if only the noise autocorrelation vanishes for $\pm T$, a condition usually consistent with the shape of the receiver filter. The condition is extended to general DPSK by having the "reference" and "data" signals independently corrupted.

Fortunately, many interferences of interest satisfy these requirements of circular symmetry and reference signal or adjacent sample independence. While numerical results are given here for several randomly phased constant-amplitude interferences only, the analysis is by no means limited to these cases. Indeed, it will be seen that the only information required about the interference to calculate the probability of error P_e are the moments of $R(kT) \sin \Psi(kT)$, i.e., the moments of the projection of the two-dimensional density onto the $\sin \omega_0 t$ (or $\sin [\omega_0 t + \alpha]$, $0 \leq \alpha \leq 2\pi$) axis.

The two-dimensional pdf of a single randomly phased sinusoid of constant amplitude a is simply

$$f(r,\psi) = (2\pi)^{-1} \delta(r - a) \tag{5}$$

which exhibits the required form (4). Now consider a sum of constant-amplitude sinusoids, whose phases ϕ_j are uniform and independent,

$$i(t) = \sum_{j=1}^{N} a_j \sin (\omega_0 t + \phi_j) \tag{6}$$

and where the a_j are arbitrary but constant. One may show that this $i(t)$ also has a circular pdf. One proof involves taking characteristic functions[2] of the components, which are themselves circularly symmetric in the transform domain. Then the characteristic function of the sum, which is a product of the characteristic functions of the components, is certainly circular and thus inverts to a circular pdf for $i(t)$.

Naturally, the radial function $f(r)$ will not be a simple impulse for a sum of sinusoids (6), but the density is of the form (4). This points up the general result that a sum of independent random variables with circular two-dimensional pdfs yields a random variable with circular symmetry.

Another example is an interfering signal returning from a random scatterer, e.g., the ionosphere, such that the phase is uniform and the amplitude is random as well, but independent of the phase. If it is possible to compute the moments of $R \cos \Psi$, then P_e may be found for this interference as well.

It is not practical, nor would space allow, to enumerate many other examples. Rather, it is better that particular cases of interference are compared with the aforementioned criteria to see if they qualify on an individual basis. In most instances extraneous analog frequency-modulation (FM) or digital PSK signals, where the interference and desired signal carrier frequencies are incommensurate, and where the baud rates are not related, will be acceptable interferences. When an interference is known to have a certain phase or frequency relationship to the signal, the P_e might be found directly; it will in general differ markedly from the P_e given here for "completely random" relationships between the interference and the desired signal.

The assumed absence of intersymbol interference is denoted by

$$g(kT) = 0, \quad k \neq 0 \tag{7}$$

and the signal amplitude is normalized to unity at the sample instant,

$$g(0) = 1. \tag{8}$$

The carrier-to-noise ratio γ is defined as the signal-to-noise power ratio at a sample time, thus

$$\gamma = g^2(0)/[2\langle n^2(kT)\rangle_{\text{av}}] = (2\sigma_n^2)^{-1} \tag{9}$$

$$\text{CNR} \triangleq 10 \log_{10} \gamma \text{ dB}. \tag{10}$$

The relationship between γ and the average signal power in the channel depends on the particular $g(t)$ used. In

[1] A polar plot of the R,Ψ variables; the locus of the signal as a low-pass function of time in this $\cos \omega_0 t$, $\sin \omega_0 t$ two space is a signal-space diagram [5].

[2] See Papoulis [6] for a good account of two-dimensional characteristic functions (Fourier transforms), and in particular the relationship of circular symmetric functions and Hankel transforms (not discussed here).

the common instances where $g(t)$ is constant (FM DPSK) or has raised cosine spectral shape, a particularly simple relationship exists.

III. Error Probability Analysis

At a sample time kT the signal occupies one of two positions, $+1$ or -1, on the in-phase ($\sin \omega_0 t$) axis. The receiver uses symmetric half-plane decision regions, which are optimum for Gaussian noise and equal a priori symbol probabilities. The error event occurs when the received composite of signal plus noise plus interference occupies the wrong half-plane at the sample instant.

For CPSK the decision boundary is fixed along the quadrature axis according to the known absolute phase of the signal carrier. For generalized DPSK the half-plane decision regions are fixed according to a derived estimate of the signal carrier phase. The estimate might be made from the phase of the last received baud (single-difference DPSK) or some average of the phases of several preceding bauds. The phase estimate may even be provided by a separate pilot signal at a different frequency (transmitted reference PSK), perhaps not even in the same channel.

In any case, generalized DPSK is characterized by a phase reference (hence decision boundary location) which suffers random perturbations.

Since the method used here is suitable for the general DPSK problem, it will be described as such in general terms, and then almost trivially specialized to the more common single-difference DPSK for which numerical results are given. The restriction for the general problem is that the reference signal is also corrupted by a circularly symmetric (in the signal space of the reference receiver) disturbance, e.g., Gaussian noise and a circular interference. Note that if the reference is sent over a channel separated in space and frequency from the data signal, the required independence of reference and data disturbances (both noise and interference) is obtained *de facto*.

Fig. 1 is a phasor diagram indicating the transmitted signal[3] as would be received in the absence of noise and interference, and a dashed line corresponding to the receiver's estimate of the signal carrier phase. This estimate is shown to be in error by α, a random variable whose pdf is certainly even about 0. The decision boundary is correspondingly misaligned by α, as shown.

The conditional probability of error, given α,k, is just the probability that $s(kT) + n(kT) + i(kT)$ lies in the wrong half-plane. It is obvious from the circular symmetry of $n + i$ that this is a function of the distance between the signal and the threshold, i.e., $\cos \alpha$, and not the orientation of the threshold. Therefore, consider defining a half-plane probability, $P[\tau]$, $\tau > 0$, which is the probability that noise plus interference lies in a half-plane τ units removed from the mean. This is extended to include negative τ by defining $P[\tau]$, $\tau \leq 0$, the probability that $n + i$ is in that half-plane which contains the mean,

[3] Assuming the zero phase change signal was sent.

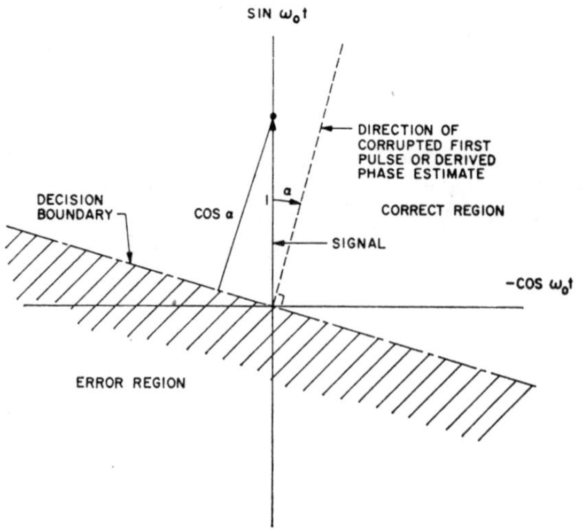

Fig. 1. Receiver signal space showing half-plane decision regions aligned according to reference phase estimate.

i.e., the opposite one. From this definition it is clear that

$$\lim_{\tau \to \infty} P[\tau] = 0$$

$$\lim_{\tau \to -\infty} P[\tau] = 1$$

$$P[0] = \tfrac{1}{2}$$

$$P[\tau] + P[-\tau] = 1. \tag{11}$$

The noise and interference are stationary of order 1 at least, so

$$P_e \mid \alpha,k = P_e \mid \alpha. \tag{12}$$

Then from the preceding definition

$$P_e \mid \alpha = P[\cos \alpha] \tag{13}$$

and from the theorem of total probability in the continuous case

$$P_e = \int_{-\pi}^{\pi} P[\cos x] f_\alpha(x)\, dx. \tag{14}$$

Up to this point, the analysis is very similar to Cahn's [7] derivation of the celebrated $\frac{1}{2} \exp(-\gamma)$ error probability expression for single-difference DPSK considering noise only.[4] The following, however, takes a different approach in an attempt to solve (14) without having $f_\alpha(x)$ directly. The first step is to integrate (14) by parts once to obtain

$$P_e = P[\cos x] F_\alpha(x) \mid_0^\pi + \int_0^\pi F_\alpha(x) P'[\cos x] \sin x\, dx \tag{15}$$

where $P' \triangleq (d/d\tau) P[\cdot]$ and

$$F_\alpha(x) \triangleq 2 \int_0^x f_\alpha(\tau)\, d\tau = \Pr\{|\alpha| \leq x\} \tag{16}$$

[4] When noise alone is present, $P[\cos x] = \frac{1}{2} \operatorname{erfc}(\gamma^{1/2} \cos x)$, $f_\alpha(x)$ is a known function, and the integral may be manipulated to yield the simple exponential term.

is the cumulative distribution function for $|\alpha|$. It follows that

$$P_e = P[-1] + \int_0^{\pi} F_\alpha(x)P'[\cos x] \sin x \, dx. \quad (17)$$

The range of integration is now bisected, and a simple change of variables in the $[\pi/2,\pi]$ range gives

$$P_e = P[-1] + \int_0^{\pi/2} F_\alpha(x)P'[\cos x] \sin x \, dx + \int_0^{\pi/2} F_\alpha(\pi - x)P'[-\cos x] \sin x \, dx \quad (18)$$

$$P_e = P[-1] + \int_0^{\pi/2} \{F_\alpha(x)P'[\cos x] + F_\alpha(\pi - x)P'[-\cos x]\} \sin x \, dx. \quad (19)$$

From the last property in (11) it can be seen that

$$P'[-x] = P'[x] \quad (20)$$

and so

$$P_e(\text{DPSK}) = P[-1] + \int_0^{\pi/2} \{F_\alpha(x) + F_\alpha(\pi - x)\} \cdot P'[\cos x] \sin x \, dx. \quad (21)$$

Fig. 2 is a phasor representation of the reference signal located on the sin $\omega_0 t$ axis, and some boundaries are drawn at $\pm A$ radians away from this axis. Consider noise and interference which has, effectively, some circular pdf in this space and corrupts the reference. Consider also a half-plane probability function, such as $P_R[\tau]$, similarly defined for this reference space. If the resultant of reference and noise plus interference has an angle α, then by equating appropriate regions one concludes that

$$\Pr\{|\alpha| \geq A\} + \Pr\{|\alpha| \geq \pi - A\} = 2P_R[\sin A], \quad A \leq \pi/2. \quad (22)$$

This can be written as

$$1 - F_\alpha(A) + 1 - F_\alpha(\pi - A) = 2P_R[\sin A] \quad (23)$$

so that

$$F_\alpha(A) + F_\alpha(\pi - A) = 2(1 - P_R[\sin A]) = 2P_R[-\sin A], \quad 0 \leq A \leq \pi/2. \quad (24)$$

Now employing (24) in (21) one has

$$P_e(\text{DPSK}) = P[-1] + 2\int_0^{\pi/2} P_R[-\sin x] \cdot P'[\cos x] \sin x \, dx \quad (25)$$

which gives the DPSK error probability solely in terms of easily computed half-plane probability functions.

However, $P[-1]$ is near 1. The integral is negative ($P' < 0$) and has magnitude slightly less than $P[-1]$, so that the expression above is in the form of a difference of two nearly equal factors. For low values of P_e, computational inaccuracies would be significant.

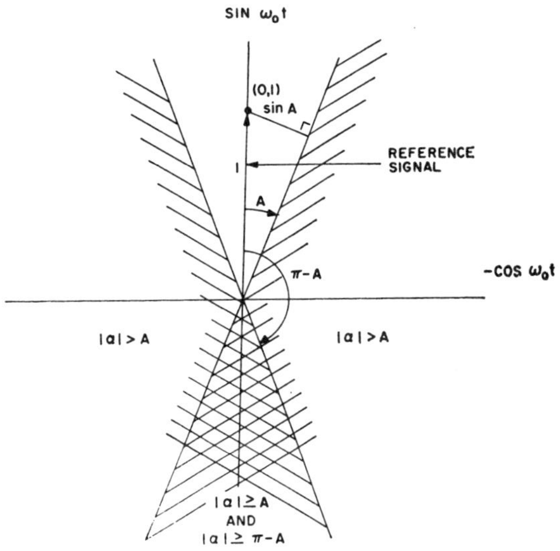

Fig. 2. Illustration of derivation of relation (22).

Therefore, one more integration by parts is required to obtain a better form. The straightforward steps lead to

$$P_e(\text{DPSK}) = P_R[1] + 2\int_0^{\pi/2} - P'_R[\sin x] \cdot P[\cos x] \cos x \, dx. \quad (26)$$

Since decoding is based on phase differences, the symmetry of the decision scheme causes the distinction between reference and signal to be an artificial but conceptually convenient one. Therefore, the titles "data" and "reference" may be transposed to obtain

$$P_e(\text{DPSK}) = P[1] + 2\int_0^{\pi/2} - P'[\sin x] \cdot P_R[\cos x] \cos x \, dx \quad (27)$$

which is immediately recognized as the CPSK error probability plus a term resulting from differential detection (imperfect phase reference).

While this P_e was derived for the zero phase change symbol, it is easily verified that P_e for the other is identical.[5] Thus (26) and (27) are the system error probabilities.

In conventional single-difference DPSK, the reference and signal use the same channel and are therefore corrupted by the same noise and interference. The solution therefore is obtained simply by setting

$$P_R[\tau] = P[\tau], \quad \text{single difference.}$$

Another example of specializing (27) to a particular problem is the transmitted reference PSK system where the reference and the signal are immersed in the Gaussian

[5] This is a result of the independence of reference and signal disturbances. In previous work [2] where the reference-signal dependence of the interference was considered, the P_e was a function of the transmitted symbol.

noise only, and have unequal CNRs, such as γ_R and γ_S. Then the half-plane probability functions are easily written as

$$P[\tau] = \tfrac{1}{2} \operatorname{erfc} [\tau(\gamma_S)^{1/2}]$$

$$P'[\tau] = -(\gamma_S/\pi)^{1/2} \exp(-\tau^2\gamma_S)$$

$$P_R[\tau] = \tfrac{1}{2} \operatorname{erfc} [\tau(\gamma_R)^{1/2}] \quad (28)$$

so that using (27) gives immediately

$$P_e = \tfrac{1}{2} \operatorname{erfc} [(\gamma_S)^{1/2}] + (\gamma_S/\pi)^{1/2} \int_0^{\pi/2} \exp(-\gamma_S \sin^2 x) \cdot \operatorname{erfc} [(\gamma_R)^{1/2} \cos x] \cos x \, dx,$$

$$\text{transmitted reference.} \quad (29)$$

This expression has also been found by Oberst [8], who obtained it by substituting the known $P_e \mid \alpha$ and f_α functions in (14), and then manipulating the resulting integral into the expression in (29). Incidentally, he demonstrated directly the symmetry property which allows interchanging, in this case, γ_R and γ_S in the expression for P_e.

IV. Numerical Results

The numerical results given here are for CPSK and single-difference (conventional) DPSK. The interference is assumed to be a sum of up to six random and independently phased equal-amplitude sinusoids. However, the equal amplitude curves may be efficiently used to bound or estimate P_e for an unequal-amplitude configuration, as will be discussed later.

In the author's previous work [2] with a single constant-amplitude interference, two important parameters were ϕ, the initial phase of the interference relative to the signal carrier, and θ, the relative phase shift in T seconds of the interference. ϕ was considered random with a uniform pdf. However, an unmodulated interference would have a constant θ which depends on the fixed frequency difference between the signal and interference. Hence the interference is not independent from baud to baud. Then P_e is a strong function of θ. Phase or frequency modulation of the interference can be viewed as a time varying θ, with P_e computed as an average with respect to θ. On the other hand, it was shown that a uniform distribution for θ gave the same P_e as one obtains with baud-interval independence of the interference.

In the present extension to multiple sinusoids, one may consider that each has an individual ϕ_j and θ_j, and that the θ_j are also uniformly distributed and independent. This situation can be shown equivalent to the one previously described, that is, overall baud-to-baud independence. It is suggestive, however, that with a large number of components the independence property is approximately met if only a few have nonuniform θ distributions.

The data are presented in ten families of curves (Figs. 3 and 4). Each family gives P_e versus CNR for a constant carrier-to-interference ratio (CIR), which is

$$-20 \log_{10} a - 10 \log_{10} N$$

for N interferences of equal amplitude a, in dB.

The curves labeled $N = 0$ give P_e for Gaussian noise alone, according to the well-known relationships

$$P_e(\text{CPSK}) = \tfrac{1}{2} \operatorname{erfc} [(\gamma)^{1/2}]$$

$$P_e(\text{DPSK}) = \tfrac{1}{2} \exp(-\gamma). \quad (30)$$

The curves labeled ∞ represent the limiting case of a sum of infinitely many components with finite total variance corresponding to the given value of CIR. This is treated as additional Gaussian noise; its variance is added to the original noise variance to obtain a new effective CNR. Thus the ∞ curves do not tend to $P_e = 0$ as CNR is increased, but approach an asymptote which is that P_e given by Gaussian noise alone (30), whose variance is that of the infinite sum.

Between these two limits are data for $N = 1, \cdots, 6$. The $N = 1$ curves were available in [2] and [3] and are reproduced here for comparison. In the case of differential detection, the interference frequency offset parameter θ was not averaged out for one interference. The $1L$ and $1U$ curves are lower and upper P_e values, respectively, as θ ranges over $[0,2\pi]$.

Before discussing the error probability curves, it should be emphasized again that these results are for an ideal channel and an ideal receiver, where the noise is truly Gaussian and baud-interval independent. Any deviations from these idealizations are naturally going to produce different error rates in a real system. In particular, most real systems exhibit an error rate "floor" which limits the achievable performance irrespective of γ. This floor, which might be caused by noise in the detector or even hardware failures, is usually well below normal regions of operation, e.g., $10^{-6} > P_e > 10^{-10}$. Nevertheless, results are provided down to 10^{-14} in order to display the analytical behavior of P_e as $\gamma \uparrow \infty$.

V. Conclusion

Perhaps the most significant conclusion to be drawn from the error probability curves is the pessimism in treating a moderate ($N < 4$) number of interferers as additional Gaussian noise, especially at a high CNR. Using the 10-dB-CIR (DPSK) data, for example [Fig. 4(d)], there is a three-decade spread between two interferences and the equal power noise conversion ($N = \infty$) at CNR = 17 dB. Using this old resort of conversion to noise, a 10-dB interference results in a P_e greater than 10^{-4} regardless of CNR, whereas the actual P_e may clearly be driven to zero with increasing CNR.

Also, the overestimation incurred by Gaussian noise conversion increases with the interference level until the interference is so large (CIR $\approx$ 10 dB) that it begins to swamp out the effect of the noise.

This datum, and others not shown, prove empirically that P_e is an increasing function of N. In addition, the

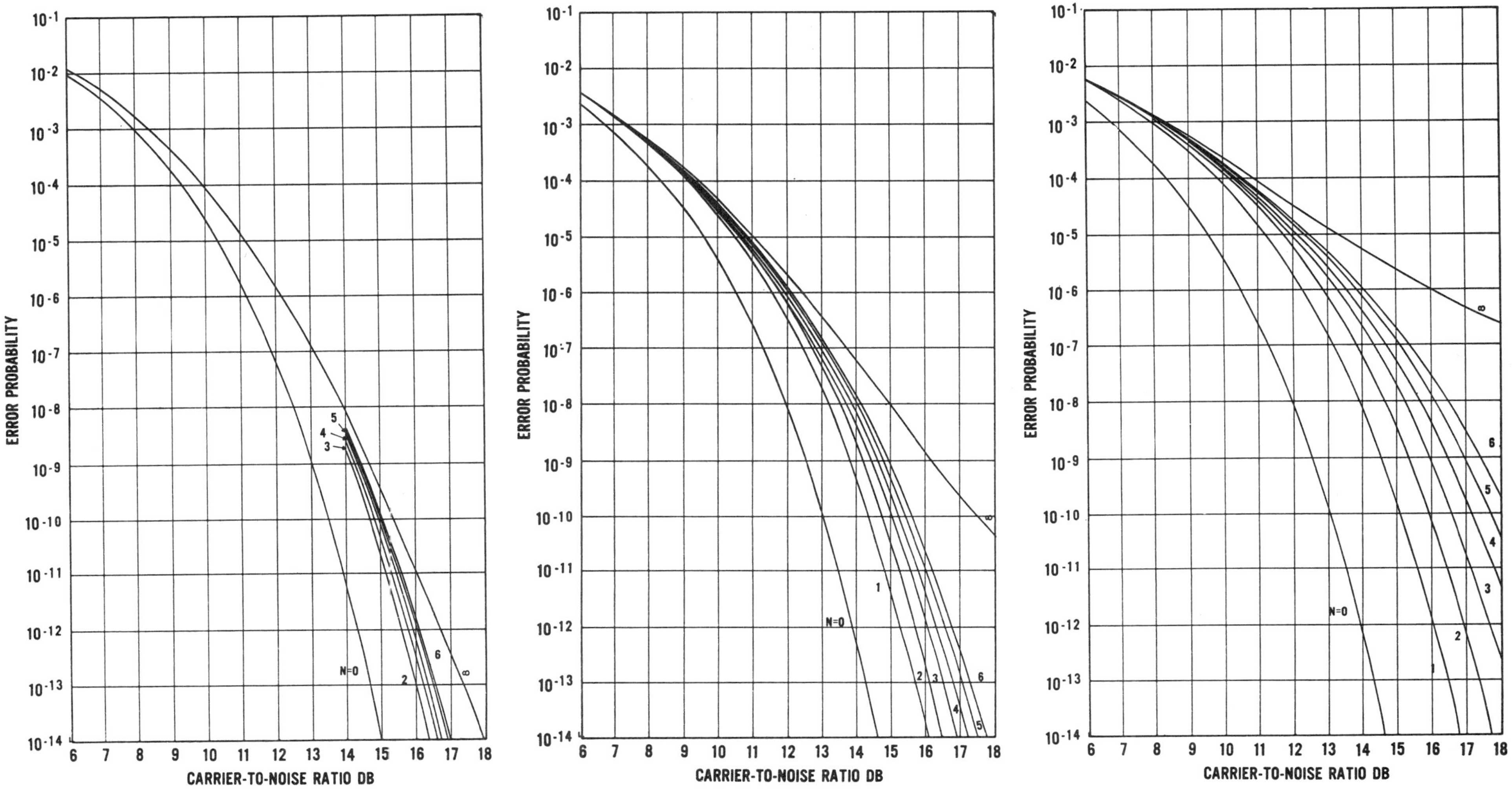
ERROR PROBABILITY
10-1
10-2
10-3
10-4
10-5
10-6
10-7
10-8
10-9
10-10
10-11
10-12
10-13
10-14
6 7 8 9 10 11 12 13 14 15 16 17 18
CARRIER-TO-NOISE RATIO DB
N=0
1
2
3
4
5
6
∞

(a) (b) (c)

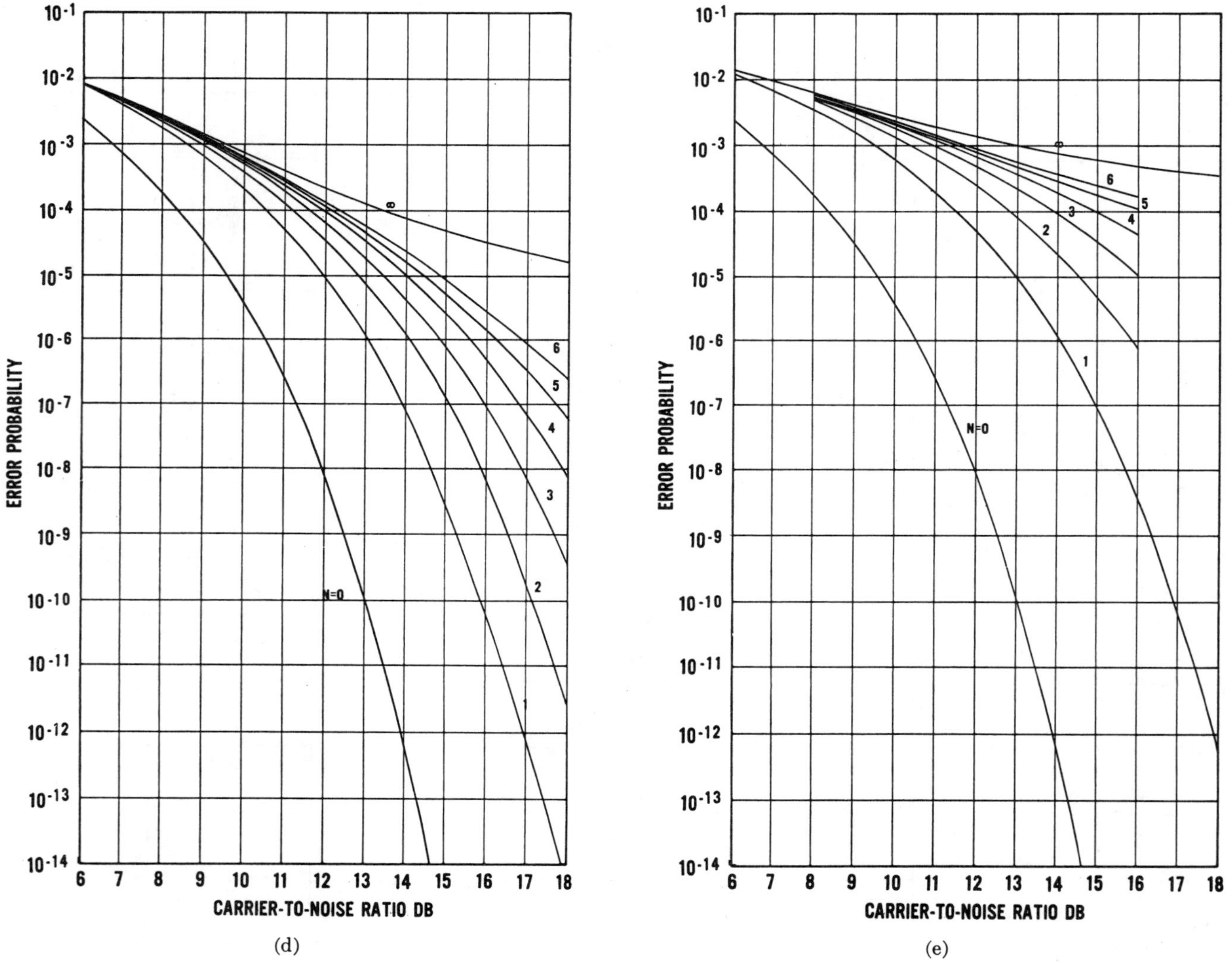

Fig. 3. CPSK. (a) CIR equal to 18 dB. (b) CIR equal to 15 dB. (c) CIR equal to 12 dB. (d) CIR equal to 10 dB. (e) CIR equal to 8 dB.

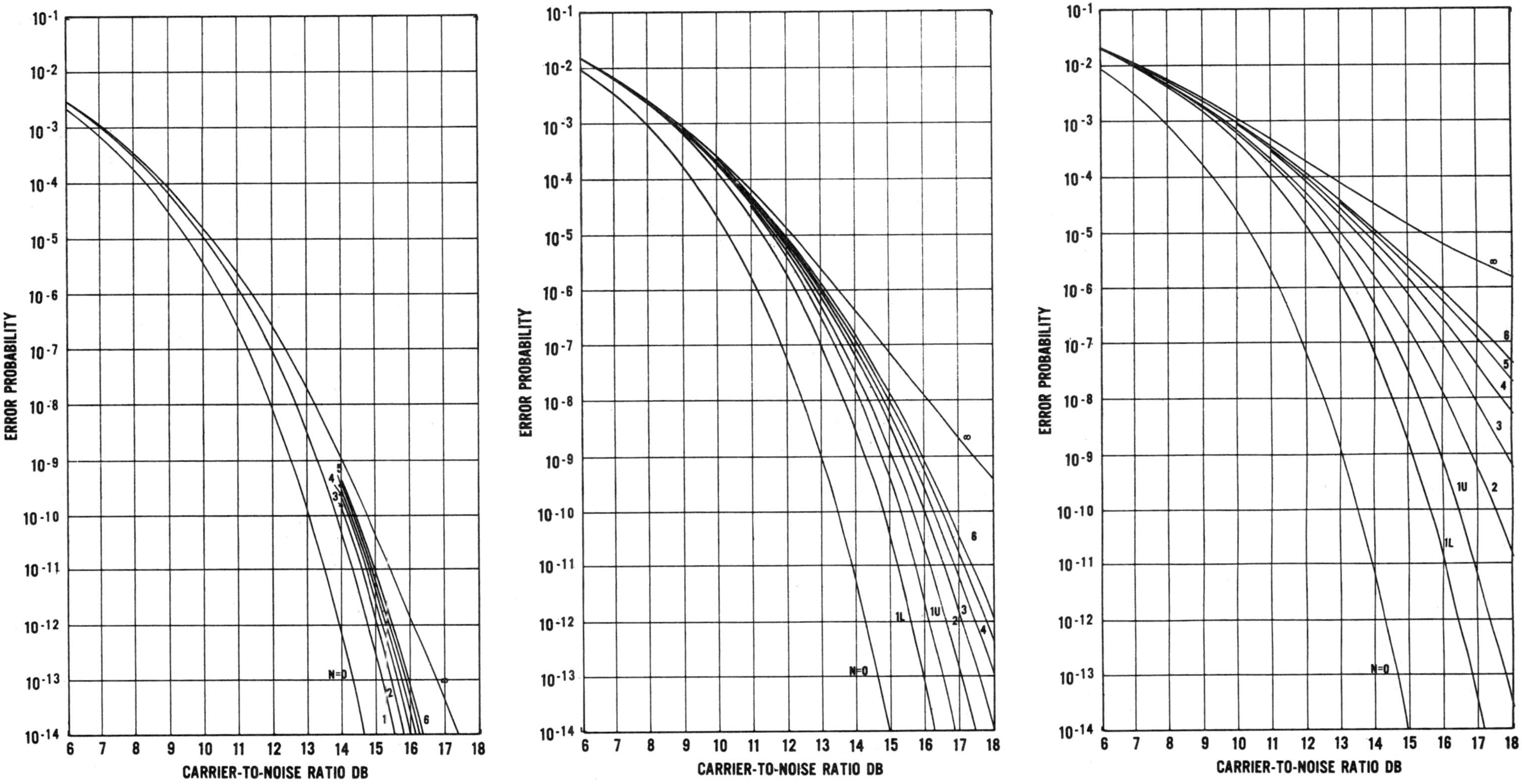
ERROR PROBABILITY
10-1
10-2
10-3
10-4
10-5
10-6
10-7
10-8
10-9
10-10
10-11
10-12
10-13
10-14
6 7 8 9 10 11 12 13 14 15 16 17 18
CARRIER-TO-NOISE RATIO DB
N=0
1
2
3
4
5
6
∞
1L
1U

(a)

(b)

(c)

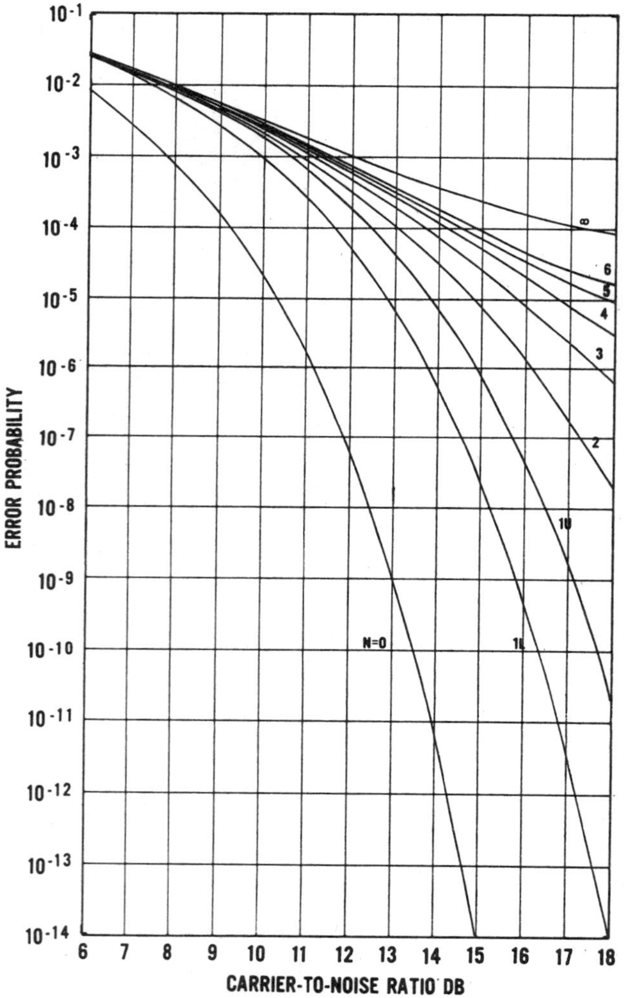

(d)

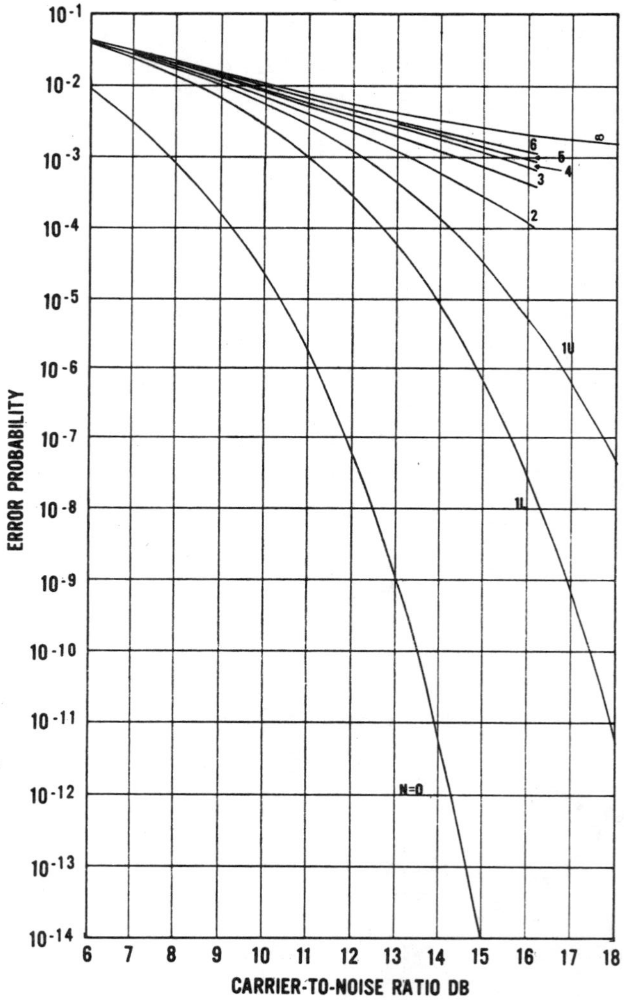

(e)

Fig. 4. DPSK. (a) CIR equal to 18 dB. (b) CIR equal to 15 dB. (c) CIR equal to 12 dB. (d) CIR equal to 10 dB. (e) CIR equal to 8 dB.

writer believes[6] that the equal amplitude ensemble gives the greatest P_e for a fixed total interference power, CIR. If true, the present curves for equal a_i will serve as upper bounds for unequal a_i configurations.

Comparing CPSK and DPSK, the former is seen to have a greater immunity from interference. When the bias inherent in the noise-only performance is removed, DPSK still has a higher P_e for the same CNR and N. Viewed another way, the increase to CNR (in dB) required to maintain a given P_e after interference is added is greater for DPSK. Thus the conclusions reached in earlier work for a single sinusoidal interference remain valid for multiple components.

It was previously suggested that these data are upper bounds to P_e for an unequal set of interference amplitudes with the same total variance, $\sum_i \frac{1}{2}a_i^2$. Short of computing the moments (see the Appendix) of the interference, one must be satisfied with this bound. However, if K components are roughly equal and the remaining $N - K$ components are relatively weak, then the following is suggested. Convert the weak ones to noise by adding their variance to the thermal noise variance, thus lowering the CNR. Then consult the curve labeled K components on the chart whose CIR corresponds to the power in the K dominant interferences. It is felt that this will give a P_e slightly higher than the true value, but considerably tighter than that for N equal amplitudes.

Another technique for estimating P_e efficiently for unequal amplitudes involves computing only the moments of the interference and eliminates calculating $P[\tau]$, $P'[\tau]$, and the integral (27). This approach, which is discussed at the end of the Appendix, is presently being investigated with favorable results.

Appendix

Some of the results pertinent to evaluating $P[\tau]$, the half-plane probability function which appears in the error probability integral, are presented. The first part deals with the series used to compute $P[\tau]$ and $P'[\tau]$. This series uses the moments of $R \sin \Psi$, and so the second part considers these moments when the interference consists of a sum of randomly phased sinusoids. Both the series expansion [12] used for $P[\tau]$ and expressions for the moments have been given before. However, an abbreviated development will be included here for the sake of completeness. Some new results are offered here, notably, closed expressions for the first five nonzero moments and a graph giving the first thirty moments for up to ten components. The Appendix concludes with a description of how this graph can be used to estimate P_e for unequal amplitude configurations.

A. Half-Plane Probability Series

Choose one of the orthogonal noise phasors perpendicular to the half-plane boundary. The projection of the interference onto this (or any) perpendicular has the statistics of $R \sin \Psi$. Call this random variable I. Then

$$P[\tau] = E\{\tfrac{1}{2} \operatorname{erfc}(\gamma^{1/2}\tau + \gamma^{1/2}I)\} \tag{31}$$

where

$$\operatorname{erfc}(\xi) \triangleq \frac{2}{\pi^{1/2}} \int_\xi^\infty \exp(-\lambda^2)\, d\lambda \tag{32}$$

is the usual complementary error function integral and $E\{\ \}$ denotes expectation. The erfc is expanded into its Taylor series about $\gamma^{1/2}\tau$ to obtain

$$P[\tau] = E\{\tfrac{1}{2} \operatorname{erfc}(\tau\gamma^{1/2}) + [\exp(-\tau^2\gamma)/\pi^{1/2}] \cdot \sum_{l=1}^{\infty} [(-1)^l/l!] H_{l-1}(\tau\gamma^{1/2})(\gamma^{1/2}I)^l\} \tag{33}$$

the $H_l(\ )$ being the Hermite polynomial of order l. Now the linearity of the expectation is employed to write

$$P[\tau] = \tfrac{1}{2} \operatorname{erfc}(\tau\gamma^{1/2}) + [\exp(-\tau^2\gamma)/\pi^{1/2}] \cdot \sum_{l=1}^{\infty} [(-1)^l/l!] H_{l-1}(\tau\gamma^{1/2}) E\{I^l\}(\gamma^{1/2})^l \tag{34}$$

and $E\{I^l\}$ is just the definition of μ_l, the lth moment of I. Since the pdf of I is clearly even about 0, the μ_l are central moments, and they vanish for odd l. Therefore,

$$P[\tau] = \tfrac{1}{2} \operatorname{erfc}(\tau\gamma^{1/2}) + [\exp(-\tau^2\gamma)/\pi^{1/2}] \cdot \sum_{l=1}^{\infty} [H_{2l-1}(\tau\gamma^{1/2})/(2l)!]\gamma^l \mu_{2l}. \tag{35}$$

The series for $P'[\tau]$ can be derived directly from (35) by differentiating term by term, and then using the recurrence relationships for the Hermite polynomials. The result is

$$P'[\tau] = -(\gamma/\pi)^{1/2} \exp(-\tau^2\gamma) \cdot \{1 + \sum_{l=1}^{\infty} [H_{2l}(\tau\gamma^{1/2})/(2l)!]\gamma^l \mu_{2l}\} \tag{36}$$

with the negative sign indicating the decreasing probability of the half-plane as it moves farther from the mean of the interference plus noise pdf.

In practice, the first 50 terms of (35) and (36) were retained. Care must be exercised, however, since adequate convergence by the 50th term is critical, and depends strongly on γ and the μ_{2l}. It is noteworthy that convergence is slower when γ (CNR) is larger and when CIR is lower. For sinusoidal interference, convergence is slower for both lower CIR and larger N. This will be evident in the following discussion of the moments for a sum of sinusoids.

B. Central Moments for a Sum of Sinusoids

Following the work of Bennett [9], the series which follows below for μ_{2l} was derived by Prabhu [3]:

$$\mu_{2l} = \frac{B^{2l}}{2l+1}\left[1 + 2\sum_{j=1}^{\infty} (-1)^j \{\prod_{i=1}^{N} J_0[a_i \pi j/B]\} \cdot \sum_{k=1}^{l} \frac{(-1)^{k+1}(2l+1)}{(2l-2k+1)!(j\pi)^{2k}}\right] \tag{37}$$

[6] The same conclusion was reached by Prabhu [3], who offered (for CPSK) an argument which is only valid for certain combinations of CNR and CIR. Therefore, a proof is still lacking, and the interested reader may wish to attempt it.

where

$$B = \sum_{i=1}^{N} a_i \tag{38}$$

is the largest possible value for I.

This works well for the lower order moments. However, the higher order moments, e.g., $l > 30$, cannot easily be machine computed from (37). Even though the summation over k has a finite number of terms, their magnitudes (for small j) become so large that the successive differences become inaccurate or grossly wrong.

To overcome this problem another series was derived. The finite summation over k represents the integral

$$(2l+1)\int_0^1 \tau^{2l} \cos(j\pi\tau)\, d\tau. \tag{39}$$

Two solutions to this integral are tabulated in [10]. One is just the finite series in (37). The other, which is in terms of confluent hypergeometric functions, can be manipulated to yield a series which converges rapidly and accurately where the first has trouble.[7] Very briefly then, with $i = (-1)^{1/2}$

$$\int = \tfrac{1}{2}[{}_1F_1(2l+1, 2l+2, ij\pi) + {}_1F_1(2l+1, 2l+2, -ij\pi)]. \tag{40}$$

Using the first Kummer transformation this becomes

$$\int = \tfrac{1}{2}[\exp(ij\pi)\,{}_1F_1(1, 2l+2, -ij\pi) + \exp(-ij\pi)\,{}_1F_1(1, 2l+2, ij\pi)]. \tag{41}$$

When the ${}_1F_1$ functions are replaced by their series, and the resulting expressions simplified, this gives

$$\int = (-1)^j \sum_{k=0}^{\infty} \frac{(-1)^k (j\pi)^{2k}}{(2l+2)_{2k}} \tag{42}$$

where $(\nu)_M$ denotes $(\nu)\cdot(\nu+1)\cdot(\nu+2)\cdots(\nu+M-1)$, $(\nu)_0 \equiv 1$.

The computer program used to compute μ_{2l} selected either (42) or the finite series in (37), depending on the current values of l and j, so that accuracy was maintained even for the high-order moments.

When the amplitudes are equal, $a_1 = a_2 = \cdots = a_N = a$, the expression for μ_{2l} takes the slightly simpler form

$$\mu_{2l} = [(Na)^{2l}/(2l+1)][1 + 2\sum_{j=1}^{\infty} (-1)^j J_0^N(j\pi/N)\cdot\Sigma]$$

where

$$\Sigma = \sum_{k=1}^{l} \frac{(-1)^{k+1}(2l-2k+2)_{2k}}{(j\pi)^{2k}} \quad \text{or} \quad \sum_{k=0}^{\infty} \frac{(-1)^k (j\pi)^{2k}}{(2l+2)_{2k}}.$$

The first few nonzero moments for equal amplitudes can be written in closed form. Since the sinusoids are independently phased, the characteristic function $\Phi(\omega)$ of their sum is the product of the individual characteristic functions:

$$\Phi(\omega) = E\{\exp(i\omega I)\} = \prod_{j=1}^{N} J_0(a_j\omega) = J_0^N(a\omega). \tag{43}$$

By the moment generating property of characteristic functions

$$(d/d\omega)^n \Phi(0) = i^n \mu_n \tag{44}$$

so that

$$\mu_{2l} = (-1)^l (d/d\omega)^{2l} J_0^N(a\omega)\,|_{\omega=0}. \tag{45}$$

Thus $2l$ successive derivatives may be taken, the last evaluated at $\omega = 0$ according to

$$J_0^{(\nu)}(0) = \begin{cases} (-1)^{\nu/2} 2^{-\nu} \binom{\nu}{\frac{1}{2}\nu}, & \nu \text{ even} \\ 0, & \nu \text{ odd} \end{cases} \tag{46}$$

and simplified to produce μ_{2l}. This becomes very tedious for $l > 3$.

However, $\Phi(\omega)$ is a composition of the two functions

$$g(\omega) = J_0(a\omega)$$
$$f(g) = g^{2l} \tag{47}$$

so that the derivatives are easily evaluated by the use of the Bell polynomials, Y_n.[8] Following the notation in Riordan [11], these polynomials have the property that

$$Y_n(fg_1, \cdots, fg_n) = (d/dt)^n f[g(t)] \tag{48}$$

where

$$f_n = (d/du)^n f(u)\,|_{g(t)} \tag{49}$$

and

$$g_n = (d/dt)^n g(t). \tag{50}$$

In this case,

$$f_n = (d/du)^n u^{2l} = (2l)_n u^{2l-n} \tag{51}$$

and

$$g_n = J_0^{(n)}(0). \tag{52}$$

At $\omega = 0$

$$u^{2l-n} = [J_0(0)]^{2l-n} = 1 \tag{53}$$

so that

$$f_n\,|_{\omega=0} = (2l)_n \tag{54}$$

and g_n is given by (46).

Using the above, and a table [11] of Bell polynomials up to Y_{10}, the following five moments were easily derived.

$$\mu_2 = (a^2/2)N$$
$$\mu_4 = (a^4/2^3)3N(2N-1)$$
$$\mu_6 = (a^6/2^5)10N(6N^2-9N+4)$$
$$\mu_8 = (a^8/2^7)35N(24N^3-72N^2+82N-33)$$
$$\mu_{10} = (a^{10}/2^9)63N(240N^4-1200N^3+2500N^2-2450N+912). \tag{55}$$

[7] A third series, obtained by expanding $\cos(j\pi\tau)$ in (39) into its Taylor series and integrating term by term, also converges very poorly.

[8] The author is grateful to S. O. Rice for calling his attention to the Bell polynomials, and suggesting their use for this problem.

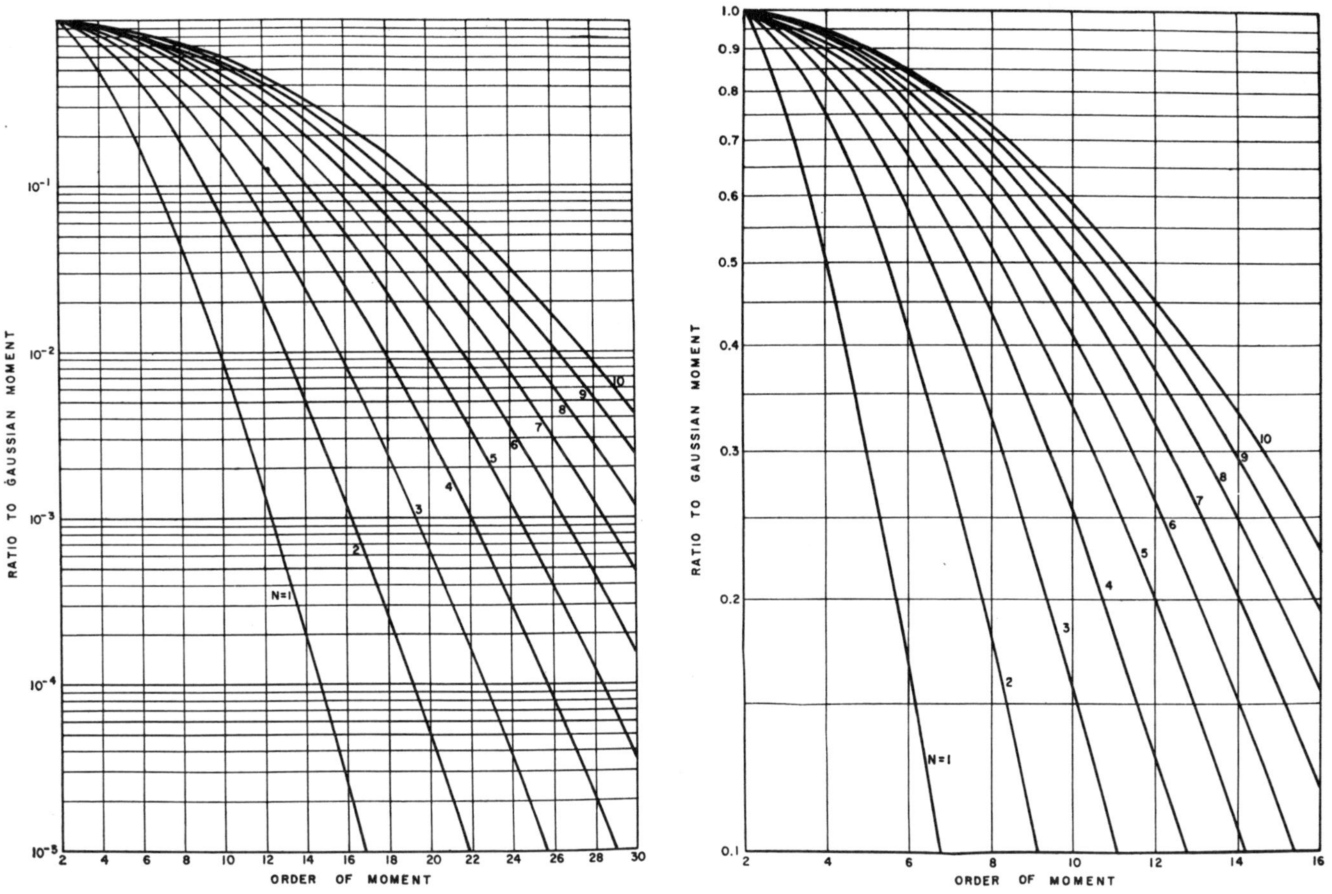

(a) (b)

Fig. 5. Ratio of μ_{2l} to corresponding moment of equal variance Gaussian.

A table was prepared giving μ_{2l} for $l = 1(1)50$ and $N = 1(1)6$, yielding approximately five places of precision. These numbers were used in series (35) and (36) to compute the error probabilities. However, since the μ_{2l} are interesting by themselves a graphical presentation is included here. In order to display the relationship between a sum of sinusoids and a Gaussian, the graphs in Fig. 5 shows the ratio of μ_{2l} (for $N \leq 10$ sinusoids) to the corresponding moment of an equal variance Gaussian.[9]

It is obvious that μ_{2l} is a monotone increasing function of N. This is consistent with the behavior of P_e versus N. Furthermore, moments were computed for many varied sets of unequal amplitudes, e.g., 1:2:3, 1:2:2:3, 1:2:3:4, etc., and plotted on Fig. 5. In each case, the graph of μ_{2l} versus l exhibited approximately the same shape as the equal amplitude curves. More important, though, is the fact that each curve for an unequal amplitude set lies below the corresponding equal amplitude curve. While it is difficult to relate exactly the error probability to the behavior of the μ_{2l} by inspection, this is certainly strong additional support to the conjecture made earlier concerning P_e for unequal amplitudes.

At the time of this writing there is insufficient data to make concrete conclusions from the graphs of μ_{2l} for unequal amplitude sinusoids or other types of interferences for that matter. On the other hand, early results strongly suggest that P_e may be fairly accurately obtained merely by plotting the interference moments on Fig. 5, and then using the equal amplitude P_e data, Figs. 3–4. For example, if a four-interference complex with amplitude ratios 3:3:4:2 were involved, one might plot μ_{2l} and see that the curve lies about two thirds of the way from the $N = 3$ to the $N = 4$ lines. Then P_e is correspondingly two-thirds from the $N = 3$ to $N = 4$ curve at the appropriate CNR; CIR is just given by the sum of the powers of the unequal variances.

Thus this relatively simple approach would avoid computing P_e for many different CNR and CIR values. That is, by relating a particular interference to an "effective" number of equal amplitude sinusoids (say $N_{\text{eff}} = 3\frac{2}{3}$ in the preceding example) much computing effort may be saved with little loss in accuracy. Work is continuing along these lines, and the validity of this idea should be determined in the near future.

[9] Several earlier works have compared the distribution (or density) functions of N sinusoids to that of a Gaussian, in order to draw conclusions as to how much larger N must be to consider the sum equivalent to the Gaussian. This view of the moments adds a different perspective to that question, since the higher moments amplify the disparity in the tails of the distributions.

References

[1] M. A. Koerner, "Effect of interference on a binary communication channel using known signals," Jet Propulsion Lab., California Inst. of Technology, Pasadena, Calif., Tech Rept. 32-1281, December 1968.

[2] A. S. Rosenbaum, "PSK error performance with Gaussian noise and interference," *Bell Sys. Tech. J.*, vol. 48, pp. 413–442, February 1968.

[3] V. K. Prabhu, "Error rate consideration for coherent phase-shift-keyed systems with co-channel interference," *Bell Sys. Tech. J.*, vol. 48, pp. 743–767, March 1969.

[4] R. J. F. Fang, "Angle-modulated signal interference on phase-shift-keyed systems," Communication Satellite Corp., Tech. Memo. CL-15-68, July 1968.

[5] J. R. Davey, "Digital data signal space diagrams," *Bell Sys. Tech. J.*, vol. 43, pp. 2973–2983, November 1964.

[6] A. Papoulis, *Systems and Transforms with Applications in Optics*. New York: McGraw-Hill, 1968, ch. 5.

[7] C. R. Cahn, "Performance of digital phase-modulation communication systems," *IRE Trans. Communications Systems*, vol. CS-7, pp. 3–6, May 1959.

[8] J. F. Oberst, "Binary phase-shift-keyed communication systems," Ph.D. dissertation, Polytechnic Inst. of Brooklyn, Brooklyn, N. Y., June 1969.

[9] W. R. Bennett, "Distribution of the sum of randomly phased components," *Quart. Appl. Math.*, vol. 5, pp. 385–393, April 1948.

[10] I. S. Gradshteyn and I. M. Ryzhik, *Tables of Integrals, Series, and Products*. New York: Academic Press, 1965, p. 421.

[11] J. Riordan, *Combinatorial Identities*. New York: Wiley, 1968, pp. 173–175.

[12] M. J. Pagones, unpublished Tech. Memo.

Multiple Error Performance of PSK Systems with Cochannel Interference and Noise

JOEL GOLDMAN, MEMBER, IEEE

Abstract—The multiple error performance of a phase-shift-keyed (PSK) communications system, when both cochannel interference (due possibly to other cochannel angle-modulated systems) and Gaussian noise additively perturb the transmitted signals, is considered. The results are fairly general: the main requirement is that the interference be circularly symmetric. All of our findings are also applicable to the case when only noise is present. The results indicate that one cannot approximate well the effect of interference on the performance of a PSK system by treating it as additional Gaussian noise. First, we derive the probability density function f_A of the phase angle of a cosinusoid plus interference and Gaussian noise. We then obtain readily computable expressions (in terms of f_A) for the probability of any number of consecutive errors in an m-phase system when either coherent or differential detection is utilized. For numerical results, the interference is assumed to be due to other cochannel angle-modulated communications systems, and the double error probability and conditional probability of error are given for 2- and 4-phase systems.

I. Introduction

THE MULTIPLE (consecutive) error performance of a phase-shift-keyed (PSK) communications system has recently received some attention in the literature. Salz and Saltzberg [1] and Oberst and Schilling [2], among others, have studied this problem under the assumption that Gaussian noise additively perturbs the transmitted signals. However, in some applications both cochannel interference (due possibly to other cochannel angle-modulated systems) and Gaussian noise are present, and this case has received little attention. In this paper we consider multiple error performance for this case with the fairly general requirement that the interference be circularly symmetric. All of our findings are also applicable to the situation when only noise is present. Our results will show that one cannot approximate well the effect of interference on the performance of a PSK system by treating it as additional Gaussian noise.

First, in Section III we derive the probability density function f_A of the phase angle of a cosinusoid plus interference and Gaussian noise. We then obtain in Section IV readily computable expressions (in terms of f_A) for the probability of any number of consecutive errors in an m-phase system when either coherent PSK (CPSK) or differential detection PSK (DPSK) is utilized. In addition we demonstrate some inequalities between the probabilities of different numbers of consecutive errors. Finally, in Section V we display numerical results for the double error probability and conditional probability of error of 2- and 4-phase systems when the interference is due to other cochannel angle-modulated communications systems.

Paper approved by the Communication Theory Committee of the IEEE Communication Technology Group for publication without oral presentation. Manuscript received January 15, 1971.
The author is with Bell Telephone Laboratories, Inc., Holmdel, N. J. 07733.

II. Model and Assumptions

In an m-phase PSK system digital information is encoded in the phase of the transmitted signal. The digit $k \in \{0,1,\cdots,m-1\}$ is sent in the ith baud of length T seconds by transmitting a signal of the form (with amplitude normalized to one):[1]

$$s_k{}^i(t) = \begin{cases} \cos(\omega_0 t + \phi_k{}^i), & t \in [(i-1)T, iT) \\ 0, & t \notin [(i-1)T, iT). \end{cases}$$

The phase angle $\phi_k{}^i = (2\pi k)/m$ when coherent detection is used, or $\phi_k{}^i = \phi_q{}^{i-1} + (2\pi k)/m$ if differential detection is utilized, where $\phi_q{}^{i-1}$ is the phase angle of the signal transmitted in the $(i-1)$th baud (digit q being sent in this baud).

Let the phase modulation $\Phi(\cdot)$ be the random[2] process defined by $\Phi(t) = \phi_k{}^i$ if $t \in [(i-1)T, iT)$ and $k \in \{0,1,\cdots,m-1\}$ is transmitted in the ith baud. The transmitted signal is then just

$$S(t) = \cos[\omega_0 t + \Phi(t)], \qquad \text{for all } t \geq 0.$$

We assume that $S(t)$ is corrupted by noise and interference so that the signal at the receiver output is

$$\hat{S}(t) = S(t) + N(t) + I(t) \tag{1}$$

where $N(t)$ and $I(t)$ are the contributions to the receiver output due to the presence of noise and interference, respectively. Suppose that $N(t)$ is the result of the passage of zero-mean white stationary Gaussian noise through a bandpass symmetrical receiver filter. Then $N(t)$ can be written as [3]

$$N(t) = N'(t)\cos\omega_0 t - N''(t)\sin\omega_0 t \tag{2}$$

where $N'(t)$ and $N''(t)$ are zero-mean independent stationary low-pass Gaussian random processes with powers equal to

$$\sigma^2 = E[N(t)]^2 = E[N'(t)]^2 = E[N''(t)]^2$$

[1] All angles will be assumed to be reduced modulo $[-\pi,\pi]$.
[2] All random quantities will be capitalized.

Reprinted from *IEEE Trans. Comm. Tech.*, vol. COM-19, pp. 420-430, Aug. 1971.

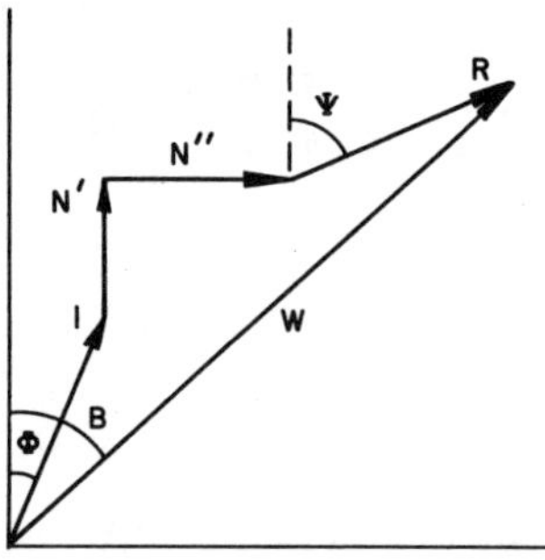

Fig. 1. Vector diagram of variables in (3).

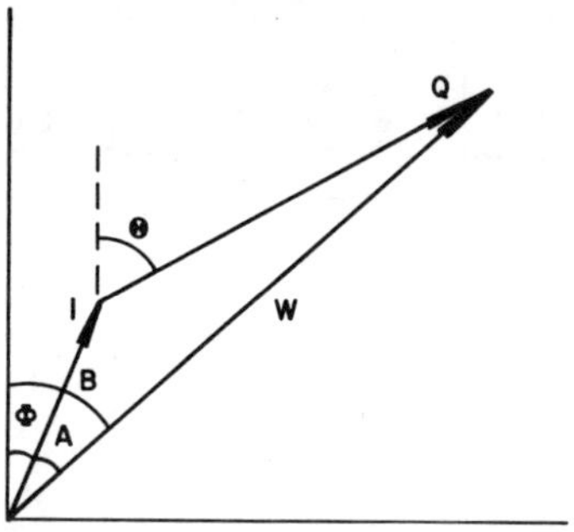

Fig. 2. Vector diagram showing angle A between transmitted and received signals.

where $E(\cdot)$ denotes expected value. Let us also conjecture that $I(t)$ is of the form

$$I(t) = R(t) \cos [\omega_0 t + \Psi(t)], \qquad R(t) \geq 0$$

where $R(t)$ and $\Psi(t)$ are stationary processes, so that the received signal is

$$\begin{aligned}\hat{S}(t) &= \cos [\omega_0 t + \Phi(t)] + N'(t) \cos \omega_0 t \\ &\quad - N''(t) \sin \omega_0 t + R(t) \cos [\omega_0 t + \Psi(t)] \\ &= W(t) \cos [\omega_0 t + B(t)]. \qquad (3)\end{aligned}$$

Another reasonable presumption is that the components of the noise and interference at the same instant of time are independent of each other and are also independent of the phase modulation at any other instant of time. That is, for any pair of (possibly different) times t_{00} and t_0, $\Phi_{00} = \Phi(t_{00})$, $R = R(t_0)$, $\Psi = \Psi(t_0)$, $N' = N'(t_0)$, and $N'' = N''(t_0)$ are independent random variables. Furthermore, suppose that Ψ is uniformly distributed on $[-\pi,\pi]$.

We say a pair (U,V) of real random variables are *circularly symmetric* [4] if and only if the joint density of U and V, f_{UV}, satisfies

$$f_{UV}(u,v) = g[(u^2 + v^2)^{1/2}]$$

for all u, v, and for some function g. In polar coordinate notation $U = \Gamma \sin \Omega$, $V = \Gamma \cos \Omega$, $\Gamma \geq 0$, $-\pi \leq \Omega \leq \pi$, this definition is equivalent to having Γ and Ω independent and Ω uniformly distributed on $[-\pi,\pi]$. We will also call the pair (Γ,Ω) circularly symmetric (polar). Note that the interference amplitude and phase (R,Ψ) are circularly symmetric (polar). It is easy to show by use of characteristic functions that if (U_1,V_1) and (U_2,V_2) are independent circularly symmetric pairs, then the sum $(U_1,V_1) + (U_2,V_2) = (U_1 + U_2, V_1 + V_2)$ is also circularly symmetric.

Finally, we also postulate that samples of the components of noise and interference are independent from baud to baud, i.e., for any time t_0, $\{N'(t_0), N''(t_0), R(t_0), \Psi(t_0)\}$ is independent of $\{N'(t_0 + T), N''(t_0 + T), R(t_0 + T), \Psi(t_0 + T)\}$. These conditions were previously used by other authors [2], [5].

Next let us set $W = W(t_0)$, $B = B(t_0)$, $\Phi = \Phi(t_0)$ and represent quantities of the form $\rho \cos (\omega_0 t_0 + \eta)$ in Euclidean 2-space by a vector from the origin of radius ρ and at an angle η from the vertical axis in a clockwise direction. Then the variables in (3) can be displayed as in Fig. 1. Let the vector sum of R, N', and N'' be a vector of radius Q at an angle Θ, and let $A = A(t_0)$ be the angle between the transmitted and received signals $S(t_0)$ and $\hat{S}(t_0)$, respectively. Fig. 1 then becomes Fig. 2.

The probability density of A, f_A, will be derived in the next section. Note that A is a function only of Q and Θ which are independent of Φ, so that A is independent of Φ, i.e., the angle between the transmitted and received signals is independent of which digit was sent.

III. Derivation of f_A

Before proceeding to the derivation of f_A let us introduce some notation. Certain special functions [6] will be used in this section. These are the mth-order $(m = 0,1,\cdots)$ Hermite polynomials H_m, the mth-order $(m = \cdots,-1,0,1,\cdots)$ parabolic cylinder functions D_m, and the mth-order $(m = 0,1,\cdots)$ (simple) Laguerre polynomials L_m. Another function H_{-2} will be defined by

$$H_{-2}(x) = \frac{1}{2} + \frac{\sqrt{\pi}}{2} x \exp (x^2) \operatorname{erfc} (-x), \qquad \forall x$$

where erfc $(\cdot)$ is the complementary error function.

The $(2k)$th moment of $R \sin \Psi$ will be denoted by

$$\mu_{2k} = E[R \sin \Psi]^{2k} < \infty.$$

Note that since R and Ψ are independent and Ψ is uniformly distributed on $[-\pi,\pi]$, we have

$$\begin{aligned}\mu_{2k} &= E[R^{2k} \sin^{2k} \Psi] \\ &= E[R^{2k}] E[\sin^{2k} \Psi] \\ &= E[R^{2k}] \frac{(2k)!}{2^{2k} (k!)^2}. \qquad (4)\end{aligned}$$

Finally, the carrier-to-noise power ratio (CNR) will be designated by $\gamma = 1/2\sigma^2$.

The result to be shown in this section is that if γ and R are such that

$$\sum_{n=0}^{\infty} \frac{\gamma^n E[R^{2n}]}{n!} < \infty$$

then

$$f_A(\alpha) = \frac{\exp (-\gamma)}{\pi} \sum_{n=0}^{\infty} \frac{\gamma^n \mu_{2n}}{(2n)!} \sum_{k=0}^{n} \binom{n}{k} \cdot H_{2n-2k}(\gamma^{1/2} \sin \alpha) H_{2k-2}(\gamma^{1/2} \cos \alpha) \qquad (5)$$

and

$$f_A(\alpha) = \frac{\exp(-\gamma/2)}{2\pi} \sum_{n=0}^{\infty} \frac{(2\gamma)^n \mu_{2n}}{(2n)!} \sum_{k=0}^{n} \binom{n}{k}$$

$$\cdot D_{2n-2k}((2\gamma)^{1/2} \sin \alpha) D_{2k-2}(-(2\gamma)^{1/2} \cos \alpha) \quad (6)$$

for all $\alpha \in [-\pi,\pi]$.

Remarks: First, the requirement that

$$\sum_{n=0}^{\infty} \frac{\gamma^n E[R^{2n}]}{n!} < \infty$$

is satisfied if, for example, the interference amplitude R is a bounded random variable. The interference model we will use for numerical calculations will satisfy this condition (see Section V).

Second, we can rewrite (5) and (6) as

$$f_A(\alpha) = p_0(\alpha) + \frac{\exp(-\gamma)}{\pi} \sum_{n=1}^{\infty} \frac{\gamma^n \mu_{2n}}{(2n)!} \sum_{k=0}^{n} \binom{n}{k}$$

$$\cdot H_{2n-2k}(\gamma^{1/2} \sin \alpha) H_{2k-2}(\gamma^{1/2} \cos \alpha) \quad (7)$$

and

$$f_A(\alpha) = p_0(\alpha) + \frac{\exp(-\gamma/2)}{2\pi} \sum_{n=1}^{\infty} \frac{(2\gamma^n) \mu_{2n}}{(2n)!} \sum_{k=0}^{n} \binom{n}{k}$$

$$\cdot D_{2n-2k}((2\gamma)^{1/2} \sin \alpha) D_{2k-2}(-(2\gamma)^{1/2} \cos \alpha) \quad (8)$$

for all $\alpha \in [-\pi,\pi]$, where

$$p_0(\alpha) = \frac{\exp(-\gamma)}{\pi} H_{-2}(\gamma^{1/2} \cos \alpha)$$

$$= \frac{\exp(-\gamma/2)}{2\pi} D_0((2\gamma)^{1/2} \sin \alpha) D_{-2}(-(2\gamma)^{1/2} \cos \alpha)$$

$$= \frac{\exp(-\gamma)}{2\pi} \exp\left(\frac{\gamma}{2} \cos^2 \alpha\right) D_{-2}(-(2\gamma)^{1/2} \cos \alpha)$$

$$= \frac{\exp(-\gamma)}{\pi} \left[\frac{1}{2} + \frac{\sqrt{\pi}}{2} \gamma^{1/2} \cos \alpha \exp(\gamma \cos^2 \alpha) \right.$$

$$\left. \cdot \operatorname{erfc}(-\gamma^{1/2} \cos \alpha)\right] \quad (9)$$

for all $\alpha \in [-\pi,\pi]$. (The last two equalities are obtained from the fact [6] that

$$D_0(x) = \exp\left(\frac{-x^2}{4}\right)$$

and

$$D_{-2}(x) = \exp\left(\frac{-x^2}{4}\right)\left[1 - \left(\frac{\pi}{2}\right)^{1/2} x \exp\left(\frac{x^2}{2}\right) \operatorname{erfc}\left(\frac{x}{\sqrt{2}}\right)\right]$$

for all x.) Equation (9) is Rice's well-known result [7] for the phase angle density function with no interference present (noise only). Thus (7) and (8) express $f_A(\alpha)$ as the sum of a term due to noise only and another due to noise and interference.

Finally, if we let

$$\hat{G}_n(\gamma,\alpha) = \frac{\exp(-\gamma)}{\pi} \cdot \frac{\gamma^n}{(2n)!} \sum_{k=0}^{n} \binom{n}{k}$$

$$\cdot H_{2n-2k}(\gamma^{1/2} \sin \alpha) H_{2k-2}(\gamma^{1/2} \cos \alpha),$$

$$\forall n = 0,1,\cdots, \quad \forall \gamma > 0, \quad \forall \alpha \in [-\pi,\pi]$$

then the density is just

$$f_A(\alpha) = \sum_{n=0}^{\infty} \mu_{2n} \hat{G}_n(\gamma,\alpha), \qquad \forall \alpha \in [-\pi,\pi].$$

This expression is quite useful for computational purposes; once we compute the set of values $\{\hat{G}_n(\gamma,\alpha)\}$, we can find f_A for various types of interference by different choices of $\{\mu_{2n}\}$.

We now prove (5) and (6). In Fig. 2, conditioned on the event $\{R = r, \Psi = \psi\}$, Q has the density function (see Rice [7])

$$f_Q(q/R = r, \Psi = \psi) = 2\gamma q \exp[-\gamma(q^2 + r^2)] I_0(2\gamma qr),$$

$$\forall q,r \geq 0, \quad \forall \psi \in [-\pi,\pi]$$

where I_0 is the zeroth-order modified Bessel function of the first kind. Taking the expectation over Ψ we get

$$f_Q(q/R = r) = 2\gamma q \exp[-\gamma(q^2 + r^2)] I_0(2\gamma qr),$$

$$\forall q,r \geq 0. \quad (10)$$

One generating function of the Laguerre polynomials is [8]

$$\exp(-u^2) I_0(2uv) = \sum_{n=0}^{\infty} \frac{(-1)^n u^{2n}}{n!} L_n(v^2), \qquad \forall u,v. \quad (11)$$

Substituting (11) into (10) we obtain

$$f_Q(q/R = r) = 2\gamma q \exp(-\gamma q^2) \sum_{n=0}^{\infty} \frac{(-1)^n \gamma^n r^{2n}}{n!} L_n(\gamma q^2),$$

$$\forall q,r \geq 0. \quad (12)$$

Taking the expectation over R and assuming for the moment that the interchange of expectation and infinite summation is valid, we find

$$f_Q(q) = 2\gamma q \exp(-\gamma q^2) \sum_{n=0}^{\infty} \frac{(-1)^n \gamma^n E[R^{2n}]}{n!} L_n(\gamma q^2),$$

$$\forall q \geq 0. \quad (13)$$

The density function of A can be expressed in terms of that of Q by using standard techniques [9]. The result is

$$f_A(\alpha) = \frac{1}{2\pi} \int_0^{\infty} z(1 + z^2 - 2z \cos \alpha)^{-1/2}$$

$$\cdot f_Q[(1 + z^2 - 2z \cos \alpha)^{1/2}] \, dz, \qquad \forall \alpha \in [-\pi,\pi].$$

$$(14)$$

Substituting (13) into (14) we see that

$$f_A(\alpha) = \frac{\gamma}{\pi}\int_0^\infty z \exp\left[-\gamma(1+z^2-2z\cos\alpha)\right]$$

$$\cdot \sum_{n=0}^{\infty} \frac{(-1)^n\gamma^n E[R^{2n}]}{n!} L_n[\gamma(1+z^2-2z\cos\alpha)]\, dz$$

$$= \frac{\gamma}{\pi}\sum_{n=0}^{\infty} \frac{(-1)^n\gamma^n E[R^{2n}]}{n!}$$

$$\cdot \int_0^\infty z \exp\left[-\gamma(1+z^2-2z\cos\alpha)\right]$$

$$\cdot L_n[\gamma(1+z^2-2z\cos\alpha)]\, dz \tag{15}$$

assuming for the moment that the interchange of integration and infinite summation is valid. In the Appendix we show that for $n = 0,1,\cdots$, for any $\gamma > 0$ and any $\alpha \in [-\pi,\pi]$:

$$\int_0^\infty z \exp\left[-\gamma(1+z^2-2z\cos\alpha)\right]$$

$$\cdot L_n[\gamma(1+z^2-2z\cos\alpha)]\, dz$$

$$= \frac{\exp(-\gamma)}{\gamma}\,\frac{(-1)^n}{n!2^{2n}}\sum_{k=0}^{n}\binom{n}{k} H_{2n-2k}(\gamma^{1/2}\sin\alpha)$$

$$\cdot H_{2k-2}(\gamma^{1/2}\cos\alpha) \tag{16}$$

$$= \frac{\exp(-\gamma/2)}{\gamma}\,\frac{(-1)^n}{n!2^{n+1}}\sum_{k=0}^{n}\binom{n}{k} D_{2n-2k}((2\gamma)^{1/2}\sin\alpha)$$

$$\cdot D_{2k-2}(-(2\gamma)^{1/2}\cos\alpha). \tag{17}$$

Using (4), (16), and (17) we obtain the desired result (5) and (6).

To complete the proof we need only to justify the interchange of expectation, integration, and infinite summation previously carried out. To do this it suffices [10] to prove that

$$\sum_{n=0}^{\infty} \frac{\gamma^n E[R^{2n}]}{n!} \mid L_n(\gamma q^2) \mid < \infty \tag{18}$$

and

$$\sum_{n=0}^{\infty} \frac{\gamma^n E[R^{2n}]}{n!} \int_0^\infty z \exp\left[-\gamma(1+z^2-2z\cos\alpha)\right]$$

$$\cdot \mid L_n[\gamma(1+z^2-2z\cos\alpha)] \mid dz < \infty. \tag{19}$$

Since [11, pp. 193–195, 205] $\mid L_n(y) \mid \leq \exp(y/2)$, for all $n,y \geq 0$, (18) and (19) follow from our hypotheses.

In Section V we will present several typical plots of f_A.

IV. Error Probabilities

A. Differential Detection

Consider $(j+1)$ consecutive bauds, without loss of generality, the 1st,2nd,$\cdots$,$(j+1)$th baud. The received signal $\hat{S}(t)$ is sampled at times $t_1,t_2,\cdots,t_{j+1}$, where $t_i = t_0 + (i-1)T$, $i = 1,2,\cdots,j+1$, and t_0 is in the first-baud interval. Let the angle between the transmitted and received signals $S(t_i)$ and $\hat{S}(t_i)$, respectively, be denoted by $A_i = A(t_i)$, $i = 1,2,\cdots,j+1$. From our assumptions and previous discussion we know that $A_1,\cdots,A_{j+1}$ are independent identically distributed (i.i.d.) random variables with probability density function f_A and are independent of the particular sequence of digits being sent. In the presence of additive Gaussian noise and assuming each of m digits is transmitted with equal probability, the differential detector which minimizes the probability of error is as follows [15].

Measure the phase difference between successively received signals $\hat{S}(t_{i+1})$ and $\hat{S}(t_i)$, quantize this angle to the nearest $(2\pi k)/m$, and then decide k was transmitted. If we use this detector when both noise *and* interference are present, we make an error in the ith digit if and only if $\mid A_{i+1} - A_i \mid$ (reduced modulo $[-\pi,\pi]$) is greater than π/m (see [15]). Thus error probabilities depend only on $A_1,\cdots,A_{j+1}$, and these probabilities are not a function of the transmitted sequence of digits. In principle error probabilities can be found by integration of the (known) joint density function of the A_i over the error region. However, using the structure of the way in which multiple errors occur, we will derive some probability of error expressions which are relatively simple to evaluate on a computer.

Let us employ the notation $\bar{z}$ to mean that the real number z is reduced modulo $[-\pi,\pi]$. Define random variables $E_1,\cdots,E_j$ by $E_i = g(A_{i+1} - A_i)$, for all $i = 1,2,\cdots,j$, where the function g satisfies, for all ξ,

$$g(\xi) = \begin{cases} 1, & \text{if } \mid \bar{\xi} \mid > \pi/m \\ 0, & \text{if } \mid \bar{\xi} \mid \leq \pi/m. \end{cases}$$

Thus an error occurs in the ith digit if and only if $E_i = 1$. The probability of j consecutive errors in an m-phase DPSK system is denoted

$$P_m{}^D(j) = P\{E_1 = 1,\cdots,E_j = 1\}.$$

We also define two other functions h and s as follows: for all $\alpha,\beta \in [-\pi,\pi]$,

$$h(\alpha) = P\{E_q = 1/A_{q+1} = \alpha\}$$

$$= P\{g(A_{q+1} - A_q) = 1/A_{q+1} = \alpha\}$$

$$= P\{g(\alpha - A') = 1\} = P\{g(A' - \alpha) = 1\}$$

$$s(\alpha,\beta) = P\{E_q = 1, E_{q+1} = 1/A_q = \alpha, A_{q+2} = \beta\}$$

$$= P\{g(A_{q+1} - A_q) = 1, g(A_{q+2} - A_{q+1}) = 1/$$

$$A_q = \alpha, A_{q+2} = \beta\}$$

$$= P\{g(A' - \alpha) = 1, g(\beta - A') = 1\}$$

where A' is a random variable with density function f_A. One can express h and s in terms of f_A, for all $\alpha,\beta \in [-\pi,$

$\pi]$, by

$$h(\alpha) = \int_{\{x:|(\alpha-x)|>\pi/m\}} f_A(x)\,dx$$

$$= 1 - \int_{\alpha-\pi/m}^{\alpha+\pi/m} \hat{f}_A(x)\,dx$$

$$= \int_{\alpha+\pi/m}^{\alpha-\pi/m+2\pi} \hat{f}_A(x)\,dx \quad (20)$$

$$s(\alpha,\beta) = \int_{\{x:|(x-\alpha)|>\pi/m,|(\beta-x)|>\pi/m\}} f_A(x)\,dx \quad (21)$$

where $\hat{f}_A$ is the periodic extension of f_A, i.e.,

$$\hat{f}_A(\xi + 2i\pi) = f_A(\xi), \qquad \xi \in [-\pi,\pi], \quad i = 0,\pm1,\pm2,\cdots.$$

We now derive the probability of j errors. If j is even, then with $f = f_A$

$$P_m{}^D(j) = \int_{-\pi}^{\pi} \cdots \int_{-\pi}^{\pi} P\{E_1 = 1, E_2 = 1, \cdots,$$
$$E_j = 1/A_2 = \alpha_2, A_4 = \alpha_4, \cdots, A_j = \alpha_j\}$$
$$\cdot f_{A_2A_4\cdots A_j}(\alpha_2,\alpha_4,\cdots,\alpha_j)\,d\alpha_2\,d\alpha_4\cdots d\alpha_j \quad (22)$$
$$= \int_{-\pi}^{\pi} \cdots \int_{-\pi}^{\pi} P\{g(A_2 - A_1) = 1,$$
$$g(A_3 - A_2) = 1, g(A_4 - A_3) = 1, \cdots,$$
$$g(A_{j-1} - A_{j-2}) = 1, g(A_j - A_{j-1}) = 1,$$
$$g(A_{j+1} - A_j) = 1/A_2 = \alpha_2, A_4 = \alpha_4, \cdots,$$
$$A_j = \alpha_j\} f(\alpha_2) f(\alpha_4) \cdots f(\alpha_j)\,d\alpha_2\,d\alpha_4 \cdots d\alpha_j$$
$$= \int_{-\pi}^{\pi} \cdots \int_{-\pi}^{\pi} P\{g(\alpha_2 - A_1) = 1, g(A_3 - \alpha_2) = 1,$$
$$g(\alpha_4 - A_3) = 1, \cdots, g(A_{j-1} - \alpha_{j-2}) = 1,$$
$$g(\alpha_j - A_{j-1}) = 1, g(A_{j+1} - \alpha_j) = 1\}$$
$$\cdot f(\alpha_2) f(\alpha_4) \cdots f(\alpha_j)\,d\alpha_2\,d\alpha_4 \cdots d\alpha_j$$
$$= \int_{-\pi}^{\pi} \cdots \int_{-\pi}^{\pi} h(\alpha_2) s(\alpha_2,\alpha_4)$$
$$\cdot s(\alpha_4,\alpha_6) \cdots s(\alpha_{j-2},\alpha_j) h(\alpha_j)$$
$$\cdot f(\alpha_2) f(\alpha_4) \cdots f(\alpha_j)\,d\alpha_2\,d\alpha_4 \cdots d\alpha_j. \quad (23)$$

Thus for $j = 2$ and $j = 4$ we have

$$P_m{}^D(2) = \int_{-\pi}^{\pi} h^2(\alpha_2) f(\alpha_2)\,d\alpha_2 \quad (24)$$

and

$$P_m{}^D(4) = \int_{-\pi}^{\pi} d\alpha_4\, h(\alpha_4) f(\alpha_4) \int_{-\pi}^{\pi} d\alpha_2\, h(\alpha_2) f(\alpha_2) s(\alpha_2,\alpha_4) \quad (25)$$

and in general for $j > 4$ and even we have

$$P_m{}^D(j) = \int_{-\pi}^{\pi} d\alpha_j\, h(\alpha_j) f(\alpha_j) \int_{-\pi}^{\pi} d\alpha_{j-2}\, f(\alpha_{j-2})$$
$$\cdot s(\alpha_{j-2},\alpha_j) \cdots \int_{-\pi}^{\pi} d\alpha_4\, f(\alpha_4) s(\alpha_4,\alpha_6)$$
$$\cdot \int_{-\pi}^{\pi} d\alpha_2\, h(\alpha_2) f(\alpha_2) s(\alpha_2,\alpha_4). \quad (26)$$

In a similar fashion, starting with

$$P_m{}^D(j) = \int_{-\pi}^{\pi} \cdots \int_{-\pi}^{\pi} P\{E_1 = 1, E_2 = 1, \cdots,$$
$$E_j = 1/A_2 = \alpha_2, A_4 = \alpha_4, \cdots, A_{j+1} = \alpha_{j+1}\}$$
$$\cdot f_{A_2A_4\cdots A_{j+1}}(\alpha_2,\alpha_4,\cdots,\alpha_{j+1})\,d\alpha_2\,d\alpha_4\cdots d\alpha_{j+1}$$

we can show that for j odd with $f = f_A$

$$P_m{}^D(j) = \int_{-\pi}^{\pi} \cdots \int_{-\pi}^{\pi} h(\alpha_2) s(\alpha_2,\alpha_4) s(\alpha_4,\alpha_6) \cdots s(\alpha_{j-1},\alpha_{j+1})$$
$$\cdot f(\alpha_2) f(\alpha_4) \cdots f(\alpha_{j+1})\,d\alpha_2\,d\alpha_4 \cdots d\alpha_{j+1}. \quad (27)$$

Thus for $j = 1$ and $j = 3$ we have

$$P_m{}^D(1) = \int_{-\pi}^{\pi} h(\alpha_2) f(\alpha_2)\,d\alpha_2 \quad (28)$$

$$P_m{}^D(3) = \int_{-\pi}^{\pi} d\alpha_4\, f(\alpha_4) \int_{-\pi}^{\pi} d\alpha_2\, h(\alpha_2) f(\alpha_2) s(\alpha_2,\alpha_4) \quad (29)$$

while in general for $j > 3$ and odd we have

$$P_m{}^D(j) = \int_{-\pi}^{\pi} d\alpha_{j+1}\, f(\alpha_{j+1}) \int_{-\pi}^{\pi} d\alpha_{j-1}\, f(\alpha_{j-1})$$
$$\cdot s(\alpha_{j-1},\alpha_{j+1}) \cdots \int_{-\pi}^{\pi} d\alpha_4\, f(\alpha_4) s(\alpha_4,\alpha_6)$$
$$\cdot \int_{-\pi}^{\pi} d\alpha_2\, h(\alpha_2) f(\alpha_2) s(\alpha_2,\alpha_4). \quad (30)$$

Of course, $P_m{}^D(1)$ can also be found by an appropriate integration of the (mod $[-\pi,\pi]$) convolution of f_A with itself. It is not hard to show that the result agrees with $P_m{}^D(1)$ given by (28) and (20).

Equations (26) and (30) are quite suitable for machine computation because they are recursive. Furthermore, the error probabilities $P_m{}^D(1), P_m{}^D(2), \cdots, P_m{}^D(j)$ can all be found in one sequence of operations, e.g., after we find

$$\eta(\alpha_4) \triangle \int_{-\pi}^{\pi} d\alpha_2\, h(\alpha_2) f(\alpha_2) s(\alpha_2,\alpha_4)$$

we can get $P_m{}^D(3)$ by the integration $\int_{-\pi}^{\pi} d\alpha_4\, f(\alpha_4)\eta(\alpha_4)$, and we can also use $\eta(\cdot)$ to find $P_m{}^D(4), P_m{}^D(5)$, etc.

B. Coherent Detection

The optimum coherent detector [15] quantizes the phase angle of the received signal to the nearest $(2\pi k)/m$ and then decides k was transmitted. Thus the probability of an error in the ith digit depends only on A_i, which is independent of $\{A_r: r \neq i\}$. Therefore, errors occur independently, and the probability of j consecutive errors in an m-phase CPSK system is just

$$P_m^C(j) = [P_m^C(1)]^j = \left[\int_{\{x:|x|>\pi/m\}} f_A(x)\,dx\right]^j$$
$$= \left[1 - \int_{-\pi/m}^{\pi/m} \hat{f}_A(x)\,dx\right]^j = \left[\int_{\pi/m}^{-\pi/m+2\pi} \hat{f}_A(x)\,dx\right]^j$$
$$= [h(0)]^j. \qquad (31)$$

C. Bounds

The derived expressions for DPSK error probabilities can be used to find an interesting collection of inequalities between the probabilities of different numbers of consecutive errors. In particular we will now show that for all $j \geq 3$ and odd

$$[P_m^D(j)]^2 \leq P_m^D(j-2)P_m^D(j+1) \qquad (32)$$

and

$$[P_m^D(1)]^2 \leq P_m^D(2). \qquad (33)$$

Equation (32) provides, for example, a lower bound for the probability of $j+1$ errors in terms of the probabilities of j errors and of $j-2$ errors. Equation (33) was also previously observed in [12].

To prove (32) we first note that for all $\alpha,\beta \in [-\pi,\pi]$, $s(\alpha,\beta) \leq h(\beta)$. Then the following sequence of inequalities is valid ($j \geq 3$ and odd):

$$[P_m^D(j)]^2 = \left[\int_{-\pi}^{\pi} \cdots \int_{-\pi}^{\pi} h(\alpha_2)s(\alpha_2,\alpha_4)s(\alpha_4,\alpha_6)\cdots \cdot s^{1/2}(\alpha_{j-1},\alpha_{j+1})s^{1/2}(\alpha_{j-1},\alpha_{j+1})f(\alpha_2)f(\alpha_4)\cdots \cdot f(\alpha_{j+1})\,d\alpha_2\,d\alpha_4\cdots d\alpha_{j+1}\right]^2$$

$$\leq \left[\int_{-\pi}^{\pi} \cdots \int_{-\pi}^{\pi} h(\alpha_2)s(\alpha_2,\alpha_4)s(\alpha_4,\alpha_6)\cdots \cdot s^{1/2}(\alpha_{j-1},\alpha_{j+1})h^{1/2}(\alpha_{j+1})f(\alpha_2)f(\alpha_4)\cdots \cdot f(\alpha_{j+1})\,d\alpha_2\,d\alpha_4\cdots d\alpha_{j+1}\right]^2$$

$$\leq \left[\int_{-\pi}^{\pi} \cdots \int_{-\pi}^{\pi} h(\alpha_2)s(\alpha_2,\alpha_4)s(\alpha_4,\alpha_6)\cdots \cdot s(\alpha_{j-1},\alpha_{j+1})h(\alpha_{j+1})f(\alpha_2)f(\alpha_4)\cdots \cdot f(\alpha_{j+1})\,d\alpha_2\,d\alpha_4\cdots d\alpha_{j+1}\right]$$
$$\cdot\left[\int_{-\pi}^{\pi} \cdots \int_{-\pi}^{\pi} h(\alpha_2)s(\alpha_2,\alpha_4)s(\alpha_4,\alpha_6)\cdots \cdot s(\alpha_{j-3},\alpha_{j-1})f(\alpha_2)f(\alpha_4)\cdots f(\alpha_{j-1}) \cdot f(\alpha_{j+1})\,d\alpha_2\,d\alpha_4\cdots d\alpha_{j-1}\,d\alpha_{j+1}\right]$$
$$= P_m^D(j+1)P_m^D(j-2)$$

where the last inequality follows from the Schwarz inequality. The proof of (33) is almost identical.

If we also assume that f_A is monotonically decreasing on $[0,\pi]$, which is the case when only noise is present or in most cases when noise and interference are present (see Section V), then we can show that for all $j \geq 3$

$$P_m^D(j) \geq P_m^C(1)P_m^D(j-1). \qquad (34)$$

To prove this inequality first note from (20) and our assumption about f_A that

$$h(\alpha) \geq h(0), \qquad \forall\alpha \in [-\pi,\pi]. \qquad (35)$$

Then, starting with

$$P_m^D(j) = \int_{-\pi}^{\pi} \cdots \int_{-\pi}^{\pi} P\{E_1 = 1, E_2 = 1, \cdots, E_j = 1/A_1 = \alpha_2, A_3 = \alpha_4, \cdots, A_{j+1} = \alpha_{j+2}\} \cdot f_{A_1A_3\cdots A_{j+1}}(\alpha_2,\alpha_4,\cdots,\alpha_{j+2})\,d\alpha_2\,d\alpha_4\cdots d\alpha_{j+2}$$

we can show by the same methods used in deriving (23) that for even j

$$P_m^D(j) = \int_{-\pi}^{\pi} \cdots \int_{-\pi}^{\pi} s(\alpha_2,\alpha_4)s(\alpha_4,\alpha_6)\cdots s(\alpha_j,\alpha_{j+2}) \cdot f(\alpha_2)f(\alpha_4)\cdots f(\alpha_{j+2})\,d\alpha_2\,d\alpha_4\cdots d\alpha_{j+2}. \qquad (36)$$

Equation (34) now follows directly from (23), (27), (31), (35), and (36). It is also easy to see that $P_m^D(j) \geq P_m^C(j)$, for all j.

V. Numerical Results and Conclusions

The interference due to several cochannel angle-modulated communications systems is modeled as the sum of M cosinusoids with constant amplitudes $a_1,\cdots,a_M$ and independent phases each uniformly distributed on $[-\pi,\pi]$. The projection of a constant amplitude uniformly phased cosinusoid onto the cos $\omega_0 t$ and sin $\omega_0 t$ axes is a circularly symmetric pair, and the sum of M of these terms is also circularly symmetric. For the numerical computations we further specialize to the case $a_1 = \cdots = a_M = a$. Then the carrier-to-interference power ratio (CIR) is

$$\text{CIR(dB)} = 10 \log_{10}(1/Ma^2),$$

and the CNR is

$$\text{CNR(dB)} = 10 \log_{10}(1/2\sigma^2),$$

where σ^2 is the noise power.

The interference moments $\{\mu_{2n}\}$ were found from expressions derived in [13]. These moments were used to obtain plots of the density function f_A (Figs. 3–6) for $M = 1$ and 2 and for typical values of CIR and CNR. The curves for $M = 1$ are in excellent agreement with those obtained earlier by Rosenbaum [14] by a different method. Note that in most situations the peak of f_A occurs at 0. However, for high CNR and relatively low CIR, the effect of the interference dominates that of the noise, and in the case of a single interferer ($M = 1$) the peak of f_A shifts away from 0.

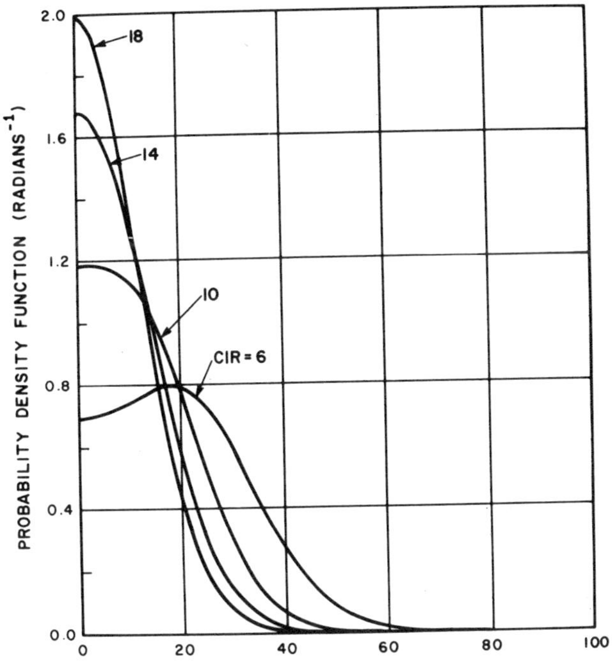

Fig. 3. Probability density function f_A for CNR = 12 and M = 1.

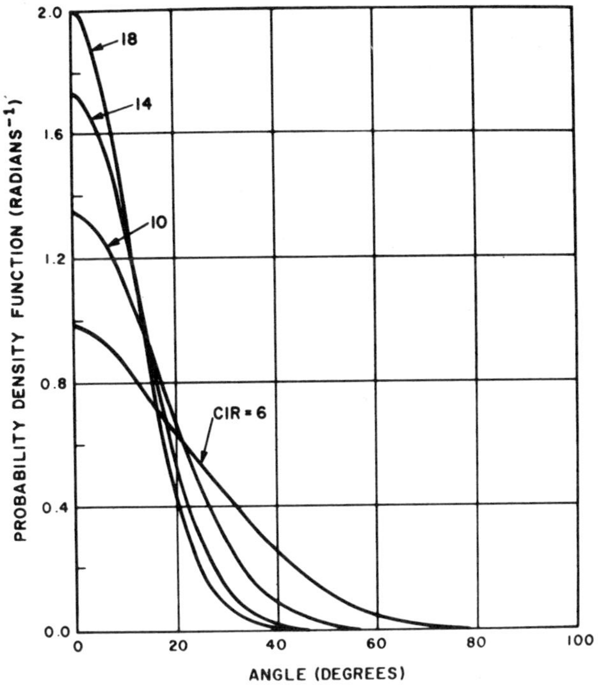

Fig. 4. Probability density function f_A for CNR = 12 and M = 2.

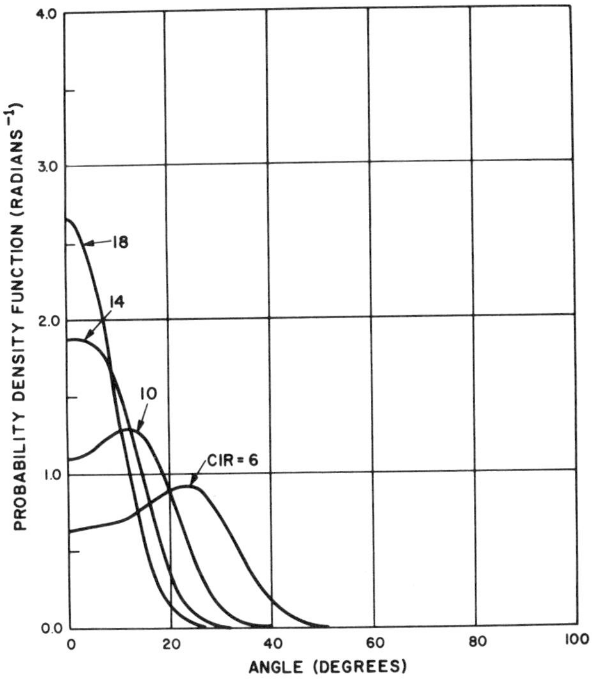

Fig. 5. Probability density function f_A for CNR = 16 and M = 1.

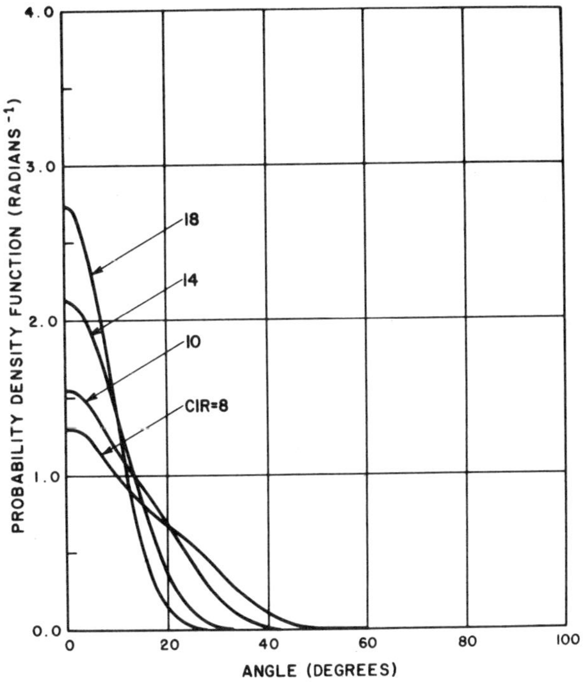

Fig. 6. Probability density function f_A for CNR = 16 and M = 2.

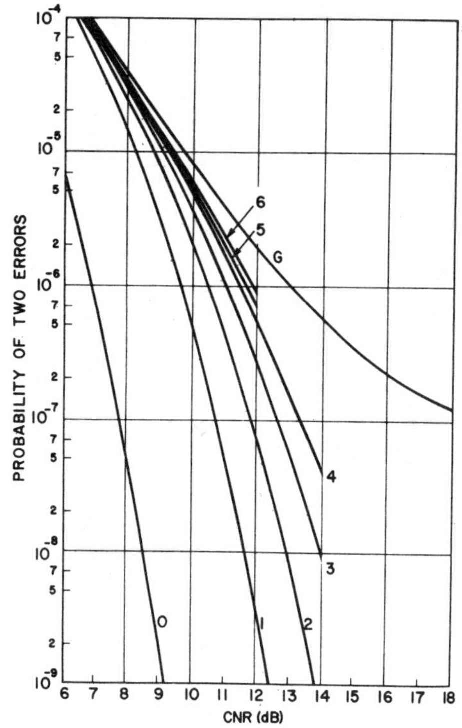

Fig. 7. Probability of two errors versus CNR for CPSK, 2-phase, CIR = 8.

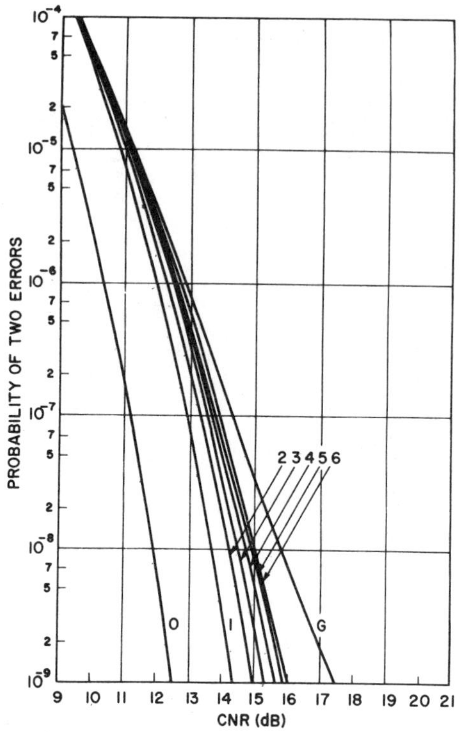

Fig. 8. Probability of two errors versus CNR for CPSK, 4-phase, CIR = 14.

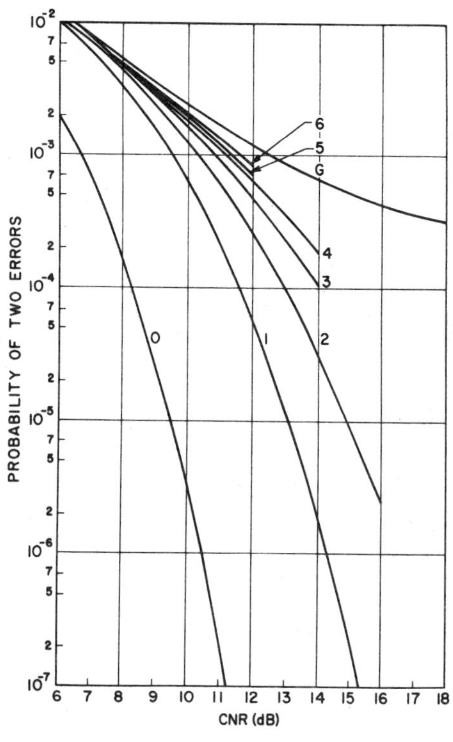

Fig. 9. Probability of two errors versus CNR for DPSK, 2-phase, CIR = 8.

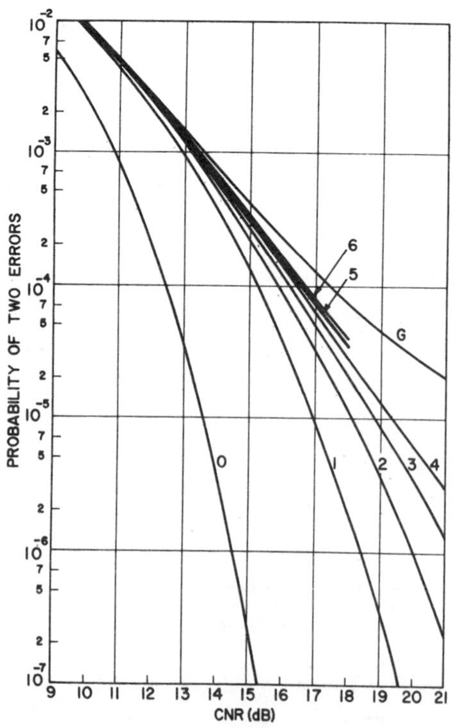

Fig. 10. Probability of two errors versus CNR for DPSK, 4-phase, CIR = 14.

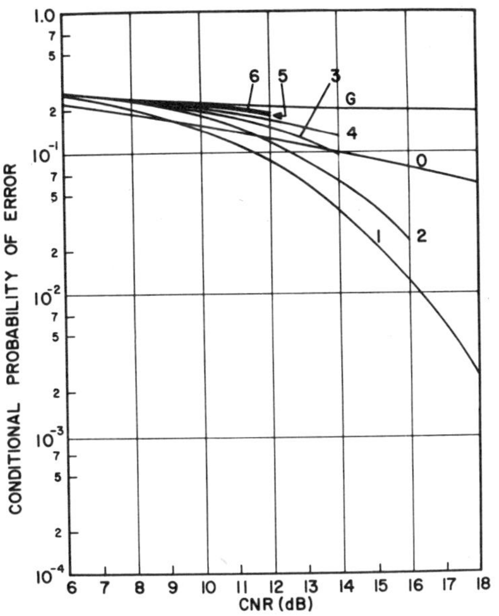

Fig. 11. Conditional probability of error versus CNR for DPSK, 2-phase, CIR = 8.

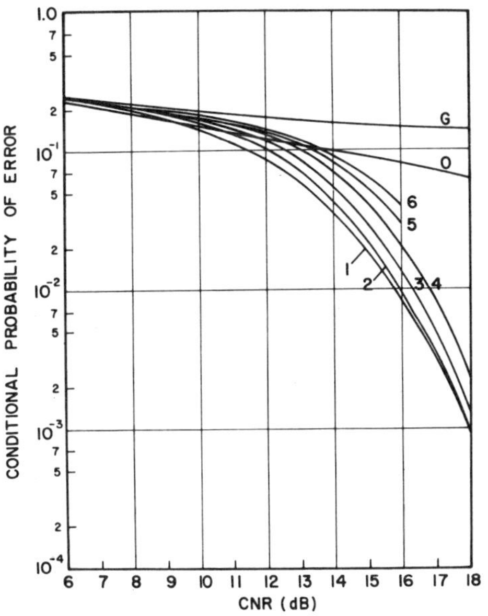

Fig. 12. Conditional probability of error versus CNR for DPSK, 2-phase, CIR = 12.

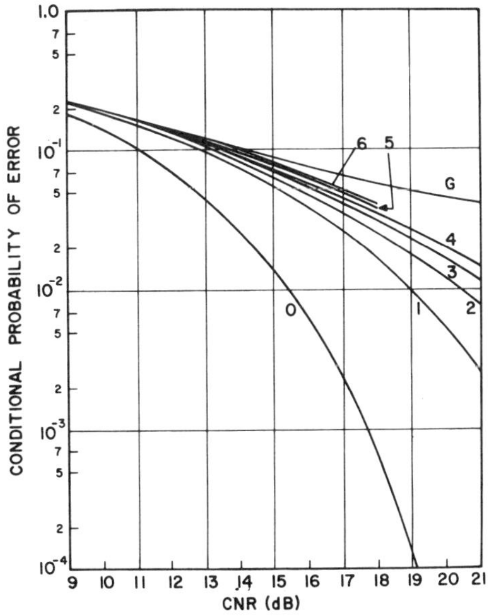

Fig. 13. Conditional probability of error versus CNR for DPSK, 4-phase, CIR = 14.

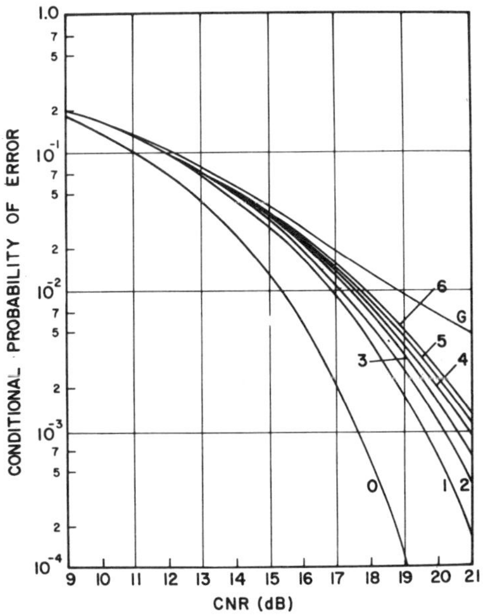

Fig. 14. Conditional probability of error versus CNR for DPSK, 4-phase, CIR = 18.

Error probabilities were calculated using the derived expressions (24), (28), and (31) with the appropriate density function f_A. Numerical integration techniques were employed to approximate the theoretical integrations. In Figs. 7–10 are typical curves of the probability of two consecutive errors as a function of CNR, for both CPSK and DPSK. In the case of DPSK the conditional probability of error $P_m^D(2)/P_m^D(1)$ is also given in Figs. 11–14 (for CPSK the conditional probability of error equals the single-error probability). Although the calculations were only performed for 2- and 4-phase systems, the derived expressions can be applied to obtain the same type of results for any m-phase system. For 2-phase PSK the curves are drawn from CNR = 6 to CNR = 18 for fixed values of CIR, while for 4-phase PSK the curves run from CNR = 9 to CNR = 21.

The curves in each set are labeled $1,2,\cdots,6$ according to the number M, of cosinusoids making up the interference. The O curve is the no interference case (noise only), while the G curve is for the case when the interference consists of additional Gaussian noise of variance $\frac{1}{2}\cdot 10^{-\mathrm{CIR}/10}$.

In general we see that the double-error curves are each monotonically decreasing as CNR increases, and each set of curves is in the increasing order $0,1,2,\cdots,6,G$. For fixed CNR and CIR, as M increases the curves approach the G curve as one would expect from the central limit theorem. Similar remarks are also valid for the conditional probabilities of error, except that for the 2-phase case the O curve is not in order with the others.

Next, consider the set of double-error curves for 2-phase CPSK. Note that at any fixed moderate to high value of CNR there is a fairly wide spread between the error probabilities for different but small values of M and also between moderate M and the G curve. In particular at CIR = 8 and CNR = 14 there is about a 2 decade "spread" between $P_2^C(2)$, for $M = 2$ and $M = 4$, and about a decade spread between $M = 4$ and G. Furthermore, as CNR increases the G curve must approach a nonzero value (equal to the error probability in the presence of Gaussian noise with fixed variance $\frac{1}{2}\cdot 10^{-\mathrm{CIR}/10}$), but the $0,1,2,\cdots,6$ curves can be driven to zero error probability. Similar observations may also be made for 2-phase DPSK as well as 4-phase CPSK and DPSK. Thus a general conclusion is that one cannot approximate well the effect of interference on the performance of a PSK system by treating it as an equivalent amount of Gaussian noise (for moderate to high CNR).

Comparing CPSK and DPSK, we note that CPSK is affected less by interference than is DPSK in the following sense. To maintain a fixed probability of error, a DPSK system requires a larger increase in CNR when interference is present over that needed for the no interference case. For example, for CIR = 14 and a double-error probability of 10^{-7}, 4-phase DPSK requires a CNR = 15.3 dB when no interference is present and 19.6 dB when $M = 1$, an increase of 4.3 dB. On the other hand, the increase for CPSK is only 12.8 − 11.2 = 1.6 dB.

Finally, we remark that numerical results for other types of interference can be obtained in the same way once the moments $\{\mu_{2n}\}$ are known. In [13] computationally simple expressions are derived for these moments when the interference is the sum of independent circularly symmetric pairs of random variables; a useful special case is the sum of independent *random* amplitude cosinusoids with independent uniformly distributed phases.

Appendix

Proof of (16) and (17)

Define

$$I_n(\gamma,\alpha) = \int_0^\infty r \exp\left[-\gamma(r^2 - 2r\cos\alpha + 1)\right] \cdot L_n[\gamma(r^2 - 2r\cos\alpha + 1)]\, dr, \quad \gamma > 0, \quad -\pi \le \alpha \le \pi, \quad n = 0,1,\cdots. \tag{37}$$

Rewrite (37) as

$$I_n(\gamma,\alpha) = \int_0^\infty r \exp\{-[\gamma(r-\cos\alpha)^2 + \gamma\sin^2\alpha]\} \cdot L_n[\gamma(r-\cos\alpha)^2 + \gamma\sin^2\alpha]\, dr$$

and set $z = \gamma^{1/2}(r - \cos\alpha)$ to obtain

$$I_n(\gamma,\alpha) = \frac{1}{\gamma}\int_{-\gamma^{1/2}\cos\alpha}^\infty (z + \gamma^{1/2}\cos\alpha) \cdot \exp\left[-(z^2 + \gamma\sin^2\alpha)\right] L_n(z^2 + \gamma\sin^2\alpha)\, dz. \tag{38}$$

One can express L_n in terms of Hermite polynomials as (see [11, p. 195, eq. 32], but note misprint: the factor $1/2^{2n}$ has been omitted):

$$L_n(x^2 + y^2) = \frac{(-1)^n}{n!2^{2n}} \sum_{k=0}^n \binom{n}{k} H_{2k}(x) H_{2n-2k}(y), \quad n = 0,1,2,\cdots, \quad \forall x,y. \tag{39}$$

Substituting for L_n in (38) one finds

$$\begin{aligned} I_n(\gamma,\alpha) &= \frac{1}{\gamma}\int_{-\gamma^{1/2}\cos\alpha}^\infty (z + \gamma^{1/2}\cos\alpha) \cdot \exp\left[-(z^2 + \gamma\sin^2\alpha)\right] \frac{(-1)^n}{n!2^{2n}} \cdot \sum_{k=0}^n \binom{n}{k} H_{2k}(z) H_{2n-2k}(\gamma^{1/2}\sin\alpha)\, dz \\ &= \frac{1}{\gamma}\frac{(-1)^n}{n!2^{2n}} \sum_{k=0}^n \binom{n}{k} H_{2n-2k}(\gamma^{1/2}\sin\alpha) \cdot \exp(-\gamma\sin^2\alpha) \int_{-\gamma^{1/2}\cos\alpha}^\infty (z + \gamma^{1/2}\cos\alpha) \cdot \exp(-z^2) H_{2k}(z)\, dz. \end{aligned} \tag{40}$$

Denote for any a, and $k \geq 0$,

$$A_k = \int_{-a}^{\infty} z \exp(-z^2) H_{2k}(z)\, dz \quad (41)$$

$$B_k = \int_{-a}^{\infty} a \exp(-z^2) H_{2k}(z)\, dz. \quad (42)$$

For $k \geq 1$, we integrate (41) by parts with $u = H_{2k}(z)$, $dv = z \exp(-z^2) dz$ and note [11] that $(d/dz) H_{2k}(z) = 2(2k) H_{2k-1}(z)$. The result is

$$A_k = -\frac{1}{2} \exp(-z^2) H_{2k}(z) \Big]_{-a}^{\infty} + 2k \int_{-a}^{\infty} \exp(-z^2) H_{2k-1}(z)\, dz$$

$$= \frac{1}{2} \exp(-a^2) H_{2k}(-a) + 2k \int_{0}^{\infty} \exp(-z^2) H_{2k-1}(z)\, dz + 2k \int_{-a}^{0} \exp(-z^2) H_{2k-1}(z)\, dz \quad (43)$$

observing that $\lim_{z \to \infty} \exp(-z^2) H_{2k}(z) = 0$.

One also knows that [11]

$$\int_{0}^{x} \exp(-z^2) H_m(z)\, dz = H_{m-1}(0) - \exp(-x^2) H_{m-1}(x). \quad (44)$$

Applying (44) to (43) it is easy to obtain

$$A_k = \tfrac{1}{2} \exp(-a^2) H_{2k}(-a) + 2k \exp(-a^2) H_{2k-2}(-a), \quad \forall k \geq 1. \quad (45)$$

Employing (44) we find similarly that

$$B_k = a \exp(-a^2) H_{2k-1}(-a), \quad \forall k \geq 1. \quad (46)$$

Thus from the recurrence relation of Hermite polynomials [11] and the fact that $H_m(-x) = (-1)^m H_m(x)$ it follows that

$$A_k + B_k = \tfrac{1}{2} \exp(-a^2) H_{2k}(-a) + 2k \exp(-a^2) \cdot H_{2k-2}(-a) + a \exp(-a^2) H_{2k-1}(-a)$$
$$= \exp(-a^2) H_{2k-2}(a), \quad \forall k \geq 1. \quad (47)$$

Furthermore, since $H_0(z) \equiv 1$,

$$A_0 + B_0 = \int_{-a}^{\infty} z \exp(-z^2)\, dz + a \int_{-a}^{\infty} \exp(-z^2)\, dz = \frac{1}{2} \exp(-a^2) + \frac{\sqrt{\pi}}{2} a \operatorname{erfc}(-a). \quad (48)$$

Then,

$$\int_{-a}^{\infty} (z + a) \exp(-z^2) H_{2k}(z)\, dz = A_k + B_k = \exp(-a^2) H_{2k-2}(a), \quad \forall k \geq 0. \quad (49)$$

Applying (49) to (40) one obtains (16). Equation (17) follows from (16) using the relations [6]

$$D_m(x) = 2^{-m/2} \exp\left(\frac{-x^2}{4}\right) H_m\left(\frac{x}{\sqrt{2}}\right), \quad \forall x, \forall m = 0,1,\cdots$$

$$D_{-2}(x) = \exp\left(\frac{-x^2}{4}\right) - \left(\frac{\pi}{2}\right)^{1/2} x \exp\left(\frac{x^2}{4}\right) \operatorname{erfc}\left(\frac{x}{\sqrt{2}}\right), \quad \forall x.$$

Acknowledgment

The author wishes to thank A. S. Rosenbaum for several interesting conversations.

References

[1] J. Salz and B. R. Saltzberg, "Double error rates in differentially coherent phase systems," *IEEE Trans. Commun. Syst.*, vol. CS-12, June 1964, pp. 202–205.

[2] J. F. Oberst and D. L. Schilling, "Double error probability in differential PSK," *Proc. IEEE* (Lett.), vol. 56, June 1968, pp. 1099–1100.

[3] D. Sakrison, *Communication Theory: Transmission of Waveforms and Digital Information.* New York: Wiley, 1968, pp. 142–148.

[4] A. Papoulis, *Systems and Transforms with Applications in Optics.* New York: McGraw-Hill, 1968, ch. 5.

[5] A. S. Rosenbaum, "Binary PSK error probabilities with multiple cochannel interferences," *IEEE Trans. Commun. Technol.*, vol. COM-18, June 1970, pp. 241–253.

[6] W. Magnus, F. Oberhettinger, and R. P. Soni, *Formulas and Theorems for the Special Functions of Mathematical Physics.* New York: Springer, 1966, pp. 239–255, 323–333.

[7] S. O. Rice, "Mathematical analysis of random noise," *Bell Syst. Tech. J.*, vol. 24, 1945, pp. 98–107.

[8] E. D. Rainville, *Special Functions.* New York: Macmillan, 1965, p. 201.

[9] A. Papoulis, *Probability, Random Variables, and Stochastic Processes.* New York: McGraw-Hill, 1965, p. 201.

[10] W. Rudin, *Real and Complex Analysis.* New York: McGraw-Hill, 1966, pp. 28–29.

[11] A. Erdélyi *et al.*, *Higher Transcendental Functions*, vol. II. New York: McGraw-Hill, 1953.

[12] A. S. Rosenbaum, unpublished memorandum.

[13] J. Goldman, "Moments of the sum of circularly symmetric random variables," to be published.

[14] A. S. Rosenbaum, "PSK error performance with Gaussian noise and interference," *Bell Syst. Tech. J.*, vol. 48, Feb. 1969, pp. 413–442.

[15] E. Arthurs and H. Dym, "On the optimum detection of digital signals in the presence of white Gaussian noise—a geometric interpretation and a study of three basic data transmission systems," *IRE Trans. Commun. Syst.*, vol. CS-10, Dec. 1962, pp. 336–372.

Spacing Limitations of Geostationary Satellites Using Multilevel Coherent PSK Signals

MICHEL C. JERUCHIM AND FLOYD E. LILLEY

Abstract—**The relationship between intersatellite spacing, error probability, and carrier-to-thermal noise ratio is studied for the case where the satellites transmit multilevel coherent phase-shift-keyed signals. Numerical results based on computer evaluation of exact expressions are presented for 2-, 4-, 8-, and 16-phase systems.**

I. Introduction

Since the establishment of geostationary communication satellites as a reality, there has been a growing recognition that the geostationary satellite orbit constitutes a unique and valuable, though unfortunately limited, natural resource. Consequently there has been rising concern that this resource be managed efficiently. The general subject of geostationary orbit utilization has been treated in detail elsewhere (see, e.g., [1], [2]). For present purposes it is sufficient merely to note the basic problem, namely that the number of satellites that can operate from a given arc of the geostationary orbit, using a common frequency band, are ultimately limited by mutual interference among the various geostationary satellite systems. Among the many factors that affect efficient use of the orbit, one of the more important is the type of modulation used by the satellites. Previous papers, e.g., [2]-[4], have dealt with the efficiency or orbit utilization for various types of analog modulation. The purpose of this paper is to examine the spacing potential of geostationary satellites using coherent phase-shift-keying (CPSK) modulation.

Numerical results have been obtained by computer evaluation of exact expressions, rather than by the approximation usually employed in the past [4]-[5] of treating the interference as an equal amount of thermal noise. Consequently, in the next section, we first review the pertinent theory for calculating error rate in CPSK systems operating in an interference environment [6]-[8]. Section III describes the problems of computer implementation in some detail as it is felt that the computational aspects and certain related results are in themselves of some interest. Finally, Section IV applies the previous results to the situation of a "homogeneous" group of CPSK satellites and presents numerical results, for 2-, 4-, 8-, and 16-phase systems, relating error probability to satellite spacing.

II. Error Probability for CPSK Systems

Only a brief development is given; more detailed derivations may be found in the references. The desired PSK signal is

Paper approved for publication by the Space Communications Committee of the IEEE Communications Society. Manuscript received June 1, 1971; revised March 20, 1972. This work was supported by the Office of Telecommunications Policy, under Contract OEP-SE-69-102.

The authors are with the Space Systems Organization, General Electric Company, Philadelphia, Pa.

Reprinted from *IEEE Trans. Comm.*, vol. COM-20, pp. 1021-1026, Oct. 1972.

given by

$$A(t) = \sqrt{2S} \cos [\omega_0 t + \theta], \tag{1}$$

where

S	average power,
$\theta = 2\pi r/M$, $1 \leqslant r \leqslant M$	carrier phase in any symbol interval,
M	possible number of symbols or carrier phases,
$f_0 = \omega_0/2\pi$	unmodulated carrier frequency.

The interference, arriving at the earth station receiver of the desired signal and arising from the other satellites[1] is assumed to consist of a number of angle-modulated carriers as follows:

$$I(t) = \sum_{j=1}^{K} \sqrt{2I_j} \cos [\omega_j t + \theta_j(t) + \mu_j], \tag{2}$$

where

I_j	average power in the jth interferer,
$\theta_j(t)$	modulation on the jth interferer,
μ_j	phase angle of the jth interferer, assumed to be a random variable uniformly distributed on $(0, 2\pi)$,
$f_j = \omega_j/2\pi$,	jth interferer carrier frequency,
K	number of interferers.

Finally, the noise in the desired receiver is assumed to be a zero-mean stationary Gaussian process with variance σ^2. The interferers are assumed to be generated independently, hence it is reasonable to assume that all the μ_j are statistically independent of one another as well as of the noise and of the $\theta_j(t)$, $1 \leqslant j \leqslant K$. Also, the $\theta_j(t)$ are assumed independent of one another. Note that in this analysis, the $\theta_j(t)$ can be arbitrary phase modulations and are not restricted to be digital PSK signals.

The signal, interference, and noise interact in the receiver, which once every symbol interval T produces an estimate of the actual transmitted phase or symbol θ. The receiver is assumed to be an ideal phase detector that measures the phase $\phi(t)$ of the input signal and whose decision mechanism is to announce that value of i for which $|\phi(t) - 2\pi i/M|$ is minimum. Equivalently, an error is made whenever $|\epsilon_s| > \pi/M$, where $\epsilon_s = \phi(t) - \theta$. The decision process can be represented graphically in terms of the familiar decision cones and is illustrated in Fig. 1 for $M = 8$ assuming, without loss of generality, that $\theta = 0^\circ$ is sent.

In terms of the geometry of Fig. 1, the error probability for an M-phase system can be generally written as

$$P_{eM} = P(AA') + P(BB') - P(A'B') \tag{3}$$

where

$P(AA')$	probability that the vector terminus lies in the upper half-plane AA',
$P(BB')$	probability that the vector terminus lies in the lower half-plane BB',
$P(A'B')$	probability that the vector terminus lies in the hatched cone $A'B'$.

Expressions for $P(A'B')$ are generally complicated and hence it is simpler to seek an upper bound to P_{eM}, which turns out to be adequately tight. From (3) it is clear that

$$P_{eM} \leqslant \hat{P}_{eM} = P(AA') + P(BB'). \tag{4}$$

[1] For simplicity, we deal only with the down-link interference. Uplink contributions can be treated in completely analogous fashion.

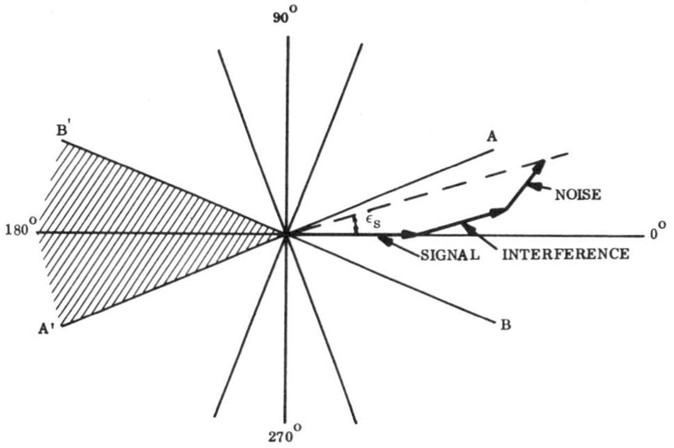

Fig. 1. Decision cones for $M = 8$.

On the other hand, we have the obvious inequality

$$P(A'B') \leqslant P_{eM} \tag{5}$$

which, along with (3) can be easily shown to yield

$$2P_{eM} \geqslant P(AA') + P(BB'), \tag{6}$$

and hence

$$0.5\,\hat{P}_{eM} \leqslant P_{eM} \leqslant \hat{P}_{eM}. \tag{7}$$

Considering the simplicity of the argument, (7) gives surprisingly tight bounds. In fact, a factor of 2 in error probability is generally considered quite small for digital systems. More elaborate lower bounds may be obtained [6], but apply only to restricted ranges of carrier-to-interference ratio. In view of (7) we deal only with the upper bound. Note that for the binary case, $P(AA') = P(BB') = P(A'B')$ and hence P_{e2} is equal to the lower bound in (7).

Referring to Fig. 1, expressions for $P(AA')$ and $P(BB')$ are easily obtained by resolving the various vectors along the lines AA' and BB', respectively. It can then be shown, [6], [9], that

$$\hat{P}_{eM} = E\,\{\text{erfc}\,(\rho \sin \pi/M + \rho\eta)\}, \tag{8}$$

where erfc is the complementary error function and

$\rho = \sqrt{S}/\sigma$	carrier-to-noise voltage ratio,
η	$= \sum_{j=1}^{K} R_j \cos \lambda_j$,
$R_j = \sqrt{I_j/S}$	interference-to-carrier voltage ratio,
λ_j	$= (\omega_j - \omega_0)t + \theta_j(t) - \theta + \mu_j$,[2]
E	expectation operator with respect to the random variable η.

Numerical evaluation of (8) turns out to be no less challenging than the original problem itself. By using the Taylor expansion of erfc (·), previous authors [6], [8] have evaluated (8) in the form

$$\hat{P}_{eM} = \text{erfc}\,(\rho \sin \pi/M) + \frac{2}{\sqrt{\pi}} \exp(-\rho^2 \sin^2 \pi/M)$$

$$\sum_{k=1}^{\infty} H_{2k-1}\,(\rho \sin \pi/M)\, \rho^{2k} \frac{E(\eta^{2k})}{(2k)!} \tag{9}$$

[2] Because of the properties of μ_j it can be shown that λ_j is also uniformly distributed on $(0, 2\pi)$. Hence, the error probability is independent of the actual carrier frequencies. This will be true in practice only if the receiver passband is wide enough to accept the interference without distortion.

where $H_n(X)$ is the Hermite polynomial of order n. Some of the difficulties in evaluating (9) are described in [8]. Our approach is based on an equivalent form of (8), namely,

$$\hat{P}_{eM} = \int_{-\Omega}^{\Omega} \operatorname{erfc}(\rho \sin \pi/M + \rho\eta)\, p(\eta)\, d\eta \tag{10}$$

where $\Omega = \Sigma R_j$ is the maximum value of η and $p(\eta)$ is the probability density function of η given by [10]

$$p(\eta) = \frac{1}{\pi} \int_0^{\infty} \cos \eta t \prod_{j=1}^{K} J_0(tR_j)\, dt \tag{11}$$

where $J_0(X)$ is the Bessel function of the first kind and order zero.

The computational aspects and some of the related considerations are deemed of sufficient interest to warrant some elaboration and are presented next.

III. Computational Considerations

Essential to any algorithm devised for the calculation of $\hat{P}_{eM}$ is a strategy for estimating and controlling both the error incurred in terminating the algorithm after a finite number of steps and the error introduced by the limited arithmetic capacity of the digital computer. It was thought that determining a suitable point of termination is more easily managed by applying numerical quadrature to (10) and (11), in spite of the improper integral in (11), than by summing the series (9). This series is actually doubly infinite, since the operator E is usually applied to the Fourier expansion of $p(\eta)$ [6], [8] and it may, for large values of ρ, have terms that threaten to exceed the arithmetic capacity of the computer before the factorial in the denominator exerts its influence as k increases. Consequently, a computer program was developed to evaluate $\hat{P}_{eM}$ in the form (10).

Our choice of numerical quadrature method is the Romberg algorithm [11], which consists of repeated applications of the trapezoidal rule for which the discretization interval h is halved each time, combined with extrapolation to the limit as $h \to 0$.

The errors that are to be estimated and controlled arise principally from four sources:

1) replacement of (11) by a proper integral (i.e., using a finite upper limit);
2) oscillation of the integrand of (11), causing the cancellation of leading digits, particularly near $\eta = \pm\Omega$;
3) tendency for erfc (X) in (10) to underflow for large X;
4) replacement of integration by a finite process.

Because of the nature of the Romberg algorithm, it is not difficult to decide with considerable confidence when the approximate integration process has produced acceptable results. When applied to (11) with a finite upper limit, this procedure will afford accurate partial integrals $p_1(\eta)$ for a number of discrete values of η. Comparison with unity of

$$\int_{-\Omega}^{\Omega} p_1(\eta)\, d\eta$$

indicates whether to extend the upper limit of (11) in order to construct a more accurate representation of $p(\eta)$. It has been our experience that control of errors 1) and 4) in this manner is not difficult. The type 3) error can be simply controlled by using the fact [12] that one form of expressing erfc (x) contains the term $\exp(-x^2)$ from which a scaling factor can be extracted thus avoiding the underflow problem.

The type 2) error is the most difficult to control and, while double precision arithmetic can be used to mitigate the effect,

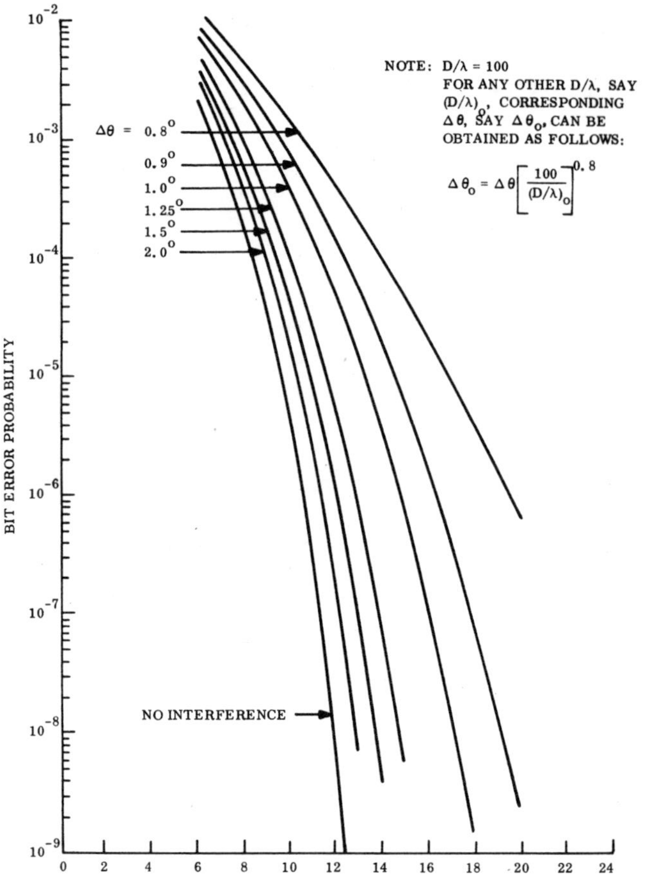

Fig. 2. Error rate for binary CPSK.

there is a region of η, (η_0, Ω) in which it cannot be controlled but only bounded. Since $p(\eta)$ is even, the comments pertaining to this interval also apply to $(-\Omega, -\eta_0)$. The number η_0 represents the largest value of η for which an acceptable approximation of $p(\eta)$ is possible using the previously described techniques. However, an upper bound on $p(\eta)$ can be found for this region and from this we can determine if in (10) the contribution of $p(\eta)$ over this interval is negligible or not. This bound depends on the fact, which can be proved straightforwardly, that

$$\int_0^{\infty} t^r e^{i\eta t} \prod_{n=1}^{K} J_0(tR_n)\, dt, \qquad r \geqslant 0 \tag{12}$$

converges uniformly for $K > 2(r+1)$, $-\infty < \eta < \infty$. This fact, along with some auxiliary theorems [13, pp. 311–312], establishes that for $K > 4$ there is a number η_1 such that $p(\eta)$ is concave upward on (η_1, Ω). The bound on $p(\eta)$ that was sought is then obtained by forming the secant line connecting $(\eta_0, p(\eta_0))$ and $(\Omega, 0)$, provided $\eta_0 \geqslant \eta_1$. The point η_1 is difficult to obtain analytically but in practice it is not difficult to find a $\eta_0 > \eta_1$. In all cases tried, the difference in computed error probability, with or without the bound, was negligible.

Finally, to ensure that the procedures for controlling and measuring the errors were applied as intelligently as possible, the computer program was designed so that the user is assigned a decision-making role via a time-sharing terminal. Besides directing operational functions such as input and output, the user controls such things as the finite upper limit of integration in the approximation of (11), the fineness of the mesh in the discretization of the integrands, the determination of η_0,

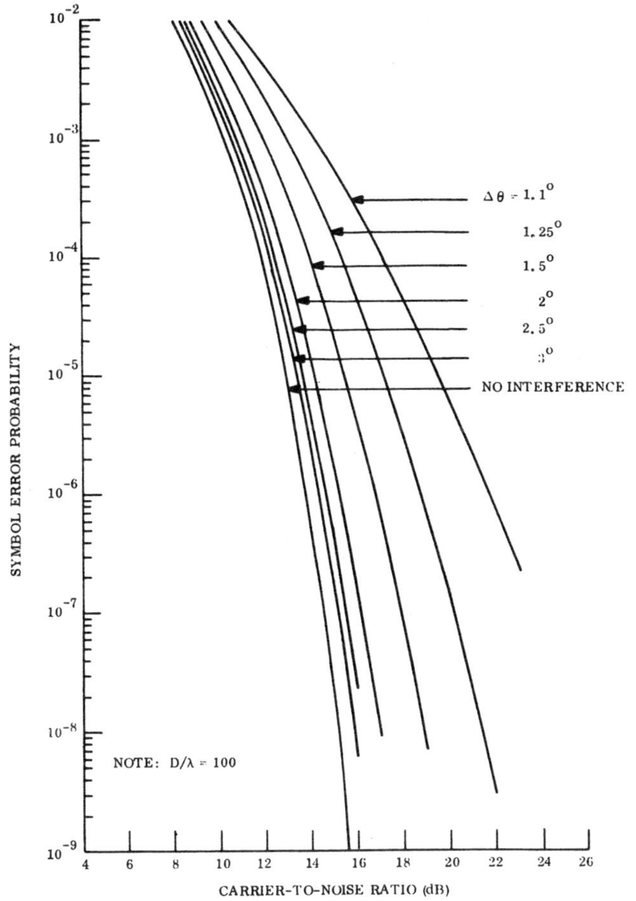

Fig. 3. Error rate for 4-phase CPSK.

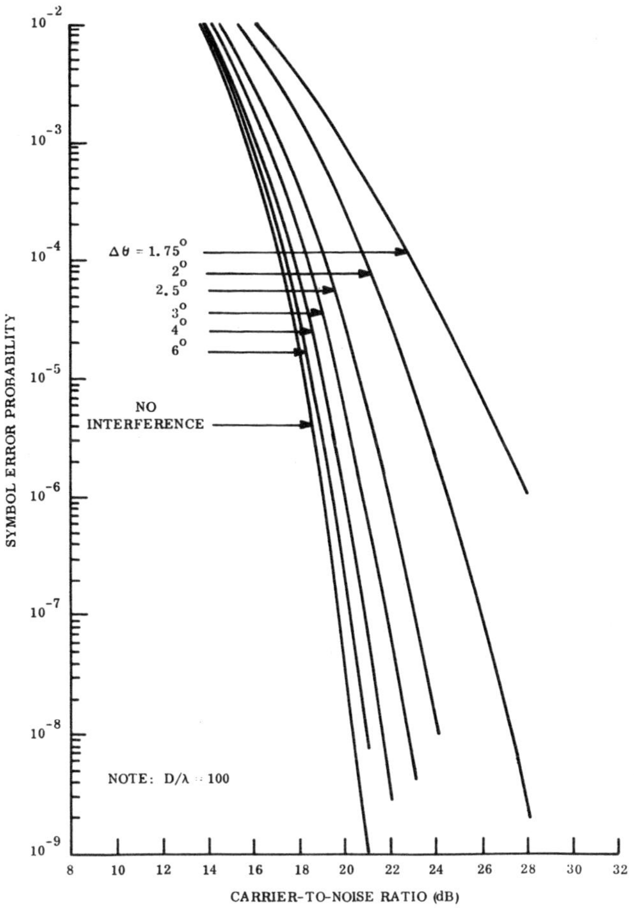

Fig. 4. Error rate for 8-phase CPSK.

and the investigation of bounds on $p(\eta)$ beyond η_0. As a consequence of having a measure of control over and awareness of the quality of intermediate calculations, there is considerable confidence in the final results. Details of program usage may be found in [9].

IV. Satellite Spacing and Error Probability

The computer program referred to in the last section was used to study the relationship between error probability and the spacing between satellites transmitting CPSK signals. For simplicity we consider down-link interference only, which is usually the limiting direction, but up-link contributions can be accounted for by straightforward extension of the analysis. To obtain numerical results it is necessary to assume a set of specific characteristics for each system. The characteristics assumed here represent a convenient reference condition and are collectively referred to as a homogeneous system model. This model makes the following assumptions: 1) there exists a set of satellites equally spaced $\Delta\theta$ degrees apart (as measured from the center of the earth) in the geostationary orbit; 2) all satellites transmit cochannel, copolarized carriers with identical equivalent isotropically radiated power (EIRP) and modulation characteristics; 3) propagation conditions are clear weather; and 4) all earth stations are identical with CCIR antenna sidelobe characteristics. The latter are given by [14]

$$G(\phi) = 32\text{–}35 \log \phi, \qquad 1^\circ \leqslant \phi \leqslant 48^\circ \tag{13}$$

where $G(\phi)$ is the gain in decibels and ϕ is the angle off-axis. In many situations, it is necessary to make interference calculations, at least of a preliminary sort, when the actual gain function of the earth station antenna may not be known, as for example during the early planning stages of a system. In cases when the precise antenna pattern is not available, for whatever reason, the pattern (13) is a useful model and is more or less universally used for interference calculations. It is generally recognized that (13) is conservative, particularly in the near-sidelobe region, in the sense that it predicts higher sidelobe levels than may be achievable with well-designed antennas. Hence, for particular antennas, it is possible to obtain closer satellite spacings than those given here.

With the previous assumptions, it is straightforward to show [9] that, to an excellent approximation, the reciprocal carrier-to-interference ratio for the jth interfering satellite east or west of the desired satellite, is given by

$$R_j = \frac{18}{|j\Delta\theta|^{1.25}\,(D/\lambda)} \tag{14}$$

where R_j is the same quantity defined following (8) and D/λ is the ratio of diameter to wavelength of the earth station antenna.

With the use of (14), the computer program was run to obtain numerical results for $\hat{P}_{eM}$ for $M = 2$, 4, 8, and 16. (For $M = 2$, the results are exact.) These results apply to a satellite flanked by five interfering satellites equally spaced on either side ($K = 10$); also, we assumed $D/\lambda = 100$ but, from (14), it is simple to convert the results to any other value of D/λ. The results are plotted in Figs. 2–5, which show error probability

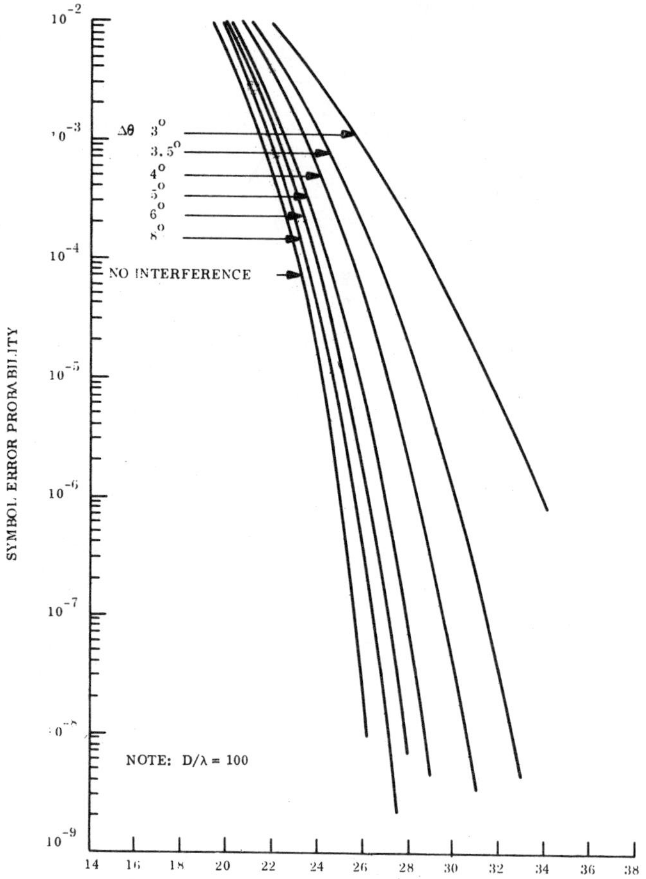

Fig. 5. Error rate for 16-phase CPSK.

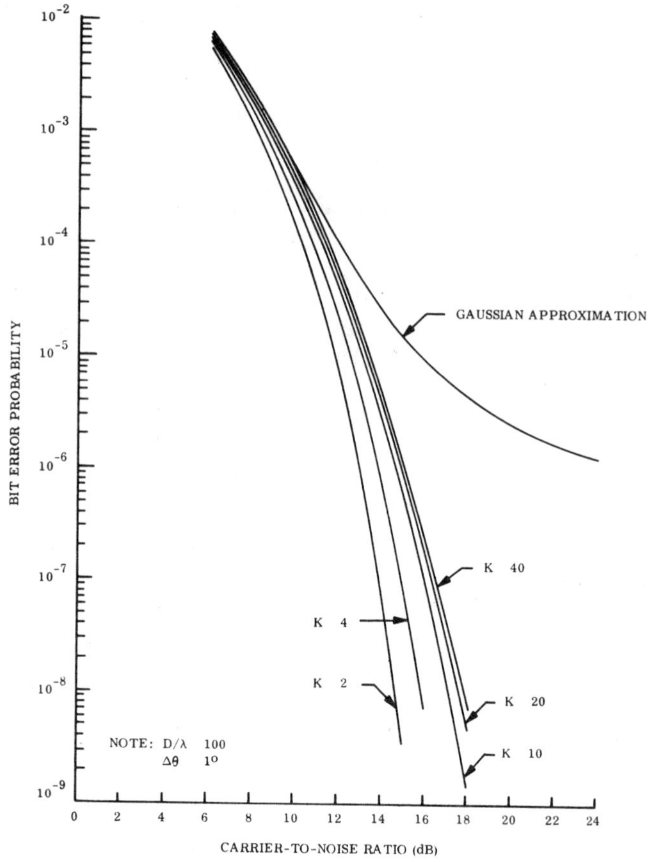

Fig. 6. Comparison of actual binary errors rates for different numbers of interfering satellites and Gaussian approximation.

versus carrier-to-noise ratio (20 log ρ) with $\Delta\theta$ as a parameter.[3] As expected, these curves show for given ρ that error probability increases with closer satellite spacing, although there exists some separation for each M beyond which the performance is negligibly different from the noise-only case. The detailed implications of these results on orbit utilization require more space than available here and are discussed elsewhere [9]; however, we may note here that four- and eight-phase systems employ the orbit/spectrum "space" quite efficiently, while FM performs about as well as, or is superior to, other values of M.

Some additional results of interest are shown in Fig. 6. A particular case was run to observe the effect of the number of satellites. As can be seen, there is relatively little difference between two and 40 satellites and this is somewhat to be expected in light of (14) since the two nearest satellites are the main contributors. The progression of the curves, from $K = 2$ to $K = 40$, also indicates that invoking a "large" number of interferers, at least in the present situation, will not justify the Gaussian approximation. This is made evident by the Gaussian approximation curve (for $K = 10$), which is obtained simply by adding the interference power to the thermal noise power. This points out the fact, reported previously [8], that the

[3] For $M = 2$ (Fig. 2) the results are exact. Note also that Fig. 2 shows two curves for $\phi < 1°$ despite the stipulation $\phi \geq 1°$ in (13). Actually the lower limit of applicability of $G(\phi)$ in (13) depends on the size and type of antenna as well as on the purpose to which such an approximation is put. The choice of 1° by the CCIR is more or less arbitrary and serves as a conservative estimate of the onset of the sidelobe region, but it can be shown that $G(\phi)$ can be meaningfully applied to smaller angles. In any case, the two curves for $\phi < 1°$ have been provided as a matter of completeness but should be used with some caution.

Gaussian approximation may lead to quite pessimistic results. On the other hand, the Gaussian approximation does yield fairly accurate results for relatively low carrier-to-noise ratios and improves as the carrier-to-interference ratio increases (or, equivalently, as the ratio of interference power to thermal noise power decreases.) However, it is evident that as ρ increases the Gaussian approximation asymptotes to a minimum error probability while the actual error rate can be reduced indefinitely. This will be true, however, only below a certain level of interference; above this level, regardless of ρ, the error probability will be interference limited to a given value.

In this context, it is interesting to note the limiting behavior of $\hat{P}_{eM}$ as $\rho \to \infty$. First we note that

$$\lim_{\rho \to \infty} \operatorname{erfc}\,[\rho(\eta + \sin \pi/M)] = \begin{cases} 0, & |\eta| < \sin \pi/M \\ 1.0, & \eta = -\sin \pi/M \\ 2.0, & \eta < -\sin \pi/M. \end{cases}$$

Denoting the limiting value by $\hat{P}_{eM}(\infty)$ and using (10) we find

$$\hat{P}_{eM}(\infty) = 2 \Pr\,[-\Omega \leq \eta < -\sin \pi/M]. \tag{15}$$

The value of $\hat{P}_{eM}(\infty)$ can be obtained by integrating $p(\eta)$ over the appropriate region and, in particular, we observe that if $\Omega \leq \sin \pi/M$, the error probability can always be made as small as desired. In the particular case when R_j is given by (14) we can obtain a simple expression for the "threshold" spacing $\Delta\theta_t$ separating the noise-limited from the interference-limited region. Specifically, since

$$\Omega = \sum R_j = \sum \frac{18}{|j\Delta\theta|^{1.25}\,(D/\lambda)} = \frac{36\,(\lambda/D)\,S(K)}{\Delta\theta^{1.25}},$$

where

$$S(K) = \sum_{j=1}^{K/2} j^{-1.25},$$

we obtain

$$\Delta\theta_t = \left\{ \frac{36\,(\lambda/D)\,S(K)}{\sin(\pi/M)} \right\}^{0.8}.$$

V. Conclusions

This paper has examined the increase in the error probability of satellite systems using CPSK signals caused by interference arising from adjacent satellites transmitting similar signals. The main results are contained in Figs. 2–5, which relate the satellite spacing to the error probability for 2-, 4-, 8-, and 16-phase systems, respectively. Inasmuch as we assumed ideal coherent reception and neglected up-link contributions, the curves of Figs. 2–5 may be taken to represent the ultimate spacing limitations for satellites using digital signals. Although the numerical results apply specifically to the CCIR antenna pattern, these results should provide useful guidance for other patterns; the technique employed, of course, is also applicable to arbitrary patterns. It was also shown that, beneath a certain interference level, the error probability can be made as small as desired by increasing the carrier-to-noise ratio. When the interference exceeds this level, it sets a minimum error probability. Thus, there exist satellite spacings beyond which the system designer can control the error probability and below which there is an irreducible minimum.

Acknowledgment

The authors are indebted to J. Kurzenknabe who performed the computer programming.

References

[1] J. L. Hult and E. E. Reinhart, "Satellite spacing and frequency sharing for communication and broadcast services," *Proc. IEEE*, vol. 59, pp. 118–128, Feb. 1971.
[2] D. M. Jansky and M. C. Jeruchim, "A technical basis for geostationary satellites to share the geostationary orbit," presented at 3rd Comm. Satellite System Conf., Los Angeles, Calif., Apr. 6–8, 1970, AIAA Paper 70-441.
[3] M. C. Jeruchim and F. D. Moore, "Orbit/spectrum utilization study," vol. 2, General Elec. Co., Valley Forge, Pa., Doc. 69SD4348, Sept. 12, 1969.
[4] J. C. Fuenzalida, "A comparative study of the utilization of the geostationary orbit," in *Proc. INTELSAT/IEE Conf. Digital Satellite Communication*, pp. 213–225, Nov. 1969.
[5] J. K. S. Jowett and A. K. Jefferis, "Ultimate communications capacity of the geostationary-satellite orbit," *Proc. Inst. Elec. Eng.* (London), vol. 116, pp. 1304–1310, Aug. 1969.
[6] V. K. Prabhu, "Error rate considerations for coherent phase-shift keyed systems with co-channel interference," *Bell Syst. Tech. J.*, vol. 48, pp. 743–767, March 1969.
[7] A. S. Rosenbaum, "PSK error performance with Gaussian noise and interference," *Bell Syst. Tech. J.*, vol. 48, pp. 413–442, Feb. 1969.
[8] A. S. Rosenbaum, "Binary PSK error probabilities with multiple cochannel interferences," *IEEE Trans. Commun. Technol.*, vol. COM-18, pp. 241–253, June 1970.
[9] M. C. Jeruchim and D. A. Kane, "Orbit/spectrum utilization study," vol. 4, General Elec. Co., Valley Forge, Pa., Doc. 70SD4293, Dec. 31, 1970.
[10] W. R. Bennett, "Distribution of the sum of randomly phased components," *Quart. Appl. Math.*, vol. 5, pp. 385–393, Apr. 1948.
[11] P. Henrici, *Elements of Numerical Analysis.* New York: Wiley, 1964.
[12] G. M. Roe, "A computer algorithm for the error integral," General Elec. Co., Rep. TIS 66-C-050, Mar. 1966.
[13] R. Courant, *Differential and Integral Calculus.* New York: Interscience, 1952.
[14] CCIR Doc. XIIth Plenary Assembly, vol. 4, pt. 2, recommendation 465, New Delhi, 1970.

Binary and Quaternary PSK Radio Systems in a Multiple-Interference Environment

C. COLAVITO AND M. SANT'AGOSTINO

Abstract—The planning of a digital radio network requires an evaluation of the effect of quite a large number of interfering sources. These in turn are dependent upon many parameters, such as the network structure, channel arrangement, antenna patterns, and so on.

This paper presents a performance analysis of a digital radio-relay system during fading periods. The possibility of attaining a required fading margin is analyzed when many co-channel and interchannel interferences are simultaneously present, both for long-haul and short-haul radio relay systems.

A comparison is presented between two possible techniques for the reusing of the same frequency band in each hop, by taking advantage of the cross-polarization discrimination (XPD). The reduction of the XPD expected during fading periods is also taken into account.

Binary and quaternary coherent PSK modulations are considered, with realizable filters (of the Butterworth type), both at the transmitter and at the receiver.

The error probability calculations were performed with a combined analytical and computer simulation approach.

The results seem to indicate the feasibility of the frequency reuse technique.

I. Introduction

The increasing demand for data communication circuits and the diffusion of PCM multiplexing are emphasizing the need for radio facilities especially designed for digital transmission, both in long-haul and short-haul systems.

In the Italian network, the exploitation of medium capacity short-haul radio-relay links is planned for the coming years [1], using a 500-MHz wide bandwidth around 13 GHz and some experimental radio-links are expected to be put soon in operation. In this system, use is made of coherent PSK at about 17 Mbd, to carry 240 or 480 telephone channels, in the binary and quaternary case, respectively.

Since the development of high capacity long-haul radio-relay systems is also under consideration, a research work has been carried out in order to analyze the performance of both systems. Both of them are assumed to be allocated in frequency bands above about 12 GHz, though this hypothesis is unavoidable only for high capacity systems. In these bands severe fadings are expected, depending on the climatic conditions, and therefore a decrease in the hop lengths will be necessary, with a subsequent increase in the number of repeaters and interference sources [9].

The various strategies proposed to avoid waste of spectrum bandwidth have been considered, and the resulting interference environment has been evaluated. Since all these strategies are based on the supposed intrinsic resistance of the digital modulations, a careful analysis of the interference effect during fading as well as during nonfading conditions is required.

It is worth noting that the results obtainable by calculating separately the effects of the various disturbances (intersymbol, co-channel, or interchannel interferences), cannot correctly be used for evaluating the overall performance of the radio-relay system.

On the other hand the criterion of substituting for many interferences a noise power corresponding to the sum of their powers, (so-called Gaussian approximation) could be justified only when the number of the disturbing sources is very large. In practical conditions it seems nearly impossible to estimate the ranges of validity of this last approximation, especially when some of the signals can give rise to much greater impairments than others.

On the contrary, the present work is based on a combined analytical and computer simulation approach that permits an estimation of the probability density function of the signal plus all the various interferences at the sampling point; the error probability is then calculated by a straightforward procedure.

The same principle was adopted in a previous work [2], but now a different computation method allows much greater speed and flexibility. A more detailed discussion concerning the accuracy of this method is outside the scope of this paper and is given elsewhere [3]. Anyway, the computed error probabilities show a satisfactory agreement with other published works [4]-[6].

Using the calculation method previously mentioned a performance analysis of digital radio-relay systems during fading

Paper approved by the Radio Communication Committee of the IEEE Communications Society after presentation at the 1972 International Conference on Communications, Philadelphia, Pa. Manuscript received October 18, 1972; revised February 8, 1973.

C. Colavito is with the SIP Società Italiana per l'Esercizio Telefonico, Rome, Italy.

M. Sant'Agostino is with the CSELT Centro Studi e Laboratori Telecomunicazioni, Turin, Italy.

Reprinted from *IEEE Trans. Comm.*, vol. COM-21, pp. 1056-1067, Sept. 1973.

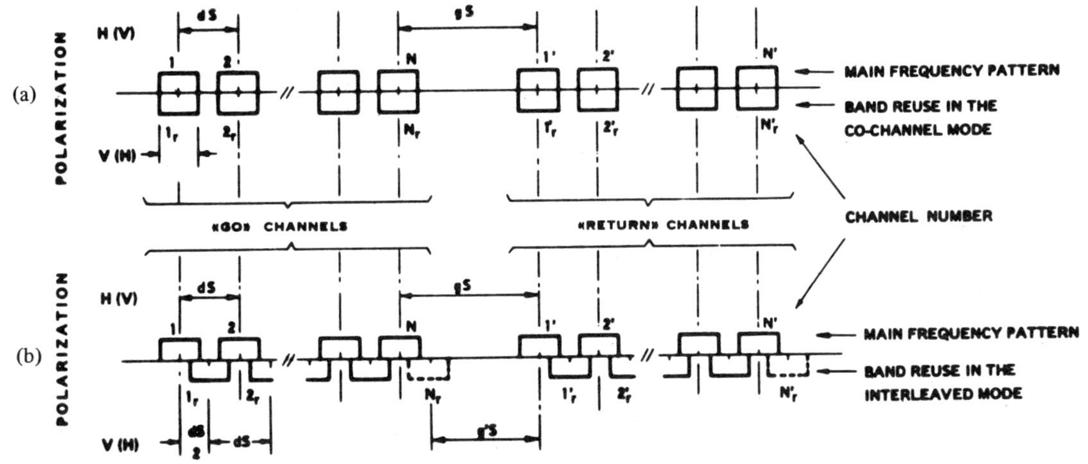

Fig. 1. Scheme of two possible techniques to double the efficiency of spectrum utilization. (a) Reutilization of the same frequencies of the main pattern. (b) Pattern reutilization with a $d/2$ shift. The two channels indicated with broken lines may not be usable.

periods has been carried out, when binary and quaternary coherent PSK modulations are used. As a part of the broader research dealing with digital transmission over radio-relay links, differentially coherent detection also was studied [7], [8]. However the results so far obtained, together with some considerations of technical feasibility, suggest the use of coherent demodulators in medium and high capacity systems.

It is worthwhile noting that this paper deals only with the case of completely digital networks. Coexistence of analog and digital radio systems would give rise to different interference problems, not considered here, and would not use all the advantages of the intrinsic interference resistance of the digital modulations.

In what follows these topics will be examined: 1) the interference environment in some different types of radio networks (Section II); 2) the computer-simulated system model, and the performance evaluation criteria (Section III); and 3) the effects of the above mentioned interferences on the performance of digital radio-relay systems (Section IV).

II. Interference Sources in Digital Radio Networks

A. *RF Channel Arrangement and Structure of the Radio Networks*

The available frequency band is assumed to be divided into two halves, one being used in transmission ("go" channels), and the other to receive ("return" channels).

Each part contains N radio channels spaced in frequency by Δf, while ΔF is the separation between the two nearest "go" and "return" channels. As the occupied bandwidth for a digitally modulated signal is directly proportional to the digit (or symbol) rate S, all the parameters related to the signal spectrum are normalized with respect to S.

In the sequel the following parameters will be extensively used:

$$d = \Delta f/S = \text{(normalized) channel spacing}$$
$$g = \Delta F/S = \text{(normalized) guard band.}$$

These parameters pertain to the main pattern [the upper patterns in Fig. 1(a) and (b)], while other RF channels [lower patterns in Fig. 1(a) and (b)] may be available to the system designer.

As a matter of fact the "reuse" of the same frequency band by exploiting the cross-polarization discrimination (XPD) is widely proposed in order to double the efficiency of the spectrum utilization.[1] It is generally agreed that this operation could be performed by means of the co-channel arrangement or with interleaved frequencies as shown in Fig. 1(a) and (b).

As it is well known that the XPD is likely to decrease during the fading periods, such a phenomenon must be considered in the performance analysis.

Though some results [10] seem to indicate that, in the frequency bands above about 12 GHz, considered in this work, the maximum XPD reduction may not occur simultaneously with the maximum attenuation of the wanted signal, a fixed relationship between the attenuation and the XPD will be assumed in Section III, for a worst case analysis.

Moreover, as the previously mentioned effects (i.e., attenuation and XPD) turn out to be different for two linearly cross-polarized signals on the same hop [11], in what follows the disturbed signal is assumed to present more attentuation and less XPD with respect to the other one.

The various possible network structures are reduced to the two types represented in Figs. 2 and 3. They correspond to the use of digital radio links, respectively, in long-haul high capacity or in short-haul mean capacity communication systems.

The last structure has the main function of connecting a large number of minor local stations to one station of higher order in the hierarchical transmission plan, and it may also represent a first step in the introduction of digital radio links in some communication networks.

In the long-haul system, a two-frequency plan is assumed, and therefore each repeater transmits all the "go" (or "return") channels on the opposite directions, with both polarizations, according to the plans of Fig. 1(a) and (b).

On this assumption, $2N$ two-way RF channels are available on each hop, with both types of channel arrangements shown in Fig. 1. The distinction between protection and working channels is not relevant in the present work.

The short-haul network is assumed to be star-shaped, and made up of many low-capacity links[2] plus one high-capacity link (hop 1 in Fig. 3), all converging into a repeater station. It

[1] It may be interesting to point out that this operation is not to be intended as a reduction of the channel spacing, which depends only on the band occupancy and hence on the spectrum of the signal.

[2] The link capacity is assumed to be high or low according to the number of the RF channels transmitted in each direction.

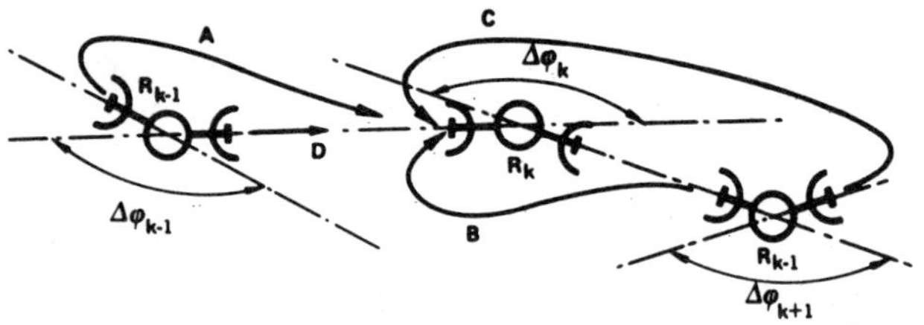

Fig. 2. Long-haul radio-relay system. Nearest interfering paths in the kth repeater. Four overreach paths are produced by the same mechanisms from the repeaters $k - 3$ and $k + 3$. R_i—regenerative repeater stations. A–D—interfering paths.

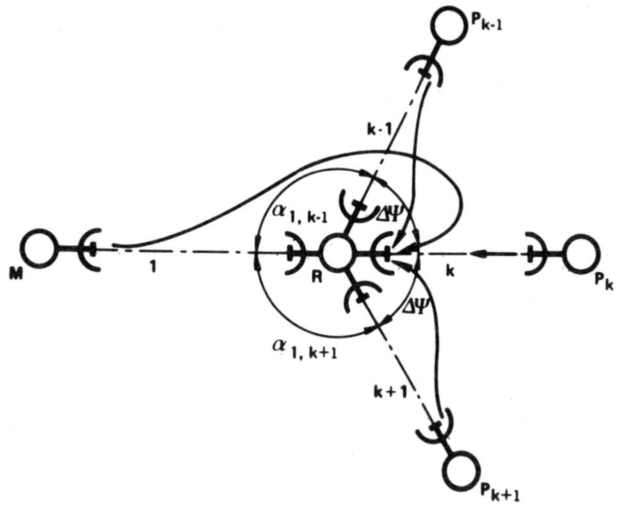

Fig. 3. Short-haul network. Interfering paths in the kth radial link. M is the main terminal station, P_i are the peripheral terminal stations, R is the repeater station, hop 1 is the main link (high capacity), hop K is a radial link (1 + 1 type), $\Delta\psi$ is the angle between the radial links, $\alpha_{1,i}$ is the angle between the main link and the ith radial link. ($\alpha_{1,i} \leq 180^\circ$).

is assumed that the required capacity in each radial link may be obtained by means of only one working plus one protection channel (1 + 1 scheme).

In addition the same RF channel is supposed to be used as working and spare channel in each radial link, with crossed polarizations, and adjacent radio channels assigned to adjacent links of the star. This frequency plan is illustrated in Fig. 4, where all the terminal stations are assumed to transmit the "go" (or "return") channels.

The same figure shows also that the amount of RF channels employed in the main link may be reduced by means of a suitable protection arrangement.

B. Interference Allocation

In this paper only some kinds of intrasystem interferences are considered, as they are likely to impair the system performance more seriously. The main interfering paths in a repeater of the long-haul or short-haul systems are shown in Figs. 2 and 3; the number and the levels of the interferences are listed in Tables I–IV, where both the polarization schemes of Fig. 1 are considered.

The interfering carrier/wanted carrier ratios evaluated at the input of the receiver, are given for nonfaded as well as for faded conditions of the wanted carrier; they are dependent on the antenna characteristics and on the fading effects (attenuation and XPD reduction).

Equal transmitted powers and equal hop lengths are assumed.

The parameters used in the tables have the following meanings.

D_a is the front-to-back ratio of the antenna pattern. In long-haul systems, when the angles $\Delta\varphi_i$ are greater than about 100°, D_a may be considered as a constant that is usually equal to the environment limit. In the star network the parameter D_a is used to indicate the decoupling between the main link and any one of the low-capacity links (sometimes called also radial links).

D'_a is the antenna decoupling between any two adjacent radial links in the star network (D'_a is assumed to be lower than D_a because the angles $\Delta\psi$ in Fig. 3 are usually lower than $\alpha_{1,k}$).

D_{po} is the XPD in normal propagation conditions between horizontally and vertically polarized signals coming from the same direction.

D_p is the XPD during fading periods between signals coming from the same direction.

A_f is the attenuation due to the fading, which is supposed to be slow and flat.

Since, in short-hop links, the attenuations due to multipath propagation are likely to have less importance than the rain attenuation, A_f is assumed to be equal, at the same time, for all the copolarized channels on the hop under examination. As a consequence, the interfering signals, which have a path common with the wanted one (i.e., the path A of Fig. 2), show no change in the interference-to-carrier ratio going from normal to fading conditions.

Moreover, to check the worst conditions, the rain attenuations are assumed not to occur simultaneously in the adjacent hops of the system.

It may be noticed that the usual values of the antenna decouplings are such that the system performance in the normal propagation conditions is not remarkably affected by the interferences, so that the checking of the system behavior is found to be necessary only under faded conditions.

Harmful interferences are due, during fading, to signals coming from paths of the type B and D. In the last column of the tables the most important disturbance sources are marked with an asterisk.

On the basis of the preceding assumptions the interference environment in each repeater of a long-haul system may be considered equal to the one reported in Tables I and II for the repeater R_k of Fig. 2 and for a radio channel in the middle of the "go" (or "return") semiband.

On the contrary in the star network the interference conditions of hop 1 are different from those of the radial hops. When the angles $\alpha_{1,i}$ are sufficiently large the situation of the main link is the same as in the long-haul system.

Furthermore, in the radial links it is assumed that no XPD is provided by the antenna with respect to the signals coming from the adjacent hops.

A more complete description of the interference paths in long-haul systems should also take into account four overreach paths produced by the same mechanisms causing the paths A–D of Fig. 2, but coming from the repeaters $k - 3$ and $k + 3$.

They are thus at a lower level than the signals coming from the corresponding shorter paths, and their effect can thus be neglected, except perhaps the analogue of D, which could carry a significant co-channel interference. However some possible natural obstacles on the route and a slight randomness of the angles $\Delta\varphi_i$ of Fig. 2 may reduce its level to a negligible value. These conditions may be less easily obtained in the very short hop systems anticipated at the higher frequencies, but are quite reasonable with medium hop lengths (hops of about 10–20 km are expected at 13 GHz).

Also the direct leakage of a transmitter into a receiver in the branching device is assumed to be maintained at a negligible amount by properly choosing the guard band g of Fig. 1.

III. Computer Simulation Model

A. General Description

A computation method for the evaluation of the error probability in the faded hop has been worked out for the case of

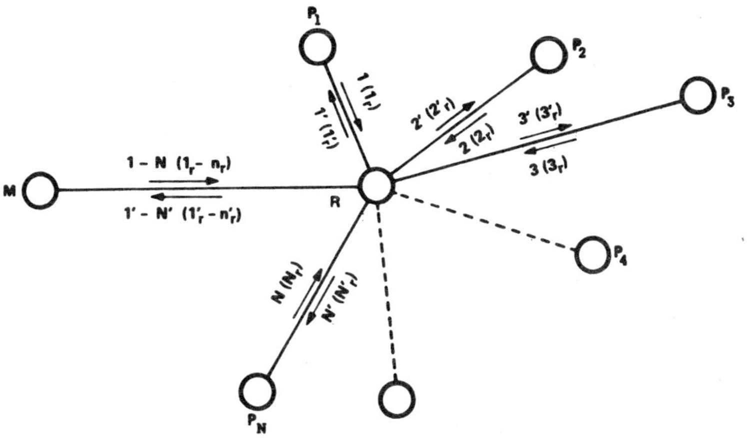

Fig. 4. Frequency plan for a typical short-haul network. M, R, and P_i have the same meaning as in Fig. 3. The frequency allocation to each link is shown by the channel number as indicated in Fig. 1. The spare channels are in brackets. In the main link the protection scheme may require a number n_r of spare channels lower than the number N of the working channels.

TABLE I
MOST SIGNIFICANT INTERFERENCES IN A REPEATER OF THE LONG-HAUL NETWORK (TYPE A CHANNEL ARRANGEMENT)

Interference Sources	Number of Interferences	[Interfering Carrier Power / Wanted Carrier Power] dB Without Fading	With Fading	
Co-Channel				
Same hop	1	$-D_{po}$	$-D_p(A_f)$	*
Path A	2	$-D_a$	$-D_a$	
Path B	2	$-D_a$	$-D_a + A_f$	*
Path C	2	$-2D_a$	$-2D_a + A_f$	
Interchannel (distance d)				
Same hop, same polarization	2	0	0	*
Same hop, crossed polarization	2	$-D_{po}$	$-D_p(A_f)$	
Path A	4	$-D_a$	$-D_a$	
Path B	4	$-D_a$	$-D_a + A_f$	
Path C	4	$-2D_a$	$-2D_a + A_f$	

TABLE II
MOST SIGNIFICANT INTERFERENCES IN A REPEATER OF THE LONG-HAUL NETWORK (TYPE B CHANNEL ARRANGEMENT)

Interference Source	Number of Interferences	[Interfering Carrier Power / Wanted Carrier Power] dB Without Fading	With Fading	
Co-Channel				
Path A	1	$-D_a$	$-D_a$	
Path B	1	$-D_a$	$-D_a + A_f$	*
Path C	1	$-2D_a$	$-2D_a + A_f$	
Interchannel (distance $d/2$)				
Same hop	2	$-D_{po}$	$-D_p(A_f)$	*
Path A	2	$-D_a$	$-D_a$	
Path B	2	$-D_a$	$-D_a + A_f$	*
Path C	2	$-2D_a$	$-2D_a + A_f$	
Interchannel (distance d)				
Same hop	2	0	0	*
Path A	2	$-D_a$	$-D_a$	
Path B	2	$-D_a$	$-D_a + A_f$	
Path C	2	$-2D_a$	$-2D_a + A_f$	

TABLE III
MOST SIGNIFICANT INTERFERENCES IN A RADIAL LINK OF THE SHORT-HAUL NETWORK (TYPE A CHANNEL ARRANGEMENT)

Interference Source	Number of Interferences	[Interfering Carrier Power / Wanted Carrier Power] dB Without Fading	With Fading	
Co-Channel				
Same hop	1[a]	$-D_{po}$	$-D_p(A_f)$	*
Hop 1	2	$-D_a$	$-D_a + A_f$	*
Interchannel (distance d)				
Hop $K + 1$	2	$-D'_a$	$-D'_a + A_f$	*
Hop $K - 1$	2	$-D'_a$	$-D'_a + A_f$	*
Hop 1	4	$-D_a$	$-D_a + A_f$	

[a]This carrier is modulated by the same PCM signal that is transmitted on the useful carrier.

TABLE IV
MOST SIGNIFICANT INTERFERENCES IN A RADIAL LINK OF THE SHORT-HAUL NETWORK (TYPE B CHANNEL ARRANGEMENT)

Interference Source	Number of Interferences	[Interfering Carrier Power / Wanted Carrier Power] dB Without Fading	With Fading	
Co-Channel				
Hop 1	1	$-D_a$	$-D_a + A_f$	*
Interchannel (distance $d/2$)				
Same hop	1[a]	$-D_{po}$	$-D_p(A_f)$	*
Hop $K - 1$	1	$-D'_a$	$-D'_a + A_f$	*
Hop 1	2	$-D_a$	$-D_a + A_f$	*
Interchannel (distance d)				
Hop $K + 1$	1	$-D'_a$	$-D'_a + A_f$	*
Hop $K - 1$	1	$-D'_a$	$-D'_a + A_f$	*
Hop 1	2	$-D_a$	$-D_a + A_f$	

[a]This carrier is modulated by the same PCM signal that is transmitted on the useful carrier.

two-phase and four-phase coherent PSK, since these modulation methods are planned to be used, in the near future, in the Italian network.

The transmitter–receiver model shown in Fig. 5 was simulated on a digital computer, with a time-domain approach, by

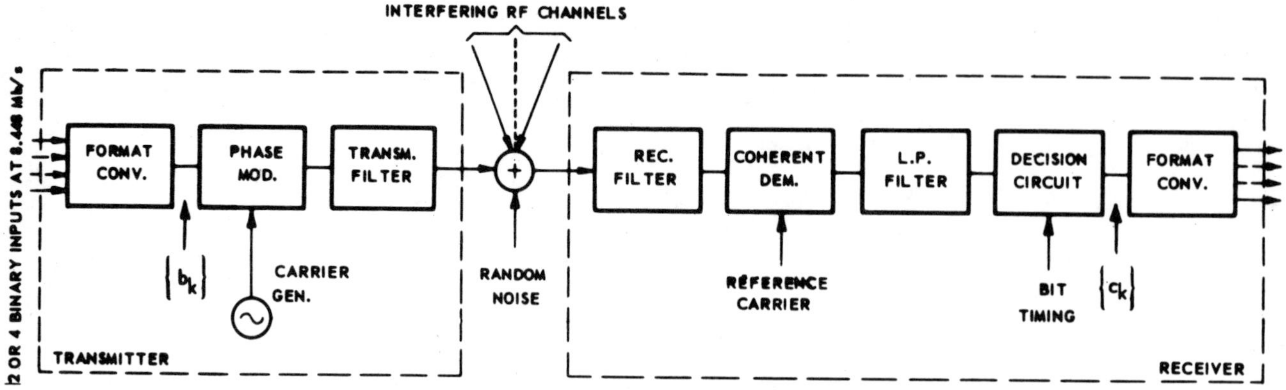

Fig. 5. Block diagram of the computer simulation model.

calculating the sampled values of the signal at the various points of the block diagram.

The carrier recovery and bit synchronization problems are not considered at the present time and perfect coherence between the wanted carrier and the reference signal is assumed. However the effect of these two impairment causes is under study and the computation method could be extended to cover these situations.

Rectangular, nonreturn to zero (NRZ) (i.e., 100 percent duty cycle) pulses are considered in transmission with equal *a priori* probabilities for the symbols 1 and 0 in the binary case.

The quaternary system corresponds to the so-called double binary scheme, in which two binary synchronous sequences modulate two quadrature carriers. Also in this case the symbols in each binary sequence are assumed to be equiprobable.

This last assumption is not usually satisfied by standard PCM line sequences, but it may be matched when scrambling–descrambling devices are used at the route ends.

A differential coder–decoder will also be required, to eliminate the phase ambiguity always present in coherent demodulators.

For two-phase demodulators (ambiguity between the sequences $\{c_k\}$ and $\{\bar{c}_k\}$) the coder operation can be written as follows:

$$b_k = a_k \oplus b_{k-1}, \tag{1}$$

where a_k is the original symbol and b_k the transmitted codified symbol, and the sign $\oplus$ is the "modulo-two sum".

To recover the original sequence, the decoder must perform the following operation:

$$d_k = c_k \oplus c_{k-1} = \bar{c}_k \oplus \bar{c}_{k-1}, \tag{2}$$

where d_k is the decoded symbol, equal to a_k in the absence of transmission errors.

In what follows the error probability P_e will be calculated between the differentially coded sequence $\{b_k\}$ and the output of the decision circuit $\{c_k\}$. The number of errors between $\{a_k\}$ and $\{d_k\}$ is actually twice the previous one; the corresponding signal-to-noise penalty is of the order of 0.5–1 dB, depending on the interference situation.

The receiver filter shown in Fig. 5 can be considered as the product of the transfer functions of the IF and RF bandpass filters, with a frequency shift due to the down-converter. The effect of the post-detection filter is included in the IF receiving filter.

The computer program calculates, by a time-domain simulation technique, the probability density function of the signal plus interference, at the decision time. Due to the linearity of the scheme of Fig. 5, the noise is still Gaussian and additive at the sampling point, so that the noise effect can be taken into account without simulating it in the computer. A description of the method is given in [2] and [3].

The calculation of the error probability for about 10–15 values of the carrier-to-noise ratio (CNR), takes, on the average, about 1 min[3] of computer time.

B. Performance Evaluation

The performance of a digital transmission system is usually expressed in terms of the error probability P_e, and described by the well-known curves of P_e versus the CNR.

However this presentation of the results is very redundant and cannot allow an immediate look at the system behavior, when the effect of many parameters is to be investigated.

Another related parameter, i.e., the value of the CNR corresponding to a given error rate P_e^*, is therefore used throughout this paper.

The normalized CNR is defined as

$$\eta = \frac{(\text{carrier energy per bit})}{(\text{noise power density})} = \frac{P}{N_0 S \log_2 L}, \tag{3}$$

where

- P received input power,
- S symbol rate,
- L number of modulation levels,
- N_0 noise power density (one-sided).

The parameter P_e^* should have the value of the maximum tolerable error rate in one hop. However this figure is dependent on the performance objectives, yet to be established by the CCITT for the whole system, and on the error/time-percentage allotment between hops.

Since it is likely that the per-hop error rate objectives will depend on the practical situations, a range of values of P_e^* sufficiently wide to cover the foreseeable needs has been chosen in this paper (from 10^{-4} to 10^{-8}).

The influence of the choice of the threshold error-rate P_e^* on the usable fading margin will be discussed in Section IV. Anyway, as far as the optimization of the system parameters is concerned, the results obtained with different values of P_e^* give the same indications[4].

C. Choice of the System Parameters

Before calculating the overall performance, the following parameters must be chosen: 1) the filtering characteristics,

[3] This figure refers to a 360/44 IBM computer.
[4] See, for example, [2, figs. 6, 7, 12, or 14], where some curves are presented of η versus a design parameter, for three values of the error probability (10^{-3}, 10^{-6}, and 10^{-9}).

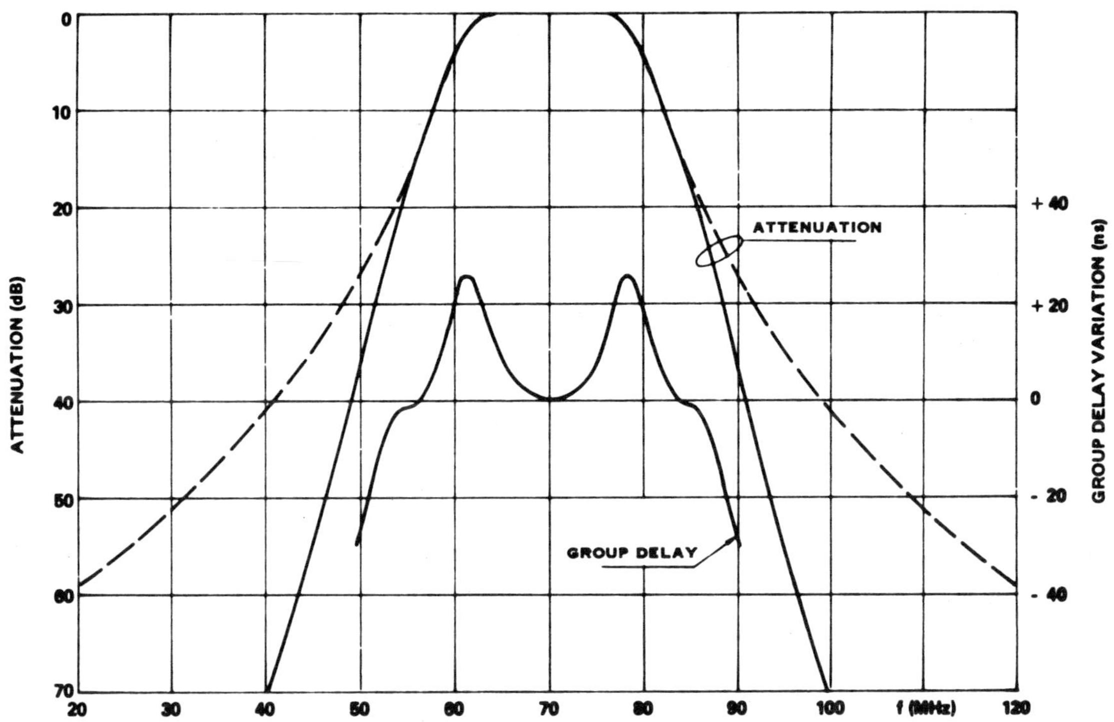

Fig. 6. Attenuation and group delay characteristic of the receive filter alone (broken line) and of the transmitter–receiver cascade (continuous lines).

both at the transmitter and at the receiver ends; 2) the normalized channel spacing d; and 3) the antenna characteristics D_a, D_a', and D_{p0}.

The choice of the filters must be made jointly for the transmitter and the receiver, and must take into account the presence of co-channel and adjacent-channel interferers in addition to thermal noise and intersymbol interference.

These impairment sources have different requirements, because adjacent-channel effects can be reduced by using narrow-band filters, while intersymbol interference phenomena would require wide-band filters.

We define the normalized transmission (reception) bandwidth $\beta_{t(r)}$ as the ratio of the filter equivalent noise bandwidth to the digit rate.

The most suitable choice of the filters within the available types (i.e., the Butterworth and the Bessel families) is made by minimizing the error probability in the presence of all the above mentioned interferences [3].

The preferable types turned out to be the following, both for two-phase and four-phase modulation.

1) Transmitting filter: 6-pole Butterworth response with normalized bandwidth $\beta_t = 2$.

2) Receiving filter: 4-pole Butterworth response with normalized bandwidth $\beta_r = 1.1$.

A plot of the filter characteristics for the two normalized bandwidths just chosen is given in Fig. 6, where, as an example, an IF carrier frequency of 70 MHz and a symbol rate S of 17.152 Mbd are assumed.

With this type of filter the interference effect due to two adjacent channels, at the same power level of the wanted signal and at a distance d, are those shown in Fig. 7, for the binary and the quaternary systems, as a function of d (continuous lines).

In the same figure, for the four-phase system only, results obtained with different types of filters are reported, in order to compare their effects. In particular the curve B was obtained with a maximally flat group-delay receiver filter, while

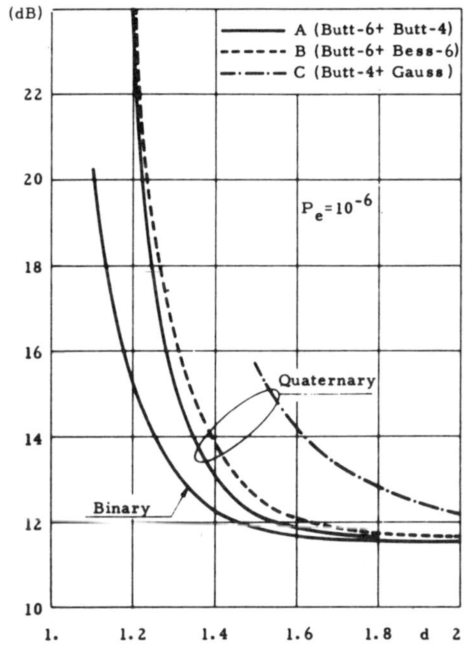

Fig. 7. Interchannel interference due to two adjacent channels at the same power level as the wanted carrier. The filter characteristics are described in Table V.

the curve C is taken from [6] and corresponds to a perfectly flat group delay Gaussian filter.

The normalized bandwidths, chosen with the above mentioned criterion, and some significant performance parameters of the filters are summarized in Table V.

The impairments due to intersymbol and interchannel interference show the convenience of the 4-pole Butterworth receiving filter with respect to the others.

TABLE V

EXAMPLES OF DIFFERENT FILTER CHOICES AND CORRESPONDING SYSTEM PERFORMANCE FOR A FOUR-PHASE PSK WITH COHERENT DETECTION

Transmitter Filter Type	β_t	Receiver Filter Type	β_r	Receiver Selectivity for $f_0 \pm 28$ MHz (dB)	Group-Delay Variation in the Nyquist Band (ns)	CNR for $P_e = 10^{-6}$ $d = \infty$ (dB)	CNR for $P_e = 10^{-6}$ $d = 1.6$ (dB)	Curve[a]
6-pole Butterworth	2	4-pole Butterworth	1.1	38.7	26.2	11.5	11.9	*A*
6-pole Butterworth	1.8	6-pole Bessel	0.9	41.7	5.5	11.6	12.1	*B*
4-pole Butterworth	1.5	Gaussian	1.0	36.4	0	11.9	14.2	*C*

[a]The Letters *A*, *B*, and *C* refer to the curves of Fig. 7.

In particular, with less selective types of filters, the presence of adjacent channels constrains the designer to choose narrower bandwidths, thus losing any possible advantage given by the flatter group delay curve.

The previous results indicate the possibility of using a channel spacing of about 1.5–1.6 times the symbol rate, both for the binary and the quaternary systems, with only a slight impairment in performance. A value of $d = 1.6$ will be used in the following [12].

The relevant antenna parameters for the calculation of the interference levels are the environmental limit D_a, the angular discrimination D_a' (which is obviously a function of $\Delta\psi$), the cross-polarization decoupling D_{p0} in the absence of fading effects.

For the usable fading margin evaluation in the long-haul system three values of D_a, i.e., 40, 50, and 60 dB are assumed. The lower values are likely to be reached in pole-mounted repeaters. In the short-haul network, when the use of existing antenna towers is foreseen, the achievable environmental limit will probably be greater. A value of $D_a = 60$ dB is adopted for this situation.

The decoupling between radial links D_a' is dependent on the angle $\Delta\psi$ and the values of 30, 40, and 50 dB are assumed.

The cross-polarization discrimination D_p is statistically dependent on the propagation phenomena but, as it was stated before, a deterministic dependence relationship between the XPD and the fading depth A_f is considered for calculation purposes.

The following linear relation is adopted:

$$D_p = D_{p0} - A_f/2 \tag{4}$$

where the parameter D_{p0} is assumed to have a value of 30 dB.

A detailed analysis concerning the effect of the relationship between D_p and A_f is now being carried out together with the acquisition of more information about the propagation behavior.

The system analysis, as it was pointed out in Sections II and III, holds for the worst behaved signal polarization, which probably is the horizontal.

IV. EVALUATION OF THE USABLE FADING MARGIN

A. Long-Haul Systems

The fading margin that has to be achieved in each hop of a digital radio link takes a paramount importance in the system design. The determination of this parameter is usually performed on the basis of the required system availability, the number of hops, the propagation characteristics in the assigned frequency band, the probability of simultaneous fadings in different hops, etc.

The choice of the *necessary* fading margin is not dealt with in the present work where, on the contrary, an analysis is carried out in order to determine the *usable* fading margin, as allowed by the interference environment.

On the other hand, the interference tables presented in Section II show that the relative levels of some unwanted signals (labeled with an asterisk) give rise to harmful disturbances increasing with the fading depth A_f. The interferences not increasing during the fading periods may be neglected, with the exception of the two adjacent channels transmitted on the same hop, whose effect comes out to be about 0.4 dB[5] for the four-phase system and with the type *A* filters of Table V.

For these reasons, the results of the error probability calculations (i.e., the values of the CNR η corresponding to an error rate $P_e^* = 10^{-4}$, in the assumed environmental situation) are reported in the following as a function of A_f.

The results obtained for the long-haul systems are plotted in Figs. 8 and 9, where both channel arrangements of Fig. 1(a) and (b) together with two-phase and four-phase modulation are considered.

The curves of Figs. 8 and 9 allow one to evaluate the usable fading margin M_f, by comparing the *necessary* CNR with that *available* at the receiver input.

Using the definition given by (3), for a symbol rate of about 17 Mbd (which corresponds to a multiple of the CCITT standard rate of 8.448 Mb/s), and a noise figure of 10 dB we obtain

$$\begin{aligned} \eta B &= P_0 - A_f + 91.5 \text{ dB} \quad &\text{two-phase PSK} \\ \eta Q &= P_0 - A_f + 88.5 \text{ dB} \quad &\text{four-phase PSK} \end{aligned} \tag{5}$$

where

P_0 received input power in normal propagation conditions;

$\eta B(Q)$ available CNR for the binary (quaternary) system.

It is remarkable that binary and quaternary systems with equal symbol rates and bandwidths are considered, and therefore the information rate is different for the two cases. This fact explains the 3-dB difference between ηB and ηQ.

To keep the error probability lower than P_e^* (10^{-4} in our example) the available CNR ηB and ηQ must be greater than the necessary η.

As shown in Fig. 8, the crossing of the straight lines representing ηB with the curves of the necessary η indicates the maximum acceptable values of A_f, i.e., the usable fading margin M_f.

It may be observed that, in order to achieve a fading margin of about 30–40 dB, an antenna front-to-back ratio of at least 50–60 dB is required. On the other hand, no great advantage is expected by improving the antenna decoupling beyond 60 dB.

The binary and quaternary performances differ by 3–5 dB while the type *B* channel arrangement gains only 1–2 dB over the type *A*.

The usable fading margins M_f are also reported in Table VI as functions of P_0 and P_e^* for the binary and quaternary systems. In the same table the increase $\Delta\eta$ of the CNR due to the effect of the fading-dependent interferences is also presented, for each value of M_f.

It may be noted that, in the considered situation, it is not convenient to have a power P_0 lower than -45 dBm or higher than -35 dBm. In the first case ($P_0 < -45$ dBm) M_f is limited by P_0 (and thus by the receiver front-end noise); in the second case ($P_0 > -35$ dBm) the errors are mainly due to the inter-

[5]This figure is for $P_e^* = 10^{-6}$, while the 0.2-dB impairment assigned to the fading-independent interferences in Section IV-*C* is for P_e^* 10^{-4}.

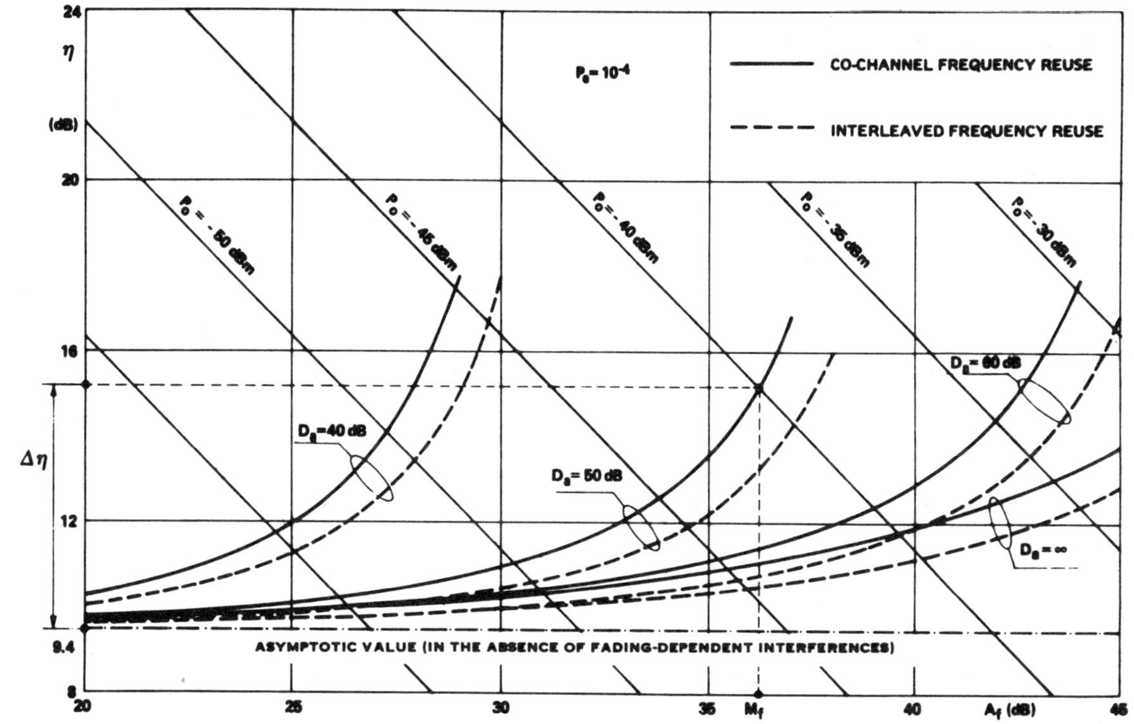

Fig. 8. Long-haul system with binary modulation. Necessary CNR as a function of the fading depth A_f, for $D_{p0} = 30$ dB, and $D_a = 40$, 50, 60, and ∞ dB. The straight lines represent the achievable values of η depending on the received power P_0.

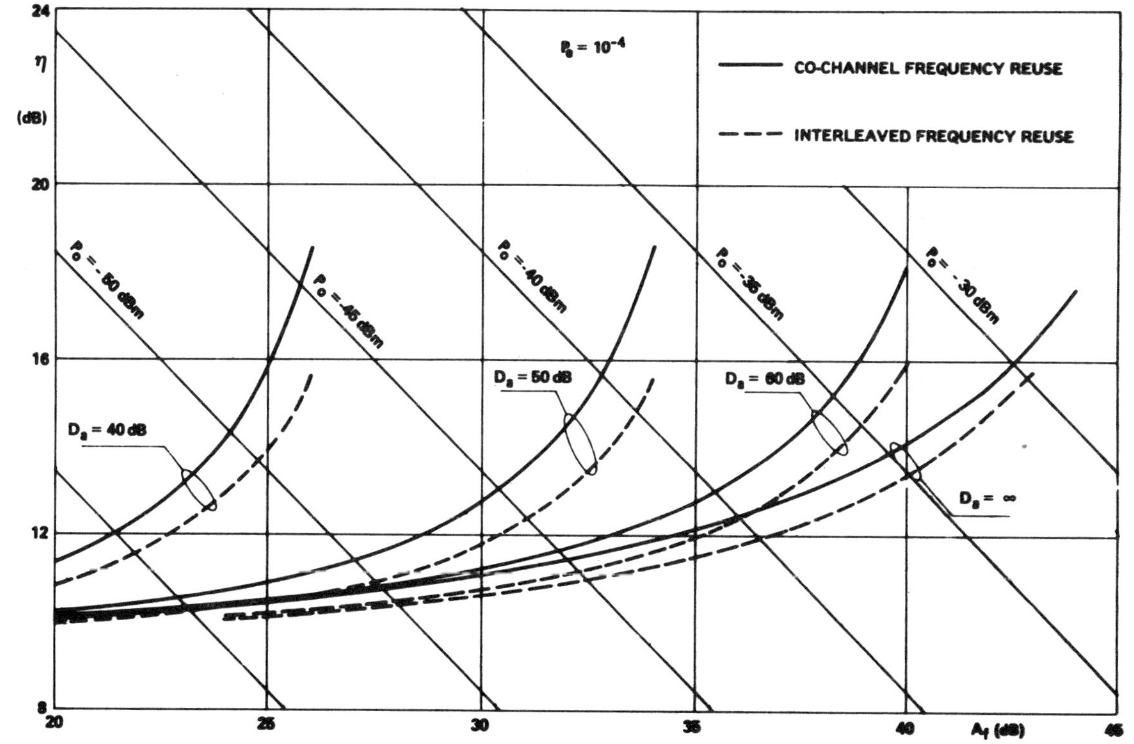

Fig. 9. Long-haul system with quaternary modulation. Necessary CNR as a function of the fading depth A_f, for $D_{p0} = 30$ dB and $D_a = 40$, 50, 60, and ∞ dB.

ferences rather than to the thermal noise, so that any increase of the available η (or of P_0) has a very slight effect on the M_f.

B. Short-Haul Systems

The same diagrams of η versus A_f that have been discussed for the long-haul system are shown in Figs. 10 and 11 for the short-haul network, with the values of the antenna parameters given in Section III-C.

In Figs. 12 and 13 the usable fading margins are plotted as functions of the power P_0; the values corresponding to $\Delta\eta = 3$ dB or 5 dB are indicated on the curves, with crosses or circles, respectively. This indication can give an idea of how close we are to the vertical asymptotes of the curves of Figs. 10 and 11.

A great difference can immediately be remarked between the performance of the type A and type B channel arrangements.

TABLE VI
LONG-HAUL SYSTEMS WITH TYPE *A* CHANNEL ARRANGEMENT[a]

	$P_e^* = 10^{-4}$		$P_e^* = 10^{-6}$		$P_e^* = 10^{-8}$	
P_0(dBm)	$\Delta\eta$(dB)	M_f(dB)	$\Delta\eta$(dB)	M_f(dB)	$\Delta\eta$(dB)	M_f(dB)
			Binary			
-30	8.1	44.0	8.1	41.9	7.9	40.5
-35	5.1	42.0	5.1	39.8	5.0	38.3
-40	3.0	39.1	3.1	36.7	3.1	35.2
-45	1.7	35.3	1.8	32.9	1.9	31.3
-50	1.0	31.1	1.1	28.6	1.2	27.0
			Quaternary			
-30	9.3	40.0	9.0	38.0	8.9	36.5
-35	5.7	38.3	5.7	35.9	5.7	34.3
-40	3.5	35.4	3.6	33.0	3.6	31.3
-45	2.1	31.8	2.3	29.3	2.3	27.7
-50	1.3	27.7	1.4	25.2	1.5	23.5

[a]Fading margins and corresponding values of Δ_η for receiver input powers ranging from -50 to -30 dBm. D_a = 60 dB and D_{p0} = 30 dB.

In fact, when the antenna decoupling D_a' between adjacent low-capacity links is not very high the interleaved mode of frequency reusing presents a very poor performance. Only for D_a' greater than about 50 dB do the two channel arrangements give about the same results.

The reason for the bad performance of the interleaved frequency plan is that each radial link suffers a strong interference from the frequency-adjacent cross-polarized channel transmitted on one of the angularly adjacent links (see Fig. 4).

A careful distribution of the channels in the various links can prevent this trouble, when a sufficient number of radio channels is available. However, to obtain such an appropriate distribution, more conditions have to be fulfilled, in addition to those imposed by site locations, so that in some cases harmful interferences could not be avoided.

C. The Influence of the Various Interfering Sources

The different disturbances considered in the previous paragraphs raise the value of the CNR necessary to obtain the required threshold error-rate P_e^*. The amount of this increase gives an indication of the influence of each disturbance.

It is worthwhile to recall that, in practice, the effects of the thermal noise, intersymbol, and interchannel interferences cannot be separated. The following figures must then be taken as representing the degradation of the transmission quality, when each impairment cause is added to the previous ones.

The CNR η corresponding to $P_e^* = 10^{-4}$ for a long-haul four-phase system is the following.

Theoretical (thermal noise only)	8.4 dB
+ Intersymbol interference	+ 0.9 dB
+ Interchannel interference (fading-independent)	+ 0.2 dB
+ Interchannel interference (fading-dependent)	+ $\Delta\eta$ dB
+ Differential encoding penalty	+ 1.0 dB
Total 10.5 + $\Delta\eta$ dB	

According to the results presented in Table VI the $\Delta\eta$ due to the fading-dependent interferences may be of about 3–6 dB, and it turns out to be the greatest contribution to the overall required CNR, apart from the thermal noise.

A further allotment of $\Delta\eta$ between the different types of fading-dependent interferences is not possible, but the harmfulness of each interference may be pointed out by means of curves like those presented in Figs. 14 and 15, where only one of the interference paths is taken into account.

The main disturbing sources considered here are two for the long-haul system (paths *B* and *D* of Fig. 2) and three for the short-haul system (coming from the same hop, the main link, and the two adjacent radial links).

V. CONCLUSIONS

On the basis of the interference situation presented in Section II, the usable fading margins have been evaluated, for two typical system structures, in connection with two different radio-frequency channeling schemes.

It has been shown that fading margins of about 30–40 dB are usable, with the usual antenna characteristics, in two-phase as well as in four-phase coherent PSK systems.

Greater values of M_f seem not to be convenient, since they could be less easily obtained and might also be reduced by second-order phenomena not considered here.

The difference between the performances of the two above mentioned modulation methods (two-phase and four-phase) has been found to be 3–5 dB, while the use of the two different types of channel arrangements proposed in this paper have shown, for the long-haul system, slight differences in performance.

In short-haul systems, on the contrary, the two frequency plans performed very differently, because of the higher sensitivity of the type *B* plan to the adjacent link interferences. The co-channel plan (type *A*) should thus be preferred in this kind of application since it requires less care in the allocation of the RF channels.

As regards the assumptions made in the present work, it is worth noticing that the XPD degradation model adopted may be considered a pessimistic one, because it assumes a complete correlation between the attenuation of the wanted carrier and the XPD. Even with this worst case assumption, the analysis concerning the effect of the various interference sources shows that the influence of the cross-polarized channel does not predominate over the other interference causes.

Further studies are required to define a more precise relationship between the path attenuation and the degradation of cross-polarization discrimination due to rainfall.

The results presented in this paper proved to be useful [12] for the solution of some design problems, relating to the choice of the channel arrangement, of the filters, etc.

ACKNOWLEDGMENT

The authors wish to thank L. Gobbo for his helpful collaboration in writing simulation programs.

REFERENCES

[1] C. Colavito, "Terrestrial radio-relay systems for digital communication networks," presented at the CEPT Sem. Digital Transmission Systems, Milan, Italy, Apr. 1971.

[2] V. Castellani and M. Sant'Agostino, "Intersymbol and interchannel interference effects in PCM signal transmission on radio links: Binary coherent phase modulation," *Alta Freq.*, vol. 40, pp. 389–397, May 1971.

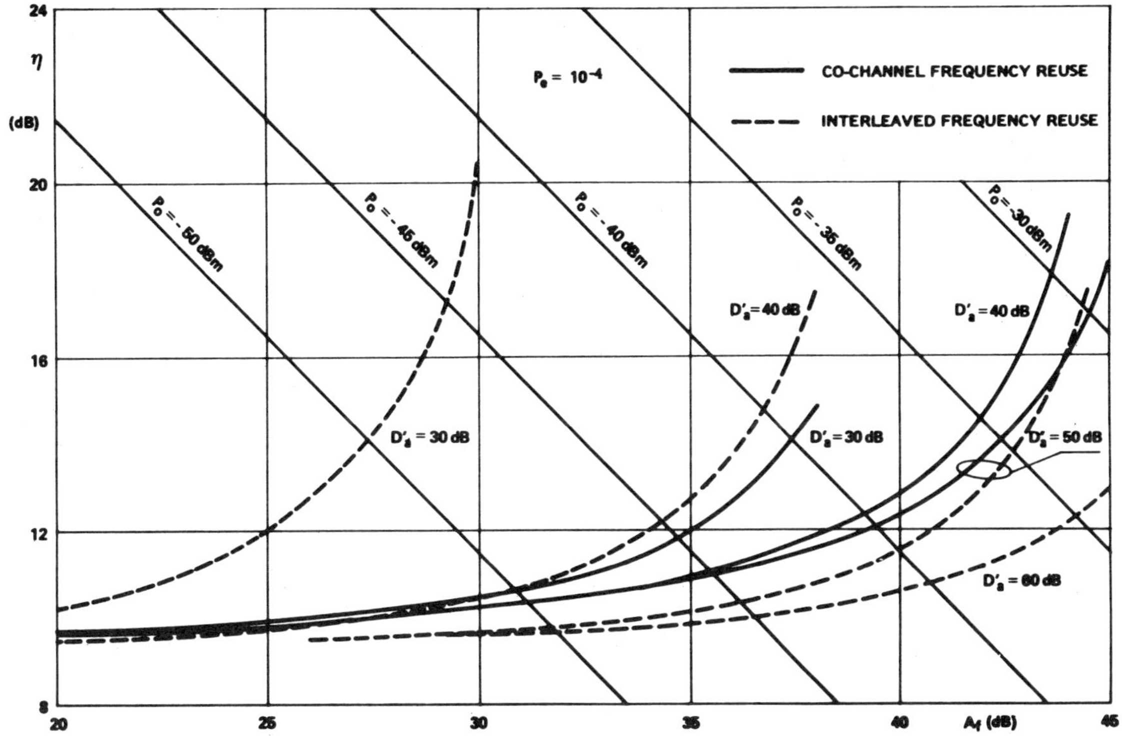

Fig. 10. Short-haul link, with binary modulation. Necessary CNR as a function of the fading depth A_f, for $D_a = 60$ dB and D'_a as indicated in the curves. The equal power lines are also given, for P_0 ranging from −50 to − 30 dBm.

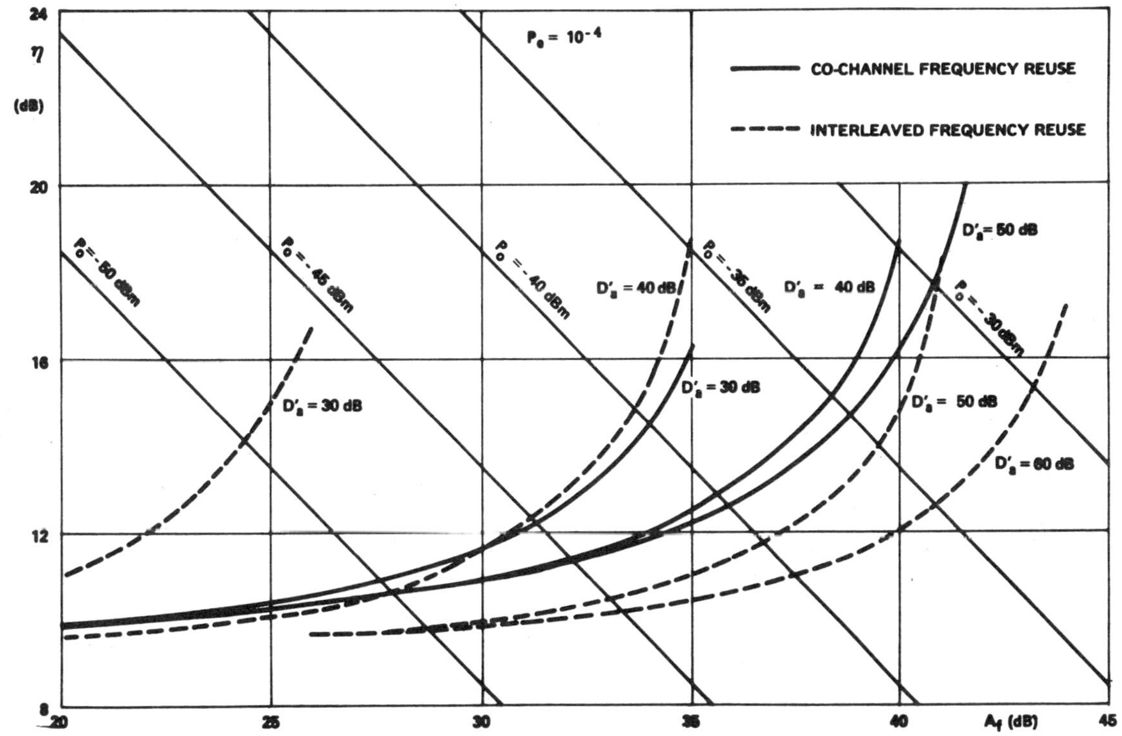

Fig. 11. Short-haul link, with quaternary modulation. Same curves as in Fig. 10.

[3] M. Sant'Agostino, "Transmission filter effects on binary and quaternary coherent PSK systems," to be published.

[4] V. K. Prabhu, "Error rate considerations for coherent phase-shift keyed systems with co-channel interference," *Bell Syst. Tech. J.*, pp. 743–767, Mar. 1969.

[5] A. S. Rosenbaum, "Binary PSK error probabilities with multiple cochannel interferences," *IEEE Trans. Commun. Technol.*, vol. COM-18, pp. 241–253, June 1970.

[6] L. Calandrino, G. Corazza, G. Crippa, and G. Immovilli, "Intersymbol, interchannel, and cochannel interferences in binary and quaternary PSK systems," *Alta Freq.*, vol. 40, pp. 407–420, May 1971.

[7] A. De Lama and A. Luvison, "State variables and numerical evaluation of intersymbol interference," *Alta Freq.*, vol. 39, pp. 953–963, Nov. 1970.

[8] E. Biglieri, R. Dogliotti, and M. Pent, "Intersymbol and interchannel interference effects in PCM signal transmission on radio links: Binary differentially coherent phase modulation," *Alta Freq.*, vol. 40, pp. 398–406, May 1971.

[9] C. L. Ruthroff and L. C. Tillotson, "Interference in a dense radio

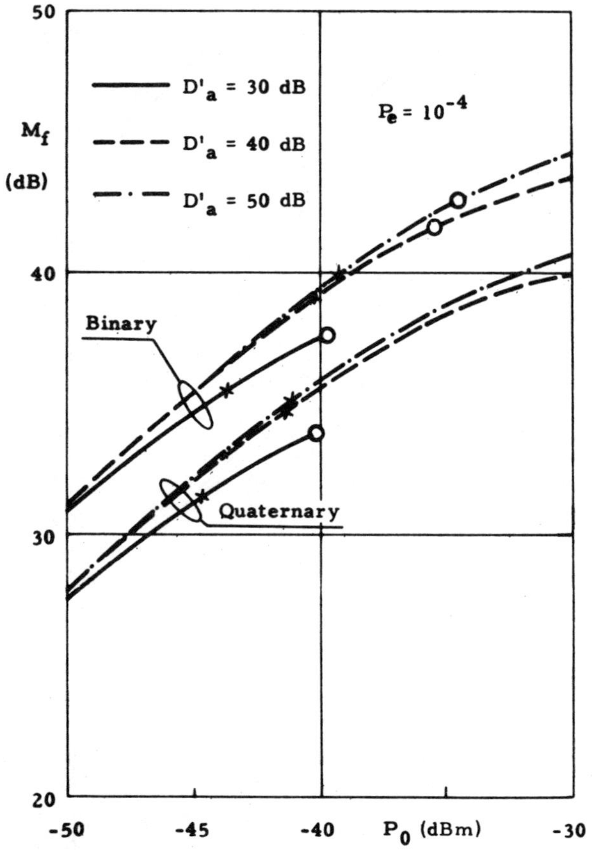

Fig. 12. Short-haul system with co-channel frequency reuse. Usable fading margin as a function of the receiver input power P_0 in normal propagation conditions. X—Values of M_f and P_0 corresponding to a 3-dB increase in the necessary η. O—Values of M_f and P_0 corresponding to a 5-dB increase in the necessary η.

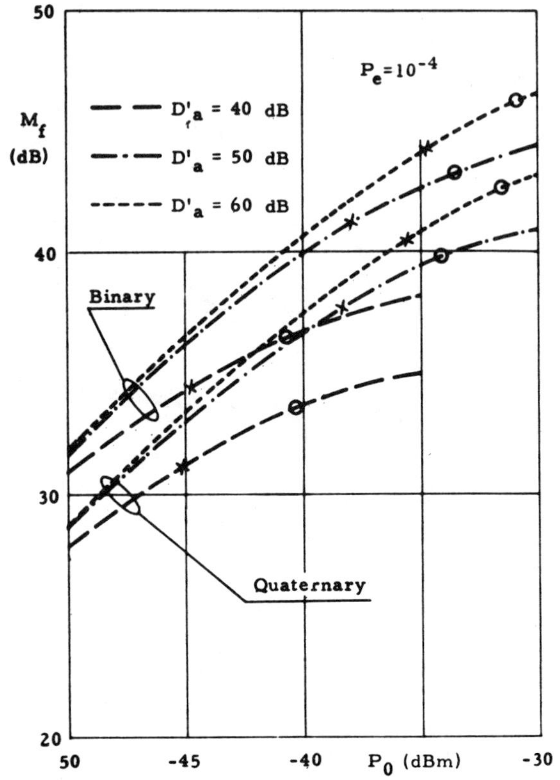

Fig. 13. Short-haul system with interleaved frequency reuse. Usable fading margin as a function of the receiver input power P_0 in normal propagation conditions. X—Values of M_f and P_0 corresponding to a 3-dB increase in the necessary η. O—Values of M_f and P_0 corresponding to a 5-dB increase in the necessary η.

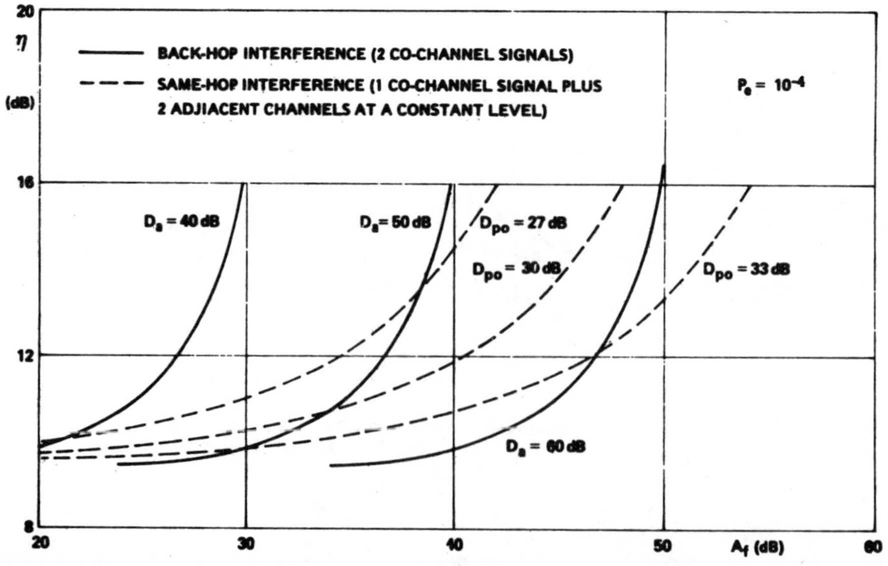

Fig. 14. Long-haul system with binary modulation and co-channel type frequency reuse. Effect of the interfering signals coming over path B (continuous lines) and path D (broken lines), for 3 different values of D_a and D_{po}.

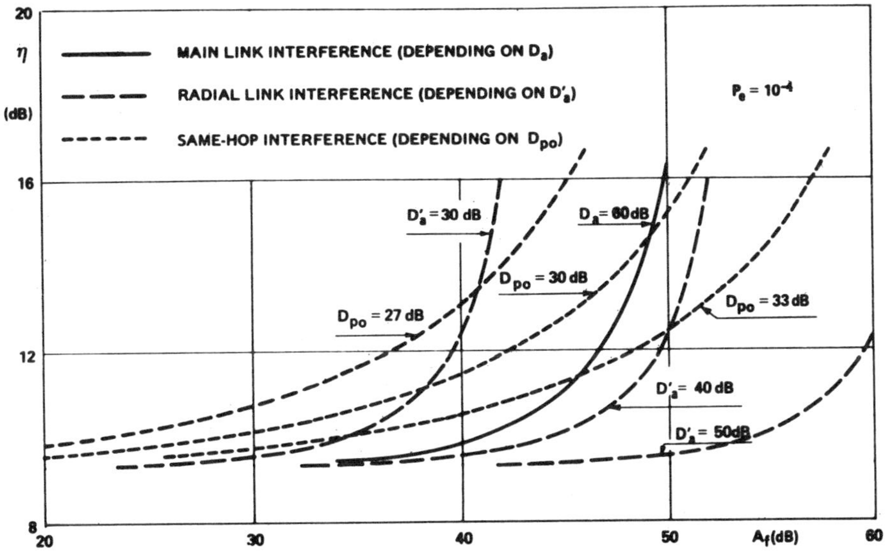

Fig. 15. Short-haul system with binary modulation and co-channel type frequency reuse. Effect of the interfering signals coming from the paths indicated in the legend for D_a = 60 dB and for 3 different values of the parameters D'_a and D_{po}.

network," *Bell Syst. Tech. J.*, vol. 48, pp. 1727–1743, July/Aug. 1969.

[10] P. A. Watson, "Attenuation and cross-polarization measurements at 11 GHz," presented at the IEEE Int. Conf. Communications, Philadelphia, Pa., June 1972.

[11] D. T. Thomas, "Cross-polarization distortion in microwave radio transmission due to rain," *Radio Sci.*, vol. 6, pp. 833–839, Oct. 1971.

[12] International Radio Consultative Committee (CCIR), Docs. 9/60 (Italy), and 9/125 (S. G. 9), 1970–1973.

Unified Analysis of a Class of Digital Systems in Additive Noise and Interference

RUSSELL FANG AND OSAMU SHIMBO

Abstract—A unified approach is presented for evaluation of the error probabilities of a class of digital communications systems in additive noise and interference. This class of systems includes coherent systems such as M-ary amplitude-shift keying (ASK), M-ary phase-shift keying (PSK), and M-ary amplitude-and-phase keying (APK); it also includes differential coherent systems such as binary differential PSK (DPSK). The noise is not necessarily Gaussian. The interference can be intersymbol interference, co-channel interference, adjacent-channel interference, any of their linear combinations, or intermodulation products at the output of some nonlinear device.

This approach essentially expands the characteristic function of the interferences into a power series so that the desired error probability can be evaluated as the sum of terms representing perturbations around the error probability due to additive noise alone. Bounds on three kinds of truncation errors, which are simple and applicable to all aforementioned digital systems, are obtained. As a result, any desired accuracy in the evaluation of the error probabilities can be achieved with this approach. In the special case in which the noise is Gaussian, explicit bounds on truncation errors are also obtained.

Examples are given to illustrate how the unified analysis can be applied to evaluate the error probabilities of various digital systems. More specifically, the combined effects of Gaussian noise, intersymbol interference, and co-channel interference on the error performance of M-ary coherent PSK and APK (MCPSK and MCAPK) systems are computed. The probability of error of a binary DPSK (BDPSK) system in the presence of Gaussian noise and intersymbol interference is analyzed. The intermodulation products at the output of a hard-limiter are also determined.

I. Introduction

In new digital communication systems design, a constant problem is that of determining the most effective way of using the available radiation power and channel bandwidth for a given data rate and an acceptable system performance, e.g., certain prescribed error probability. If the channel is Gaussian, there are many theoretical results that can serve as design guidelines. However, Gaussian channels rarely exist in practical applications, even in modern satellite communications. (Of course, because of severe power limitations on some older generation satellites, Gaussian noise was once a primary concern. Nevertheless, since high-power satellites are now readily available, the limitation on channel bandwidth due to hardware or legal constraints becomes increasingly important for future satellite communications. Consequently, other disturbances such as intersymbol interference and adjacent-channel interference are just as important as Gaussian noise.) Thus, all major disturbances as well as additive noise must be considered in the design. Since in many situations the system design margin is quite costly, an accurate performance analysis of various candidate digital communication systems to be operated over such non-Gaussian channels is necessary so that the most effective system can be chosen and the design margin can be minimized.

In this paper, symbol error probability is chosen as the criterion for evaluating system performance.

This paper is organized into two major parts: Sections II and III form the theoretical part while Sections IV-VII form the application part. Those readers interested only in the applications part may go through Sections II and III quickly, and directly get into the applications.

Section II poses the problem of calculating the probability of an n-dimensional random vector, defined as $z = S + N + \sum_{k=1}^{L} a_k$, falling into some region D in R^n,[1] i.e., $\Pr[z \in D]$, where S is an n-dimensional known signal vector, while N and a_k are n-dimensional random vectors representing noise and interference, respectively. This additive noise is not necessarily Gaussian, but it is assumed to have a probability density function in R^n satisfying certain minor regularity conditions. The interferences $\{a_k\}$ characterize disturbances such as those caused by intersymbol interference, co-channel interference, adjacent-channel interference, or any of their combinations, intermodulation products at the output of some nonlinear device. For simplicity, it is assumed that $N, a_1, a_2, \cdots$, and a_L are statistically independent, and that $\{a_k\}$ are ordered according to their expected Euclidian norms, i.e., $E[\|a_{k+1}\|] \leqslant E[\|a_k\|]$ for $k = 1, 2, \cdots, L$. The number of interferences L may be infinite, for example, when $\{a_k\}$ represents the intersymbol interference due to all past symbols in the message sequence.

In general, since the probability density function of interference may not exist [1] or is just too difficult to obtain [2], [3], the probability density function of z is practically impossible to find. Therefore, the problem of evaluating $\Pr[z \in D]$ is by no means simple. However, since for almost all practical cases the characteristic functions of $\{a_k\}$ do exist and can be found without much difficulty, $\Pr[z \in D]$ can be expressed in terms of the characteristic functions of N and $\{a_k\}$. The characteristic function of the interference $\sum_{k=1}^{L} a_k$ is then expanded into a power series in the hope that $\Pr[z \in D]$ can be evaluated as the sum of terms representing perturba-

[1] The meaning of the dimensionality n can be clearly seen from the examples in the applications part of this paper.

Paper approved by the Associate Editor for Communication Theory of the IEEE Communications Society for publication after presentation at the IEEE International Symposium on Information Theory, Pacific Grove, Calif., January 31–February 4, 1972. Manuscript received February 12, 1973; revised May 22, 1973. This work was performed at COMSAT Laboratories under the joint sponsorship of the Communications Satellite Corporation and the International Telecommunications Consortium (INTELSAT). Views expressed are not necessarily those of INTELSAT.

The authors are with COMSAT Laboratories, Clarksburg, Md. 20734.

Reprinted from *IEEE Trans. Comm.*, vol. COM-21, pp. 1075–1091, Oct. 1973.

tions around $\Pr[z \in D \mid \Sigma_{k=1}^{L} a_k = 0]$. In what follows it is shown that this can be effectively done under some minor assumptions on $\{a_k\}$ and N.

In Section III, three kinds of truncation errors are present. The first kind is caused by integration over a finite region instead of over D, which is sometimes infinite. The second kind is caused by considering some finite number less than the total number of interferences. The third is caused by the termination of the power series. With these truncation errors it is shown that the power series expansion method presented here can achieve any desired accuracy in the evaluation of $\Pr[z \in D]$. Explicit bounds on these truncation errors are also obtained for the most usual case in which the additive noise is Gaussian.

Sections IV–VII then demonstrate how the results in the theoretical part, Sections II and III, can be applied to the problems of calculating the probabilities of error of a number of digital communications systems subject to both additive noise and interference. These systems may employ modulation techniques such as M-ary coherent amplitude-shift keying (MCASK), M-ary coherent phase-shift keying (MCPSK), and M-ary coherent amplitude-and-phase-keying (MCAPK) [7]-[9].

For systems subjected to Gaussian noise and intersymbol interference only, several methods have been previously proposed for calculating the error probabilities of digital systems such as MCASK [10]-[15], MCPSK [16]-[18], and binary differentially-coherent PSK (BDPSK) [19]-[21]. Methods for computing the probability of error for an MCPSK system with only Gaussian noise and co-channel interference have also been recently suggested [22]-[28]. Nevertheless, all of these methods are somewhat limited in their scope of applicability, whereas the method presented here is most general and very powerful, especially when numerical results are desired, as in most practical situations.

For illustrative purposes, the combined effects of Gaussian noise, intersymbol interference, and co-channel interference on the error performance of MCPSK and MCAPK systems will be analyzed in Section IV and Section V, respectively. The error probability of a binary differential PSK (DPSK) system in the presence of Gaussian noise and intersymbol interference will be determined in Section VI. Then, the intermodulation products due to hard-limiting multiple signals and noise are treated in Section VII. Finally, Section VIII gives the conclusion for this paper and includes some future research problems.

II. Problem Statement and Method of Solution

Problem Statement

Let

$$z \equiv S + N + \sum_{k=1}^{L} a_k, \tag{1}$$

where S, N, and a_k are n-dimensional real vectors. S is deterministic, while N and a_k are random with associated characteristic functions denoted by ϕ_0 and ϕ_k, respectively. That is,

$$\phi_0(u) = \int \exp[jx \cdot u]\, dF_0(x) \tag{2}$$

$$\phi_k(u) = \int \exp[jx \cdot u]\, dF_k(x), \qquad \text{for } k = 1, 2, \cdots, L, \tag{3}$$

where $x \cdot u$ denotes the scalar product of two vectors x and u, and F_0 and F_k are probability distribution functions of N and a_k, respectively. Assume that ϕ_0 is absolutely integrable over R^n so that the probability density function of N, denoted by p_N, exists and let

$$\phi_0(u) = \int \exp[jx \cdot u]\, p_N(x)\, dx. \tag{4}$$

Also assume the following.

Assumption 1) $N, a_1, a_2, \cdots$, and a_L are zero-mean and statistically independent.

Assumption 2) For simplicity, ϕ_k for $k = 1, 2, \cdots, L$, are real; i.e., all odd moments of a_k are zero.

Assumption 3) $\Sigma_{k=1}^{L} E[\|a_k\|^2] = P < \infty$; when $L = \infty$, this condition guarantees that $\Pi_{k=1}^{\infty} \phi_k$ exists [4].

Assumption 4) $\sup_k \|a_k\| = A_0 < \infty$.

Assumption 5) $\int \|u\|^{2l} \, |\phi_0(u)|\, du < \infty$ for $l = 1, 2, \cdots$.

Then, the problem is to effectively evaluate $\Pr[z \in D]$ for some $D \subset R^n$ within any desired accuracy.

Method of Solution

Observe that the probability density function p_z of z exists, since the probability density function p_N of N and the probability distribution function G_L of $\Sigma_{k=1}^{L} a_k$ exist. [If $L = \infty$, then Assumption 3 implies the existence of G_∞]. Hence,

$$\begin{aligned}\Pr[z \in D] &= E\left[\int_D p_z\left(x \,\middle|\, \sum_{k=1}^{L} a_k = t\right) dx\right] \\ &= \iint_D p_N(x - S - t)\, dx\, dG_L(t) \\ &= \iint_D p_N(x - S - t)\, dG_L(t)\, dx, \end{aligned} \tag{5}$$

where $E[\cdot]$ denotes the ensemble average over $\Sigma_{k=1}^{L} a_k$.

Let the characteristic function of $\Sigma_{k=1}^{K} a_k$ be represented by

$$\Phi_K(u) = \int \exp(jt \cdot u)\, dG_K(t), \tag{6}$$

where G_K is the probability distribution function of $\Sigma_{k=1}^{K} a_k$. Then, because $\{a_k\}$ are statistically independent,

$$\Phi_K(u) = \prod_{k=1}^{K} \phi_k(u). \tag{7}$$

Now, application of Bochner's theorem [6] to (5) results in

$$\Pr[z \in D] = (2\pi)^{-n} \int_D dx \int_{R^n} \phi_0(u)\, \Phi_L(u) \cdot \exp[-j(x - S) \cdot u]\, du. \tag{8}$$

Computation of (8) is extremely difficult, if not impossible, especially when $L = \infty$. A power series expansion method is proposed to evaluate this integral as follows. Construct an n-dimensional cube D_s of length a on each side, centered at the vector S, as illustrated in Fig. 1. Let

$$P(D \cap D_s; K) \equiv (2\pi)^{-n} \int_{D \cap D_s} dx \int_{R^n} du\, \phi_0(u)\, \Phi_K(u) \cdot \exp[-j(x - S) \cdot u] \quad (9)$$

be the probability that the random variable $z_K \equiv S + N + \sum_{k=1}^{K} a_k$ falls into the set $D \cap D_s$. Also, define

$$P(D \cap D_s; K; R) \equiv (2\pi)^{-n} \int_{D \cap D_s} dx \int_{R^n} du\, \phi_0(u) \cdot \sum_{l=0}^{R-1} \frac{(-1)^l}{(2l)!} \left[\left(\frac{\partial}{\partial S} \cdot \frac{\partial}{\partial t}\right)^{2l} \Phi_K(t)\right]_{t=0} \exp[-j(x - S) \cdot u], \quad (10)$$

where

$$\frac{\partial}{\partial y} \equiv \left(\frac{\partial}{\partial y_1}, \frac{\partial}{\partial y_2}, \cdots, \frac{\partial}{\partial y_n}\right)$$

and $\{y_k\}$ are the n components of the vector y. Then, by triangular inequality

$$\begin{aligned} |\Pr[z \in D] - P(D \cap D_s; K; R)| &\leq |\Pr[z \in D] - P(D \cap D_s; K)| \\ &\quad + |P(D \cap D_s; K) - P(D \cap D_s; K; R)| \\ &\leq P(D \cap D_s^c; L) + |P(D \cap D_s; L) - P(D \cap D_s; K)| \\ &\quad + |P(D \cap D_s; K) - P(D \cap D_s; K; R)|, \end{aligned} \quad (11)$$

where D_s^c is the complement of D_s.

Since, as will be shown in (19), (26), and (33), respectively, for any $\epsilon > 0$

$$P(D \cap D_s^c; L) \leq P(D_s^c; L) \equiv E_0(a) < \epsilon/3 \quad (12a)$$

for sufficiently large a;

$$|P(D \cap D_s; L) - P(D \cap D_s; K)| < \epsilon/3 \quad (12b)$$

for sufficiently large K; and

$$|P(D \cap D_s; K) - P(D \cap D_s; K; R)| < \epsilon/3 \quad (12c)$$

for sufficiently large R, it follows by substituting (12a-c) into (11) that

$$|\Pr[z \in D] - P(D \cap D_s; K; R)| < \epsilon. \quad (13)$$

Thus, $P(D \cap D_s; K; R)$ can be used to approximate $\Pr[z \in D]$ within any desired accuracy. The question is simply how to compute $P(D \cap D_s; K; R)$ effectively. Before proceeding to the problem of computing $P(D \cap D_s; K; R)$, the following observations should be made.

The first term on the right-hand side (RHS) of (11) represents the truncation error caused by the confinement of D in a cube D_s centered at the vector S. The second term on the RHS of (11) represents the truncation error due to the approximation of $\Phi_L(u)$ by $\Phi_K(u)$ with $K \leq L$. This approximation is sometimes necessary, because $P(D \cap D_s; L)$ may

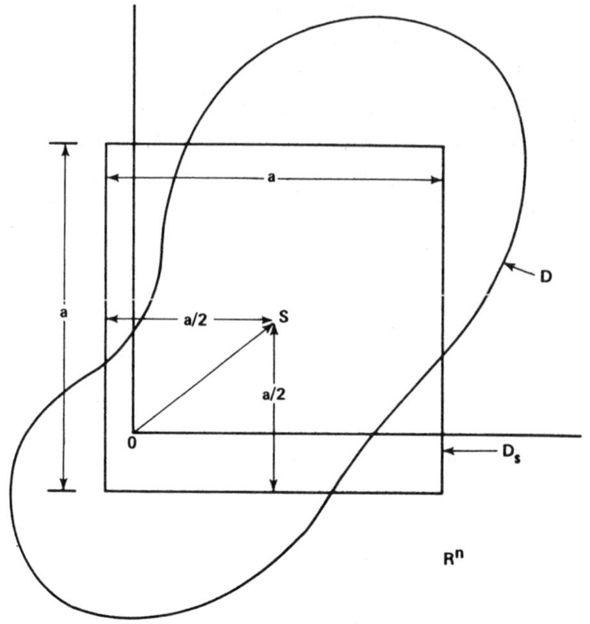

Fig. 1. Definition of D and D_S.

still be difficult to evaluate, especially if $L = \infty$. The third term on the RHS of (11) represents the error resultant from approximating $P(D \cap D_s; K)$ by $P(D \cap D_s; K; R)$. Since, from Lemma 1 of Appendix A, $\Phi_K(u)$ can be expanded into the power series

$$\Phi_K(u) = \sum_{l=0}^{\infty} \frac{(-1)^l}{(2l)!} \left[\left(u \cdot \frac{\partial}{\partial t}\right)^{2l} \Phi_K(t)\right]_{t=0}, \quad (14)$$

it follows by substituting (14) into (9) that

$$P(D \cap D_s; K) = (2\pi)^{-n} \int_{D \cap D_s} dx \int_{R^n} du\, \phi_0(u) \cdot \sum_{l=0}^{\infty} \frac{(-1)^l}{(2l)!} \left[\left(\frac{\partial}{\partial S} \cdot \frac{\partial}{\partial t}\right)^{2l} \Phi_K(t)\right]_{t=0} \exp[-j(x - S) \cdot u]. \quad (15)$$

Consequently, a comparison of (10) and (15) reveals that the third term on the RHS of (11) is just the truncation error caused by terminating the power series in (14) at the $(R - 1)$th term.

Note that the individual integrals in (10) do exist under Assumptions 4 and 5; according to Lemma 2 in Appendix B the summation and the integration in (10) can be interchanged. Hence,

$$\begin{aligned} P(D \cap D_s; K; R) &\equiv \sum_{l=0}^{R-1} \frac{(-1)^l}{(2l)!} (2\pi)^{-n} \int_{D \cap D_s} dx \int_{R^n} du\, \phi_0(u) \cdot \left[\left(\frac{\partial}{\partial S} \cdot \frac{\partial}{\partial t}\right)^{2l} \Phi_K(t)\right]_{t=0} \exp[-j(x - S) \cdot u] \\ &= \sum_{l=0}^{R-1} \frac{(-1)^l}{(2l)!} \left[\left(\frac{\partial}{\partial S} \cdot \frac{\partial}{\partial t}\right)^{2l} \Phi_K(t)\right]_{t=0} \cdot \int_{D \cap D_s} p_N(x - S)\, dx, \end{aligned} \quad (16)$$

where

$$\int_{D \cap D_s} p_N(x - S) \equiv (2\pi)^{-n} \int_{D \cap D_s} dx \int_{R^n} du\, \phi_0(u) \cdot \exp\left[-j(x - S) \cdot u\right] \quad (17)$$

is the probability that $z \in D \cap D_s$ given $\Sigma_{k=1}^{L} a_k = 0$.

Therefore, if $\int_{D \cap D_s} p_N(x - S)\, dx$ and its derivatives with respect to S can be found by some means and if the coefficients of the power series expansion of $\Phi_K(u)$ can be found, e.g., through the recurrence method, then $\Pr[z \in D]$ can be evaluated within any desired accuracy by using (16), (17).

III. Truncation Error Bounds

There are three kinds of truncation errors. The first kind is due to the confinement of D in a cube D_s centered at the vector S. Since p_z is a probability density function that is by definition nonnegative and integrable over the whole R^n, it follows that the first kind of error bound as defined in (12a), $E_0(a)$, is a monotonically decreasing function of the volume of D_s and, consequently, a monotonically decreasing function of a. Let p_{z_i} be the ith marginal density of z. Then, from the definition of $P(D_s^c; L)$,

$$\begin{aligned} E_0(a) &\equiv P(D_s^c; L) \\ &= \int_{D_s^c} p_z(x - S)\, dx \\ &\leqslant \sum_{i=1}^{n} \left(\int_{-\infty}^{-a/2} p_{z_i}(x_i)\, dx_i + \int_{a/2}^{\infty} p_{z_i}(x_i)\, dx_i \right). \end{aligned} \quad (18)$$

Thus, for any $\epsilon > 0$,

$$E_0(a) < \epsilon/3 \quad (19)$$

for sufficiently large value of a.

The second kind of truncation error, which is caused by considering only K instead of L interference terms, can be bounded from above as follows:

$$\begin{aligned} &|P(D \cap D_s; L) - P(D \cap D_s; K)| \\ &= \left| (2\pi)^{-n} \int_{D \cap D_s} dx \int_{R^n} du\, \phi_0(u)\, \Phi_K(u) \cdot \left[1 - \prod_{k=K+1}^{L} \phi_k(u)\right] \exp\left[-j(x - S) \cdot u\right] \right| \\ &\leqslant A(2\pi)^{-n} \int_{R^n} |\phi_0(u)| \left| 1 - \prod_{k=K+1}^{L} \phi_k(u) \right| du, \end{aligned} \quad (20)$$

where

$$A \equiv \int_{D \cap D_s} dx \leqslant \int_{D_s} dx = a^n < \infty. \quad (21)$$

However,

$$\prod_{k=K+1}^{L} \phi_k(u) \equiv E\left[\exp ju \cdot \left(\sum_{k=K+1}^{L} a_k \right)\right]. \quad (22)$$

Since the exponential term can be expanded into a Taylor series with an integral reminder [5, p. 309],

$$\exp\left(ju \cdot \sum_{k=1}^{L} a_k\right) = 1 + ju \cdot \sum_{k=K+1}^{L} a_k - \frac{1}{2}\left(u \cdot \sum_{k=K+1}^{L} a_k\right)^2 \int_0^1 (1 - t) \exp\left(jtu \cdot \sum_{k=K+1}^{L} a_k\right) dt, \quad (23)$$

and since a_k has zero mean for all k, we can combine (22) and (23) to yield

$$\begin{aligned} &\left| 1 - \prod_{k=K+1}^{L} \phi_k(u) \right| \\ &= \frac{1}{2} \left| E\left[\int_0^1 \left(u \cdot \sum_{k=K+1}^{L} a_k\right)^2 \exp\left(jtu \cdot \sum_{k=K+1}^{L} a_k\right) (1 - t)\, dt \right] \right| \\ &\leqslant \frac{1}{2} E\left[\left(u \cdot \sum_{k=K+1}^{L} a_k\right)^2\right] \int_0^1 (1 - t)\, dt \\ &\leqslant \frac{1}{4} \|u\|^2 \sum_{k=K+1}^{L} E[\|a_k\|^2] < \infty. \end{aligned} \quad (24)$$

Substituting (24) into (20) the following theorem is obtained.

Theorem 1: If p_N exists and if Assumptions 1-3 and 5 are satisfied, then

$$\begin{aligned} &|P(D \cap D_s; L) - P(D \cap D_s; K)| \\ &\leqslant \frac{A}{4(2\pi)^n} \int_{R^n} \|u\|^2\, |\phi_0(u)|\, du \sum_{k=K+1}^{L} E[\|a_k\|^2] \\ &\equiv E_1(K) < \infty, \end{aligned} \quad (25)$$

and $E_1(K)$ is obviously decreasing monotonically. Thus, for any $\epsilon > 0$, K can be selected large enough so that

$$E_1(K) < \epsilon/3. \quad (26)$$

It should be noted that this bound does not depend upon whether or not a power series expansion method is used, but it does depend upon K, the number of interference terms used.

The third kind of truncation error is caused by the termination of the power series expansion of $\Phi_K(u)$. In other words, it is due to the approximation of $P(D \cap D_s; K)$ by $P(D \cap D_s; K; R)$. According to Taylor's theorem [5, p. 245], the series in (14) can also be represented by only R terms plus a remainder; namely,

$$\Phi_K(u) = \sum_{l=0}^{R-1} \frac{1}{(2l)!} \left[\left(u \cdot \frac{\partial}{\partial t}\right)^{2l} \Phi_K(t)\right]_{t=0} + \frac{1}{(2R)!} \left[\left(u \cdot \frac{\partial}{\partial t}\right)^{2l} \Phi_K(t)\right]_{t=u_0}, \quad (27)$$

where u_0 is a vector on the line joining 0 and u. From (27), (15), and (10), it is evident that

$$|P(D \cap D_s; K) - P(D \cap D_s; K; R)| = \left| (2\pi)^{-n} \int_{D \cap D_s} dx \int_{R^n} du\, \phi_0(u) \frac{1}{(2R)!} \cdot \left[\left(u \cdot \frac{\partial}{\partial t} \right)^{2R} \Phi_K(t) \right]_{t=u_0} \exp\,[-j(x-S)\cdot u] \right|. \quad (28)$$

However,

$$\left| \left[\left(u \cdot \frac{\partial}{\partial t} \right)^{2R} \Phi_K(t) \right]_{t=u_0} \right| = \left| \int \left(u \cdot \frac{\partial}{\partial t} \right)^{2R} \exp\,[jx\cdot t]\, dG_K(x) \right|_{t=u_0}$$

$$= \left| \int (u \cdot x)^{2R} \exp\,[jx \cdot t]\, dG_K(x) \right|_{t=u_0}$$

$$\leqslant \left| \int (u \cdot x)^{2R}\, dG_K(x) \right|_{t=u_0}$$

$$\leqslant \|u\|^{2R} \int \|x\|^{2R}\, dG_K(x) \quad \text{(Schwartz inequality)}$$

$$\equiv \|u\|^{2R}\, E\left[\left\| \sum_{k=1}^{K} a_k \right\|^{2R} \right], \quad (29)$$

which is obviously finite according to Assumption 4. Thus,

$$\left| \frac{1}{(2R)!} \left[\left(u \cdot \frac{\partial}{\partial t} \right)^{2R} \Phi_K(t) \right]_{t=u_0} \right| \leqslant \|u\|^{2R}\, E\left[\left\| \sum_{k=1}^{K} a_k \right\|^{2R} \right] \Big/ (2R)!. \quad (30)$$

The RHS of (30) approaches zero as R increases to ∞ since all a_k are bounded.

Substituting (30) into (28) and applying Assumption 5 results in

$$|P(D \cap D_s; K) - P(D \cap D_s; K; R)| \leqslant \frac{A}{(2\pi)^n (2R)!} E\left[\left\| \sum_{k=1}^{K} a_k \right\|^{2R} \right] \int_{R^n} |\phi_0(u)|\, \|u\|^{2R}\, du \equiv E_2(R). \quad (31)$$

Whenever the condition

$$\frac{A}{(2\pi)^n (2R)!} E\left[\left\| \sum_{k=1}^{K} a_k \right\|^{2R} \right] \int_{R^n} |\phi_0(u)|\, \|u\|^{2R}\, du \searrow 0 \quad \text{as } R \nearrow \infty \quad (32)$$

is satisfied, the truncation error bound $E_2(R)$ in (31) can be made arbitrarily small if R is made sufficiently large. Therefore, the following theorem can be stated.

Theorem 2: If Assumptions 1-5 and relation (32) are true, then, for any $\epsilon > 0$, R can be chosen sufficiently large so that

$$|P(D \cap D_s; K) - P(D \cap D_s; K; R)| \leqslant E_2(R) < \epsilon/3. \quad (33)$$

Gaussian Noise Case

For many practical applications, the special case in which the noise vector N is Gaussian is of primary interest. Without loss of generality, N can be assumed to have a zero mean and a covariance matrix $\sigma_n^2 I$, where I is an $n \times n$ identity matrix. That is,

$$\phi_0(u) = \exp\,(-\tfrac{1}{2}\sigma_n^2 \|u\|^2). \quad (34)$$

Substituting (34) into (16) and (17) yields

$$P(D \cap D_s; K; R) = \sum_{l=0}^{R-1} \frac{(-1)^l}{(2l)!} \left[\left(\frac{\partial}{\partial S} \cdot \frac{\partial}{\partial t} \right)^{2l} \Phi_K(t) \right]_{t=0} \cdot \int_{D \cap D_s} (2\pi\sigma_n^2)^{-n/2} \exp\left(- \frac{\|x - S\|^2}{2\sigma_n^2} \right) dx. \quad (35)$$

Now, represent $\Phi_K(u)$ by the power series

$$\Phi_K(u) \equiv \sum_{k_1, k_2, \cdots, k_n} b_{k_1, k_2, \cdots, k_n} u_1^{k_1} u_2^{k_2} \cdots u_n^{k_n}. \quad (36)$$

And, for $n = 0, 1, 2, \cdots$, define

$$\lambda_n(x) = (2\pi)^{-1/2} H_n(x) \exp\,(-\tfrac{1}{2}x^2)$$

$$\lambda_n'(x) = -\lambda_{n+1}(x)$$

$$\lambda_{n+1}(x) = x\,\lambda_n(x) - n\,\lambda_{n-1}(x)$$

$$\lambda_{-1}(x) \equiv (2\pi)^{-1/2} \int_x^\infty \exp\left(-\frac{1}{2}t^2 \right) dt \equiv \frac{1}{2}\,\text{erfc}\,(x/\sqrt{2}), \quad (37)$$

where H_n are the Hermite polynomials of degree n, and λ_n are the associated Hermite functions. Then (35) can be expressed as

$$P(D \cap D_s; K; R) = \sum_{k_1 + k_2 + \cdots + k_n = 0}^{R-1} b_{k_1, k_2, \cdots, k_n} (-j)^{\Sigma_i k_i} \cdot \int_{D \cap D_s} \prod_{i=1}^{n} \lambda_{k_i} [(x_i - S_i)/\sigma_n]\, dx. \quad (38)$$

Since each term of (38) contains an integration of a product of n Hermite functions, the desired approximation $P(D \cap D_s; K; R)$ of $\Pr\,[z \in D]$ is simply a weighted sum of such integrals. The weighting factors $b_{k_1, k_2, \cdots, k_n}$ can be obtained by developing recurrence relations. For specific problems the recurrence relations may be developed differentially. This subject will be postponed until Sections IV-VII.

Here the attention will be restricted to investigation of the error bounds in (25) and (31) for the special case in which the noise is Gaussian. Clearly,

$$E_1(K) = \frac{A}{4(2\pi)^n} \int_{R^n} \|u\|^2 \exp\left(-\frac{1}{2}\sigma_n^2 \|u\|^2 \right) du \cdot \sum_{k=K+1}^{L} E[\|a_k\|^2] = \frac{a^n n \sigma_n^2}{4(2\pi\sigma_n^2)^{n/2}} \sum_{k=K+1}^{L} E[\|a_k\|^2], \quad (39)$$

which decreases monotonically with K. Substituting (34) into (31) yields

$$E_2(R) = \frac{a^n}{(2\pi)^n (2R)!} E\left[\left\|\sum_1^K a_k\right\|^{2R}\right] \cdot \int_{R^n} \exp\left[-\frac{1}{2}\sigma_n^2 \|u\|^2\right] \|u\|^{2R}\, du$$

$$= \frac{a^n}{(2\pi)^{n/2} (2R)!} E\left[\left\|\sum_1^K a_k\right\|^{2R}\right] \cdot \frac{1}{2^{n/2-1}\Gamma(n/2)\sigma_n^{2R+n}} \int_0^\infty r^{2R+n+1} \exp\left[-\frac{1}{2}r^2\right] dr$$

$$= \frac{\Gamma(R+n/2)\,2^R}{\Gamma(n/2)(2R)!(2\pi)^{n/2}} \left(\frac{a}{\sigma_n}\right)^n E\left\{\left[\left\|\sum_{k=1}^K a_k\right\|^2 \Big/ \sigma_n^2\right]^R\right\}. \quad (40)$$

The RHS of (40) is a monotonically decreasing function of $R(R >> n/2)$ when R is sufficiently large. In the case that $E[\|\Sigma_{k=1}^K a_k\|^{2R}] < \sigma_n^{2R}$, the error bound $E_2(R)$ decreases faster than

$$E_2(R) \leqslant \frac{\Gamma(R+n/2)\,2^R}{(2\pi)^{n/2}\,\Gamma(n/2)\,(2R)!} \left(\frac{a}{\sigma_n}\right)^n. \quad (41)$$

Hence, when the noise is Gaussian, the truncation error bounds can be made more explicit, as illustrated in (39) and (41).

The applications part of this paper follows. It demonstrates how the above two theoretical sections can be applied to solve practical problems.

IV. Combined Effects of Gaussian Noise, Intersymbol Interference, and Co-Channel Interference on the Error Performance of MCPSK Systems

Let the M-ary coherent PSK system be depicted by Fig. 2, in which $W(t)$, a trapezoidal gating function, is defined as

$$W(t) = \begin{cases} 1, & |t| \leqslant (T-\tau_0)/2 \\ [(T+\tau_0)/2 - |t|]/\tau_0, & (T-\tau_0)/2 < |t| \leqslant (T+\tau_0)/2 \\ 0, & \text{elsewhere}, \end{cases} \quad (42)$$

where T is the inverse of symbol rate and τ_0 represents both the finite rise and delay time of the pulse. For convenience, M is assumed to be even. The transmitted symbol $\hat{\phi}_k$ is assumed to be selected with equal probability from the set $\{\pi/M + i\Omega,\ i = 0, 1, \cdots, (M-1)\}$ in which $\Omega \equiv 2\pi/M$. The receiver bandpass filter, represented by the complex envelope $q(t)$ of its impulse response, is usually matched to the transmitted signal

$$S(t) = \frac{1}{2}\sum_k \int_{-\infty}^{\infty} p(\tau)\,W(t-\tau-kT)\,d\tau \exp(j\hat{\phi}_k),$$

where $p(t)$ is the complex envelope of the impulse response of the transmit bandpass filter.

The receiver front-end noise is assumed to be zero-mean, white, stationary, and Gaussian with a two-sided power spectral density of $N_0/2$. This noise is represented by its complex envelope $N(t)$.

The lth co-channel interference can be represented by its complex envelope $A_l(t) \exp[j\theta_l(t) + j\xi_l]$, in which $A_l(t)$, $\theta_l(t)$, and ξ_l are amplitude, phase modulation, and random carrier phase, respectively. For many practical applications, the random phase ξ_l may be assumed to have a uniform density of $[0, 2\pi]$.

The noise and co-channel interference terms at the input to the decision box can be represented by the two-dimensional vectors

$$N \equiv \begin{pmatrix} n_1 \\ n_2 \end{pmatrix} \equiv \begin{Bmatrix} \mathrm{Re} \\ \mathrm{Im} \end{Bmatrix} \int_{-\infty}^{\infty} q(t_0-\tau)\,N(\tau)\,d\tau$$

$$\sum_{l=1}^{L_1} a_l \equiv \sum_{l=1}^{L_1} \begin{pmatrix} a_{l1} \\ a_{l2} \end{pmatrix} \equiv \begin{Bmatrix} \mathrm{Re} \\ \mathrm{Im} \end{Bmatrix} \int_{-\infty}^{\infty} \sum_{l=1}^{L_1} A_l(\tau) \exp[j\theta_l(\tau) + j\xi_l] \cdot q(t_0-\tau)\,d\tau, \quad (43)$$

respectively, where t_0 is the sampling time, and L_1 is the number of co-channel interferers at the input to the receiver. The first and second components of N and a_l denote the in-phase and quadrature components, respectively.

Let

$$\begin{pmatrix} R_k \\ I_k \end{pmatrix} \equiv \begin{pmatrix} R_k(t_0) \\ I_k(t_0) \end{pmatrix} \equiv \begin{Bmatrix} \mathrm{Re} \\ \mathrm{Im} \end{Bmatrix} \int_{-\infty}^{\infty}\int_{-\infty}^{\infty} p(t_0 - kT - \sigma - \tau) \cdot q(\tau)\,W(\sigma)\,d\tau\,d\sigma \quad (44)$$

$$\alpha_k \equiv \tfrac{1}{2}(R_k \cos\hat{\phi}_k - I_k \sin\hat{\phi}_k)$$
$$\beta_k \equiv \tfrac{1}{2}(R_k \sin\hat{\phi}_k + I_k \cos\hat{\phi}_k). \quad (45)$$

Then, the signal and intersymbol interference terms at the input to the decision box are

$$S \equiv \begin{pmatrix} \alpha_0 \\ \beta_0 \end{pmatrix} \quad (46)$$

$$\sum_k{}' a_k^* \equiv \sum_k{}' \begin{pmatrix} \alpha_k \\ \beta_k \end{pmatrix}, \quad (47)$$

respectively, where Σ_k' denotes summation over all integers k except $k = 0$.

Without loss of generality, the order of summation in (47) can be rearranged so that the summation is taken over all positive integers greater than L_1 and the intersymbol interference vectors (α_k, β_k) are ordered in the descending order of their magnitudes. (Note that, from (45), $\alpha_k^2 + \beta_k^2 = \frac{1}{4}(R_k^2 + I_k^2)$, $\forall\, k$. This reordering of (α_k, β_k) can obviously be performed). Let $L = L_1 + L_2$, where L_2 is the total number of intersymbol interference terms and may be ∞. Then,

$$\sum_k{}' \begin{pmatrix} \alpha_k \\ \beta_k \end{pmatrix} \equiv \sum_{l=L_1+1}^{L} \begin{pmatrix} a_{l1} \\ a_{l2} \end{pmatrix} \equiv \sum_{l=L_1+1}^{L} a_l. \quad (48)$$

Therefore, the following decision statistics can be formed at the decision box:

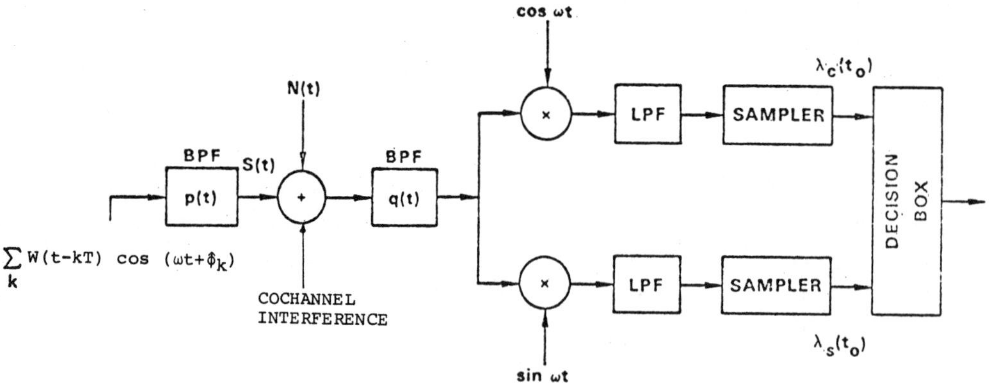

Fig. 2. Block diagram of MCPSK receiver.

$$\begin{pmatrix}\lambda_c\\ \lambda_s\end{pmatrix} \equiv z \equiv S + N + \sum_{k=1}^{L} a_k, \tag{49}$$

which is in the same form as (1) with $n = 2$.

The receiver decides that the ith symbol was sent if and only if

$$i\Omega \leqslant \underline{/(\lambda_c + j\lambda_s)} < (i+1)\,\Omega. \tag{50}$$

For illustrative purposes, assume that the co-channel interferences are random carrier waves (CW's) or angle-modulated signals with bandwidths sufficiently narrow that the receiving filter $q(t)$ has negligible distortions on them.[2] Then, $A_l(\tau) = A_l$, $\forall l \leqslant L_1$, and the kth interference signal at the output of the receiving filter $q(t)$ has a uniform phase and an amplitude

$$|a_k| = A_k \left| \int_{-\infty}^{\infty} q(t_0 - \tau)\, d\tau \right| < \infty, \qquad \forall 1 \leqslant k \leqslant L_1. \tag{51}$$

It can easily be shown that the characteristic functions associated with a_k are

$$\begin{aligned}\phi_k(u) &= J_0(a_k\sqrt{u_1^2 + u_2^2}), \qquad \text{for } 1 \leqslant k \leqslant L_1\\ &= \frac{2}{M}\sum_{i=0}^{M/2-1} \cos\left[\frac{1}{2}(R_k u_1 + I_k u_2)\cos(\pi/M + i\Omega)\right.\\ &\quad \left. + \frac{1}{2}(R_k u_2 - I_k u_1)\sin(\pi/M + i\Omega)\right], \qquad \text{for } L_1 < k \leqslant L,\end{aligned} \tag{52}$$

where J_0 is the 0th order Bessel function of the first kind; M is even; and u is a column vector whose corresponding transpose is $u^+ = (u_1, u_2)$.

Since for almost all practical systems, the filter impulse responses are bounded and also have finite energy,

$$\sup_k \|a_k\| = \max\left\{\sup_k \left(\frac{1}{2}\sqrt{R_k^2 + I_k^2}\right), \sup_l A_l\right\} < \infty$$

$$\sum_{k=1}^{L} E[\|a_k\|^2] = \frac{1}{2}\sum_{l=1}^{L_1} A_l^2 + \frac{1}{4}\sum_k{}' (R_k^2 + I_k^2) < \infty. \tag{53}$$

[2]For ASK or AM co-channel interference on M-ary coherent PSK systems, see [31, Appendix D].

Assumptions 2-4 in Section II are clearly satisfied. If the decision region D is identified as the wedge of angle Ω in Fig. 3, and if the characteristic functions ϕ_k are further equated to those in (52), then the correct decision probability can be obviously put into the form of (8); namely

$$\Pr[z \in D] = (2\pi)^{-2}\int_D dx \int_{R^2} \phi_0(u)\,\Phi_L(u) \cdot \exp[-j(x - S)\cdot u]\, du, \tag{54}$$

where

$$\Phi_L(u) = \prod_{k=1}^{L} \phi_k(u)$$

$$\phi_0(u) = \exp\left(-\frac{1}{2}\sigma_n^2 \|u\|^2\right)$$

$$\sigma_n^2 = N_0/2 \int_{-\infty}^{\infty} |q(\tau)|^2\, d\tau.$$

To compute the error probability, or equivalently, the correct decision probability, it is sometimes sufficient to consider only L_1' (instead of L_1) of the more significant co-channel interference terms and only L_2' (instead of L_2) of the more significant intersymbol interference terms. Let $K \equiv L_1' + L_2'$, and

$$\begin{aligned}\sigma^2 &\equiv \sigma_n^2 + \frac{\Delta_1}{2}\sum_{k=1}^{L_1'} A_k^2 + \frac{\Delta_2}{8}\sum_{l=1}^{L_2'} (R_l^2 + I_l^2)\\ \lambda^2 &\equiv \frac{\Delta_1}{2\sigma^2}\sum_{k=1}^{L_1'} A_k^2 + \frac{\Delta_2}{8\sigma^2}\sum_{k=1}^{L_2'} (R_k^2 + I_k^2)\\ A_{k1} &\equiv A_k/\sigma, \qquad R_{k1} \equiv R_k/\sigma, \qquad I_{k1} \equiv I_k/\sigma\\ \alpha' &\equiv \alpha_0/\sigma, \qquad \beta' \equiv \beta_0/\sigma,\end{aligned} \tag{55}$$

where $\Delta_i \in [0, 1]$ for $i = 1$ and 2 are weighting factors controlling the fraction of total interference power that can be effectively regarded as an equivalent Gaussian noise. (These factors Δ_i could be determined either by making educated guesses based upon specific problems, or by trial-and-error [16]). Moreover, construct a rectangle D_s centered at (α_0, β_0) as shown in Fig. 3. Then, the following approximation of (54)

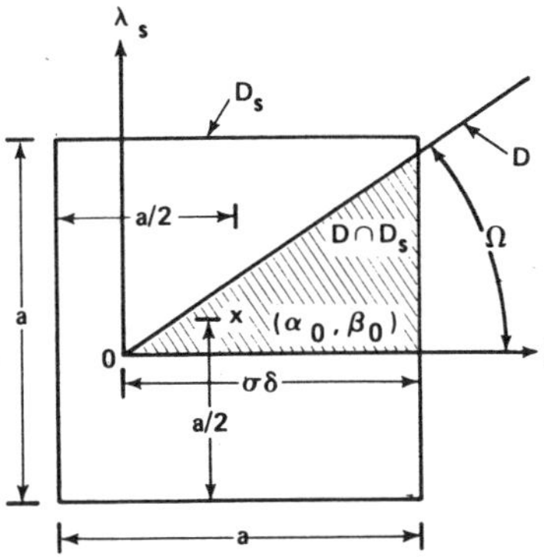

Fig. 3. Decision region for an MCPSK system.

is obtained:

$$P(D \cap D_s; K) = (2\pi)^{-2} \int_{-\infty}^{\infty} \int_{(D \cap D_s)/\sigma^2} \exp\left[-\frac{1}{2}(u_1^2 + u_2^2)\right] \cdot \Phi_{K1}(u_1, u_2) \cdot \exp\left[\frac{1}{2}\lambda^2(u_1^2 + u_2^2)\right] \cdot \exp\left[-j(x_1 - \alpha')u_1 - j(x_2 - \beta')u_2\right] dx_2\, dx_1\, du_2\, du_1, \tag{56}$$

where

$$\Phi_{K1}(u_1, u_2) \equiv \prod_{k=1}^{K} f^{(k)}(u_1, u_2), \tag{57}$$

in which

$$f^{(k)}(u_1, u_2) \equiv J_0(A_k \sqrt{u_1^2 + u_2^2}) \tag{58}$$

for $k = 1, 2, \cdots, L_1$, and

$$f^{(k)}(u_1, u_2) \equiv \frac{2}{M} \sum_{i=0}^{M/2-1} \cos(\xi_{ki} u_1 + \eta_{ki} u_2) \tag{59}$$

$$\xi_{ki} \equiv \frac{1}{2}[R_{(k-L_1')1} \cos(\pi/M + i\Omega) - I_{(k-L_1')1} \sin(\pi/M + i\Omega)] \tag{60}$$

$$\eta_{ki} \equiv \frac{1}{2}[I_{(k-L_1')1} \cos(\pi/M + i\Omega) + R_{(k-L_1')1} \sin(\pi/M + i\Omega)] \tag{61}$$

for $k = L_1' + 1, L_1' + 2, \cdots, K$.

Note that $\Phi_{K1}(u_1, u_2) \exp[\frac{1}{2}\lambda^2(u_1^2 + u_2^2)]$ is unchanged if the signs of u_1 and u_2 are changed simultaneously. Thus, the product can be expressed by

$$\Phi_{K1}(u_1, u_2) \exp\left[\frac{1}{2}\lambda^2(u_1^2 + u_2^2)\right] = \sum_{k_1, k_2, 2} b_{k_1, k_2} u_1^{k_1} u_2^{k_2}, \tag{62}$$

where $\Sigma_{k_1, k_2, 2}$ denotes the summation over all nonnegative k_1 and k_2 for $k_1 + k_2$ even. If $f^{(k)}(u_1, u_2)$ is represented by the power series

$$f^{(k)}(u_1, u_2) = \sum_{m,n,2} \tau_{m,n}^{(k)} u_1^m u_2^n, \tag{63}$$

it is easy to verify

$$\tau_{0,0}^{(k)} = 1 \quad \forall k$$

$$\tau_{2m,2n+1}^{(k)} = \tau_{2m+1,2n}^{(k)} = 0 \quad \forall k$$

$$\tau_{2m,2n}^{(k)} = \frac{(-1)^{m+n} A_{k1}^{2(m+n)}}{[(m+n)!\, 2^{m+n}]^2} \frac{(m+n)!}{m!\, n!}, \quad \text{for } k = 1, 2, \cdots, L_1'$$

$$\tau_{2m+1,2n+1}^{(k)} = 0, \quad \text{for } k = 1, 2, \cdots, L_1'$$

$$\tau_{m,n}^{(k)} = (-1)^{(m+n)/2} \frac{2}{M} \cdot \frac{1}{m!\, n!} \sum_{i=0}^{M/2-1} \xi_{ki}^m \eta_{ki}^n, \quad \text{for } k = L_1' + 1, L_1' + 2, \cdots, K. \tag{64}$$

Now, by employing the results in [16, Section V] we can find b_{k_1, k_2} according to the following recurrence relations:

$$\alpha_{0,0}^{(k)} = 1 \quad \forall k \neq 0$$

$$\alpha_{m,n}^{(k)} = \sum_{p=0}^{m} \sum_{q=0}^{n} \tau_{p,q}^{(k)} \tau_{m-q,n-q}^{(k)}, \quad \text{for even } m + n \tag{65}$$

$$\beta_{0,0}^{(k)} = 0 \quad \forall k \neq 0$$

$$\beta_{m,0}^{(k)} = \tau_{m,0}^{(k)} - \sum_{p=1}^{m} (1 - p/m) \tau_{p,0}^{(k)} \beta_{m-p,0}^{(k)}, \quad \text{for even } m > 0 \tag{66}$$

$$\beta_{0,n}^{(k)} = \tau_{0,n}^{(k)} - \sum_{q=1}^{n} (1 - q/n) \tau_{0,q}^{(k)} \tau_{0,n-q}^{(k)}, \quad \text{for even } n > 0 \tag{67}$$

$$\beta_{m,n}^{(k)} = \frac{1}{mn}\left[\lambda_{m,n}^{(k)} - \sum_{r=0}^{m} \sum_{t=0}^{n} \epsilon_{r,t} \alpha_{r,t}^{(k)} \beta_{m-r,n-t}^{(k)} (m-r)(n-t)\right], \quad \text{for } m > 0, n > 0 \text{ and even } m + n > 0, \tag{68}$$

where $\epsilon_{r,t} = 0$ if $r = t = 0$ and $\epsilon_{r,t} = 1$ if otherwise; and

$$\lambda_{m,n}^{(k)} = \sum_{p=0}^{m} \sum_{q=0}^{n} (2p - m) q \tau_{p,q}^{(k)} \tau_{m-p,n-q}^{(k)} \tag{69}$$

$$\sigma_{2,0} = \frac{1}{2}\lambda^2 + \sum_{k=1}^{K} \beta_{2,0}^{(k)}$$

$$\sigma_{0,2} = \frac{1}{2}\lambda^2 + \sum_{k=1}^{K} \beta_{0,2}^{(k)}$$

$$\sigma_{m,n} = \sum_{k=0}^{K} \beta_{m,n}^{(k)}, \quad \text{for } (m,n) \neq (2,0) \text{ or } (m,n) \neq (0,2) \tag{70}$$

$$b_{0,0} = 1$$

$$b_{k_1,0} = \sum_{m=0}^{k_1-1} \left(1 - \frac{m}{k_1}\right) \sigma_{k_1-m,0}\, b_{m,0}, \qquad \text{for even } k_1 > 0 \tag{71}$$

$$b_{0,k_2} = \sum_{n=0}^{k_2-1} \left(1 - \frac{n}{k_2}\right) \sigma_{0,k_2-n}\, b_{0,n}, \qquad \text{for even } k_2 > 0 \tag{72}$$

$$b_{k_1,k_2} = \frac{1}{k_1} \sum_{m=0}^{k_1} \sum_{n=0}^{k_2} \frac{1}{2} [1 + (-1)^{m+n}]\, m\, b_{k_1-m,k_2-n}\, \sigma_{m,n}. \tag{73}$$

Now, the truncated version $P(D \cap D_s; K; R)$ of $P(D \cap D_s; K)$ can be evaluated as follows.

$$P(D \cap D_s; K; R) = \sum_{k_1,k_2,2}^{R-1} b_{k_1,k_2} (-1)^{(k_1+k_2)/2} \cdot \frac{1}{2\pi} \iint_{(D\cap D_s)/\sigma^2} H_{k_1}(x_1 - \alpha') H_{k_2}(x_2 - \beta') \cdot \exp\left[-\frac{1}{2}\{(x_1 - \alpha')^2 + (x_2 - \beta')^2\}\right] dx_2\, dx_1$$

$$= \sum_{k_1,k_2,2}^{R-1} (-1)^{(k_1+k_2)/2} b_{k_1,k_2} \int_0^{\delta} dx_1 \cdot \lambda_{k_1}(x_1 - \alpha') \int_0^{\Gamma x_1} dx_2\, \lambda_{k_2}(x_2 - \beta'),$$

where

$$\Gamma \equiv \tan \Omega, \qquad \delta \equiv \alpha' + a/2\sigma, \tag{74}$$

and λ_n are Hermite functions. Let

$$I_{k_1,k_2}(\alpha', \beta', \rho, \delta) \equiv \int_0^{\delta} \lambda_{k_1}(x - \alpha')\, \lambda_{k_2}(\rho x - \beta')\, dx. \tag{75}$$

Then,

$$P(D \cap D_s; K; R) = \int_0^{\delta} \int_0^{\Gamma x_1} \lambda_0(x_1 - \alpha')\, \lambda_0(x_2 - \beta')\, dx_2\, dx_1 + \sum_{k_1,k_2,2}^{R-1}{}' (-1)^{(k_1+k_2)/2} b_{k_1,k_2} \cdot [I_{k_1,k_2-1}(\alpha', \beta', 0, \delta) - I_{k_1,k_2-1}(\alpha', \beta', \Gamma, \delta)]. \tag{76}$$

The prime after the summation sign in (76) denotes the exclusion of the term with $k_1 = k_2 = 0$.

The first term of (76) can be evaluated by direct integration, i.e.,

$$\int_0^{\delta} \int_0^{\Gamma x_1} \lambda_0(x_1 - \alpha')\, \lambda_0(x_2 - \beta')\, dx_2\, dx_1 = (2\pi)^{-1/2} \int_0^{\delta} \exp\left[-\frac{1}{2}(x_1 - \alpha')^2\right] \frac{1}{2} \{\text{erfc}(-\beta'/\sqrt{2}) - \text{erfc}[(\Gamma x_1 - \beta')/\sqrt{2}]\}\, dx_1. \tag{77}$$

Since, as mentioned before, the coefficients b_{k_1,k_2} can be computed by using the results in [16], the summation terms in (76) can thus be evaluated if $I_{k_1,k_2}(\alpha', \beta', \rho, \delta)$ are known.

$$I_{k,l}(\alpha', \beta', \rho, \delta) = \lambda_{k-1}(-\alpha')\, \lambda_l(-\beta') - \lambda_{k-1}(\delta - \alpha')\, \lambda_l(\rho\delta - \beta') - \Gamma I_{k-1,l+1}(\alpha', \beta', \rho, \delta). \tag{78}$$

Hence, if $I_{0,l}(\alpha', \beta', \rho, \delta)$ are known for $l = 1, 2, \cdots, R-1$, then all $I_{k,l}$ for $0 \leqslant k, l \leqslant R-1$ can be computed from the recurrence relationships in (78).

Since

$$I_{0,0}(\alpha', \beta', \rho, \delta) \equiv \int_0^{\delta} \lambda_0(x - \alpha')\, \lambda_0(\rho x - \beta')\, dx = (2\pi)^{-1} \int_0^{\delta} \exp\left[-\frac{1}{2}(x - \alpha')^2 - \frac{1}{2}(\rho x - \beta')^2\right] dx$$

$$= (2\pi)^{-1/2} (1 + \rho^2)^{-1/2} \exp[-(\alpha'\rho - \beta')^2/2(1 + m^2)] \cdot \{\lambda_{-1}[-(\alpha' + \beta'\rho)/\sqrt{1 + \rho^2}] - \lambda_{-1}[\delta\sqrt{1 + \rho^2} - (\alpha' + \beta'\rho)/\sqrt{1 + \rho^2}]\}, \tag{79}$$

it follows, by differentiating (79) l times with respect to β',

$$I_{0,l}(\alpha', \beta', \rho, \delta) \equiv \int_0^{\delta} \lambda_0(x - \alpha')\, \lambda_l(\rho x - \beta)\, dx = \sum_{p=0}^{l} \binom{l}{p} \lambda_p\left(\frac{\alpha'\rho - \beta'}{\sqrt{1 + \rho^2}}\right) \left\{\lambda_{l-p-1}\left(-\frac{\alpha' + \beta'\rho}{\sqrt{1 + \rho^2}}\right) - \lambda_{l-p-1}\left(\delta\sqrt{1 + \rho^2} - \frac{\alpha' + \beta'\rho}{\sqrt{1 + \rho^2}}\right)\right\} \frac{m^{l-p}}{(1 + \rho^2)^{(l+1)/2}}. \tag{80}$$

Indeed, $I_{0,l}(\alpha', \beta', \rho, \delta)$, for $l = 0, 1, \cdots, R-1$, can all be computed by (79) and (80) and, consequently, $I_{k,l}(\alpha', \beta', \rho, \delta)$ can be obtained from (78).

Therefore, $P(D \cap D_s; K; R)$ can be evaluated from (76), which in turn gives the desired error probability

$$P_e = 1 - \Pr(z \in D) \doteqdot 1 - P(D \cap D_s; K; R). \tag{81}$$

It seems appropriate to note that, when $L_1 = 0$, namely, when there is no co-channel interference at all, the results in this section obviously reduce to those derived for Gaussian noise and intersymbol interference only [16]. On the other hand, when there is no intersymbol interference (i.e., $L_2 = 0$), the results in this section can be simplified to those results obtained for Gaussian noise and co-channel interference only [22]. In general, the combined effects of Gaussian noise, intersymbol interference, and co-channel interference on the error probability of an MCPSK system are quite different from their separate effects, e.g., measured in terms of the degradation in signal-to-noise ratio needed to achieve a certain error performance.

It also seems appropriate to make the following remarks. It has been shown in [16] that, if the channel is symmetric, the error probability of a four-phase coherent PSK system can be derived from that of a binary coherent PSK system over the same channel, even though there is intersymbol interference in addition to Gaussian noise. The reason is simple. When $M = 4$ and when the channel is symmetric, the characteristic function $\Phi_L(u_1, u_2)$, corresponding to Gaussian noise and all intersymbol interferences can be factorized into a product of two functions. One is a function of u_1 only and another is a function of u_2 only. However, whenever there is random CW co-channel interference, the error probability of a four-phase PSK system can no longer be obtained from that of a two-phase PSK system because, in the presence of such co-channel interference, the characteristic function $\Phi_L(u_1, u_2)$ contains such factors as $J_0(A_l\sqrt{u_1^2 + u_1^2})$ for $l = 1, 2, \cdots, L_1$, which obviously cannot be factorized into products of two single-variable functions of u_1 and u_2. Hence it can be asserted that whenever there is random CW or narrow-band angle-modulated interference, the in-phase and the quadrature-phase components of a four-phase coherent PSK system are always correlated even though the channel is symmetric.

Next is a numerical example that is computed by employing the recurrence relationships given in (55), (60), (61), (64)–(74), (76)–(81).

Example 1. Four-Phase PSK/Butterworth Channel

The channel is assumed to be characterized by a three-pole Butterworth RF filter. The complex envelope of the impulse response of such a filter can be expressed by

$$\begin{aligned} h_3(t) &= 4\pi B\,[\exp(-2\pi Bt) \\ &\quad + 3^{-1/2}\exp(-\pi Bt)\sin(\sqrt{3}\,\pi Bt - \pi/3)], \qquad \text{for } t \geqslant 0 \\ &= 0, \qquad \text{for } t < 0. \end{aligned}$$

The symbol rates used in this example for the four-phase PSK signals are $2B/1.25$ and $2B$ symbols/s, respectively. Two cases of co-channel interferences are investigated: one has only a single interfering random CW($L_1 = 1$), and the other has four equal-amplitude interfering random CW's($L_1 = 4; A_l = A$; $\forall l = 1, 2, 3, 4$).

For five values of the signal-to-total co-channel interference-power ratios, S/I(= 10, 15, 20, 25, and 40 dB, respectively), the results in this section have been used to evaluate the symbol error probabilities as functions of SNR's at both signaling rates.[3] In this example the following parameters have been chosen: $K \equiv L_1' + L_2' = L_1 + L_2 = 4 + 6 = 10$, $R = 7$, $\Delta_2 = 4\Delta_1 = \Delta = 0.5$, and $a = 2\sqrt{2}$. The computed results are shown in Fig. 4(a) and (b). It can be observed from Fig. 4 that, for same S/I, four equal-amplitude random CW's degrade the system performance more than a single random CW, in agreement with the conclusions of [22]. Also, note that at high S/I (for which the co-channel interference is negligible), the error probability curves agree well with the results in [16] as expected.

[3] S/I is defined as $(R_0^2 + I_0^2)/4(\Sigma_k A_k^2)$ and SNR is defined as $(R_0^2 + I_0^2)/8\sigma_n^2$.

V. Combined Effects of Gaussian Noise, Intersymbol Interference, and Co-Channel Interference on the Error Performance of MCAPK Systems

The block diagram of the MCAPK system is identical to Fig. 2, which is for an MCPSK system. However, the M message symbols, instead of being placed on a unit circle in the in-phase and quadrature-phase plane, are now located anywhere in this plane. That is, they are now represented by the M points $(\hat{a}_k, \hat{b}_k)$ instead of by the points $(\cos\hat{\phi}_k, \sin\hat{\phi}_k)$. Of course, if $\hat{a}_k = \cos\hat{\phi}_k$, and $\hat{b}_k = \sin\hat{\phi}_k$, MCPSK can be treated as a special case of MCAPK systems. Fig. 5(a) shows the "rectangular" design of a 32-ary APK signal set, whereas Fig. 5(b), (c), and (d) illustrate the rectangular, the (4, 4) circular, and the triangular design of the 8-ary APK signal sets, respectively. Obviously, if a_k(or $\hat{b}_k$) is set equal to zero for all k and $\hat{b}_k$ (or $\hat{a}_k$) is set equal to any of a prescribed M levels, the M-ary coherent ASK [or pulse-amplitude modulation (PAM)] signal set can be treated as a special case of the MCAPK signal set. Thus, the MCAPK signals, in theory, are both amplitude-and-phase keyed (APK).

The receiver forms the same kind of decision statistics as in (49), but the $(\hat{a}_k, \hat{b}_k)$ are no longer $(\cos\hat{\phi}_k, \sin\hat{\phi}_k)$; instead they are determined by the particular APK signal set of interest. The decision regions for high SNR's generally have linear decision boundaries such as those shown in Fig. 5(a)–(d). The receiver's decision rule is simply to decide that the ith symbol was sent if and only if z falls into the ith decision region, Ω_i. Thus, a symbol error occurs if z falls outside the decision region corresponding to the correct symbol.

The average symbol error probability can be written as

$$P_e = \frac{1}{M}\sum_{i=1}^{M}(1 - P_r[\Omega_i]), \tag{82}$$

where the MCAPK message symbols are assumed to be equally likely, and $P_r[\Omega_i]$ denotes the correct decision probability for the ith symbol given that the ith symbol was sent.

Assume that there are L_1 interfering random CW's. Obviously, the characteristic function, $\psi_c(u_1, u_2)$, corresponding to these CW's is

$$\psi_c(u_1, u_2) \equiv \prod_{k=1}^{L_1}\phi_k(u_1, u_2) = \prod_{l=1}^{L_1}J_0(A_l\sqrt{u_1^2 + u_2^2}), \tag{83}$$

where $\phi_k(u_1, u_2)$ is the characteristic function of the kth co-channel interference.

For illustrative purposes, consider the 32-ary APK system with the rectangular design shown in Fig. 5(a). Since the in-phase and quadrature components of the reordered lth intersymbol interference term are

$$\begin{aligned} \alpha_l &\equiv \tfrac{1}{2}(\hat{a}_l R_l - \hat{b}_l I_l) \\ \beta_l &\equiv \tfrac{1}{2}(\hat{a}_l I_l + \hat{b}_l R_l), \end{aligned} \tag{84}$$

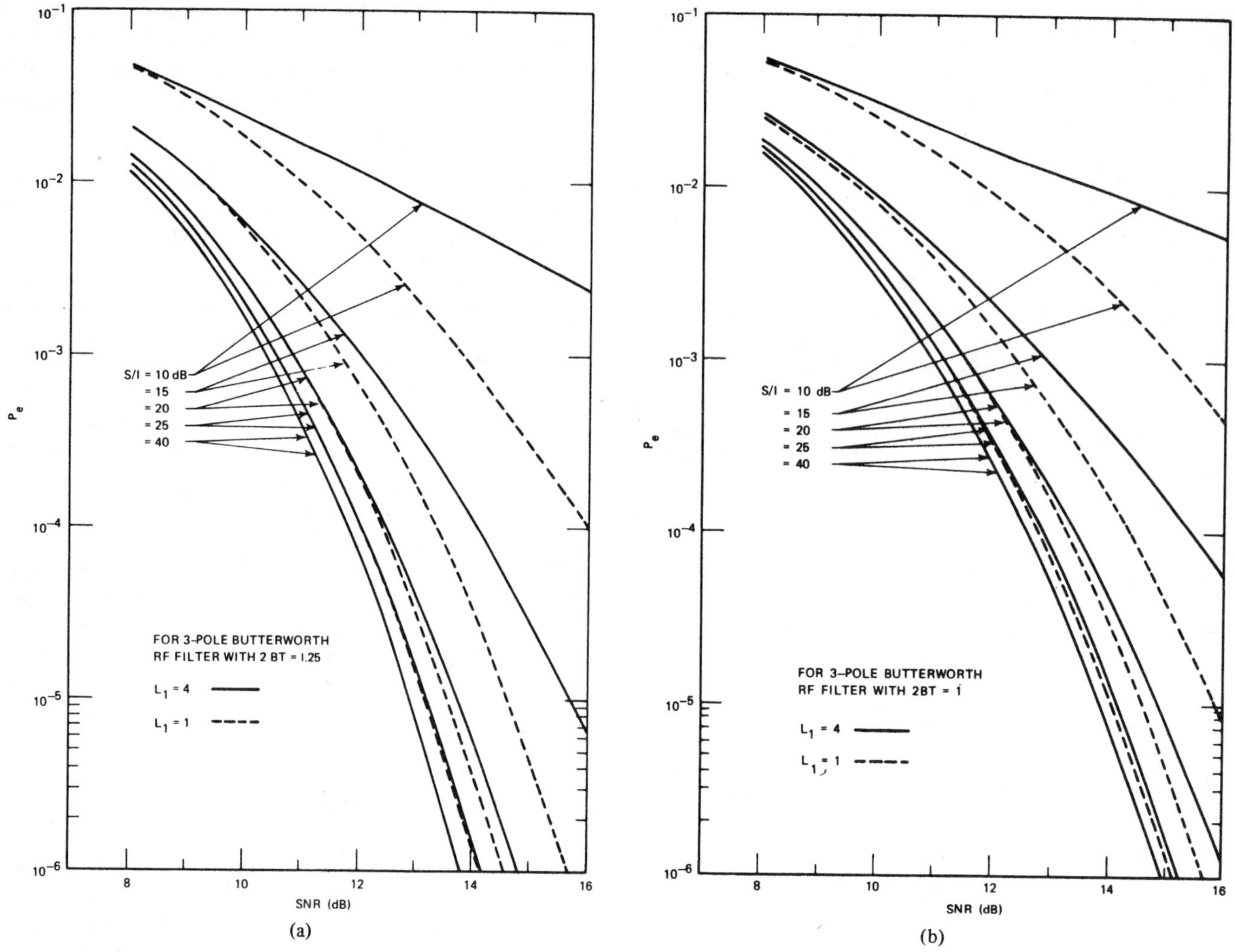

Fig. 4. (a) Probability of symbol error versus SNR, 2 BT = 1.25. (b) Probability of symbol error versus SNR, 2 BT = 1.

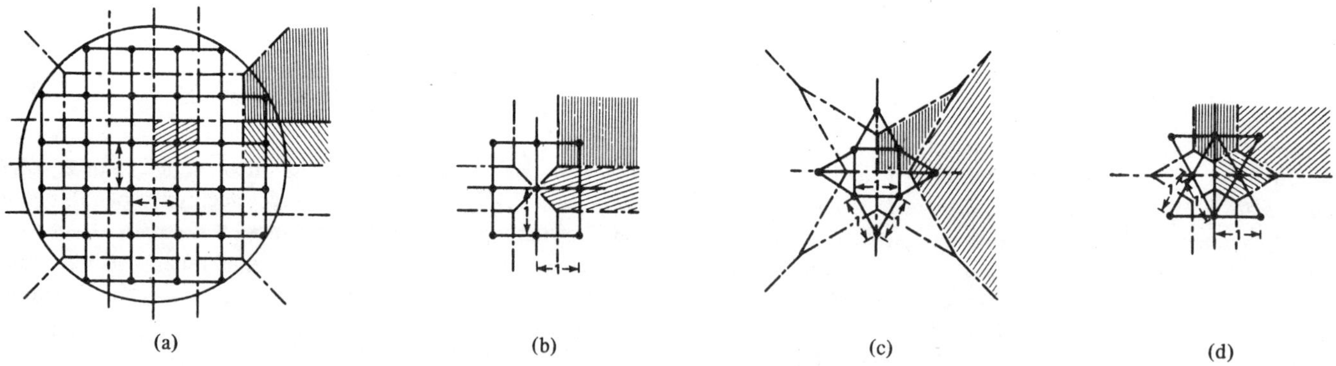

Fig. 5. APK signal sets. (a) 32-ary rectangular design, $\bar{A} = \sqrt{5}$. (b) 8-ary rectangular design, $\bar{A} = \sqrt{3}/2$. (c) 8-ary (4, 4) circular design, $\bar{A} = (7/8 + \sqrt{3}/4)^{1/2}$. (d) 8-ary triangular design, $\bar{A} = \sqrt{7}/2$.

respectively, it follows

$$E[\alpha_l^2 + \beta_l^2] = (R_l^2 + I_l^2)\bar{A}^2/4 \tag{85}$$

where

$$\bar{A} = \{E[\hat{a}_l^2 + \hat{b}_l^2]\}^{1/2} = 5^{1/2} \tag{86}$$

is the rms amplitude of the signals. It can be easily verified that the characteristic function $\psi_I(u_1, u_2)$ corresponding to the intersymbol interference is

$$\psi_I(u_1, u_2) = \prod_{k=L_1+1}^{L_1+L_2} \phi_k(u_1, u_2), \tag{87}$$

where, for $l = 1, 2, \cdots, L_2$,

$$\phi_{L_1+l}(u_1, u_2) \equiv \frac{1}{16} \sum_{i=0}^{15} \cos [u_1(\pm X_i R_l - Y_i I_l) + u_2(\pm X_i I_l + Y_i R_l)] \tag{88}$$

is the characteristic function of the lth intersymbol interference. In (88), the summation is taken over the 16 pairs of $(\pm X_i, Y_i)$: $\{(\pm 1, 1), (\pm 1, 3), (\pm 3, 1), (\pm 3, 3), (\pm 1, 5), (\pm 5, 1), (\pm 3, 5), (\pm 5, 3)\}/4\bar{A}$.

Redefine α^2, λ^2, α', β', and $\tau_{m,n}^{(k)}$ as follows instead of as in (55) and (64):

$$\sigma^2 \equiv \sigma_n^2 + \frac{\Delta_1}{2} \sum_{k=1}^{L_1'} A_k^2 + \frac{\Delta_2}{8} \bar{A}^2 \sum_{k=1}^{L_1'} (R_k^2 + I_k^2) \qquad (89)$$

$$\lambda^2 \equiv \frac{\Delta_1}{2\sigma^2} \sum_{k=1}^{L_1'} A_k^2 + \frac{\Delta_2}{8\sigma^2} \bar{A}^2 \sum_{l=1}^{L_2'} (R_l^2 + I_l^2) \qquad (90)$$

$$\alpha' \equiv (\hat{a}_0 R_0 - \hat{b}_0 I_0)/2\sigma\bar{A} \qquad (91)$$

$$\beta' \equiv (\hat{b}_0 R_0 + \hat{a}_0 I_0)/2\sigma\bar{A} \qquad (92)$$

$$\tau_{m,n}^{(k)} \equiv (-1)^{(m+n)/2} \frac{1}{16} \frac{1}{m!\,n!} \sum_{i=0}^{15} \xi_{ki}^m \eta_{ki}^n, \qquad (93)$$

for $k = L_1' + 1, L_1' + 2, \cdots, K$, where the 16 pairs of (ξ_{ki}, η_{ki}) for every k are defined by

$$(\xi_{ki}, \eta_{ki}) \equiv \{(\pm X_i R_{(k-L_1')1} - Y_i I_{(k-L_1')1}), (\pm X_i I_{(k-L_1')1} + Y_i R_{(k-L_1')1})\} \quad \text{for } i = 0, 1, \cdots, 15. \qquad (94)$$

Then the coefficients b_{k_1,k_2} can be computed with the same recurrence relationships as in (55), (60), (61), (64)-(74), (76)-(81).

Since the decision regions are assumed to have linear boundaries, they must be polygons, which of course can be partitioned into a direct sum of disjoint triangles. Through a proper choice of coordinates, the decision polygons can all be partitioned into a direct sum of triangles defined by the lines $y = \Gamma x$, $y = 0$, and $x = \delta$ in the x-y plane. The probabilities that z falls into such triangles given that the signal is at (α', β') can then be evaluated by using recurrence relationships (74) and (77)-(80). From these probabilities, the desired error probability can therefore be obtained according to (82).

Example 2. 32-ary APK System with Rectangular Signal Set

Because of the speed of logic elements in the hardware implementation, a contemplated 33.3-Msymbol/s 32-ary APK system has the following window function corresponding to a 10 percent rise time and a 10 percent delay time:

$$W(t) = \begin{cases} 1, & \text{for } |t|/T \leqslant 9/20 \\ 11/2 - 10\,|t|/T, & \text{for } 9/20 < |t|/T \leqslant 11/20 \\ 0, & \text{otherwise.} \end{cases} \qquad (95)$$

The amplitude and group-delay characteristics of the transmit and receive filters are given in Fig. 6(a) and (b). The computed average symbol error probabilities are shown in Fig. 7(a) and (b), respectively, for one-co-channel interference and for four equal-amplitude co-channel interference terms. From these curves, it can be seen that, for a given signal-to-total co-channel interference power ratio, S/I,[4] four co-channel interference terms cause more degradation in the error probabilities then a single co-channel interference term.

[4] S/I is defined as $\bar{A}^2(R_0^2 + I_0^2)/4(\Sigma_k A_k^2)$ and SNR is defined as $\bar{A}^2(R_0^2 + I_0^2)/8\sigma_n^2$.

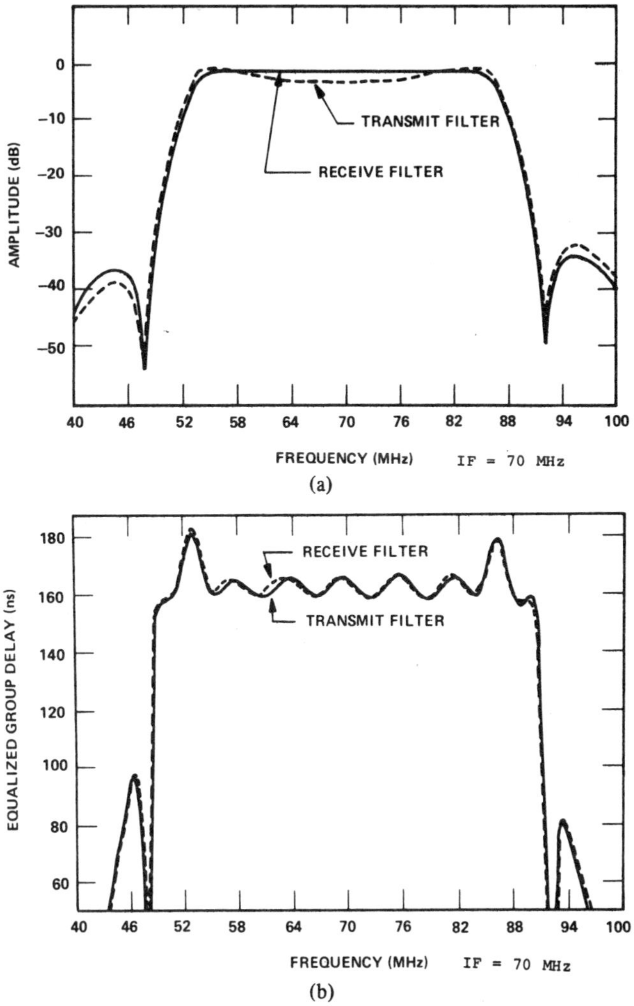

Fig. 6. (a) Amplitude characteristics of modem (elliptic) filters (adjusted to 33.3 Mbd). (b) Modem-filter group delay response (equalized).

It is an easy matter to obtain the error probabilities as functions of SNR for this APK system in the absence of intersymbol interferences by simply setting R_{k1} and I_{k1} in (95) equal to zero for all $k \neq 0$ and by setting $R_0 = 1$ and $I_0 = 0$ in (91) and (92). On the other hand, if only the effects of Gaussian noise and intersymbol interference are desired, then the co-channel interference A_k can be simply set equal to zero for all k in the computation of the coefficients b_{k_1,k_2}. These matters are trivial, however. What is important is that the error performance of APK systems in the presence of Gaussian noise, intersymbol interference, and co-channel interference can be calculated accurately by using the general approach presented here.

If the ϕ_k in (83) are replaced by the ϕ_k in [31, D-8] and the definitions given in [31, D-10–D-12] are used for σ^2, $f^{(k)}$, and $\tau_{2m,2n}^{(k)}$, respectively, then it is also possible to compute the error probabilities of APK systems in the presence of Gaussian noise, intersymbol interference, and AM co-channel interference. Of course, other extensions are just as possible, but the key is that the characteristic function of the co-channel interference terms at the sampler outputs must be obtainable.

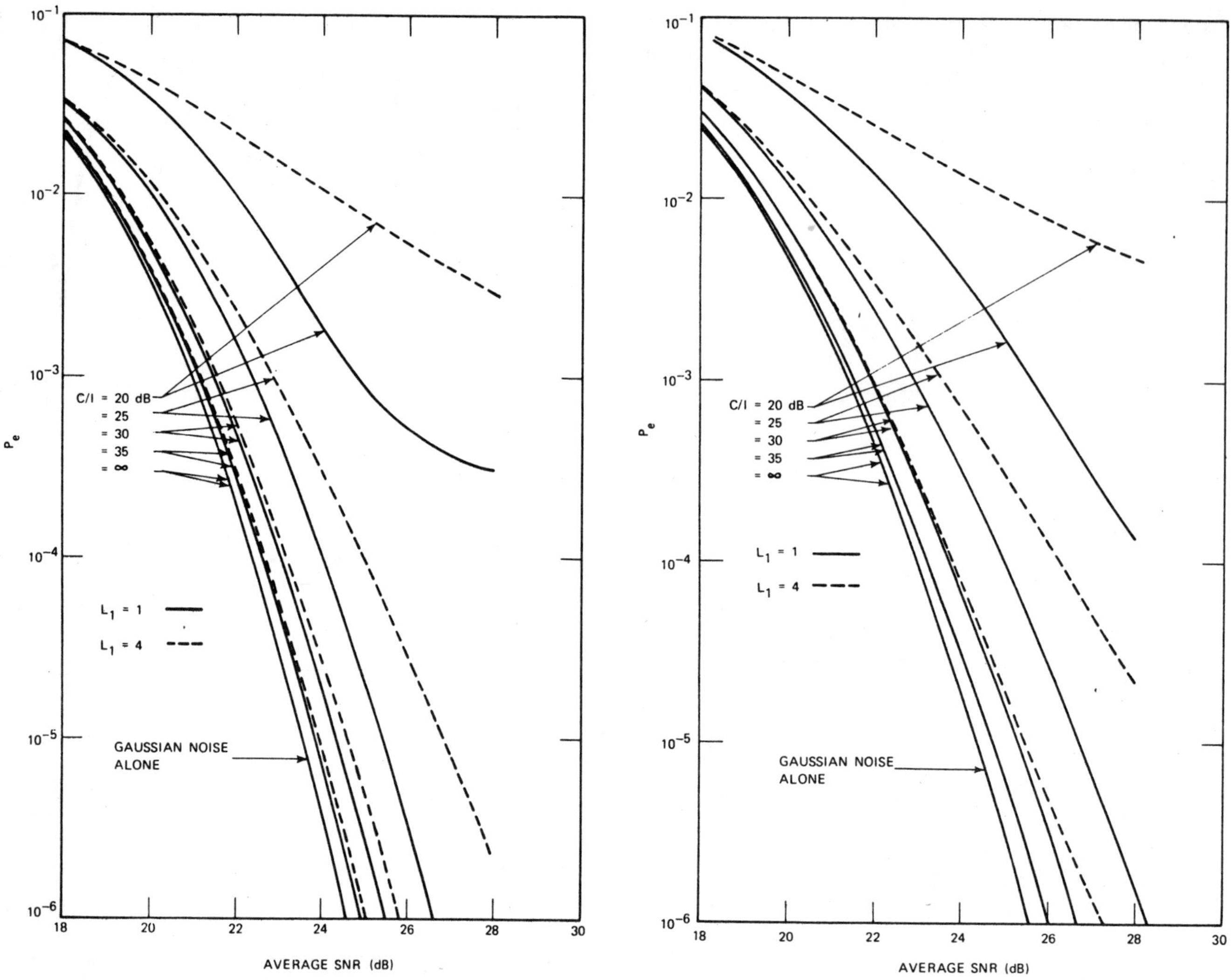

(a) (b)

Fig. 7. (a) Symbol error probability versus average SNR for a 32-ary rectangular APK signal without intersymbol interference but with cochannel interference. (b) Symbol error probability versus average SNR for a 32-ary APK signal with intersymbol interference and cochannel interference.

VI. Error Probability of a Binary DPSK System in the Presence of Gaussian Noise and Intersymbol Interferences

From [19, eq. (36)], the error probability of a binary DPSK system in the presence of Gaussian noise and intersymbol interference is

$$P_e = (1 - P_2)/2, \tag{96}$$

where

$$P_2 \equiv \frac{1}{\pi^2}\int_{-\infty}^{\infty}\int_{-\infty}^{\infty} \exp\left[-\frac{1}{2}(u_1^2 + u_2^2)\right] \{\hat{K}_1(u_1 u_2) - \rho\hat{K}_0(u_1 u_2)\} \sin(h_0 u_1 + h_1 u_2)\sin(h_{-1} u_1 + h_0 u_2) \cdot \prod_{k \neq 0,1} \cos(h_k u_1 + h_{k+1} u_2)\, du_1\, du_2. \tag{97}$$

In (97), $\hat{K}_0(x) \equiv K_0(|x|)$ and $\hat{K}_1(x) \equiv K_1(x)$ if $x \geqslant 0$, and $\hat{K}_1(x) \equiv -K_1(|x|)$ if $x < 0$; K_0 and K_1 are the zero-order and first-order modified Bessel functions of the second kind, respectively; ρ is the correlation of the circularly symmetric narrow-band, zero-mean Gaussian noise at the front end of the receiver; $h_k \equiv h_k(t_0)/\sigma$, where t_0 is the sampling time, σ is the standard deviation of the Gaussian noise; and

$$h_k(t) \equiv \int_{-\infty}^{\infty} W(\tau - kT)\, h(t - \tau)\, d\tau. \tag{98}$$

In (98) W is the window function defined in (42) with $\tau_0 = 0$, and h is half of the complex envelope of the impulse response of the symmetric bandpass channel filter.

With $\sin\alpha \sin\beta = \frac{1}{2}[\cos(\alpha - \beta) - \cos(\alpha + \beta)]$, (97) can be rewritten as

$$P_2 = \frac{1}{2\pi^2}\sum_{l=0}^{1}(-1)^l \int_{-\infty}^{\infty}\int_{-\infty}^{\infty} \exp[-(u_1^2 + u_2^2)/2]\ \{\hat{K}_1(u_1 u_2) - \rho\hat{K}_0(u_1 u_2)\} \prod_{k \neq 0,1} \cos(h_k u_1 + h_{k+1} u_2) \cdot [\cos x_l u_1 \cos y_l u_2 - \sin x_l u_1 \sin x_l u_2]\, du_1\, du_2, \tag{99}$$

where

$$x_0 \equiv h_0 - h_{-1}, \quad y_0 \equiv -h_0 + h_1$$
$$x_1 \equiv h_0 + h_{-1}, \quad y_0 \equiv h_0 + h_1.$$

Suppose that it is possible to expand

$$\prod_{k \neq 0,1} \cos(h_k u_1 + h_{k+1} u_2) = \sum_{k_1=0}^{\infty} \sum_{k_2=0}^{\infty} b_{k_1,k_2} u_1^{k_1} u_2^{k_2}, \tag{100}$$

as will be performed later. Then, since $\hat{K}_0$ is even and $\hat{K}_1$ is odd, from (99) and (100),

$$P_2 = \frac{2}{\pi^2} \sum_{p=0}^{1} (-1)^p \Bigg\{ \sum_{k_1} \sum_{k_2} b_{2k_1,2k_2} \int_0^{\infty} \int_0^{\infty} u_1^{2k_1} u_2^{2k_2} \cdot \exp[-(u_1^2+u_2^2)/2]\ [-K_1(uv) \sin x_p u_1 \sin y_p u_2 - \rho K_0(u_1 u_2) \cos x_p u_1 \cos y_p u_2]\ du_1\, du_2$$
$$+ \sum_{k_1} \sum_{k_2} b_{2k_1+1,2k_2+1} \int_0^{\infty} \int_0^{\infty} u_1^{2k+1} u_2^{2l+1} \cdot \exp[-(u_1^2+u_2^2)/2]\ [K_1(u_1 u_2) \cos x_p u_1 \cos u_p u_2 + \rho K_0(u_1 u_2) \sin x_p u_1 \sin u_p u_2]\ du_1\, du_2 \Bigg\}, \tag{101}$$

where term-wise integration can be justified by showing that the integrated series is actually absolutely convergent.

Let

$$\alpha_{0,0}(x,y) \equiv \int_0^{\infty} \int_0^{\infty} K_0(u_1 u_2) \exp[-(u_1^2+u_2^2)/2] \cdot \cos xu_1 \cos yu_2\, du_1\, du_2 \tag{102}$$

$$\beta_{0,0}(x,y) \equiv \int_0^{\infty} \int_0^{\infty} K_1(u_1 u_2) \exp[-(u_1^2+u_2^2)/2] \cdot \sin xu_1 \sin yu_2\, du_1\, du_2 \tag{103}$$

$$\alpha_{k_1,k_2}(x,y) \equiv \frac{\partial^{k_1+k_2}}{\partial_x^{k_1} \partial_y^{k_2}} \alpha_{0,0}(x,y) \tag{104}$$

$$\beta_{k_1,k_2}(x,y) \equiv \frac{\partial^{k_1+k_2}}{\partial_x^{k_1} \partial_y^{k_2}} \beta_{0,0}(x,y). \tag{105}$$

Then,

$$P_2 = \frac{2}{\pi^2} \sum_{p=0}^{1} (-1)^{p+1} \sum_{k_1,k_2,2} (-1)^{(k_1+k_2)/2}\, b_{k_1,k_2} \cdot [\beta_{k_1,k_2}(x_p, y_p) + \rho \alpha_{k_1,k_2}(x_p, y_p)], \tag{106}$$

in which α_{k_1,k_2} and β_{k_1,k_2} can be computed by using (56), (60), and [19, eq. (61)]. Hence, the question is how to compute b_{k_1,k_2}.

Without loss of generality, the product in (100) can be reordered so that the left-hand side (LHS) of (100) is a product of

$$f^{(k)}(u_1,u_2) \equiv \cos(\xi_k u_1 + \eta_k u_2) \equiv \sum_{m,n,2} \tau_{m,n}^{(k)} u_1^m u_2^n, \quad \text{for } k = 0, 1, 2, \cdots. \tag{107}$$

The reordering may be performed according to the sequence $k = -1, 2, -2, 3, -3, 4, -4, \cdots$, in (100). Obviously, then

$$\tau_{0,0}^{(k)} = 1\ \forall k$$
$$\tau_{2m,2n+1}^{(k)} = \tau_{2m+1,2n}^{(k)} = 0\ \forall k$$
$$\tau_{m,n}^{(k)} = (-1)^{(m+n)/2} \frac{1}{m!\,n!} \xi_k^m \eta_k^n, \quad \text{for all even } m+n > 0. \tag{108}$$

The recurrence relationships in (65)-(73) can thus be used to obtain the coefficients b_{k_1,k_2}.

Therefore, the error probability of a binary DPSK system in the presence of Gaussian noise and intersymbol interference can be computed.

VII. Intermodulations due to Hard-Limiting of L Signals and Noise

Let the hard-limiter be represented by

$$y = \frac{2}{\pi} \int_0^{\infty} \frac{\sin xu}{u}\, du = \begin{cases} 1, & \text{if } x > 0 \\ -1, & \text{if } x \leq 0. \end{cases} \tag{109}$$

The input x to this hard-limiter is a sum of multi-carriers plus noise:

$$x(t) = S_i(t) + n(t) \tag{110}$$

where

$$S_i(t) = \sum_{k=0}^{L} a_k(t) \cos[(\omega_c + \omega_k)t + \tilde{\phi}_k(t) + \theta_k] \tag{111}$$

$$n(t) = r(t) \cos(\omega_c t + \phi(t)) \equiv n_c(t) \cos \omega_c t - n_s(t) \sin \omega_c t. \tag{112}$$

In the above equations, a_k, $\tilde{\phi}_k$, and $\omega_c + \omega_k$ are the amplitude modulation, the phase modulation, and the carrier frequency of the kth carrier, respectively. The noise $n(t)$ is a narrowband stationary Gaussian random process with mean zero and variance σ_n^2. That is, for any t, $r(t)$ and $\phi(t)$ have a Rayleigh and a uniform density, respectively. The phase angles θ_k, for $k = 0, 1, 2, \cdots, L$, are assumed to be independent identically distributed random variables, each uniformly distributed on $[0, 2\pi]$.

Jain has shown [29] that the $s \times s$ term at the hard-limiter output is

$$S_0(t) = \frac{1}{2} \sum_{k_0,k_1,k_2,\cdots,k_L=-\infty}^{\infty} A_{k_0,k_1,\cdots,k_L} \cdot \sin\left[\sum_{l=0}^{L} k_l \left\{(\omega_c + \omega_l)t + \tilde{\phi}_l(t) + \theta_l + \frac{\pi}{2}\right\}\right], \tag{113}$$

where

$$A_{k_0,k_1,\cdots,k_L} \equiv \frac{4}{\pi} \int_0^{\infty} \frac{1}{u} \prod_{l=0}^{L} J_{k_l}(ua_l) \exp(-u^2 \sigma_n^2/2)\, du \tag{114}$$

are the amplitudes of the output signal and intermodulation components.

In (113), only those terms that fall into the channel bandwidth of the hard-limiter output filter centered at ω_c are usually of interest. Let the ideally filtered output be $S(t)$. Then,

$$S(t) = \frac{1}{2} \sum_{k_0, k_1, \cdots, k_L = -\infty}^{\infty}{}^{*} A_{k_0, k_1, \cdots, k_L} \cdot \cos\left[\omega_c t + \sum_{l=0}^{L} k_l(\omega_l t + \tilde{\phi}_l(t) + \theta_l)\right], \quad (115)$$

where the asterisk * after the summation sign denotes that the summation is taken under the condition $\Sigma_{l=1}^{L} k_l = 1$.

Suppose that $a_0(t) \cos [(\omega_c + \omega_0) t + \tilde{\phi}_0(t) + \theta_0]$ is the desired signal at the input of the limiter. Then

$$I(t) \equiv \sum_{k=1}^{L} a_k(t) \cos [(\omega_c + \omega_k) t + \tilde{\phi}_k(t) + \theta_k] + n(t) \quad (116)$$

is simply the undesired signal. As a result of hard-limiting, the term $A_{1,0,0,\cdots,0}$ represents the distortion on the pure signal $a_0(t) \cos [(\omega_c + \omega_0) t + \tilde{\phi}_0(t) + \theta_0]$ due to the presence of the undesired signal "interference."

$$A_{1,0,\cdots,0} = \frac{4}{\pi} \int_0^{\infty} \frac{1}{u} J_1(a_0 u) \left[\prod_{l=1}^{L} J_0(a_l u)\right] \exp(-u^2 \sigma_n^2/2)\, du. \quad (117)$$

However, the factor $[\Pi_{l=1}^{L} J_0(a_l u)] \exp(-u^2 \sigma_n^2/2)$ in the integral of (117) is just the characteristic function of the undesired signal $I(t)$ in (116). What it amounts to is that the hard-limiter simply "transposes" the undesired carriers $\Sigma_{k=1}^{L} \cdot a_k(t) \cos [(\omega_c + \omega_k) t + \tilde{\phi}_k(t) + \theta_k]$ onto the desired carrier, and effectively results in co-channel interference. Therefore, it should not be surprising to see that the integral in (117) also appears in the error rate analysis of the CPSK system in the presence of Gaussian noise and co-channel interference.

Consider the evaluation of (114) or the evaluation of (117) in particular. If L is a small integer (e.g., $\leqslant 3$) then after some manipulations with Bessel functions these integrals may be computed by numerical means. If L is a moderate or a large number, however, those integrals become extremely difficult to evaluate even with high-speed digital computers. Although, for example, the integral in (117) may be expressed as a generalized hyper-geometric function of L variables, this hyper-geometric function has no tabulation yet and its series expansion involves $(L + 1)$-fold summation, which is of course impractical for numerical computation. On the other hand, accurate bounds on (117) are not easy to obtain. Hence, power-series expansion method presented here will be used to integrate (117).

$$\sigma^2 \equiv \sigma_n^2 + \frac{\Delta}{2} \sum_{l=1}^{L} a_l^2$$

$$a_l' \equiv a_l/\sigma \;\; \forall l$$

$$\lambda^2 \equiv \frac{\Delta}{2} \sum_{l=1}^{L} a_l'^2 < 1. \quad (118)$$

Then,

$$A_{1,0,\cdots,0} = \frac{4}{\pi} \int_0^{\infty} \frac{1}{v} J_1(a_0' v) \left[\prod_{l=1}^{L} J_0(a_l' v)\right] \cdot \exp(\lambda^2 v^2/2) \exp(-v^2/2)\, dv. \quad (119)$$

Since $[\Pi_{l=1}^{L} J_0(a_l' v)] \exp(\lambda^2 v^2/2)$ is even in v, it can be expanded into a power series:

$$\Psi(v) \equiv \exp(\lambda^2 v^2/2) \prod_{l=1}^{L} J_0(a_l' v) = \sum_{m=0}^{\infty} b_{2m} v^{2m}. \quad (120)$$

Substitute (120) into (119) and note that [30]

$$\int_0^{\infty} x^{\mu} \exp(-\alpha x^2) J_\nu(\beta x)\, dx = \frac{\Gamma[(\mu + \nu + 1)/2]}{\beta \alpha^{\mu/2}\, \Gamma(\nu + 1)} \exp(-\beta^2/8\alpha) \cdot M_{\mu/2, \nu/2}(\beta^2/4\alpha), \quad \text{for Re } \alpha > 0 \text{ and Re } (\mu + \nu) > -1, \quad (121)$$

in which $M_{m,n}$ is the Whittaker function, yields

$$A_{100\cdots0} = \sqrt{2/\pi}\, a_0' \exp(-a_0'^2/4) [I_0(a_0'^2/4) + I_1(a_0'^2/4)] + \sum_{m=1}^{\infty} b_{2m} \pi^{-1} 2^{m-3/2} a_0'^{-1} \Gamma(m + 1/2) \cdot \exp(-a_0'^2/4) M_{m-1/2, 1/2}(a_0'^2/2). \quad (122)$$

The first term in (121) represents the amplitude distortion on the pure signal component when the L intermodulation or "interference" signals in (116) are regarded as an equivalent Gaussian noise with power equal to $\Delta/2\, \Sigma_{l=1}^{L} a_l^2$. The rest of (122) thus represents the corrections that are necessary to obtain an accurate $A_{100\cdots0}$. Hence, it is only necessary to find an effective way of computing b_{2m}.

Parallel to the derivation of the recurrence relations for $b_{k,l}$ in [16], it can be shown [31] that b_{2n} admits the following one-dimensional version of similar recurrence relations:

$$b_{2n} = \sum_{m=0}^{n-1} (1 - m/n)\, b_{2m}\, \sigma_{2(n-m)}, \quad (123)$$

where

$$\sigma_2 = \lambda^2/2 + \sum_{l=1}^{L} \beta_2^{(l)}$$

$$\sigma_{2m} = \sum_{l=1}^{L} \beta_{2m}^{(l)}, \qquad \text{for all } m \neq 1$$

$$\beta_{2m}^{(l)} = \left[\tau_{2m}^{(l)} - \sum_{n=0}^{m-1} n\, \tau_{2(m-n)}^{(l)}\, \beta_{2n}\right] \Big/ \tau_0^{(l)}$$

$$\tau_{2n}^{(l)} \equiv \frac{(-1)^n a_l'^{2n}}{(n!\, 2^n)^2}.$$

Similarly, to compute other intermodulation products in (114), the relationship in (123) can be used. Therefore, the problem of evaluating intermodulation products after hard-limiting is nothing but some form of interference and of course can be handled by our approach.

VIII. CONCLUSION

In this paper it has been shown that the wide range of interference or intermodulation problems in digital communications systems can be formulated on a general setting, and that a power series expansion method can be effectively applied to solve these problems. This is the basis of our unified analysis.

Truncation error bounds have been found. Although they are not claimed to be tight, they do show that the method presented here is theoretically rigorous. Of course, to find tighter truncation error bounds may be a significant future research subject.

The power of the unified approach has been demonstrated with examples, for which there has been no other method to provide effective solutions. More specifically, the combined effects of Gaussian noise, intersymbol interference, and cochannel interference on the error performance of MCPSK and MCAPK systems which includes the MCASK (or PAM) systems, have been evaluated. The error probability of a BDPSK system in the presence of Gaussian noise and intersymbol interference has been computed. Also, the intermodulation products at the output of a hard-limiter have been analyzed.

Finally, it should be pointed out that the Chernoff-type bound may be found for the general problem formulated in this paper. However, a really tight or "optimized" Chernoff bound may be extremely difficult to find. The effort required to find an optimized Chernoff bound may be even greater than required to implement the power series expansion approach. After all, the approach presented here provides accurate answers for almost all practical applications, as can be seen from Sections IV–VII of this paper.

APPENDIX A. EXISTENCE OF TAYLOR EXPANSION FOR $\Pi_{k=1}^{K} \phi_k(u)$

Lemma 1

If Assumptions 1–4 (see Section II of the text) are true, then the Taylor series of $\Pi_{k=1}^{K} \phi_k(u)$ exists for any finite $K \leqslant L$; i.e.,

$$\Phi_K(u) \equiv \prod_{k=1}^{K} \phi_k(u)$$

$$= \sum_{l=0}^{\infty} \frac{(-1)^l}{(2l)!} E\left[\left(u \cdot \sum_{k=1}^{K} a_k\right)^{2l}\right]$$

$$= \sum_{l=0}^{\infty} \frac{(-1)^l}{(2l)!} \left(u \cdot \frac{\partial}{\partial t}\right)^{2l} \Phi_K(t)\Big|_{t=0}. \qquad \text{(A-1)}$$

Proof: See [31, Appendix B].

APPENDIX B

Lemma 2

The integral

$$\Lambda_l \equiv \frac{1}{(2\pi)^n} \int_{D \cap D_s} dx \int_{R^n} du\, \phi_0(u) \left[\left(\frac{\partial}{\partial S} \cdot \frac{\partial}{\partial t}\right)^{2l} \Phi_K(t)\right]_{t=0} \cdot \exp\left[-j(x - S) \cdot u\right] \qquad \text{(B-1)}$$

exists for $l = 0, 1, 2, \cdots, R - 1$, and

$$\Lambda_l = \left[\left(\frac{\partial}{\partial S} \cdot \frac{\partial}{\partial t}\right)^{2l} \Phi_K(t)\right]_{t=0} \int_{D \cap D_s} p_N(x - S)\, dx, \qquad \text{(B-2)}$$

where the integral in (B-2) is as defined in (17).

Proof: See [31, Appendix C].

ACKNOWLEDGMENT

The authors would like to thank Dr. K. Bhatnagar for his computer support.

REFERENCES

[1] O. Shimbo and M. I. Celebiler, "The probability of error due to intersymbol interference and Gaussian noise in digital communication systems," *IEEE Trans. Commun. Technol.*, vol. COM-19, pp. 113–119, Apr. 1971.
[2] B. R. Saltzberg, "Intersymbol interference error bounds with application to ideal bandlimited signaling," *IEEE Trans. Inform. Theory*, vol. IT-14, pp. 563–568, July 1968.
[3] R. W. Lucky, J. Salz, and E. J. Weldon, Jr., *Principle of Data Communication*. New York: McGraw-Hill, 1968, p. 65.
[4] W. Feller, *An Introduction to Probability Theory and Its Applications*, vol. II. New York: Wiley, 1966, p. 259.
[5] R. G. Bartle, *The Elements of Real Analysis*. New York: Wiley, 1964, pp. 245, 309.
[6] S. Bochner, *Lectures on Fourier Integrals*. Princeton, N.J.: Princeton University Press, 1959, p. 318.
[7] C. R. Cahn, "Combined digital phase and amplitude modulation communication systems," *IRE Trans. Commun. Syst.*, vol. CS-8, pp. 150–155, Sept. 1960.
[8] R. W. Lucky and J. C. Hancock, "On the optimum performance of *N*-ary systems having two degrees of freedom," *IRE Trans. Commun. Syst.*, vol. CS-10, pp. 185–192, June 1962.
[9] C. N. Campopiano and B. G. Glazer, "A coherent digital amplitude and phase modulation scheme," *IRE Trans. Commun. Syst.*, vol. CS-10, pp. 90–95, Mar. 1962.
[10] E. Y. Ho and Y. S. Yeh, "Error probability of a multilevel digital system with intersymbol interference and Gaussian noise," *Bell Syst. Tech. J.*, vol. 50, pp. 1017–1023, Mar. 1971.
[11] —, "A new approach for evaluating the error probability in the presence of intersymbol interference and additive Gaussian noise," *Bell Syst. Tech. J.*, vol. 49, pp. 2249–2265, Nov. 1970.
[12] M. R. Aaron and D. W. Tufts, "Intersymbol interference and error probability," *IEEE Trans. Inform. Theory*, vol. IT-12, pp. 26–34, Jan. 1966.
[13] R. Lugannani, "Intersymbol interference and probability of error in digital systems," *IEEE Trans. Inform. Theory*, vol. IT-15, pp. 682–688, Nov. 1969.
[14] J. M. Aein and J. C. Hancock, "Reducing the effects of intersymbol interference with correlation receivers," *IEEE Trans. Inform. Theory*, vol. IT-9, pp. 167–175, July 1963.
[15] F. E. Glave, "An upper bound on the probability of error due to intersymbol interference for correlated digital signals," *IEEE Trans. Inform. Theory*, vol. IT-18, pp. 356–363, May 1972.
[16] O. Shimbo, R. J. Fang, and M. I. Celebiler, "Performance of *M*-ary PSK systems in Gaussian noise and intersymbol interference," *IEEE Trans. Inform. Theory*, vol. IT-19, pp. 44–58, Jan. 1973.
—, "Correction to 'Performance of *M*-ary PSK systems in Gaussian noise and intersymbol interference,'" *IEEE Trans. Inform. Theory* (Corresp.), vol. IT-19, p. 365, May 1973.
[17] V. K. Prabhu, "Performance of coherent phase-shift-keyed systems with intersymbol interference," *IEEE Trans. Inform. Theory*, vol. IT-17, pp. 418–431, July 1971.
[18] —, "Some consideration of error bounds in digital systems," *Bell Syst. Tech. J.*, vol. 50, pp. 3127–3151, Dec. 1971.
[19] O. Shimbo, M. I. Celebiler, and R. Fang, "Performance analysis of DPSK systems in both thermal noise and intersymbol interference," *IEEE Trans. Commun. Technol.*, vol. COM-19, pp. 1179–1188, Dec. 1971.
[20] W. M. Hubbard, "The effect of intersymbol interference on error rate in binary differentially coherent phase-shift-keyed system," *Bell Syst. Tech. J.*, vol. 46, pp. 1149–1172, July–Aug. 1967.
[21] J. J. Bussgang and M. Leiter, "Error performance of differential phase-shift transmission over a telephone line," *IEEE Trans. Commun. Technol.*, vol. COM-16, pp. 411–419, June 1968.

[22] O. Shimbo and R. Fang, "Effects of cochannel interference and Gaussian noise in *M*-ary PSK systems," *COMSAT Tech. Rev.*, vol. 3, pp. 183–207, Spring 1973.

[23] A. S. Rosenbaum, "PSK error performance with Gaussian noise and interference," *Bell Syst. Tech. J.*, vol. 48, pp. 413–442, Feb. 1969.

[24] —, "Binary PSK error probabilities with multiple cochannel interferences," *IEEE Trans. Commun. Technol.*, vol. COM-18, pp. 241–253, June 1970.

[25] V. K. Prabhu, "Error rate consideration for coherent phase-shift keyed systems with cochannel interference," *Bell Syst. Tech. J.*, vol. 48, pp. 743–767, Mar. 1969.

[26] —, "Error-probability upper bound for coherently detected PSK signals with cochannel interference," *Electron. Lett.*, vol. 5, Aug. 7, 1969.

[27] J. Goldman, "Multiple error performance of PSK systems with cochannel interference and noise," *IEEE Trans. Commun. Technol.*, vol. COM-19, pp. 420–430, Aug. 1971.

[28] J. M. Aein, "On the effects of undesired signal interference to a coherent digital carrier," Institute for Defense Analysis, Paper p-812, IDA Log HQ71-13729, Feb. 1972.

[29] P. C. Jain, "Limiting of signals in random noise," *IEEE Trans. Inform. Theory*, vol. IT-18, pp. 332–340, May 1972.

[30] I. S. Gradshteyn and I. M. Ryzhik, *Table of Integrals, Series and Products* (4th ed. prepared by Yu. V. Geronimus and M. Yu. Tseytlin). New York: Academic, 1965.

[31] R. Fang and O. Shimbo, "Unified analysis of a class of digital systems in additive noise and interferences," Comsat Tech. Memorandum, CL-57-72, Dec. 18, 1972.

Index: telecommunications, modulation, phase-shift keying, random noise, electromagnetic interference, probable error.

Effects of cochannel interference and Gaussian noise in M-ary PSK systems

O. SHIMBO AND R. FANG

Abstract

The power series expansion technique can be used to develop an effective computational procedure for analyzing the combined effects of Gaussian noise and cochannel interference in M-ary coherent PSK systems. This procedure can be easily implemented on the computer. As numerical examples, the error probabilities of 4-, 8-, and 16-phase PSK systems as a result of impairments caused by Gaussian noise and one, two, three, and four equal-strength interferences are evaluated. Tradeoffs in system design between carrier-to-noise ratio and carrier-to-interference ratio can thus be made to achieve a given error probability performance for a given number of interferences.

Introduction

In some satellite communications environments, the effect of cochannel interference in addition to that of Gaussian noise can be quite significant. For instance, in the geostationary orbit, the effect of cochannel interference has been an important factor in efficient utilization of the available "parking space" in the "parking window." Also, for the case in which an earth station of a satellite system operates near some terrestrial communications facilities that share the same frequency spectrum, the resulting performance degradation in either facility must be accurately estimated. In this paper, the combined effects of Gaussian noise and cochannel interference on the error probability performance of M-ary coherent PSK systems are to be evaluated.

This problem has been studied previously by many investigators [1]–[7]. However, all of them either attack a less general problem or are unable to provide accurate numerical results for the general M-ary (e.g., 8- and 16-phase) PSK systems. Modification of the results in Reference 8 indicates that the combined effects of Gaussian noise and cochannel interference on the probability of error of M-ary PSK systems can be obtained rather easily.

First, the necessary modifications to the analyses in Reference 8 are adapted to solve the problem presented here. Then a modified computational procedure, which is implemented on COMSAT's computer, is described. Typical results are presented in figures which show the combined effects of Gaussian noise and one, two, three, and four equal-strength cochannel interference entries on the error probabilities of 4-, 8-, and 16-phase systems.*

Analysis

It is assumed that the receiver is an ideal PSK receiver as shown in Figure 1. It is further assumed that the filter in Figure 1 does not distort either the main signal or the cochannel interference. The case in which the filter produces intersymbol interference on the desired signal is outside the scope of the present paper and will be reported elsewhere.

The output signal of the filter in Figure 1 can be represented by

$$R(t) = A \sin (\omega_c t + \theta) + N_c(t) \cos \omega_c t + N_s(t) \sin \omega_c t + \sum_{l=1}^{H} B_l \sin (\omega_l t + \phi_l + \lambda_l) \quad (1)$$

This paper is based upon work performed at COMSAT Laboratories under the sponsorship of the International Telecommunications Satellite Organization (INTELSAT). Views expressed in this paper are not necessarily those of INTELSAT.

* For convenience, the cochannel interferences are assumed to have equal strength in all illustrations, but this assumption is by no means necessary, as can be seen from the analysis in the next section.

Reprinted with permission from *COMSAT Tech. Rev.*, vol. 3, pp. 183–206, Spring 1973.

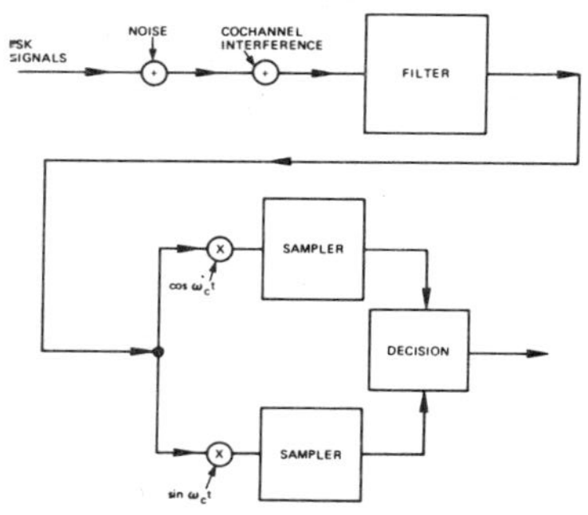

Figure 1. *Receiving Scheme of PSK Systems*

where
A = amplitude of the desired signal
ω_c = angular carrier frequency of the desired signal
θ = modulating angle of the desired signal
N_c = in-phase component of noise
N_s = quadrature component of noise
B_l = amplitude of the lth carrier
ω_l = angular frequency of the lth carrier
ϕ_l = modulating angle of the lth carrier
λ_l = random phase of the lth carrier
H = number of cochannel interferences.

If it is assumed that θ is equally likely to be any of the M phases, the probability with which the received signal point P falls outside the decision cone D in Figure 2 represents the desired error probability. According to

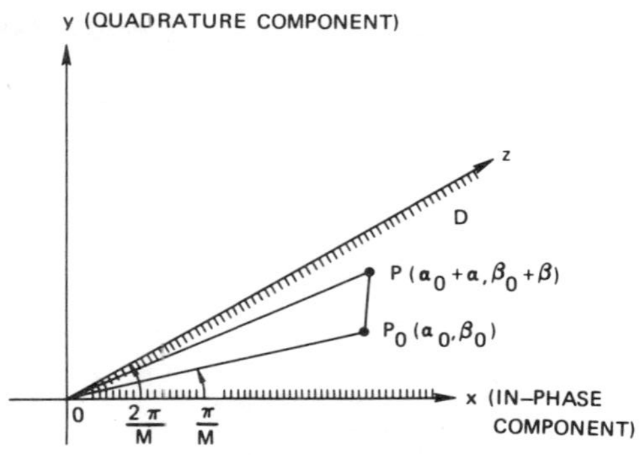

Figure 2. *Signal Space of PSK Systems*

the analysis given in Reference 8, the probability with which P falls inside D is

$$P_c = (2\pi\sigma_n^2)^{-1} \underset{\alpha,\beta}{E} \left[\int_0^\infty \int_0^{\Gamma x} \exp\left\{ -\frac{1}{2\sigma_n^2} [(x - \alpha_0 - \alpha)^2 + (y - \beta_0 - \beta)^2] \right\} dx\, dy \right] \quad (2)$$

where

σ_n^2 = baseband power of $N_c(t)$ or $N_s(t)$

$$\alpha_0 = A \cos\left(\frac{\pi}{M}\right)$$

$$\beta_0 = A \sin\left(\frac{\pi}{M}\right)$$

$$\Gamma = \tan\left(\frac{2\pi}{M}\right)$$

$$\alpha = \sum_{l=1}^{H} B_l \cos [(\omega_l - \omega_c)\, t + \phi_l + \lambda_l]$$

$$\beta = \sum_{l=1}^{H} B_l \sin [(\omega_l - \omega_c)\, t + \phi_l + \lambda_l]$$

$\underset{\alpha,\beta}{E}$ = averaging with respect to α and β.

Rewriting equation (2) in terms of the characteristic function Φ_0 of the cochannel interference yields

$$P_c = (2\pi)^{-2} \int_{-\infty}^{\infty} \int_{-\infty}^{\infty} \int_0^{\infty} \int_0^{\Gamma x} e^{-j(x-\alpha_0)u - j(y-\beta_0)v} \cdot e^{-1/2\, \sigma_n^2(u^2+v^2)}\, \Phi_0(u, v)\; du\, dv\, dx\, dy \quad (3)$$

where

$$\Phi_0(u, v) \equiv \underset{\alpha,\beta}{E} \left\{ e^{ju\alpha + jv\beta} \right\} \quad (4)$$

If it is assumed that the random phases of the cochannel interference ϕ_l are mutually independent and uniformly distributed in $(0,2\pi)$, then $\Phi_0(u,v)$ can be expressed as

$$\Phi_0(u, v) = \prod_{l=1}^{H} J_0 \left(B_l \sqrt{u^2 + v^2} \right) \quad (5)$$

Note that the form of equation (3) is identical to that of equation (16) in Reference 8 except that $\Phi_0(u,v)$ is now given by equation (5) instead of by equations (17) and (18) in Reference 8.

To achieve faster convergence in the evaluation of error probabilities via the power series expansion method in Reference 8, define

$$\sigma^2 = \sigma_n^2 + \frac{\Delta}{2} \sum_{l=1}^{H} B_l^2 \tag{6}$$

Hence, σ^2 can be interpreted as the sum of the RF noise power and a part of the RF cochannel interference power (with the fraction Δ). With this definition, equation (3) can be modified as follows:

$$P_e = (2\pi)^{-2} \int_{-\infty}^{\infty} \int_{-\infty}^{\infty} \int_0^{\infty} \int_0^{\Gamma x} e^{-1/2\,(u^2+v^2)} \cdot e^{-j(x-\alpha_1)u - j(y-\beta_1)v}\, \Phi_1(u, v) \cdot e^{\lambda^2/2\,(u^2+v^2)}\, du\, dv\, dx\, dy \tag{7}$$

where

$$\alpha_1 = \frac{\alpha_0}{\sigma} = \frac{A}{\sigma} \cos\left(\frac{\pi}{M}\right) = \frac{A \cos(\pi/M)}{\sqrt{\sigma_n^2 + \frac{\Delta}{2} \sum_{l=1}^{H} B_l^2}} \tag{8a}$$

$$\beta_1 = \frac{\beta_0}{\sigma} = \frac{A}{\sigma} \sin\left(\frac{\pi}{M}\right) = \frac{A \sin(\pi/M)}{\sqrt{\sigma_n^2 + \frac{\Delta}{2} \sum_{l=1}^{H} B_l^2}} \tag{8b}$$

$$\lambda^2 = \frac{\frac{\Delta}{2} \sum_{l=1}^{H} B_l^2}{\sigma^2} = \frac{\frac{\Delta}{2} \sum_{l=1}^{H} B_l^2}{\sigma_n^2 + \frac{\Delta}{2} \sum_{l=1}^{H} B_l^2} \tag{8c}$$

$$\phi_1(u, v) = \prod_{l=1}^{H} J_0 \left(\frac{B_l}{\sigma} \sqrt{u^2 + v^2}\right) \tag{9}$$

Now expand

$$\Phi_1(u, v)\, e^{\lambda^2/2(u^2+v^2)}$$

into a power series:

$$\phi_1(u, v)\, e^{(\lambda^2/2)(u^2+v^2)} = \sum_{m=0}^{\infty} \sum_{n=0}^{\infty} b_{2m,\,2n}\, u^{2m}\, v^{2n} \tag{10}$$

Define the Hermite functions $\phi_n(x)$ as

$$\phi_n(x) = (2\pi)^{-1/2} H_n(x) \exp\left(\frac{-x^2}{2}\right) \tag{11}$$

for $n = 0, 1, 2, \ldots$, where $H_n(x)$ is the Hermite polynomial of degree n. These Hermite functions admit the following recurrence relationships:

$$\phi_n'(x) = -\phi_{n+1}(x) \tag{12a}$$

$$\phi_{n+1}(x) = x\phi_n(x) - n\phi_{n-1}(x) \tag{12b}$$

where the prime denotes the differentiation with respect to x and

$$\phi_{-1}(x) \equiv (2\pi)^{-1/2} \int_x^{\infty} \exp\left(\frac{-t^2}{2}\right) dt \equiv \frac{1}{2} \operatorname{erfc}\left(\frac{x}{\sqrt{2}}\right) \tag{12c}$$

Hence, substitution of equations (10)–(12) into equation (7) yields

$$P_e = \int_0^{\infty} \int_0^{\Gamma x} \phi_0(x - \alpha_1)\, \phi_0(y - \beta_1)\, dy\, dx + \sum_{m,n}{}' (-1)^{m+n}\, b_{2m,\,2n} [\phi_{2m-1}(-\alpha_1)\, \phi_{2n-1}(-\beta_1) - I_{2m,\,2n-1}(\alpha_1, \beta_1)] \tag{13}$$

where the prime after the Σ denotes the exclusion of the term with $m = n = 0$, where $b_{0,0} = 1$, as can easily be seen from equations (9) and (10), and where

$$I_{m,n}(\alpha_1, \beta_1) \equiv \int_0^{\infty} \phi_m(x - \alpha_1)\, \phi_n(\Gamma x - \beta_1)\, dx \tag{14}$$

The quantity in equation (14) can be evaluated by using the recurrence method. The recurrence relationships for computing $I_{m,n}(\alpha_1,\beta_1)$ have been derived in Reference 8 and are as follows:

$$I_{m,n}(\alpha_1, \beta_1) = \phi_{m-1}(-\alpha_1)\, \phi_l(-\beta_1) - \Gamma\, I_{m-1,n+1}(\alpha_1, \beta_1) \tag{15}$$

for $M > 4$, and

$$I_{0,n}(\alpha_1, \beta_1) = \sum_{l=0}^{n} \binom{n}{l} \phi_l \left(\alpha_1 \sin \frac{2\pi}{M} - \beta_1 \cos \frac{2\pi}{M}\right) \left(\cos \frac{2\pi}{M}\right)^{l+1}$$
$$\cdot \left[\delta_{n,l} - \phi_{n-l-1} \left(\alpha_1 \cos \frac{2\pi}{M} + \beta_1 \sin \frac{2\pi}{M}\right)\right.$$
$$\left.\cdot \left(-\sin \frac{2\pi}{M}\right)^{n-l}\right] \tag{16}$$

in which

$$\delta_{n,l} = 1 \text{ if } n = l$$
$$= 0 \text{ if } n \neq l$$

If $M = 4$, then the term $I_{2m,2n-1}(\alpha_1,\beta_1)$ in equation (13) should be set equal to zero and computation of any $I_{m,n}(\alpha_1,\beta_1)$ is unnecessary.

Computation of the coefficients $b_{2m,2n}$ is slightly different from the computation in Reference 8. The new computational procedure will be provided in the next section so that the Σ' term in equation (13) can be evaluated. The first term in equation (13) can be computed by direct integration or by the following approximation derived in Reference 8:

$$\int_0^\infty \int_0^{\Gamma x} \phi_0(x - \alpha_1)\, \phi_0(y - \beta_1)\; dy\, dx$$
$$= \frac{1}{2} \operatorname{erfc}[2^{-1/2}\, C \cos \theta_2] + \frac{1}{2} \operatorname{erfc}[-2^{-1/2}\, C \cos \theta_1]$$
$$+ \exp\left(-\frac{1}{2} C^2\right) \Big\{(4\pi d)^{-1}\, C^2 \sin(\theta_1 - \theta_2) \cos(\theta_1 + \theta_2)$$
$$- (2Md)^{-1} C^2 + (4\pi d)^{-1} C\, [\cos \theta_2 (2d + C^2 \sin^2 \theta_2)^{1/2}$$
$$- \cos \theta_1 (2d + C^2 \sin^2 \theta_1)^{1/2}] + (4\pi d)^{-1} (C^2 + 2d)$$
$$\cdot \left[\sin^{-1}\left\{\frac{C \cos \theta_2}{(C^2 + 2d)^{1/2}}\right\} - \sin^{-1}\left\{\frac{C \cos \theta_1}{(C^2 + 2d)^{1/2}}\right\}\right]\Big\} \tag{17}$$

where $2 \geq d \geq 4/\pi$, and

$$C \equiv (\alpha_1^2 + \beta_1^2)^{1/2} \qquad \theta_1 = \frac{\pi}{M} + \frac{2\pi}{M} - \theta_0$$
$$\theta_0 = \tan^{-1}\left(\frac{\beta_1}{\alpha_1}\right) \qquad \theta_2 = \frac{\pi}{2} - \theta_0 \quad .$$

Therefore, the error probability

$$P_e \equiv 1 - P_c \tag{18}$$

can be numerically computed from equations (18) and (13).

Recurrence relationship for $b_{2m,\,2n}$

Denote the total RF-interference-to-RF-noise power ratio as k, and define L as the ratio of the carrier power to the sum of the RF noise power and the total RF interference power. Namely,

$$k = \sum_{l=1}^{H} \frac{B_l^2}{2\sigma_n^2} \tag{19a}$$

$$L \equiv \frac{A^2}{2\left(\sigma_n^2 + \frac{1}{2}\sum_{l=1}^{H} B_l^2\right)} = \frac{C/N}{1+k} \quad . \tag{19b}$$

Then, equation (8) can be rewritten as

$$\alpha_1 = \left[\frac{2L(1+k)}{1+\Delta k}\right]^{1/2} \cos\left(\frac{\pi}{M}\right) \tag{20a}$$

$$\beta_1 = \left[\frac{2L(1+k)}{1+\Delta k}\right]^{1/2} \sin\left(\frac{\pi}{M}\right) \tag{20b}$$

$$\lambda^2 = \frac{\Delta k}{1+\Delta k} \quad . \tag{20c}$$

Assuming all interferences to be equal (otherwise only eqs. 27a and b would be modified),

$$B = B_1 = B_2 = \ldots = B_H \quad . \tag{21}$$

Thus, equations (9) and (10) indicate that

$$\Phi_1(u, v) \exp\left[\frac{1}{2}\lambda^2(u^2 + v^2)\right]$$
$$= \left[J_0\left\{\left[\frac{2k}{H(1+\Delta k)}\right]^{1/2} [u^2 + v^2]^{1/2}\right\}\right]^H \exp\left[\frac{1}{2}\lambda^2(u^2 + v^2)\right]$$
$$= \sum_{m,n} b_{2m,2n}\, u^{2m}\, v^{2n} \quad . \tag{22}$$

Two methods can be employed to obtain $b_{2m,2n}$: the recurrence method and the convolution method.

Recurrence method

Define

$$\gamma = (u^2 + v^2)^{1/2}$$

and

$$J_0\left\{\left[\frac{2k}{H(1+\Delta k)}\right]^{1/2}\gamma\right\} = \sum_{m=0}^{\infty} C_{2m}\,\gamma^{2m} \quad (23)$$

where

$$C_{2m} = (-1)^m\,(2^m\,m!)^{-2}\left[\frac{2k}{H(1+\Delta k)}\right]^m \quad (24)$$

Then, equation (22) can be put into the following form:

$$\Phi_1(u, v)\exp\left[\frac{1}{2}\lambda^2(u^2+v^2)\right]$$

$$= \left[J_0\left\{\left[\frac{2k}{H(1+\Delta k)}\right]^{1/2}\gamma\right\}\right]^H \exp\left(\frac{1}{2}\lambda^2\gamma^2\right) \equiv \sum_{m=0}^{\infty} d_{2m}\,\gamma^{2m} \quad (25)$$

Since $\exp[(1/2)\,\lambda^2\gamma^2]$ can be expanded into the power series

$$\exp\left(\frac{1}{2}\lambda^2\gamma^2\right) = \sum_{l=0}^{\infty}\frac{1}{l!}\left(\frac{1}{2}\lambda^2\right)^l\gamma^{2l} \quad (26)$$

the coefficients d_{2m} can be obtained by comparing the coefficients of γ^{2m} on both sides of equation (25) after substituting equations (23) and (26) into equation (25). Hence, for $i \neq 0$,

$$d_{2i} = \frac{1}{2i}\sum_{p=0}^{i-1}[2H(i-p)\;C_{2i-2p} + \lambda^2 C_{2i-2p-2} - (2p)\;C_{2i-2p}]\;d_{2p} \quad (27a)$$

$$d_0 = 1 \quad . \quad (27b)$$

Now, equation (25) can be written as

$$\Phi_1(u, v)\exp\left[\frac{1}{2}\lambda^2(u^2+v^2)\right] = \sum_{m=0}^{\infty} d_{2m}\,\gamma^{2m}$$

$$= \sum_{p=0}^{\infty} d_{2p}(u^2+v^2)^p$$

$$= \sum_{m,n} b_{2m,2n}\,u^{2m}\,v^{2n} \quad (28)$$

Using binominal expansion on $(u^2 + v^2)^p$ in equation (28) and comparing coefficients yields

$$b_{2m,2n} = d_{2(m+n)}\,\frac{(m+n)!}{m!n!} \quad (29)$$

Convolution method

From equations (22), (23), and (25), it is obvious that d_{2m} can be computed by convolving C_{2m} H times and then convolving with the coefficients in equation (26). That is,

$$C_{2k}^{(1)} = \sum_{m=0}^{k} C_{2m}\,C_{2k-2m} \quad (30a)$$

$$C_{2k}^{(2)} = \sum_{m=0}^{k} C_{2m}\,C_{2k-2m}^{(1)} \quad (30b)$$

$$C_{2k}^{(H-1)} = \sum_{m=0}^{k} C_{2m}\,C_{2k-2m}^{(H-2)} \quad (30c)$$

$$d_{2i} = \sum_{m=0}^{i} C_{2i-2m}^{(H-1)}\left(\frac{\lambda^2}{2}\right)^m\frac{1}{m!} \quad . \quad (30d)$$

Computational procedure

The following is a modified computational procedure which is implemented on COMSAT's computer to determine the effects of Gaussian noise and up to four equal-strength cochannel interferences on the error probabilities of 4-, 8-, and 16-phase PSK systems:

a. Decide the values of M, H, k, L, and Δ. (To choose Δ, see Reference 8.)
b. Compute α_1, β_1, and λ^2.

c. Compute $1 - \int_0^\infty \int_0^{\Gamma x} \phi_0(x - \alpha_1)\ \phi_0(y - \beta_1)\ dy\ dx$ from equation (17).

d. Compute $b_{2m,2n}$ from equations (24), (27), and (29) or from equations (29) and (30).

e. Compute $\phi_n(-\alpha_1)$ and $\phi_n(-\beta_1)$ from the recurrence relationship in equation (12).

f. Compute $I_{m,n}(\alpha_1,\beta_1)$ from equation (15) if $M > 4$. (If $M = 4$, this computation is unnecessary.)

g. Compute $P_e = 1 - P_c$ from equation (13).

h. If the convergence in the computation of equation (13) is not fast enough, adjust the value of Δ to obtain a faster convergence (see Reference 8).

Numerical results and discussion

Figures 3–5 show the numerical results for one, two, three, and four cochannel interferences in 4-, 8-, and 16-phase PSK systems, respectively. In all figures, the vertical axes represent the error probabilities, the horizontal axes represent L in dB, and the parameters represent k in dB. Since k, as defined in equation (19a), is the total RF-interference-to-RF-noise power ratio, the two extrema with $k = +\infty$ dB and $-\infty$ dB correspond to noise-free and interference-free situations, respectively. Thus, the curves associated with $k = -\infty$ dB in all figures should be identical to those error probabilities resulting from Gaussian noise alone. Consequently, the curves with $k = -\infty$ dB in Figure 3 are identical, as are the curves with $k = -\infty$ in Figure 4 and Figure 5.

On the other hand, for the noise-free cases in which $k = +\infty$ dB, the error probabilities are always equal to zero for L above certain thresholds, since the amplitudes of these interferences are bounded. (When the carrier-to-total-interference power ratio exceeds a certain threshold, the received signal will always lie in the correct decision region and hence the error probability will be zero.) These thresholds are determined by the number of cochannel interferences for a given M-ary PSK system. For instance, it can be seen from Figure 3 that, for the 4-phase PSK system, the thresholds are 3, 6, 8, and 9 dB, respectively, as a result of one, two, three, and four cochannel interferences.

Also, from these sets of figures, it can be deduced that, for a fixed number of phases, M, a fixed interference-to-noise power ratio, k, and a fixed

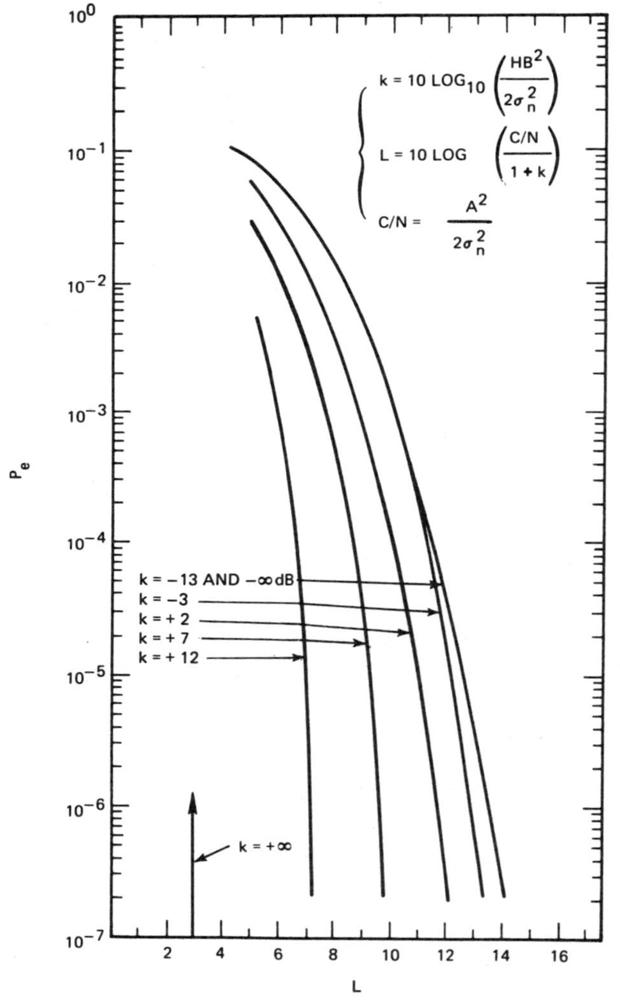

Figure 3a. *Error Probabilities of 4-Phase PSK in the Presence of Gaussian Noise and One Cochannel Interference ($H = 1, 4\phi$)*

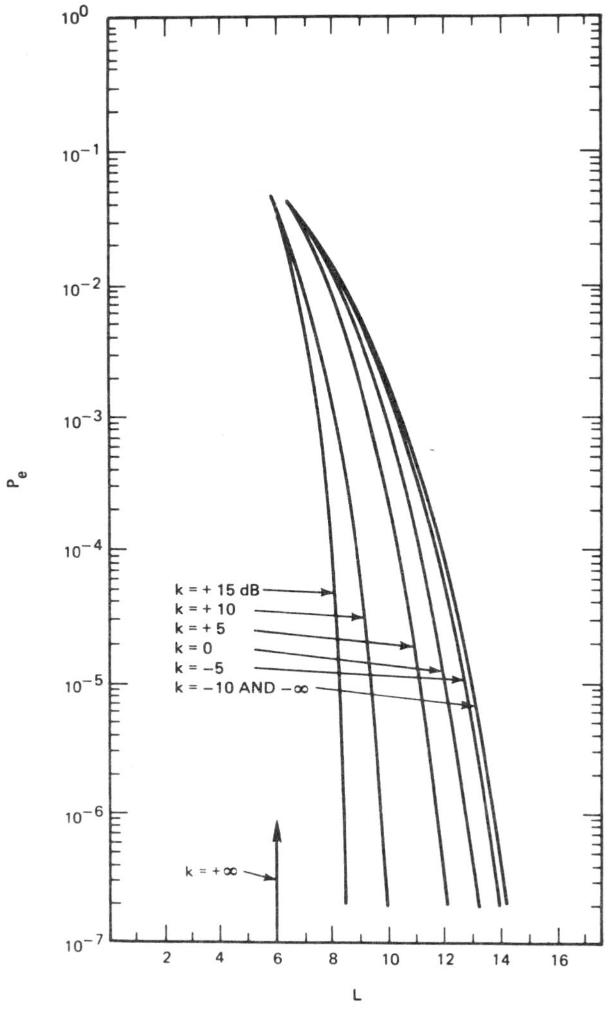

Figure 3b. *Error Probabilities of 4-Phase PSK in the Presence of Gaussian Noise and Two Cochannel Interferences* ($H = 2, 4\phi$)

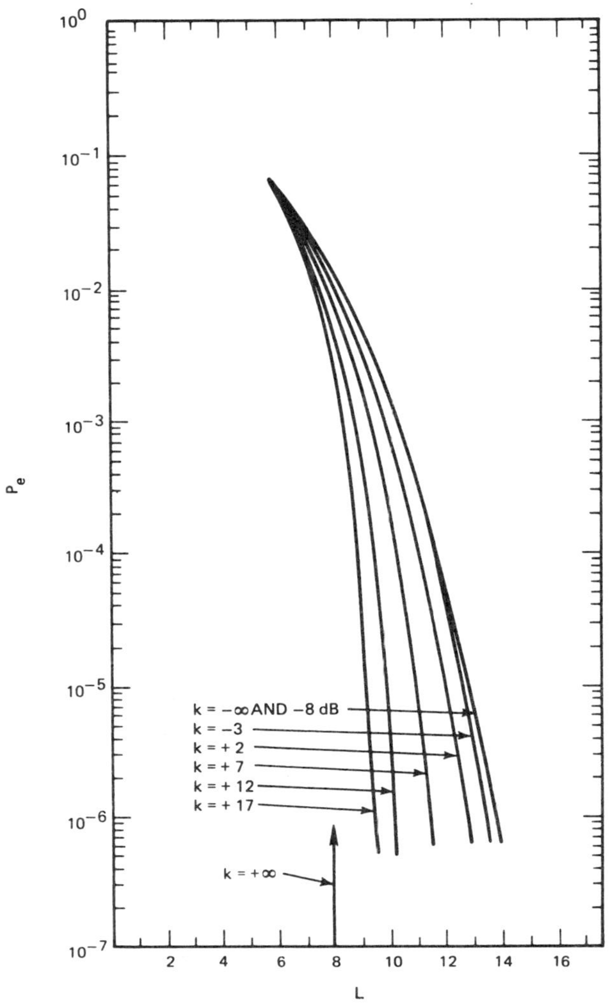

Figure 3c. *Error Probabilities of 4-Phase PSK in the Presence of Gaussian Noise and Three Cochannel Interferences* ($H = 3, 4\phi$)

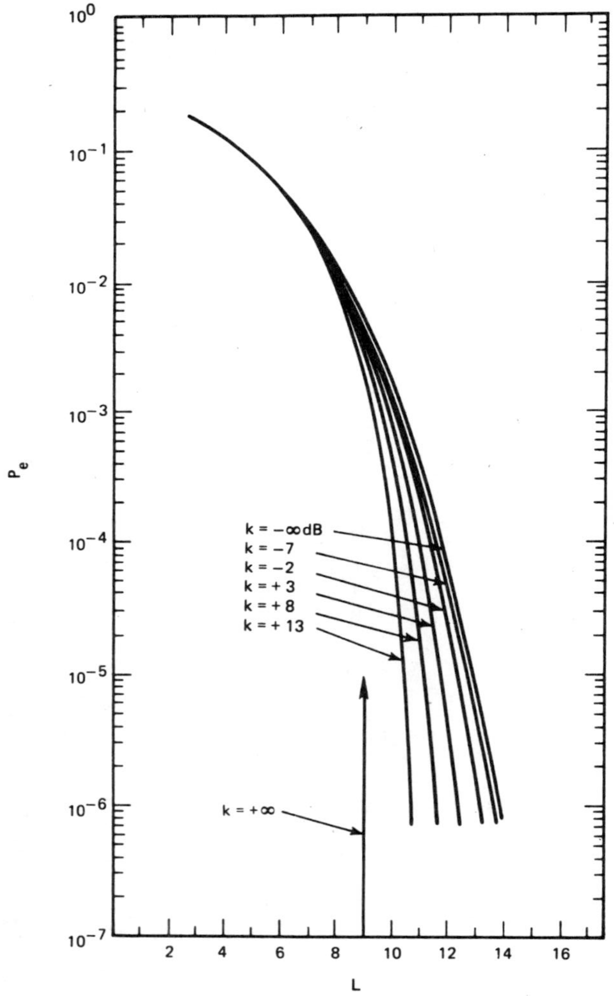

Figure 3d. *Error Probabilities of 4-Phase PSK in the Presence of Gaussian Noise and Four Cochannel Interferences* ($H = 4, 4\phi$)

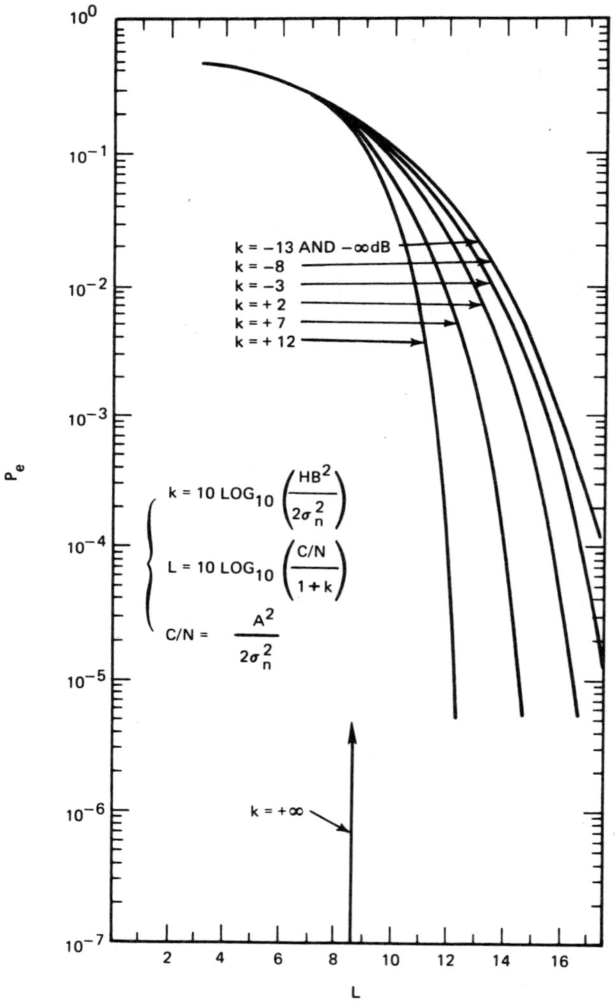

Figure 4a. *Error Probabilities of 8-Phase PSK in the Presence of Gaussian Noise and One Cochannel Interference* ($H = 1, 8\phi$)

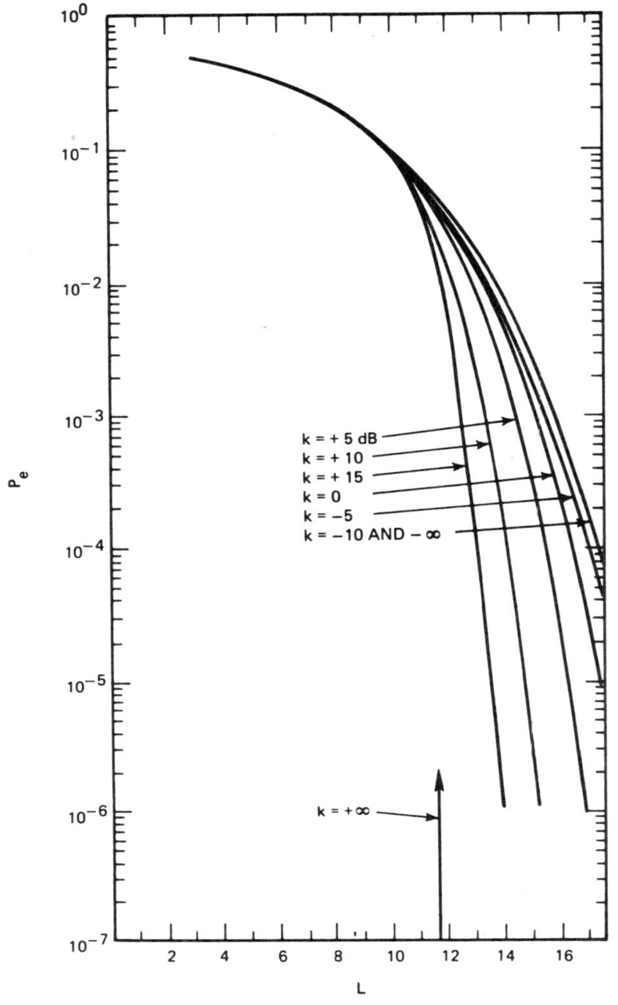

Figure 4b. *Error Probabilities of 8-Phase PSK in the Presence of Gaussian Noise and Two Cochannel Interferences* ($H = 2, 8\phi$)

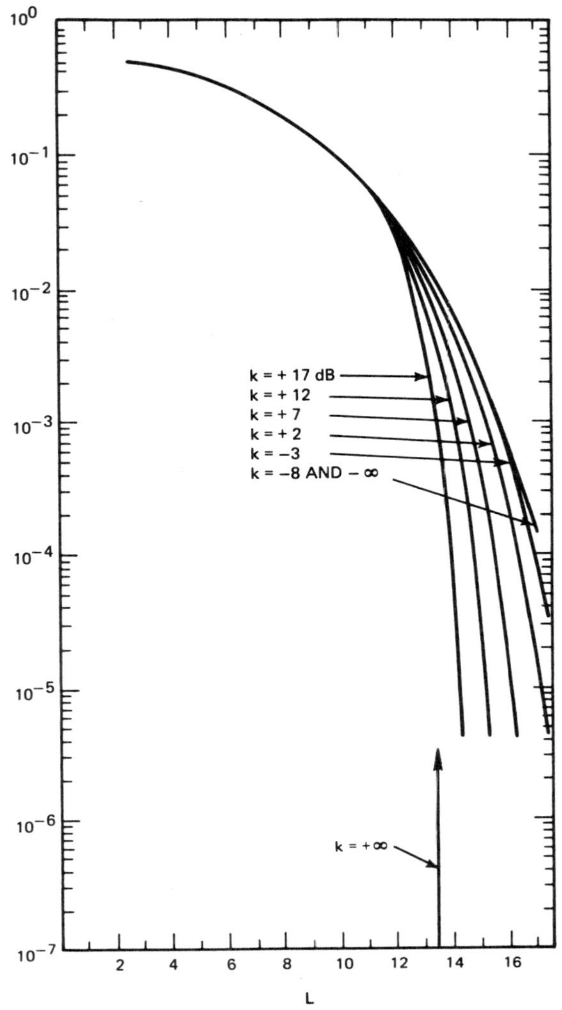

Figure 4c. *Error Probabilities of 8-Phase PSK in the Presence of Gaussian Noise and Three Cochannel Interferences* ($H = 3, 8\phi$)

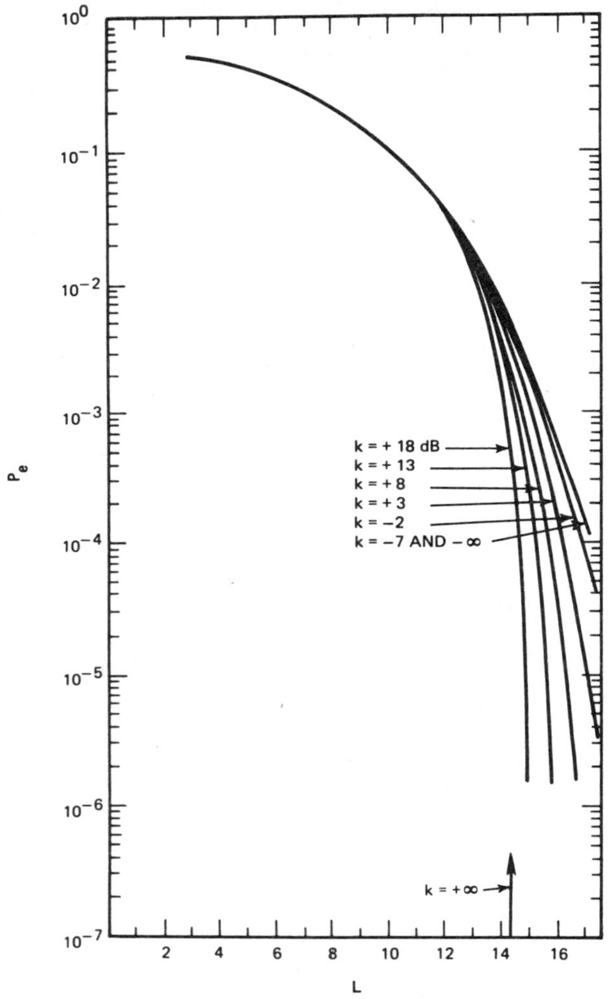

Figure 4d. *Error Probabilities of 8-Phase PSK in the Presence of Gaussian Noise and Four Cochannel Interferences* ($H = 4, 8\phi$)

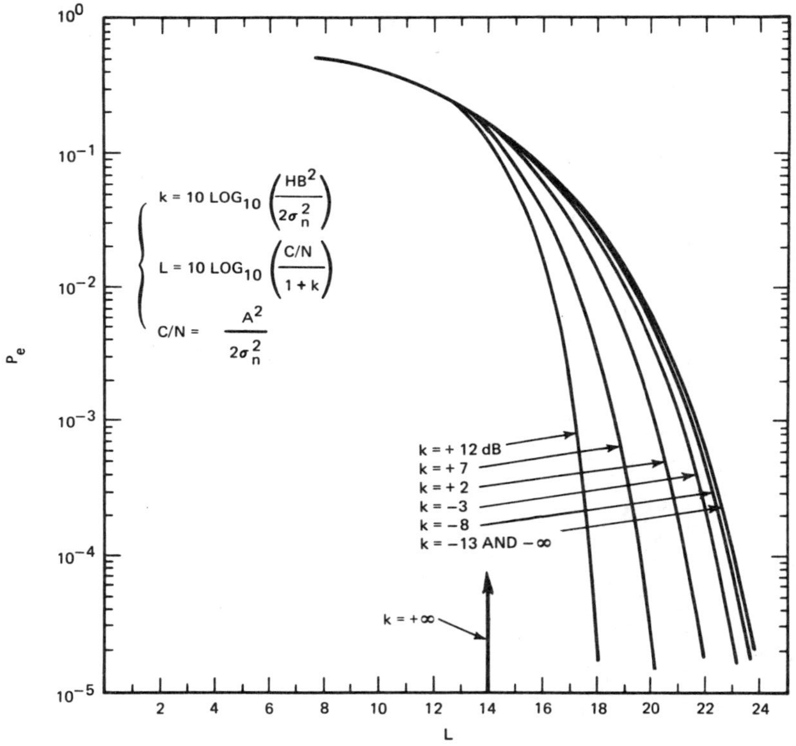

Figure 5a. *Error Probabilities of 16-Phase PSK in the Presence of Gaussian Noise and One Cochannel Interference* ($H = 1, 16\phi$)

carrier-to-total-interference-and-noise power ratio, L, the error probability increases as the number of interferences, H, increases. This is quite reasonable. That is, as H increases to $+\infty$, the statistics of the sum of these H independent interferences become more close to Gaussian. Thus, the error probability should approach that caused by Gaussian noise with the same amount of power.

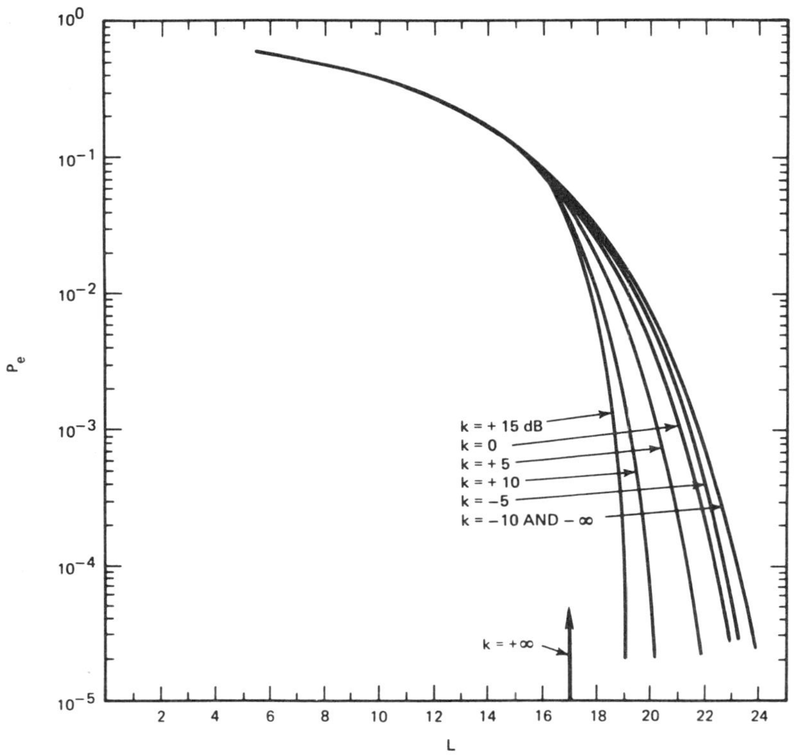

Figure 5b. *Error Probabilities of 16-Phase PSK in the Presence of Gaussian Noise and Two Cochannel Interferences* ($H = 2, 16\phi$)

Conclusion

Adaptation of the results obtained in Reference 8 to analyze the effects of cochannel interference in M-ary PSK systems has been demonstrated. The necessary modifications are few. The computational procedure described in this paper has been programmed on COMSAT's computer to

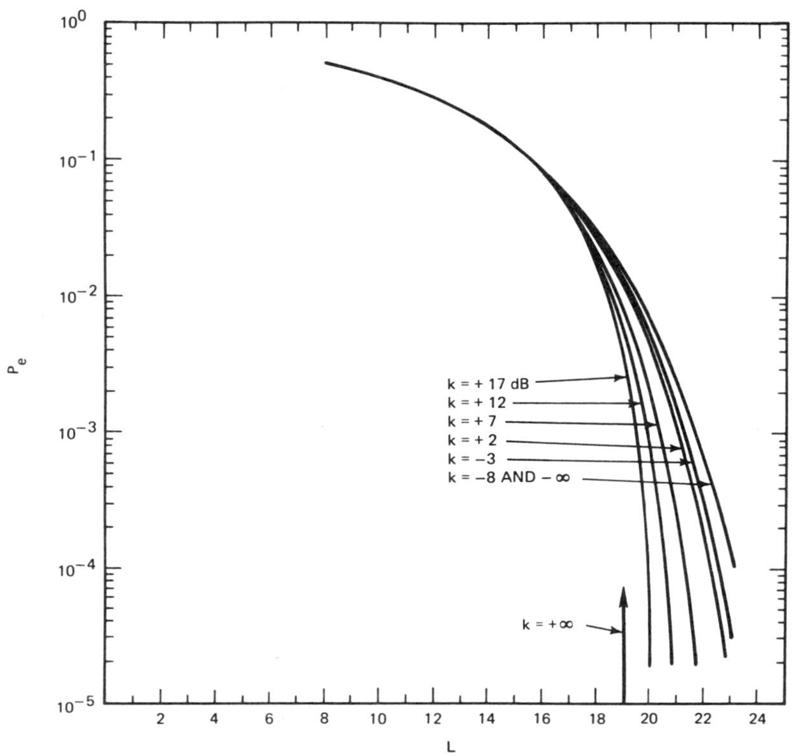

Figure 5c. *Error Probabilities of 16-Phase PSK in the Presence of Gaussian Noise and Three Cochannel Interferences* ($H = 3, 16\phi$)

yield numerical results. These numerical results can also provide a tradeoff in system design between the required carrier-to-noise power ratio and carrier-to-interference power ratio to achieve a given error performance for a given number of interferences. For example, from Figure 3a for the case of 4-phase PSK and a single cochannel interference, to achieve a 10^{-4} error probability, L must equal 8.5 and 11.2 dB for $k = 7$ and -3 dB, respec-

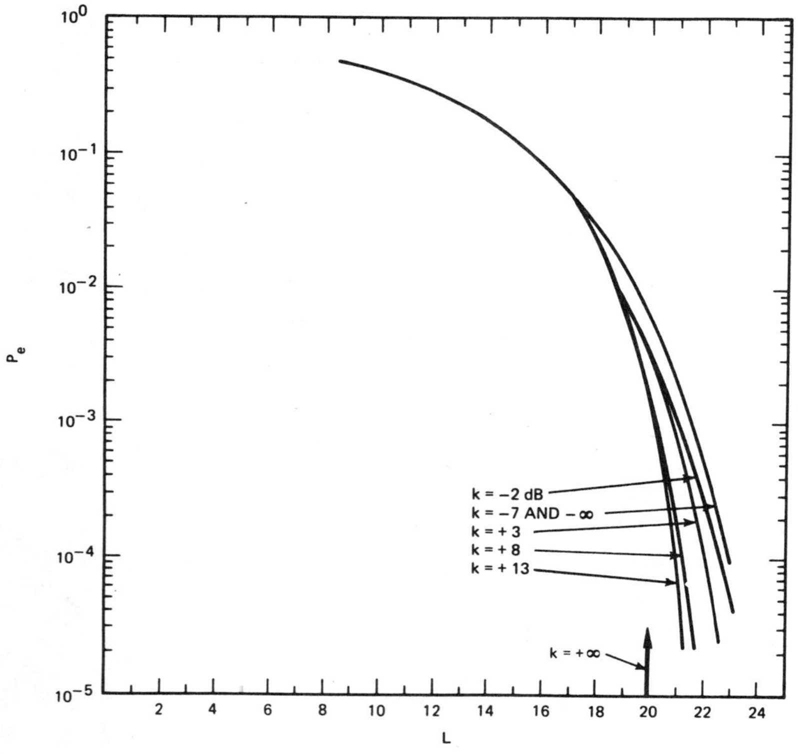

Figure 5d. *Error Probabilities of 16-Phase PSK in the Presence of Gaussian Noise and Four Cochannel Interferences* ($H = 4, 16\phi$)

tively. Therefore, to increase the carrier-to-noise-and-interference power ratio, L, from 8.5 to 11.2 dB, the interference-to-noise power ratio can be reduced from 7 dB to -3 dB. Many other tradeoffs can be also made.

It should be noted here that intersymbol interference has been assumed to be negligible in this paper. The tradeoff between noise, cochannel interference, and intersymbol interference will be presented elsewhere.

Acknowledgment

The authors wish to thank Dr. K. Bhatnagar for his contributions to the computation of all curves given in this paper.

References

[1] A. S. Rosenbaum, "PSK Error Performance with Gaussian Noise and Interference," *Bell System Technical Journal*, Vol. 48, February 1969, pp. 413–442.

[2] A. S. Rosenbaum, "Binary PSK Error Probabilities with Multiple Co-channel Interferences," *IEEE Transactions on Communications Technology*, COM-18, June 1970, pp. 241–253.

[3] V. K. Prabhu, "Error Rate Consideration for Coherent Phase-Shift Keyed Systems with Co-channel Interference," *Bell System Technical Journal*, Vol. 48, March 1969, pp. 743–767.

[4] V. K. Prabhu, "Error-probability Upper Bound for Coherently Detected Signals with Co-channel Interference," *Electronics Letters*, Vol. 5, No. 16, August 7, 1969.

[5] M. C. Jeruchim and D. A. Kane, "Orbit/Spectrum Utilization Study," Vol. IV, Document No. 70SD4293, General Electric Company, December 31, 1970.

[6] J. Goldman, "Multiple Error Performance of PSK Systems with Co-channel Interference and Noise," *IEEE Transactions on Communications Technology*, COM-19, No. 4, August 1971, pp. 420–430.

[7] J. M. Aein, "On the Effects of Undesired Signal Interference to a Coherent Digital Carrier," Institute for Defense Analysis, Paper p-812, IDA Log No. HQ71-13729, February 1972.

[8] O. Shimbo, R. Fang, and M. Celebiler, "Performance of M-ary PSK Systems in Gaussian Noise and Intersymbol Interference," *IEEE Transactions on Information Theory*, IT-19, No. 1, January 1973, pp. 44–58, IT-19, No. 3, May, 1973, p. 365 (corrections).

Combined Effects of Intersymbol, Interchannel, and Co-Channel Interferences in M-ary CPSK Systems

SERGIO BENEDETTO, EZIO BIGLIERI, AND VALENTINO CASTELLANI

Abstract—An expression for the error probability of multilevel coherent phase-shift-keyed (CPSK) systems in the presence of intersymbol, interchannel, and co-channel interferences and additive Gaussian noise is derived. An exact expression is given for the binary CPSK system, whereas upper and lower bounds are presented for the multilevel systems. The approach proposed in this paper overcomes the difficulties of the exhaustive methods and allows accurate and fast evaluation of the character error probability. Only the impulse responses of the overall channels are needed, and the computational labor is reduced to the numerical evaluation of one particular type of integral, whose computation is based upon nonclassical Gaussian quadrature rules.

The model of the system is general enough to allow the choice of a rectangular or otherwise-shaped modulating pulse and a constant or shaped envelope of the modulating carrier. Phase incoherence between adjacent carriers is also assumed, and a random misalignment among the modulating bit streams can be taken into account, if necessary.

Extensive numerical results are presented for 2- and 4-level systems. The presentation of the results stresses a possible utilization of them in system design problems.

I. Introduction

THE PERFORMANCE of coherent phase-shift-keyed (CPSK) systems has been investigated by many authors [1]-[14] and digital phase modulation is assuming increased interest as a modulation scheme for transmitting digital information. As a matter of fact, digital phase modulation is an efficient technique for trading bandwidth for signal-to-noise ratio (SNR).

The performance of CPSK systems was first investigated in the presence of disturbance due only to additive Gaussian noise [1]-[3].

Binary and quaternary CPSK modulations in the presence of both intersymbol interference and Gaussian noise were investigated by Calandrino *et al.* [4]. Their approach is based on the direct enumeration of all possible interfering sequences, so that it becomes computationally impractical if many intersymbol interference terms have to be considered. The first possibility of overcoming this drawback is to look for upper bounds that are easy to compute. This approach was investigated by Prabhu [5], who used the Chernoff bound to bound the exact expression of the error probability for the binary case, whereas for the quaternary case he used the same technique to bound an upper bound derived with the geometric arguments used by Arthurs and Dym in the case of no intersymbol interference [2]. Prabhu's bounds are not always very tight and in any case his technique does not allow any absolute comparison between different channels. One relevant feature, however, is that only the impulse response of the CPSK system is needed for the evaluation of the bound.

Benedetto and Biglieri [6] have recently proposed a computational technique that allows the evaluation of the exact error probability in the binary case within any desired accuracy; for M-ary CPSK systems the same technique has been applied to get an accurate and fast computation of the upper bound that Prabhu [5] simply bounded using the Chernoff bound. Even a lower bound has been proposed, and the upper bound came out to be a good approximation to the true value of the character error probability.

Rosenbaum [7] has investigated the case of one co-channel interfering modulated signal and Gaussian noise in M-ary systems. Error probabilities are given as convergent series.

The case considered by Prabhu [8] takes into account multiple co-channel interference and Gaussian noise; exact expressions for the error probability are given in the form of convergent power series for binary and quaternary systems, whereas upper and lower bounds are presented for the M-ary case. The same author [9] used the Chernoff bound to get a simpler upper bound for M-ary systems.

The binary system was further investigated by Rosenbaum [10] in the case of multiple co-channel interferences and Gaussian noise; comprehensive numerical results for the CPSK system were presented.

The most realistic model of a transmission link has to deal therefore with the following degradation effects: intersymbol interference due to the transmission and reception filters; co-channel and interchannel interferences coming from the radio network in which the useful channel is embedded, and additive Gaussian noise. Such a general model is a little tricky to analyze and exact expressions for the character error probability are rather difficult to derive unless effective computational methods are available.

The first conceivable approach is, in fact, an exhaustive computation of all the conditional error probabilities followed by an average over the conditioning random variables (RV's). This approach was pursued using either computer simulation techniques [11], [13], [14] or direct enumeration of the interfering sequences [12]. Results are available for binary and quaternary CPSK systems. In spite of some particular subtleties that can eventually be devised, this approach suffers an in-

Paper approved by the Data Communications Committee of the IEEE Communications Society for publication without oral presentation. Manuscript received September 12, 1972.

The authors are with the Istituto di Elettronica e Telecomunicazioni del Politecnico di Torino, Turin, Italy.

Reprinted from *IEEE Trans. Comm.*, vol. COM-21, pp. 997-1008, Sept. 1973.

herent limitation that cannot be avoided. When the number of phase-modulating levels M and/or the number of interfering symbols is only moderately high, the computer time becomes unacceptable unless the channel impulse response is truncated to a very limited number of samples, thus causing a rough approximation in the error-probability evaluation.

In this paper a new method is proposed to analyze the CPSK system performance under the combined influence of intersymbol interference, interchannel and co-channel interferences, and additive Gaussian noise. The proposed approach overcomes the difficulties of the exhaustive methods and allows accurate and fast evaluation of the character error probability. An exact expression is derived for the binary CPSK systems and evaluated within any preassigned accuracy. A generalization of the bounding technique used by Arthurs and Dym [2] is applied to the M-ary case and an accurate evaluation of upper and lower bounds is performed; the upper bound is conjectured to be a good approximation to the true value of the error probability. The model is general enough to allow the choice of a rectangular or otherwise-shaped modulating pulse and a constant or shaped envelope of the modulated carrier. Moreover, phase incoherence between adjacent carriers and random misalignment of the modulating bit streams can be assumed.

Only the impulse responses of the overall channels are needed to obtain the results. The computational labor is reduced to the numerical evaluation of only one particular type of integral whose computation is based upon nonclassical Gaussian quadrature rules.

One relevant feature that results in a significant saving of computer time is that the statistical averages over the various RV's are performed, once for all, in the computation of the moments of a suitably defined RV.

The problem to be solved will be formally stated in Section II, and expressions for the error probability will be given in Section III. The computational procedure for the evaluation of the results is described in Section IV, whereas Section V is devoted to a brief recapitulation of the whole method. In Section VI some numerical results for 2- and 4-level CPSK are presented.

II. Statement of the Problem

An M-ary CPSK system is considered in the presence of intersymbol and interchannel interference and additive Gaussian noise. The model of the system is given in Fig. 1. The transmitter filters are assumed to have an equivalent bandwidth larger than the receiving one, and to be equal except for the frequency location. In other words, let

$$H_i(\omega) = H_0(\omega - i\omega_d) \tag{1}$$

where ω_d is the angular frequency spacing between two adjacent channels.

Provided that the center angular frequency ω_c of the useful channel is assumed to be much larger than the modulating signal bandwidth, the complex envelope of the useful modulated signal can be written as

$$s(t) = \sum_{n=-\infty}^{\infty} f(t - nT) \exp\,[j\alpha_n g(t - nT)], \tag{2}$$

where

$f(t)$ takes into account a possible amplitude shaping of the modulated signal;

$g(t)$ is the shape of the phase-modulating pulse; and

α_n are assumed to be statistically independent and equally likely RV's. Their range is the set

$$\mathfrak{A} = \{(2r + 1)\pi/M\} \tag{3}$$

with the integer r ranging as

$$-M/2 \leqslant r < M/2, \qquad M \text{ equals a power of 2.}$$

The ith interfering signal $s_i(t)$ is assumed to have a complex envelope similar to (2) and can be written as

$$s_i(t) = \sum_{m_i=-\infty}^{\infty} f(t - \tau_i - m_iT) \exp\,[j\alpha_{m_i} g(t - \tau_i - m_iT)] \cdot \exp\,[j(i\omega_d t + \varphi_i)] \tag{4}$$

where the meaning of the symbols is as follows.

τ_i takes into account the possible misalignment of the signaling intervals of the different channels. For instance, τ_i is equal to zero when the ith channel is aligned with the useful one; otherwise, τ_i may be a RV ranging in the interval $(0, T)$.

φ_i is a random phase uniformly distributed in the interval $(0, 2\pi)$, and represents the lack of coherence among the different carriers.

α_{m_i} are the information symbols pertaining to the ith channel.

A complex envelope can also be defined for the impulse response of the transmitting filters as follows:

$$\tilde{h}_i(t) = h_0(t) \exp\,(ji\omega_d t) \tag{5}$$

where $h_0(t)$ is the impulse response corresponding to the low-pass equivalent transfer function $H_0(\omega + \omega_c)$ of the central filter, which is assumed to be symmetric and strictly band-limited. This simplified situation makes $h_0(t)$ a real function of time and allows neater results. The extension to the more general case in which $H_0(\omega)$ is not symmetric and the transmitting filters are not equal is straightforward, and will not be pursued here. Notice, however, that in many practical cases our assumptions are justified.

Suppose now that there are N_L interfering channels at angular frequencies lower than ω_c, and N_R at angular frequencies higher than ω_c; then the complex envelope $y(t)$ (see Fig. 1) is given by

$$y(t) = x(t) + \sum_{i=-N_L}^{N_R} V_i x_i(t). \tag{6}$$

The term with $i = 0$ in the summation of (6) represents a co-channel interference at the same frequency as the useful signal, and the coefficients V_i take into account possible attenuations on the different interfering channels.

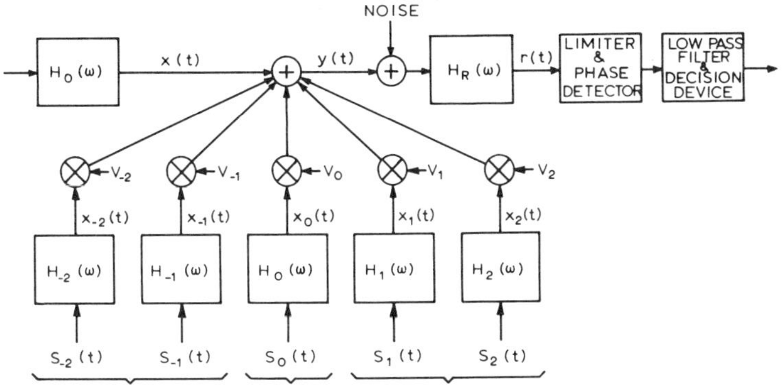

Fig. 1. Model of the system.

Assume now that the receiver filter is also symmetric, and let $h_R(t)$ be the low-pass equivalent impulse response. The complex envelope of the received signal can be written as

$$r(t) = y(t) * h_R(t) + n(t) \tag{7}$$

where $n(t)$ is a complex noise process.

After the receiving filter, an ideal limiter removes all the amplitude variations from the received signal, which is then presented to an ideal phase detector. At the input of the decision device the detected phase signal is sampled every T s at the instants $t = t_0 + kT, -\infty < k < \infty$. Thus, the sampled phase angle determined by the receiver is

$$\varphi_0 = tg^{-1} \frac{r_{0Q}}{r_{0P}} \tag{8}$$

where r_{0P} and r_{0Q} are the real and imaginary parts of $r(t_0)$, respectively.

Let us define now the following random processes for the useful signal:

$$a(t; \alpha_n) = \{f(t) \cos [\alpha_n g(t)]\} * h_0(t) * h_R(t) \tag{9a}$$

$$b(t; \alpha_n) = \{f(t) \sin [\alpha_n g(t)]\} * h_0(t) * h_R(t) \tag{9b}$$

and for the interfering signals

$$a^{(i)}(t; \alpha_{m_i}, \varphi_i) = [w_{cc}^{(i)}(t; \alpha_{m_i}) - w_{ss}^{(i)}(t; \alpha_{m_i})] \cos \varphi_i - [w_{cs}^{(i)}(t; \alpha_{m_i}) + w_{sc}^{(i)}(t; \alpha_{m_i})] \sin \varphi_i \tag{10a}$$

$$b^{(i)}(t; \alpha_{m_i}, \varphi_i) = [w_{cs}^{(i)}(t; \alpha_{m_i}) + w_{sc}^{(i)}(t; \alpha_{m_i})] \cos \varphi_i + [w_{cc}^{(i)}(t; \alpha_{m_i}) - w_{ss}^{(i)}(t; \alpha_{m_i})] \sin \varphi_i \tag{10b}$$

where

$$w_{cc}^{(i)}(t; \alpha_{m_i}) \triangleq \{[f(t) \cos [\alpha_{m_i} g(t)] * h_0(t)] \cdot \cos (i\omega_d t)\} * h_R(t) \tag{11a}$$

$$w_{cs}^{(i)}(t; \alpha_{m_i}) \triangleq \{[f(t) \cos [\alpha_{m_i} g(t)] * h_0(t)] \cdot \sin (i\omega_d t)\} * h_R(t) \tag{11b}$$

$$w_{sc}^{(i)}(t; \alpha_{m_i}) \triangleq \{[f(t) \sin [\alpha_{m_i} g(t)] * h_0(t)] \cdot \cos (i\omega_d t)\} * h_R(t) \tag{11c}$$

$$w_{ss}^{(i)}(t; \alpha_{m_i}) \triangleq \{[f(t) \sin [\alpha_{m_i} g(t)] * h_0(t)] \cdot \sin (i\omega_d t)\} * h_R(t). \tag{11d}$$

With these definitions, and introducing the RV $\epsilon_i = \tau_i/T$, the sampled values of the components of the received signal can be given the following form:

$$r_{0P} = a_0(\alpha_0) + \sum_n{}' a_n(\alpha_n) + \sum_i \sum_{m_i} V_i a_{m_i+\epsilon_i}^{(i)}(\alpha_{m_i}, \varphi_i) + n_P \tag{12}$$

$$r_{0Q} = b_0(\alpha_0) + \sum_n{}' b_n(\alpha_n) + \sum_i \sum_{m_i} V_i b_{m_i+\epsilon_i}^{(i)}(\alpha_{m_i}, \varphi_i) + n_Q \tag{13}$$

where n_P and n_Q are the real and imaginary parts of the noise sample $n(t_0)$. The first terms in (12) and (13) represent the information bearing sample; the simple summation is the intersymbol interference disturbance, whereas the double summation represents the contribution due to both interchannel and co-channel interference. The significance of the heavy but necessary notation can be summarized as follows: the subscripts stand for the sampling time, the arguments in parentheses indicate the RV's upon which the sample is dependent, and the superscripts refer to the interfering channel. So the random sample

$$a_{m_i+\epsilon_i}^{(i)}(\alpha_{m_i}, \varphi_i)$$

comes from the ith interfering channel sampled at time $t = t_0 + (m_i + \epsilon_i)T$ with ϵ_i previously defined, and is dependent upon the random phase φ_i and the modulating symbol α_{m_i}.

If the two RV's X and Y are introduced for simplicity,

$$X \triangleq \sum_n{}' a_n(\alpha_n) + \sum_i \sum_{m_i} V_i a_{m_i+\epsilon_i}^{(i)}(\alpha_{m_i}, \varphi_i) \tag{14}$$

$$Y \triangleq \sum_n{}' b_n(\alpha_n) + \sum_i \sum_{m_i} V_i b_{m_i+\epsilon_i}^{(i)}(\alpha_{m_i}, \varphi_i), \tag{15}$$

then the phase angle determined by the receiver and defined in (8) becomes

$$\varphi_0 = tg^{-1}\frac{b_0(\alpha_0)+Y+n_Q}{a_0(\alpha_0)+X+n_P}. \qquad (16)$$

III. The Probability of Error

Under the assumption that the detector threshold levels $\varphi_T(\alpha_0)$ are given by

$$\varphi_T(\alpha_0) = \alpha_0 \pm \frac{\pi}{M}, \qquad (17)$$

the character error probability for M equally likely signals can be expressed as

$$P_M(e) = \frac{1}{M}\sum_{\alpha_0 \in \mathcal{A}} \Pr\{e|\alpha_0\}$$

$$= 1 - \frac{1}{M}\sum_{\alpha_0 \in \mathcal{A}} \Pr\left\{\alpha_0 - \frac{\pi}{M} < \varphi_0 < \alpha_0 + \frac{\pi}{M}\ \Big|\ \alpha_0\right\}, \qquad (18)$$

where φ_0 is the sampled received phase angle defined in (16).

The computation of the probabilities appearing in the summation of (18) requires the knowledge of the statistics of the RV's X and Y defined in (14) and (15), and of the noise process $n(t)$. The problems concerning X and Y will be examined in greater detail in a later section. The noise process $n(t)$ is assumed to be Gaussian, even if the authors feel that the method can be extended to other noise statistics without too much labor.

For the case of $M = 2$ an exact evaluation of (18) can be pursued. For the M-ary case, the direct evaluation of (18) has been avoided because of the messy computational difficulties; upper and lower bounds are proposed instead, which were found using a bounding technique first proposed by Arthurs and Dym [2] for the case of no interference.

Since the analytical procedure is an extension of results presented in another paper [6], the interested reader can eventually refer to it. Here, only the final expressions are given with some brief comments.

In the binary case, it is immediately verified that there is an error if the imaginary part r_{0Q} of the sampled received signal has a sign opposite to the transmitted information symbol α_0. Equation (18) gives the result:

$$P_2(e) = \frac{1}{4}\int\left[\operatorname{erfc}\left(\frac{Y + b_0(\pi/2)}{\sqrt{2}\,\sigma_n}\right) + \operatorname{erfc}\left(\frac{-Y - b_0(-\pi/2)}{\sqrt{2}\,\sigma_n}\right)\right] dF(Y) \qquad (19)$$

where $F(Y)$ is the distribution function of the RV Y, and the integration is performed over the range of Y.

For the M-ary case, let us first define the RV's

$$\lambda(\beta) \triangleq X \sin\beta - Y\cos\beta \qquad (20)$$

$$\chi(\alpha_0, \beta) \triangleq a_0(\alpha_0)\sin\beta - b_0(\alpha_0)\cos\beta \qquad (21)$$

and the two integrals

$$I_1(\alpha_0) \triangleq \frac{1}{2}\int \operatorname{erfc}\left\{\frac{\lambda(\alpha_0 + \pi/M) + \chi(\alpha_0, \alpha_0 + \pi/M)}{\sqrt{2}\,\sigma_n}\right\} \cdot dF[\lambda(\alpha_0 + \pi/M)] \qquad (22)$$

$$I_2(\alpha_0) \triangleq \frac{1}{2}\int \operatorname{erfc}\left\{\frac{-\lambda(\alpha_0 - \pi/M) - \chi(\alpha_0, \alpha_0 - \pi/M)}{\sqrt{2}\,\sigma_n}\right\} \cdot dF[\lambda(\alpha_0 - \pi/M)] \qquad (23)$$

where $F[\lambda(\cdot)]$ is the distribution function of the RV $\lambda(\cdot)$, and the integrals are performed over the range of $\lambda(\cdot)$.

Then the error probability $P_M(e)$ can be bounded as follows:

$$\frac{1}{M}\sum_{\alpha_0 \in \mathcal{A}} \max[I_1(\alpha_0), I_2(\alpha_0)] \leqslant P_M(e)$$

$$\leqslant \frac{1}{M}\sum_{\alpha_0 \in \mathcal{A}} [I_1(\alpha_0) + I_2(\alpha_0)]. \qquad (24)$$

It is easy to see that the lower bound is minimized when

$$I_1(\alpha_0) = I_2(\alpha_0) \qquad (25)$$

and in this case it differs from the upper bound by a factor of 2. As a consequence, the exact computation of $I_1(\alpha_0)$ and $I_2(\alpha_0)$ allows us to know the error probability $P_M(e)$ within an interval whose bounds differ by no more than a factor of 2.

Notice, however, that the upper bound, when particularized to the case of no interference (i.e., $\lambda = 0$), assumes exactly the same expression obtained by Cahn [1] in a different manner. He showed that this upper bound gives an excellent approximation to the true value of $P_M(e)$, and therefore it is reasonably expected that this statement holds true also when interferences are present.

As a final comment, notice that the definition (20) is the key point of the derivation: in fact it allows us to average the conditional error probability over a single RV, thus avoiding the presence of a two-fold integral. The RV $\lambda(\cdot)$ is a linear combination of the RV's X and Y and its statistics will be discussed later on.

It can be finally observed that the evaluation of the error probability is reduced to the numerical computation of integrals of the same type. This problem is the subject of Section IV.

IV. Computation of the Probability of Error

As we have seen in the last section, the evaluation of (19), (22), and (23) requires always the numerical computation of integrals of the following form:

$$I = \int \operatorname{erfc}\left(\frac{\xi + c}{\sqrt{2}\,\sigma_n}\right) dF(\xi) \qquad (26)$$

where the distribution function $F(\xi)$ is not known explicitly.

Our main result here is to reduce the problem of the error probability in CPSK multilevel systems in the presence of intersymbol and interchannel interference to the numerical computation of (26). This approach depends significantly on the fact that we are able to compute (26), thus allowing an exact evaluation of $P_2(e)$ and of the bounds to $P_M(e)$.

The problem of the computation of integrals such as (26) is not new in the study of digital transmission systems. As a matter of fact, it has been given considerable attention in connection with the evaluation of error probability in binary or multilevel pulse-amplitude modulation (PAM) systems [15]-[19].

Our method is based on the knowledge of the first $2L + 1$ moments of the RV ξ. In this case, in fact, a Gaussian quadrature rule [20] can be used to approximate the integral (26) as follows:

$$I \cong \sum_{j=1}^{L} w_j \operatorname{erfc}\left(\frac{\xi_j + c}{\sqrt{2}\sigma_n}\right) \tag{27}$$

where the ξ_i are called the abscissas of the formula and the w_i the weights. The set of numbers $\{w_i, \xi_i\}_{i=1}^{L}$ is called a quadrature rule and can be obtained from the moments of the RV ξ.

This method of computation is described in great detail in [21]. The integral (26) can be evaluated within any desired accuracy since the error involved in the approximation (27) can be upper bounded by a quantity that tends to zero as the number L increases [21].

The interested reader will also find in [21] a critical comparison with other approaches for the approximation of the integral (26). As a matter of fact, it was observed that the method proposed here assures accurate results and very satisfactory performance even when other methods fail. Also the computer time required is shorter, especially when the error probability is computed for many values of the SNR.

The last step to be performed is therefore the evaluation of the moments of the RV $\lambda(\cdot)$ defined in (20), which appears as the RV in the integrals such as (26).

This is the last key point of our approach, since an accurate and fast computation of the moments of $\lambda(\cdot)$ allows us to obtain results for $P_2(e)$ and $P_M(e)$ with a high degree of accuracy and short computer time.

Evaluation of the Moments of $\lambda(\cdot)$

Recalling definitions (14), (15), and (20), the RV $\lambda(\beta)$ can be given the following expression:

$$\lambda(\beta) = \sum_n{}' \left[a_n(\alpha_n) \sin\beta - b_n(\alpha_n)\cos\beta\right] + \sum_i \sum_{m_i} V_i \left[a^{(i)}_{m_i+\epsilon_i}(\alpha_{m_i}, \varphi_i)\sin\beta - b^{(i)}_{m_i+\epsilon_i}(\alpha_{m_i}, \varphi_i)\cos\beta\right]. \tag{28}$$

The terms appearing in (28) have been previously defined in (9)-(13).

The RV $\lambda(\beta)$ is the sum of a finite number of terms: the first summation comes from the intersymbol interference on the useful channel and the number R of interfering samples that significantly contribute to $P_M(e)$ can be chosen following the method described in the Appendix; the double summation comes from the interchannel and co-channel interference, and we can assume that there are R_i terms for the ith channel, R_i being chosen again with the same procedure used for the choice of R.

Notice that for each interfering channel the subscript m_i ranges from a lower value $-m_{iL}$ to an upper value m_{iU} such that

$$R_i = m_{iL} + m_{iU} + 1.$$

The terms appearing in (28) can be suitably renumbered and $\lambda(\beta)$ put in the form

$$\lambda(\beta) = \sum_{k=1}^{K} \lambda_k(\beta) \tag{29}$$

where

$$K = R + \sum_i R_i \tag{30a}$$

and, moreover, for $k > R$ it is

$$k = R + m_{iL} + m_i + 1 + \sum_{j=-N_L}^{i} R_{j-1}, \qquad R_{-N_L-1} \equiv 0. \tag{30b}$$

There are different methods [18], [22] that allow a recurrent evaluation of the moments when the RV's $\lambda_k(\beta)$ are statistically independent; if this is the case, the method proposed by Prabhu [22] turns out to be very convenient.

Define the partial sum

$$\lambda^{(n)}(\beta) \triangleq \sum_{k=1}^{n} \lambda_k(\beta). \tag{31}$$

The jth moment of $\lambda(\beta)$ can then be written as

$$E\{[\lambda(\beta)]^j\} = E\{[\lambda^{(K)}(\beta)]^j\}. \tag{32}$$

But due to the statistical independence just noticed, the following recurrence relation holds:

$$E\{[\lambda^{(n+1)}(\beta)]^j\} = E\{[\lambda^{(n)}(\beta) + \lambda_{n+1}(\beta)]^j\} = \sum_{h=0}^{j} \binom{j}{h} E\{[\lambda^{(n)}(\beta)]^h\} E\{[\lambda_{n+1}(\beta)]^{j-h}\}. \tag{33}$$

So the moment in (32) can be computed recurrently, provided that we can evaluate the moments of each term appearing in the sum. And this is the last problem to be solved to get the solution.

1) Moments of the terms of intersymbol interference: The first R terms in (29) have the following form:

$$\lambda_k(\beta) = a_k(\alpha_k)\sin\beta - b_k(\alpha_k)\cos\beta \tag{34}$$

and it is easily seen that they are statistically independent.

Averaging over α_k, which can take M equally likely values, we get

$$E\{[\lambda_k(\beta)]^j\} = \frac{1}{M} \sum_{\alpha_k \in \mathcal{A}} [a_k(\alpha_k)\sin\beta - b_k(\alpha_k)\cos\beta]^j. \tag{35}$$

2) Moments of the terms of interchannel interference: The K-R terms from $R + 1$ to K in the sum (29) have the following expression:

$$\lambda_k(\beta) = V_i\{a^{(i)}_{m_i+\epsilon_i}(\alpha_k, \varphi_i)\sin\beta - b^{(i)}_{m_i+\epsilon_i}(\alpha_k, \varphi_i)\cos\beta\}. \tag{36}$$

With the following definitions

$$u^{(i)}_{m_i+\epsilon_i}(\alpha_k) \triangleq [w^{(i)}_{cc}(t;\alpha_k) - w^{(i)}_{ss}(t;\alpha_k)]\,\Big|_{t=t_0+(m_i+\epsilon_i)T} \tag{37}$$

$$v^{(i)}_{m_i+\epsilon_i}(\alpha_k) \triangleq [w^{(i)}_{cs}(t;\alpha_k) + w^{(i)}_{sc}(t;\alpha_k)]\,\Big|_{t=t_0+(m_i+\epsilon_i)T} \tag{38}$$

(36) can be rewritten as

$$\lambda_k(\beta) = V_i \{u^{(i)}_{m_i+\epsilon_i}(\alpha_k) \sin \psi_i - v^{(i)}_{m_i+\epsilon_i}(\alpha_k) \cos \psi_i\} \quad (39)$$

where, for each value of β, $\psi_i = \beta - \varphi_i$ is seen to be an RV, uniformly distributed in the interval $(0, 2\pi)$.

Notice that for these terms, the right-hand side of (39) is no more a function of β and depends only on the RV's α_k, ϵ_i, and ψ_i.

Due to the presence of the same random phase φ_i, the terms λ_k of interchannel interference coming from the ith channel are not in general statistically independent; however they become statistically independent for given values of φ_i and ϵ_i. So the aforementioned recurrent procedure can be used in the computation of the conditioned moments:

$$E\left\{\left[\sum_{k=-m_{iL}}^{m_{iu}} \lambda_k(\beta)\right]^j \Bigg| \psi_i, \epsilon_i\right\}. \quad (40)$$

For the set of terms pertaining to each interfering channel, standard quadrature formulas allow us to compute the moments:

$$E\left\{\left[\sum_{k=-m_{iL}}^{m_{iu}} \lambda_k(\beta)\right]^j\right\}$$

$$= \iint E\left\{\left[\sum_{k=-m_{iL}}^{m_{iu}} \lambda_k(\beta)\right]^j \Bigg| \psi_i, \epsilon_i\right\} dF(\epsilon_i)\, dF(\psi_i). \quad (41)$$

Once the moments have been computed for each term of the useful channel (35) and for the sums of terms (41) coming from each of the interfering channels, the same recursive method is used to get the moments of the RV $\lambda(\beta)$, defined in (29).

One final comment to this section. Notice the simplification that is made possible if the averages over ψ_i, α_k, and ϵ_i are performed when computing the moments. In this case, in fact, these calculations are made once for all the values of SNR.

In the direct enumeration method, on the contrary, the conditional error probability must be computed first for a given value of SNR; then the averages are performed both over the RV's ψ_i and ϵ_i and on the set of possible symbols sequences. As a result, there is also a drastic increase of computer time.

V. Recapitulation of the Computational Procedure

The model of the system is given when the number M of modulating levels, the number $N_L + N_R + 1$ of interfering signals, the normalized distance $D = \omega_d T/2\pi$, and the filters are assigned; the waveforms $f(t)$ and $g(t)$ are also given.

The computational procedure is then organized in the following steps.

Step 1: Obtain the in-phase and quadrature components of the overall impulse response of the useful and interfering channels. For the useful channel, $a(t; \alpha_n)$ and $b(t; \alpha_n)$ must be derived using (a) and (b); there are M of such functions, one for each value of the information phase α_n. Then, for each interfering signal (i.e., one value of the index i), the four functions $w^{(i)}_{cc}$, $w^{(i)}_{cs}$, $w^{(i)}_{ss}$, $w^{(i)}_{sc}$ are derived using (11a)–(11d); there are M of such functions too.

The number of samples to be taken into account on the useful and interfering channels can be chosen according to the method described in the Appendix. Notice however that on each interfering signal a higher number of samples is needed if the random misalignment ϵ_i has to be taken into account. Notice also that a change of the parameter D affects only the computation of (11a)–(11d).

It must be emphasized, however, that only the impulse response of the overall CPSK system needs to be known for the derivation of the results.

Step 2: The samples of the functions derived in Step 1 are used for the computations described in Section IV. Equation (35) allows us to compute the moments of the single terms $\lambda_k(\beta)$ pertaining to the useful channel; the moments of the sums of RV's of each interfering channel are computed using (40) and (41), and finally the recurrence relationship (33) is applied to get the moments of $\lambda(\beta)$ defined in (29). In the cases we have considered, no more than seventeen moments were needed.

Step 3: Construct the L abscissas and weights of the Gaussian quadrature rule [21] starting from the $2L + 1$ moments derived in Step 2.

Step 4: Apply the Gaussian quadrature rule (27) to evaluate the integrals of the type (26) appearing in the error probability expressions (19) for $M = 2$ or (24) for $M > 2$. It is important to notice that only this step must be repeated for different values of SNR to get a complete curve of error probability.

VI. Numerical Examples

The examples presented in this section refer to a special case that has always been given considerable attention either because it is easier to analyze or because it has enough practical importance. Assume that M is even, and that $f(t)$ and $g(t)$ are of the type

$$f(t) = g(t) = u(t) - u(t - T), \quad (42)$$

$u(t)$ being the unit step function. We have therefore a constant-envelope modulated signal with rectangular-shaped modulating pulses. With these assumptions, the formulas previously derived for the more general case particularize to more simple and manageable expressions.

Going again through the computational steps summarized in Section V, the following simplifications become possible.

Step 1: Instead of $2M$ time responses, only one is needed for the useful channel; for each of the interfering channels only two instead of $4M$ are sufficient.

Step 2: All the moments of the elementary terms in the sum (28) come out to be independent of the threshold angle β[6], and furthermore the moments of the interchannel interference terms are also independent of the number of phase levels M.

Steps 3–4: The RV λ has a symmetrical distribution, and therefore only one integral must be computed in evaluating (19). Moreover, the summations in (24) disappear because the integrals do not depend on the transmitted information symbol α_0.

The examples presented here refer to both cases of interchannel and co-channel interference.

The filters at the transmitting and receiving end of the system are assumed to have a low-pass equivalent transfer function of the Butterworth type.

The parameters used to define the system completely are the following: 1) the normalized equivalent noise bandwidth $(B_{eq}T)_T$ and the number of poles n_T of the transmitter filter; 2) the normalized equivalent noise bandwidth $(B_{eq}T)_R$ and the number of poles n_R of the receiving filter; 3) the normalized distance $D = \omega_d T/2\pi$ between two adjacent channels; and 4) the SNR η, defined through the relation

$$\eta \triangleq \frac{T}{2G_0} \tag{43}$$

where G_0 is the double-sided noise spectral density. So η is the ratio between the energy of a constant-envelope signal of unity amplitude over a time interval T and the noise spectral density. The variance σ_n^2 of the noise process appearing in all the expressions of the error probability is related to G_0 by the equation

$$\sigma_n^2 = G_0 B_{eq} \tag{44}$$

where B_{eq} is the equivalent noise bandwidth of the receiving filter. Therefore η and σ_n^2 can be related as follows:

$$\eta = \frac{(B_{eq}T)_R}{2\sigma_n^2}. \tag{45}$$

For any given value of η, (45) gives the corresponding value of σ_n^2 to be used for the computation of the curves of the error probability. The sampling time is chosen at the maximum of the impulse response of the overall system.

A. Interchannel Interference

Two symmetrically located interfering channels are present at the same power level as the interfered one; the bit intervals in the three channels are first assumed synchronous and aligned (i.e., $\epsilon_i = 0$ for both the interfering channels). However, some results are also given for the case of random misalignment between signaling periods.

1) 2-level systems: For this case, an exact expression for the error probability $P_2(e)$ was derived. A lot of results were obtained for this binary system in order to show the flexibility of the computational approach. The computer time required to perform the four steps described in Section V for 20 values of SNR, when 10 samples have to be considered in each channel, is about 20 s (of an IBM 360/44 computer).

First of all, the normalized equivalent noise bandwidth at the receiver $(B_{eq}T)_R$ is chosen to be 1.1; this value was found to minimize the error probability in the presence of only inter-

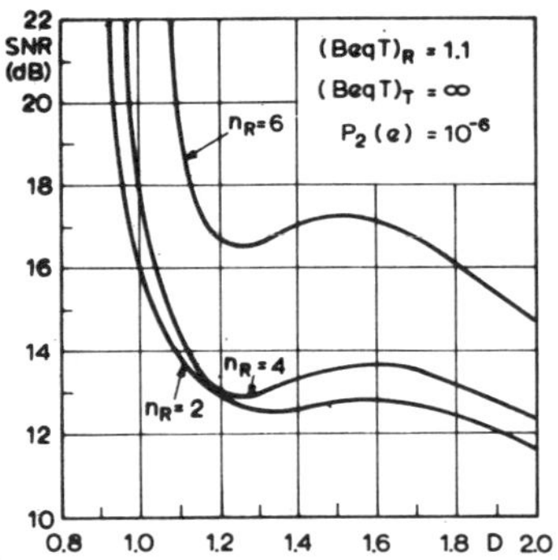

Fig. 2. SNR in binary CPSK system versus the normalized frequency displacement of two symmetrically interfering channels for a probability of error of 10^{-6}. The interfering channels are at the same power level as the interfered one and the bit streams are synchronous and aligned. No transmission filter is present; the parameter n_R is the number of poles of the Butterworth receiving filter. Rectangular modulating pulses are assumed.

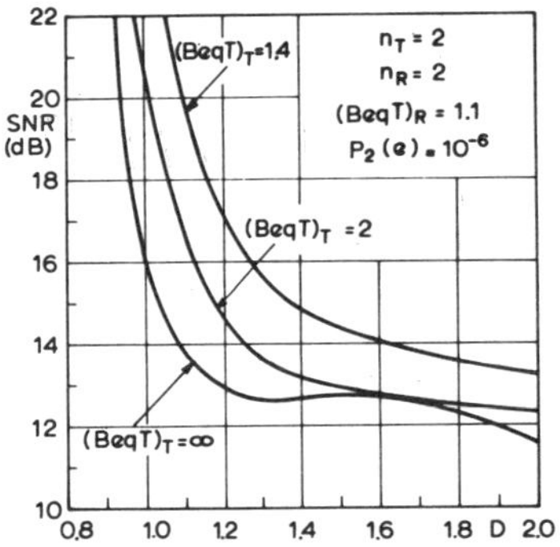

Fig. 3. Binary CPSK system: same situation as in Fig. 2, except for the presence of 2-poles transmission filters with equivalent bandwidth $(B_{eq}T)_T$ labelled on each curve. The receiving filter is a 2-poles Butterworth.

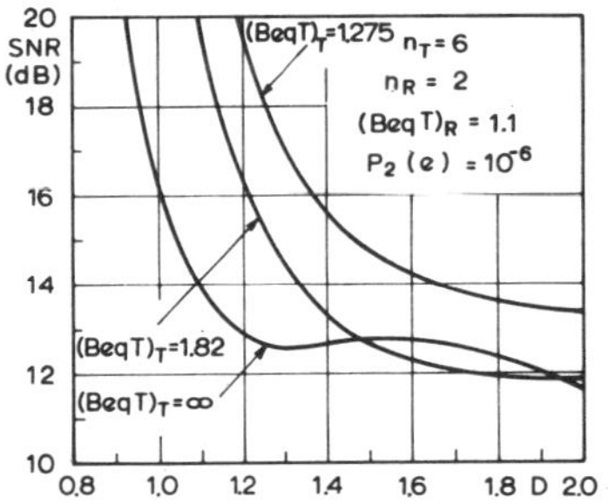

Fig. 4. Binary CPSK system: same situation as in Fig. 2, except for the presence of 6-poles transmission filters.

symbol interference [11], and it is probably a satisfactory choice also for this more general case. In Fig. 2 the SNR necessary to achieve a value of the error probability of 10^{-6} is plotted versus the normalized distance of the interfering channels. The transmission filter is not present and the parameter is the number of poles n_R of the receiving filter. It is seen that the filter with six poles introduces too much distortion, and therefore $n_R = 2$ and $n_R = 4$ will be used in the following. Curves such as those of Fig. 2 can therefore guide the choice of the receiving filter. Figs. 3 and 4 are intended to suggest the choice of the transmitter filter. In these curves the receiving filter is assumed to have two poles and $(B_{eq}T)_R = 1.1$; the error

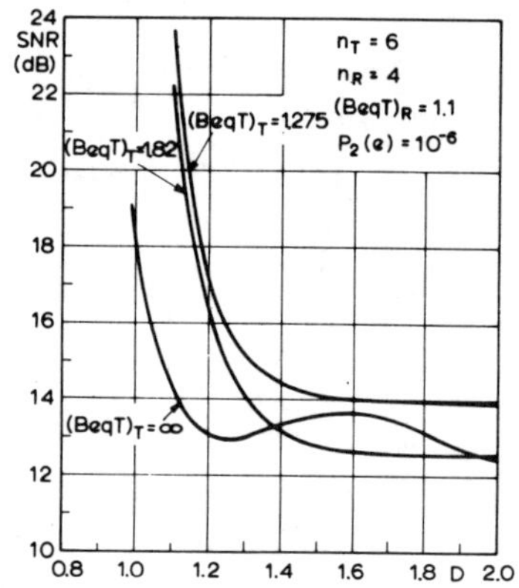

Fig. 5. Same situation as in Fig. 4, except for a 4-poles receiving filter.

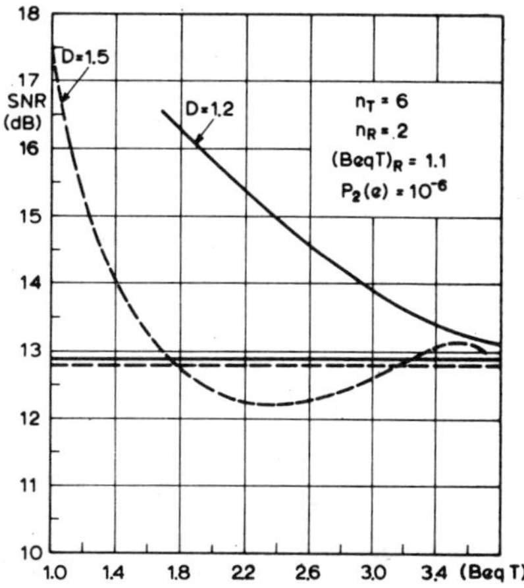

Fig. 6. Binary CPSK system: SNR that assures an error rate of 10^{-6} versus the equivalent bandwidth of the transmission filter. The receiving filter has two poles and the parameter is the normalized frequency displacement of the interfering channels.

probability is still fixed to 10^{-6} and the difference is the number of poles of the transmitter filter. It is seen from Fig. 3 that a transmitter filter with two poles does not improve the performances with respect to the case of no filter (curve labelled $(B_{eq}T)_T = \infty$), whereas for the case of six poles an improvement is possible, for a given range of distances D, provided that the equivalent bandwidth is large enough. This result checks the intuition that the transmitter filter bandwidth has to be large enough to avoid intersymbol interference and steep enough to reduce interchannel interference; only at these conditions its presence causes an improvement of performances.

The case of a 4-poles receiving filter is presented in Fig. 5, which has the same meaning as Fig. 4. Fig. 6 should allow an optimization of the equivalent bandwidth of the transmitter filter. It refers to the case of a 2-poles receiving filter. It is easily seen that with $D = 1.2$ there is no improvement by using the transmitter filter; on the contrary, when $D = 1.5$, a transmitter filter of equivalent bandwidth around 2.4 allows a saving of about 1 dB to get the same error probability. Finally Fig. 7 checks the choice that was made for $(B_{eq}T)_R$. For the 2-poles filter this choice remains the best, whereas for the 4-poles filter the value should be slightly greater than 1 to be optimum.

This presentation of results should outline a possible use of them for the optimum choice of the parameters in order to carefully design the system.

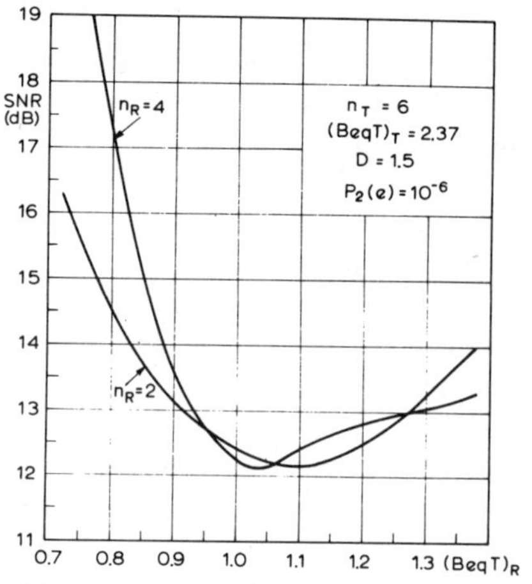

Fig. 7. Binary CPSK system: SNR versus the equivalent bandwidth of the receiving filter.

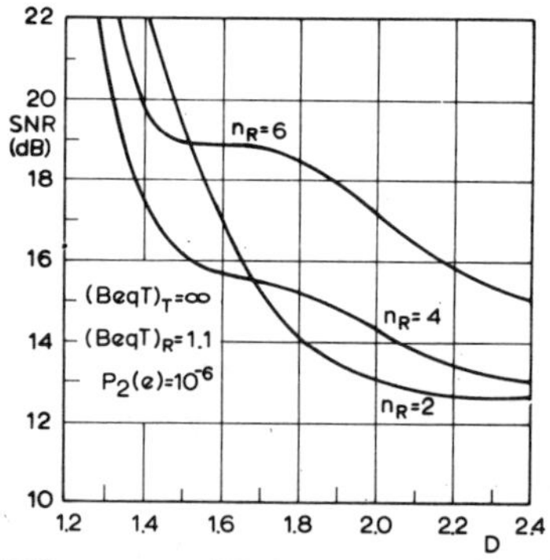

Fig. 8. Binary CPSK system: SNR versus the normalized frequency displacement of two symmetrically interfering channels. The interfering channels are modulated by randomly misaligned bit streams. The average over the ϵ_i was performed using standard quadrature formulas with ten points.

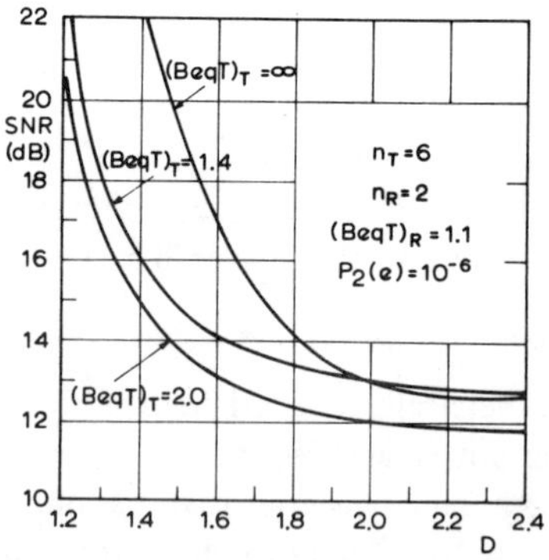

Fig. 9. Binary CPSK system: same situation as in Fig. 8, except for the presence of 6-poles transmission filters.

Results were also obtained on the assumption that the signaling periods of the interfered and interfering channels are randomly misaligned, i.e., when ϵ_i are RV's uniformly distributed in the interval (0, 1).

Fig. 8 gives the results for the same case described in Fig. 2; it can be seen that the average over the ϵ_i shifts the curves towards the right, and therefore a higher distance between adjacent channels is required to get the same SNR. Fig. 9 reproduces, for this case, the situation of Fig. 4; a significant difference now is that the presence of a transmission filter seems to be more effective for the case of aligned signaling

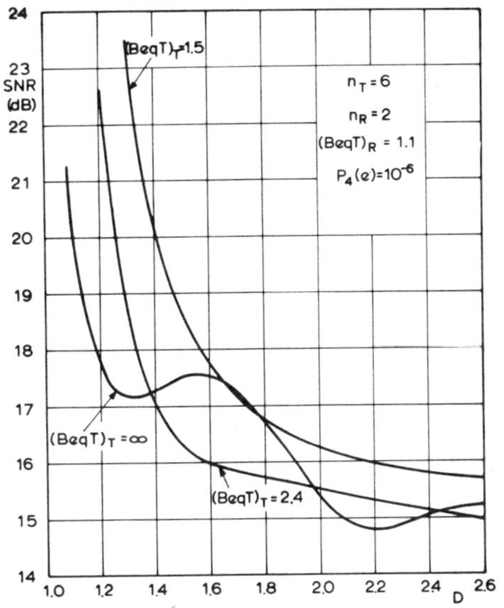

Fig. 10. Quaternary CPSK system: SNR versus the normalized frequency displacement of the two symmetrically located interfering channels. The interfering signals are at the same power level as the interfered one and are modulated by synchronous and aligned bit streams. The modulating pulses are rectangular. The error probability is now the upper bound (24).

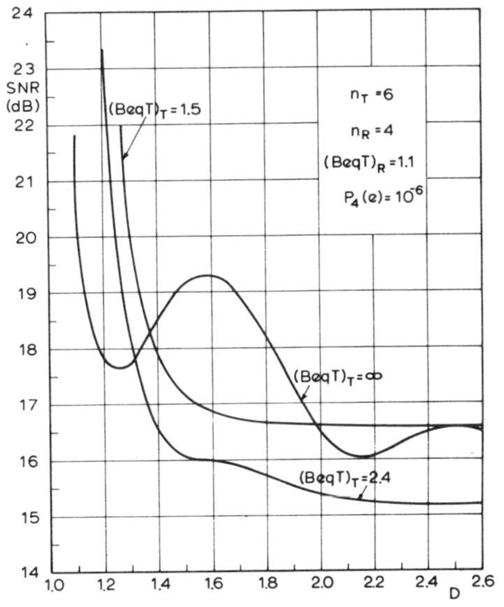

Fig. 11. Same situation as in Fig. 10, except for the number of poles of the receiving filter.

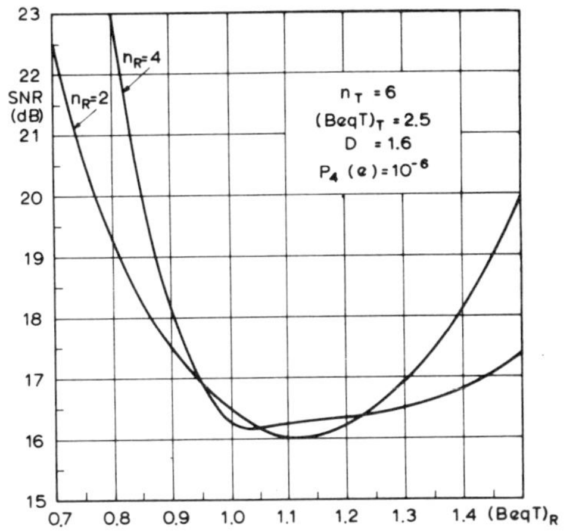

Fig. 12. Quaternary CPSK system: SNR versus the equivalent bandwidth of the receiving filter with a given error probability.

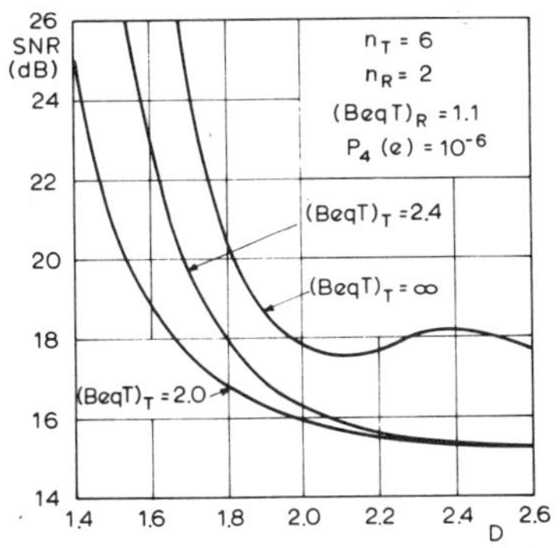

Fig. 13. Quaternary CPSK system: same situation as in Fig. 10, except for the random misalignment among the modulating bit streams of the interfering and interfered signals. The average over the ϵ_i was performed with standard quadrature formulas with ten points.

intervals on the various channels. It was also noticed that a receiver filter with four poles has better performances if compared with the 2-poles filter.

2) 4-level systems: The results presented here are of the same type as those obtained for the binary case. The only difference is that now the error probability $P_4(e)$ is an upper bound to the true value. We have indeed verified, by simulation of some points, that this upper bound is an excellent approximation to the true value, in the sense that the simulation points are practically indistinguishable from the plotted curves. In any case, the knowledge of a lower bound assures the knowledge of the error probability within an interval whose end values differ by no more than a factor of 2. In terms of SNR, this interval is of the order of 0.25 dB for error probabilities around 10^{-6}. In Figs. 10 and 11 the transmission filter is assumed to have six poles and the receiver filter, two or four poles, respectively; both figures plot the SNR necessary to achieve an error probability of 10^{-6} as a function of the normalized distance D, having the equivalent noise bandwidth as a parameter. The possible improvement due to the presence of the transmission filter is now greater than for the binary case (see Figs. 2 and 4 for comparison).

The choice of the bandwidth of the receiver filter is checked finally in Fig. 12 and it is seen again that the 4-poles filter requires a bandwidth slightly narrower. All these curves refer to the case of constant-envelope signals, rectangular modulating pulses, and synchronous and aligned interfering channels.

A situation of no alignment between different modulating bit streams is instead plotted in Fig. 13; this figure must be compared with Fig. 10 to see the effects of the lack of alignment. The most evident result is that the presence of the transmitter filter is more effective and useful when no alignment between channels is provided.

B. Co-Channel Interference

The presence of only one interfering channel at the same frequency as the useful one was considered. The modulating bit stream on the interfering channel is supposed to have an independent timing. The parameters of the system are indicated in Fig. 14 where results are given for $M = 2$, 4, and 8. The SNR for an error rate of 10^{-6} is plotted as a function of the

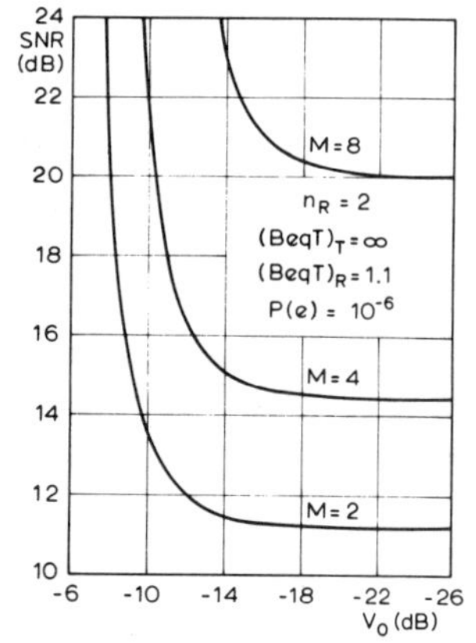

Fig. 14. Co-channel interference effects in binary, quaternary, and octonary CPSK systems. SNR versus the attenuation of the interfering channel. The interference is at the same frequency of the useful channel; the modulating bit streams are assumed to be randomly misaligned.

attenuation of the interfering channel. It is seen that the attenuation has to be of the order of 14, 16, or 20 dB for $M = 2$, 4, and 8, respectively, in order to assure a negligible degradation of performance as compared with the case of no interference.

VII. Conclusions

A very general approach has been presented for the computation of the character error probability in M-ary CPSK systems in the presence of intersymbol, interchannel, and co-channel interferences and additive Gaussian noise.

The results presented take into account a shaping of the modulating signal and of the envelope of the modulated carrier; these results have been particularized to the usual case of rectangular modulating pulses and constant-envelope modulating signals. Phase incoherence between adjacent carriers was assumed and, if necessary, the lack of alignment among the modulating bit streams can also be considered.

The computational procedure is based on the numerical evaluation of only one particular type of integrals, and the method requires the knowledge of only the samples of the impulse responses of the overall channels.

An exact expression is given for binary systems, whereas for $M > 2$ upper and lower bounds to the character error probability are presented. Moreover, the upper bound is conjectured to be a good approximation of the true value of the error probability.

Due to the computational method, the procedure is fast and accurate; some remarks on the accuracy achieved are given in the Appendix.

A good many numerical examples are also presented for 2-, 4-, and 8-level systems. These results should show, in the authors' opinion, the possible utilization of this procedure for system design.

Appendix

The error probability $P_2(e)$ and the upper and lower bound for $P_M(e)$ derived in Section III suffer two types of truncation errors. The first, due to the truncation of Gaussian quadrature rule, has been analyzed in detail in [21]. The second is due to the truncation of the impulse responses of the useful and interfering channels. Although the criterion of increasing the number of samples until $P(e)$ stops changing seems to be sufficient from a practical viewpoint, in this Appendix some results recently obtained in the literature [22], [23] will be extended to the case studied in this paper. The purpose here is not to give a complete solution to the problem, but rather to use the available results and get a rough estimate of the number of samples of the impulse response to be taken into account for each channel.

Since the numerical examples presented in Section VI refer to the case of a rectangular-shaped modulating pulse, only this case will be considered here.

The RV λ defined in (28) is indeed an infinite sum of terms; let us partition it into two disjoint sums of RV's λ^* and λ^{**} as follows:

$$\lambda^* \triangleq \sum_{k=1}^{R} \lambda_k + \sum_i \sum_{k_i=1}^{R_i} \lambda_{k_i} \tag{46}$$

$$\lambda^{**} \triangleq \sum_{k>R} \lambda_k + \sum_i \sum_{k_i>R_i} \lambda_{k_i}. \tag{47}$$

λ^* is the partial sum of the first $R + \Sigma_i R_i$ terms of λ, and λ^{**} represents the sum of the remaining terms. In both definitions (46) and (47), the first sum comes from the intersymbol interference, whereas the double summation is due to the interchannel interference.

Let us consider the integral

$$\int \operatorname{erfc}\left(\frac{\lambda + z_0 \sin \pi/M}{\sqrt{2}\,\sigma_n}\right) dF(\lambda) \tag{48}$$

that arises in the computation of error probability in the special case considered in this Appendix; following the calculations of [22] it is easily seen that such an integral can be bounded as follows:

$$\begin{aligned}
&\int \operatorname{erfc}\left(\frac{\lambda^* + z_0 \sin \pi/M}{\sqrt{2}\,\sigma_n}\right) dF(\lambda^*) \\
&\leqslant \int \operatorname{erfc}\left(\frac{\lambda + z_0 \sin \pi/M}{\sqrt{2}\,\sigma_n}\right) dF(\lambda) \\
&\leqslant [1 - (\sigma_r/\sigma_n)^2]^{-1/2} \\
&\cdot \int \operatorname{erfc}\left\{\frac{\lambda^* + z_0 \sin \pi/M}{\sqrt{2}\,\sigma_n [1 - (\sigma_r/\sigma_n)^2]^{-1/2}}\right\} dF(\lambda^*)
\end{aligned} \tag{49}$$

where σ_r^2 is a quantity such that

$$E\{\exp(A\lambda^{**})\} \leqslant \exp\left(\frac{A^2}{2}\sigma_r^2\right) \qquad \forall A. \tag{50}$$

Notice that the inequalities (49) hold true only when the eye pattern is open (i.e., the peak distortion is less than one), and the RV λ has a symmetric distribution; both these assumptions are verified in the examples presented in Section VI.

Let us first consider the conditional average

$$E\{\exp(A\lambda^{**})\,|\,\epsilon_1, \epsilon_2, \cdots; \varphi_1, \varphi_2, \cdots\}$$

$$= \prod_{k>R} E\{\exp(A\lambda_k)\} \prod_i \prod_{k_i>R_i} E\{\exp(A\lambda_{k_i})|\epsilon_i, \varphi_i\}. \quad (51)$$

To compute an upper bound to the right-hand side of (51), let us start with the terms coming from intersymbol interference, which in this case have the expression

$$\lambda_k = z_k \sin\theta_k, \qquad \theta_k \in \mathfrak{A} \quad (52)$$

where the z_k are the samples of the impulse response of the interfered channel. Salzberg [24] has shown that

$$E\{\exp(A\lambda_k)\} \leqslant \exp\left(\frac{A^2}{2}E\{\lambda_k^2\}\right) = \exp\left[\frac{A^2}{2}f(M)z_k^2\right] \quad (53)$$

where

$$f(M) = \frac{1}{M}\sum_{\theta_j \in \mathfrak{A}} \sin^2\theta_j. \quad (54)$$

For the terms of interchannel interference, recalling (36), it is easy to show that we have

$$\lambda_{k_i} = V_i \zeta_{k_i}(\epsilon_i)\cos(\alpha_{k_i} + \varphi_i + \chi_{k_i}) \quad (55)$$

where

$$\zeta_{k_i}(\epsilon_i) = [(\mu^{(i)}_{k_i+\epsilon_i})^2 + (\nu^{(i)}_{k_i+\epsilon_i})^2]^{1/2}$$

$$\chi_{k_i} = tg^{-1}\frac{\nu^{(i)}_{k_i+\epsilon_i}}{\mu^{(i)}_{k_i+\epsilon_i}} \quad (56)$$

where $\mu^{(i)}_{k_i+\epsilon_i}$ and $\nu^{(i)}_{k_i+\epsilon_i}$ can be obtained from (36), and are equal to

$$\mu^{(i)}_{k_i+\epsilon_i} \triangleq \{[f(t) * h_0(t)]\cos(i\omega_d t)\} * h_R(t)\big|_{t=t_0+(k_i+\epsilon_i)T} \quad (57)$$

$$\nu^{(i)}_{k_i+\epsilon_i} \triangleq \{[f(t) * h_0(t]\sin(i\omega_d t)\} * h_R(t)\big|_{t=t_0+(k_i+\epsilon_i)T}\cdot \quad (58)$$

For all these terms, we get

$$E\{\exp(A\lambda_{k_i})|\epsilon_i, \varphi_i\} \leqslant \exp\left(\frac{A^2}{2}\sigma^2_{k_i}\right). \quad (59)$$

If the two sides of the inequality in (59) are expanded in a Taylor series and the two series are compared term by term, it can be verified that the inequality is satisfied if $\sigma^2_{k_i}$ is given by

$$\sigma^2_{k_i} = V_i^2\zeta^2_{k_i}(\epsilon_i)F_{k_i}(\varphi_i)$$

where

$$F_{k_i}(\varphi_i) = \sum_{\alpha_{k_i}\in\mathfrak{A}} \cos^2(\alpha_{k_i} + \varphi_i + \chi_{k_i}). \quad (60)$$

Combining (53) and (60) we have

$$E\{\exp(A\lambda^{**})|\epsilon_1, \epsilon_2, \cdots; \varphi_1, \varphi_2, \cdots\}$$

$$\leqslant \exp\left\{\frac{A^2}{2}\left[f(M)\sum_{k>R} z_k^2 + \sum_i V_i \sum_{k_i>R_i}\zeta^2_{k_i}(\epsilon_i)F_{k_i}(\varphi_i)\right]\right\} \quad (61)$$

and

$$E\{\exp(A\lambda^{**})\}$$

$$\leqslant \exp\left\{\frac{A^2}{2}\left[f(M)\sum_{k>R} z_k^2 + \sum_i V_i^2 \max_{\epsilon_i,\varphi_i}\sum_{k_i>R_i}\zeta^2_{k_i}(\epsilon_i)F_{k_i}(\varphi_i)\right]\right\} \quad (62)$$

so that finally we get

$$\sigma_r^2 = f(M)\sum_{k>R} z_k^2 + \sum_i V_i^2 \max_{\epsilon_i,\varphi_i}\sum_{k_i>R_i}\zeta^2_{k_i}(\epsilon_i)F_{k_i}(\varphi_i). \quad (63)$$

The difference between the upper and lower bound for the integral (48) is a monotone decreasing function of the number of samples taken into account, so that this integral can be computed within any desired accuracy.

We made some checks on the results outlined in this Appendix in order to have a rough indication on the number of samples that must be used in the computation.

If the value of $(\sigma_r/\sigma_n)^2$ is chosen to be of the order of 10^{-2} the integral (48) will be bounded within an interval of about 0.05 dB, as can be seen computing the bounds in (49).

This accuracy requires a value of σ_r^2 not greater than $5 \cdot 10^{-5}$ when the SNR is 20 dB. It was verified that no more than ten samples on each channel are sufficient to assure this accuracy in all the cases considered in the numerical examples.

ACKNOWLEDGMENT

The authors wish to thank one reviewer, in particular, for his comments leading to improvements of this paper, and their colleague, M. Pent, for his critical reading of the manuscript.

REFERENCES

[1] C. R. Cahn, "Performance of digital phase-modulation communication systems," *IRE Trans. Commun. Syst.*, vol. CS-7, pp. 3–6, May 1959.
[2] R. Arthurs and H. Dym, "On the optimum detection of digital signals in the presence of white Gaussian noise—A geometric interpretation and a study of three basic data transmission systems," *IRE Trans. Commun. Syst.*, vol. CS-10, pp. 336–372, Dec. 1962.
[3] V. K. Prabhu, "Error rate considerations for digital phase modulation systems," *IEEE Trans. Commun. Technol.*, vol. COM-17, pp. 33–42, Feb. 1969.
[4] L. Calandrino, G. Crippa, and G. Immovilli, "Intersymbol interference in binary and quaternary PSK and DCPSK systems," *Alta Freq.*, part I, vol. 38, pp. 337–344, May 1969; *Alta Freq.*, part II, vol. 38, pp. 562–569, Aug. 1969.
[5] V. K. Prabhu, "Performance of coherent phase-shift-keyed systems with intersymbol interference," *IEEE Trans. Inform. Theory*, vol. IT-17, pp. 418–431, July 1971.
[6] S. Benedetto and E. Biglieri, "Performance of *M*-ary PSK systems in the presence of intersymbol interference and additive noise," *Alta Freq.*, vol. 41, pp. 225–239, Apr. 1972.
[7] A. S. Rosenbaum, "PSK error performance with Gaussian noise and interference," *Bell Syst. Tech. J.*, vol. 47, pp. 413–442, Feb. 1968.
[8] V. K. Prabhu, "Error rate consideration for coherent phase-shift-

keyed systems with co-channel interference," *Bell Syst. Tech. J.*, vol. 48, pp. 743–767, Mar. 1969.

[9] ——, "Error probability upper bound for coherently detected PSK signals with co-channel interference," *Electron. Lett.*, vol. 5, pp. 383–385, Aug. 1969.

[10] A. S. Rosenbaum, "Binary PSK error probabilities with multiple cochannel interferences," *IEEE Trans. Commun. Technol.*, vol. COM-18, pp. 241–253, June 1970.

[11] V. Castellani and M. Sant'Agostino, "Intersymbol and interchannel interference effects in PCM signal transmission on radio links: Binary coherent phase modulation," *Alta Freq.*, vol. 40, pp. 389–397, May 1971.

[12] L. Calandrino, G. Corazza, G. Crippa, and G. Immovilli, "Intersymbol, interchannel and cochannel interferences in binary and quaternary PSK systems," *Alta Freq.*, vol. 40, pp. 407–420, May 1971.

[13] E. Biglieri, V. Castellani, R. Dogliotti, M. Pent, and M. Sant'Agostino, "Interchannel interference in PCM signal transmission," in *Conf. Rec. IEEE Int. Conf. Communications*, Montreal, Canada, June 1971.

[14] C. Colavito and M. Sant'Agostino, "Multiple co-channel and interchannel interference effects in binary and quaternary PSK radio systems," in *Conf. Rec. IEEE Int. Conf. Communications*, Philadelphia, Pa., June 1972.

[15] E. Y. Ho and Y. S. Yeh, "A new approach for evaluating the error probability in the presence of intersymbol interference and additive Gaussian noise," *Bell Syst. Tech. J.*, vol. 49, pp. 2249–2265, Nov. 1970.

[16] ——, "Error probability of a multilevel digital system with intersymbol interference and Gaussian noise," *Bell Syst. Tech. J.*, vol. 50, pp. 1017–1023, Mar. 1971.

[17] M. I. Celebiler and O. Shimbo, "The probability of error due to intersymbol interference and Gaussian noise in digital communication systems," COMSAT, Tech. Memo., May 5, 1970.

[18] S. Benedetto, E. Biglieri, and R. Dogliotti, "Probabilità di errore per trasmissione numerica a più livelli e codificazione lineare," *Alta Freq.*, vol. 40, pp. 725–732, Sept. 1971.

[19] O. Shimbo and M. I. Celebiler, "The probability of error due to intersymbol interference and Gaussian noise in digital communication systems," *IEEE Trans. Commun. Technol.*, vol. COM-19, pp. 113–119, Apr. 1971.

[20] V. I. Krylov, *Approximate Calculation of Integrals.* New York: McMillan, 1962.

[21] S. Benedetto, G. De Vincentiis, and A. Luvison, "Error probability in the presence of intersymbol interference and additive noise for multilevel digital signals," *IEEE Trans. Commun.*, vol. COM-21, pp. 181–188, Mar. 1973.

[22] V. K. Prabhu, "Some considerations of error bounds in digital systems," *Bell Syst. Tech. J.*, vol. 52, pp. 3127–3151, Dec. 1971.

[23] Y. S. Yeh and E. Y. Ho, "Improved intersymbol interference error bounds in digital systems," *Bell Syst. Tech. J.*, pp. 2585–2598, Oct. 1971.

[24] B. R. Saltzberg, "Intersymbol interference error bounds with application to ideal band-limited signaling," *IEEE Trans. Inform. Theory*, vol. IT-14, pp. 563–568, July 1968.

An Error-Probability Upper Bound for Coherent Phase-Shift Keying with Peak-Limited Interference

ARNOLD S. ROSENBAUM, MEMBER, IEEE, AND FREDERICK E. GLAVE, MEMBER, IEEE

Abstract—An upper bound on the probability of error for coherent phase-shift-keyed (CPSK) systems operating in the presence of peak-limited interference is derived. Such a bound is particularly useful in analyzing the effect of co-channel and adjacent channel interference on CPSK systems. The bound is often tight enough to be employed as a system design tool, and is computationally simple to apply.

I. Introduction

SEVERAL techniques are now available for calculating the error probability (Pe) of a coherently detected phase-shift-keyed (CPSK) transmission link when the signal is subjected to certain significant degradations. Broadly speaking, we have analysis for CPSK where the signal is additively disturbed by a combination of noise[1] and interference. An interference might be another PSK signal on the same nominal carrier frequency, termed co-channel interference, or an adjacent-channel interference, where the undesired signal is primarily outside of the passband of the wanted signal. There is also the self-generated interference caused by lingering effects of past pulses which is called intersymbol interference.

The cornerstone of any analysis that leads to the exact value of Pe is an adequate probabilistic description of all the interferences present. In the case of CPSK reception (implemented by an ideal phase discriminator followed by an ideal quantizer) an adequate description consists of the two-dimensional first-order probability density function (pdf) of the interference complex envelope in the IF passband of the receiver.

Perhaps the pdf itself is not directly used, but some set of data, for example the infinite set of moments of all orders, which completely specifies the pdf, is required for an exact analysis. When less than complete knowledge of such statistics of the particular interference is available, we must of course be content with approximating or bounding types of analyses.

Specifically, an exact statistical description of either co-channel or adjacent-channel interference in actual CPSK systems is practically impossible to obtain. However, it is usually true that these interferences are circular symmetric (CS) [a property that will be defined in Section II], and moreover one can usually determine empirically the rms and peak values of the interference envelope. We present in this paper an upper-bound analysis for CPSK that depends solely on these two key parameters of the interference. In order to emphasize the role of the peak envelope information, it will be called the limited peak bound.

It is seen to be a marked improvement over both the crude "eye diagram" bound (which also uses the peak envelope value) and the noise conversion approximation (which uses only the rms envelope) especially in the important range of large carrier-to-noise ratio (CNR). Noise conversion refers to the old resort of treating a statistically unknown interference as if it were additional noise of the same average power. Of course, the mean-square envelope, which is twice the power of the interference, is a strong factor, but it alone cannot be used to give reliable performance estimates in many cases. However, it, in conjunction with the peak envelope value, is sufficient information to provide a bound that is surprisingly tight over a wide range in CNR.

The concept of the limited-peak bound is quite simple. It is to determine, of all admissible interference envelope probability distribution functions that have the given rms and peak values, that one which would produce the largest Pe. Clearly, if the supremum exists and can be found, it is a bound on the Pe associated with any other admissible interference with the same rms and peak envelope. It turns out that a maximizing distribution (or, equivalently, a maximizing pdf) can be found, and that it is simple enough so that Pe is easily evaluated for that distribution to produce the bound.

The "maximizing probability distribution" approach was successfully used by Glave [1] in bounding Pe due to intersymbol interference for baseband pulse-transmission systems. We have here the natural extension of that bound to two-dimensional signal space with noise and CS interference. Actually, the CS property of the interference effectively reduces the problem to one dimension, but the functions involved are more complicated. The proof of the maximizing distribution is similar to that used by Glave, and is based on the work of Smith [2].

II. Signal, Noise, and Interference Representations

We are concerned with phase detection of an M-ary PSK signal such as [3, p. 233]

$$s(t) = \mathrm{Re} \sum_{k=-\infty}^{\infty} e^{j(\phi_k + \omega_0 t)} p(t - kT), \tag{1}$$

$$\phi_k \in \left\{0, \frac{2\pi}{M}, \cdots, \frac{M-1}{M} 2\pi\right\}, \tag{2}$$

Paper approved by the Associate Editor for Communication Theory of the IEEE Communications Society for publication after presentation at the International Conference on Communications, Seattle, Wash., June 1973. Manuscript received March 22, 1973; revised August 10, 1973.

A. S. Rosenbaum is with Bell Laboratories, Holmdel, N.J. 07733.
F. E. Glave is with Bell-Northern Research, Ottawa, Ont. Canada.

[1] Throughout, noise will mean a Gaussian-distributed, bandpass random process that is the model for thermally generated internal noise.

Reprinted from *IEEE Trans. Comm.*, vol. COM-22, pp. 6-16, Jan. 1974.

which has been corrupted by the addition of interference

$$i(t) = \mathrm{Re}\, r(t)\, e^{j[\xi(t)+\omega_0 t]} \tag{3}$$

and Gaussian noise $n(t)$. It is convenient to deal with the complex envelopes of these quantities. With ω_0 the radian carrier frequency, these are

$$\overline{S} = \sum_{k=-\infty}^{\infty} p(t - kT)\, e^{j\phi_k}, \tag{4}$$

$$\overline{I} = r(t)\, e^{j\xi(t)}, \tag{5}$$

and

$$\overline{N} = \{n(t) + j\hat{n}(t)\} e^{-j\omega_0 t}. \tag{6}$$

We assume $p(t)$ is an ideal Nyquist pulse function (any intersymbol interference that meets the criteria of the bound may be included in the interference term) and so the signal term is simply

$$\overline{S} = p(0)\, e^{j\phi_k}, \qquad t_k = kT \tag{7}$$

for integer k.

A random process is defined to be circular symmetric (CS) if its complex envelope modulus is independent of the phase, and the phase, modulo 2π, is uniformly distributed. For example, the interference random process will be CS if $r(t_0)$ and $\xi(t_0)$ are independent random variables for any t_0, and $\xi(t_0)$ modulo 2π is uniformly distributed. Furthermore, the interference will be peak limited, with peak envelope R, if the pdf of $r(t_0)$ has support with finite upper end point R. We note that a CS process and its envelope modulus have the same peak limit. Also, it is clear that Gaussian noise is CS, but is not peak limited. One may usually assume that the interference and the wanted signal carrier differ by a randomly varying phase angle that is modulo 2π uniform. Possible exceptions are, of course, intersymbol interference and constant delay echoes. On the other hand, adjacent channel and most co-channel interferences will have the requisite random phase relationship to cause them to be CS.

III. Error Probability for Noise and CS Interference

We will now recall a well-known form for the exact value of Pe for binary CPSK with the only signal impairments being a combination of noise and CS interference. This result will then be used in the bound derivation given in the following section. Conditioned on a fixed interference modulus and angle, the input to the phase angle discriminator is

$$\overline{\Psi} = p(0)\, e^{j\phi_k} + \overline{N} + re^{j\xi} \tag{8}$$

at a sample time $t = kT$. We assume that ϕ_k can be either 0 or π, so that the decision regions are the right and left half complex planes,[2] and an error will occur if $\phi_k = 0$ and $\overline{\Psi}$ is in the left-half plane. Therefore we may restrict attention to the real axis projections of the components involved, and an error occurs whenever $p(0) + \nu + r \cos \xi < 0$, where $\nu = \mathrm{Re}\, \overline{N}$ is a zero mean Gaussian random variable with variance σ_n^2. The probability of error is written as

$$\Pr\{\mathrm{Re}\, \overline{\Psi} < 0\} = \Pr\{p(0) + \nu + r\cos\xi < 0\} = \frac{1}{2}\,\mathrm{erfc}\left(\frac{p(0) + r\cos\xi}{\sqrt{2}\,\sigma_n}\right). \tag{9}$$

The unconditioned probability is the expected value of this quantity with respect to both r and ξ. To remove the ξ conditioning, expand the complementary error function into its Taylor series about $(p(0)/\sqrt{2}\,\sigma_n) \equiv \sqrt{\gamma}$. We first get

$$\frac{1}{2}\,\mathrm{erfc}\left(\sqrt{\gamma} + \sqrt{\gamma}\frac{r}{p(0)}\cos\xi\right) = \frac{1}{2}\,\mathrm{erfc}(\sqrt{\gamma}) + \frac{e^{-\gamma}}{\sqrt{\pi}} \sum_{l=1}^{\infty} \frac{(-1)^l H_{l-1}(\sqrt{\gamma})}{l!} \sqrt{\gamma}^{\,l} \left[\frac{r}{p(0)}\right]^l \cos^l \xi \tag{10}$$

where H is the Hermite polynomial, related to derivatives of $\exp(-x^2)$.

Since

$$E\{\cos^l \xi\} = \frac{1}{2\pi}\int_0^{2\pi} \cos^l \xi \, d\xi = \begin{cases} 2^{-l}\begin{pmatrix} l \\ \frac{1}{2}l \end{pmatrix}, & l \text{ even} \\ 0, & l \text{ odd} \end{cases}, \tag{11}$$

and expectation is linear, we then have

$$\mathrm{Pe} \mid r = \frac{1}{2}\,\mathrm{erfc}(\sqrt{\gamma}) + \frac{e^{-\gamma}}{\sqrt{\pi}} \sum_{l=1}^{\infty} \frac{H_{2l-1}(\sqrt{\gamma})\,\gamma^l}{2^{2l}\, l!\, l!} \left[\frac{r}{p(0)}\right]^{2l} \tag{12}$$

when $\phi_k = 0$. By symmetry this is also the expression that would result if $\phi_k = \pi$, and so the above is $\mathrm{Pe} \mid r$.

The unconditioned value of Pe is given by

$$\mathrm{Pe} = \int_0^{\infty} (\mathrm{Pe} \mid r)\, dF(r), \tag{13}$$

where the Stieltjes integral is used to admit distribution functions $F(r)$ that are not absolutely continuous. For brevity, we will denote $\mathrm{Pe} \mid r$ as $q(r)$, where the dependence of this function on γ will be implicit.

In what follows, we will bound

$$\mathrm{Pe} = \int_0^{\infty} q(r)\, dF(r), \tag{14}$$

subject to the second-moment constraint

$$\int_0^{\infty} r^2\, dF(r) \leq \sigma_r^2 \tag{15a}$$

and also the peak interference envelope constraint $r \leq R$, i.e.,

$$\int_{R^+}^{\infty} dF(r) = 0. \tag{15b}$$

[2] The complex plane obviously coincides in this case with the two-dimensional signal space [4] on which bandpass processes are usually represented.

It can easily be shown that the second moment of the complex envelope modulus is twice the average power in the real signal, and for a CS process the real signal is evenly distributed about zero, which makes the power equal to the variance.

IV. Optimization Theory

We begin by defining $\mathcal{F}$ to be the infinite set of all distribution functions F whose support is contained in $[0, R]$. Equivalently, $\mathcal{F}$ is the class of functions that are nondecreasing on $[0, R]$, equal to 0 for $r < 0$, and equal to 1 for $r > R$. Consider

$$\text{Pe}(F) = \int_0^\infty q(r)\, dF(r). \tag{16}$$

as a functional, mapping $\mathcal{F}$ to the (positive) real line. We seek the maximum of this functional, if it exists, and the corresponding maximizing distribution $F_0 \in \mathcal{F}$ subject to the second moment constraint (15a).

This maximization problem might be approached by classical variational calculus techniques. However, through the application of some relatively simple—yet elegant—optimization theory, a set of necessary and sufficient conditions for a maximizing distribution can be obtained. The sufficiency criterion will then prove that a certain distribution, conjectured to produce maximum Pe, is indeed an extremal. The problem of finding extremals on a set of bounded support distribution functions under a continuous functional mapping has been studied by Smith [2]. His results that bear on our problem are summarized in the following two theorems.

Theorem 1: Let

$$C = \sup \text{Pe}(F), \tag{17}$$

where the supremum is taken over all $F \in \mathcal{F}$ for which

$$\int_0^\infty r^2\, dF(r) \leqslant \sigma_r^2.$$

If Pe is a convex cap, weakly differentiable continuous mapping from $\mathcal{F}$ into the reals, then C is attained with a random variable having distribution function $F_0 \in \mathcal{F}$. A necessary and sufficient condition for F_0 to produce C is that for some constant $\lambda \geqslant 0$,

$$\int_0^\infty [q(r) - \lambda r^2]\, dF(r) \leqslant \text{Pe}(F_0) - \lambda \sigma_r^2, \qquad \forall F \in \mathcal{F}. \tag{18}$$

Corollary 1: Let F_0 be an arbitrary distribution function in $\mathcal{F}$ such that

$$\int_0^\infty r^2\, dF_0(r) \leqslant \sigma_r^2. \tag{19}$$

Let E_0 denote the points of increase of F_0 on $[0, R]$. Then F_0 is an extremal if and only if for some $\lambda \geqslant 0$

$$q(r) \leqslant \text{Pe}(F_0) + \lambda(r^2 - \sigma_r^2), \qquad \forall r \in [0, R] \tag{20a}$$

$$q(r) = \text{Pe}(F_0) + \lambda(r^2 - \sigma_r^2), \qquad \forall r \in E_0. \tag{20b}$$

Theorem 2: If $q(r)$ is analytic on $[0, R]$, then E_0 is a finite set of points.

The functional Pe $(\cdot)$ and the space $\mathcal{F}$ for the present problem are similar to those considered by Smith. In our case one may easily show that $\mathcal{F}$ is convex, and compact in the Levy metric [8, p. 215], and since Pe $(\cdot)$ is a linear functional it is trivially convex cap and weakly differentiable. The continuity of Pe $(\cdot)$ follows immediately from the Helly-Bray Theorem [8, p. 182], since $q(r)$ is bounded on $[0, R]$. Also, it is straightforward to verify the analyticity of $q(r)$ required in Theorem 2. The reader is referred to [2] for definitions and a complete proof of the theorems.

The importance of these results lies in the fact that the maximizing distribution we seek is simply a composition of a finite number of step functions, and the corresponding pdf is zero except for a finite number of impulses, or mass points. As will be seen later, propitious interpretation of a certain function involving $q(r)$ leads to the correct discrete pdf. At that point it is only a candidate for a maximizing density, but Corollary 1 then provides a simple means of proving the conjecture that it is indeed maximizing, and thereby gives an upper bound.

V. The Limited-Peak Bound

Let U be the unit step function

$$U(r) = \begin{Bmatrix} 0, & r < 0 \\ 1, & r \geqslant 0 \end{Bmatrix},$$

and define $\rho \triangleq (\sigma_r^2/R^2)$. We will now show that

$$F_0(r) = (1 - \rho)\, U(r) + \rho\, U(r - R) \tag{21}$$

is a maximizing distribution for a large, nontrivial region in the R, γ plane. The corresponding upper bound is easily found from (14) and (12) to be

$$\text{Pe} \leqslant \text{Pe}(F_0) = \frac{1}{2}\, \text{erfc}(\sqrt{\gamma}) + \rho \frac{e^{-\gamma}}{\sqrt{\pi}} \sum_{l=1}^\infty \frac{H_{2l-1}(\sqrt{\gamma})}{2^{2l} l!\, l!} \cdot [\sqrt{\gamma} R/p(0)]^{2l}. \tag{22}$$

First, however, let us see why the two-impulse pdf, i.e.,

$$f_0(r) = \frac{dF_0}{dr} = (1 - \rho)\, \delta(r) + \rho\, \delta(r - R), \tag{23}$$

is a plausible maximizing density. The functional kernel $q(r)$ can be shown to be nondecreasing in r, with $q(0) = \frac{1}{2}$ erfc $(\sqrt{\gamma})$, $q(\infty) = \frac{1}{2}$. This is true for any $\gamma \geqslant 0$, furthermore q is continuous in r if $\gamma < \infty$. Since $q(r)$ increases with r, the maximizing density should in general have the greatest mass possible concentrated at the largest values of r. In the absence of the second-moment constraint, this would obviously be accomplished by placing unit mass at $r = R$, i.e., $f(r) = \delta(r - R)$. This is not possible, however, if the rms modulus is required to be strictly less than the peak. For values of ρ less than unity, then, consider placing as much mass at $r = R$ as is consistent with the second-moment constraint. The remaining mass must then be placed at $r = 0$ where it does not contribute to the second moment. Thus we would get an impulse at $r = R$ whose value is ρ (which contributes σ_r^2 to the second moment), and the remaining $1 - \rho$ of the mass is accounted for by an impulse at $r = 0$. One might expect this to be correct, i.e., yield a maximizing density, if $q(r)$ rises steeply enough over the interval $[0, R]$ so that any tradeoff, taking mass away from the impulse at R and redistributing it in $[0, R)$ in such a way as to keep the second moment constant, results in no greater value of Pe. Indeed, we will see that there is a range of R, with the upper limit $R_{\max}$ depending on σ_n and $p(0)$, where that statement is true.

We will henceforth denote the second term in (12) as the function $\Sigma(r, \gamma)$, whereupon

$$q(r) = \frac{1}{2} \operatorname{erfc}(\sqrt{\gamma}) + \Sigma(r, \gamma). \tag{24}$$

We do this primarily to save writing out the summation; however the convenience of this abbreviation will be needed throughout. The bound (22) is given in this notation as

$$\mathrm{Pe} \leqslant \mathrm{Pe}(F_0) = \frac{1}{2} \operatorname{erfc}(\sqrt{\gamma}) + \rho \Sigma(R, \gamma). \tag{25}$$

The necessary and sufficient conditions, given by the corollary (20), for (21) to be a maximizing distribution are that a positive constant λ exists such that

$$\frac{1}{2} \operatorname{erfc}(\sqrt{\gamma}) + \Sigma(r, \gamma)$$
$$\leqslant \frac{1}{2} \operatorname{erfc}(\sqrt{\gamma}) + \rho \Sigma(R, \gamma) + \lambda(r^2 - \sigma_r^2), \quad \forall r \in [0, R], \tag{26}$$

with equality to hold at $r = 0, R$. We choose λ by forcing the equality at either point. Thus, setting $r = 0$ in the above, and using $\Sigma(0, \gamma) = 0$, we find

$$\lambda = \frac{1}{R^2} \Sigma(R, \gamma). \tag{27}$$

Substituting this expression into (26) gives

$$\Sigma(r, \gamma) \overset{?}{\leqslant} \frac{r^2}{R^2} \Sigma(R, \gamma), \quad \forall r \in [0, R]. \tag{28}$$

Now equality for $r = 0, R$ is achieved, and it remains to verify the inequality for $0 < r < R$.

If we divide both sides of (28) by r^2, and let

$$g(r) \triangleq \frac{1}{r^2} \Sigma(r, \gamma), \tag{29}$$

then (28) becomes

$$g(r) \overset{?}{\leqslant} g(R), \quad 0 < r < R. \tag{30}$$

This is obviously true if g is a nondecreasing function on $[0, R]$. We will now show that the inequality holds for any R, where $0 \leqslant R \leqslant R_{\max}$, if and only if g is nondecreasing on $[0, R_{\max}]$.

The sufficiency of this property is again obvious, and we need only consider the necessity of a nondecreasing g. Assume g is decreasing on some interval in $[0, R_{\max}]$. Then there exists $r_1, r_2 \in [0, R_{\max}]$ such that $0 < r_1 < r_2 \leqslant R_{\max}$ and $g(r_1) > g(r_2)$. Then with $r_2 = R$, we have

$$g(r_1) > g(R), \quad 0 < r_1 < R,$$

but this violates the inequality of the corollary and so necessity is proved. Thus $R_{\max}$ is the upper limit of the interval $[0, R_{\max}]$ of monotonicity of g.

We shall see later that for most practical cases of interest $R_{\max} \geqslant R$ and (25) is thus generally applicable. Nevertheless, it is interesting to determine the limited-peak upper bound when $R > R_{\max}$. To this end the following interpretation of $g(r)$ is useful.

We may think of $g(r)$ as the incremental contribution to error probability per unit of second moment used up by mass placed in the interference envelope pdf at point r. For example, if $g(r_2) > g(r_1)$, then mass placed at point r_2 adds more to Pe than mass placed at r_1 for equal expenditure of envelope second moment. Thus if we find a point $R_{\sup}$ such that

$$g(R_{\sup}) = \sup_{0 \leqslant r \leqslant R} g(r) \tag{31}$$

then, according to the above, we should construct a maximizing distribution by exhausting the second moment available (i.e., σ_r^2) with mass concentrated solely at the point $R_{\sup}$. If less than unit mass is accounted for thereby, the remainder should be placed at $r = 0$ where it does not add to the second moment.

For example, we know from the monotonicity of g on $[0, R_{\max}]$ that when $R \leqslant R_{\max}$, then $R_{\sup} = R$ satisfies (31). The previous reasoning is therefore entirely consistent with the maximizing distribution, and associated upper bound, already established for the case $R \leqslant R_{\max}$.

Now consider the case $R > R_{\max}$. We proceed again by constructing a maximizing distribution according to the method given in the foregoing heuristic discussion. Suppose that $\sigma_r < R_{\sup}$, so that it is possible to exhaust the second-moment allowance σ_r^2 by concentrating $(\sigma_r/R_{\sup})^2$ mass at $R_{\sup}$ and still have $1 - (\sigma_r/R_{\sup})^2$ mass left over. Thus we would get the distribution

$$F_0(r) = (1 - \rho^*) U(r) + \rho^* U(r - R_{\sup}) \tag{32}$$

where

$$\rho^* \triangleq (\sigma_r/R_{\sup})^2. \tag{33}$$

Next, we verify the corresponding bound

$$\mathrm{Pe} \leqslant \mathrm{Pe}(F_0) = \frac{1}{2} \operatorname{erfc}(\sqrt{\gamma}) + \rho^* \Sigma(R_{\sup}, \gamma), \quad \sigma_r \leqslant R_{\sup}, \tag{34}$$

by applying the sufficiency criteria, (20). We must find a constant $\lambda > 0$ such that

$$\frac{1}{2} \operatorname{erfc}(\sqrt{\gamma}) + \Sigma(r, \gamma) \leqslant \frac{1}{2} \operatorname{erfc}(\sqrt{\gamma}) + \rho^* \Sigma(R_{\sup}, \gamma)$$
$$+ \lambda(r^2 - \sigma_r^2), \quad \forall r \in [0, R], \tag{35}$$

with the equality to hold at $r = 0, R_{\sup}$. The steps that would follow are obviously identical to (26) through (28) if $R_{\sup}$ is substituted for R, and therefore lead to

$$\Sigma(r, \gamma) \leqslant \frac{r^2}{R_{\sup}^2} \Sigma(R_{\sup}, \gamma), \quad \forall r \in [0, R]. \tag{36}$$

The equality required at $r = 0$, R_{sup} is satisfied, and furthermore the inequality everywhere else is an immediate consequence of the definition of R_{sup}. Thus (34) is established. Furthermore, (34) is the general form of the bound when $\sigma_r \leqslant R_{sup}$, and (25) is simply the special case that obtains when $R \leqslant R_{max}$, i.e., when $R_{sup} \equiv R$.

Finally, when $\sigma_r > R_{sup}$, it is obviously impossible to obtain a valid pdf with mass place solely at R_{sup} and 0 since both points are less than σ_r, and hence the resulting distribution would yield a second moment strictly less than σ_r^2. To jointly satisfy the unity total mass and σ_r^2 second-moment requirements of the pdf there are only two possibilities: 1) some mass must be placed at points that lie both above and below σ_r; or 2) unit mass must be concentrated at σ_r. First, consider case 1. If a certain fraction of the total pdf mass is to be located at points $r \geqslant \sigma_r$, our previous argument concerning $g(r)$ demands, for a maximizing distribution, that this mass must be concentrated at R'_{sup}, where

$$g(R'_{sup}) = \sup_{\sigma_r \leqslant r \leqslant R} g(r), \tag{37}$$

and that any mass below σ_r should be concentrated at R_{sup} and/or 0. The proper maximizing division of mass among R'_{sup}, R_{sup}, and 0 depends further on the nature of $g(r)$.

In our case it so happens that $g(r)$ is unimodal. Consequently, when $R_{sup} < \sigma_r$, it follows that $g(r)$ is strictly monotone decreasing on $[\sigma_r, R]$, and hence $R'_{sup} = \sigma_r$. This leads to the conclusion that case 2 is the maximizing distribution, i.e.,

$$F_0(r) = U(r - \sigma_r), \qquad \sigma_r > R_{sup}, \tag{38}$$

and the associated upper bound is given by

$$\text{Pe} \leqslant \text{Pe}(F_0) = \frac{1}{2}\text{erfc}(\sqrt{\gamma}) + \Sigma(\sigma_r, \gamma), \qquad \sigma_r > R_{sup}. \tag{39}$$

The proof that $g(r)$ is unimodal, and (surprisingly) the application of the sufficiency criterion to establish (35), is rather involved. Since the case $\sigma_r > R_{sup}$ is of negligible practical significance, these proofs have not been included. In fact, it turns out that R_{max} is sufficiently large for γ values of interest that the bound for the special case $R \leqslant R_{max}$ applies in nearly all practical interference situations. For example, if $\gamma = 2$, which corresponds to a CNR of only 3 dB, the value of R_{max} is approximately $0.9p(0)$, and furthermore $R_{max} > 0.9p(0)$ if $\gamma > 2$. A fairly complete description of R_{max} is given by the graphs in Figs. 1 and 2. In Fig. 1, the abscissa is $\sqrt{\gamma}$ and the ordinate is $R_{max}/p(0)$. Fig. 2 gives the same information in decibel equivalents, wherein the abscissa is

$$\text{CNR} \triangleq 10 \log_{10} \gamma, \tag{40}$$

and the ordinate represents the carrier-to-R_{max} ratio

$$\text{CRR} \triangleq 20 \log_{10} p(0)/R_{max}. \tag{41}$$

R_{max} could not be determined analytically because $g(r)$ is not yet expressible in closed form. Resort was made to a digital computer solution, using standard techniques. The heart of the program is a routine that rapidly and efficiently evaluates the series for $\Sigma(r, \gamma)$, a series that is also central to the evaluation of the bound itself. A listing and brief description of a PL-1 routine for computing the bound is given in Appendix A.

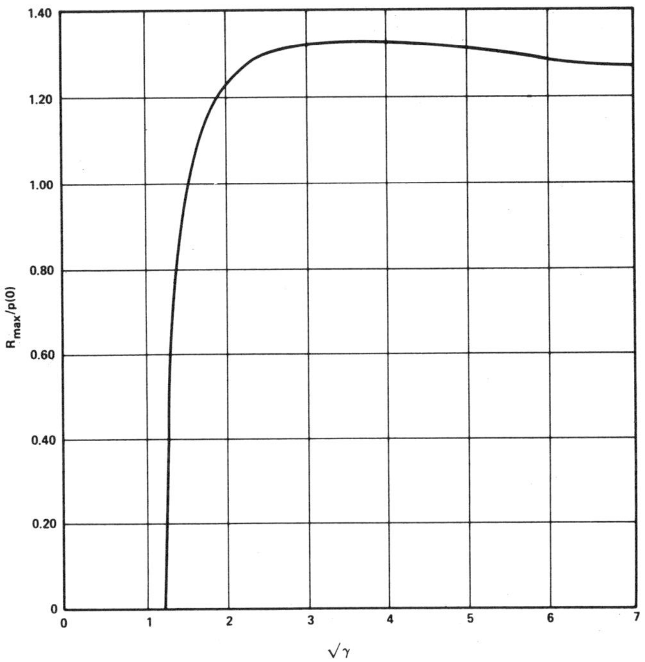

Fig. 1. Graph of R_{max} versus $\sqrt{\gamma} = [p(0)/\sqrt{2}\,\sigma_n]$. Data were obtained numerically by digital computer.

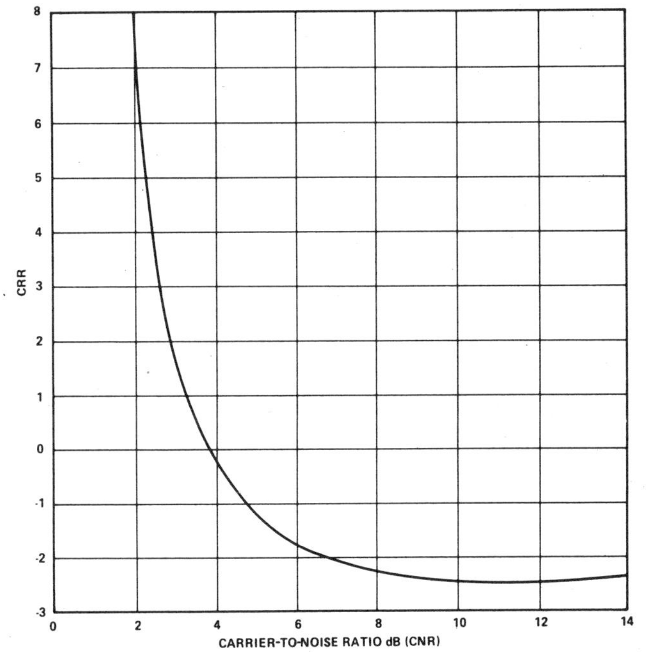

Fig. 2. Graph of R_{max} versus $\sqrt{\gamma}$ in decibel units. CNR = $10 \log_{10}\gamma$. CRR = $20 \log_{10} p(0)/R_{max}$.

The two end points of the R_{max} versus γ characteristics are obtainable analytically, however. We show in Appendix B that $R_{max} = 0$ if and only if $\gamma \leqslant 3/2$, and also that $\lim_{\gamma \uparrow \infty} R_{max} = 1.26\,p(0)$.

It will be observed in Fig. 1 that the limiting value for R_{max} as $\gamma \uparrow \infty$ is actually less than the largest value of R_{max} for finite γ, which is approximately $1.32p(0)$. No major significance is attributed to the apparent lack of monotonicity of R_{max} with γ, since R_{max} is a function of the shape of $g(r)$, and that shape is a very complicated and dynamic function of γ. On the other hand, it is encouraging that the limiting analysis for unbounded γ gives a value for R_{max} so close to that which appears to be the limiting value based on the numerical evaluation for finite γ, as may be seen in Fig. 1.

A comprehensive evaluation of the limited-peak bound for $\sigma_r \leqslant R_{sup}$ is presented in graphical form in Fig. 3(a)-(f). The set of curves in each figure are a family of constant carrier-to-interference-ratio (CIR) bounds, that is, the parameter

$$\text{CIR} \triangleq -20 \log_{10} \frac{\sigma_r}{p(0)} \tag{42}$$

is common to each. The curve parameter is the peak factor (PF),

$$\text{PF} \triangleq 20 \log_{10} \frac{R}{\sigma_r} = -10 \log_{10} \rho. \tag{43}$$

As it stands, this definition for PF is consistent with the common situation wherein $R \leqslant R_{max}$. However, in order to utilize such curves for the case $R \geqslant R_{max}$ (if $\sigma_r \leqslant R_{sup}$), one would merely substitute R_{sup} and ρ^* for R and ρ, respectively, in (43). PF would then still be the peak factor of the maximizing distribution, but would no longer be the peak factor of the actual interference envelope.

The bound itself is plotted logarithmically versus CNR. For the ordinate, only the exponent of 10 is shown at each decade. Dotted lines between decades are the 2 and 5 lines. For example, the second dotted line above the solid line labeled -7 represents 5×10^{-7}.

The heavy dashed line running through each family is the so-called Gaussian noise conversion approximation to Pe. This is the value of Pe obtained by pretending that the interference behaves precisely as additional independent Gaussian noise whose power is equal to the interference power. This will be discussed in Section VII.

Combining (41), (42), and (43), we find that the condition $R \leqslant R_{max}$ is equivalently given by

$$\text{CIR} - \text{PF} \geqslant \text{CRR}, \tag{44}$$

where CRR depends on CNR. The CNR range of the curves is 6 to 21 dB, and we see from Fig. 2 that the largest value that CRR attains in this interval is approximately -1.75 dB; this occurring at CNR = 6 dB. Therefore, with the exception of the small CNR portion of the PF = 7 dB curve of Fig. 3(a) (CIR = 5 dB), the CIR and PF parameters of all the bound curves presented satisfy (44), and hence all points on the curves pertain to $R \leqslant R_{max}$.

VI. Upper Bound for *M*-ary CPSK

The upper bound for binary CPSK can be used to bound Pe for *M*-ary CPSK as well. Quite simply, the binary bound is treated as a general half-plane probability bound in the complex plane of the receiver signal space. Then the well-known union bound, which upper bounds *M*-ary Pe for CPSK by twice a half-plane probability, is applied in conjunction with the binary bound to give a general *M*-ary bound. Everything goes through, except that the range of parameters, (e.g., CIR, CNR, PF), where (25) [as opposed to (34)] applies differs slightly in the *M*-ary case.

The error region in the complex envelope plane for *M*-ary CPSK is the union of two symmetric half planes, each located at a perpendicular distance $p(0) \sin \pi/M$ from the signal, $p(0)e^{j\phi k}$. Because of the CS property of the interference and noise, the probability that the total received complex envelope $\overline{\Psi}$ is found in one of those half planes is equal to the binary CPSK error probability, where the signal modulus would instead be $p(0) \sin \pi/M$. In other words, each half plane has occupancy probability equal to the left half-plane probability in the binary case with a foreshortened signal, all other parameters remaining the same. Therefore, the complete *M*-ary error region has probability that is upper bounded by twice the binary CPSK Pe, where the values of CNR and CIR each would be reduced by the amount

$$\text{CF} \triangleq -20 \log_{10} \sin \frac{\pi}{M}. \tag{45}$$

Finally, the limited peak bound for binary Pe, being a half-plane probability bound, is therefore also an upper bound to half the *M*-ary Pe when the appropriate adjustments to CNR and CIR are made. For example, to use the binary charts, Fig. 3(a)-(f), for *M*-ary CPSK simply add the amount CF to the abscissa CNR labels, add CF to each CIR label, and then double the ordinate identifications. The latter is quite easily accomplished by mentally shifting the decade and 2X labels to the next lower line.

The effective reduction in $p(0)$ must also be taken into account when determining the proper form of the bound for a given *M*-ary case, inasmuch as the boundary $R < R_{max}$ must similarly be readjusted. The simplest way to accomplish this is to add CF to both the abscissa and ordinate labels of Fig. 2. However, since CRR and CIR were each adjusted by CF in the same direction, we see that (44) remains unchanged for the *M*-ary case. As a consequence, points on the binary bound curves that correspond to $R \leqslant R_{max}$, do so also for any $M > 2$.

For example, consider $M = 4$, quaternary CPSK. We find that CF = 3 dB. Relabeling abscissas, the bound curves now cover the CNR range 9-24 dB. The maximum value of CRR, read from an adjusted Fig. 2, is +1.25 dB (-1.75 + 3, which corresponds to an adjusted CNR = 9 dB). Fig. 3(a) now represents CIR = 8 dB, so that (44) would read

$$\text{PF} \leqslant 8 - \text{CRR} \Longleftrightarrow R \leqslant R_{max}.$$

Thus, just as we found before for the binary case, all curves in Fig. 3(a) except part of the PF = 7 dB curve (the excluded part runs only from CNR = 9 to CNR = 10 dB, since CRR > 1 dB for CNR > 10 dB), and, obviously, all the bound curves in the remaining figures correspond to $R \leqslant R_{max}$.

VII. Applications and Comparisons

We will assume in the following discussion that all interferences under consideration have been determined to satisfy the requirements for being CS, and for practical purposes that the important parameters σ_r^2 and R are known or can be mea-

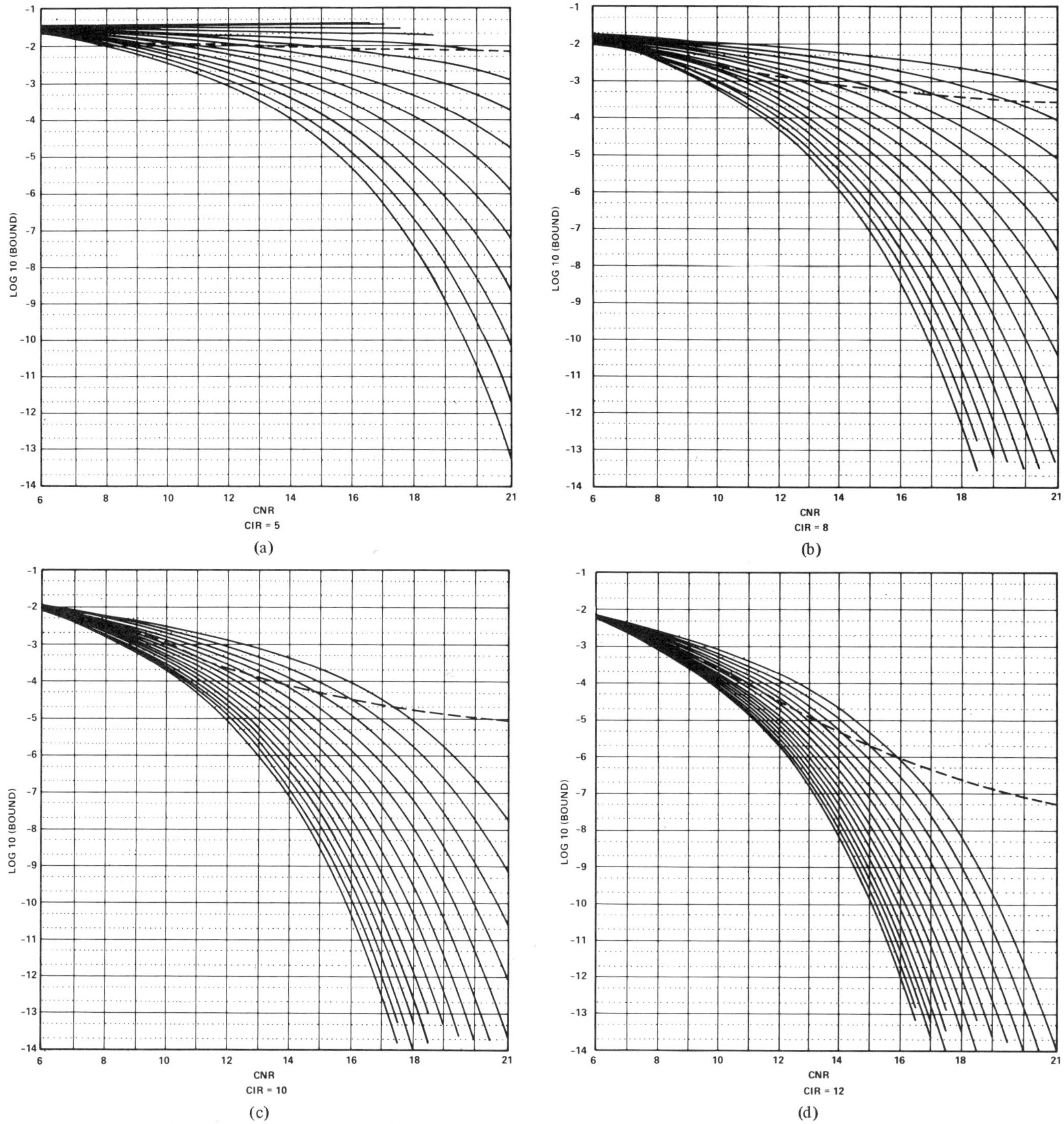

Fig. 3. Peak-limited bound for binary CPSK in families of constant CIR. Lowermost curve is PF = 0 dB; uppermost curve is PF = 7 dB. Intermediate curves are at every $\frac{1}{2}$ dB [every 1 dB in (f)]. Dashed line represents noise conversion approximation.

sured. Furthermore, we will limit discussion to binary CPSK, with the knowledge that the M-ary case is a straightforward extension through the use of the half-plane union bound.

The co-channel interference arising from an independent but similar PSK source has been evaluated extensively [7], and so provides a convenient basis of comparison. The simplest case is a single interferer that is modeled by a randomly phased, but constant modulus signal. The interference complex envelope pdf is a circular line mass, and the modulus pdf is just a single impulse. The peak equals the rms modulus, so that $\rho = 1$. The limited-peak bound then coincides precisely with the well-known result for a single constant-amplitude component. Thus for PF = 0 dB, the bound is exact, and identical with the single, constant amplitude co-channel interference error probability.

Incidentally, a reasonably good approximation to the value of the limited-peak bound can be obtained from the curves in [7, fig. 3] (note, however, that [7, figs. 3(a) and 4(a)] were inadvertently transposed). Since the $N = 1$ curves of [7, fig. 3] give the aforementioned single constant-amplitude error probability, and the $N = 0$ curve (corresponding to no interferers) is just $\frac{1}{2}$ erfc $(\sqrt{\gamma})$, the limited-peak bound is a simple linear interpolation with proportionality constant ρ, i.e., limited-

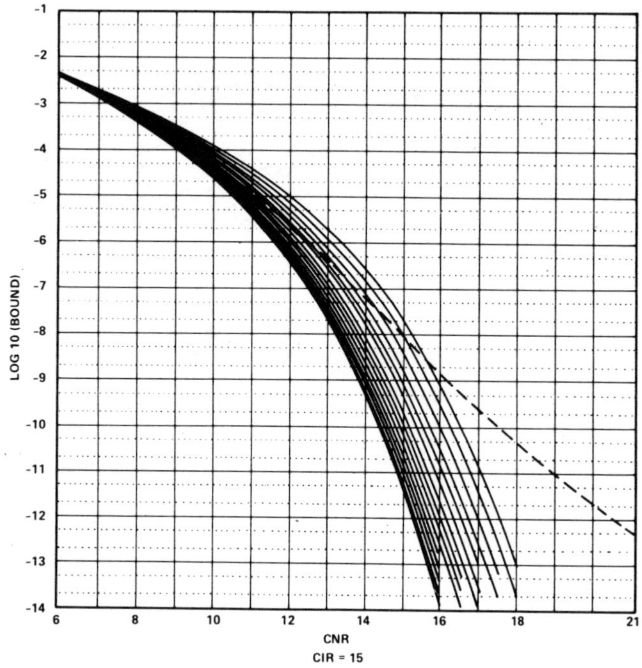

(e)

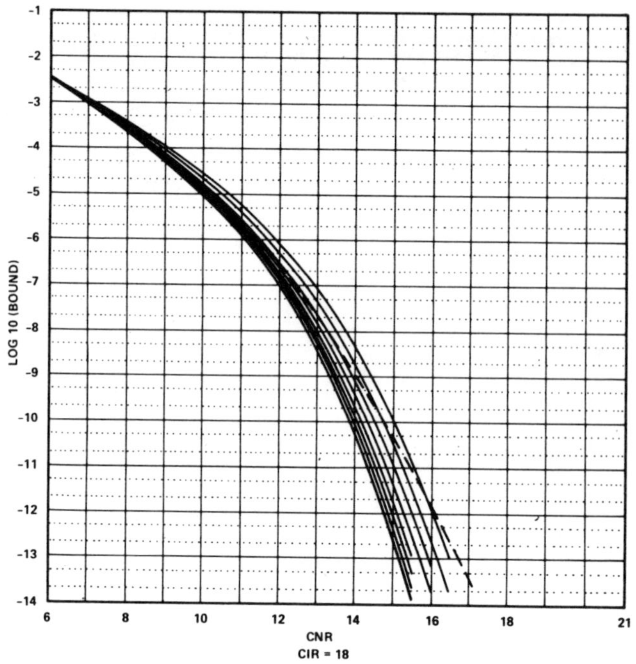

(f)

Fig. 3. (*Continued*).

peak bound = $\rho \cdot (N = 1 \text{ curve}) + (1 - \rho) \cdot (N = 0 \text{ curve})$.

The proper CIR family in [7, fig. 3] is selected so that the single component amplitude is made equal to R or R_{sup}. Assuming we want R, this is accomplished by using the CIR′ curves in [7], where

$$\text{CIR}' = \text{CIR} + \text{PF}.$$

Of course, an "eyeball" interpolation between the $N = 0$ and $N = 1$ curves of [7, fig. 3] must be done carefully because of the logarithmic ordinate scale, unless ρ is close to either 0 or 1.

Another set of curves that also may be used to estimate the limited-peak bound are found in [5, fig. 4], where the interpolation is made between the $S/I = \infty$ curve (corresponds to $N = 0$) and the appropriate $S/I < \infty$ curve. This figure exhibits the curious property that if the ordinate scale is chosen to make the no-interference curve ($\frac{1}{2}$ erfc $(\sqrt{\gamma})$) become a straight line, then the single constant-amplitude interference curves will be nearly straight parallel lines over a wide range in CNR and CIR. The interpolation curve would certainly also be approximately a straight line, and therefore easy to sketch on that figure.

The curves of [7, fig. 3] for $N > 1$ also provide an opportunity to compare the limited-peak bound with exact error probabilities arising from fairly realistic peak-limited interference envelope distributions. For example, the $N = 2$ curves are the exact value of Pe for two independent CS interferers with equal constant (nonrandom) envelopes. The envelope modulus of the composite interference is, however, random and it may be shown to have the pdf

$$f(r) = (\pi a)^{-1} \left(1 - \left(\frac{r}{2a}\right)^2\right)^{-1/2}, \quad 0 \leq r \leq 2a, \tag{46}$$

where a is the common envelope amplitude of the two components. This density is sketched in Fig. 4(a). Furthermore, the phase is uniform and independent of the modulus, so that the composite is CS. Last, it is a consequence of the independence that $\sigma_r^2 = 2a^2$, and of course $R = 2a$, so that $\rho = \frac{1}{2}$ for the composite interference distribution. Assuming $2a \leq R_{max}$, the maximizing distribution is given by

$$F_0(r) = \frac{1}{2} U(r) + \frac{1}{2} U(r - 2a),$$

and the corresponding pdf is sketched in Fig. 4(b). Comparing the actual and the bounding densities in Fig. 4, it is not obvious that the bound will be particularly good, or even that it actually is a bound.

Suppose that σ_r^2 is such that CIR = 12 dB. The exact value of Pe is given by the $N = 2$ curve of [7, fig. 3(d)]. The limited peak bound is the PF = 3 dB curve of Fig. 3(d) of this paper. The two curves are reproduced here in Fig. 5. The bound is seen to be approximately 1/4 dB away throughout the CNR range. Also shown in Fig. 5 is the same comparison for three equal-amplitude components, where $\rho = 1/3$, PF = 4.76 dB. Here the bound departs from the exact Pe curve by $\frac{1}{2}$ to $\frac{2}{3}$ dB.

Except at very low CNR values where the Gaussian noise completely dominates, the tightness of the bound is obviously governed by how "close" the actual envelope distribution is to the maximizing distribution. When the interference comprises several independent similar-amplitude components, the envelope distribution tends to be spread out having a long tail, and PF becomes correspondingly large. In this case the actual distribution is not as well approximated by F_0 as in the low PF case where the interference is such that a significant portion of the pdf mass is near the peak. As a trend, therefore, the bound is tightest for distributions with low PF, and is least

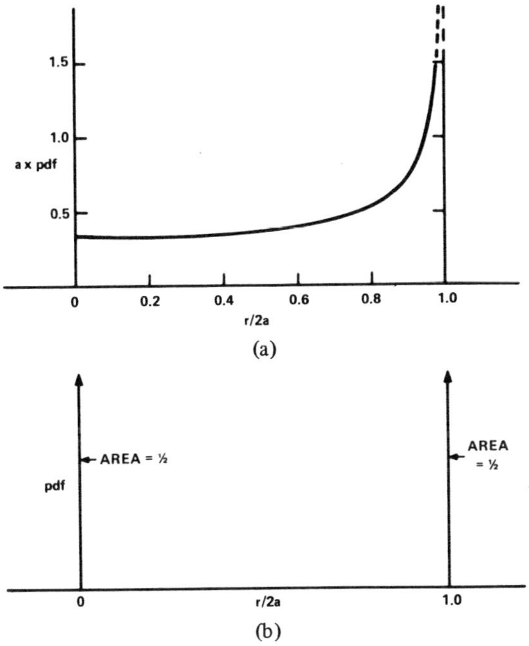

Fig. 4. (a) Actual envelope pdf for two CS components with equal (constant) amplitudes a. (b) Maximizing pdf for $R = 2a$, $\rho = \frac{1}{2}$, which produces the corresponding Pe upper bound.

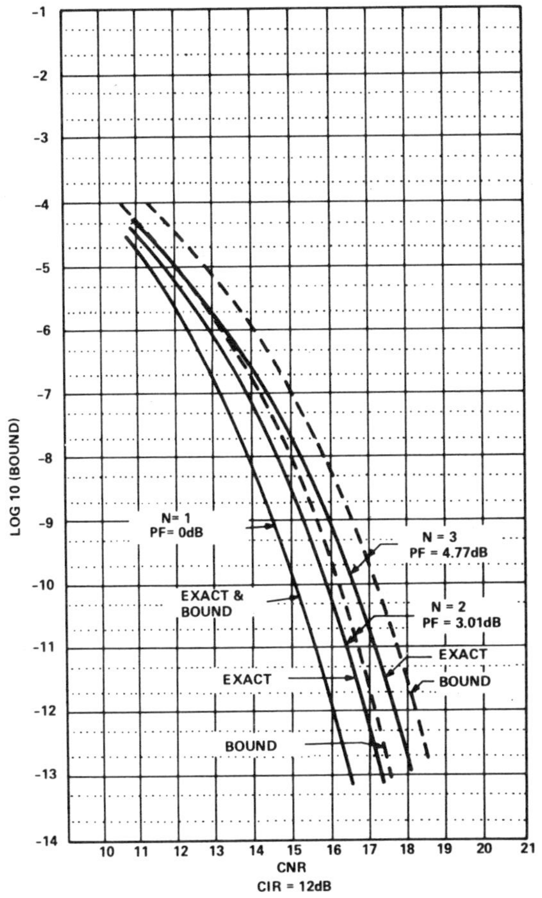

Fig. 5. Comparison of bound with exact Pe for a sum of independent CS interferences with equal constant envelopes. CIR = 12 dB.

tight for highly composite distributions with large PF.

Nevertheless, the bound is applicable to composite interference situations,[3] and should give satisfactory results in the high CNR range for PF $\leqslant 7$ dB. To apply the bound, we use the easily proven fact that the sum of independent CS components is CS, with

$$R = \sum R_i, \quad \sigma_r^2 = \sum \sigma_{r_i}^2. \tag{47}$$

Thus for multiple interferences we sum the individual envelope limits, and sum the individual envelope second moments, to obtain the respective two parameters of the CS resultant. As in the previous example, given N equal constant-amplitude interferences with $R_i = a$, then $R = Na$, $\sigma_r^2 = Na^2$, and $\rho = N^{-1}$.

The case of one or more adjacent channel interferences is treated the same, except each component will now have PF $>$ 0. For the symmetrical case of N interferences with equal PF_i and equal R_i (or equal σ_{r_i}), the resultant has

$$\text{PF} = \text{PF}_i + 10 \log_{10} N, \tag{48}$$

while

$$\text{CIR} = \text{CIR}_i - 10 \log_{10} N. \tag{49}$$

A significant property of the bound is that it inherently exploits the salient feature of the interference—its peak-limited modulus—by giving the correct exponential behaviour as $\gamma \uparrow \infty$. An important corollary property for an "open eye" situation (i.e., $R < p(0)$) is that the bound will approach 0, as it should, in the limit $\gamma \uparrow \infty$. We note that this is in marked contrast with certain other approaches, notably the Gaussian noise conversion approximation.

On the other hand, the so-called "eye diagram" bound, which in our notation would be $\frac{1}{2}$ erfc $[\sqrt{\gamma}(1 - R/p(0))]$, also tends toward zero for large γ if $R < p(0)$. However, we see immediately the gross pessimism in that bound. It does not account for any averaging over phase or modulus variations, and ignores second-moment information.

A little better than the crude eye diagram bound for moderate γ in certain cases is a bound [6] derived from the Chernoff bound, which turns out to be in some sense a mixture of the noise conversion and eye diagram philosophies. The limited-peak bound has been found to be tighter, however, for interferences that are not highly composite, i.e., for the lower values of PF, and especially in the large-CNR, low-Pe range.

It will be observed in the bound curves in Fig. 3 that the limited-peak bound exceeds the noise conversion result in a low-CNR region for certain combinations of CIR and PF. This is somewhat surprising, for in previous work it was found empirically that "practical" interference configurations, e.g., multiple co-channel interferences with arbitrary relative amplitudes, had an exact Pe that always was less than the noise conversion value. On that basis it was conjectured that noise conversion is an upper bound, for the multiple co-channel

[3] The fact that a given total interference is composite, being the sum of distinct, independent CS components with individual parameters σ_{r_i}, R_i, is important additional information, beyond σ_r and R, which is not utilized by the present limited-peak bound. However, work is being completed on an improved bound that optimally incorporates this information to give an improved result for composite interferences.

```
conversion from decibel   { X=10**(CNR/20);
input parameters          { TWOX=2*X;
                          { PWRTRM=0.25*10**((CNR-CIR+PF)/10);
recurrence                { TERM2=PWRTRM;
initialization            { TERM3=TWOX*TERM2;
                            SUM=TERM3;
                   LOOP:
                            DO L=2 TO 400;
                 | 1st iteration { TERM1=TERM2;
recurrence to    |               { TERM2=TERM3;
generate         |               { TERM3=TWOX*TERM2-(4*L-6)*TERM1;
series terms     | 2nd iteration { TERM1=TERM2*PWRTRM/(L*L);
                 |               { TERM2=TERM3*PWRTRM/(L*L);
                 |               { TERM3=TWOX*TERM2-(4*L-4)*TERM1;
                                   SUM=SUM+TERM3;
convergence test . . . . . . . . . IF ABS(TERM3/SUM)<=1.E-9 THEN GO TO QUIT;
                            END LOOP;
                   QUIT:
bound computation . . . . . BOUND=.56418958*EXP(-X*X)*10**(-PF/10)*SUM+.5*ERFC(X);
```

Fig. 6. PL-1 program statements for bound computation.

interference model at least. Nevertheless, the limited-peak bound, which we see can exceed noise conversion, is certainly derived from a pdf that satisfies the requirements of a peak envelope constrained interference, and so would seem to admit the possibility that noise conversion is not always an upper bound for the multiple co-channel interference model.

We may ask, however, if the maximizing distribution F_0 is a physically realizable interference distribution, i.e., whether or not F_0 is a legitimate distribution for the interference envelope modulus at the input to the detector. Such a distribution, in addition to satisfying the peak and mean-square constraints, must also represent the complex envelope modulus of a narrow-band random process. It is still possible, thus, that noise conversion remains an upper bound within the restricted class of narrow-band interferences for a certain range of CNR.

The limited-peak bound might be improved by maximizing Pe over a restricted class of constrained-amplitude, spectrally realistic modulus distributions. We feel that this is, perhaps, a more fruitful way to "fine tune" the bound than the natural extension of using additional first-order statistics, such as third moment, etc. Regardless, the present limited-peak bound is still significantly better than noise conversion over much of the practically important region of CNR, CIR, PF parameter space, and furthermore the bound gives the proper exponential behavior of Pe as CNR grows large.

Appendix A

The utility of the limited-peak bound for CPSK owes much to our ability to rapidly and accurately compute it. In particular, the ease and precision with which we can evaluate the series

$$\sum_{l=1}^{\infty} \frac{H_{2l-1}(\sqrt{\gamma})}{2^{2l} l! \, l!} \gamma^l \left(\frac{r}{p(0)}\right)^{2l} \tag{50}$$

on a digital computer is an important factor. We include here, therefore, the core of a program written in PL-1 (its Fortran equivalent is obvious) that will enable the reader to compute the bound with a minimum of initial setup effort.

The central problem, of course, is the evaluation of the Hermite polynomial in each term. This, and certain other difficulties having to do with exponent overflow, loss of precision, efficiency, etc., are fairly well overcome by generating successive terms in the series through recurrence relationships. In this case, we use the well-known recurrence

$$H_{n+1}(x) = 2xH_n(x) - 2nH_{n-1}(x) \tag{51}$$

for the Hermite polynomial, and the obvious single-term recurrences for the remaining factorials and power terms.

An annotated listing of the PL-1 statements is given in Fig. 6. Notice that two iterations of the Hermite recurrence are needed for each pass through the DO loop, since successive series terms involve Hermite polynomials differing in index by two. On the second iteration, the power term

$$\text{PWRTRM} \equiv \frac{\gamma r^2}{4p^2(0)} \tag{52}$$

and L ** $2 \equiv l \cdot l$, for the factorial are combined with the Hermite polynomial recurrence to achieve the desired new series term, TERM3, which then augments SUM. The convergence test used is merely ad-hoc, but one that seems to work well.

In the initialization statements, we have assumed that the input parameters are in decibel form, i.e., CNR, CIR, and PF, which were defined in (40), (42), and (43), respectively. It is straightforward to verify from these three definitions that the computation for PWRTRM agrees with (52).

The general problem of evaluating various probability distribution function expressions associated with a Gaussian plus a sinusoid has been addressed recently by Esposito and Wilson [9]. Their basic computational approach parallels ours, and the reader is referred to [9] for a detailed account. While Esposito and Wilson do discuss some programming considerations, unfortunately no listings of programs were presented in their paper.

Appendix B

While an analytical expression for R_{max} was not found, we are able to show in a fairly straightforward way that

$$R_{max} = 0, \quad \gamma < \frac{3}{2} \tag{53}$$

$$\lim_{\gamma \uparrow \infty} R_{max} = 1.26\, p(0). \tag{54}$$

To obtain (53), we first write out the series for $g(r)$. Omitting the positive constant factor $e^{-\gamma}/\sqrt{\pi}$, we get

$$g(r) \triangleq \frac{1}{r^2} \Sigma(r, \gamma) = \frac{H_1(\sqrt{\gamma})\gamma}{4} + \frac{H_3(\sqrt{\gamma})\gamma^2}{64} r^2 + (\text{terms in higher powers of } r^2). \tag{55}$$

From this, it is clear that

$$\frac{1}{r} \frac{\partial g}{\partial r} = \frac{H_3(\sqrt{\gamma})\gamma^2}{32} + o(r), (r \to 0). \tag{56}$$

For any fixed γ, a neighborhood of $r = 0$ exists where the right side (hence the left) has the same sign as does $H_3(\sqrt{\gamma})$. Furthermore, since $H_3(x) = 8x^3 - 12x$, we see that $H_3(\sqrt{\gamma})$ will be negative for $0 < \gamma < 3/2$. If $H_3(\sqrt{\gamma})$ is negative, then there exists a neighborhood of $r = 0$ wherein $(1/r)(\partial g/\partial r)$ is negative, and in particular there exists an interval $0 < r < R_{max}$ for any $R_{max} > 0$ wherein $(\partial g/\partial r)$ is negative and $g(r)$ is decreasing. This shows that when $\gamma < 3/2$, $g(r)$ cannot be nondecreasing on $[0, R_{max}]$ for any $R_{max} > 0$. Hence we have $R_{max} = 0$.

In order to derive (54), we first consider the limiting behavior of $q(r)$ for $\gamma \uparrow \infty$. That is, keeping $p(0)$ fixed, we examine the limit of binary Pe for a CS interference with constant modulus r, as the noise variance decreases to zero. We have

$$\text{Pe} = \Pr\{r \cos \xi + \nu \to p(0)\}, \tag{57}$$

and since ν is zero mean, $\nu \xrightarrow{\text{ms}} 0$ as $\sigma_n^2 \to 0$. This implies that $r \cos \xi + \nu \xrightarrow{p} r \cos \xi$ (for all points since the distribution function of $r \cos \xi$ is continuous). Therefore we have the intuitively obvious result,

$$\lim_{\sigma_n \downarrow 0} \text{Pe} = \Pr\{r \cos \xi > p(0)\}, \tag{58}$$

that the limit is just the value of Pe in the absence of noise. From the well-known pdf

$$f_{r \cos \xi}(x) = \begin{cases} (\pi\sqrt{r^2 - x^2})^{-1}, & |x| \leqslant r \\ 0, & |x| > r \end{cases} \tag{59}$$

of a sinusoid with uniform phase, we easily find by integrating the density between the limits $p(0), \infty$ that

$$q^*(r) \triangleq \lim_{\gamma \uparrow \infty} q(r) = \begin{cases} \dfrac{1}{\pi} \cos^{-1}\left(\dfrac{p(0)}{r}\right), & r \geqslant p(0) \\ 0, & r < p(0) \end{cases}. \tag{60}$$

This result may also be easily found by considering the more intuitive geometric interpretation. Heuristically speaking, the limiting pdf of $\overline{I} + \overline{N}$ in the complex plane is just the pdf of $\overline{I}$ alone, which is a circular line mass (of uniform weight) centered at $(p(0), 0)$ with radius r. $q^*(r)$ is the proportion of the circular arc lying in the left half plane. Simple trigonometry gives at once that arc length to be $2r \cos^{-1}(p(0)/r)$, and dividing this by the total length $2\pi r$ produces the desired result. Of course, when $r < p(0)$, the circle lies wholly within the right half plane, and so $q^*(r) = 0$. This corresponds to the so-called "open eye" situation.

Assuming the maximizing distribution in (21), and applying the conditions of Corollary 1 with $q^*(r)$, we would find that R_{max} is simply the point where the function

$$g^*(r) \triangleq \frac{q^*(r)}{r^2}, \qquad r \geqslant 0 \tag{61}$$

ceases to be nondecreasing. Since q^* is identically zero on $(0, p(0))$, equivalently we seek the first solution of

$$\frac{dg^*}{dr} = 0, \qquad r > p(0), \tag{62}$$

a transcendental equation that was solved numerically to yield $R_{max} = 1.26\, p(0)$.

REFERENCES

[1] F. E. Glave, "An upper bound on the probability of error due to intersymbol interference for correlated digital signals," *IEEE Trans. Inform. Theory*, vol. IT-18, pp. 356-363, May 1972.

[2] J. G. Smith, "The information capacity of amplitude- and variance-constrained scalar Gaussian channels," *Inform. Contr.*, vol. 18, pp. 203-219, Apr. 1971.

[3] W. R. Bennett and J. R. Davey, *Data Transmission*. New York: McGraw-Hill, 1965, ch. 10.

[4] J. R. Davey, "Digital data signal space diagrams," *Bell Syst. Tech. J.*, vol. 43, pp. 2973-2983, Nov. 1964.

[5] V. K. Prabhu, "Error rate considerations for coherent phase-shift keyed systems with cochannel interference," *Bell Syst. Tech. J.*, vol. 48, pp. 743-767, Mar. 1969.

[6] —, "Error-probability upper bound for coherently detected PSK signals with cochannel interference," *Electron. Lett.*, vol. 5, pp. 383-385, Aug. 7, 1969.

[7] A. S. Rosenbaum, "Binary PSK error probabilities with multiple cochannel interferences," *IEEE Trans. Commun. Technol.*, vol. COM-18, pp. 241-253, June 1970.

[8] M. Loeve, *Probability Theory*. New York: Van Nostrand, 1955.

[9] R. Esposito and L. R. Wilson, "Statistical properties of two sine waves in Gaussian noise," *IEEE Trans. Inform. Theory*, vol. IT-19, pp. 176-183, Mar. 1973.

An Upper Bound Analysis for Coherent Phase-Shift Keying with Cochannel, Adjacent-Channel, and Intersymbol Interference

FREDERICK E. GLAVE, MEMBER, IEEE, AND ARNOLD S. ROSENBAUM, MEMBER, IEEE

***Abstract*—An error-probability upper-bound is developed for coherent phase-shift keying (CPSK) considering the combined effects of Gaussian noise, intersymbol interference (ISI), and several peak-limited circular-symmetric (CS) components such as adjacent-channel interference (ACI) and cochannel interference (CCI). In an earlier paper [1], an upper bound was introduced, based on the "maximizing probability density function" concept, for CPSK with Gaussian noise and a single CS interference. The natural extension of this type of bound to multiple mixed interferences—such as are encountered in practice—is given in this sequel. This multiple interferer bound, which is computationally simple to use, requires only the peak and mean-square values of the several interferences involved, and is optimal in the sense that no other bound, based solely on the same information, will be lower than it.**

Through examples we draw comparisons with previously available results, which show that the multiple-interference limited-peak bound, when applied to composite interference situations, can be an effective tool for system design and analysis.

I. INTRODUCTION

SEVERAL techniques are presently available for calculating the error probability of a coherently detected phase-shift-keyed (CPSK) transmission link when the signal is subjected to certain significant degradations [2]–[9]. Specifically, analyses are available for CPSK where the signal is disturbed by additive noise and intersymbol interference (ISI), or additive noise and cochannel interference (CCI). That is, the presently available analyses basically consider the presence of only one or at most two types of interference, in addition to Gaussian noise.

However, the system designer is faced with the problem of determining the performance of a CPSK link operating in the presence of a multiplicity of interferers, the most common and important of which besides Gaussian noise are ISI, CCI, and adjacent-channel interference (ACI). The joint effect of all such signal impairments must be taken into account when deciding upon key system parameters, such as required carrier-to-noise ratio (CNR), frequency plan, filter selectivity, tolerable interference levels, etc.

One approach to evaluation of the error probability in the multiple-interference environment is direct machine computation. This can take the form of either a computer simulation using a Monte Carlo technique, or more straightforwardly, an averaging of the conditional probability of error over all the interference random variables. An example of the former is [10] and quite recently Benedetto *et al.* [11] have taken the latter approach in an analysis which depends upon calculating the moments of the interferences in order to perform the necessary averaging operation. However, in most cases a sufficient (for exact results) statistical description of the various interferences—such as the infinite set of moments of all orders—is difficult or impossible to obtain. Furthermore, while these techniques can provide close approximations to the error probability, they are usually time consuming and expensive, and also do not readily yield insight into the nature of the effect of a particular interference on the error performance of a system.

The bounds we describe stem from the concept of a maximizing probability density function (pdf) associated with the given interference and depend only on the peak and rms values of the interference. This technique was first applied to the problem of ISI in pulse-amplitude modulation systems having Gaussian noise by Glave [12], and later by Matthews [13] who also obtained lower bounds. Recently, the authors [1] have applied it to the problem of a single interference in CPSK systems.

In this sequel, we indicate how the "maximizing probability density" concept is applied to optimally bound the error probability in the case of a composite interference comprised of several distinct components, including CCI, ACI, ISI, each having a known peak and rms value. This bound is optimal in the sense that no other bound which also depends solely on the peak and rms values of the interferences will be lower. Even though those parameters would seem to be meager statistical information indeed, in fact the bound is reasonably tight in many cases of practical importance. This, coupled with its computational simplicity, makes it an effective tool for system design and analysis, especially in those many cases where the statistics of the interference cannot be fully enough described to employ other methods.

II. CPSK SYSTEM MODEL

We are concerned with detection of an M-ary PSK signal such as

$$s(t) = \mathrm{Re} \sum_{k=-\infty}^{\infty} \exp\left[j(\phi_k + \omega_0 t)\right] p(t - kT) \qquad (1)$$

Paper approved by the Associate Editor for Communication Theory of the IEEE Communications Society for publication without oral presentation. Manuscript received June 28, 1974; revised December 13, 1974.

F. E. Glave is with Bell-Northern Research, Ottawa, Ont., Canada.
A. S. Rosenbaum is with Bell Laboratories, Holmdel, N. J. 07733.

Reprinted from *IEEE Trans. Comm.*, vol. COM-23, pp. 586-597, June 1975.

which has been corrupted by the addition of one or more interference terms of the form

$$i(t) = \text{Re}\, r(t) \exp(j[\beta(t) + \omega_0 t]) \tag{2}$$

and Gaussian noise $n(t)$.

It is convenient to deal with the complex envelopes of these quantities. With ω_0 the radian carrier frequency, the complex envelopes have the form

$$\bar{S} = \sum_{k=-\infty}^{\infty} p(t - kT) \exp[j\phi_k] \tag{3a}$$

$$\bar{I} = r(t) \exp[j\beta(t)] \tag{3b}$$

and $\bar{N} = \nu \exp[j\alpha]$ is the filtered Gaussian noise envelope. Let σ_n^2 be the mean-squared value of the real noise voltage which is, of course, the common variance of the independent jointly Gaussian components of any orthogonal decomposition.

The function $p(t)$ is the overall response of the channel including transmit and receive filters. At a sample instant $t = 0$, the signal complex envelope is represented as

$$\bar{S} = p(0) \exp[j\phi_0] + \sum_{k=-\infty;\, k\neq 0}^{\infty} p(kT) \exp[j\phi_k]. \tag{4}$$

The first term is the desired signal component and the terms in the summation represent the ISI. Henceforth, we shall denote the envelope of the ISI term as

$$\bar{Z} = \sum_{k=-\infty}^{\infty}{}' \; p(kT) \exp[j\phi_k], \tag{5}$$

where the prime indicates that the $k = 0$ term has been deleted from the summation. In most practical situations an exact statistical description of $\bar{Z}$ is impossible to obtain. However, it is usually true that $\bar{Z}$ is peak-limited in that there exists some Z such that $|\bar{Z}| \leq Z$. Furthermore, one can usually compute the mean-square value of $|\bar{Z}|$, σ_z^2.

Interference terms of the form given in (2) are assumed to be circular symmetric (CS) random processes, whereupon the complex envelope modulus is independent of the phase, and the phase, modulo 2π, is uniformly distributed. For example, the interference process given in (2) will be CS if $r(t)$ and $\beta(t)$ are independent random variables for any t, and $\beta(t)$ is uniform, modulo 2π. Note that the noise process is CS. In PSK systems, ACI and CCI will be CS provided there is no phase coherence between the wanted and interfering carriers.

Also, just as in the case of ISI, practically all common forms of CS interference are peak-limited with peak envelope R, in that $r(t) \leq R$. Finally, CS interference has mean-square modulus

$$\sigma_r^2 = E\{r^2(t)\}, \tag{6}$$

which can easily be shown to be twice the mean-square value of the real signal $i(t)$.

Thus the input to the phase angle detector in the presence of intersymbol and CS interference is a signal whose complex envelope at the sample instant is

$$\bar{\psi} = p(0) \exp[j\phi_0] + \bar{N} + \bar{Z} + \sum_{i=1}^{N} r_i \exp[j\beta_i]. \tag{7}$$

For binary signaling, where ϕ_k is either 0 or π, the receiver signal space decision regions are the right- and left-half complex planes. The conditional probability of error, given the values of the interference terms, when $\phi_k = 0$ or π, is given by

$$\text{Pe} = \tfrac{1}{2} \,\text{erfc} \left\{ \frac{p(0) + z + \sum_{i=1}^{N} r_i \cos \beta_i}{\sqrt{2}\sigma_n} \right\}, \tag{8}$$

where z is the in-phase component of the ISI,

$$z = \sum_{k=-\infty}^{\infty}{}' \; p(kT) \cos(\phi_k). \tag{9}$$

The overall probability of error is then obtained by averaging (8) with respect to the joint distribution of the random variables z, $\{r_i\}_{i=1}^{N}$ and $\{\beta_i\}_{i=1}^{N}$. It is precisely this process of averaging over the interference distribution that is so difficult and makes an exact computation of Pe practically prohibitive.

However, it is possible to derive useful upper bounds on Pe that require no statistical information concerning the various interference random variables, other than the previously mentioned peak and mean-square values.

We begin by extending the bounding technique for a single CS interferer alone [1] to the analysis of two independent peak-limited CS interferers (for example a co-channel and an adjacent-channel interferer). We then demonstrate how to extend the bound further to include both CS and intersymbol interferers simultaneously. We shall concentrate primarily on the derivation and examination of the two-CS bound. All of the salient features and advantages of the bounding technique are clearly in evidence in the two-CS bound and the subsequent inclusion of ISI is then but a formal extension.

III. BOUND FOR TWO INDEPENDENT PEAK-LIMITED CS INTERFERENCES

Suppose in the model of Section II that the interference is comprised of two independent peak-limited CS components. In terms of complex envelopes,

$$\bar{I} = \bar{I}_1 + \bar{I}_2 = r_1 \exp[j\beta_1] + r_2 \exp[j\beta_2], \tag{10}$$

where $r_1, r_2, \beta_1, \beta_2$ are independent random variables at any given sample time.

From (8), the probability of error, conditioned on these four variables, in the absence of ISI, is simply

$$\text{Pe} \mid r_1, r_2, \beta_1, \beta_2 = \tfrac{1}{2} \,\text{erfc} \left(\frac{p(0) + r_1 \cos \beta_1 + r_2 \cos \beta_2}{\sqrt{2}\sigma_n} \right). \tag{11}$$

It can be shown [14] that the expected value of Pe, with respect to the interference phase angles β_1 and β_2, may be expressed as

$$\mathrm{Pe}\mid r_1,r_2 = \tfrac{1}{2}\,\mathrm{erfc}\,(\gamma^{1/2}) + \frac{\exp(-\gamma)}{\pi^{1/2}} \sum_{l=1}^{\infty} \frac{H_{2l-1}(\gamma^{1/2})\gamma^l}{2^{2l}l!l!} \cdot \left(\frac{r_1}{p(0)}\right)^{2l} G_l\left[\left(\frac{r_2}{r_1}\right)^2\right], \quad (12)$$

where

$$\gamma^{1/2} \triangleq \frac{p(0)}{\sqrt{2}\sigma_n}, \quad (13a)$$

and

$$G_l(x^2) \triangleq \sum_{k=0}^{l} \binom{l}{k}^2 x^{2k}, \quad (13b)$$

is the factor which relates the (even order) moment of a sinusoid to the same order moment of the sum of two randomly phased sinusoids with amplitude ratio x. Actually, (12) above is a variant of the expression given by Esposito and Wilson [14], who refer to $G_l(x)$ as the Gaussian hypergeometric function, since indeed

$$G_l(x) = {}_2F_1(-l,-l;1,x). \quad (14)$$

A natural extension of the notation used in the single CS bound [1, eq. (24)] is to write (12) in the abbreviated form

$$\mathrm{Pe}\mid r_1,r_2 = \tfrac{1}{2}\,\mathrm{erfc}\,(\gamma^{1/2}) + \Sigma(r_1,r_2,\gamma). \quad (15)$$

The error probability for two independent CS interferences may thus be expressed as (integrating out the r_1 and r_2 dependence)

$$\mathrm{Pe} = \tfrac{1}{2}\,\mathrm{erfc}\,(\gamma^{1/2}) + \int_0^{R_2}\int_0^{R_1} \Sigma(r_1,r_2,\gamma)\, dF_1(r_1)\, dF_2(r_2). \quad (16)$$

Note that the interference peak envelopes R_1,R_2 need not be equal, and in general they will be unequal.

Let $\mathfrak{F}_1$ and $\mathfrak{F}_2$ be spaces of distribution functions with support contained in $[0,R_1]$ and $[0,R_2]$, respectively. The upper bound we seek is given by the supremum of (16) taken over all distribution function pairs (F_1,F_2) in the region of the product space $\mathfrak{F}_1 \times \mathfrak{F}_2$, where the respective envelope second-moment constraints are satisfied. The extremal element in this region, i.e., the jointly maximizing constrained second-moment distribution pair (F_{01},F_{02}), cannot be found as simply as in the one-dimensional single-component case. The theory does not seem to carry over directly. However, by a judicious conditioning argument, coupled with the symmetry of $\Sigma(\cdot,\cdot,\gamma)$ in the envelope variables, the theory for the single-interference bound can be used to find and establish F_{01} and F_{02}, and therefore to obtain the desired bound for two independent CS interferences.

Consider the expression for Pe conditioned on only one of the interference envelopes, say r_2. We get

$$\mathrm{Pe}\mid r_2 = \tfrac{1}{2}\,\mathrm{erfc}\,(\gamma^{1/2}) + \int_0^{R_1} \Sigma(r_1,r_2,\gamma)\, dF_1(r_1). \quad (17)$$

Now let us maximize $\mathrm{Pe}\mid r_2$ over the second-moment constrained region in $\mathfrak{F}_1$. Clearly, this problem is formally identical with the single interference maximization problem, and differs only in the functional kernel $\Sigma(r_1,r_2,\gamma)$ which is in place of $\Sigma(r,\gamma)$ used in [1]. We may therefore retrace the steps which led to the correct single-interference maximizing distribution F_0, and we should obtain the conditional maximizing distribution $F_{01}\mid r_2$.

From physical considerations, if not by inspection of (12), it is clear that

$$\Sigma(0,u,\gamma) = \Sigma(u,0,\gamma) \equiv \Sigma(u,\gamma), \quad (18)$$

and also that $F_{01}\mid r_2 = 0$ is precisely F_0 for a single interference $r_1 \cos\beta_1$. We would therefore not expect a drastic change in $F_{01}\mid r_2$ if now r_2 is made positive, but small (because the basic theory always demands a discrete distribution for F_0). Therefore, let us hypothesize, initially for r_2 small, that

$$F_{01}\mid r_2 \equiv F_0 = (1-\rho_1)U(r) + \rho_1 U(r-R_1), \quad (19)$$

where

$$\rho_1 \triangleq \sigma_{r_1}^2/R_1^2 \quad (20)$$

is the mean-square to peak-square ratio for interference one. Substituting the candidate maximizing distribution into (17) gives

$$\mathrm{Pe}\,(F_{01}\mid r_2) = \tfrac{1}{2}\,\mathrm{erfc}\,(\gamma^{1/2}) + \rho_1[\Sigma(R_1,r_2,\gamma) - \Sigma(0,r_2,\gamma)] + \Sigma(0,r_2,\gamma). \quad (21)$$

This is indeed a conditional upper bound, that is $\mathrm{Pe}\mid r_2 \le \mathrm{Pe}\,(F_{01}\mid r_2)$ for all admissible r_1 distributions, if (and only if) conditions similar to (26) in [1] are satisfied. Namely, $F_{01}\mid r_2$ above is maximizing if there exists $\lambda > 0$ such that

$$\mathrm{Pe}\mid r_1,r_2 \le \mathrm{Pe}\,(F_{01}\mid r_2) + \lambda(r_1^2 - \sigma_{r_1}^2), \qquad \forall\, r_1 \in [0,R_1] \quad (22a)$$

and

$$\mathrm{Pe}\mid r_1,r_2 = \mathrm{Pe}\,(F_{01}\mid r_2) + \lambda(r_1^2 - \sigma_{r_1}^2), \qquad \forall\, r_1 \in E_{01} \quad (22b)$$

where E_{01} is the set of points of increase of F_{01}. Making the appropriate substitutions, the test becomes

$$\tfrac{1}{2}\,\mathrm{erfc}\,(\gamma^{1/2}) + \Sigma(r_1,r_2,\gamma) - \Sigma(0,r_2,\gamma) \le \tfrac{1}{2}\,\mathrm{erfc}\,(\gamma^{1/2}) + \rho_1[\Sigma(R_1,r_2,\gamma) - \Sigma(0,r_2,\gamma)] + \lambda(r_1^2 - \sigma_{r_1}^2), \qquad \forall\, r_1 \in [0,R_1], \quad (23)$$

with equality to hold on the set E_{01}, which by our hypothesis consists of the two points $0,R_1$. We choose λ by forcing the equality at $r_1 = 0$, and so we get

$$\lambda = (1/R_1^2)[\Sigma(R_1,r_2,\gamma) - \Sigma(0,r_2,\gamma)]. \quad (24)$$

Substituting this value for λ, the test becomes

$$\Sigma(r_1,r_2,\gamma) - \Sigma(0,r_2,\gamma) \leq (r_1^2/R_1^2)[\Sigma(R_1,r_2,\gamma) - \Sigma(0,r_2,\gamma)], \qquad \forall\, r_1 \in [0,R_1]. \tag{25}$$

We have, by this choice of λ, achieved the required equality on the points $r_1 = 0,R_1$, and we now need to establish the inequality for r_1 on $(0,R_1)$. Analogous to what was done previously in [1], define the function

$$g_1(r_1,r_2) = \frac{\Sigma(r_1,r_2,\gamma) - \Sigma(0,r_2,\gamma)}{r_1^2}, \tag{26}$$

and note that the inequality is certainly satisfied if $g_1(r_1,r_2)$ is a nondecreasing function of r_1 on the interval $[0,R_1]$.

Let $R_{\max 1}(r_2)$ be the point where $g_1(r_1,r_2)$, as a function of r_1, ceases to be nondecreasing away from zero. Then it is easy to show that the necessary and sufficient conditions for $F_{01}(r_1) = (1-\rho_1)U(r_1) + \rho_1 U(r_1 - R_1)$ to be a conditional maximizing distribution for a given r_2 will be satisfied for any $R_1 \leq R_{\max 1}(r_2)$.

In summary of what we have shown so far, the probability of error conditioned on a given value r_2 for the envelope modulus of interference two is maximized when the modulus of interference one possesses the two impulse pdf

$$\frac{dF_{01}}{dr_1} = (1-\rho_1)\delta(r_1) + \rho_1\delta(r_1 - R_1), \tag{27}$$

whenever $R_1 \leq R_{\max 1}(r_2)$.[1] Now suppose that

$$R_1 < \inf_{0 \leq r_2 \leq R_2} R_{\max 1}(r_2),$$

i.e., as r_2 varies over its given interval $[0,R_2]$, the value of $R_{\max 1}(r_2)$ is never exceeded by R_1. This means that, for every possible value of r_2, the maximizing distribution F_{01} remains the same, as given above. Wc conclude, therefore, that if $R_1 \leq R_{\inf}(R_2)$, where

$$R_{\inf}(l) \triangleq \inf_{0 \leq x \leq l} R_{\max 1}(x), \tag{28}$$

then the conditional maximizing distribution for interference one, since it remains the same for each realizable value of r_2, must actually be independent of the distribution for interference two, and in fact is given by (19) above.

Now we condition on r_1, and look for the conditional maximizing distribution F_{02} for interference two. However, by the symmetry of the functional kernel $\Sigma(r_1,r_2,\gamma)$ in the variables r_1,r_2, and by the previous result for F_{01}, we see that

$$F_{02}(r_2) = (1-\rho_2)U(r_2) + \rho_2 U(r_2 - R_2), \tag{29}$$

(with ρ_2 defined analogously to ρ_1) is the correct maximizing distribution, independent of the distribution assumed by interference one, if conversely

$$R_2 \leq \inf_{0 \leq r_1 \leq R_1} R_{\max 2}(r_1) \equiv R_{\inf}(R_1).$$

Therefore, whenever we have jointly

$$R_1 \leq R_{\inf}(R_2), \tag{30a}$$

$$R_2 \leq R_{\inf}(R_1), \tag{30b}$$

the error probability is maximized by taking F_1 and F_2, respectively, to be the conditionally maximizing two-step distributions F_{01} and F_{02} as indicated in (19) and (29). Finally, the corresponding two-interference limited-peak bound is easily written in terms of Σ functions by inserting these maximizing distributions into (16), to get

$$\text{Pe} \leq \text{Pe}\,(F_{01},F_{02}) = \tfrac{1}{2}\,\text{erfc}\,(\gamma^{1/2}) + \rho_1(1-\rho_2)\Sigma(R_1,0,\gamma) + \rho_2(1-\rho_1)\Sigma(0,R_2,\gamma) + \rho_1\rho_2\Sigma(R_1,R_2,\gamma). \tag{31}$$

The joint conditions given by (30) on the individual peak envelopes describe a region in the $R_1 \times R_2$ plane where the bound (31) has been shown to be valid. The above inequalities are only a sufficient condition, however, and the complete validity region is actually somewhat larger than is given by them.

Fig. 1 illustrates precisely how the validity region is formed as a result of (30a) and (30b). The shaded regions below each curve satisfy the respective inequality, and therefore the doubly shaded intersection region contains those points (R_1,R_2) that jointly satisfy both conditions. This, of course, gives the validity region only for a fixed value of γ, which in the present example corresponds to a carrier-to-noise ratio of 10 dB.

We may also easily superimpose the validity region appropriate to the sum of peaks, sum of powers bound,[2] which is just

$$R_1 + R_2 \leq R_{\max}, \tag{32}$$

where we know that

$$R_{\max} \equiv R_{\max 1}(0) = R_{\inf}(0). \tag{33}$$

Thus, we get the straight-line boundary as shown. The dependence of the validity region on CNR is shown in Fig. 2. Again, the accompanying straight (dashed) lines describe the corresponding single-interference bound regions.

Because there are too many individual parameters involved, it is not possible to display comprehensively the two-interference limited-peak bound in just a few charts, such as was done in [1] for the single-interference bound. Nevertheless, by making a few reasonable assumptions about the interference, we can present some computations, and draw meaningful comparisons with other techniques.

[1] Bear in mind that $R_{\max 1}(r_2)$, and hence the indicated R_1 range. is implicitly a function of γ.

[2] This bound is obtained by recognizing that the sum of two independent CS interferers is itself CS (with peak and rms values equal to $R_1 + R_2$ and $\sigma_{r_1}^2 + \sigma_{r_2}^2$) and then applying the corresponding single-CS bound.

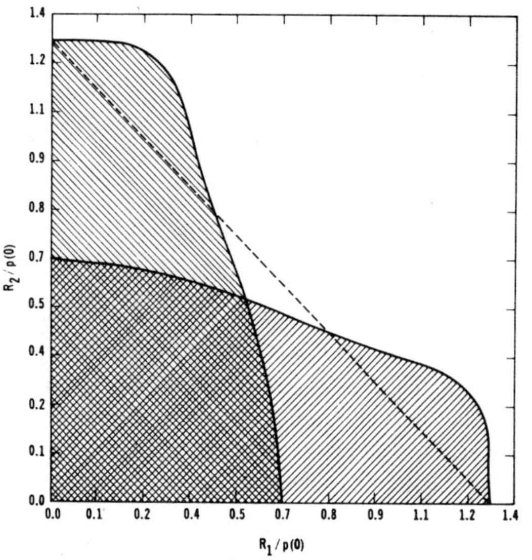

Fig. 1. Formation of bound validity regions for two CS interferers.

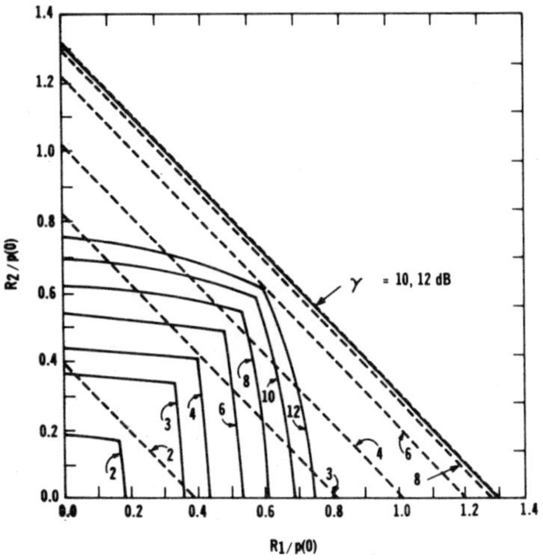

Fig. 2. Bound validity regions for several values of CNR.

Before continuing, we need to define the following (decibel) system parameters: the CNR,

$$\text{CNR} \triangleq 10 \log_{10} \gamma, \tag{34}$$

the carrier-to-interference ratio (CIR) for the ith interferer,

$$\text{CIR}_i \triangleq 20 \log_{10} \frac{p(0)}{\sigma_{r_i}}, \tag{35}$$

and its peak factor (PF),

$$\text{PF}_i \triangleq 20 \log_{10} \frac{R_i}{\sigma_{r_1}}. \tag{36}$$

For the first example, consider a CS interference for which CIR = 10 dB and PF = 6 dB. Suppose, furthermore, that the given interference is composite, and is actually the sum of two independent but statistically equivalent interferers. By equivalent, we mean that the constituent interferers have equal powers and equal PF's. It follows that the individual interference parameters are

$$\text{CIR}_1 = \text{CIR}_2 = 10 + 3 \text{ dB} \tag{37}$$

and

$$\text{PF}_1 = \text{PF}_2 = 6 - 3 \text{ dB}. \tag{38}$$

The two-component limited peak bound for this example is plotted in Fig. 3. Also plotted on this figure for comparison are the results of three other techniques: 1) the eye-diagram bound; 2) the Gaussian noise conversion approximation (this assumes that the interference behaves as if it were a Gaussian process with the same power); and 3) the sum-of-peaks, sum-of-powers bound referred to in the previous section.

We see that the two-interference bound is a significant improvement over the single-interference bound, being approximately a full decade lower at the higher values of CNR. Furthermore, the two-interference bound appears to approach Gaussian conversion asymptotically as CNR decreases, whereas the single-interference bound crosses over and exceeds Gaussian conversion for low CNR.

The second example is that of two similar interferences with individual PF's of only 1 dB. Thus, everything is the same as in the first example, except that the total interference has PF = 4 dB. CIR is still 10 dB, and the individual parameters CIR_1, CIR_2 remain at 13 dB. The resulting bounds are shown in Fig. 4. Note that the noise conversion curve is unchanged from Fig. 3, since it depends only on CIR. The three bounds, however, all depend on the ratio $R/p(0)$, so they have changed, reflecting the new value of PF. The same general behavior that was seen for PF = 6 dB is evident at PF = 4 dB, but the lower PF results in somewhat smaller differences between the bounds.

As a broad generalization, error-probability analysis for Gaussian noise and peak-limited interference that depend on interference power alone will be reasonably good for very low CNR, and those that make use of the peak value alone may be acceptable for very high CNR (where the curves are so steep that the separation in Pe may remain very great). The limited peak bounds discussed here and in [1], by utilizing optimally the rms and peak-envelope data of the interference components, bridge the gap and give good error-probability performance indication over the entire CNR range of practical interest.

IV. BOUND FOR COMBINED INTERSYMBOL INTERFERENCE AND CS INTERFERENCES

Most CPSK systems are subject to a plurality of different types of interferences. Almost all systems must operate in the presence of some ISI. In addition, there will usually be some CS interference present arising from either an adjacent-channel or a cochannel signal operating in the same band as the wanted signal. The two-CS bound of the previous sections carries over almost directly to provide a bound for the other, more important two-interference case: ISI plus one-CS interferer. The only difference is that because the ISI is not CS, the random

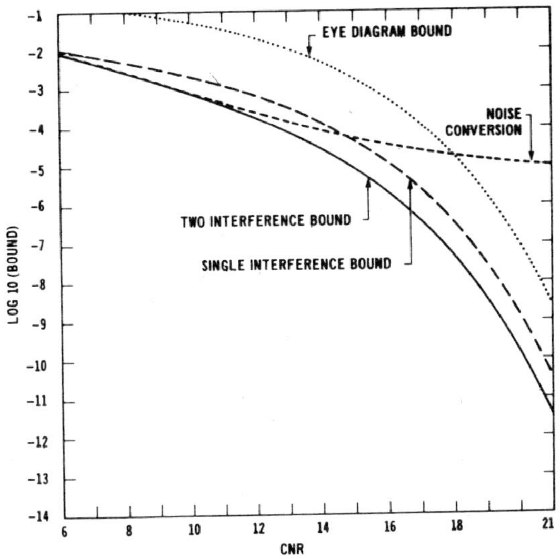

Fig. 3. Comparison of results for interference comprising two similar CS components, $PF_1 = PF_2 = 3$ dB, $CIR_1 = CIR_2 = 13$ dB.

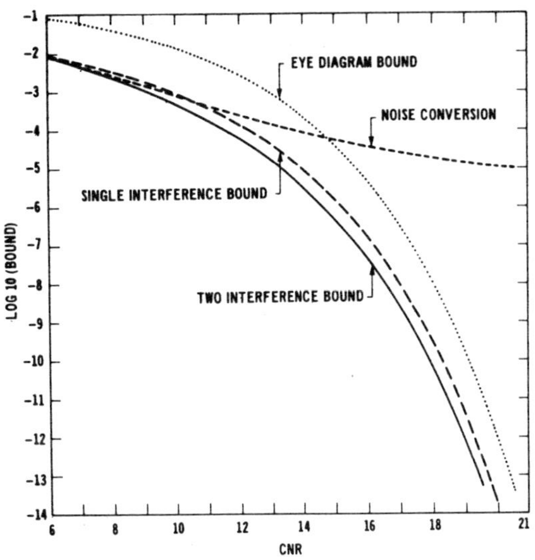

Fig. 4. Comparison of results for interference comprising two similar CS components, $PF_1 = PF_2 = 1$ dB, $CIR_1 = CIR_2 = 13$ dB.

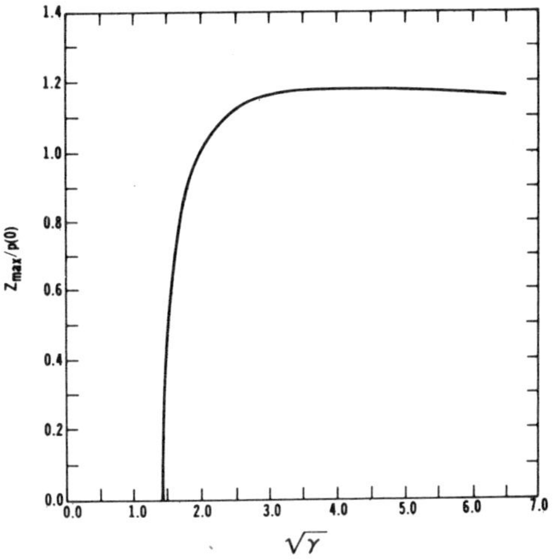

Fig. 5. Validity region—ISI.

variable we deal with is not simply the modulus or envelope of the interference, but is instead the actual interference component in the "direction" of the wanted signal vector. That is, with a binary signal we deal with $\bar{z} = \mathrm{Re}\,\bar{Z}$. As first shown by Glave [12], this results in a maximizing pdf for ISI having 3 mass points, at $\pm Z$ and 0. It can thus take on both positive and negative values.

For example, the bound for ISI alone is obtained with the following maximizing distribution function:

$$F_0(z) = \frac{\rho_z}{2}\left[U(z-Z) + U(z+Z)\right] + (1-\rho_z)U(z), \tag{39}$$

where $\rho_z = \sigma_z^2/Z^2$, and provided $Z \leq Z_{\max}(\gamma)$. $Z_{\max}(\gamma)$ is that point where the function

$$g(z) = \frac{\frac{1}{4}\,\mathrm{erfc}\,[\gamma^{1/2}(1 + z/p(0))] + \frac{1}{4}\,\mathrm{erfc}\,[\gamma^{1/2}(1 - z/p(0))] - \frac{1}{2}\,\mathrm{erfc}\,[\gamma^{1/2}]}{z^2} \tag{40}$$

ceases to be nondecreasing away from zero. The function $Z_{\max}(\gamma)$ is plotted in Fig. 5. If the peak ISI Z is larger than $Z_{\max}(\gamma)$, then the distribution function given by (39) is no longer maximizing. Matthews [13] has shown (at least for the ISI case) that even under this condition a maximizing distribution still exists. However, for most practical cases of interest, the distribution given by (39) applies.

The corresponding upper bound is easily obtained:

$$\mathrm{Pe} \leq \frac{\rho_z}{4}\left\{\mathrm{erfc}\left[\gamma^{1/2}\left(1 - \frac{Z}{p(0)}\right)\right] + \mathrm{erfc}\left[\gamma^{1/2}\left(1 + \frac{Z}{p(0)}\right)\right]\right\} + \tfrac{1}{2}(1-\rho_z)\,\mathrm{erfc}\,(\gamma^{1/2}). \tag{41}$$

Now consider the combined presence of ISI and one CS interferer. Using (8) and averaging over the random phase β, the probability of error in the binary case is

$$\mathrm{Pe} = \iint [q(r,z,\gamma)]\, dF(z)\, dF(r), \tag{42a}$$

where now

$$q(r,z,\gamma) = \tfrac{1}{4}\operatorname{erfc}\left[\gamma^{1/2}\left(1 - \frac{z}{p(0)}\right)\right] + \tfrac{1}{4}\operatorname{erfc}\left[\gamma^{1/2}\left(1 + \frac{z}{p(0)}\right)\right] + \frac{1}{2}\frac{\exp[-\gamma(1 - z/p(0))^2]}{\pi^{1/2}} \cdot \sum_{l=1}^{\infty} \frac{H_{2l-1}[\gamma^{1/2}(1 - z/p(0))]}{(l!)^2}\left[\frac{r\gamma^{1/2}}{2p(0)}\right]^{2l} + \frac{1}{2}\frac{\exp[-\gamma(1 + z/p(0))^2]}{\pi^{1/2}} \cdot \sum_{l=1}^{\infty} \frac{H_{2l-1}[\gamma^{1/2}(1 + z/p(0))]}{(l!)^2}\left[\frac{r\gamma^{1/2}}{2p(0)}\right]^{2l}. \tag{42b}$$

The maximization uses exactly the same conditioning argument as in Section III, where now the kernel $\Sigma(r_1,r_2,\gamma)$ is replaced by $q(r,z,\gamma)$. The resulting bound is

$$\mathrm{Pe} \le \rho_z[\rho_r q(R,Z,\gamma) + (1 - \rho_r)q(0,Z,\gamma)] + (1 - \rho_z)[\rho_r q(R,0,\gamma) + (1 - \rho_r)q(0,0,\gamma)] \tag{43}$$

where, as before, $\rho_z = \sigma_z^2/Z^2$, $\rho_r = \sigma_r^2/R^2$, and the bound is valid provided

$$Z \le \inf_{0 \le r \le R} Z_{\max}(r,\gamma) \tag{44a}$$

$$R \le \inf_{0 \le z \le Z} R_{\max}(z,\gamma). \tag{44b}$$

Here, $Z_{\max}(r,\gamma)$ is the point where

$$\frac{q(r,z,\gamma) - q(r,0,\gamma)}{z^2}, \tag{45a}$$

as a function of z, ceases to be nondecreasing away from zero. Similarly, $R_{\max}(z,\gamma)$ is the point where

$$\frac{q(r,z,\gamma) - q(0,z,r)}{r^2}, \tag{45b}$$

as a function of r, ceases to be nondecreasing away from zero. This validity region is plotted in Fig. 6 for a range of CNR values. The extension to three interferences (ISI plus two-CS) is similarly quite straightforward. The probability of error in the binary case is

$$\mathrm{Pe} = \iiint [q(r_1,r_2,z,\gamma)]\, dF(r_1)\, dF(r_2)\, dF(z), \tag{46}$$

where [compare with (12) and (42b)]

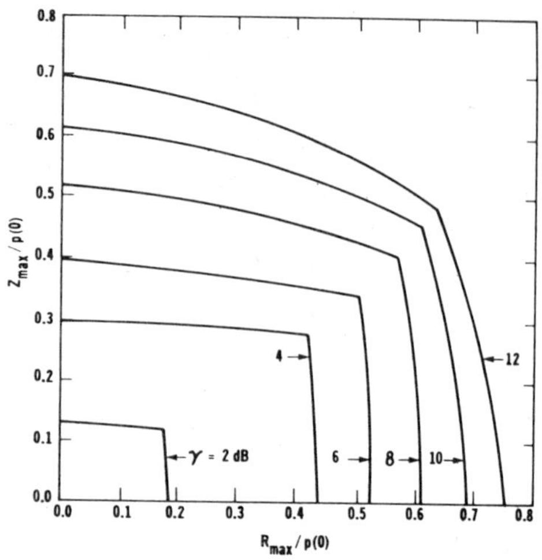

Fig. 6. Validity region—ISI + one CS.

$$q(r_1,r_2,z,\gamma) = \tfrac{1}{4}\operatorname{erfc}\left[\gamma^{1/2}\left(1 - \frac{z}{p(0)}\right)\right] + \tfrac{1}{4}\operatorname{erfc}\left[\gamma^{1/2}\left(1 + \frac{z}{p(0)}\right)\right] + \tfrac{1}{2}\pi^{-1/2}\exp\left[-\gamma\left(1 - \frac{z}{p(0)}\right)^2\right] \cdot \sum_{l=1}^{\infty} \frac{H_{2l-1}\{\gamma^{1/2}(1 - z/p(0))\}}{2^{2l}l!l!}\gamma^l\left[\frac{r_1}{p(0)}\right]^{2l} \cdot G_l\left[\left(\frac{r_2}{r_1}\right)^2\right] + \tfrac{1}{2}\pi^{-1/2}\exp\left[-\gamma\left(1 + \frac{z}{p(0)}\right)^2\right] \cdot \sum_{l=1}^{\infty} \frac{H_{2l-1}\{\gamma^{1/2}(1 + z/p(0))\}}{2^{2l}l!l!}\gamma^l\left[\frac{r_1}{p(0)}\right]^{2l} \cdot G_l\left[\left(\frac{r_2}{r_1}\right)^2\right]. \tag{47}$$

Extending the maximization argument in Section III to a third variable, and this time replacing the kernel $q(r,z,\gamma)$ with $q(r_1,r_2,z,\gamma)$, the resulting bound is

$$\mathrm{Pe} \le \rho_z\{\rho_{r1}[\rho_{r2}q(R_1,R_2,Z,\gamma) + (1 - \rho_{r2})q(R_1,0,Z,\gamma)] + (1 - \rho_{r1})[\rho_{r2}q(0,R_2,Z,\gamma) + (1 - \rho_{r2})q(0,0,Z,\gamma)]\} + (1 - \rho_z)\{\rho_{r1}[\rho_{r2}q(R_1,R_2,0,\gamma) + (1 - \rho_{r2})q(R_1,0,0,\gamma)] + (1 - \rho_{r1}) \cdot [\rho_{r2}q(0,R_2,0,\gamma) + (1 - \rho_{r2})q(0,0,0,\gamma)]\}. \tag{48}$$

Naturally, $\rho_z = \sigma_z^2/Z^2$, $\rho_{r1} = \sigma_{r1}^2/R_1^2$, $\rho_{r2} = \sigma_{r2}^2/R_2^2$, and the bound is valid provided that

$$Z \leq \inf_{0 \leq r_1 \leq R_1; 0 \leq r_2 \leq R_2} Z_{\max}(r_1,r_2,\gamma) \tag{49a}$$

$$R_1 \leq \inf_{-Z \leq z \leq Z; 0 \leq r_2 \leq R_2} R_{\max 1}(r_2,z,\gamma) \tag{49b}$$

$$R_2 \leq \inf_{-Z \leq z \leq Z; 0 \leq r_1 \leq R_1} R_{\max 2}(r_1,z,\gamma). \tag{49c}$$

Again, $Z_{\max}(r_1,r_2,\gamma)$ is the point where

$$\frac{q(r_1,r_2,z,\gamma) - q(r_1,r_2,0,\gamma)}{z^2}, \tag{50a}$$

as a function of z, ceases to be nondecreasing away from zero. Similarly, $R_{\max 1}(r_2,z,\gamma)$ and $R_{\max 2}(r_1,z,\gamma)$ are the points where the functions

$$\frac{q(r_1,r_2,z,\gamma) - q(0,r_2,z,\gamma)}{r_1^2} \tag{50b}$$

and

$$\frac{q(r_1,r_2,z,\gamma) - q(r_1,0,z,\gamma)}{r_2^2}, \tag{50c}$$

respectively, cease to be monotonically increasing away from zero.

A typical plot of the validity region given by (49) and (50) is presented in Fig. 7 for CNR = 10 dB, using Z as a parameter in the $R_1 \times R_2$ plane.

Apparently complicated expressions, such as (47) and (48) above for the ISI plus two-CS interference case, would seem to contradict the previously stated computational simplicity of the bound. Closer inspection, however, or better still some operating familiarity, should convince the reader of the correctness of our contention that the bounding technique is, indeed, relatively simple to apply. First, the bound itself is just a combination of similar terms, either Σ functions (for CS interference) or q functions (for CS and ISI), and the coefficients involve only the PF's ρ of the interferences. Furthermore, there is a definite order, or pattern, to the terms, that is, to the argument lists of the Σ or q functions and their ρ coefficients, so that with a little familiarity the bound could be written out from memory.

Second, and most important, the series which Σ or q represent are efficiently evaluated on a digital computer. Evaluation of the basic series, that for $\Sigma(r,\gamma)$ which pertains to one CS interference, was covered in detail in [1], wherein a PL-1 program was included. The extension to two CS components, for which $\Sigma(r_1,r_2,\gamma)$ is to be evaluated, can be accomplished with a simple augmentation of the aforementioned program. Basically, the $G_l(\cdot)$ function (14) which multiplies each term of the series possesses a simple recurrence relationship (easily found by writing G_l as the appropriate Legendre polynomial), and this recurrence can be included in the main recurrence (which generates the Hermite polynomial, factorial, etc.). This augmented program is seen in Fig. 10, which should help the interested reader to easily implement a bound evaluation program on his own computer. Finally, it is

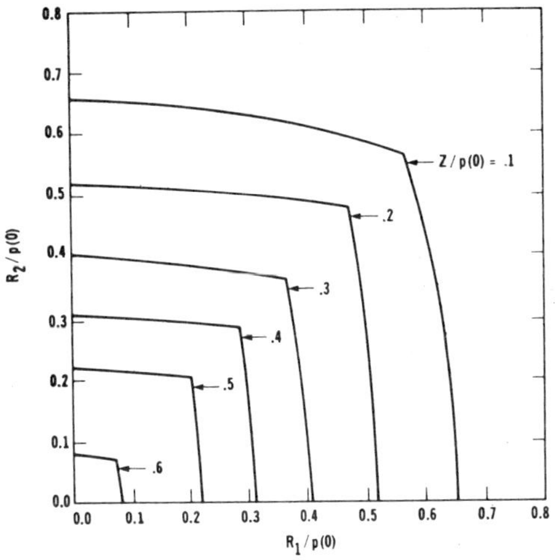

Fig. 7. Validity region—ISI + two CS.

seen that each of the two series which comprise a q function is a trivial variant of a Σ function series, wherein γ has been modified in both the exponential term and the Hermite polynomial argument. (To accordingly modify the program, merely insert $X = X * (1 + Z)$ between the first and second lines.) Thus, using the program code of Fig. 10 as a basic subroutine, bound expressions such as (48) are easily evaluated.

V. APPLICATIONS OF THE INTERFERENCE BOUND TO COMBINED ISI AND CS

In this section two examples are presented to illustrate the use of the ISI plus single-CS and ISI plus two-CS bounds for a quaternary CPSK system. The extension from the binary case presented above to the quaternary CPSK case is made by treating the binary bound as a general bound on the probability that the received signal occupies a given half plane in the signal space.

Specifically, the error region in the complex signal space for quaternary CPSK is the union of two half planes, each located a distance $p(0)/\sqrt{2}$ from the received signal point $p(0)\exp[j\phi_k]$. From (5) the in-phase and quadrature ISI components are also each reduced by a factor $1/\sqrt{2}$. Finally, because both the Gaussian noise and the CS interferences are CS, the probability that the total received complex envelope, $\tilde{\psi}$ in (7), lies in one of these half planes is equal to the binary CPSK error probability where both the signal and intersymbol interference terms are reduced by $1/\sqrt{2}$, all other terms remaining the same, i.e.,

$$\tfrac{1}{2}\,\mathrm{erfc}\left\{\frac{p(0) + z + \sqrt{2}\sum_{i=1}^{N} r_i \cos\beta_i}{\sqrt{2}\sigma_n}\right\}.$$

Hence, the complete quaternary error probability is upper bounded by twice the binary CPSK Pe when the values of σ_n and $\{r_i\}_{i=1}^{N}$ are all increased by a factor of $\sqrt{2}$ (equivalently, when CNR and $\{\mathrm{CIR}_i\}_{i=1}^{N}$ are all reduced by 3 dB).

Thus by using the above ISI and CS peak-limited bounds on these "adjusted" binary Pe values, we obtain the desired upper bound on the quaternary Pe.

In the examples we consider a quaternary CPSK system where the transmitted signal is of the form

$$s(t) = \mathrm{Re}\exp[j\omega_0 t]\int_{-\infty}^{\infty} g(\tau) \cdot \Big[\sum_{k=-\infty}^{\infty} w(t-\tau-kT)\exp(j\phi_k)\,d\tau\Big], \quad (51)$$

where ϕ_k is the transmitted symbol, selected with equal probability from the set $\{\pm\pi/4, \pm 3\pi/4\}$, $g(t)$ is the complex envelope of the impulse response of the transmitting bandpass filter, and

$$w(t) = \begin{cases} 1, & |t| < T/2 \\ 0, & \text{otherwise.} \end{cases} \quad (52)$$

The received signal (in the absence of noise and interference) may then be expressed as

$$p(t) = \mathrm{Re}\sum_{k=-\infty}^{\infty} \exp[j(\phi_k + w_0 t)]p(t-kT), \quad (53)$$

where $p(t)$ is the overall rectangular pulse response of the channel including transmit and receive filters

$$p(t) = \frac{1}{2}\int_{-\infty}^{\infty}\int_{-\infty}^{\infty} h(\tau)g(t-\tau-\sigma)w(\sigma)\,d\tau\,d\sigma, \quad (54)$$

where $h(t)$ is the complex envelope of the impulse response of the receive filter.

As a first example we assume that the transmit filter is ideal, and the receive filter is characterized by a three-pole Butterworth RF filter with time-bandwidth product $2BT = 1$. The complex envelope of the impulse response of the receive filter is given by

$$h(t) = \begin{cases} 4\pi B[\exp(-2\pi Bt) + (1/\sqrt{3})\exp(-\pi Bt) \cdot \sin(3\pi Bt - \pi/3)], & t \geq 0 \\ 0, & t < 0. \end{cases} \quad (55)$$

Let the CCI be a single interfering sinusoid with CIR = 12 dB. That is

$$10\log_{10}\frac{p^2(0)}{\sigma_r'^2} = 12. \quad (56)$$

There is no ACI.

The peak and mean-square values of the ISI are given by

$$Z = \sum_{i=-\infty}^{\infty}{}' \, |p(iT)| \quad (57a)$$

$$\sigma_z^2 = \sum_{i=-\infty}^{\infty}{}' \, p^2(iT). \quad (57b)$$

These ISI parameters can be computed simply and quickly by using a straightforward finite sum truncation approximation.

Then, substituting (57), together with the "adjusted" CNR and CIR values into (43), and then multiplying by two, we obtain the desired upper bound. This is plotted in Fig. 8 and then compared with the approximation result of Fang and Shimbo, which is within a few percent of the exact value. The bound is reasonably tight, being within $\frac{1}{3}$ dB of the approximation result.

As a second example we consider a very common interference environment for CPSK: the presence of two randomly phased independent adjacent-channel interferers. If we assume no carrier coherence between the adjacent and wanted channels, each of these adjacent interferences will be CS. We also assume no bit synchronism between the adjacent and wanted channels. Finally, the presence of nonideal transmit and receive filters will inevitably introduce some ISI.

Specifically, we consider a quaternary PSK system that uses Butterworth transmit and receive filters. The transmit filter is a six-pole type with $2BT = 2$ and the receive filter is four-pole, with $2BT = 1$. The peak and mean-square values of the resulting ISI are computed exactly as they were in the previous example.

We assume that on either side of the wanted signal are identical PSK signals with carriers displaced by amounts $\pm\omega_d$ from the wanted carrier ω_0 and randomly phased with respect to the wanted carrier.

At the receiver, each of these interferers will be of the form

$$i(t) = \mathrm{Re}\sum_{k=-\infty}^{\infty} \hat{p}(t+t'-kT) \cdot \exp(j[\phi_k + \omega_0(t+t') + \theta]) \quad (58)$$

where

$$\hat{p}(t) = \hat{p}_c(t) + j\hat{p}_s(t) \quad (59)$$

$$\hat{p}_c(t) = \tfrac{1}{2}\mathrm{Re}\int_{-\infty}^{\infty}\int_{-\infty}^{\infty} h(\tau)g(t-\tau-\sigma)w(\sigma) \cdot \exp[j(\omega_0 \pm \omega_d)\sigma]\,d\tau\,d\sigma \quad (60a)$$

$$\hat{p}_s(t) = \tfrac{1}{2}\mathrm{Im}\int_{-\infty}^{\infty}\int_{-\infty}^{\infty} h(\tau)g(t-\tau-\sigma)w(\sigma) \cdot \exp[j(\omega_0 \pm \omega_d)\sigma]\,d\tau\,d\sigma. \quad (60b)$$

θ is the random (uniform) phase offset between the wanted and interfering carriers, and t' is a random variable uniform on $[-T/2, T/2]$ representing the lack of bit synchronism between the wanted and interfering carriers.

Each of these adjacent channel interferers can be expressed in the form

$$i(t) = \mathrm{Re}\, r(t,t')\exp(j[\omega_0(t+t') + \theta]) \quad (61)$$

where

$$r(t,t') = \sum_{k=-\infty}^{\infty} \hat{p}(t+t'-kT)\exp[j\phi_k]. \quad (62)$$

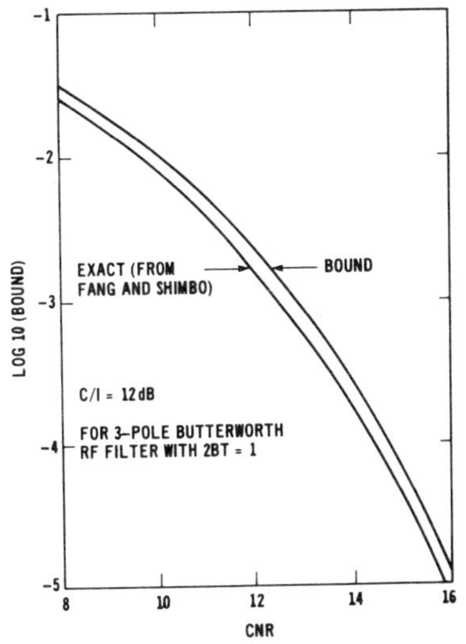

Fig. 8. Comparison of the peak-constrained upper bound with Fang and Shimbo's approximation for ISI and one sinusoidal interferer.

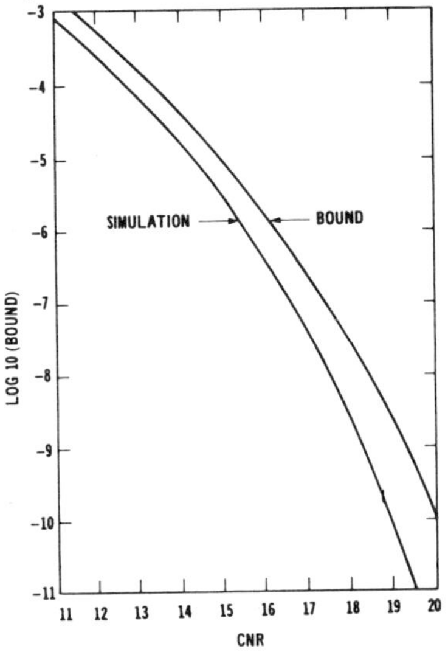

Fig. 9. Comparison of the peak constrained upper bound with a computer simulation approximation for ISI and two similar ACI interferers.

In quaternary CPSK, each of the ϕ_k independently takes on one of the four possible values $\{\pi/4 + i\pi/2\}_{i=1}^4$. The mean-square value of the ACI envelope (at $t = 0$ and for fixed t') is thus given by

$$\sigma_i^2(t') = \sum_{k=-\infty}^{\infty} \hat{p}_c^2(t' - kT) + \sum_{k=-\infty}^{\infty} \hat{p}_s^2(t' - kT). \quad (63)$$

We describe in the Appendix an algorithm for computing the peak ACI envelope.

Both the peak and mean-square values of the ACI envelope are computed using fast Fourier transform techniques and truncating the sums involved when the percentage change decreases below a fixed value.

Assuming that $\omega_d = 1.4T$ and that t' is uniformly distributed on $[-T/2,T/2]$, a further computation to average over this random variable yields R_i and σ_i^2 for $i = 1$ and 2.

Finally, substitution of these peak and mean-square values for all three interferers into (48) yields the desired upper bound. [It is readily verified that the peak values of each of the interferers satisfies (49)].

This bound on Pe versus CNR is plotted in Fig. 9 and compared with a computer-simulation approximation of the same system [15]. Again, even in the presence of multiple interference, the bound remains tight enough to be useful as a design tool.

VI. CONCLUSION

An upper bound on Pe has been determined for coherent PSK transmission subject to Gaussian noise and multiple interferences. All that is required to know is the power and peak envelopes of the interferences.

The bound is conceptually simple, and what is more important, it is simple computationally. The major part of the programming required to compute the bound has been supplied here in Fig. 10.

For the case of two CS interferers, the bound presented here represents a significant improvement, in many important cases, over the sum of peaks, sum of powers, application of the single-interference bound introduced in [1].

The extension of the bound to a mixture of CS and non-CS components (in particular, to a combination of ACI and ISI) is a natural one and has been shown to yield results that are tight enough to render the bound a useful design tool.

Refinements of the bound are to be had through the inclusion of additional statistical information (which further restricts the range of allowed envelope distributions). Perhaps the most effective improvement is to be found in relating the easily measured spectral properties of the interference to equivalent restrictions on the envelope pdf. It seems intuitive, for example, that a stationary process cannot have an envelope pdf which contains only isolated impulses (such as the maximizing pdf's F_{01} and F_{02}) if that process is also narrow band. We need to make such notions precise, and to find out just how the general band limiting of an interference imposes limitations on its envelope pdf, and thereby, in turn, serves (in conjunction with other data such as rms and peak values) to bound the error probability.

In conclusion, it is felt that the limited-peak bounds in their present form provide an effective and convenient solution to the problem of PSK error performance estimation in situations where the interference cannot be described statistically beyond a few rudimentary attributes.

APPENDIX

We wish to compute the peak value of the modulus of

$$r(t,t') = \sum_{k=-\infty}^{\infty} \hat{p}(t + t' - kT) \exp[j\phi_k], \quad (64)$$

```
                                  X = 10**(CNR/20);
main                              TWOX = 2 * X;
initialization . . . .            LAMBDA = 0.25 * 10**((CNR - CIR1 + PF1)/10);
                                  TERM2 = LAMBDA;
                                  TERM3 = TWOX * LAMBDA;
                                  ARG = 10**((PF2 - CIR2 - PF1 + CIR1)/10);
Gℓ recurrence                     C = ARG + 1;
initialization . . . .            D = (1-ARG)**2;
                                  OLD = 1;
                                  NEW = C;
first term in series . . .        SUM = TERM3 * C;
                             LOOP:
                                  DO I = 2 TO 400;
                                        TERM1 = TERM2;
main recurrence,                        TERM2 = TERM3;
including Hermite,                      TERM3 = TWOX * TERM2 - TERM1 * (4 * I - 6);
factorial, and        . .               TERM1 = TERM2 * LAMBDA / I**2;
geometric terms                         TERM2 = TERM3 * LAMBDA / I**2;
                                        TERM3 = TWOX * TERM2 - TERM1 * (4 * I - 4);
                                        OLDEST = OLD;
Gℓ recurrence . . . .                   OLD = NEW;
                                        NEW = ((2*I-1) * C * OLD - (I-1) * D * OLDEST) / I;
                                        SUM = SUM + TERM3 * NEW;
convergence test . . . . . . .          IF ABS(TERM3*NEW) / ABS(SUM) <= .1E-09 THEN GO TO QUIT;
                                  END LOOP;
                             QUIT:
```

Fig. 10. PL-1 program to compute series $\Sigma(r_1, r_2, \gamma)$.

for fixed t and t', by appropriately selecting each of the ϕ_k from the set $\{\pi/4 + i\pi/2\}_{i=1}{}^4$, where $\hat{p}(\cdot)$ is the overall complex channel response.

With t and t' fixed we may rewrite (64) in the simplified notation

$$r = \sum_{k=-\infty}^{\infty} \hat{p}_k \exp[j\phi_k]. \tag{65}$$

Let r_0 be the maximizing resultant vector we seek, that is

$$\|r_0\| = \max_{\{\phi_k\}} \|\sum_k \hat{p}_k \exp[j\phi_k]\| \tag{66}$$

where $\|\cdot\|$ denotes the norm, or modulus, of a vector.

We assert that in order for r to achieve this maximum r_0, all of the component vectors $\hat{p}_k \exp[j\phi_k]$ must have arguments within $\pi/4$ radians of r_0 and hence within $\pi/2$ radians of each other. That is, the maximizing vector is such that all of its components lie within some 90° quadrant. The following is an informal proof of this intuitively appealing notion.

Suppose we have found the maximum length vector r_0 and one of the components, say $\hat{p}_m \exp[j\phi_m]$, does not lie within $\pi/4$ radians of r_0. Then we could form a new resultant r_1 by selecting a different value for ϕ_m. This is equivalent to rotating $\hat{p}_m \exp[j\phi_m]$ to one of the other three possible positions, such that its new orientation now lies within $\pi/4$ radians of r_0. That is,

$$r_1 = r_0 + \hat{p}_m \exp[j(\phi_m + \pi)] + \hat{p}_m \exp[j(\phi_m \pm k\pi/2)] \tag{67}$$

where $k = 1$ or 2 as necessary to place $\hat{p}_m \exp[j(\phi_m \pm k\pi/2)]$ within $\pi/4$ radians of the argument of r_1.

Now since the angle between r_0 and $p_m \exp[j\phi_m]$ was originally assumed to be greater than $\pi/4$, the angle between $p_m \exp[j(\phi_m + \pi)]$ and r_0 must be less than $3\pi/4$. Further, since $p_m \exp[j(\phi_m \pm k\pi/2)]$ now lies within $\pi/4$ of r_0, the vector sum of the last two equal length vectors, call it $\tilde{r}$, in (68) must lie within $\pi/2$ of r_0.

Then using dot product notation,

$$\|r_1\|^2 = \|r_0\|^2 + \|\tilde{r}\|^2 + 2r_0 \cdot \tilde{r}. \tag{68}$$

But the angle between r_0 and $\tilde{r}$ is less than $\pi/2$ and hence the last term in (67) is positive. Thus $\|r_1\| > \|r_0\|$, which contradicts the assumption that r_0 was the maximizing vector, and so proves the assertion that all components must be within the same quadrant.

The above proof also suggests an algorithm for finding r_0, and thereby computing the peak envelope. The choice of the angular coefficients $\{\phi_n\}$, which yields the maximizing vector, must be such that all of the components $\hat{p}_n \exp[j\phi_n]$ lie within some 90° quadrant. By symmetry, then, we can reduce the range of admissible data coefficients $\{\phi_n\}$ to that subset which places all of the vectors $\{\hat{p}_n \exp[j\phi_n]\}$ in some half plane. We then need to search for the maximizing vector over all those distinct choices of the set of $\{\phi_n\}_{n=1}{}^{\infty}$ such that the component vectors of the corresponding set of $\{\hat{p}_n \exp[j\phi_n]\}$ all lie within some 90° quadrant in the chosen half plane. Note that in general it will be a different 90° quadrant that is associated with each candidate set of $\{\hat{p}_n \exp[j\phi_n]\}$. This is equivalent to searching over all possible orientations in any half plane of a 90° quadrant each containing all the component vectors.

By truncating the vector sum to N terms, there are only N different possible sets of orientations of the component vectors (i.e., N different orientations of the 90° quadrant). For each of these orientations we compute the modulus of the resultant and select the maximum, which is the desired result.

Specifically, if we wish to find approximately the maximum modulus of

$$r = \sum_{n=1}^{\infty} \hat{p}_n \exp[j\phi_n] \tag{69}$$

we use the following algorithm. Select a finite subset of components, by choosing those having the largest $|\hat{p}_n|$.

Then order these components by argument such that

$$-\pi/2 \leq \angle \hat{p}_m \exp[j\phi_m] < \angle \hat{p}_n \exp[j\phi_n] \leq 0,$$
$$0 \leq m \leq n \leq N.$$

Compute the modulus of the resultant; call it $R^{(1)}$. Rotate $\hat{p}_1 \exp[j\phi_1]$ counterclockwise through $\pi/2$ radians and compute the modulus of the new resultant; call it $R^{(2)}$. Continue with each vector in turn until each component has been rotated through $\pi/2$ radians. The peak modulus R is equal to max $\{R^{(i)}\}_{i=1}{}^{N}$.

By increasing N the peak modulus of the infinite sum may be computed to any desired accuracy.

REFERENCES

[1] A. S. Rosenbaum and F. E. Glave, "An error probability upper bound for coherent phase-shift keying with peak-limited interference," *IEEE Trans. Commun.*, vol. COM-22, pp. 6–16, Jan. 1974.

[2] O. Shimbo, R. J. Fang, and M. Celebiler, "Performance of *M*-ary PSK systems in Gaussian noise and intersymbol interference," *IEEE Trans. Inform. Theory*, vol. IT-19, pp. 44–58, Jan. 1973.

[3] V. K. Prabhu, "Performance of coherent phase-shift-keyed systems with intersymbol interference," *IEEE Trans. Inform. Theory*, vol. IT-17, pp. 418–431, July 1971.

[4] ——, "Some considerations of error bounds in digital systems," *Bell Syst. Tech. J.*, vol. 50, pp. 3127–3151, Dec. 1971.

[5] A. S. Rosenbaum, "PSK error performance with Gaussian noise and interference," *Bell Syst. Tech. J.*, vol. 48, pp. 413-442, Feb. 1969.

[6] ——, "Binary PSK error probabilities with multiple cochannel interferences," *IEEE Trans. Commun. Technol.*, vol. COM-18, pp. 241–253, June 1970.

[7] V. K. Prabhu, "Error rate considerations for coherent phase-shift-keyed systems with cochannel interference," *Bell Syst. Tech. J.*, vol. 48, pp. 743–767, Mar. 1969.

[8] J. Goldman, "Multiple error performance of PSK systems with cochannel interference and noise," *IEEE Trans. Commun. Technol.*, vol. COM-19, pp. 420–430, Aug. 1971.

[9] R. Fang and O. Shimbo, "Unified analysis of a class of digital systems in additive noise and interference," *IEEE Trans. Commun.*, vol. COM-21, pp. 1075–1091, Oct. 1973.

[10] C. Colavito and M. Sant' Agostino, "Multiple co-channel and interchannel interference effects in binary and quaternary PSK radio systems," presented at IEEE Int. Conf. Communications, Philadelphia, Pa., 1972.

[11] S. Benedetto, E. Biglieri, and V. Castellani, "Combined effects of intersymbol, interchannel, and co-channel interferences in *M*-ary CPSK systems," *IEEE Trans. Commun.*, vol. COM-21, pp. 997–1008, Sept. 1973.

[12] F. E. Glave, "An upper bound on the probability of error due to intersymbol interference for correlated digital signals," *IEEE Trans. Inform. Theory*, vol. IT-18, pp. 356–363, May 1972.

[13] J. W. Matthews, "Sharp error bounds for intersymbol interference," *IEEE Trans. Inform. Theory*, vol. IT-19, pp. 440–447, July 1973.

[14] R. Esposito and L. R. Wilson, "Statistical properties of two sine waves in Gaussian noise," *IEEE Trans. Inform. Theory*, vol. IT-19, pp. 176–183, Mar. 1973.

[15] P. Whalen, private communication, results from Bell-Northern Research PSK simulation program.

TV cochannel interference on a PCM-PSK SCPC system

D. KURJAN AND M. WACHS

(Manuscript received May 12, 1976)

Measurements have been made while investigating the effects of cochannel TV frequency modulating an interfering carrier on a multicarrier single-channel-per-carrier (SCPC) system with PCM-PSK modulation. This note presents results which deal with a single voice channel.

Experimental setup

The SCPC system under consideration would consist of several hundred 64-kbps 4-phase PSK channel slots in a single transponder. Since it would be impractical to generate that many carriers, multicarrier operation was simulated by using six 32-kbps 2-phase PSK carriers independently PSK modulated and clocked. The total power in these six carriers was adjusted so that the operating point of the traveling wave tube amplifier (TWTA) was at an input backoff of 10 dB. The SCPC carrier under test was located at a frequency corresponding to a carrier-to-intermodulation (C/I) ratio of approximately 20 dB.

Figure 1 is a schematic of the experimental setup. An SCPC carrier is generated and up-converted to 4 GHz along with the six PSK cotransponder carriers. The up-converter is linear, providing a 2-carrier C/I greater than 40 dB. After amplification by the TWT, the composite is down-converted to 70 MHz. A common local oscillator is used for up- and down-conversion to eliminate translation oscillator instability. The TV interference and thermal noise are added at IF on the down-link. The TV interference is centered on the desired SCPC channel frequency and is represented as a frequency modulated carrier driven by a triangular dispersal wave with a peak-to-peak deviation of 1 MHz.

Test results

The performance of the SCPC modem in a back-to-back configuration without added interference is shown in the bit-error rate (BER) vs carrier-to-noise (C/N) ratio curve of Figure 2. It can be seen that an error probability (P_e) of 10^{-4} occurs at a C/N of 11.65 dB. The corresponding ratio of energy per bit to noise density (E_b/N_o) is computed as follows:

$$E_b/N_o = (C/N)\, BT_b$$

where B = channel bandwidth = 38 kHz

$$T_b = \text{information bit period} = \frac{1}{64 \text{ kbps}}$$

$$= 1.56 \times 10^{-5} \text{ seconds.}$$

Then,

$$(E_b/N_o)_{\text{meas}} = 11.65 - 2.26 = 9.4 \text{ dB}$$

while theoretical calculations yield

$$(E_b/N_o)_{\text{th}} = 8.4 \text{ dB}$$

The performance of the SCPC channel through the INTELSAT IV satellite simulator without external (*i.e.*, TV) interference is shown in Figure 3. The BER vs C/N curve for a C/I of 30 dB is the same as that for the back-to-back case (Figure 2). For C/I ratios of 20 and 15 dB, a P_e of 10^{-4}

This note is based upon work performed at COMSAT Laboratories under the sponsorship of the International Telecommunications Satellite Organization (INTELSAT). Views expressed in this note are not necessarily those of INTELSAT.

Mr. Kurjan is a member of the technical staff in the Systems Applications and Simulation Department of the Transmissions Systems Laboratory, COMSAT *Laboratories.*

Mr. Wachs is Manager, Systems Applications and Simulation Department, Transmission Systems Laboratory, COMSAT *Laboratories.*

Reprinted with permission from *COMSAT Tech. Rev.*, vol. 6, pp. 413-424, Fall 1976.

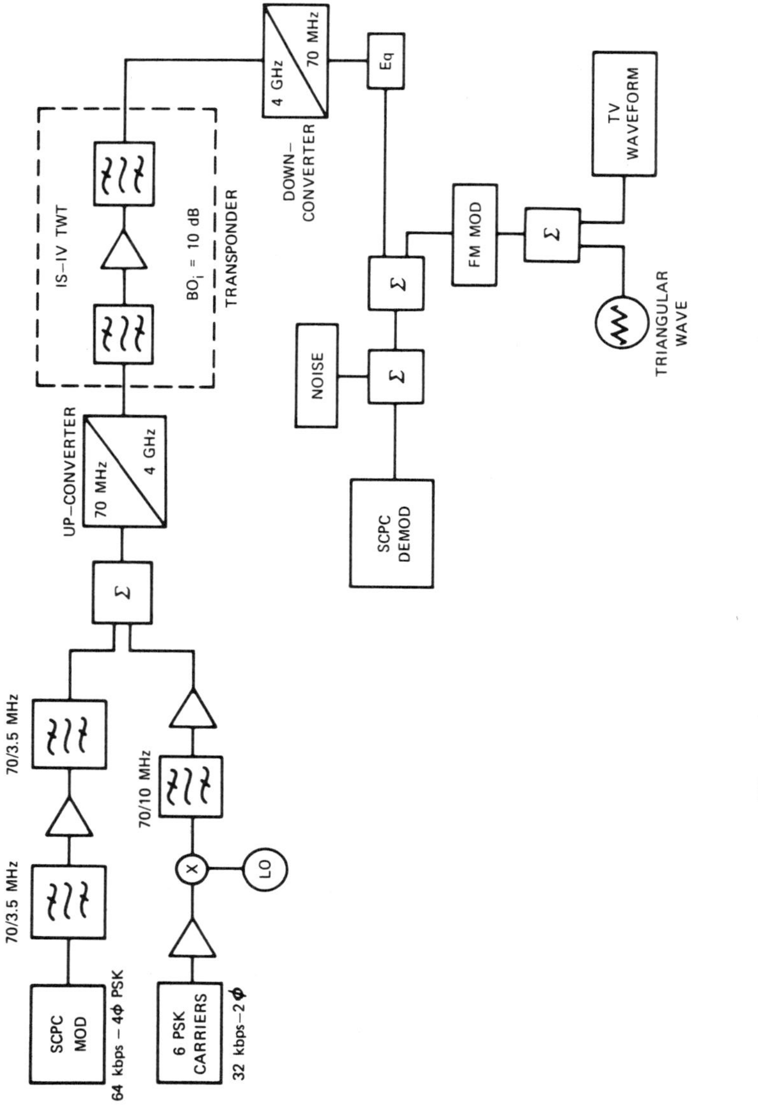

Figure 1. *Block Diagram of Experimental Setup*

requires C/N ratios of 12.45 and 16.6 dB, respectively. It should be noted that C/I refers to the modulated carrier power relative to the intermodulation power measured after the demodulator input filter without added thermal noise or interference.

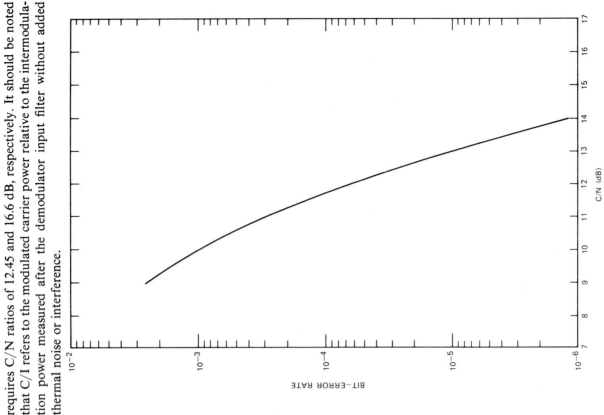

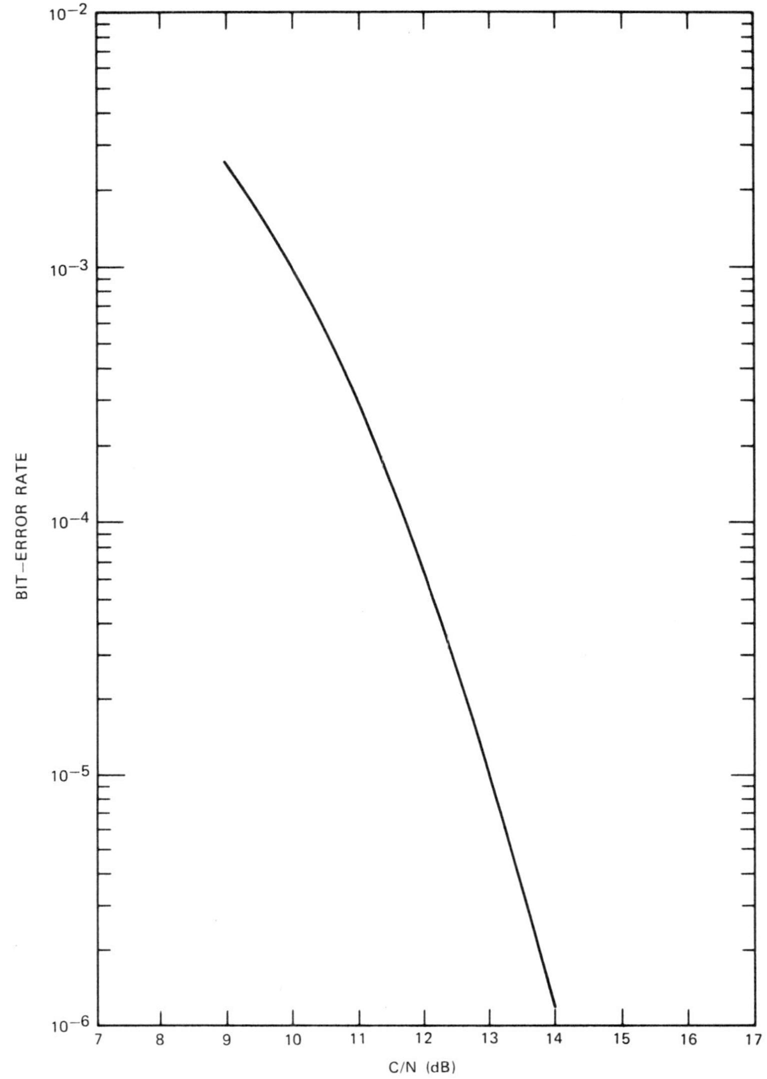

Figure 2. *SCPC Modem Back-to-Back Performance*

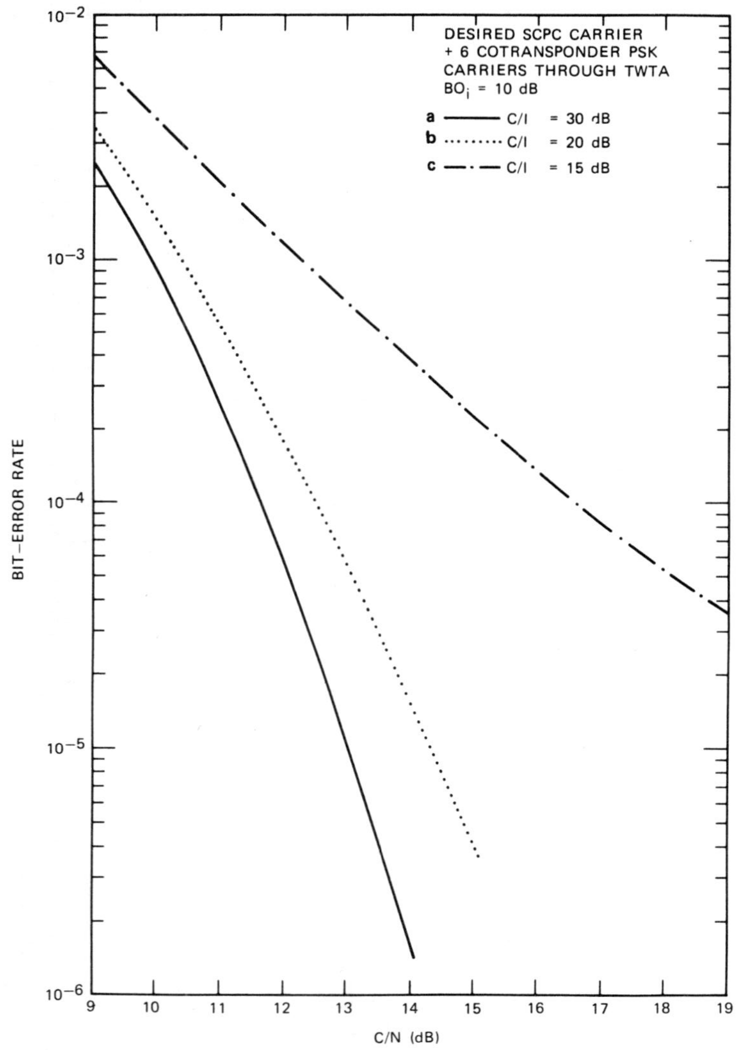

Figure 3. *SCPC Performance Through a Transponder*

The carrier-to-interference (C/INT) ratio is defined as the ratio of modulated carrier power to the power of an unmodulated external interference source at the demodulator input without added thermal noise (or intermodulation, in the case of transponder operation). The SCPC performance in a back-to-back mode with a frequency dispersal (TV) interference source added is shown in curves *a–c* of Figure 4. It can be seen that a C/INT of 10 dB yields an increase in the required C/N of less than 0.75 dB for a 10^{-4} bit-error probability. The average modulated (sweeping) interference power appearing after the demodulator input filter is 14.2 dB less than the power of the unmodulated interference source, since the swept interference source is in the channel bandwidth 3.8 percent of the time in these tests. Therefore, for a C/INT of 10 dB, the average power ratio of modulated carrier to modulated interference source at the channel filter output is 24.2 dB.

Curve *d* in Figure 4 shows the result of an unmodulated interference source at 70 MHz with the same C/INT (24 dB). In this case, performance is seen to be better than for the modulated interference source, even though the average power levels of the interference source at the channel filter output are the same. This performance difference is a result of transient effects in the demodulator caused by the spreading wave sweeping through the channel rather than being a stationary carrier within the band. Performance through the transponder with the spreading wave interference source added is shown in Figure 5, which considers the case of a 20-dB C/I for a range of interference levels. As can be seen from the figure, the presence of the swept interference source results in a further performance degradation of 0.5 dB for a C/INT of 15 dB, while a C/INT of 10 dB yields a 1.2-dB degradation for a P_e of 10^{-4}.

The effect of the modulating frequency (f_m) of the interference source is shown in Figure 6. It can be seen that the average BER does not vary significantly from 20 to 1,200 Hz. Even though the average BER is almost constant, the actual number of errors per interference sweep will vary directly with the time during which the interference source is in the channel bandwidth. No periodic effects were noticed in the quality of the channel over the entire range of sweep frequencies.

A "window" error rate, which corresponds to the error rate in the period during which the interference source sweeps through the channel bandwidth, can be computed from Figures 3 and 5 by eliminating the error rate due to noise and intermodulation alone. The resulting window error rate curves are shown in Figure 7. The fact that these error rates are not constant indicates that the additional errors in this period are not independent of the thermal noise.

For a typical operating condition of C/N = 15 dB and C/I ~ 20 dB,

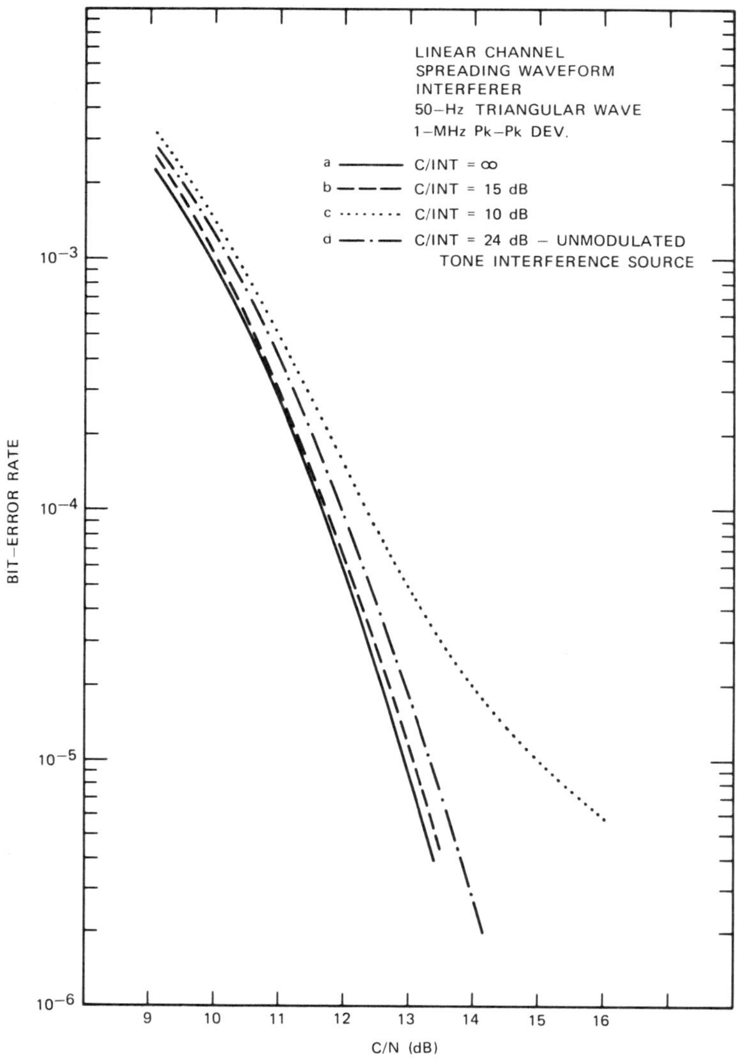

Figure 4. *SCPC Performance with Spreading Wave (back to back)*

C/INT was varied to determine the point at which the synchronization became unreliable. It was found that at C/INT = 9 dB occasional bursts

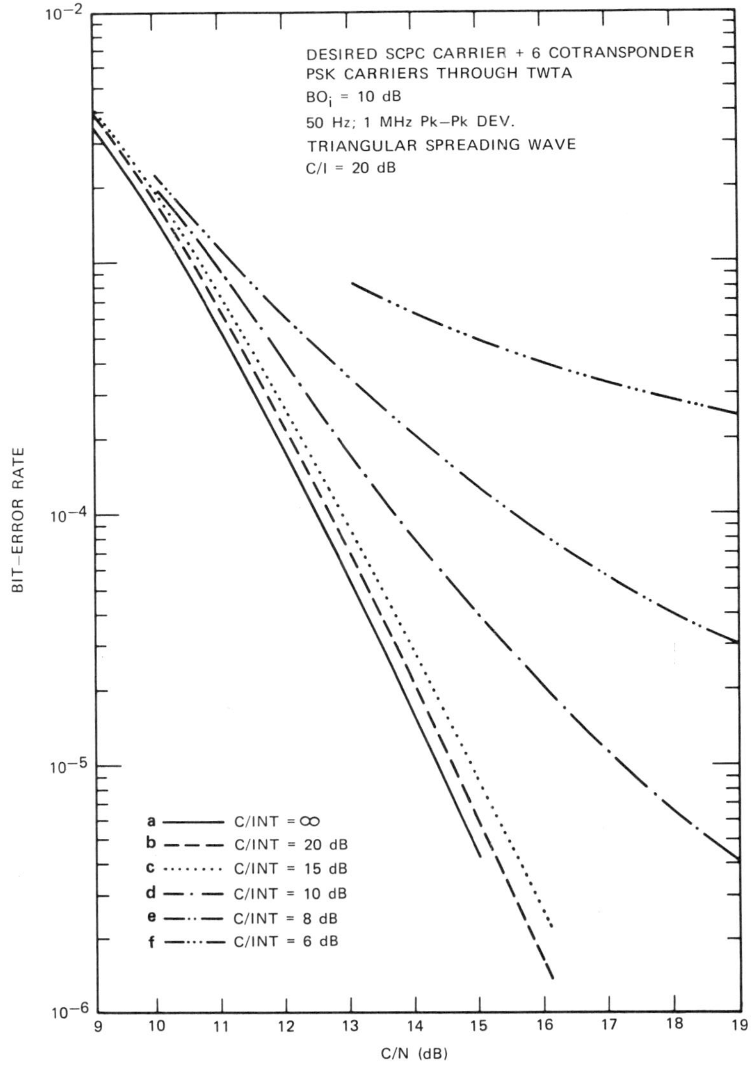

Figure 5. *SCPC Performance Through a Transponder Plus Spreading Wave Interference*

of approximately 100 errors occurred, corresponding to the loss of start-of-message synchronization.

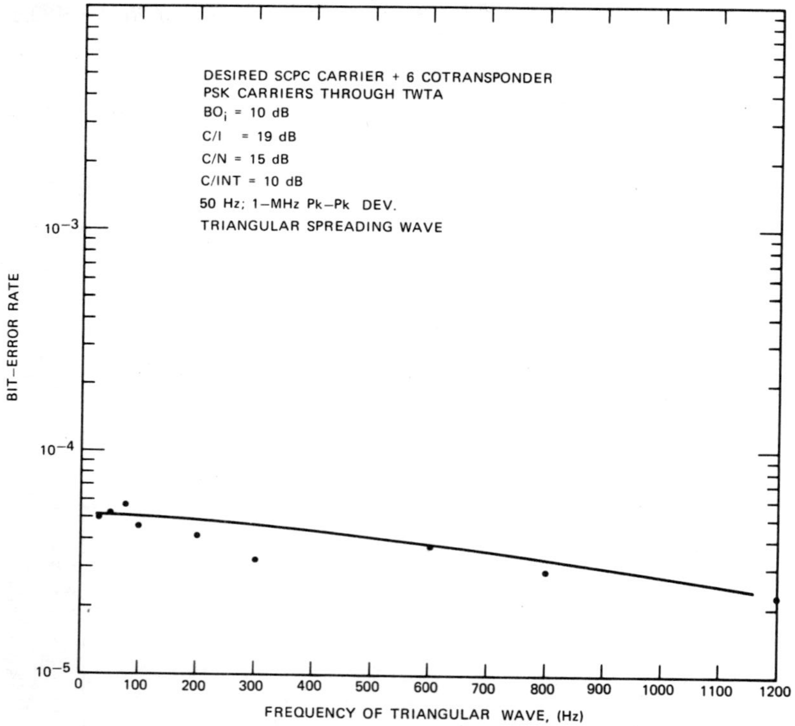

Figure 6. *Effect of Interference Source Modulating Frequency*

The SCPC channel performance through the simulator with both the spreading wave and TV signal (black field plus sync pulse and color burst) is shown in curves *a* and *b* of Figure 8. There is no significant difference between the corresponding curves of Figures 5 and 8 to within measurement accuracy. In this case, a spreading wave alone yields slightly higher error rates. Curve *c* in Figure 8 shows that error rates are lower for color bars than for a black field.

Conclusions

Based on the experimental results presented in this note, the following general conclusions can be drawn. In the PCM-PSK SCPC system, the error rate increase caused by a wideband sweeping inference source whose average power within a particular channel slot is at a certain level cannot

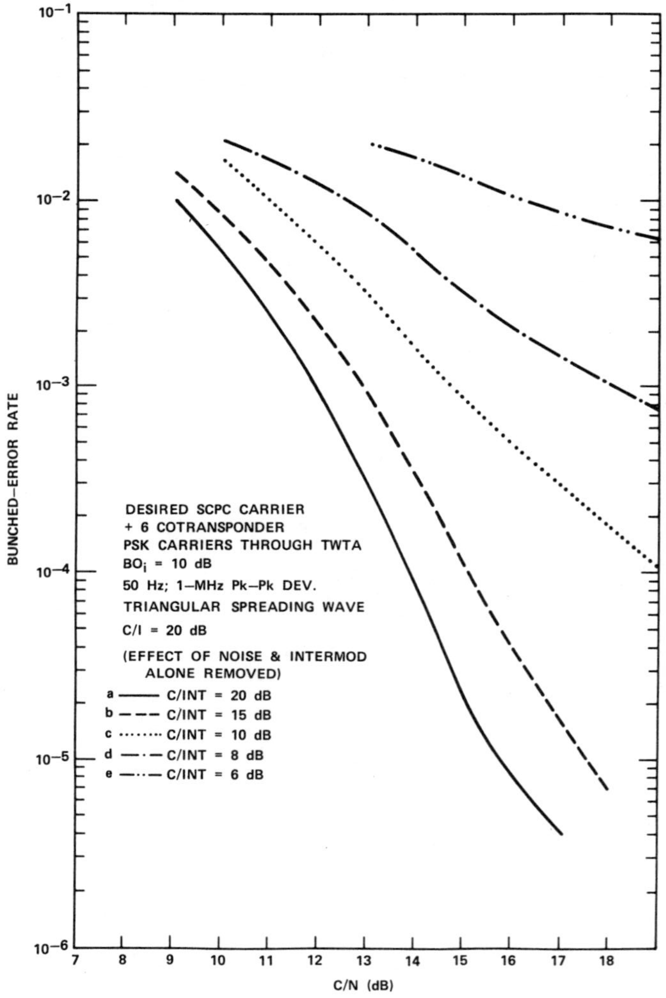

Figure 7. *Bunched Error Rate vs C/N*

be duplicated by a cochannel CW interference source with the same average in-band power.

The average BER for SCPC operation including intermodulation, AM/PM, and interference is given in Table 1. The increase in the average BER is

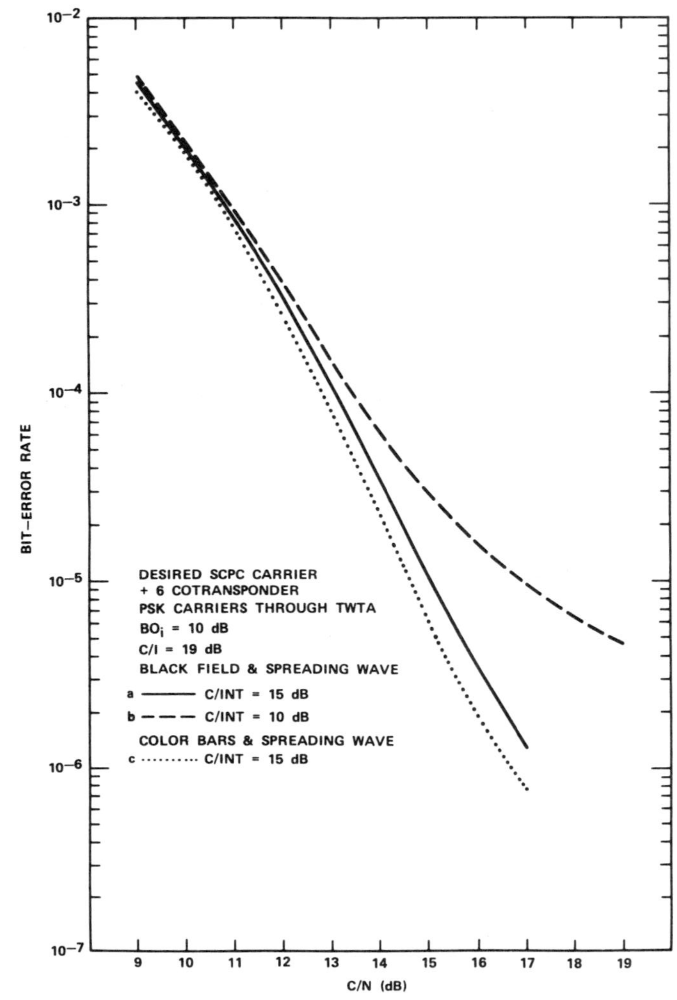

Figure 8. *SCPC Performance with TV Signal and Spreading Wave*

due to the occurrence of interference-induced errors; at times these errors bunch as the sweeping interference source impinges on the SCPC demodulator passband. The bunched or window error rate, which is the rate occurring when the interference energy is within the demodulator input filter bandwidth, is given in Table 2.

TABLE 1. SCPC AVERAGE BER

C/N (dB)	C/I (dB)	C/INT (dB)	Average BER
15	20	∞	4.2×10^{-6}
15	20	20	5.1×10^{-6}
15	20	15	8.5×10^{-6}
15	20	10	4.0×10^{-5}

TABLE 2. SCPC WINDOW ERROR RATE

C/N (dB)	C/I (dB)	C/INT (dB)	Bunched Rate	Average Errors per Pass*
15	20	∞	4.2×10^{-6}	0.000167
15	20	20	2.4×10^{-5}	0.0138
15	20	15	1.1×10^{-4}	0.045
15	20	10	9.4×10^{-4}	0.383

* Computed for 30-Hz sweep rate.

Based on a 30-Hz 1-MHz peak-to-peak interference sweep, the interference carrier is within the receiver's 38-kHz passband for 633 μs (or approximately 40 bits) per pass. The effect of the impulsive nature of the interference on telephone channel quality has been examined. Based on 3-minute testing periods, Table 3 summarizes the audible click occurrences vs operating conditions.

TABLE 3. AUDIBLE CLICK OCCURRENCES VS OPERATING CONDITIONS

C/N (dB)	C/I (dB)	C/INT (dB)	Average Audible Clicks in 3 min.
18	18.5	15	4
18	18.5	20	2.25
18	18.5	25	2
18	18.5	30	1

Acknowledgment

The authors wish to thank S. Fry for his assistance with the measurements.

An Approximate Method to Estimate an Upper Bound on the Effect of Multipath Delay Distortion on Digital Transmission

WILLIAM C. JAKES, JR., FELLOW, IEEE

***Abstract*—Microwave transmission on line-of-sight paths occasionally experiences severe fading associated with multiple transmission paths of differing propagation delays. An expression for the envelope delay distortion caused by two such paths is presented. Assuming that envelope delay distortions comparable to the symbol length of a digital signal could be intolerable, an expression is derived for critical pair-values of fade depth and path delay for a digital signal transmitted over a two-path medium. Good agreement of the principle was obtained with laboratory measurements made on 4- and 8-level systems. Curves are given showing the relationship between critical fade depths and relative delays, and estimates are made of channel outages based on assumed distributions of fades and time delays. The outage times are extemely sensitive to hop length, and exceed a commonly used objective for hops more than 25 mi long.**

I. INTRODUCTION

IT IS well known that microwave transmission over line-of-sight paths can occasionally experience periods of rapid and deep fading. This effect is ascribed to the formation of tropospheric inversion layers that permit additional paths whose vector sum makes up the received signal. It has been directly demonstrated[(1)] that these separate "rays" may arrive at slightly different times, with relative delay differences typically on the order of a few nanoseconds. Transmission through such a medium shows appreciable variation with frequency because of these delays. If two rays differ in path length by one foot, for example, the transmission delay difference is approximately one nanosecond, and as the transmitted frequency is varied the two rays will interfere destructively at frequencies separated by 1000 MHz.

In the case of analog radio operating over the relatively narrow channels in the 4 and 6 GHz bands the frequency dispersive nature of the fading is generally of secondary importance compared to the average value of the faded signal across the channel. If the average value of the signal falls below a specified level, a switch is usually made to a different channel (frequency diversity) or antenna (space diversity) for protection against the fade. In contrast, high-speed digital signals occupy the entire channel bandwidth and are much more sensitive to the time delays associated with the multiple paths; it is the purpose of this study to estimate some upper bounds of this sensitivity.

Section II describes some of the estimates available to date of the delay differences to be expected, and Section III presents expressions for the delay distortions. The value of the delay distortion required to cause excessive errors in a digital system will be estimated in Section IV and compared with some measured results. Estimates of channel outage will then be presented in Section V.

II. MULTIPATH DELAYS

In the early 1950's a number of transmission experiments were conducted which resulted in estimates of path length differences.[(2,3)] These were done by measuring the transmitted amplitude across a 500 MHz band at 4000 MHz and attempting to fit the observed frequency response by a combination of a small number of paths with empirically adjusted amplitudes and delays. Delay differences of as much as 12 ns were required.

Guided by these early tests, C. L. Ruthroff[(4)] used an atmospheric model containing a layer of air having an anomalous refractive index to predict the path length differences one might expect. His analysis shows that a single layer can result in two additional rays, at most, besides the direct path. The maximum value of delay difference he obtained for this model is:

$$\tau_m = 3.7 \left(\frac{D}{20}\right)^3 \text{ ns}, \qquad (1)$$

where D is the path length in miles.

This equation predicts a maximum delay of 5.5 ns on a 22.8 mi path; DeLange[(1)] observed 7 ns. On a 30.8 mi path we would compute 13.5 ns; Kaylor[(3)] observed 12.4 ns. Thus the few available observations tend to corroborate Ruthroff's approach, even though it is probable that the actual atmospheric conditions during the reported experiments were at times more complex than Ruthroff's model. Thus Eq. (1) should be useful in estimating the maximum delays to be expected on a given hop. Very little is known, however, about how often particular delays can occur, i.e., the statistical distribution of the delay differences.

III. DELAY DISTORTION

In the general case a number of separate rays are required to completely describe the amplitude and phase response of

Paper approved by the Editor for Communication Theory of the IEEE Communications Society for publication after presentation at the International Conference on Communications, Toronto, Ont., Canada, June 1978. Manuscript received February 27, 1978; revised July 6, 1978.

The author is with the Radio Transmission Laboratory, Bell Laboratories, North Andover, MA 01845.

Reprinted from *IEEE Trans. Comm.*, vol. COM-27, pp. 76–81, Jan. 1979.

the channel during a multipath fading episode. However, in most cases a two-ray model can serve as a good approximation for the relatively narrow (20-40 MHz) channel bandwidths of interest, and hence will be adopted here to simplify the analysis. We assume that the direct ray has unity amplitude, and that the secondary ray has amplitude $r(<1)$ and propagation delay τ relative to the direct ray. It can then be easily shown that the envelope delay, T, for this model is given by

$$T(\omega) = r\tau \frac{r + \cos \omega\tau}{1 + 2r \cos \omega\tau + r^2}. \tag{2}$$

The most negative value of T occurs when $\omega\tau = \pi$ and is given by:

$$T_{\min} = -\frac{r\tau}{1 - r}. \tag{3}$$

Here we see the amplification of the path delay difference τ by the fading factor $r(1 - r)^{-1}$; i.e. a 20 dB fade results when $r = 0.9$, and in this case $T_{\min} = -9\tau$.

Correspondingly the most positive value of T occurs when $\omega\tau = 0$:

$$T_{\max} = \frac{r\tau}{1 + r}. \tag{4}$$

Typical delay shapes are shown in Figure 1 for 5 dB and 20 dB fades. Note that the peaks are quite sharp and repeat for a frequency separation of $1/\tau$. Also note that the delay varies non-linearly with frequency and thus can be classified as delay distortion.

IV. EFFECT ON DIGITAL TRANSMISSION

Envelope delay distortion (EDD) causes serious intersymbol interference in a digital signal, and cannot be alleviated by increasing the transmitted power. The effect of thermal noise on a digital signal is well-known, of course, but here it is assumed that the radio system has been designed to provide adequate margin against "flat" fades so that thermal noise only becomes significant at deep fade minima of 30 dB or more. Since these minima are generally very narrow in frequency and occur only rarely, the contribution of thermal noise to system outage is small in comparison to that due to delay distortion and will be neglected in this analysis.

We are interested here in determining when the digital radio system "breaks" completely because of excessive delay distortion. It seems reasonable to assume this would occur at the point where the total peak-to-peak envelope delay distortion contained within the channel bandwidth is comparable to the symbol length of the digital signal being passed through this bandwidth. "Breaking" in this case means error rates of 10^{-2} or greater; no attempt will be made to associate BER quantitatively with envelope delay distortion before the break point is reached. The above assumption will be tested by expressing it analytically and comparing the results with experimental measurements.

It might be noted in passing that the amplitude shape of the multipath signal may also have a strong influence on the digital signal, but there does not appear to be as intuitively attractive a criterion to define the break point as there is for EDD and its relationship to symbol length, as stated above.

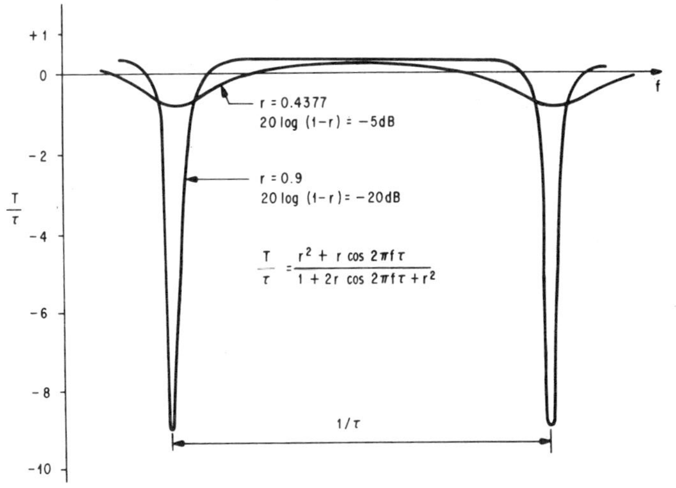

Figure 1 Two-ray envelope delay for fades of 5 dB and 20 dB

To express the above criterion analytically, we assume the maximum negative peak of the envelope delay falls at one edge of the channel bandwidth B, since this places the greatest amount of *p-p* EDD within the band. In Eq. (2) for EDD we let $\omega_1 = \pi/\tau$ correspond to one edge of the band, and $\omega_2 = \pi/\tau + 2\pi B$ to the other edge. The *p-p* EDD in band is then:

$$\Delta T \equiv T(\omega_2) - T(\omega_1)$$

$$= r\tau \frac{1 + r}{1 - r} \frac{1 - \cos 2\pi B\tau}{1 - 2r \cos 2\pi B\tau + r^2}, \quad B\tau \leqslant 0.5. \tag{5}$$

For $B\tau > 0.5$ the maximum possible value of ΔT is contained in the band, so we set

$$\Delta T = \frac{2r\tau}{1 - r^2}, \qquad B\tau > 0.5. \tag{6}$$

By the above criterion we assume the error rate becomes excessive when $\Delta T = k\tau_s$, where k is some constant near unity and τ_s is the symbol length of the digital signal. But, in general, for digital radio systems the bandwidth required by the Nyquist criterion is $B \geqslant 1/\tau_s$. The critical condition for high error rate is then defined by the equations:

$$r \frac{1 + r}{1 - r} \frac{1 - \cos 2\pi B\tau}{1 - 2r \cos 2\pi B\tau + r^2} = \frac{k}{B\tau}, \quad B\tau \leqslant 0.5; \tag{7}$$

or

$$\frac{2r}{1 - r^2} = \frac{k}{B\tau}, \qquad B\tau > 0.5. \tag{8}$$

Note in passing that we expect the parameter k to depend somewhat on the format of the digital signal, particularly on the number of levels, and also on the modulation method and filter shapes.

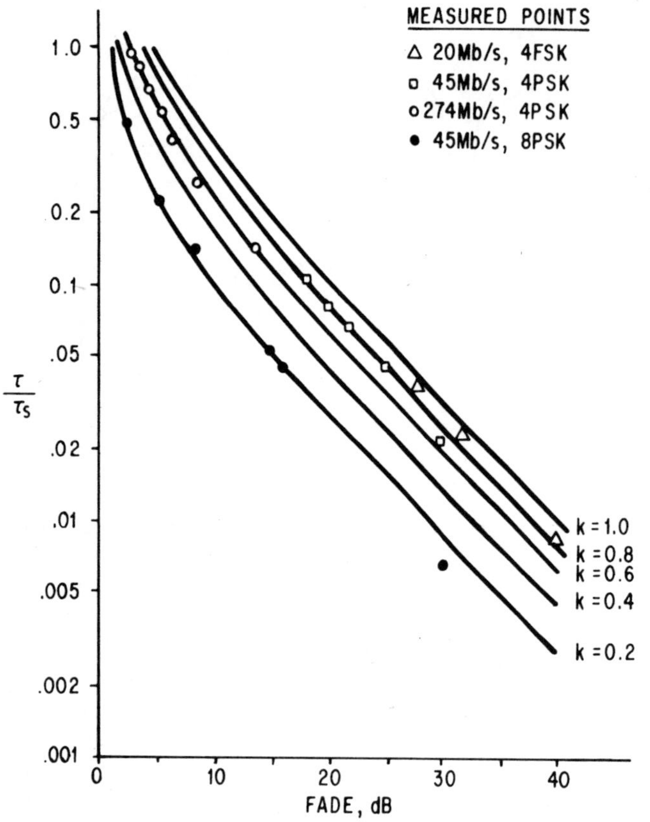

Figure 2 Values of τ/τ_s and fade depth to cause failure assuming 2-ray fade with minimum at band edge

Equations (7) and (8) are to be solved for $B\tau = \tau/\tau_s$ as a function of r. The solution is shown in Figure 2, where τ/τ_s is plotted as a function of the fade depth $= -20 \log (1 - r)$ for five values of k. For small values of τ/τ_s the solution of Equations (7) and (8) may be expressed to good approximation by:

$$1 - r = \frac{\tau}{\tau_s} k^{-0.85} . \tag{9}$$

Some laboratory experiments have been made recently where digital radio transmitters and receivers were connected by means of a waveguide medium to simulate a two-path situation. The relative phase, delay, and amplitude of the secondary path could be varied and the effect on error rate observed. Tests were made on four systems, having the parameters of interest listed below:

Modulation	Bit rate, Mbits/s	Symbol length, τ_s, ns
4 FSK	20	100
4 PSK	44.7	44.7
4 PSK	274.2	7.29
8 PSK	44.7	67.1

In each case a number of measurements were made whereby the fade depth required to reach the break point was determined for various values of relative delay. These observations have also been plotted on Figure 2. It can be seen that the results for all of the four-level systems lie close to the curve for $k = 0.7$, while those for the eight-level system are close to the curve $k = 0.2$. The fact that the observations support the shape of the theoretical curves is taken as justification of the approach, and the values of k thus empirically determined will be assumed to characterize 4- and 8-level PSK systems in subsequent estimates of outage.

In conjunction with Equation (9) these values of k lead to convenient rules for estimation of critical values of delay, τ_c, which, if exceeded can cause a system to break, given the maximum fade depth $(1 - r)$ and symbol length:

$$\begin{aligned} &\text{In general: } \tau_c = k^{0.85}(1 - r)\tau_s \\ &\text{Four-level systems: } \tau_c \doteq \tfrac{3}{4}(1 - r)\tau_s \\ &\text{Eight-level systems: } \tau_c \doteq \tfrac{1}{4}(1 - r)\tau_s . \end{aligned} \tag{10}$$

For example, a 90 Mbit/s 8 PSK system has a symbol length $\tau_s = 33.3$ ns. If a maximum fade of 20 dB ($r = 0.9$) occurred at band edge the system would break if the delay exceeded 0.8 ns.

V. SYSTEM OUTAGE

Having established a critical relationship between multipath delay and fade depth for digital channel failure it is tempting to try to relate this to the known statistical properties of multipath fading and thus determine an estimate of channel failure probability. To do this, one needs the joint probability of maximum fade depth and multipath delay. Much is known about the distribution of single-frequency fade depths,[5] but until now there has been little need to investigate the statistics of the associated delays. We will assume that the fade depths and time delays are statistically independent so that their joint probability density function may be written:

$$p(x, \tau) = p(x)p(\tau) , \tag{11}$$

where the maximum fade depth is 20 log x and τ is the time delay random variable. (We have substituted $x = 1 - r$ in this section.)

Analysis of a small amount of fading data taken a few years ago on a 26 mi path at 6 GHz suggests that an exponential distribution of delays is appropriate, so that $p(\tau)$ is expressed:

$$p(\tau) = \frac{1}{\tau_0} e^{-\tau/\tau_0} , \tag{12}$$

where τ_0 is the mean value of delay, $\langle\tau\rangle$. We choose τ_0 to be such that the probability of τ exceeding the maximum value predicted by Ruthroff in Eq. (1) is very small:

$$P[\tau \geqslant \tau_m] = e^{-\tau_m/\tau_0} \ll 1 \tag{13}$$

or:

$$\tau_0 = \mu 3.7 \left(\frac{D}{20}\right)^3 , \quad \text{where } \mu \ll 1. \tag{14}$$

A value of $\mu = 0.07$ satisfies this criterion and also yields $\tau_0 = 0.6$ ns for a 26-mi hop, comparable to an experimental value of 0.4 ns suggested by the measurements of amplitude shapes on this path at 6 GHz.

Greenstein and Prabhu[6] have assumed an exponential form for $p(x)$:

$$p(x) = \frac{\alpha}{1 - e^{-\alpha}} e^{-\alpha x}, \qquad 0 \leqslant x \leqslant 1 . \tag{15}$$

This expression is intuitively attractive and easy to analyze. A value of $\alpha = 10$ has been proposed and will be used here; this places 99.6% of the maximum fade depths below −5 dB, which represents the situation during the fading events of interest. Equations (11), (12), and (15) then define the joint probability density $p(x, \tau)$:

$$p(x, \tau) = \frac{\alpha}{\tau_0} \frac{1}{1 - e^{-\alpha}} e^{-(\tau/\tau_0 + \alpha x)} . \tag{16}$$

We are now in a position to proceed with an outage estimate. In the preceding section, critical values of fade depth and delay were established which, if exceeded, would cause channel failure. An assumption used in obtaining these results was that the fading minimum be located at one edge of the channel band. In the actual case, the frequency for a minimum varies, being controlled by the phase difference between the two rays. This phase difference is random, and may be thought of as a small random perturbation of the delay difference which results in any phase occurring with equal likelihood between 0 and 2π. The minimum thus resides at band edge only a small fraction of the time. If the minimum falls within about $0.2B$ of either band edge, most of the envelope delay distortion is contained within the band, and we assume this condition must be met before a failure can occur. Now the frequency separation between minima is $1/\tau$, thus the fraction of time $q(\tau)$ we might expect the minimum to lie within $0.2B$ of either band edge is, for a given τ:

$$q(\tau) = 2 \frac{0.2B}{1/\tau} = 0.4 \frac{\tau}{\tau_s} . \tag{17}$$

The conditional outage probability, given that multipath is occurring, may now be expressed as the probability of finding τ and x in the failure region and a minimum within $0.2B$ of either band edge:

$$P_c = \int_0^1 dx \int_{\tau_c}^{\infty} q(\tau) p(x, \tau)\, d\tau . \tag{18}$$

Note that this is a worst-case estimate since it is assumed that all fades are two-ray fades with delay difference equal to τ.

Equation (10) in the preceding section gives a simple, approximate expression for τ_c, valid for $20 \log \chi \leqslant -10$ dB:

$$\tau_c = \tau_s k^{0.85} x . \tag{19}$$

To accommodate the range of validity for this expression the upper limit of integration on x in Equation (18) must be changed to $x_0 = 10^{-1/2}$, but there is little loss of accuracy since $p(x)$ is essentially zero for $x > x_0$.

The average fraction of time that fading occurs in a fading month is a function of carrier frequency, location, and path length. Greenstein[6] has shown that this fraction is given by $(4R/\alpha)(1 - e^{-\alpha})$, where R is the multipath occurrence factor defined by Barnett.[5] R is related to carrier frequency, hop length, and location as follows:

$$R = 2.5 \times 10^{-6}\, C f D^3 , \tag{20}$$

where D is the hop length in miles, f is the carrier frequency in GHz, and C is a terrain-related factor between $\frac{1}{4}$ and 4.

The unconditional outage probability is then

$$P_u = \frac{4R}{\alpha} (1 - e^{-\alpha}) P_c , \tag{21}$$

and the total outage time in a month is given by the product of P_u and the number of seconds in a month, $T_m = 2.63 \times 10^6$:

$$P = T_m P_u = \frac{4RT_m}{\alpha} (1 - e^{-\alpha}) P_c . \tag{22}$$

Substituting Equations (16)-(19) into Equation (22) and carrying out the integration we get:

$$P = \frac{1.6RT_m k^{0.85}}{uv} \left[1 + \frac{u}{v} - e^{-vx_0} \left(1 + \frac{u}{v} + ux_0 \right) \right] , \tag{23}$$

where $u = (\tau_s/\tau_0)k^{0.85}$, $v = \alpha + u$. It turns out that for all cases of interest, the exponential term may be neglected, thus

$$P \doteq \frac{1.6mk^{0.85}}{uv} \left(1 + \frac{u}{v} \right) ,$$

where

$$m = RT_m = 6.575 C f D^3 . \tag{24}$$

This expression gives an upper bound on the time during the worst fading month we can expect channel failure, and has been evaluated for four cases of interest assuming average terrain ($C = 1$):

Bit rate, Mbits/s	No. Levels	Frequency GHz
20	4	4
45	8	4
78	8	6
90	8	11

Results of these calculations are shown in Figure 3; for comparison the current Bell System short-haul one-way outage objective[7] of 0.005% annually is also shown, assuming a 250 mi route and that fading occurs three months out of the year.*

* It should be noted that the short-haul outage objective used was established for non-solid state analog FM systems and may not be applicable to digital radio systems. Since new objectives are not available, the existing objectives are used as a benchmark.

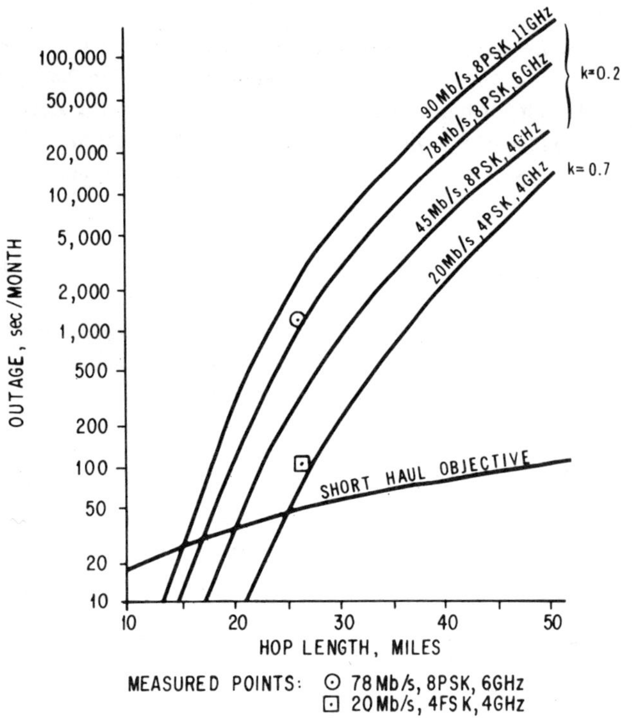

Figure 3 Channel outage for worst fading month versus hop length. Measured points: ⊙ 78 Mbit/s, 8 PSK, 6 GHz; ⊡ 20 Mbit/s, 4 FSK, 4 GHz.

We observe from the curves that the channel failure time increases with operating frequency and symbol rate, and the higher speed systems may be in trouble for hops longer than 20 mi.

Several experiments have been performed recently which, although not exhaustive, provide interesting comparisons to the outages predicted above.

1. 4 GHz: During August 1976, a 20 Mbit/s 4-level FSK signal was transmitted over an unprotected 8-hop route using non-regenerating FM radio repeaters. The route was looped over four hops (out and back) using different channels in the 4 GHz band. Hop lengths are shown below, along with the expected outage calculated from Equation (24):

Hop Length, Miles		Outage calculated for one fading month, sec
23.6		25
30.4		228
34.0		588
21.6		12
	Total	853

Doubling this value to get the 8-hop outage gives 1706 s; the observed value was 2670 s, 1.6 times larger than predicted. It was observed during the course of the experiment that a number of interfering 4 GHz signals were present along the route, and these may have increased the outages since FSK is somewhat less resistant to interference. Flat fades to the noise threshold would also increase the outage.

Another test of 20 Mbit/s 4-level FSK was conducted during the summer of 1977 on a one-hop looped-around arrangement using equipment similar to that of the first experiment, but in another geographical area 900 mi away. The hop was 26.4 mi long, giving a calculated outage for the two-hop loop of 135 s. The observed value was 204 s, 1.5 times as large. Again it would be expected that interference and flat fades contributed additional outage.

2. 6 GHz: Test transmissions of 78 Mbit/s, 8-level PSK were also carried out during the summer of 1977 on the same 26.4 mi hop. One fading month caused an observed total outage of 1008 s; 1125 s would be calculated.

The measured results for the 1977 tests are plotted in Figure 4.** They are in fair agreement with the theory, notwithstanding the short duration of the tests and the presence of other factors such as interference and flat fading.

V. DISCUSSION AND CONCLUSIONS

A number of assumptions have been made in obtaining the results presented in this paper, generally identified in the course of the analysis, but it may be worthwhile to summarize them here.

1. Independent statistics were assumed for the fade depth and time delay distributions. This is intuitively satisfying but needs to be verified.

2. The distribution of delays was assumed to be exponential; much more data need to be gathered and examined to determine the functional form and mean values.

3. It may be possible to define more precisely the parameter $k = \Delta T/\tau_s$ for which the channel fails, particularly as a function of the digital signal format.

4. Multipath transmission through the atmosphere is assumed to be the only source of delay distortion. On actual installations with long waveguide runs and multi-band channel dropping networks, very large delay distortions (tens or even hundreds of nanoseconds) have been occasionally observed. These would have to be equalized.

5. No attention was paid to filter shapes in the present analysis. This factor could be expected to have a bearing on the critical curves established in IV, and may partially explain the slight differences in the apparent value of k for the three 4-level systems measured.

For short hops with D less than about 15 mi u becomes very large and approximately equal to v. In this case Equation (24) becomes

$$P = \frac{3.2mk^{0.85}}{u^2} = 2.27 \times 10^{-8} \frac{Cf}{\tau_s^2} k^{-0.85} D^9. \qquad (25)$$

Thus control of the hop length offers the strongest possibility for controlling the outage without including other system features such as use of a separate protection channel, adaptive equalization, or special coding techniques. The symbol length is the next strongest controller, followed by the modulation form (parameter k) and the operating frequency as the weaker contributors.

Space diversity would be a very effective means of controlling system outage. This is applied on a per-hop basis, of

** For the 20 Mbit/s, 4 FSK measurement the observed value of 204 s was halved to get the outage for one hop.

course, thus the equipment at a repeater station must be capable of recognizing the need for a switch to the space diversity antenna. This may be difficult or expensive; since the total power in the digital signal is a very weak function of a narrow fade capable of failing the channel the common practice of using the AGC voltage to initiate a switch is not sensitive enough. Complete demodulation of the digital signal and monitoring its error rate would require significant equipment not essential for the basic transmission process.

Much remains to be done, but the results presented here may be useful for rough worst-case estimations of digital channel performance bounds, until such time as the analytical methods are sharpened and a more detailed propagation data base is available.

ACKNOWLEDGMENT

Discussions with many members of Bell Labs have been most helpful in providing insight into the various factors involved in the analysis. I am also grateful to Messrs. S. H. Lee, G. H. Lentz, A Hamori, and R. C. Mathews for making available the results of their laboratory measurements on specific digital radio systems, and to Mr. W. T. Barnett for the outage results observed at 4 and 6 GHz.

REFERENCES

1. O. E. DeLange, "Propagation Studies at Microwave Frequencies by Means of Very Short Pulses", *BSTJ,* January, 1952, p. 91.
2. A. B. Crawford, W. C. Jakes, Jr., "Selective Fading of Microwaves", *BSTJ,* January, 1952, p. 68.
3. R. L. Kaylor, "A Statistical Study of Fading of Super-High Frequency Radio Signals", *BSTJ,* September, 1953, p. 1187.
4. C. L. Ruthroff, "Multiple-Path Fading on Line-of-Sight Microwave Radio Systems as a Function of Path Length and Frequency", *BSTJ* September, 1971, p. 2375.
5. W. T. Barnett, "Multipath Propagation at 4, 6, and 11 GHz", *BSTJ,* February, 1972, p. 321.
6. L. J. Greenstein, V. K. Prabhu, "Analysis of Multipath Outage with Applications to 90 Mbit/s PSK Systems at 6 and 11 GHz", this issue, pp. 68–75.
7. A. Vigants, "Space Diversity Engineering", *BSTJ,* January, 1975, p. 103.

Imperfect Carrier Recovery Effect on Filtered PSK Signals

V.K. PRABHU
Bell Laboratories

Abstract

Coherent demodulation of a PSK signal requires the generation of a local carrier phase reference. Methods are given to determine the detection loss caused by noisy phase recovery and its use in the coherent detection of filtered BPSK and QPSK signals. It is assumed that the phase noise can have a static part and a random component with a Tikhonov-type distribution. The static part is mostly due to offset frequency tracking of the PLL used to recover the carrier, while the random component is due to thermal noise present in the carrier recovery loop and is also due to the random nature of the phase modulation. It is shown that the probability of error of BPSK and QPSK can be expressed as a finite sum of a set of strictly alternating converging series when the number of ISI terms is finite. Upper and lower bounds on the probability of error have been derived when this number becomes infinite and we show how this error rate can be computed with any desired accuracy. Numerical results are presented for various values of static error and phase noise variance when the transmit and receive filters are 4-pole Butterworth filters. For filtered PSK signals and for a bit error rate of 10^{-6}, our results show that the additional degradation in present-day receiver systems due to imperfect carrier recovery can be less than 0.1 dB for BPSK and less than 1 dB for QPSK.

Manuscript received June 10, 1977.

Author's address: Bell Laboratories, Box 400, Holindel, NJ 07733,

I. Introduction

Coherent demodulation of a phase-shift keyed (PSK) signal requires the generation or extraction of a local carrier phase reference. Frequency instabilities, inherent in transmitter and receiver systems, do not allow the bandwidth of the carrier recovery circuit to be made arbitrarily small in order to reduce the noise jitter of the phase reference to negligible levels. Methods are given in [1-3] to determine the performance loss caused in unfiltered binary phase-shift keyed (BPSK) and quaternary phase-shift keyed (QPSK) systems by imperfect carrier phase recovery. In [3] it is assumed that the phase noise associated with the recovered carrier can have a static component and a Gaussian or a Tikhonov-type distribution.

Filters are almost always used in practical systems to increase the spectral efficiency of transmission and to limit the thermal noise in the receiver. In radio systems, they are also used to satisfy the out-of-band emission requirements imposed on the radio signal. For such filtered BPSK and QPSK systems, a method is presented in this paper to determine the detection performance loss caused by imperfect carrier recovery when the carrier phase noise is assumed to have a Tikhonov-type distribution. The phase noise can have a nonzero mean due to offset frequency tracking of the phase-locked loop (PLL) used to recover the carrier or may include an equivalent phase error produced by the comparator dead zone [3, 5].

II. PSK System Description

The unfiltered M-ary PSK signal $s(t)$ is assumed to be a constant-amplitude (we normalize the amplitude to unity) sinusoidal carrier of frequency f_c, phase-modulated by a message sequence $\{\alpha_k\}$ at the baud rate $1/T$. Assuming that the carrier frequency is much larger than the modulation signal bandwidth and that $g(t)$ is the signaling pulse, it is convenient to represent $s(t)$ in terms of its complex envelope $x(t)$ where

$$x(t) = \exp\{j[\Sigma\alpha_k g(t - kT)\ \mathrm{rect}\{(t - kT)/T\}]\} \tag{1}$$

$$s(t) = \mathrm{Re}\{x(t) \exp(j2\pi f_c t)\}. \tag{2}$$

In this paper we also assume that the M phase amplitudes are uniformly distributed between $(-\pi, \pi]$, that they are equally probable, and that the data phases in different time slots are statistically independent.

Under the hypothesis that a "conventional" PSK receiver is used in detecting the symbols and that, besides thermal noise, imperfect carrier recovery and distortion produced by filters are the only other sources of degradation, the system we analyze is shown in Fig. 1. All RF (or IF) filters are considered to be linear and time invariant and are specified by their equivalent low-pass characteristics. The complex envelope $y(t)$ at the output of the receive filter may, therefore, be written as

Reprinted from *IEEE Trans. Aerosp. Electron. Syst.*, vol. AES-14, pp. 608-615, July 1978.

$$y(t) = x(t)*h_T(t)*h_C(t)*h_R(t) + n(t) + j\hat{n}(t) \tag{3}$$

where $h_T(t)$, $h_C(t)$, and $h_R(t)$ are, respectively, the impulse responses of the transmit filter, the channel, and the receive filter, while the $n(t) + j\hat{n}(t)$ is the complex envelope of the Gaussian noise passed through the receive filter. The symbol $*$ denotes convolution. The noise is assumed to be white, Gaussian, and of one-sided spectral density N_0. Since we assume that the receive filter used in the PSK system is symmetric, n and $\hat{n}$ are independently and identically distributed zero-mean Gaussian random variables with variance

$$\sigma^2 = 2N_0 B_R \tag{4}$$

where

$$B_R = \int_{-\infty}^{\infty} |H_R(f)|^2 \, df \tag{5}$$

is the noise bandwidth and $H_R(f)$ the transfer function of the receive filter.

First we consider binary systems.

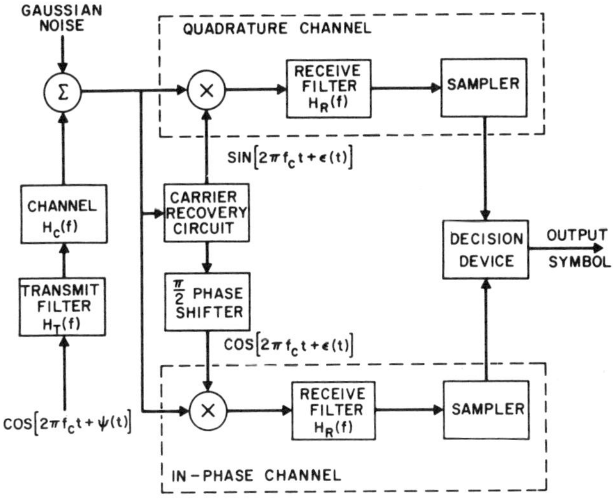

Fig. 1. M-ary PSK receiver, $M > 2$. (*Note*: for BPSK only quadrature channel need be present and all decisions are made based on its output. The output of the in-phase channel can be disregarded for BPSK.)

III. Error Rate for Binary System

Let us first consider the case in which the data sequence $\{\alpha_k\}$ is finite and of length $2N + 1$, $k = -N, \ldots, -1, 0, 1, \ldots, N$. For a binary system, the conditional probability of error $P_2\,[\mathbf{E}|\pi/2, N]$, given that the zeroth modulating symbol is $\pi/2$, can be shown [3, 6] to be

$$P_2(\mathbf{E}|\pi/2, N) = \tfrac{1}{2}\Big\langle \mathrm{erfc}\Big\{\Big[\,[\cos\{\epsilon(t_0)\}\,\mathrm{Im}\{x(t_0)*h(t_0)\} - \sin\{\epsilon(t_0)\}\,\mathrm{Re}\{x(t_0)*h(t_0)\}]\Big]_{\alpha_0=\pi/2}\Big\}\Big\rangle_{\{\alpha_k\}',\,\epsilon(t)} \tag{6}$$

where t_0 is the sampling time, $\langle\cdot\rangle_z$ denotes an average over z, $\epsilon(t)$ represents a very slowly varying phase noise associated with the recovered carrier,

$$h(t) \triangleq h_T(t)*h_C(t)*h_R(t) \tag{7}$$

$$\mathrm{erfc}(z) \triangleq (2/\sqrt{\pi}) \int_z^{\infty} \exp(-t^2)\, dt \tag{8}$$

and the sequence $\{\alpha_k\}'$ contains all elements of $\{\alpha_k\}$ except α_0.

For a given sampling time t_0 and for a given N, there are 2^{2N} (M^{2N} for an M-ary system) possible data sequences and

$$P_2(\mathbf{E}|\pi/2, N) = \tfrac{1}{2}(1/2^{2N}) \sum_n \Big\langle \mathrm{erfc}\Big\{\Big[\,[\cos\{\epsilon(t_0)\}\,\mathrm{Im}\{x_n(t_0)*h(t_0)\} - \sin\{\epsilon(t_0)\}\,\mathrm{Re}\{x_n(t_0)*h(t_0)\}]/[\sqrt{2}\,\sigma]\Big]\Big\}\Big\rangle_{\epsilon(t_0)} \tag{9}$$

where $x_n(t)$ is the nth data sequence for $\alpha_0 = \pi/2$. Using the notation

$$a_k + jb_k = h(t_0)*\{[\exp\{j\alpha_k g(t_0 - kT)\}]\ \mathrm{rect}\{(t_0 - kT)/T\}\},\ k \neq 0 \tag{10}$$

$$\beta_c + j\beta_s = h(t_0)*\{[\exp\{j(\pi/2)g(t_0)\}]\ \mathrm{rect}\{t_0/T\}\}\Big| \tag{11}$$

we can write

$$P_2(\mathbf{E}|\pi/2, N) = \tfrac{1}{2}(1/2^{2N}) \sum_n \Big\langle \mathrm{erfc}\Big\{\Big[[\cos\{\epsilon(t_0)\}\{\beta_s + \Sigma b_{kn}\} - \sin\{\epsilon(t_0)\}\{\beta_c + \Sigma a_{kn}\}]\Big]\Big\}\Big\rangle_{\epsilon(t_0)} \tag{12}$$

where a_{kn} and b_{kn} denote the values of a_k and b_k for the nth data sequence. From [3]

$$\tfrac{1}{2}\,\mathrm{erfc}(\rho\cos\theta) = \tfrac{1}{2} + \sum_{l=0}^{\infty} (-1)^{l+1} H_l \cos\{(2l+1)\,\theta\} \tag{13}$$

where

$$H_l = \{\rho\exp(-\rho^2/2)/[\sqrt{\pi}\,(2l+1)]\}\{I_l(\rho^2/2) + I_{l+1}(\rho^2/2)\} \geqslant 0 \tag{14}$$

and $I_l(x)$ is the modified Bessel function of the first kind and of order l.

Regardless of how the carrier is extracted, it will contain some phase noise ϵ, with a mean $\bar{\epsilon}$ and a standard deviation σ_ϵ. The static error $\bar{\epsilon}$ is primarily due to carrier frequency offset, but can also be assumed to include the comparator dead zone effects [3, 6]. The random component of ϵ is due to additive noise present in the carrier recovery circuit and due to random phase modulation. In general, ϵ is a function of time; but in cases of practical interest, the variation is slow enough compared with the modulation rate that ϵ can be considered to be a constant over the interval of interest.

In [3] two different models are assumed for the probability distribution of ϵ. It is also shown in [3] that both models essentially give the same results for small values of $\bar{\epsilon}$ and σ_ϵ. Since we are primarily interested in receiver systems with small values of $\bar{\epsilon}$ and σ_ϵ, in this paper we assume that the probability density function $f_\epsilon(\Phi)$ of ϵ is Tikhonov type and is given by [3]

$$f_\epsilon(\Phi) = \{\exp[\alpha \cos(\Phi - \bar{\epsilon})]\}/\{2\pi I_0(\alpha)\}, \quad |\Phi| \leqslant \pi \tag{15}$$

where

$$\sigma_\epsilon^2 = (\pi^2/2) + 4 \sum_{l=1}^{\infty} (-1)^l I_l(\alpha)/\{l^2 I_0(\alpha)\}. \tag{16}$$

It may be noted that (15) is suggested by the case of a first-order PLL acting on an unmodulated carrier in Gaussian noise [7] and

$$\alpha = 2G/(N_p B_L) \tag{17}$$

where G is the power in the carrier, N_p is the single-sided noise spectral density, and B_L is the noise bandwidth of the linearized PLL. For a second-order PLL and for a first-order PLL with modulation, it can be shown [7] that the probability density of ϵ is approximately given by (15).

From (12), (13), and (15) we can show [3] that

$$P_2(\mathbf{E}|\pi/2, N) = (1/2^{2N}) \cdot \sum_n \Big\{\tfrac{1}{2}\operatorname{erfc}(\rho_n) + \sum_{l_n=1}^{\infty} (-1)^l H_{l_n}(\rho_n) \cdot \{1 - [I_{2l_n+1}(\alpha)/I_0(\alpha)]\} \cos\{(2l_n+1)(\theta_n + \bar{\epsilon})\}\Big\} \tag{18}$$

where

$$\rho_n = \{(\beta_s + \Sigma b_{kn})^2 + (\beta_c + \Sigma a_{kn})^2\}^{1/2}/(\sqrt{2}\,\sigma) \tag{19}$$

$$\rho_n \cos(\theta_n) = \beta_s + \Sigma b_{kn} \tag{20}$$

$$\rho_n \sin(\theta_n) = \beta_c + \Sigma a_{kn}. \tag{21}$$

As proven in [3], the series in (18) can be shown to be strictly alternating for all values of α and for all values of $\sigma_\epsilon \neq 0$. It can also be demonstrated that an m_n can be found such that the $(l_n + 1)$th term is numerically less than or equal to the l_nth term, $l_n \geqslant m_n$, and $H_{l_n} \to 0$ as $l_n \to \infty$. The inifinite series in (18), therefore, converges and the magnitude of the truncation error $E_2(L_n)$, if L_n terms, $L_n \geqslant m_n$, are included in the partial sum to compute the infinite series, can be bounded by

$$|E_2(L_n)| \leqslant H_k\{1 + [I_{2k+1}(\alpha)/I_0(\alpha)]\}, \quad k = L_n + 1; L_n \geqslant m_n. \tag{22}$$

Since the summation in (18) with respect to n is finite, it therefore follows that $P_2(\mathbf{E}|\pi/2, N)$ can be computed with any desired accuracy.

Let us now assume that the data sequence $\{\alpha_k\}$ in (6) is infinite and that Ω is a finite subset containing $2N + 1$ consecutive elements of $\{\alpha_k\}$. It follows that

$$\begin{aligned} P_2(\mathbf{E}|\pi/2) &= \lim_{k\to\infty} P_2(\mathbf{E}|\pi/2, k) \\ &= \Pr[\cos\{\epsilon(t_0)\}\{\beta_s + \Sigma' b_k\} \\ &\quad - \sin\{\epsilon(t_0)\}\{\beta_c + \Sigma' a_k\} + \eta(t) < 0] \end{aligned} \tag{23}$$

where Σ' does not include the term $k = 0$, and $\eta(t)$ is a Gaussian random variable with mean zero and variance σ^2. It is very difficult to evaluate (23) when $k \to \infty$. Because of the presence of sine and cosine terms in (23), it can be shown that no useful bounds can even be derived by using the very general results given in [8] and [9] under the assumptions that intersymbol interference (ISI) is peak limited and the eye is open.

However, it is shown in the Appendix that

$$\begin{aligned} &\tfrac{1}{2}\Big\langle \operatorname{erfc}\Big\{[\cos\{\epsilon(t_0)\}\{\beta_s + \sum_\Omega b_k\} - \sin\{\epsilon(t_0)\}\{\beta_c + \sum_\Omega a_k\}]/\sqrt{2}\,\sigma\Big\}\Big\rangle - [2/(2\pi)^{1/2}](R_N/\sigma) \\ &\leqslant P_2(\mathbf{E}|\pi/2) \\ &\leqslant \tfrac{1}{2}\Big\langle \operatorname{erfc}\Big\{[\cos\{\epsilon(t_0)\}\{\beta_s + \sum_\Omega b_k\} - \sin\{\epsilon(t_0)\}\{\beta_c + \sum_\Omega a_k\}]/\sqrt{2}\,\sigma\Big\}\Big\rangle \\ &\quad + [2/(2\pi)^{1/2}](R_N/\sigma) \end{aligned} \tag{24}$$

where R_N is a non-negative number such that

$$0 \leqslant \sum_{\Omega^c} |b_k| < R_N \tag{25}$$

$$0 \leqslant \sum_{\Omega^c} |a_k| < R_N. \tag{26}$$

The complement of the set Ω is denoted by Ω^c.

Methods of evaluating the expectation appearing in (24) when the set is finite are given elsewhere in this section. For any $g(t)$ in (1), and $h(t)$ in (7), we choose a sufficiently small N and calculate upper and lower bounds on $P_2(\mathbf{E}|\pi/2)$ from (24). The value of N is gradually increased until the desired accuracy is attained. The monotonically decreasing nature of R_N in (25) and (26) assures the convergence of this procedure.

Instead of using the rigorous method given in (24), for most filters used in practice, the criterion of increasing the number of ISI terms so that the value of $P_2(E|\pi/2, N)$ calculated stops changing appreciably seems to be sufficient from a practical viewpoint. We similarly compute $P_2(E|-\pi/2)$ and then determine the average probability of error P_2.

IV. Error Rate for Quaternary System

For the QPSK system shown in Fig. 1, the probability of bit error [3] $P_{4B}(E|\pi/4, N)$ can be shown to be

$$P_{4B}(E|\pi/4, N) = 1/4^{2N} \sum_n \tfrac{1}{2}\,\mathrm{erfc}(\rho_n)$$

$$+ \sum_{l_n=1}^{\infty} (-1)^{l} H_{l_n}(\rho_n)\{1 - [I_{2l_n+1}(\alpha)/I_0(\alpha)]$$

$$\cdot \cos\{(2l_n + 1)(\pi/4)\} \cos\{(2l_n + 1)[\bar{\epsilon} + \theta_n - (\pi/4)]\}\} \tag{27}$$

where ρ_n and θ_n are given by (19) through (21). In (27)

$$\beta_c + j\beta_s \triangleq h(t_0) * \{[\exp\{j(\pi/4)g(t_0)\}]\ \mathrm{rect}(t_0/T)\}. \tag{28}$$

If the data sequence $\{\alpha_k\}$ is infinite, upper and lower bounds on $P_{4B}(E|\pi/4)$ can be derived by methods very similar to those in Section III. Since no insight in methodology would be gained, we do not give these bounds here. We similarly determine the probability of error for other values of α_0 and then obtain the average, P_{4B}.

V. Numerical Computation of P_2 and P_{4B}

For various values of $\bar{\epsilon}$ and σ_ϵ, for a Tikhonov-type distribution of ϵ, and for rectangular signaling, P_2 and P_{4B} have been calculated when a 4-pole Butterworth transmit and a 4-pole Butterworth receive filter are used in the system. The sampling time t_0 and the bandwidth of the receive filters are chosen so that P_2 and P_{4B} are minimum for a given thermal noise density N_0. The results are shown in Figs. 2 through 5. The carrier noise ratio (CNR) in decibels is defined as $10 \log_{10}$ (CNR),

$$\mathrm{CNR} \triangleq T/2N_0 = (1/2)/N_0(1/T) = \text{unmodulated carrier}$$

power/noise power in a band equal to the baud rate. (29)

For BPSK and QPSK systems and for bit error rates of 10^{-4}, 10^{-6}, and 10^{-8}, we list in Tables I and II the increase in CNR needed to maintain the same error rate. Power penalties associated with ISI and with imperfect carrier recovery when their effects are treated separately are also summarized in these tables.

For small values of $(\bar{\epsilon}, \sigma_\epsilon)$ and for values of $1/(B_T T)$ not too large, the combined effect due to ISI and imperfect ϵ is the sum of their individual effects treated separately. This is especially true for BPSK but is approximately valid for QPSK. For large values of $(\bar{\epsilon}, \sigma_\epsilon)$ and/or large values of $1/(B_T T)$, the penalty due to the combined effects is larger than the sum of the individual penalties, necessitating the complex analysis given in this paper.

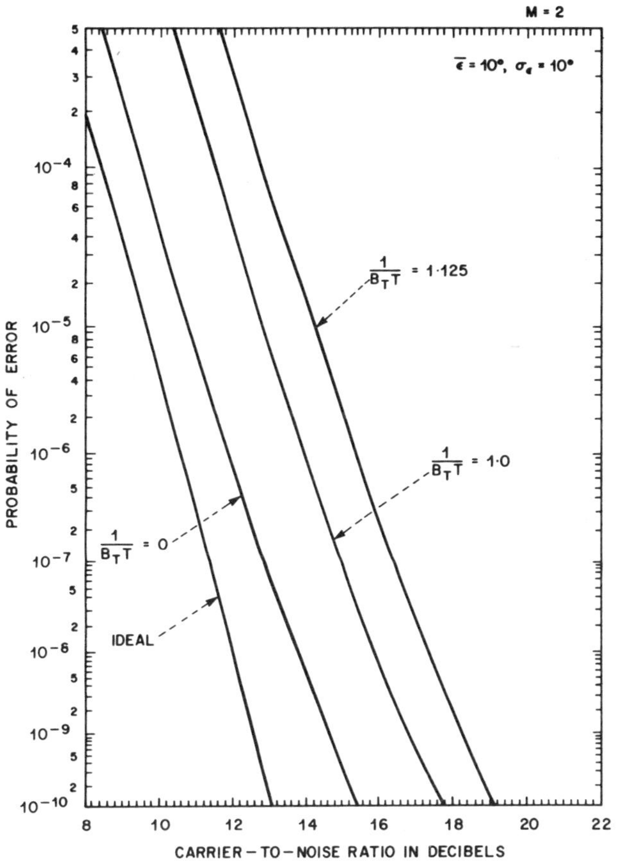

Fig. 2. Probability of bit error for BPSK with rectangular signaling, noisy carrier phase recovery, and 4-pole Butterworth transmit and 4-pole Butterworth receive filter. (*Note*: Noise associated with the recovered carrier is assumed to have a mean $\bar{\epsilon}$ and a Tikhonov-type distribution with variance σ_ϵ^2. The CNR in decibels is defined as $10 \log_{10} [T/2N_0]$, where N_0 is the single-sided noise spectral density and the received carrier power without distortion is assumed to be $\frac{1}{2}$. B_T denotes the two-sided 3-dB bandwidth of the transmit filter. The receive filter bandwidth and the sampling time are chosen so that the probability of bit error, calculated to an accuracy of 5 percent, is minimum.)

VII. Summary and Conclusions

Methods are given to determine the detection performance loss caused by noisy phase recovery and its use in the coherent detection of filtered BPSK and QPSK signals. It is assumed that the phase noise can have a static component and a random component with Tikhonov-type distribution. The static component is mostly due to offset frequency tracking of the PLL used to recover the carrier, and the random component is due to thermal noise present in the recovery loop and to the random phase modulation.

It is shown that the probability of error of both BPSK

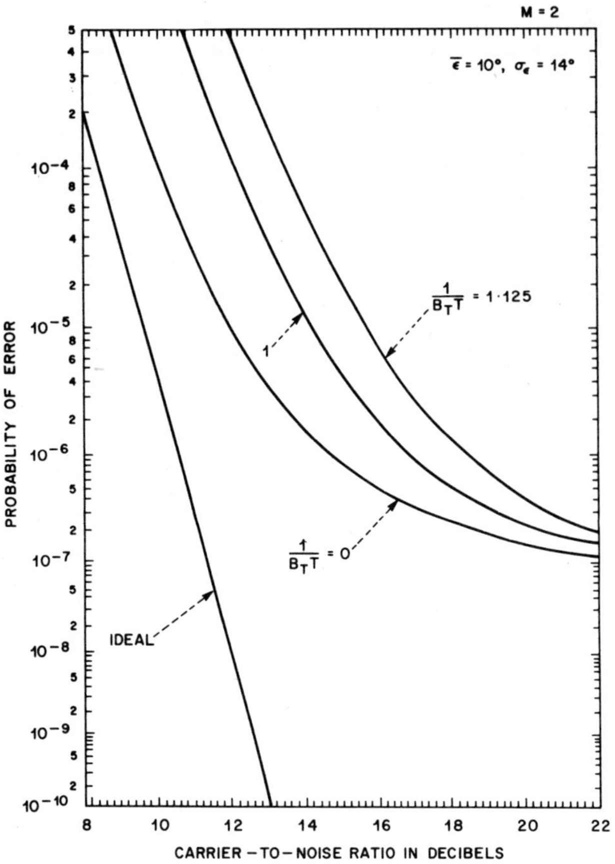

Fig. 3. Probability of bit error for BPSK. (*Note*: Other assumptions as in Fig. 2.)

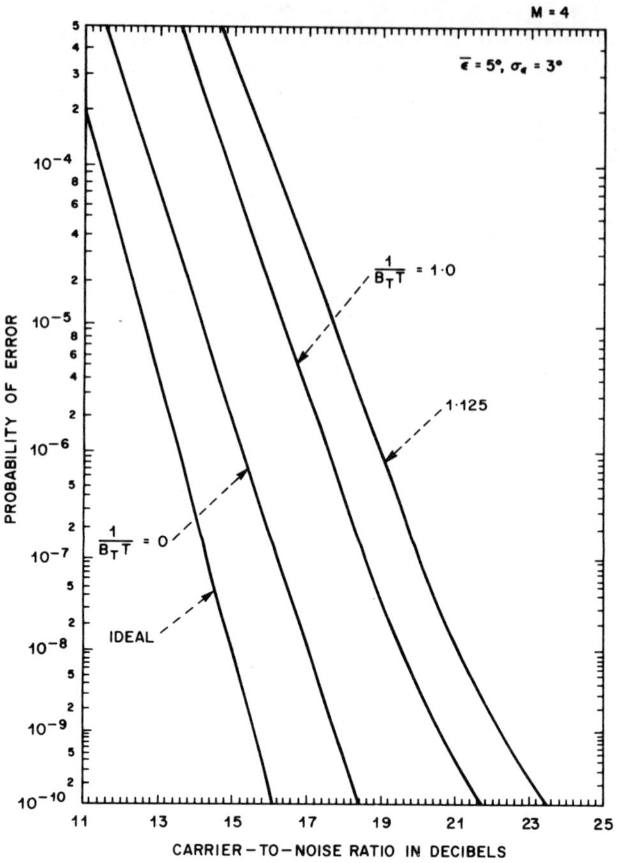

Fig. 4. Probability of bit error for QPSK. (*Note*: It is assumed that the Gray code described in [3] is used in assigning pairs of bits in the original message. Other assumptions as in Fig. 2.)

and QPSK can be expressed as a finite sum of a set of strictly alternating converging series when the number of ISI terms is finite. Upper and lower bounds on the probability of error have been derived when this number becomes infinite and we show how this error rate can be computed with any desired accuracy.

Numerical results for both BPSK and QPSK have been presented for various values of static error and phase noise variance when the transmit and receive filters are 4-pole Butterworth filters.

Appendix

Since [10]

$$\Pr[z_N < \rho - \Delta] - \Pr[z_R \geqslant \Delta] \leqslant \Pr[z_N + z_R < \rho] \leqslant \Pr[z_N < \rho + \Delta] + \Pr[z_R \leqslant -\Delta] \tag{30}$$

where z_N, z_R are any two arbitrary random variables and Δ is any arbitrary number, one can show from (23) that

$$\begin{aligned}
&\Pr[\cos\{\epsilon(t_0)\}\{\beta_s + \sum_{\Omega} b_k\} - \sin\{\epsilon(t_0)\}\{\beta_c + \sum_{\Omega} a_k\} + \eta(t) < -\Delta] \\
&\quad - \Pr[\cos\{\epsilon(t_0)\}\{\sum_{\Omega^c} b_k\} \geqslant \Delta/2] \\
&\quad - \Pr[-\sin\{\epsilon(t_0)\}\{\sum_{\Omega^c} a_k\} \geqslant \Delta/2] \\
&\leqslant P_2(\mathrm{E} \mid \pi/2) \leqslant \Pr[\cos\{\epsilon(t_0)\}\{\beta_s + \sum_{\Omega} b_k\} \\
&\quad - \sin\{\epsilon(t_0)\}\{\beta_c + \sum_{\Omega} a_k\} + \eta(t) < \Delta] \\
&\quad + \Pr[\cos\{\epsilon(t_0)\}\{\sum_{\Omega^c} b_k\} \leqslant -\Delta/2] \\
&\quad + \Pr[-\sin\{\epsilon(t_0)\}\{\sum_{\Omega^c} a_k\} \leqslant -\Delta/2]
\end{aligned} \tag{31}$$

or

$$\begin{aligned}
&\tfrac{1}{2}\left\langle \mathrm{erfc}\left\{[\cos\{\epsilon(t_0)\}\{\beta_s + \sum_{\Omega} b_k\} - \sin\{\epsilon(t_0)\}\{\beta_c + \sum_{\Omega} a_k\} + \Delta]/\sqrt{2}\,\sigma\right\}\right\rangle - \lambda_l \\
&\leqslant P_2(\mathrm{E} \mid \pi/2) \\
&\leqslant \tfrac{1}{2}\left\langle \mathrm{erfc}\left\{[\cos\{\epsilon(t_0)\}\{\beta_s + \sum_{\Omega} b_k\} - \sin\{\epsilon(t_0)\}\{\beta_c + \sum_{\Omega} a_k\} - \Delta]/\sqrt{2}\,\sigma\right\}\right\rangle + \lambda_u
\end{aligned} \tag{32}$$

where

$$\lambda_l \triangleq \Pr[\cos\{\epsilon(t_0)\}\{\sum_{\Omega^c} b_k\} > \Delta/2] + \Pr[-\sin\{\epsilon(t_0)\}\{\sum_{\Omega^c} a_k\} > \Delta/2] \quad (33)$$

and

$$\lambda_u \triangleq \Pr[\cos\{\epsilon(t_0)\}\{\sum_{\Omega^c} b_k\} < -\Delta/2] + \Pr[-\sin\{\epsilon(t_0)\}\{\sum_{\Omega^c} a_k\} < -\Delta/2]. \quad (34)$$

Since it can be shown (see Fig. 6 for a simple proof) that, for any $a > 0$, and for any two arbitrary random variables x and y,

$$\Pr[z = xy > a] \leqslant \Pr[|x| > p] + \Pr[|y| > a/p], \quad p > 0 \text{ but arbitrary} \quad (35)$$

we have

$$\lambda_l \leqslant \lambda \triangleq \Pr[|\sum_{\Omega^c} b_k| > \Delta/2] + \Pr[|\sum_{\Omega^c} a_k| > \Delta/2], \quad \Delta > 0 \quad (36)$$

$$\lambda_u \leqslant \lambda. \quad (37)$$

From (32), (36), and (37),

$$\tfrac{1}{2}\left\langle \operatorname{erfc}\left\{[\cos\{\epsilon(t_0)\}\{\beta_s + \sum_{\Omega} b_k\} - \sin\{\epsilon(t_0)\}\{\beta_c + \sum_{\Omega} a_k\} + \Delta]/\sqrt{2}\,\sigma\right\}\right\rangle - \lambda \leqslant P_2(\mathbf{E} \mid \pi/2) \leqslant \tfrac{1}{2}\left\langle \operatorname{erfc}\left\{[\cos\{\epsilon(t_0)\}\{\beta_s + \sum_{\Omega} b_k\} - \sin\{\epsilon(t_0)\}\{\beta_c + \sum_{\Omega} a_k\} - \Delta]/\sqrt{2}\,\sigma\right\}\right\rangle + \lambda. \quad (38)$$

For filters considered in this paper the ISI is assumed to be bounded and a number R_N can be found such that (25) and (26) are satisfied.

From (36) through (38) and (25) and (26) it follows that

$$\tfrac{1}{2}\left\langle \operatorname{erfc}\left\{[\cos\{\epsilon(t_0)\}\{\beta_s + \sum_{\Omega} b_k\} - \sin\{\epsilon(t_0)\}\{\beta_c + \sum_{\Omega} a_k\} + 2R_N]/\sqrt{2}\,\sigma\right\}\right\rangle \leqslant P_2(\mathbf{E} \mid \pi/2) \leqslant \tfrac{1}{2}\left\langle \operatorname{erfc}\left\{[\cos\{\epsilon(t_0)\}\{\beta_s + \sum_{\Omega} b_k\} - \sin\{\epsilon(t_0)\}\{\beta_c + \sum_{\Omega} a_k\} - 2R_N]/\sqrt{2}\,\sigma\right\}\right\rangle. \quad (39)$$

Since R_N is a monotone decreasing function of N, R_N can be made arbitrarily small. By using one of the mean value theorems

$$f(a + z) = f(a) + zf'(a + \mu z), \quad 0 \leqslant \mu \leqslant 1 \quad (40)$$

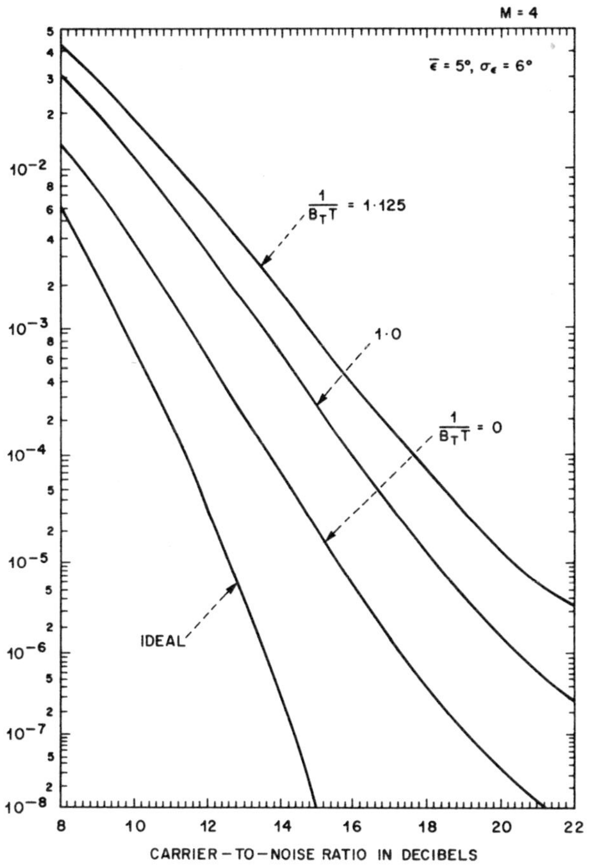

Fig. 5. Probability of bit error for QPSK. (*Note*: Other assumptions as in Fig. 4.)

Fig. 6. Upper bound on distribution function $F_z(a) \triangleq \Pr[z = xy > a]$. (*Note*: It is assumed that $a > 0$ and $p > 0$.)

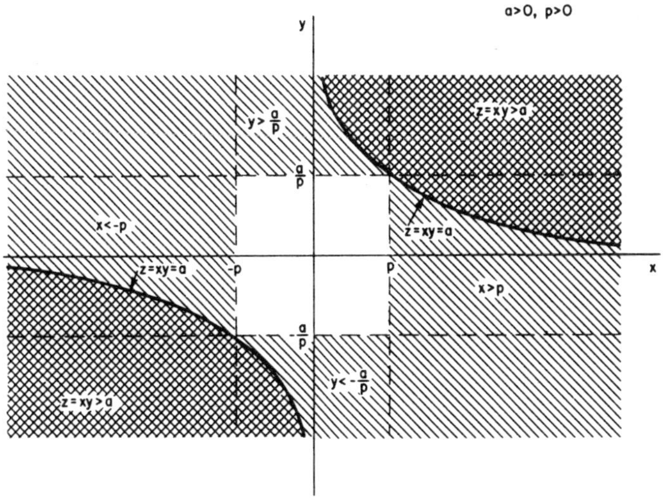

and the inequality

$$\tfrac{1}{2}\operatorname{erfc}'(z) = -(1/\sqrt{\pi})\exp(-z^2) \geqslant -1/\sqrt{\pi} \quad (41)$$

we can derive (24) from (39).

TABLE I

Power Penalties for BPSK

			Power Penalty in DB Due to Noise and				
	Ideal CNR in DB			Imperfect ϵ Only		ISI and Imperfect ϵ	
P_2		$1/B_T T$	ISI Only	$\bar{\epsilon} = 10°$ $\sigma_\epsilon = 10°$	$\bar{\epsilon} = 10°$ $\sigma_\epsilon = 14°$	$\bar{\epsilon} = 10°$ $\sigma_\epsilon = 10°$	$\bar{\epsilon} = 10°$ $\sigma_\epsilon = 14°$
		0	0.69	0.27	0.85	1.11	1.69
10^{-4}	8.40	1	2.61	0.27	0.85	3.05	3.71
		1.125	3.96	0.27	0.85	4.41	5.02
		0	0.73	0.45	3.40	1.31	4.17
10^{-6}	10.53	1	2.80	0.45	3.40	3.40	6.25
		1.125	4.31	0.45	3.40	4.94	7.88
		0	0.75	0.91		1.73	
10^{-8}	11.97	1	2.88	0.91		3.87	
		1.125	4.36	0.91		5.36	

Note: B_T is the two-sided 3-dB bandwidth of the transmit filter.

TABLE II

Power Penalties for QPSK

			Power Penalty in DB Due to Noise and				
	Ideal CNR in DB			Imperfect ϵ Only		ISI and Imperfect ϵ	
P_{4B}		$1/B_T T$	ISI Only	$\bar{\epsilon} = 5°$ $\sigma_\epsilon = 3°$	$\bar{\epsilon} = 5°$ $\sigma_\epsilon = 6°$	$\bar{\epsilon} = 5°$ $\sigma_\epsilon = 3°$	$\bar{\epsilon} = 5°$ $\sigma_\epsilon = 6°$
		0	0.70	0.66	1.55	1.42	2.27
10^{-4}	11.40	1	2.62	0.66	1.55	3.44	4.58
		1.125	3.97	0.66	1.55	4.71	6.21
		0	0.73	0.97	2.89	1.73	3.81
10^{-6}	13.53	1	2.81	0.97	2.89	4.03	5.98
		1.125	4.32	0.97	2.89	5.42	
		0	0.75	1.22	5.18	2.08	6.15
10^{-8}	14.97	1	2.88	1.22	5.18	4.58	
		1.125	4.36	1.22	5.18	6.07	

Note: B_T is the two-sided 3-dB bandwidth of the transmit filter.

References

[1] W.C. Lindsey, "Phase-shift-keyed signal detection with noisy reference signals," *IEEE Trans. Aerosp. Electron. Syst.*, vol. AES-2, pp. 393-401, July 1966.

[2] S.A. Rhodes, "Effect of noisy phase reference on coherent detection of offset-QPSK signals," *IEEE Trans. Commun.*, pp. 1046-1054, Aug. 1974.

[3] V.K. Prabhu, "Effect of imperfect carrier phase recovery on the performance of PSK systems," *IEEE Trans. Aerosp. Electron. Syst.*, vol. AES-12, pp. 275-285, Mar. 1976.

[4] G.J. Foschini, R.D. Gitlin, and S.B. Weinstein, "On the selection of a two-dimensional signal constellation in the presence of phase jitter and Gaussian noise," *Bell Syst. Tech. J.*, vol. 52, pp. 927-965, July-Aug. 1973.

[5] V.K. Prabhu, "Error-rate considerations for digital phase-modulation systems," *IEEE Trans. Commun.*, vol. COM-17, pp. 33-42, Feb. 1969.

[6] ———, "PSK-type modulation with overlapping baseband pulses," *IEEE Trans. Commun.* pp. 950-990, Sept. 1977.

[7] W.C. Lindsey, *Synchronization Systems in Communication and Control.* Englewood Cliffs, N.J.: Prentice-Hall, 1972, pp. 488-489.

[8] F.E. Glave, "An upper bound on the probability of error due to intersymbol interference for correlated digital signals," *IEEE Trans. Inform. Theory*, vol. IT-19, pp. 440-447, July 1973.

[9] J.W. Matthews, "Sharp error bounds for intersymbol interference," *IEEE Trans. Inform. Theory*, vol. IT-19, pp. 440-447, July 1973.

[10] V.K. Prabhu, "Some considerations of error bounds in digital systems," *Bell Syst. Tech. J.*, vol. 50, pp. 3127-3151, Dec. 1971.

Phase-Shift-Keyed Signal Detection with Noisy Reference Signals

W. C. LINDSEY, Member, IEEE
Jet Propulsion Laboratory
Pasadena, Calif.

Abstract

This paper derives and graphically illustrates the performance characteristics of Phase-Shift-Keyed communication systems where the receiver's phase reference is noisy and derived from the observed waveform by means of a narrow-band tracking filter (a phase-locked loop). In particular, two phase measurement methods are considered. One method requires the transmission of an auxiliary carrier (in practice, this signal is usually referred to as the sync subcarrier). This carrier is tracked at the receiver by means of a phase-locked loop, and the output of this loop is used as a reference signal for performing a coherent detection. The second method is self-synchronizing in that the reference signal is derived from the modulated data signal by means of a squaring-loop.

The statistics (and their properties) of the differenced-correlator outputs are derived and graphically illustrated as a function of the signal-to-noise ratio existing in the tracking filter's loop bandwidth and the signal-to-noise ratio in the data channel. Conclusions of these results as well as design trends are presented.

Key Words—Communication, keyed, noise, phase lock, phase shift, signal, signal-to-noise, system.

Manuscript received October 1, 1965.
This paper presents the results of one phase of research carried out at the Jet Propulsion Laboratory, California Institute of Technology, Pasadena, under Contract NAS 7-100, sponsored by the National Aeronatucis and Space Administration.

Introduction

The detection of deterministic signals, i.e, signals whose form is *exactly* known at the receiver, in white Gaussian noise has been studied by many authors—notably North, Van Vleck, Middleton, and Woodward. In fact, the principle operation which must be performed at the receiver so as to minimize the error probability is one of cross-correlation, matched filtering, or correlation detection. The receiver cross-correlates the received signal, say $y(t)$, with each possible transmitted signal, $w_k(t)=x_k(t)\cos\omega t$, $k=1, 2$, and makes a decision as to what was transmitted by comparing the result with a preset threshold. One element of this operation may be mechanized as indicated in Fig. 1.

Basically, the cross-correlation operation is performed by a multiplier-integrator combination, with readout at time $t=T$, where T is the duration of the kth signal $x_k(t)$, or by a matched filter with readout at $t=T$. For the case of a perfectly matched filter, the output statistics at the sample instances, their properties, and the performance of the detector as a function of the signal-to-noise ratio are well known and documented. In practice, however, the readout instances as well as the phase of the reference signal $r(t)$ are usually not known exactly. The reason, of course, is that this synchronizing information is usually derived at the receiver in the presence of noise. Consequently, the receiver is unable to maintain a perfect "match" between what is transmitted and what is received; i.e., uncertainties in timing are introduced. This inability to maintain a perfect "match" between reference models of the transmitted waveforms and those stored at the receiver is due to random disturbances which may be referred to as *timing noise.*

In space communications, one source of timing noise is generated at the receiver; e.g., the carrier reference $r(t)$ is usually derived at the receiver by means of a tracking device commonly known as a phase-locked loop.[1] In this situation the timing noise appears as phase jitter on the reference carrier and, consequently, a randomly varying mismatch between the stored waveforms and what is received is produced. There is another source of timing noise which also introduces a similar randomly varying mismatch. This has to do with the time instances at which the correlator outputs (see Fig. 1) are sampled. Such timing information is usually referred to as *bit synchronization.* It is not the purpose of this paper to consider bit-sync jitter. However, this concept could easily be explored using similar analytical methods. (In fact, the degradation on system performance caused by bit-sync jitter errors may be interpolated from the results which follow.)

[1] This reference signal is usually referred to as a subcarrier signal. However, in this paper it is convenient to refer to the reference signal as the carrier reference. The practical implications of this terminology are that in communication systems which utilize data and sync subcarriers to modulate the radio frequency (RF) carrier, the demodulation of these subcarriers from the RF carrier is done with zero timing error. If such is not the case, there will be further degradation in the performance of the detector due to this additional component of timing noise. The reader is referred to [1] for a treatment of radio-frequency carrier-reference timing noise.

Reprinted from *IEEE Trans. Aerosp. Electron. Syst.*, vol. AES-2, pp. 393-401, July 1966.

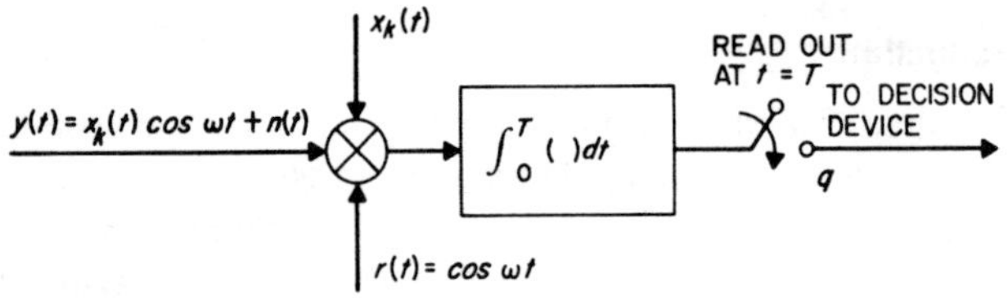

Fig. 1. Coherent detector; no mismatching.

Fig. 2. Coherent detector; mismatching of phases by $\theta - \theta$ radians.

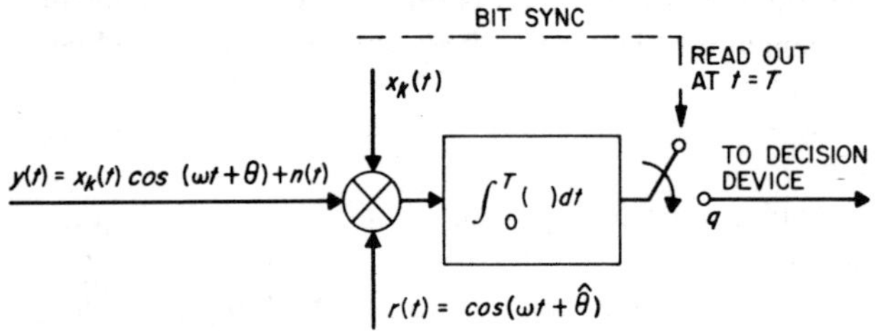

A functional diagram of a correlation detector which uses a noisy carrier replica to perform the detection is illustrated in Fig. 2. The detector shown is of the "integrate-and-dump" type and may be mechanized with an operational amplifier of sufficient stability. The difference between the correlation detector of Fig. 1 and that of Fig. 2 is exhibited by the relationship between the received and stored carrier phases. In this case the phase difference is $\phi = \theta - \hat{\theta}$. To perform a coherent detection, we must have $\theta = \hat{\theta}$ which is tantamount to saying that $\phi = 0$. In practice, however, this is rarely the case since the channel changes the phase of the transmitted signal. Thus to perform a coherent detection, one usually derives the reference signal $r(t)$ at the receiver and uses the result in the detection process as though it were noise free. For the present we will assume that the receiver detects or estimates the channel phase characteristic by means of a phase-locked loop which is tracking an auxiliary carrier of average power P watts.[2]

There are obvious disadvantages of this measurement technique. For example, the auxiliary carrier requires energy which could be used for information transmission, or the channel may affect the auxiliary carrier in a way which is different from that of the information bearing signal. For these reasons, measuring schemes which allow one to detect the phase θ by operating on the information bearing signal are of interest. One which immediately comes to mind and one which may be mechanized in practice is the squaring-loop device (see Fig. 3). In essence, the phase-shift-keyed modulation is removed by prefiltering, squaring the result, and tracking the double-frequency component by means of a phase-locked loop. The phase-locked loop output is frequency divided by two and the result is used as the reference signal $r(t)$ for demodulation.

In this paper we treat the problem of specifying the sample statistics (and their properties) of the differenced-correlator outputs when the modulation is known exactly but the receiver has stored a noisy version of the transmitted carrier; i.e., we are given *bit synchronization* exactly, but only an estimate of the carrier phase is available for performing the detection.[3] We further access the probability that a binary decision will be in error for the two different carrier-phase-measuring methods. These results are useful for comparison with experimental investigations and for use in carrying out a particular design.

Probability Distribution for the Phase-Error

The probability distribution of the phase error is of importance in specifying the differenced-correlator output statistics and the performance of the receiver. Consequently, the purpose of this section is to present the probability distribution for the phase-error $(\theta - \hat{\theta})$ and discuss some of its properties.

It is well known that the response distribution of a phased-locked loop is highly dependent on the filter which is placed in the loop [2] and [3]. In particular, the nonlinear response of a phase-locked loop, tracking a sinusoid in the presence of white Gaussian noise, is known for only a few special cases, viz., the so-called first-order loop [2] and [3] and the loop which contains an RC-circuit loop filter [3].

In practice, the loop filter of greatest interest for carrier tracking is the proportional plus-integral-control type; however, the response distribution for the loop phase error is known only in an approximate form [4]. Furthermore, experimental evidence shows that the phase-error distribution obtained using the proportional-plus-integral control type loop filter is closely related (for signal-to-noise ratios where phase-locked loops are generally expected to work) to the response distribution given in [2] and [3]. The resemblance, of course, is due to the fact that the region of operation is usually in the vicinity where linear phase-locked loop theory applies, viz., the region where the variance of the phase-error is approximately the reciprocal of the signal-to-noise ratio existing in the bandwidth of the tracking loop. Consequently, in what follows, we base the receiver phase-error distribution on the results which are available in [2] and [3]. Thus for the probability density of the phase error $\phi - \hat{\theta}$ we postulate

$$p(\phi) = \frac{\exp(\alpha \cos \phi)}{2\pi I_0(\alpha)}; \quad |\phi| \leq \pi \tag{1}$$

where $I_0(x)$ is the imaginary Bessel function of zero order and argument x. The parameter α specifies the performance of the auxiliary carrier-tracking loop. Furthermore, α represents the signal-to-noise ratio existing in the tracking loop; viz.,

$$\alpha = \frac{2P}{N_0 B_L} \tag{2}$$

[2] In practice this auxiliary carrier is usually referred to as the sync subcarrier.

[3] By differenced-correlator outputs we mean the value obtained by sampling (at the conclusion of the signaling interval) the output of the correlator which is matched to $w_1(t)$ and the correlator which is matched to $w_2(t)$ and differencing the result. If, however, $w_1(t) = -w_2(t)$, only one correlator is required.

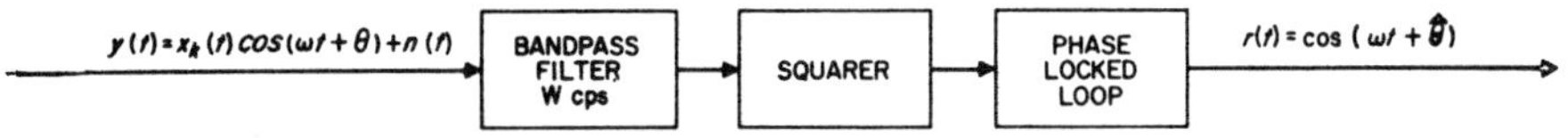

Fig. 3. Squaring-loop mechanization.

where P is the power in the auxiliary carrier, N_0 is the single-sided noise spectral density, and B_L is the bandwidth of the carrier tracking loop defined from the linear phase-locked loop theory; i.e.,

$$B_L = \frac{1}{2\pi j}\int_{-j\infty}^{j\infty} |H(s)|^2 ds$$

where $H(s)$ is the closed-loop transfer function. Further, α gives some indication of the degree of coherence between the observed carrier phase and the estimated carrier phase. If, for example, $\alpha=0$, the probability distribution of the phase measurement is uniformly distributed. At the other extreme (i.e., $\alpha=\infty$), the measurement is perfect and the probability distribution of the phase measurement is a delta function centered at the true value of observed carrier phase. This corresponds to the perfectly synchronized case. For sufficiently large signal strength the tracking loop may be linearized and the distribution $p(\phi)$ becomes Gaussian distributed with zero mean and variance α^{-1}. Thus we have postulated a phase-error distribution which checks with linear phase-locked loop theory at one extreme and is intuitively satisfying at the other extreme of weak signal conditions.[4]

Differenced Cross-Correlator Output Statistics[3]

Before we present the derivation of the differenced cross-correlator output statistics, $p(q)$, we define several basic communication parameters which serve to characterize the distribution $p(q)$ and the performance of the receiver, i.e., the probability of error. First, there is the signal-to-noise ratio $R=E/N_0=ST/N_0$. This parameter represents the ratio of the received signal energy to the one-sided spectral density of the noise. The signal cross-correlation coefficient λ is a measure of the degree of sameness of the two signals $w_1(t)$ and $w_2(t)$; i.e.,

$$\lambda = \frac{1}{E}\int_0^T w_1(t)w_2(t)dt$$

where

$$E = \int_0^T w_k^2(t)dt = ST; \qquad k = 1, 2. \tag{3}$$

For the case of phase-shift keying, $w_1(t) = -w_2(t)$ and $\lambda=-1$. Another convenient parameter, which arises in the analysis, is given by $\rho=2(1-\lambda)R$.

In order to minimize the length of the subsequent material, we shall draw heavily upon previous results. The approach which we shall take is to make use of the method of conditional probabilities; i.e., we shall develop expressions for the particular parameter, distribution, or moment of interest conditioned on the fact that the system phase error is fixed at ϕ radians. Thus by averaging over the condition, which is random, we easily determine the behavior of interest as the system phase error takes on all possible values.

We begin our analysis by specifying the differenced-statistics at the output of the cross-correlators given a fixed value of ϕ; i.e., we are interested in determining the probability density of the difference

$$q = \int_0^T y(t)[w_1(t) - w_2(t)]dt, \tag{4}$$

for it is this difference upon which a decision is made. If $q>0$, we announce w_1 and if $q<0$ we announce w_2.

Since, for a fixed ϕ, the input to the cross-correlator is Gaussian, the output of the cross-correlator is also Gaussian. In fact, the normalized-conditional output statistics of the cross-correlator differences, contingent upon the phase error being ϕ radians and the signal $x_1(t)$ having been transmitted, are given by (see Appendix I)

$$p(q_1|\phi) = \frac{1}{\sqrt{2\pi}}\exp\left[-\frac{(q_1-\sqrt{\rho}\cos\phi)^2}{2}\right]. \tag{5}$$

In the presence of noise only, the normalized conditional output statistics are given by (see Appendix I)

$$p(q_2|\phi) = \frac{1}{\sqrt{2\pi}}\exp(-q_2^2/2) = p(q_2) \tag{6}$$

and we see that the output statistics become independent of the system phase error. This result is physically comprehensible.

The output statistics $p(q_1)$ may be obtained by averaging (5) over the distribution of the phase error given by (1). Carrying out this averaging yields (see Appendix I)

$$p(q_1) = \int_{-\pi}^{\pi} p(\phi)p(q_1|\phi)d\phi$$

$$= \frac{\exp\left[-\frac{2q_1^2+\rho}{4}\right]}{\sqrt{2\pi}2\pi I_0(\alpha)} \sum_{k=-\infty}^{\infty} I_k(\rho/4)I_{2k}(\alpha+\sqrt{\rho}q_1)\exp(j\pi k) \tag{7}$$

where $I_k(x)$ is the modified Bessel function of order k and of argument x.

Illustrated in Figs. 4 and 5 are the differenced-correlator output statistics for various values of the parameter α and for two values of ρ. The values of ρ were chosen such that the error rate is 10^{-3} (typical for telemetry systems) and

[4] For $\alpha \geq 2$dB, the form of this distribution has been shown to agree with measurements obtained using the proportional-plus-integral-control type loop filter. In practice, this filter is of greatest interest in carrier tracking applications.

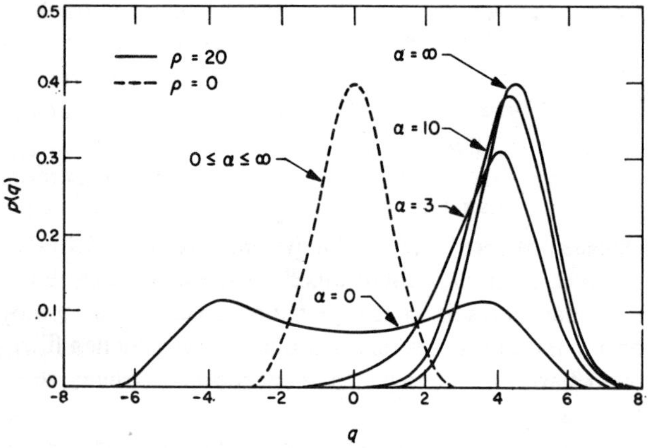

Fig. 4. Output statistics of decision variable.

Fig. 5. Output statistics of decision variable.

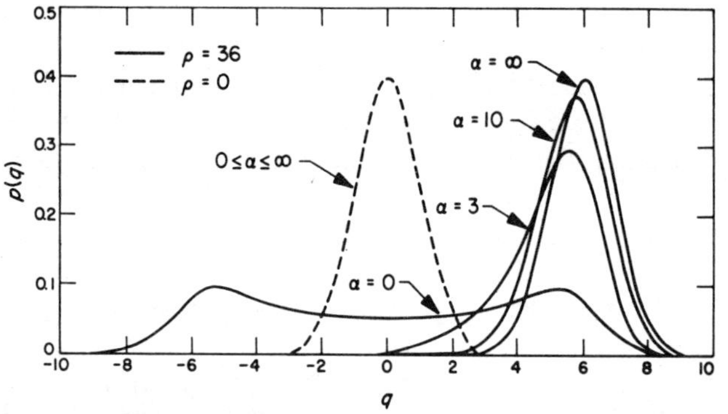

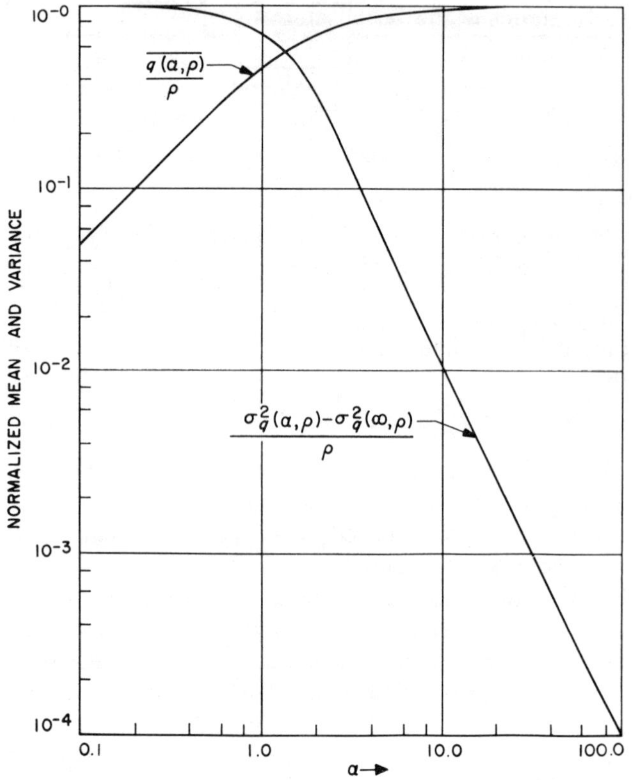

Fig. 6. Normalized mean and variance vs. SNR in tracking loop.

10^{-5} (typical for command systems) when the timing noise is zero, i.e., perfect phase synchronization. From either of these figures we notice the following results.

1) The distribution $p(q)$ does not depend on the parameter α if the input to the detector is only noise. In this case $p(q)$ is always Gaussian.
2) Under the hypothesis that signal plus noise is present we see that $p(q)$ is highly dependent on the value of α or equivalently the reference signal timing error.
3) The distribution $p(q)$ is bimodal for small α, i.e., large timing errors. This property may be attributed to the mixing of two probability distributions.
4) The distribution $p(q)$ is asymmetric for large values of α, i.e., small timing errors.
5) For $\alpha = \infty$, $p(q)$ is Gaussian with mean $\sqrt{\rho}$. This corresponds to the condition of zero timing noise or perfect time synchronization.
6) For $0 < \alpha < \infty$, the mean of $p(q)$ is less than $\sqrt{\rho}$.

In practice, the moments of this distribution are of interest. In particular, the mean of the differenced-cross-correlator output may be computed using (1) and (5) and the fact that the mean is given by

$$\bar{q}_1 = \int_{-\infty}^{\infty}\int_{-\pi}^{\pi} q_1 p(q_1 \mid \phi) p(\phi) d\phi dq_1. \tag{8}$$

Substituting (1) and (5) into (8) and carrying out the necessary integration gives (see Appendix I)

$$\bar{q}_1 = \sqrt{\rho}\,\frac{I_1(\alpha)}{I_0(\alpha)}\cdot \tag{9}$$

The mean-squared value of the differenced-cross-correlator output may be determined from

$$\bar{q}_1^2 = \int_{-\infty}^{\infty}\int_{-\pi}^{\pi} q_1^2 p(\phi) p(q_1 \mid \phi) d\phi dq_1. \tag{10}$$

Substitution of (1) and (5) into (10) and carrying out the necessary integration gives (see Appendix I)

$$\bar{q}_1^2 = 1 + \frac{\rho}{2} + \frac{\rho}{2}\left(\frac{I_2(\alpha)}{I_0(\alpha)}\right). \tag{11}$$

Thus the variance of the distribution $p(q_1)$ may be obtained from (9) and (11); i.e.,

$$\sigma_{q_1}^2 = 1 + \frac{\rho}{2} + \frac{\rho}{2}\left(\frac{I_2(\alpha)}{I_0(\alpha)}\right) - \rho\left[\frac{I_1(\alpha)}{I_0(\alpha)}\right]^2. \tag{12}$$

In practice, the expressions for the moments may be used in the laboratory to check a particular design.

Figure 6 shows the effects on the normalized mean $\bar{q}_1$ and the variance $\sigma_{q_1}^2$ of the distribution $p(q_1)$ as a function of the tracking-loop signal-to-noise ratio α. The normalization is with respect to the values of the mean and variance of $p(q_1)$ when there is zero timing error. For signal-to-noise ratios in the tracking loop greater than $\alpha = 10$ dB the mean and variance of $p(q_1)$ are relatively unaffected. This result may be used in setting power levels when designing phase-measuring tracking loops.

Performance of the Data Detector

We now turn to the problem of specifying the probability that a set of cross-correlators will err in making a decision. The method which we shall employ is that of computing the conditional error probability, the condition being that the system phase error is fixed to ϕ radians. Assuming, without loss of generality, that $x_1(t)$ was transmitted, the probability of error is one minus the probability that the decision made is correct. The probability that the decision is correct is the probability that the output of that filter matched to signal $x_2(t)$ is less than the output of the filter matched to $x_1(t)$. This probability is given by

$$P_E(\phi) = 1 - \int_{-\infty}^{\infty} p(q_1 \mid \phi) dq_1 \int_{-\infty}^{q_1} p(q_2 \mid \phi) dq_2. \quad (13)$$

Substituting (5) and (6) into (13) and simplifying the result yields

$$P_E(\phi) = \frac{1}{\sqrt{2\pi}} \int_{\sqrt{\beta}\cos\phi}^{\infty} \exp\left[-\frac{x^2}{2}\right] dx$$

$$= E\,rfc[\sqrt{\beta}\cos\phi] \quad (14)$$

where $\beta = \rho/2 = R(1-\lambda)$. The average error rate is easily obtained by averaging (14) over the phase-error distribution $p(\phi)$ given in (1); i.e.,

$$P_E = \frac{1}{\pi} \int_0^{\pi} \frac{\exp(\alpha\cos\phi)}{I_0(\alpha)} E\,rfc\,(\sqrt{\beta}\cos\phi) d\phi. \quad (15)$$

The integrals in (15) may be evaluated (see Appendix II); however, the desire to place the result into closed closed form seems, at this point, to be extremely difficult to realize. In any case, it is possible to show (see Appendix II) that (15) integrates to

$$P_E = \frac{1}{2}\Bigg[1 - \sqrt{\frac{2\beta}{\pi}} \exp\left(-\frac{\beta}{4}\right) \cdot \sum_{k=0}^{\infty} (-1)^k \epsilon_k b_{2k+1} \frac{({}_kI\beta/4)}{(1-4k^2)}\Bigg] \quad (16)$$

where

$$b_{2k+1} = \frac{I_k(\alpha)}{I_0(\alpha)}$$

and $\epsilon_k = 1$ if $k=0$ and $\epsilon_k = 2$ if $k \neq 0$. Although this expression is extremely difficult to evaluate on a digital computer, it does indicate several important results. For example, if we let ρ approach infinity, then (16) reduces to

$$P_E = \frac{1}{2}\left[1 - \frac{2}{\pi} \sum_{k=0}^{\infty} \frac{(-1)^k \epsilon_k b_{2k+1}}{1-4k^2}\right] \quad (17)$$

which shows that the performance of the receiver exhibits an irreducible error probability which depends on the signal-to-noise ratio in the auxiliary carrier tracking loop. In the limit as α approaches infinity, i.e., perfect phase measurement, we note that P_E approaches zero.

Plotted in Fig. 7 is the probability that the detector will err vs. R for different values of α. Notice the "bottoming" of the error rate as R increases. This says that as more energy is placed into the information bearing signal the ultimate performance of the system is governed by the signal-to-noise ratio in the auxiliary carrier tracking loop. Thus the effects on system performance caused by timing errors, inaccurate reference models, or mismatched filters are clearly manifested. In a particular design, these results may be used to access the degradation due to the noisy phase reference and to compensate for the loss in performance.

Performance of the Data Detector when a Squaring-Tracking-Loop Is Used

A convenient and attractive way to derive the phase of the carrier reference $r(t)$ is to make use of the squaring loop.[5] The basic operations which must be performed are best illustrated in Fig. 3. The observed data, $y(t)$, is first filtered by a band-pass filter of W c/s, and this result is then squared. The double frequency term is then tracked by means of a phase-locked loop whose voltage control oscillator (VCO) output is frequency divided by two and used as an estimate of the carrier reference $r(t)$. Stiffler [6] and Van Trees [7] were among the first to consider this measurement technique.

In order to carry out the analysis, we shall make several assumptions pertaining to the nature of the band-pass filter, squarer, and phase-locked loop combination. First, we assume that the bandwidth of the band-pass filter is sufficiently wide to pass without distortion the signal component in the observed data $y(t)$. We further assume that the ratio W/B_L is sufficiently large such that the output of the squaring device is, to a good approximation, Gaussian, and that the bandwidth of the phase-locked loop is sufficiently small so that the spectral density of the input noise is flat with a spectral density determined by the value of the output spectrum at $f_s = 2f$. If one assumes that the output z of the squarer is related to its input x through $z = ax^2$, then it may be shown ([5], p. 262) that the value of the output spectrum (exclusive of the sine wave component at $f_s = 2f$) is given by

$$N_0' = a^2\left[N_0 S + 2\left(\frac{N_0}{2}\right)^2 W\right] \quad (18)$$

while the intensity of the double-frequency sine-wave component occurring at the input to the phase-locked loop is given by

$$S_i = \frac{a^2}{2} S. \quad (19)$$

Thus the variance of the phase error is, using linear phase-locked loop theory, given by

[5] In this section we shall assume that $\lambda = -1$; i.e., $x_1(t) = -x_2(t)$.

$$\sigma_\phi^2 = \frac{1}{\alpha} = \frac{N_0}{ST_b}\left[\left(1 + \frac{N_0 W}{2S}\right)\frac{B_L}{\mathcal{R}}\right] = (\delta R)^{-1} \quad (20)$$

where $\mathcal{R} = T_b^{-1}$ is the data rate and

$$\delta = \frac{2\mathcal{R}}{B_L}\left[\left(1 + \frac{N_0 W}{2S}\right)\right]^{-1}. \quad (21)$$

Thus the parameter δ depends on the data rate, the bandwidth of the phase-locked loop, and the signal-to-noise ratio existing at the output of the band-pass filter. It will subsequently become apparent that the design engineer should adjust the parameters of the squaring loop so as to maximize the parameter δ.

Proceeding as before, we have that the error probability is given by

$$P_E = \frac{1}{\pi}\int_0^\pi \frac{\exp(\delta R \cos\phi)}{I_0(\delta R)}\,d\phi \int_{\sqrt{2R}\cos\phi}^\infty \frac{1}{\sqrt{2\pi}} \cdot \exp\left(-\frac{x^2}{2}\right) dx. \quad (22)$$

This integral is shown, in Appendix II, to reduce to

$$P_E = \frac{1}{2}\left[1 - \sqrt{\frac{4R}{\pi}}\exp\left(-\frac{R}{2}\right) \cdot \sum_{k=0}^{\infty} (-1)^k \epsilon_k b_{2k+1} \frac{I_k\left(\frac{R}{2}\right)}{(1 - 4k^2)}\right] \quad (23)$$

where

$$b_{2k+1} = \frac{I_k(\delta R)}{I_0(\delta R)}.$$

In this case the irreducible error probability does not exist. Figure 8 illustrates the performance of the detector for various values of the parameter δ. The parameter δ gives a good measure of how effective the squaring-loop is in estimating the phase of the signal component in $y(t)$. In practice, these results may be used to carry out a particular design or to access the performance of a particular design.

Conclusions

The foregoing investigation has developed a number of interesting points in connection with Phase-Shift-Keyed communication systems. In particular, the effects introduced by timing errors on the detector output statistics are noted to be asymmetric and in some situations bimodal. The variations in the mean and variance of these distributions due to timing noise is graphically illustrated. For signal-to-noise ratios which are greater than 10 dB in the carrier-tracking loop, the mean and variance of the detector output statistics are essentially unaffected by timing errors.

Phase-Shift-Keyed communication systems which em-

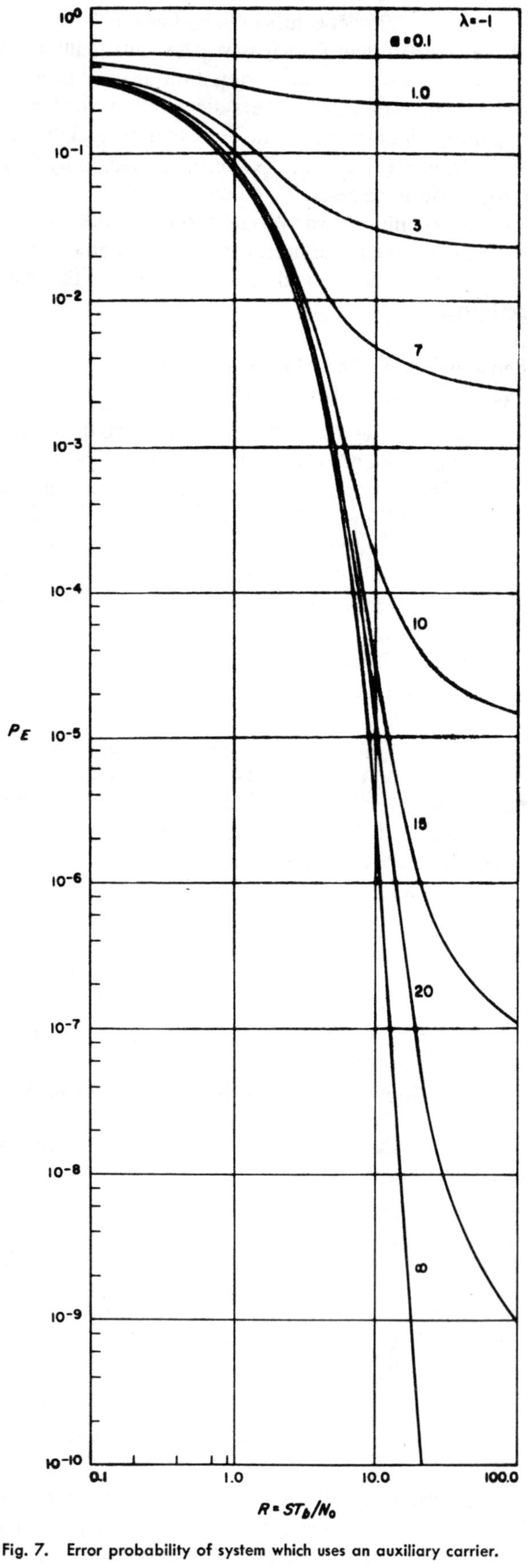

Fig. 7. Error probability of system which uses an auxiliary carrier.

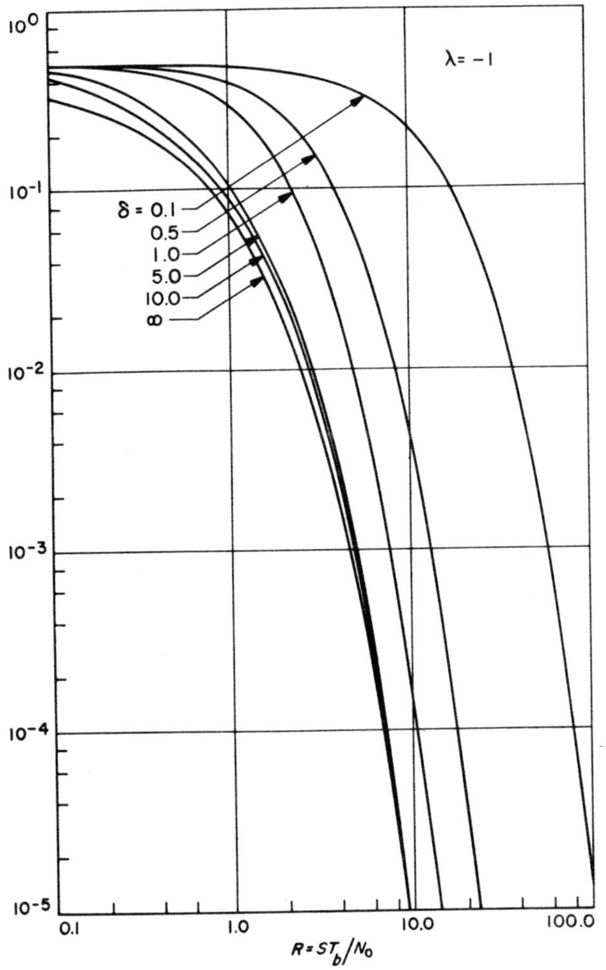

Fig. 8. Error probability of system which uses a squaring loop.

ploy auxiliary channels for establishing the cross-correlators reference signals inherently contain an irreducible error probability. The irreducible error probability is ultimately determined by the amount of timing noise which is introduced into the data detection process by the auxiliary carrier-tracking loop. The performance characteristic for such a system is graphically illustrated. These results may be used for design purposes.

If a squaring-loop is used as the synchronizing device it is shown that, for $\delta > 5$, the carrier reference is, for all practical purposes, perfect. The parameter δ is determined by the bandwidth of the band-pass filter, the data rate, the bandwidth of the phase-locked loop, the signal power, and the spectral density of the noise. The design of the squaring-loop should be aimed at maximizing δ.

Appendix I

First, we wish to derive (5) and (6). We begin the analysis by fixing the phase error to ϕ radians and deriving the sample statistics $p(q|\phi)$ at the output of the cross-correlators. We also assume that x_1 was transmitted. Since the input to the demodulator is Gaussian and the operations which the cross-correlator must perform are linear, the decision variable q (for a fixed ϕ) is also Gaussian. To specify the distribution $p(q|\phi)$, we need only compute the conditional mean and variance of q and properly normalize the result; i.e., (4), which is easily shown to be (assuming x_1 is transmitted)

$$\mu_1 = ST(1 - \lambda) \cos \phi = E(1 - \lambda) \cos \phi \quad (24)$$

while the conditional variance is easily shown to be

$$\sigma_1^2 = EN_0(1 - \lambda). \quad (25)$$

Using the subscript one to denote that x_1 was transmitted we have

$$p(q_1 | \phi) = \frac{\exp\left[-\frac{(q_1 - \mu_1)^2}{2\sigma_1^2}\right]}{\sqrt{2\pi\sigma_1^2}} \quad (26)$$

while letting $q_1' = \sigma_1 q_1$ and making use of the Jacobian of the transformation, (26) becomes

$$p(q_1' | \phi) = \frac{\exp\left[-\frac{(q_1' - \mu_1/\sigma_1)^2}{2}\right]}{\sqrt{2\pi}}. \quad (27)$$

Under the hypothesis that x_2 was transmitted, the conditional mean of the decision variable, i.e., (4), is easily shown to be

$$\mu_2 = -\mu_1 \quad (28)$$

while the conditional variance is

$$\sigma_2^2 = EN_0(1 - \lambda) = \sigma_1^2. \quad (29)$$

Thus as before

$$p(q_2' | \phi) = \frac{\exp\left[-\frac{(q_2' + \mu_2/\sigma_2)^2}{2}\right]}{\sqrt{2\pi}} \quad (30)$$

and the primes on the q's may be dropped with no loss in generality. Using (26) and (30) it is easy to show that system performance may be studied by assuming that the decision variable has mean $\sqrt{2}\mu_1/\sigma_1$ and unit variance under the hypothesis that signal plus noise is present and zero mean and unit variance in the presence of noise only. Thus, for a fixed ϕ the statistics in the presence of signal plus noise are given by

$$p(q_1 | \phi) = \frac{\exp\left[-(q_1 - \sqrt{2}\mu_1/\sigma_1)^2\right]}{\sqrt{2\pi}} \quad (31)$$

and if we let $\rho_n \cos \phi = 2\mu_n^2/\sigma_n^2$ we have (5) where

$$\rho = 2R(1 - \lambda). \quad (32)$$

In the presence of noise only we set $\rho_n = 0$ and obtain (6).

To obtain (7) we begin with (5) and (1). The product $p(\phi)p(q|\phi)$ is

$$p(q, \phi) = \frac{\exp\left[-\frac{2q^2+\rho}{4}\right] \exp\left[(\alpha+\sqrt{\rho q}) \cos \phi + \frac{\rho}{4} \cos 2\phi\right].}{\sqrt{2\pi} 2\pi I_0(\alpha)} \quad (33)$$

Now from Watson [8] we have

$$\exp(x \cos \theta) = \sum_{k=0}^{\infty} \epsilon_k I_k(x) \cos k\theta \quad (34)$$

where $\epsilon_k = 1$ if $k=0$ and $\epsilon_k = 2$ if $k \neq 0$. Thus (33) may be written

$$p(q, \phi) = \frac{\exp\left[-\frac{2q^2 + \rho}{4}\right]}{\sqrt{2\pi} 2\pi I_0(\alpha)} \sum_{k=0}^{\infty} \epsilon_k I_{2k}\left(\frac{\rho}{4}\right)$$

$$\cdot \exp\left[(\alpha + \sqrt{\rho q}) \cos \phi\right] \cos 2k\phi. \quad (35)$$

If one integrates (35) with respect to ϕ from $-\pi$ to π, uses the definition [8]

$$I_k(x) = \frac{1}{2\pi} \int_{-\pi}^{\pi} \cos k\theta \exp(x \cos \theta) d\theta, \quad (36)$$

and notices that ([8], p. 80)

$$I_k[z \exp(km\pi i)] = \exp(mk\pi i) I_k(z), \quad (37)$$

he immediately obtains (7).

The mean and variance of q may be derived using the normalized statistics of (7). First we note ([8], p. 31) that

$$\exp(\pi n i) I_{2n}(z) = \frac{1}{2\pi} \int_{-\pi}^{\pi} \exp\left[i(2n\theta - iz \sin \theta)\right] d\theta. \quad (38)$$

Using (38) in (7) gives

$$p(q) = \frac{\exp\left[-\frac{2q^2 + \rho}{4}\right]}{2\pi\sqrt{2\pi} I_0(\alpha)} \int_{-\pi}^{\pi} \sum_{k=0}^{\infty} I_k\left(\frac{\rho}{4}\right)$$

$$\cdot \exp(2ik\theta) \exp\left[(\alpha + \sqrt{\rho q}) \sin \theta\right] d\theta. \quad (39)$$

Applying formula (34) to (39) we have

$$p(q) = \frac{\exp\left[-\frac{2q^2 + \rho}{4}\right]}{2\pi\sqrt{2\pi} I_0(\alpha)} \int_{-\pi}^{\pi}$$

$$\cdot \exp\left[\frac{\rho}{4} \cos 2\theta + (\alpha + \sqrt{\rho q}) \sin \theta\right] d\theta. \quad (40)$$

To obtain the mean we multiply $p(q)$ by q and integrate from minus infinity to plus infinity. Thus

$$\bar{q} = \frac{\exp\left(-\frac{\rho}{2}\right)}{\sqrt{2\pi} 2\pi I_0(\alpha)} \int_{-\pi}^{\pi} \exp\left[\alpha \sin \theta + \frac{\rho}{2} \cos^2 \theta\right] d\theta$$

$$\cdot \int_{-\infty}^{\infty} q \exp\left[-\frac{q^2}{2} - \sqrt{\rho}\, q \sin \theta\right] dq. \quad (41)$$

Performing the integration with respect to q yields

$$\bar{q} = \frac{\sqrt{\rho}}{I_0(\alpha)} \frac{1}{2\pi} \int_{-\pi}^{\pi} \sin \theta \exp(\alpha \sin \theta) d\theta. \quad (42)$$

Using (36) we have (9).

Similarly for the variance $\bar{q}^2$ we must evaluate

$$\bar{q}^2 = \frac{\exp\left(-\frac{\rho}{2}\right)}{\sqrt{2\pi} 2\pi I_0(\alpha)} \int_{-\pi}^{\pi} \exp\left(\alpha \sin \theta + \frac{\rho}{2} \cos^2 \theta\right) d\theta$$

$$\cdot \int_{-\infty}^{\infty} q^2 \exp\left[-\left(\frac{q^2}{2} - \sqrt{\rho}\, q \sin \theta\right)\right] dq. \quad (43)$$

Integration with respect to q yields

$$\bar{q}^2 = \frac{1}{2\pi I_0(\alpha)} \int_{-\pi}^{\pi} (1 + \rho \sin^2 \theta) \exp(\alpha \sin \theta) d\theta$$

$$= 1 + \frac{\rho}{I_0(\alpha)} \frac{\partial}{\partial \alpha} \int_{-\pi}^{\pi} \sin \theta \exp(\alpha \sin \theta) d\theta. \quad (44)$$

Carrying out the integration with respect to θ and then differentiating the result with respect to α yields (11).

Appendix II

To establish the validity of (9) and (11) we make use of a result given by Luke ([9], p. 174); viz.,

$$Erf(z \sin \theta) = \frac{1}{2} z \sum_{k=0}^{\infty} \epsilon_k$$

$$\cdot \left[\frac{\sin(2k+1)\theta}{2k+1} + \frac{\sin(2k-1)\theta}{2k-1}\right]$$

$$\cdot \exp(-z^2/2) I_k(z^2/2) \quad (45)$$

where

$$Erf(x) = \int_0^x \exp(-t^2) dt. \quad (46)$$

If one introduces the change of variable $\theta = \theta' + \pi/2$ and uses the fact that $\cos \theta = \sin(\theta + \pi/2)$, it is easy to show that

$$Erf(z \cos \theta) = \frac{z \exp(-z^2/2)}{2} \sum_{k=0}^{\infty} (-1)^k \epsilon_k$$

$$\cdot \left[\frac{\cos(2k+1)\theta}{2k+1} - \frac{\cos(2k-1)\theta}{2k-1}\right]$$

$$\cdot I_k(z^2/2). \quad (47)$$

Also we note that

$$\frac{Erf(z)}{\sqrt{\pi}} = \frac{1}{2} - \int_{\sqrt{2}\, z}^{\infty} \frac{1}{\sqrt{2\pi}} \exp(-t^2/2) dt. \quad (48)$$

From (15) we write ($\phi = \phi_E$)

$$P_E(n) = \int_{-\pi}^{\pi} p(\phi) d\phi \int_{\sqrt{2x} \cos \phi}^{\infty} \exp(-t^2/2) dt \quad (49)$$

where $x=R(1-\lambda)/2$. Equation (49) may be written as

$$P_E(n) = \frac{1}{2} - \frac{1}{\sqrt{\pi}} \int_{-\pi}^{+\pi} p(\phi) d\phi \cdot \int_0^{\sqrt{x} \cos \phi} \exp(-w^2/2) dw$$

$$= \frac{1}{2} - \frac{1}{\sqrt{\pi}} \int_{-\pi}^{\pi} Erf(\sqrt{x} \cos \phi) p(\phi) d\phi$$

$$= \frac{1}{2} - \frac{1}{\sqrt{\pi}} \int_{-\pi}^{\pi} G(\phi) d\phi \qquad (50)$$

where $G(\phi)=p(\phi)\ Erf(\sqrt{x} \cos \phi)$. From Watson [8] we have

$$\exp[\alpha \cos \phi] = \sum_{k=0}^{\infty} \epsilon_k I_k(\alpha) \cos k\phi. \qquad (51)$$

Thus the distribution $p(\phi)$, given by (1), becomes

$$p(\phi) = \sum_{k=0}^{\infty} \epsilon_k a_k \cos k\phi \qquad (52)$$

where

$$a_k = \frac{I_k(\alpha_1)}{2\pi I_0(\alpha_1)} \qquad (53)$$

and the function $G(\phi)$ reduces to

$$G(\phi) = \frac{\sqrt{x}}{2} \exp(-x/2) \sum_{j=0}^{\infty} \sum_{k=0}^{\infty} (-1)^k \epsilon_j \epsilon_k a_j \cdot \left[\frac{\cos(2k+1)\phi}{2k+1} - \frac{\cos(2k-1)\phi}{2k-1}\right] \cdot \cos j\phi I_k(x/2). \qquad (54)$$

Integrating (54) from $-\pi$ to π using the facts

$$\int_{-\pi}^{\pi} \cos j\theta \cos(2k+1)\theta d\theta = \pi \delta_{j,2k+1}$$

$$\int_{-\pi}^{\pi} \cos j\theta \cos(2k-1)\theta d\theta = \pi \delta_{j,2k-1} \qquad (55)$$

gives

$$\int_{-\pi}^{\pi} G(\phi) d\phi = \sqrt{x} \exp(-x/2) \cdot \sum_{k=0}^{\infty} (-1)^k b_{2k+1} \frac{I_k(x/2)}{1-4k^2} \qquad (56)$$

where $\delta_{j,k}=0$ if $j \neq k$ and $\delta_{jk}=1$ if $j=k$ and $b_{2k+1}=2\pi a_{2k+1}$. If we substitute (56) into (50) and let $x=\beta/2$, we have (16).

To arrive at (17) we let x approach infinity in (56). Since [8]

$$I_k(x) \sim \frac{\exp(x)}{\sqrt{2\pi x}} \qquad (57)$$

for large x, (56) becomes

$$\int_{-\pi}^{\pi} G(\phi) d\phi = \frac{1}{\sqrt{\pi}} \sum_{k=0}^{\infty} (-1)^k \epsilon_k b_{2k+1} \left(\frac{1}{1-4k^2}\right). \qquad (58)$$

Using this result in (50) gives (17).

REFERENCES

[1] W. C. Lindsey, "Optimal design of one-way and two-way coherent communication links," *IEEE Trans. on Communication Technology*, to be published, August 1966.

[2] A. J. Viterbi, "Phase-locked loop dynamics in the presence of noise by Fokker-Planck techniques," *Proc. IEEE*, vol. 51, pp. 1737–1753, December 1963.

[3] V. I. Tikhonov, "Influence of noise on phase-locked oscillator operation," *Avtomat. i Telemekh.*, Akad. Nauk SSSR, vol. 20, September 1959. Also "Phase-lock automatic frequency control operation in the presence of noise," *Avtomat. i Telemekh.*, Akad. Nauk SSSR, vol. 21, p. 301, March 1960.

[4] W. C. Lindsey, "Investigation of second-order phase-locked loops by Fokker-Planck methods," Pasadena, Calif., JPL Space Programs Summary 37–30, vol. IV, pp. 262–268, Nobember 1964.

[5] W. B. Davenport and W. L. Root, *An Introduction to the Theory of Random Signals and Noise*. New York: McGraw-Hill, 1958.

[6] J. P. Stiffler, "Bit and subcarrier synchronization in a binary PSK communication system," *Proc. Nat'l Telemetering Conf.*, Los Angeles, Calif., June 2–4, 1964.

[7] S. W. Golomb, J. R. Davey, I. S. Reed, H. L. Van Trees, and J. J. Stiffler, "Synchronization," *IEEE Trans. on Communication Systems*, vol. CS-11, pp. 481–491, December 1963.

[8] G. N. Watson, *A Treatise on the Theory of Bessel Functions*. New York: Cambridge, 1958.

[9] Y. L. Luke, *Integrals of Bessel Functions*. New York: McGraw-Hill, 1962.

Effect of Noisy Phase Reference on Coherent Detection of Offset-QPSK Signals

SMITH A. RHODES, MEMBER, IEEE

Abstract—**Imperfect carrier synchronization causes a performance loss for coherent phase-shift-keyed (PSK) communications. This detection loss is greater for quaternary signaling (QPSK) than for the binary case (BPSK). The use of an offset form of QPSK, also known as double-biphase modulation, is shown to yield a probability of bit error in detection that is equal to the average of the detection performances for BPSK and conventional QPSK. Because of frequency instabilities in communications systems, it is sometimes difficult to obtain carrier synchronization with sufficiently low jitter to preclude significant detection losses. The use of offset QPSK in lieu of conventional QPSK modulation is shown to lower by almost 3 dB the required SNR of the synchronizer phase reference for satisfying a specified value of allowable detection loss.**

I. INTRODUCTION

PHASE-SHIFT keying (PSK) is an efficient modulation technique that is used extensively in digital communications. When coherent detection is employed, there is a loss [1], [2] of detection efficiency associated with the use of an imperfect or noisy phase reference. This loss is greater [3] for quaternary signaling (QPSK) than for the binary case (BPSK). The use of an offset form of QPSK, also termed double-biphase [4], [5] modulation, is shown to reduce the detection sensitivity to phase synchronization for quadriphase communications. Although offset QPSK has been known to possess other properties [6] that are superior to those for conventional QPSK, the effect of a noisy phase reference upon the detection of offset QPSK signals apparently has not been investigated previously.

Gray [7], [8] mapping of the source onto the signal vectors produces QPSK transmissions that are equivalent to the sum of two binary sources modulated onto quadrature components of the carrier. In conventional QPSK modulation, the two binary modulation waveforms are time coincident, whereas the bit alignments are staggered for offset QPSK modulation. Under ideal conditions, the two binary components of either form of QPSK signal may be detected orthogonally. Thus, the absence of cross-coupling between the binary components allows both forms of QPSK to provide detection performances that are identical to the BPSK case. When the orthogonality between the two binary channels of a QPSK signal is destroyed by nonideal conditions, however, the resulting interchannel interference (cross coupling) causes the detection performance to deteriorate more than for BPSK signaling. The staggered alignment for offset QPSK allows the cross coupling from one binary channel to change state at the midpoint of the bit interval for the other channel. Because of the coincident alignment for conventional QPSK signals, the interchannel interference is always constant during a bit interval. Therefore, the detection performances for the two forms of quadriphase modulation are degraded differently when cross coupling occurs.

Imperfect carrier synchronization causes a correlation loss of $\cos\phi$ for the demodulated signal voltage, where ϕ is the error in the phase estimate supplied by the synchronizer. In BPSK signaling, there is no further degradation as the result of a synchronization error. For QPSK transmissions, however, interchannel interference is introduced by cross coupling between the binary signal components that is proportional to $\sin\phi$. Because of the staggered alignment of the binary modulation components of offset QPSK signals, the cross coupling can change state at midbit, such that the interference during the first half of the bit is cancelled by interference of opposite polarity during the second half of the bit interval. Consequently, offset QPSK signaling will have a detection performance identical to BPSK when the cross coupling changes state. When the cross coupling remains constant over the entire bit interval, the detection performance will be the same as for conventional QPSK communications. With the assumption of equally-likely binary states of $+A$ or $-A$ that are statistically independent for each source bit, the probability of a change in the cross-coupling polarity from a bit transition is one half. It follows that for offset QPSK, the probability of decision error per detected bit is equal to the average of the detection performances for BPSK and conventional QPSK.

Because of frequency instabilities [9], [10] in communications systems, the bandwidth of the carrier synchronizer cannot be lowered indefinitely in order to reduce the thermal noise jitter of the phase reference to negligible levels. In fact, it is difficult [11] in some systems to obtain a coherent phase reference with a signal-to-noise power ratio (SNR) that is sufficiently high to prevent significant losses in detection performance. Hence, the improved immunity to noisy phase synchronization that results from the offset bit alignment is an important factor in achieving efficient quadriphase communications. It is shown that

Paper approved by the Associate Editor for Communication Theory of the IEEE Communications Society for publication after presentation at the 1973 National Telecommunications Conference, Atlanta, Ga., November 26–28, 1973. Manuscript received October 23, 1973; revised February 22, 1974. The analyses and results presented in this paper were obtained while performing studies for the Defense Communications Agency under Contract DCA-100-73-C-0008.

The author is with the Computer Sciences Corporation, Falls Church, Va. 22046.

Reprinted from *IEEE Trans. Comm.*, vol. COM-22, pp. 1046–1055, Aug. 1974.

the use of offset QPSK in lieu of conventional quadriphase modulation reduces by almost 3 dB the SNR requirement for carrier phase synchronization to provide a specified detection efficiency.

Only two system imperfections, additive white Gaussian noise and imperfect carrier synchronization, are taken into account in the detection analyses. The modulation waveforms are restricted to rectangular shapes, and only uncoded PSK communications are treated. The detection results are plotted as graphs of probability of error per binary decision versus E_b/N_0, the received ratio of signal energy per bit to noise power density, with the SNR of the carrier phase reference as a parameter. BPSK detection curves are shown for comparison with the two forms of quadriphase signaling. Additional curves for BPSK and the two forms of QPSK illustrate the detection losses as a function of α, the SNR of the noisy phase reference. Also, for certain fixed losses in detection, the required values of α are plotted versus the detection probability of error for the three types of PSK that are analyzed. These latter curves indicate clearly the advantage of the offset alignment in reducing the SNR requirements of the phase reference for coherent quadriphase communications. In all of the graphs, the detection performances are based upon the use of a simple phase-lock loop (PLL) to derive the coherent phase reference from a noisy unmodulated auxiliary carrier.

In most communications systems, it is desired to limit the detection loss to a very small value, such as 0.1 dB. It is difficult to read the detection curves accurately when the loss is that low. However, theoretical approximations have been obtained for the detection performance that are valid when α is large, which corresponds to the low-loss case. These analytical expressions indicate that for a given loss specification, the use of the offset bit alignment for QPSK signals allows almost a 3-dB relaxation of the requirement on α, the SNR of the phase reference.

II. ANALYSIS

For the most efficient BPSK communications with respect to power utilization, the signal vectors are displaced by π rad. Thus the binary signal vectors are of opposite polarity or antipodal, and the BPSK transmission may be described as illustrated in Fig. 1(a) by

$$s_B(t) = C_k(2S)^{1/2} \cos(\omega_c t - \theta). \tag{1}$$

In this expression, ω_c represents the carrier frequency and θ is some arbitrary constant. S denotes the received signal power. The polarity of the kth bit is given by C_k, which independently takes on a value of either $+1$ or -1 with equal probability for each bit of the transmission.

In QPSK communications, the most efficient signaling space results in a transmission that is equivalent to the sum of two binary-antipodal signals which are in phase quadrature, that is, separated by $\pi/2$ rad. Hence, this QPSK signal may be formed [see Fig. 1(b)] by two binary source sequences $\{A_k\}$ and $\{B_k\}$ as follows:

$$s_Q(t) = A_k(2S)^{1/2} \cos(\omega_c t - \theta) + B_k(2S)^{1/2} \sin(\omega_c t - \theta). \tag{2}$$

Both A_k and B_k are independently distributed and take on $+1$ or -1 values with equal probabilities of one half. For consistency with the BPSK notation, S denotes the received signal power per binary digit of the QPSK transmission. Thus, the total power is $2S$ for the quaternary case. The distinguishing feature for the two forms of QPSK is the relative epoch alignments for the two binary modulation sequences $\{A_k\}$ and $\{B_k\}$. As illustrated in Fig. 2, the two bit sequences are time coincident for conventional QPSK modulation, whereas the bit sequences have their epochs relatively shifted by one half of a bit duration for offset QPSK modulation.

Coherent demodulation is performed by correlating the received signal with a phase reference that is derived by the carrier synchronizer from a noisy unmodulated residual carrier or auxiliary transmission. Integrate-and-dump (I&D) circuits are employed for matched filtering of the binary modulation waveforms prior to the polarity decisions on the transmitted bits. Perfect bit synchronization is assumed, and all other system functions are considered to be ideal except for signal fluctuations caused by thermal noise and jitter of the carrier phase reference. Under the assumption that the thermal disturbance is additive white Gaussian noise, the probability of decision error on each binary digit of either a BPSK or a QPSK transmission is given by

$$p(\phi) = Q[(2E_b/N_0)^{1/2}\rho(\phi)] \tag{3}$$

where E_b is the received signal energy per binary digit and N_0 is the single-sided density of the additive noise. Q denotes the normalized inverse cumulative Gaussian distribution function as defined by

$$Q[v] \triangleq \frac{1}{(2\pi)^{1/2}} \int_v^\infty \exp(-\lambda^2/2)\, d\lambda. \tag{4}$$

The factor $\rho(\phi)$ accounts for any decrease in the demodulated signal voltage that results from a carrier-synchronization error of ϕ rad.

In the absence of a carrier-tracking error, $\rho = 1$ and the detected distance between binary-antipodal vectors is $2(E_b)^{1/2}$. The presence of a synchronization error ϕ introduces a correlation loss [12] in signal demodulation that is equal to $\cos \phi$. Hence, for the BPSK case,

$$\rho_B(\phi) = \cos \phi. \tag{5}$$

Therefore, for BPSK detection,

$$p_B(\phi) = Q[(2E_b/N_0)^{1/2} \cos \phi]. \tag{6}$$

When carrier synchronization is perfect, the two binary components of a QPSK signal may be detected orthogonally. Hence, the QPSK detection performance is the same as for BPSK in this ideal coherent case ($\phi = 0$). When $\phi \neq 0$, however, there is not only a correlation loss of $\cos \phi$, as in the binary case, but orthogonality be-

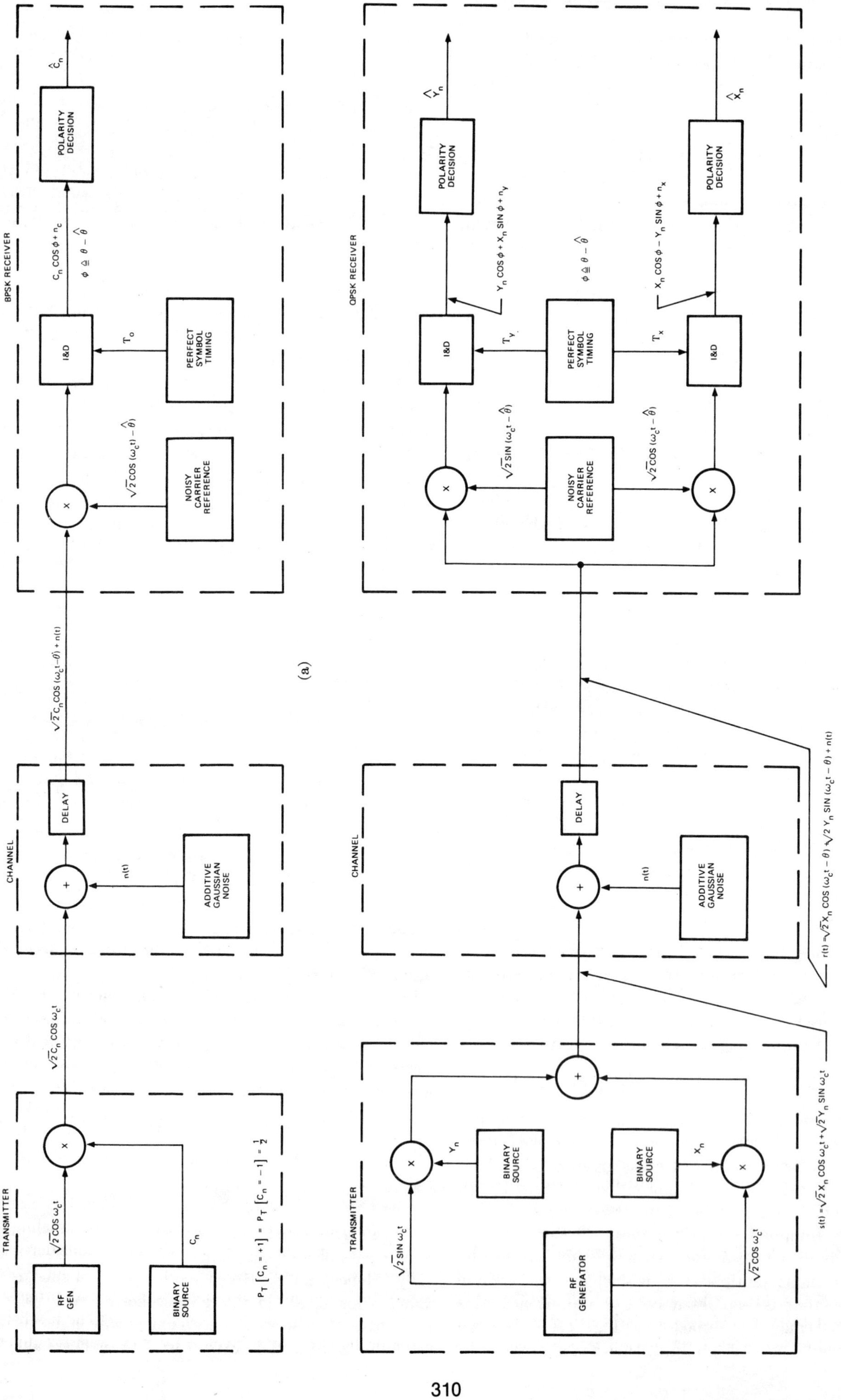

Fig. 1. Communications models for (a) BPSK and (b) QPSK signaling.

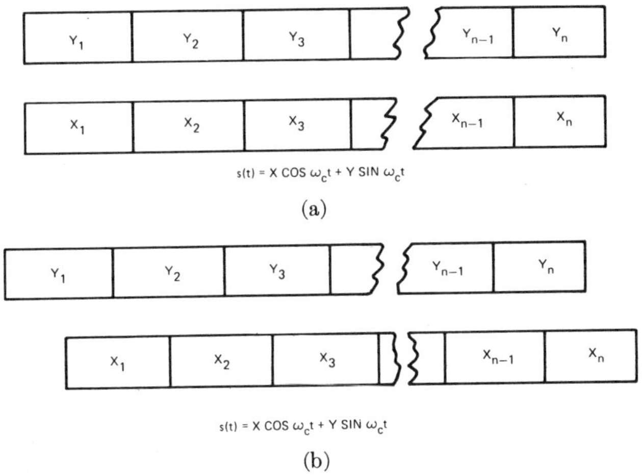

Fig. 2. Alignment of binary modulation sequences for (a) conventional and (b) offset forms of QPSK.

tween the two binary channels of a QPSK signal is lost. The mutual interference (cross coupling) between the two binary components is proportional to $\pm\sin\phi$. Because of the interchannel interference, QPSK detection is degraded more by imperfect carrier synchronization than is BPSK communications. For a given phase error ϕ in carrier synchronization, the resulting distances between the demodulated voltages for the binary-antipodal vectors of a QPSK transmission are reduced in accordance with

$$\rho_Q(\phi) = \cos\phi \pm \sin\phi. \tag{7}$$

In the determination of the conditional probability of error per binary decision for a given ϕ, the two forms of QPSK must be treated separately. For conventional QPSK, during each quaternary symbol interval one binary component suffers destructive cross-coupling interference, while the other binary component experiences constructive interference. Therefore, the conditional probability of decision error per binary digit is the arithmetic mean of the detection performances for the cases of destructive and constructive cross coupling.

$$p_{CQ}(\phi) = \tfrac{1}{2}[p_+(\phi) + p_-(\phi)] \tag{8}$$

where

$$p_+(\phi) = Q[(2E_b/N_0)^{1/2}(\cos\phi + \sin\phi)]$$
$$p_-(\phi) = Q[(2E_b/N_0)^{1/2}(\cos\phi - \sin\phi)]. \tag{9}$$

In offset QPSK signaling, the bit transitions for one binary channel occur at the middle of the bit intervals for the other channel. Under the assumption of independent equally-likely choices of positive or negative polarities for each bit, the probability of transition is one half. When a transition occurs, the cross coupling changes polarity at midbit of the other binary channel, and the interchannel interference during the first half of the bit interval is consequently cancelled by the interference of opposite polarity during the second half of the interval. Hence, the detection performance is identical to that for BPSK signaling when a bit transition takes place. If no transition occurs, then the cross-coupling interference remains constant during the entire interval of the detected bit. This cross coupling has equal probabilities of being constructive or destructive, corresponding to equally-likely polarities (plus and minus) for the $\sin\phi$ cross-coupling coefficient. Thus, when there is no bit transition, the detection performance is identical to that for conventional QPSK signaling. It follows that the detection performance for offset QPSK is equal to the average of the performances for the BPSK and conventional QPSK cases. Thus,

$$p_{OQ}(\phi) = \tfrac{1}{2}[p_B(\phi) + p_{CQ}(\phi)]. \tag{10}$$

The probability density function for the carrier-tracking error ϕ is dependent upon the technique that is utilized for carrier synchronization. For a phase reference that is derived by a first-order PLL from an unmodulated residual component of the PSK transmission or from an auxiliary carrier transmission, the density function [13] of the tracking error is

$$f_\Phi(\phi) = \frac{\exp(\alpha\cos\phi)}{2\pi I_0(\alpha)}, \qquad -\pi \le \phi \le +\pi \tag{11}$$

where α is the SNR of the phase reference, which is approximately equal to the inverse of the variance of ϕ. I_0 refers to a zero-order modified Bessel function [14] of the first kind.

If carrier synchronization is derived from a suppressed-carrier PSK transmission, then a nonlinear operation must be used for modulation removal. The squaring operation that is required to remove biphase modulation introduces a loss [15] in the SNR α of the phase reference. In the case of suppressed-carrier QPSK transmission, the quadrupling operation that is necessary to remove the modulation so that a phase reference may be obtained causes a greater loss [16] in α than does squaring. All forms of modulation removal for M-ary PSK transmissions introduce an m-fold phase ambiguity [17] in that the synchronizer will have m stable points. The two-state ambiguity resulting from BPSK removal and the four-state ambiguity associated with QPSK removal necessitate differential encoding and decoding [18] or some other technique in order to provide nonambiguous communications. The present analysis is restricted to the relative detection performances of offset and conventional forms of QPSK for specified qualities of the noisy phase reference without further considerations that relate to synchronization techniques. Thus, it is assumed in all cases that the carrier phase reference is obtained from an unmodulated transmission that is degraded only by additive Gaussian noise.

The bandwidth of the carrier synchronizer is assumed to be small in comparison with the digital signaling rate so that the synchronization error may be considered as a constant during each bit interval. Thus, for all types of coherent communications, the unconditional probability of decision error is obtained by averaging the conditional error probability for a given ϕ over the density function

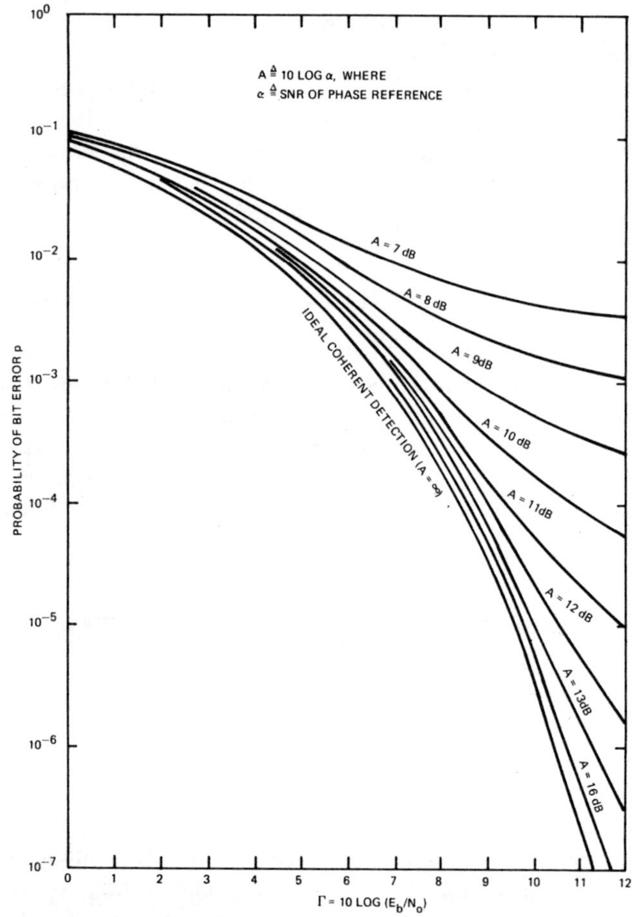

Fig. 3. Coherent detection performance for BPSK.

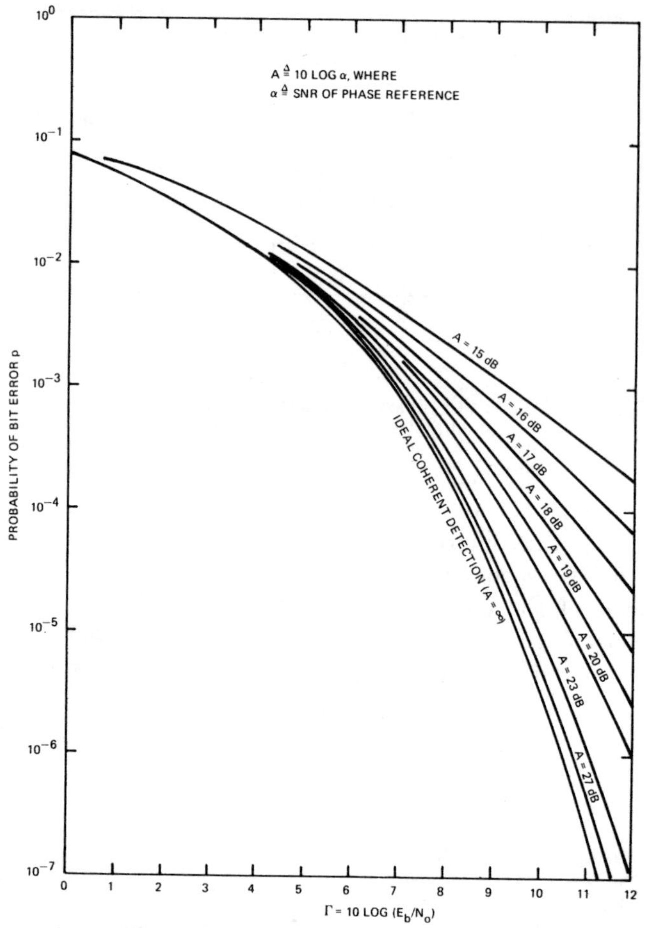

Fig. 4. Coherent detection performance for conventional QPSK.

of the synchronization error ϕ. It follows for BPSK, conventional QPSK, and offset QPSK communications that the average detection performances are given by

$$p_B = \int_{-\pi}^{+\pi} p_B(\phi) f_\Phi(\phi)\, d\phi$$

$$p_{CQ} = \int_{-\pi}^{+\pi} p_{CQ}(\phi) f_\Phi(\phi)\, d\phi$$

$$p_{OQ} = \int_{-\pi}^{+\pi} p_{OQ}(\phi) f_\Phi(\phi)\, d\phi = \frac{p_B + p_{CQ}}{2}. \quad (12)$$

Numerical evaluation of the three integrals yields the detection functionals that are plotted in Figs. 3, 4, and 5 versus E_b/N_0 and parametric in $A = 10 \log \alpha$, the SNR of the phase reference expressed in decibels. Note the improved detection performance for offset QPSK relative to the conventional quadriphase modulation technique. This advantage in detection for offset QPSK modulation is achieved directly as a result of the staggered bit alignment, which allows the cross coupling between binary components to be cancelled for a detected bit whenever a bit transition takes place in the other binary component.

The loss in detection efficiency that is associated with a noisy phase reference is defined as the required increase in E_b/N_0 relative to ideal coherent detection for maintaining a given probability p of error per binary decision. Figs. 3, 4, and 5 may be used to determine the detection losses for the three forms of PSK signaling when p and α are specified. However, the curves cannot be plotted or read with sufficient accuracy when the loss L is very small to compare the required values of α for given values of p and L. It is usually desirable to limit the loss L in detection efficiency to only one or two tenths of a decibel in practical communications systems. In order to assess the detection performance when L is small, corresponding to large α, theoretical approximations for the detection functionals will be derived that are valid for the low-loss case, when α is 20 or larger. These analytical expressions are very useful in providing estimates of the required values of α for different forms of PSK communications when L and p are specified.

When the SNR α of the phase reference is sufficiently large to prevent the detection loss L from exceeding a few tenths of a decibel, the phase error ϕ in carrier synchronization will usually be small. Hence, for a nominal operating point p_0 on the detection curve, only small excursions about p_0 will take place with significant probability as a result of imperfect carrier synchronization. Therefore, the loss in detection efficiency associated with a noisy phase

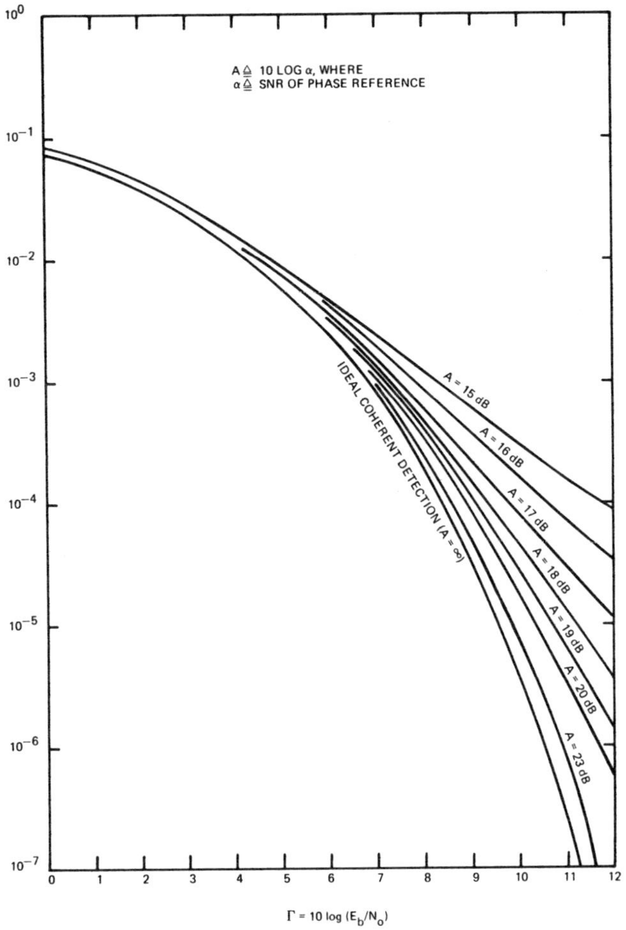

Fig. 5. Coherent detection performance for offset QPSK.

reference can be determined from approximations of the bit detection probability of bit error p about p_0 when the variation in ϕ is small. As derived in [19], the Q function that describes the detection curves for coherent PSK communications is bounded as follows:

$$\left(1-\frac{1}{u^2}\right)\frac{1}{(2\pi)^{1/2}u}\exp\left(-\frac{u^2}{2}\right) < Q[u] < \frac{1}{(2\pi)^{1/2}u} \cdot \exp\left(-\frac{u^2}{2}\right). \quad (13)$$

The upper bound on the Q function is very tight for other than small values of the argument; in fact, this bound is reasonably accurate for error probabilities p as high as 0.1. For p less than 10^{-3}, which includes most cases of interest in communications systems, the difference between the Q function and its approximation by the upper bound is less than the resolution of most detection plots.

Define the effective value of E_b/N_0 as γ. Thus

$$\gamma = \rho^2 E_b/N_0 \quad (14)$$

where ρ^2 denotes the effective loss in E_b/N_0 caused by a synchronization error ϕ. Note that $\rho^2 = 1$ when carrier synchronization is perfect. Any discrepancy in the predicted value of $\gamma = \gamma_0$ required to achieve a nominal detection performance of $p = p_0$ can be compensated for with a suitable multiplying factor C_1, which may be lumped along with other constants into a new constant C. Then, the approximation for the detection curve is

$$p = Q[(2\gamma)^{1/2}] \approx \frac{C_1}{(2\pi)^{1/2}(2\gamma)^{1/2}} \exp(-\gamma) = \frac{C}{\gamma^{1/2}} \exp(-\gamma). \quad (15)$$

When (p,γ) are confined to a nearby region of the nominal operating points (p_0,γ_0), the following relationship may be obtained:

$$p = p_0\left(\frac{\gamma}{\gamma_0}\right)^{-0.5} \exp(-(\gamma-\gamma_0)) \approx p_0 \exp(-h(\gamma-\gamma_0)) \quad (16)$$

where

$$h \triangleq 1 + \frac{0.5}{\gamma}.$$

Let $p = p_0$ designate the desired detection probability of bit error, which may be obtained with an E_b/N_0 value of γ_0 when carrier synchronization is perfect. When the phase reference has a variable phase error ϕ as the result of noise, E_b/N_0 must be raised to a value of γ_1 in order to maintain an average detection performance of $p = p_0$. Thus, the detection efficiency η drops below unity when carrier synchronization is imperfect, where

$$\eta \triangleq \frac{\gamma_0}{\gamma_1}. \quad (17)$$

Now define the value of E_b/N_0 in decibels by

$$\Gamma = 10 \log \gamma. \quad (18)$$

The loss in detection efficiency is expressed in decibels as

$$L\text{ (dB)} \triangleq \Gamma_1 - \Gamma_0 = -10 \log \eta. \quad (19)$$

Define as ϵ the small decrease in detection efficiency as a result of noisy phase synchronization:

$$\epsilon \triangleq 1 - \eta = 1 - \frac{\gamma_0}{\gamma_1} = \frac{\gamma_1 - \gamma_0}{\gamma_1}. \quad (20)$$

Therefore, the performance loss for coherent detection is

$$L\text{ (dB)} = -10 \log(1-\epsilon) \approx 4.34\epsilon\left(1+\frac{\epsilon}{2}\right). \quad (21)$$

The performance analyses may thus be obtained by the evaluation of ϵ for the different forms of PSK communications.

In the BPSK case, the effective value of E_b/N_0 is reduced by a correlation factor of $\cos^2 \phi$.

$$\gamma = \gamma_1 \cos^2 \phi = \gamma_1(1 - \sin^2 \phi). \quad (22)$$

For the low-loss case, ϕ is usually very small, such that $\sin \phi \approx \phi$. Therefore, the conditional probability of bit error for coherent BPSK detection may be approximated as

$$p_B(\phi) \approx p_0 \exp(-h(\gamma - \gamma_0)) = p_0 \exp(-h(\gamma_1 - \gamma_0)) \exp(+h\gamma_1\phi^2). \quad (23)$$

Thus the average detection performance is given by

$$p_B = \mathcal{E}[p_B(\phi)] = p_0 \exp(-h(\gamma_1 - \gamma_0))\mathcal{E}[\exp(+h(\gamma_1\phi^2))] = p_0 \frac{\exp(-h(\gamma_1 - \gamma_0))}{(1 - 2h\gamma_1\sigma^2)^{1/2}}. \quad (24)$$

But γ_1 is chosen to yield an average detection performance of p_0 when the phase reference is noisy. With p_B equated to p_0, the decrease in efficiency for BPSK detection is given by

$$\epsilon_B = \frac{\gamma_1 - \gamma_0}{\gamma_1} = \sigma^2[1 + h\gamma_1\sigma^2] = \sigma^2[1 + (\gamma_1 + 0.5)\sigma^2]. \quad (25)$$

At the high SNR values of α required for the phase reference to provide efficient detection, the density function of ϕ is approximately Gaussian with zero mean and a variance of $\sigma^2 = 1/\alpha$. Consequently, the detection loss for coherent BPSK communications is given by

$$L_B \text{ (dB)} \approx 4.34\epsilon\left(1 + \frac{\epsilon}{2}\right) \approx \frac{4.34}{\alpha}\left(1 + \frac{\gamma_1 + 1}{\alpha}\right). \quad (26)$$

For coherent QPSK communications, the operating point γ is different for the two bits of each quaternary symbol as a result of cross coupling of opposite polarities.

$$\gamma = \gamma_1(\cos \phi \pm \sin \phi)^2 = \gamma_1(1 \pm 2 \sin \phi \cos \phi) = \gamma_1(1 \pm \sin 2\phi). \quad (27)$$

Thus,

$$\gamma - \gamma_0 = (\gamma_1 - \gamma_0) \pm \gamma_1 \sin 2\phi \approx (\gamma_1 - \gamma_0) \pm 2\gamma_1\phi \quad (28)$$

and the conditional probability of bit error for a given ϕ is

$$p_{CQ}(\phi) = \tfrac{1}{2}[p_+(\phi) + p_-(\phi)] = p_0 \exp(-h(\gamma_1 - \gamma_0)) \times \frac{\exp(+2h\gamma_1\phi) + \exp(-2h\gamma_1\phi)}{2} = p_0 \exp(-h(\gamma_1 - \gamma_0)) \cosh(2h\gamma_1\phi). \quad (29)$$

With ϕ assumed to be a Gaussian random variable, the average error rate in detection is

$$p_{CQ} = \mathcal{E}[p_{CQ}(\phi)] = p_0 \exp(-h(\gamma_1 - \gamma_0))\mathcal{E}[\cosh 2h\gamma_1\phi] = p_0 \exp(-h(\gamma_1 - \gamma_0))\exp(2h^2\gamma_1^2\sigma^2). \quad (30)$$

When $\gamma_1 = E_b/N_0$ is adjusted to yield an average error probability of p_0, it is seen that

$$\epsilon_Q = \frac{\gamma_1 - \gamma_0}{\gamma_1} = \frac{1 + 2\gamma_1}{\alpha}. \quad (31)$$

Therefore, the detection loss for coherent QPSK communications of the conventional type is approximated by

$$L_{CQ} \text{ (dB)} \approx 4.34\epsilon\left(1 + \frac{\epsilon}{2}\right) \approx \frac{4.34}{\alpha}(1 + 2\gamma_1)\left[1 + \frac{0.5}{\alpha}(1 + 2\gamma_1)\right]. \quad (32)$$

For coherent detection of offset QPSK signals, the probability of bit error was shown to equal the mean of the error rates for the BPSK and conventional QPSK cases. Thus,

$$p_{OQ} = \frac{p_B + p_{CQ}}{2} \approx p_0 \exp(-h(\gamma_1 - \gamma_0)) \cdot \left[\frac{(1 - 2h\gamma_1\sigma^2)^{-0.5} + \exp(+2h^2\gamma_1^2\sigma^2)}{2}\right]. \quad (33)$$

The two quantities within the brackets may be expanded and then added. In the low-loss situation of interest, only the leading terms need to be considered. Thus,

$$p_{OQ} \approx p_0 \exp(-h(\gamma_1 - \gamma_0))[1 + \gamma_1 h\sigma^2(0.5 + \gamma_1 h)] \approx p_0 \exp(-h(\gamma_1 - \gamma_0)) \exp(\gamma_1 h\sigma^2(0.5 + \gamma_1 h)). \quad (34)$$

With γ_1 set to yield an average detection performance of p_0, the small drop in detection efficiency caused by the use of a noisy phase reference with high SNR α is given by

$$\epsilon_{OQ} = \frac{\gamma_1 - \gamma_0}{\gamma_1} \approx \sigma^2(1 + \gamma_1) \approx \frac{1 + \gamma_1}{\alpha}. \quad (35)$$

It follows that the performance loss for coherent detection of offset-QPSK signals may be approximated as

$$L_{OQ} \text{ (dB)} \approx 4.34\epsilon\left(1 + \frac{\epsilon}{2}\right) \approx \frac{4.34}{\alpha}(1 + \gamma_1)\left[1 + \frac{0.5}{\alpha}(1 + \gamma_1)\right]. \quad (36)$$

This loss for offset-QPSK detection is considerably greater than for the BPSK case, but only about one half the degradation associated with coherent detection of conventional QPSK transmissions.

III. SUMMARY

Imperfect carrier synchronization causes a performance loss for coherent PSK communications. For BPSK signaling, a carrier-synchronization error of ϕ rad results in a correlation loss of $\cos\phi$. A QPSK signal may be considered as two BPSK channels that are in phase quadrature. If perfect carrier synchronization exists ($\phi = 0$), then the two binary signals may be detected orthogonally. When $\phi = 0$, however, the orthogonality is lost because of cross coupling between the binary channels that is proportional to $\sin\phi$. Because of the additonal performance degradation that is related to the cross-coupling interference, QPSK losses from imperfect carrier synchronization exceed those for BPSK. Whereas the bit alignment for the two channels is coincident for conventional QPSK signaling, the alignments are staggered for the offset form of QPSK. The staggered or offset alignment allows the integral of the cross-coupling interference to vanish over any decoded bit interval during which the other channel experiences a bit transition. For a transition probability of one half, the detection performance for offset QPSK is thus equal to the average of the performances for BPSK and conventional QPSK.

Figs. 3, 4, and 5 are graphs of the detection performance for BPSK, conventional QPSK, and offset QPSK, respectively, as obtained by numerical computation. The probability of error per bit decision is plotted parametrically in $A = 10 \log \alpha$, where α is the SNR of the noisy phase reference supplied by carrier synchronization. The density of the phase error in synchronization is taken as the steady-state tracking performance of a simple first-order PLL when the reference is derived from any unmodulated or residual carrier component. Perfect frequency synchronization is assumed. Also, the results are based upon ideal bit timing for the I&D circuits, which serve as matched filters for the rectangular waveforms of the binary modulation sequences. In the computation of probability of bit error, the carrier-tracking error ϕ is treated as a constant over each bit interval, which is virtually true for synchronizer bandwidths that do not exceed one tenth of the digital signaling rate.

As seen from Figs. 3 and 4, the required SNR of the carrier phase reference to prevent significant losses in detection efficiency is much greater for quaternary signaling than for the binary case. In Fig. 5 the detection curves for offset QPSK indicate some reduction in the required SNR α of the phase reference for the quaternary case. In most communications systems it is desired to limit the detection loss to small values, such as 0.1 dB. The computed detection curves have poor resolution for the determination of small detection losses. However, theoretical results have been obtained that yield approximate expressions for the detection performances which are valid for the low-loss case. These analytical expressions indicate that for a given loss specification, the use of the offset bit alignment for QPSK signaling allows almost a 3-dB relaxation in the requirement on the SNR of the phase reference. The derived approximations for the loss in detection efficiency are

$$L\ (\mathrm{dB}) \approx \begin{cases} \dfrac{4.34}{\alpha}\left(1+\dfrac{1+\gamma}{\alpha}\right), & \text{BPSK} \\[2ex] \dfrac{4.34}{\alpha}(1+2\gamma)\left(1+\dfrac{1+2\gamma}{2\alpha}\right), & \text{conventional QPSK} \\[2ex] \dfrac{4.34}{\alpha}(1+\gamma)\left(1+\dfrac{1+\gamma}{2\alpha}\right), & \text{offset QPSK} \end{cases} \tag{37}$$

where α is the SNR of the phase reference and γ is the value of E_b/N_0 at the receiver. In these three equations, only the loss L is expressed in decibels.

In general, the detection curves of Figs. 3, 4, and 5 are quite accurate for other than the low-loss case, while the derived analytical expressions are accurate only when the detection loss is small. Therefore, an accurate measure of detection performance can be obtained from the detection curves and loss expressions combined. In Fig. 6, the loss in detection efficiency is compared for the three forms of PSK signaling as a function of the SNR of the phase reference. The losses are plotted for two operating points in detection, probabilities of bit error of $p = 10^{-3}$ and $p = 10^{-5}$. These curves illustrate the increase in detection loss for quaternary signaling relative to the BPSK case. Although the loss for offset QPSK is considerably larger than that for BPSK, there is a significant decrease in detection loss relative to the conventional form of QPSK signaling. When the SNR of the phase reference is sufficiently high to limit the loss to a few tenths of a decibel, the detection loss in decibels for offset QPSK is only slightly greater than half of the corresponding loss for conventional QPSK.

Halving a small loss in detection efficiency is not of great importance. The significant feature of the detection properties of offset QPSK is that it allows the same detection efficiency to be achieved with a smaller value of SNR of the phase reference than that required for conventional QPSK. Although the reduction in the SNR requirements of $A = 10 \log \alpha$ can be determined from the graphs of Fig. 6, the effect is dramatized in Fig. 7 by plotting the required A for the different forms of PSK signaling versus the detection operating point p. The results are portrayed for three specified values of detection loss: 0.05, 0.10, and 0.20 dB. Whereas the SNR requirement A for the phase reference is fairly insensitive to the operating point in detection for BPSK communications, the required values of A to maintain a fixed detection loss increase for QPSK as the operating point is shifted to a lower probability of bit error. It is evident from the curves of Fig. 7 that the use of offset QPSK in lieu of conventional quadriphase modulation reduces by almost 3 dB the SNR requirement

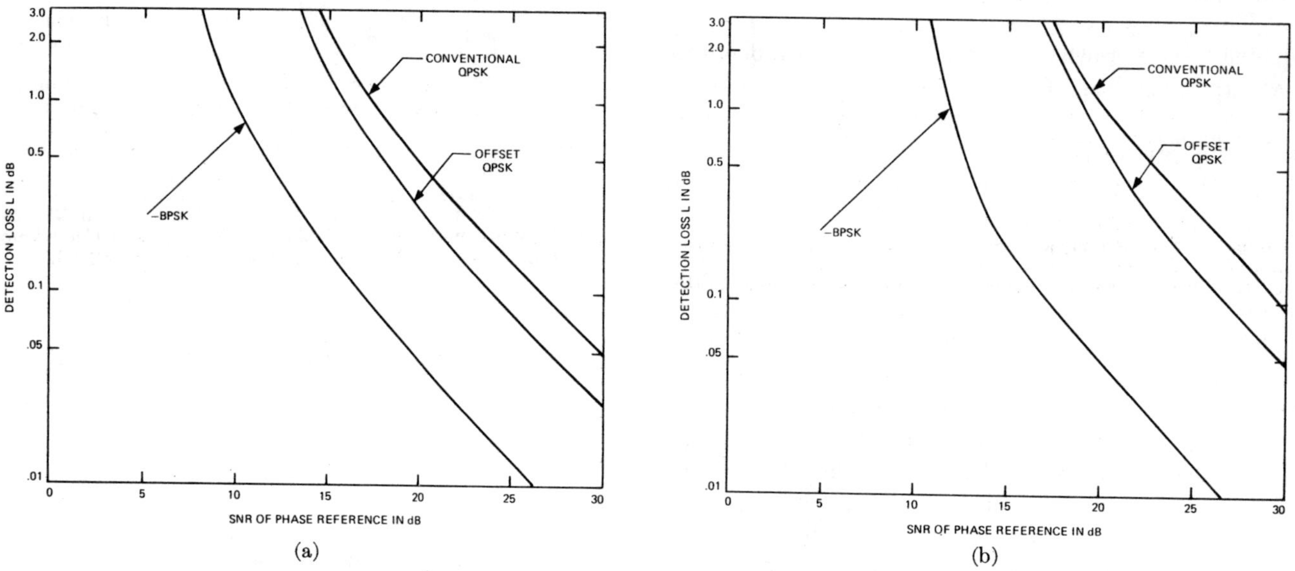

Fig. 6. Detection losses as a function of SNR of phase reference (a) Evaluated at $P_0 = 10^{-3}$. (b) Evaluated at $P_0 = 10^{-5}$.

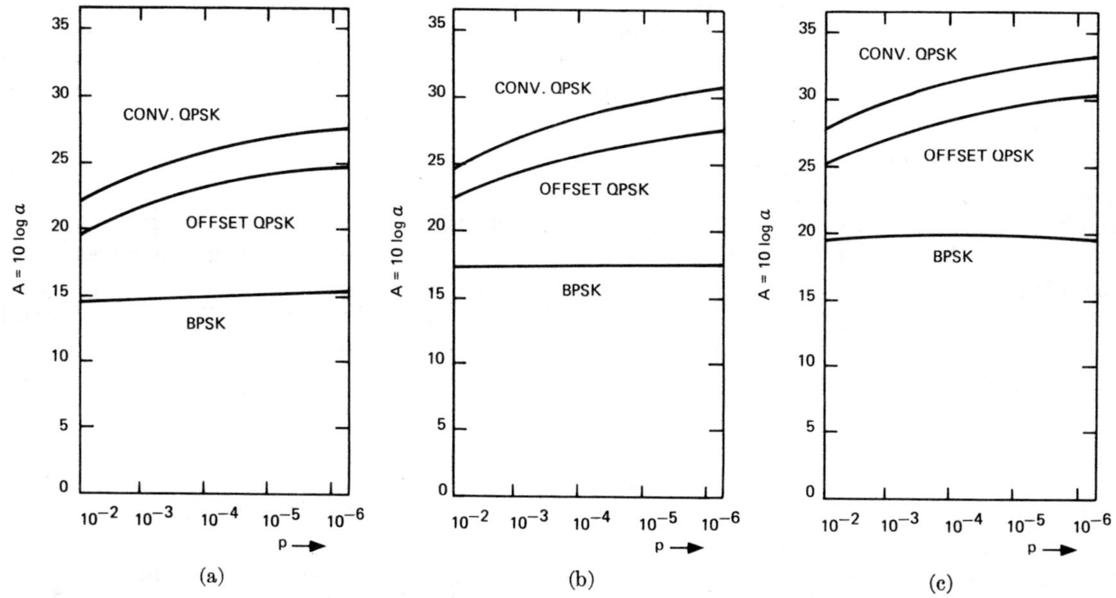

Fig. 7. Required SNR of phase reference as a function of probability of bit error when detection loss is specified. (a) $L = 0.20$ dB. (b) $L = 0.10$ dB. (c) $L = 0.05$ dB.

A on carrier phase synchronization for providing a specified detection efficiency. Because of short-term frequency instabilities in communications systems, the tracking bandwidth of the carrier synchronizer cannot be lowered indefinitely in order to obtain higher values of A; in fact, it is difficult in some systems to obtain values of A that are sufficiently high to prevent large losses in detection efficiency. Hence, the improved immunity to noisy carrier synchronization that results from the staggered bit alignment of offset QPSK can be valuable in achieving efficient quadriphase communications.

REFERENCES

[1] W. C. Lindsey, "Phase-shift-keyed signal detection with noisy reference signals," *IEEE Trans. Aerosp. Electron. Syst.*, vol. AES-2, pp. 393–401, July 1966.

[2] A. J. Viterbi, *Principles of Coherent Communication.* New York: McGraw-Hill, 1966, pp. 198–204.

[3] Final Rep. on "Conceptual design of a TDMA system," prepared under Contract DAAB07-69-0392 for U.S. Army Satellite Commun. Agency, Ft. Monmouth, N. J., by Communications Satellite Corp., Dec. 1, 1970, Appendixes 4-2 and 4-3.

[4] R. K. Kwan, "The effects of filtering and limiting a double-binary PSK signal," *IEEE Trans. Aerosp. Electron. Syst.*, vol. AES-5, pp. 589–594, July 1969.

[5] C. J. Wolejsza, A. M. Walker, and A. M. Werth, "PSK modems for satellite communications," in *Proc. 1st Intelsat Int. Conf. Digital Satellite Commun.*, London, England, Nov. 1969.

[6] S. A. Rhodes, "Effects of hardlimiting on bandlimited transmissions with conventional and offset QPSK modulation," in *1st IEEE Nat. Conf. Telecommun., Houston, Tex.*, Dec. 1972.

[7] W. R. Bennett and J. R. Davey, *Data Transmission.* New York: McGraw-Hill, 1965, p. 219.

[8] J. J. Stiffler, *Theory of Synchronous Communications.* Englewood Cliffs, N. J.: Prentice-Hall, 1971, p. 106.

[9] N. Brienza, "Phase noise theory-specification," Defense Commun. Rep. R-241303-2-3, prepared under Contract DCA100-70-C-0009 by Computer Sciences Corp., Falls Church, Va., Aug. 1970.

[10] L. C. Palmer and S. A. Rhodes, "Computer simulation of a digital satellite communications system utilizing TDMA and coherent quadriphase signaling," in *Proc. 1972 IEEE Int. Conf. Commun.*, Philadelphia, Pa.
[11] N. Brienza and P. Trafton, "Phase noise specification for digital satellite communications," in *Proc. 1972 IEEE Int. Conf. Commun.*, Philadelphia, Pa.
[12] S. Stein and J. Jones, *Modern Communications Principles.* New York: McGraw-Hill, 1967, p. 248.
[13] A. J. Viterbi, "Phase-locked loop dynamics in the presence of noise by Fokker-Planck techniques," *Proc. IEEE*, vol. 51, pp. 1737–1753, Dec. 1963.
[14] C. R. Wylie, *Advanced Engineering Mathematics.* New York: McGraw-Hill, 1969, pp. 357–358.
[15] R. L. Didday and W. C. Lindsey, "Subcarrier tracking methods and communication system design," *IEEE Trans. Commun. Technol.*, vol. COM-16, pp. 541–550, Aug. 1968.
[16] R. J. Sherman, "Quadriphase-shift-keyed signal detection with noisy reference signal," in *1969 EASCON Rec.*, pp. 45–52.
[17] W. C. Lindsey and M. K. Simon, "Carrier synchronization and detection of polyphase signals," *IEEE Trans. Commun.*, Part I, pp. 441–454, June 1972.
[18] R. W. Lucky, J. Salz, and E. J. Weldon, *Principles of Data Communication.* New York: McGraw-Hill, 1968, pp. 247–248.
[19] J. N. Wozencraft and I. M. Jacobs, *Principles of Communication Engineering.* New York: Wiley, 1967, pp. 82–83.

Co-Channel Interference of Spread Spectrum Systems in a Multiple User Environment

SAMUEL A. MUSA, SENIOR MEMBER, IEEE, AND WASYL WASYLKIWSKYJ, SENIOR MEMBER, IEEE

Abstract—This paper considers the mutual interference problem of several users employing the same spread spectrum technique in selected multiple user environments. The spread spectrum techniques consist of pseudo noise (PN), time division multiple access/PN, synchronous and asynchronous frequency hopping (FH). The environment consists of a desired transmitter-receiver pair located in an area where there are M interfering users distributed in accordance with a specified probability density function. Both coherent phase-shift-keyed and noncoherent frequency-shift-keyed modulations are considered. The general relationship between the probability of bit error of PN and FH systems is derived which is independent of the signal modulation and distribution of users. The degradation of the communication system performance (average probability of bit error) of the desired link as a function of the total number of interfering users within the considered area is investigated. The analysis shows that the mutual interference problem is less severe with users employing synchronous FH than with the other spread spectrum techniques. The comparison between asynchronous FH and PN is highly dependent on the relative location of interferers to the desired link and the time duty factor of the hopping.

I. INTRODUCTION

THE problem of maintaining radio communication in the presence of interfering signals originating from nearby transmitters competing for the same range of the RF spectrum is gaining in importance both in military and civilian applications. Aside from the obvious case of military tactical ground-based communication networks, examples in the civilian sector include communications in the urban environment, receiving stations serviced by a common satellite transponder as well as high density communication nets for air traffic control.

While a detailed analysis of degradation in communication system performance requires consideration of a host of factors endemic to the particular application, there are several questions of principle transcending the particular system implementation and which at the same time appear to be susceptible to relatively straightforward analytical formulations. For example, taking the overall bandwidth that must be shared among M users situated within a given perimeter as one of the constraints, differences in fidelity of information transmission attainable by any one member of the communication links can be investigated in general terms for a variety of so called spread spectrum modulation techniques: pseudorandom noise (PN), asynchronous and synchronous "frequency hopping" (FH) as well as time division multiple access PN (TDMA/PN). Of course, as in any analytical study (as opposed to a direct numerical system simulation), certain idealizations in respect to signal statistics and propagation modeling are unavoidable if tractable results are to be obtained.

The idealized model investigated in this paper is the following. A receiver is located at the center of a planar circular region within which there are $M+1$ transmitters, only one of which is in communication with the receiver. The degree of the resulting RF interference by the other M users (transmitter-receiver pairs) of the RF band will depend on a) the RF power radiated by each interferer, b) the number of interferers (M) within the communications perimeter, c) the distribution of the interferers within the circular region in question, d) the law of propagation of the electromagnetic waves within the perimeter, and finally, e) the type of spread spectrum modulation employed by each of the users of the RF band in question.

The co-channel interference problem for coherent reception of phase shift keying (CPSK) has been investigated in Refs. 1, 2, and 3, without considering the geometrical distribution of the interferers, whereas in this paper the effect of the distribution of the interferers is of primary concern. The main objective of this paper is a quantitative comparison of the effectiveness of the various spread spectrum techniques in combating mutual interference. Consequently, the effects of alternate propagation models are not investigated. For definiteness, $1/r^4$ (power) propagation law is assumed which corresponds, approximately, to ground wave propagation at VHF (Ref. 4). Therefore, the numerical results obtained apply specifically to this case, although the technique can be extended to other propagation models as well. Perhaps of greater importance than the propagation model per se is the distribution of interferers within the circular communications perimeter. Since the intent here is to obtain results applicable to as wide a range of geometrical arrangements of interferers as possible, the distribution of the interferers is treated statistically. In this connection an assumption had to be made with regard to the temporal variation of the interferer location: it is assumed that the interferer locations do not change within the period of a typical message length. While this constraint may render the results inapplicable to certain cases involving sufficiently rapid changes in the interferer scenario (e.g., when the interfering signals originate from moving aircraft) the relative comparison of performance attained by using different spread spectrum techniques (e.g., PN vs. FH) would still apply in a qualitative "average" sense.

The analytical development leading to expressions for bit error probability for PN, (slow) FH and TDMA/PN is presented in Section II. The special case of interferers distributed in

Paper approved by the Editor for Communication Theory of the IEEE Communications Society for publication after presentation at the International Union of Radio Science Symposium, Palo Alto, CA, June 1977. Manuscript received November 8, 1977; revised June 2, 1978.

The authors are with the Institute for Defense Analyses, Arlington, VA 22202.

Reprinted from *IEEE Trans. Comm.*, vol. COM-26, pp. 1405-1412, Oct. 1978.

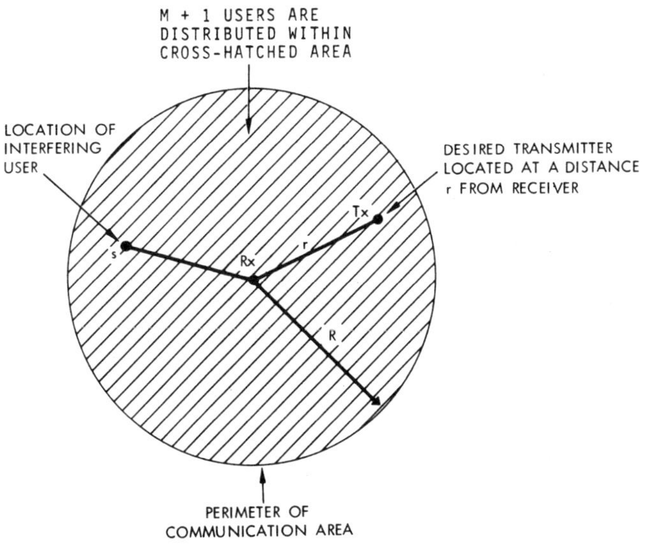

Figure 1. Geometric Model for User Distribution

accordance with the "bell" shaped distribution is analyzed for users employing CPSK in background noise, and CPSK and noncoherent FSK channel with Gaussian interferers. Numerical results showing the sensitivity of the probability of bit error of the spread spectrum systems to the number of users and background noise are given in Section III.

II. MATHEMATICAL FORMULATION

A. General Relationships for Bit Error Probability of PN and FH Systems

Consider M interferers located within the circular region shown in Fig. 1. The interferers are distributed with equal probability along the circumference of any circle of radius s, and the probability that an interferer is located within the annulus with radius s_l and $s_l + ds_l$ is $p(s_l)ds_l$. For a typical interferer located at distance s_l from the receiver, the ratio of interference power to desired power at the receiver is given by:

$$x_l = \frac{(P_I G_I G_R)\dfrac{h_I^2 h_R^2}{s_l^4}}{(P_T G_T G_R)\dfrac{h_T^2 h_R^2 G}{r^4}}, \qquad 1 \leq l \leq M \tag{1}$$

where P_I, G_I, h_I are the transmitted power, antenna gain and antenna height of the interfering user respectively. P_T, G_T, h_T are the transmitted power, antenna gain and antenna height of the desired transmitter, respectively. G_R is the receiver antenna gain and h_R is the receiver antenna height. G is the processing gain between transmitter and receiver (PN system), and for FH system $G = 1$. The effect of the ground results in the received power being inversely proportional to the fourth power of the distance for VHF ground-to-ground communication instead of the second power as expected in free space propagation (Ref. 4). For the sake of simplicity, assume that the desired transmitter as well as the interferers have equal transmitted powers, antenna gains, and antenna heights. With this assumption, Eq. 1 reduces to

$$x_l = \frac{r^4}{G s_l^4}. \tag{2}$$

Denoting the probability density of the interference to signal ratio x_l at the receiver by $f(x_l)$, the probability density governing the location of an interferer along the radial distance s_l is:

$$p(s_l) = \begin{cases} \dfrac{4r^4}{G s_l^5} f\left(\dfrac{r^4}{G s_l^4}\right); & s_l > 0, \\ 0; & s_l < 0, \qquad 1 \leq l \leq M. \end{cases} \tag{3}$$

Thus, $f(x_l)$ is determined once the probability density governing the location of an interferer is specified. If one assumes that the location of transmitters giving rise to the interference signals are mutually independent, the interference signals arriving at the receiver are likewise statistically independent. Hence the joint density function for M interference to signal ratios $x_1, x_2, \cdots, x_M$ is simply

$$F_G(x_M) \equiv F_G(x_1, x_2, \cdots, x_M) = \prod_{l=1}^{M} f_G(x_l), \tag{4}$$

where the subscript G is employed to emphasize the fact that there is an implicit dependence on processing gain G. Suppose each of the $M+1$ transmitter receiver pairs employs its own PN code with processing gain G. Since each of the interferers is incoherent with respect to the desired signal transmission, the probability of bit error for a specified modulation format is a function of the normalized intensities $x_1, x_2, \cdots, x_M$. We denote this probability of bit error by

$$P(x_1, x_2, x_3, \cdots, x_M). \tag{5}$$

For a specified spatial distribution of interferers this probability of bit error is a fixed number that can be computed provided, of course, that the law of propagation of electromagnetic energy is also specified. In our case, we assume an $(r)^{-4}$ dependence. In principle, one could perform such a calculation for each distribution. We prefer to treat the distribution of interferers as a statistical ensemble, in which case $P(x_1, x_2, \cdots, x_M)$ is a random variable.

One reasonable measure of systems performance would then be the probability that the bit error falls within specified limits. Instead of finding the complete statistical description of P we only consider its mean. Such a more limited description appears adequate since our principle goal is to obtain relative measures of performance for various cases of PN and FH operation. Accordingly, we define the average bit error rate by the statistical average $\langle P \rangle$, taken over all possible interferer locations consistent with (4). For a PN system it will be convenient to denote $\langle P \rangle$ by $P_M(G)$, highlighting the fact that this average depends on the processing gain G and the number of interferers. The average probability of bit error is therefore

$$P_M(G) = \int_0^\infty \cdots \int_0^\infty P(x_1, x_2, \cdots, x_M) \cdot \prod_{l=1}^{M} f_G(x_l)\, dx_1 \cdots dx_M. \tag{6}$$

An alternative interpretation of $P(x_1, x_2, \cdots, x_M)$ is as a conditional probability of bit error (conditional on the intensities $x_1, x_2, \cdots, x_M$) which is then averaged over $x_1, x_2, \cdots, x_M$ to obtain the unconditional probability of error $P_M(G)$. If instead of PN the $M+1$ users employ asynchronous frequency hopping over N channels with a duty factor b, then the probability density $q(x_l)$ of x_l for a typical interferer signal at the receiver is

$$q(x_l) = \left(1 - \frac{b}{N}\right)\delta(x_l) + \frac{b}{N} f_1(x_l), \qquad 1 \leqslant l \leqslant M, \tag{7}$$

where $\delta(x_l)$ is the Dirac Delta function and $f_1(x_l)$, $1 \leqslant l \leqslant M$, is the conditional probability density of x_l given that the channel is selected during the appropriate time to interfere with the desired transmission. This conditional probability density is, of course, the same as the $f_G(x_l)$, $1 \leqslant l \leqslant M$, in (4) with the processing gain G set equal to unity. The average probability of bit error Q_M for asynchronous FH is then given by:

$$Q_M = \int_0^\infty \cdots \int_0^\infty P(x_1, x_2, \cdots, x_M) \cdot \prod_{l=1}^{M} \left[\left(1 - \frac{b}{N}\right)\delta(x_l) + \frac{b}{N} f_1(x_l)\right] dx_1 \cdots dx_M. \tag{8}$$

If the 2^M sums that result after the M-fold multiplication of the binominal in Eq. 8 are written out explicitly, one obtains

$$Q_M = \int_0^\infty \cdots \int_0^\infty P(x_1, x_2, \cdots, x_M) \cdot \sum_{l=0}^{M} \left[\left(1 - \frac{b}{N}\right)^{M-l} \left(\frac{b}{N}\right)^l \underbrace{\sum_{k_1} \cdots \sum_{k_M}}_{\text{all combinations}} f(x_{k_1}) \cdot f(x_{k_2}) \cdots f(x_{k_l}) \delta(x_{k_{l+1}}) \cdot \delta(x_{k_{l+2}}) \cdots \delta(x_{k_M})\right] dx_1 dx_2 \cdots dx_M. \tag{9}$$

The functional dependence of the conditional bit error probability P on the individual interferer to signal ratio x_l, $1 \leqslant l \leqslant M$, depends on the particular modulation employed. A not too restrictive assumption is that the interferer powers are additive, in which case $P(x_1, x_2, \cdots, x_M)$ remains unchanged if the x_l ($l = 1, 2, \cdots, M$) are permuted. Equation 9 can then be reduced to a series of only $M + 1$ terms. This simplification results if one notes that for each l there are $\binom{M}{l}$ terms ($l = 0, 1, \cdots, M$). However, from the assumed invariance of P under the permutation of the subscripts of the x_k, the integration over the $\binom{M}{l}$ delta function products yields $\binom{M}{l}$ identical terms. The final result can therefore be written in the following form:

$$Q_M = \sum_{l=0}^{M} \left(1 - \frac{b}{N}\right)^{M-l} \left(\frac{b}{N}\right)^l \binom{M}{l} P_l(1) \tag{10a}$$

where

$$P_l(1) = \int_0^\infty \cdots \int_0^\infty P(x_1, x_2, \cdots, x_l) \cdot \prod_{j=1}^{l} f_1(x_j)\, dx_1 dx_2 \cdots dx_l \qquad \text{for } l \neq 0 \tag{10b}$$

and

$$P_0(1) = P(0). \tag{10c}$$

Equation 10 lends itself to an interesting interpretation. The quantity $P_l(1)$ entering into the sum and defined by (10b) is, in accordance with the notation used in (6), just the average probability of bit error for PN operation with l interferers and unity processing gain. For the special case $l = 0$ (no interferers) $P_0(1)$ in (10c) is the intrinsic bit error rate as determined from the ambient signal to noise considerations. Equation 10a immediately yields two general bounds on the average bit error probability for asynchronous FH. First, since

$$P_l(1) \leqslant \frac{1}{2},$$

one finds, upon summing the terms in (10a) for $l > 0$ and a constant $P_l(1)$, that

$$P_1(1)\left[1 - \left(1 - \frac{b}{N}\right)^M\right] \leqslant Q_M - P_0(1)\left(1 - \frac{b}{N}\right)^M \leqslant \frac{1}{2}\left[1 - \left(1 - \frac{b}{N}\right)^M\right]. \tag{11}$$

Also, by considering the single term $l = M$ of the sum in (10a) one readily obtains

$$Q_M \geqslant \left(\frac{b}{N}\right)^M P_M(1). \tag{12}$$

In the next section, the average probability of bit error is calculated for PN and asynchronous FH in an environment with M interfering users distributed according to a bell-shaped probability density function.

B. Interference Distributed in Accordance with a Bell-Shaped Probability Density Function

The preceding analytical formulation will now be applied to the special case of interferers distributed within a ring-shaped region surrounding the given receiver. Direct numerical computations show that for unimodal probability density functions (i.e., with a single peak at $s = s_o$) the exact functional form of $p(s)$ is not very important. It is, therefore, reasonable to choose for $p(s)$ a unimodal function that yields analytically tractable results. Motivated by these considerations, we chose

$$p(s) = \begin{cases} \dfrac{s_0{}^4}{s^5} \exp\left[-\dfrac{5}{4}\left(\dfrac{s_0}{s}\right)^4\right]; & s > 0 \\ 0; & s < 0. \end{cases} \tag{13}$$

This density function is plotted in Fig. 2, and has the desirable characteristic of very rapid decay for small s, in agreement with the obvious physical requirement that finding an interferer collocated with the given receiver should be extremely unlikely. The most likely location of interferers is the neighborhood of the ring of radius s_o. Using (3), one finds that the interference to signal ratio is governed by the probability density function

$$f(x_l) = \begin{cases} \alpha e^{-\alpha x_l}; & x_l > 0, \\ 0; & x_l < 0, \end{cases} \tag{14}$$

with x_l given by (2) and the parameter α by

$$\alpha = \frac{5}{4}\left(\frac{s_0}{r}\right)^4 G. \tag{15}$$

In order to calculate the average probability of bit error for a coherent BPSK channel with M BPSK interfering signals, the probability of bit error P needs to be calculated. It may be shown that

$$P = \frac{1}{2\pi}\int_{-\infty}^{-1}\int_{-\infty}^{\infty} e^{-i\omega z_M - (\gamma\omega^2)/2} \prod_{l=1}^{M} J_0(\omega a_l)\, d\omega dz_M$$

where $\gamma = N_o/A^2T$ is the signal-to-noise ratio at the receiver. Eqs. 6 and 13 with $x_j = a_j^2$ are used in conjunction with P to yield:

$$P_M(G) = \int_0^{\infty} \cdots \int_0^{\infty} \prod_{j=1}^{M} 2a_j f(a_j{}^2)\, da_j P(a_1{}^2, a_2{}^2, \cdots, a_M{}^2)$$

$$= \int_{-\infty}^{-1} dz_M \frac{1}{2\pi}\int_{-\infty}^{\infty} d\omega e^{-i\omega z_M - (\gamma\omega^2)/2}$$

$$\cdot \prod_{j=1}^{M} \int_0^{\infty} 2a_j \alpha e^{-\alpha a_j{}^2} J_0(\omega a_j)\, da_j. \tag{16}$$

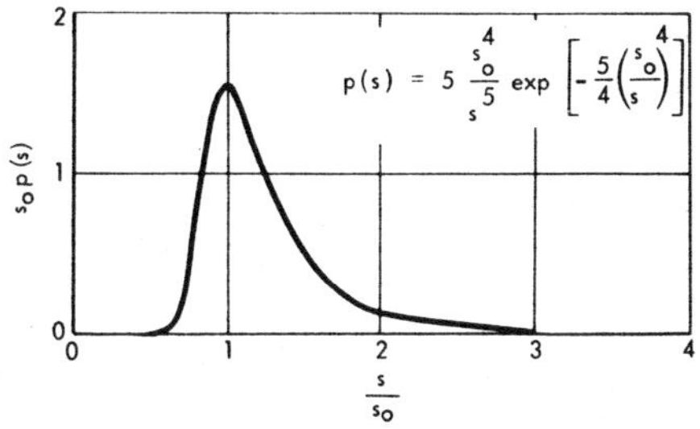

Figure 2. Probability Density of an Interferer

Employing the well known formulas (Ref. 5)

$$\int_0^{\infty} a_j e^{-\alpha a_j{}^2} J_0(a_j\omega)\, da_j = \frac{1}{2\alpha} e^{-\frac{\omega^2}{4\alpha}} \tag{17}$$

$$\frac{1}{2\pi}\int_{-\infty}^{\infty} e^{-i\omega z_M - (k\omega^2)/\alpha}\, d\omega = \sqrt{\frac{\alpha}{4\pi k}} \exp\left[-\frac{\alpha z_M{}^2}{4k}\right]; \quad k \neq 0 \tag{18}$$

equation 16 reduces to

$$P_M(G) = \frac{1}{2}\,\text{erfc}\left[\frac{M}{\alpha} + 2\gamma\right]^{-1/2}, \tag{19}$$

where the complementary error function erfc (y) is defined by $(2/\sqrt{\pi}) \int_y^{\infty} e^{-t^2} dt$. Equation 19 represents the average probability of bit error for M CPSK interfering signals in a PN system with processing gain G. Since $M/\alpha = (4Mr^4)/(5s_o{}^4G)$, it is evident that the average probability of bit error increases with increasing M/G.

Gaussian Interferers

Although the main concern here is interferers having the same modulation format as the desired signal, for purposes of comparison, it is also of interest to compute the bit error probability for the case in which the M interferers are Gaussian random processes. In this case, the bit error probability is a function of the sum of the powers of the interferers. We denote this fact symbolically by writing

$$P(x_1, x_2, \cdots, x_M) \equiv P\left(\sum_{l=1}^{M} x_l\right). \tag{20}$$

With this simplification together with Eq. 14, one can write Eq. 6 in the following form:

$$P_M(G) = \alpha^M \int_0^{\infty} \cdots \int_0^{\infty} P\left(\sum_{l=1}^{M} x_l\right) e^{-\alpha \sum_{l=1}^{M} x_l} dx_1 dx_2 \cdots dx_M$$

$$= \frac{\alpha^M}{(M-1)!}\int_0^{\infty} P(x) x^{M-1} e^{-\alpha x}\, dx. \tag{21}$$

The last expression follows from a change of variables of integration to $\xi_j = \Sigma_{l=1}^{j} x_l$; $1 \leqslant j \leqslant M-1$ and $\xi_M = x$, and then performing the $M-1$ fold integration with respect to $\xi_1, \xi_2, \cdots, \xi_{M-1}$. For incoherent FSK and Gaussian noise the probability of bit error is $(1/2)\exp[-S/2I]$ (Ref. 6), where S is the signal and I is the variance of the Gaussian noise. In our case

$$\frac{I}{S} = x = \sum_{l=1}^{M} x_l \qquad (22)$$

so that in Eq. 21, $P(x) = (1/2)\exp[-1/(2x)]$. Substituting this into Eq. 21 and changing variables of integration to y with $y = \ln x$, one obtains

$$P_M(G) = \frac{1}{2}\frac{\alpha^M}{(M-1)!}\int_{-\infty}^{\infty} e^{-\sqrt{2\alpha}\cosh(y+z)+My}\,dy \qquad (23)$$

where $\tanh z = (\alpha - 1/2)/(\alpha + 1/2)$.

Since $K_M(\sqrt{2\alpha}) = \int_0^\infty e^{-\sqrt{2\alpha}\cosh t}\cosh(Mt)\,dt$ (Ref. 5), where K_M is the modified Bessel function of the second kind (Kelvin function of order M), (23) with $y + z = t$ and $z = \ln\sqrt{2\alpha}$ reduces to

$$P_M(G) = \frac{(\sqrt{2\alpha})^M K_M(\sqrt{2\alpha})}{2^M(M-1)!}. \qquad (24)$$

Equation 24 represents the probability of bit error for a PN system with M interferers. For $\sqrt{2\alpha} \gg M$, an approximation to Eq. 24 is

$$P_M(G) \sim \sqrt{\frac{\pi}{2\sqrt{2\alpha}}}\, e^{-\sqrt{2\alpha}}\, \frac{(\sqrt{2\alpha})^M}{2^M(M-1)!}. \qquad (25)$$

In the case of one interferer ($M = 1$), Eq. 25 reduces to

$$P_1(G) \sim \frac{1}{2}\sqrt{\frac{\pi}{2}}\, e^{-\sqrt{2\alpha}}(2\alpha)^{1/4}.$$

Since α is proportional to G, increasing the processing gain yields exponential improvement in the probability of bit error for a PN system. From (10a), the probability of bit error for an FH system with a single interferer is given by $Q_1 = (b/N)P_1(1)$. Thus, increasing the number of channels in an FH system yields linear improvement in the probability of bit error.

Similarly, the probability of bit error for PSK channel with M Gaussian interferers may be calculated to give:

$$P_M(G) = \frac{\alpha^M}{2(M-1)!}\int_0^\infty x^{M-1}\,\mathrm{erfc}\left(\frac{1}{\sqrt{2x}}\right)e^{-\alpha x}\,dx$$

$$= \frac{(-1)^{M-1}\alpha^M}{2(M-1)!}\frac{d^M}{d\alpha^M}\left(\frac{1}{\alpha}e^{-\sqrt{2\alpha}}\right). \qquad (26)$$

III. NUMERICAL RESULTS

A comparison of the performance of PN and asynchronous FH systems with M BPSK interferers is illustrated in Fig. 3 for $N = 100$ frequencies, $G = 20$ dB, $b = 1$, $(N_o/S) = 0$ and various values of s_o/r. When the interferers are far removed from the receiver (e.g., $s_o/r = \sqrt{2}$), PN is superior to asynchronous FH. In particular, for a 10^{-4} probability of bit error seven times as many users can be tolerated with PN as with asynchronous FH. When the distance of interferers to receiver is comparable to that of the desired receiver, there is a crossover point corresponding to about 20 interferers, where the performance of PN becomes better than that of asynchronous FH. For close-in interferers (e.g., $\sqrt{2}s_o/r = 0.5$) the performance of asynchronous FH is superior to that of PN. The latter result is intuitively obvious, since as the total interference power increases, there is always a finite probability of selecting a slot with few interferers, so that $P_{PN} > P_{FH}$.

When the interferers are far removed from the receiver and there is background noise, the relationship between the average probability of bit error and the number of interferers is plotted in Fig. 4. As the level of noise power increases, the difference between PN and FH becomes insignificant.

Another interesting case is shown in Fig. 5, where the number of interferers is fixed and the relationship between the average probability of bit error and the signal-to-background noise ratio is examined. It is clear from Fig. (5) that the interference power is dominant when the S/N_o is greater than 30 dB, while for $S/N_o < 30$ dB the receiver noise becomes significant.

A comparison of four spread spectrum techniques, PN, asynchronous FH, synchronous FH and TDMA/PN is illustrated in Fig. (6), where $N = 100$ frequencies, processing gain $G = 20$ dB, duty factor $b = 1$ and $\sqrt{2}(s_o/r) = 1$. The curves of bit error probability for PN and asynchronous FH are extracted from Fig. (3). In the case of synchronous FH, the average probability of bit error is zero for 100 users employing 100 frequencies. The average bit error probability of synchronous FH with $100 + n$ users corresponds to asynchronous FH with n users. The probability of bit error for TDMA/PN with N time slots is likewise derived from the probability of bit error of a PN system. In particular, the probability of bit error of TDMA/PN with 8 to 16 users is equivalent to that of PN with a single interferer. The eight additional users are distributed among the eight time slots and hence act as a single interferer on the system. In general, the average probability of bit error of TDMA/PN associated with jN to $(j+1)N$ users corresponds to the probability of bit error of PN associated with j users. If the data rate is kept the same for both PN and TDMA/PN, TDMA/PN would require eight times the bandwidth of PN. In order to provide a common basis for comparison of PN and TDMA/PN, the TDMA/PN probability of bit error curve is derived from a PN probability of bit error curve with 1/8th the processing gain (i.e., $G = 12.5$).

From Fig. (6) it is clear that synchronous FH could tolerate the maximum number of interfering users for fixed probability of bit error. In particular, for a 0.1 average probability of bit error, synchronous FH can accommodate almost three times

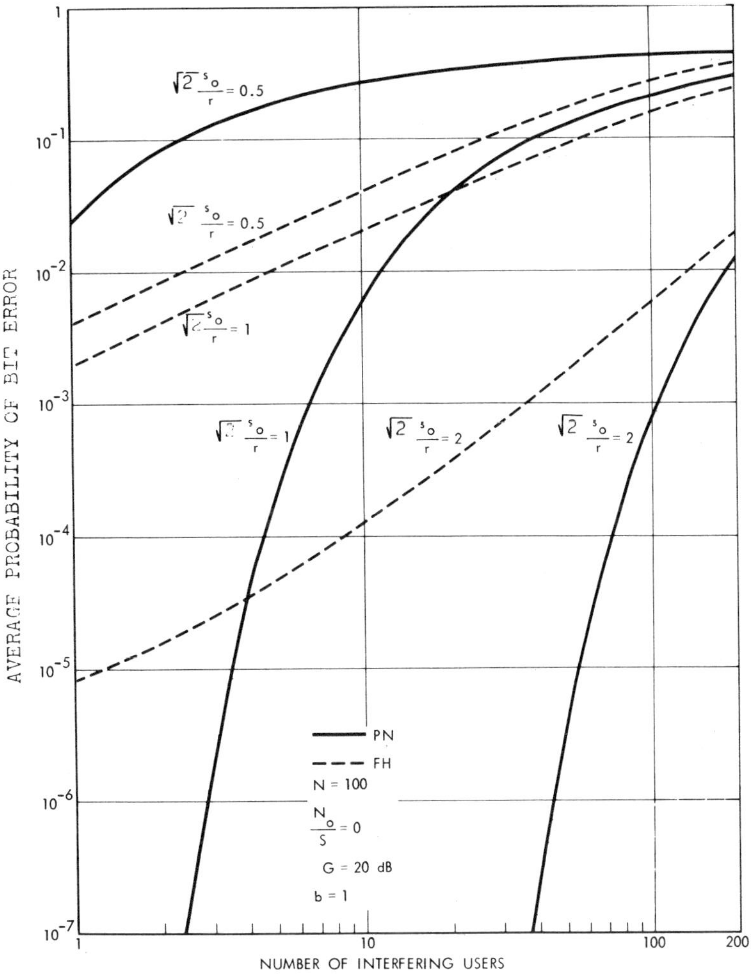

Figure 3. Comparison of Spread-Spectrum Techniques for Different Location of Interferers

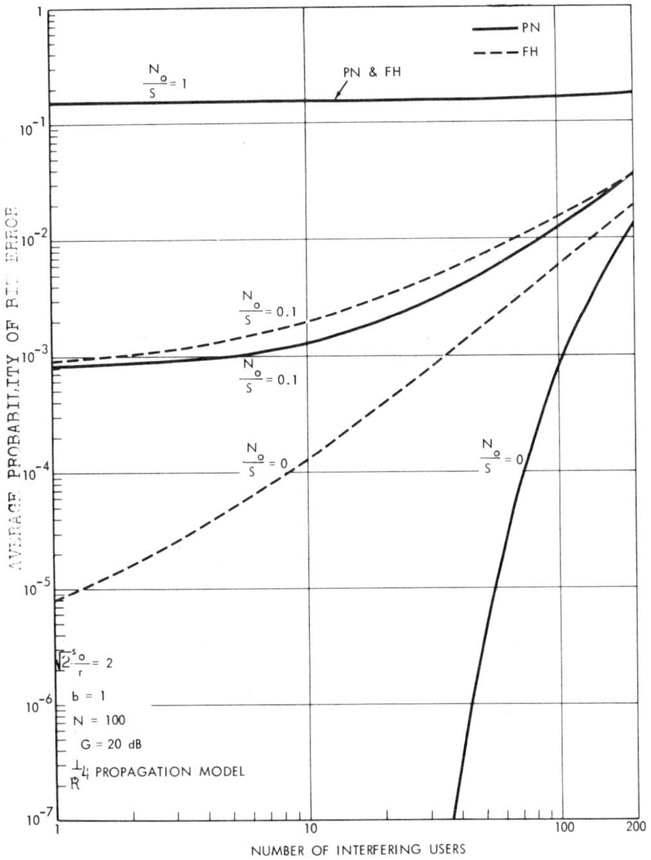

Figure 4. Comparison of Spread-Spectrum Techniques with Different Background Noise Levels

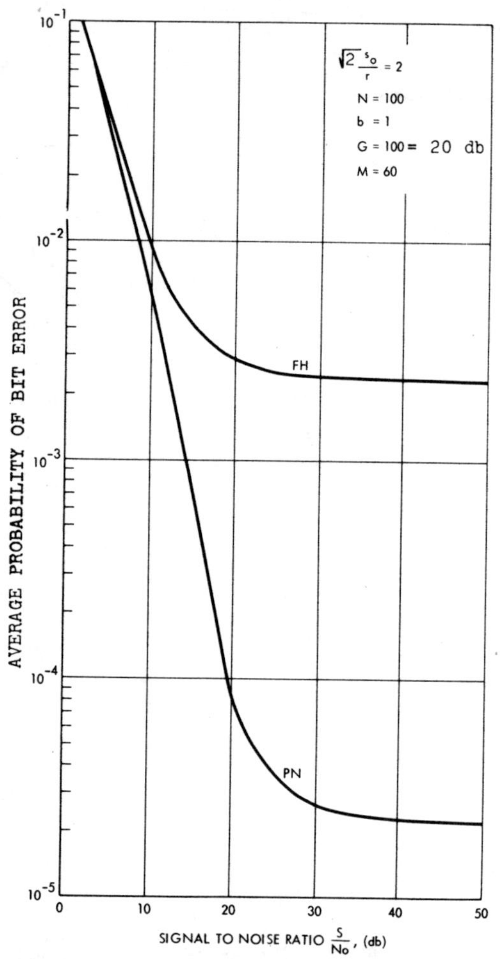

Figure 5. Error Rate for CPSK (60 Interferers)

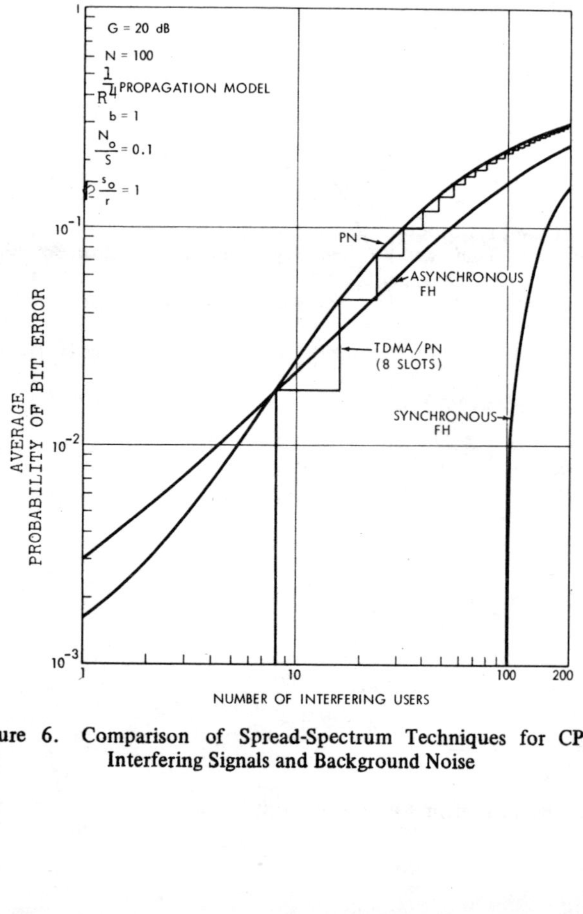

Figure 6. Comparison of Spread-Spectrum Techniques for CPSK Interfering Signals and Background Noise

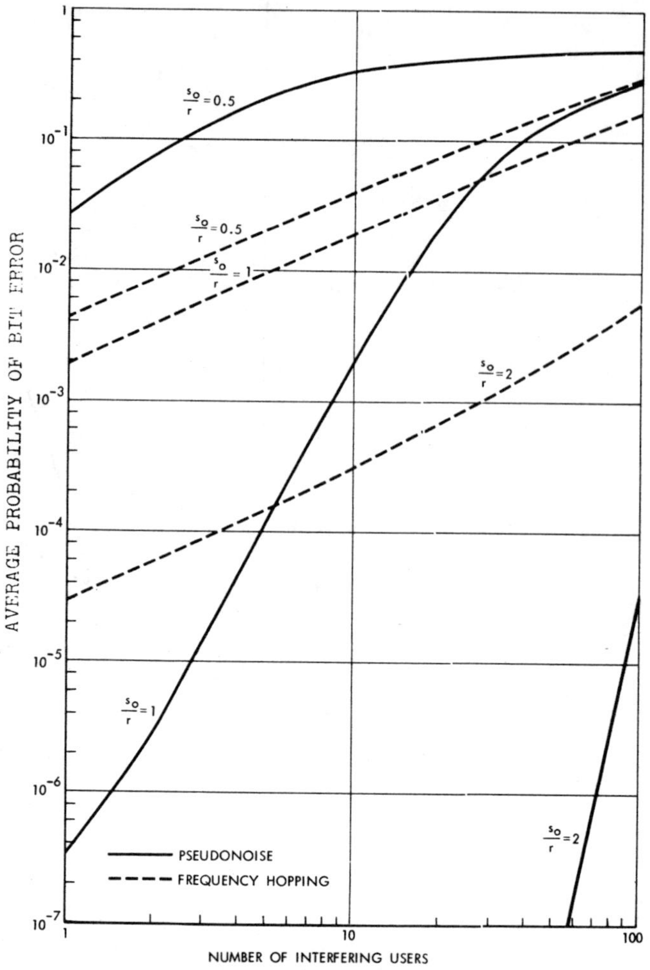

Figure 7. Comparison of Spread-Spectrum Techniques for Bell-Shaped Distribution of Users (Noncoherent FSK Gaussian Interferers)

the number of users employing asynchronous FH and five times the users employing PN and TDMA/PN. The relative performance of asynchronous FH and PN is highly dependent on the location of the interferers from the receiver as compared to the location of the desired transmitter, as well as on the time duty factor of the hopping. Furthermore, it should be emphasized that in the analysis of the FH system there is a tacit assumption that the hopping rate is slow. Thus, the results shown in Fig. 6 are representative of the performance of slow frequency hopping systems, i.e., the frequency splatter into adjacent channels is minimal.

Comparable results are obtained in the case of noncoherent FSK with M Gaussian interfering users. Figs. (7) and (8) show results for the bell-shaped distribution discussed in the preceding and for a distribution of users on a ring of radius R. In the latter case, the average probability of bit error simplifies to $P_M = (1/2) \exp [-(G/2M)(R/r)^4]$.

A comparison of the four spread spectrum techniques for bell-shaped distribution of Gaussian interferers and noncoherent FSK modulation is given in Fig. (9). It is evident from Fig. (9) that the system performance attained for each of the four techniques with noncoherent FSK Gaussian interferers is quite comparable to that with coherent PSK interferers.

IV. CONCLUSIONS

The results of the analysis show that any conclusions to be drawn from a quantitative comparison of the effectiveness of

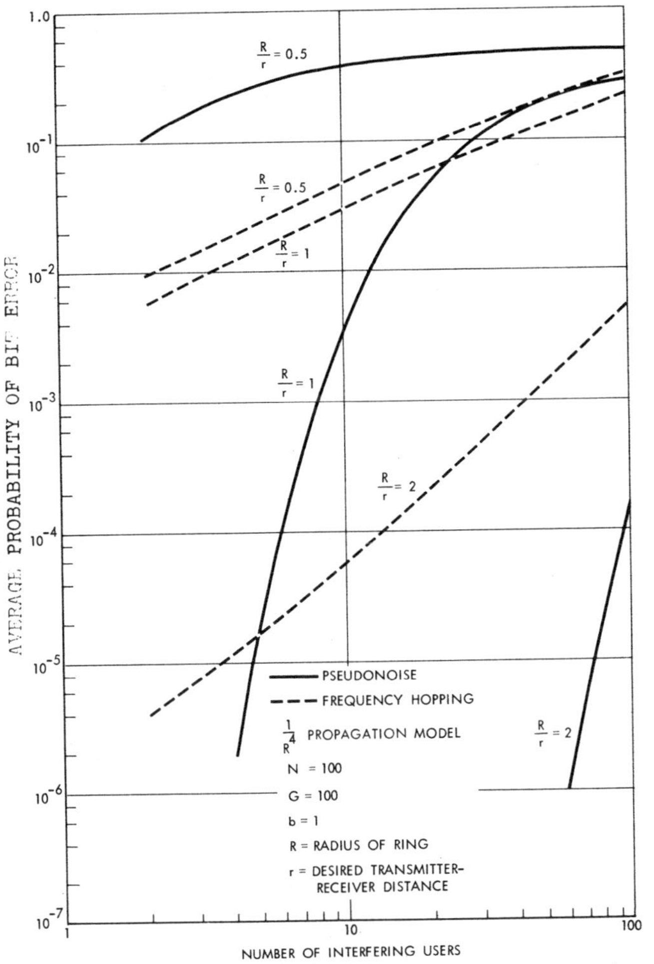

Figure 8. Comparison of Spread-Spectrum Techniques for Ring Distribution of Users (Noncoherent FSK Gaussian Interferers)

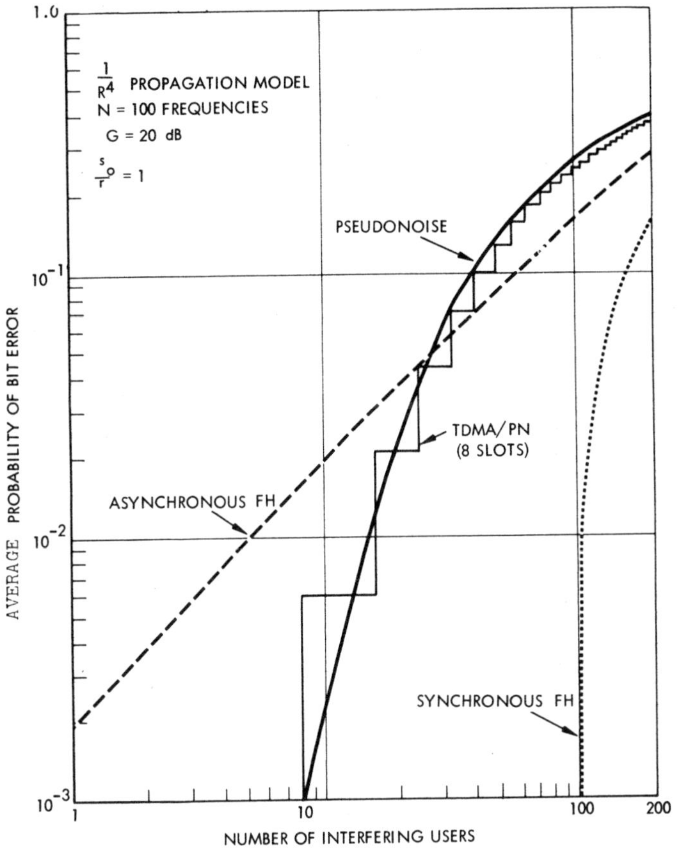

Figure 9. Comparison of Spread-Spectrum Techniques for Bell-Shaped Distribution of Users (Noncoherent FSK Gaussian Interferers)

asynchronous FH and PN are highly dependent on the relative location of interferers and the desired transmitter, the number of interferers, the time duty factor of hopping and ambient signal-to-noise ratio. For high signal-to-noise ratio, PN is superior to asynchronous FH for interferers far removed from the receiver. The reverse holds true for close interferers. Moreover, the difference between PN and asynchronous FH is more significant for small number of interferers.

The comparison of the mutual interference of spread spectrum techniques (PN, TDMA/PN, synchronous and asynchronous FH) shows that synchronous FH can tolerate the maximum number of interfering users. However, the penalty of such a scheme is an increased complexity of system timing and net management.

REFERENCES

1. V. K. Prabhu, "Error Rate Considerations for Coherent Phase-Shift Keyed Systems with Co-Channel Interference," *Bell System Technical Journal*, Vol. 48, pp. 743-767, March 1969.
2. J. M. Aein and R. D. Turner, "Effect of Co-Channel Interference on CPSK Carriers," *IEEE Transactions on Communications*, Vol. COM-21, No. 7, pp. 783-790, July 1973.
3. R. N. Tobin and K. Yao, "Upper and Lower Error Bounds for Coherent Phase-Shift Keyed (CPSK) Systems with Co-Channel Interference," *IEEE Transactions on Communications*, Vol. COM-25, No. 2, pp. 281-287, February 1977.
4. J. J. Egli, "Radio Propagation Above 40 Mc over Irregular Terrain," *Proceedings of the IRE*, October 1957.
5. *National Bureau of Standards, Handbook of Mathematical Functions*, AMS 55, June 1964.
6. W. C. Lindsey and M. K. Simon, *Telecommunication Systems Engineering*, Prentice Hall, Inc., 1973, p. 250.

Fading Effects on the Performance of a Spread Spectrum Multiple Access Communication System

CHESTER S. GARDNER, SENIOR MEMBER, IEEE, AND JOHN A. ORR, MEMBER, IEEE

Abstract—**The effects of fading on the performance of a spread spectrum multiple access communication system using direct sequence modulation are investigated. Expressions for the system signal-to-noise ratio and for the mean and variance of the desired and interfering users are derived in terms of the spreading code and fading correlation functions. The system error probability is also evaluated for the special case of slow Rayleigh fading. The results suggest that under conditions of fading and interference, long code sequences may be required to maintain low error rates.**

I. INTRODUCTION

CODE-DIVISION multiple access communication techniques are presently being considered as attractive alternatives to conventional time and frequency division multiplexing.[1,2] Code-division multiplexing is commonly implemented as a spread spectrum multiple access (SSMA) communication system in which each user is assigned a unique pseudorandom code sequence. Each data bit for a particular user is encoded into many code bits, resulting in a substantial increase in the signal bandwidth. In addition to providing multiple access capability, the large bandwidth is expected to reduce the effects of selective fading and multipath distortion.[3]

The performance of SSMA systems is affected by the usual channel and receiver noise and by interference from other users. The effects of user interference depend on the structure of the codes employed by the system and in general are very difficult to evaluate analytically. Yao has developed a moment space technique for bounding the error probabilities in terms of various moments of the interfering signal.[4] This approach has been used to evaluate the error probability of an SSMA system operating under conditions of user interference and Gaussian noise.

In this paper, we consider the effects of fading on the performance of SSMA communication systems. Expressions for the system signal-to-interference-plus-noise ratio and for the means and variances of the desired and interfering signals are derived in terms of the fading statistics and the correlation properties of the user codes. The results are then used to calculate the error probability for a system whose signals are contaminated by slow Rayleigh fading and additive white Gaussian noise.

Paper approved by the Editor for Space Communication of the IEEE Communications Society for publication without oral presentation. Manuscript received December 16, 1977; revised July 16, 1978. This work was supported in part by the Naval Telecommunication System Command under Contract N00014-67-A-03-5-0002.

J. A. Orr was with the Department of Electrical Engineering, University of Illinois, Urbana, IL 61801. He is now with the Department of Electrical Engineering, Worcester Polytechnic Institute, Worcester, MA.

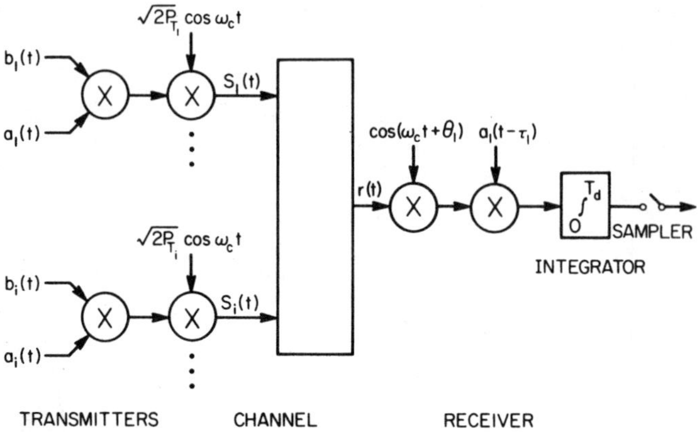

Figure 1. SSMA communication system diagram. $b(t)$ is the data sequence and $a(t)$ is the spreading code sequence.

II. SYSTEM EQUATIONS

The spread spectrum multiple access (SSMA) communication system to be analyzed in this paper uses direct sequence modulation. The system block diagram is illustrated in Fig. 1. At the transmitter, the binary data sequence on the ith channel, $b_i(t)$, is multiplied by a synchronous pseudorandom spreading code, $a_i(t)$ (also called the code sequence). The resultant binary sequence is then used to phase modulate a sinusoidal carrier (sin $\omega_c t$). In both the data sequence and spreading code, logical zero is represented by -1 and logical one is represented by $+1$. Because the code-induced phase shifts are $\pm\pi/2$ radians, the transmitted signal can be represented as an amplitude modulated carrier

$$S_i(t) = \sqrt{2P_{Ti}}a_i(t)b_i(t)\cos\omega_c t \tag{1}$$

where P_{Ti} is the transmitted power and ω_c is the carrier frequency. The spreading code is synchronized to the data sequence and consists of l chips of duration T_c so that the data bit duration T_d is given by

$$T_d = lT_c. \tag{2}$$

The channel model is illustrated in Fig. 2. The channel introduces fading and noise. f_i is the random fading envelope of the ith signal and n is the additive white Gaussian noise. The

* Time dependence is not explicitly indicated to simplify notation.

Reprinted from *IEEE Trans. Comm.*, vol. COM-27, pp. 143-149, Jan. 1979.

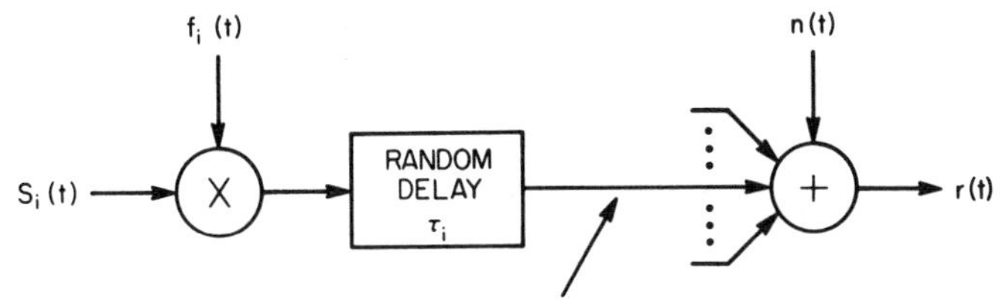

Figure 2. Channel model. $f_i(t)$ is the random fading envelope, τ_i is the random propagation delay, θ_i is the random carrier phase shift and $n(t)$ is the additive white Gaussian noise.

fading envelopes for each channel are assumed to be mutually independent and identically distributed. The transmitters are assumed to be operating asynchronously so that the time delays τ_i and phase angles θ_i are also random processes.* The random processes τ_i are independent and uniform on $[0, T_d]$, while the processes θ_i are independent and uniform on $[0, 2\pi]$.

The total received signal is

$$r(t) = \sum_{i=1}^{K} \sqrt{2P_i} f_i(t-\tau_i) a_i(t-\tau_i) b_i(t-\tau_i) \cdot \cos(\omega_c t + \theta_i) + n(t) \tag{3}$$

where P_i is the effective received power for the ith channel. The receiver is assumed to be frequency, phase and code synchronized to the first signal which we will hereafter refer to as the desired channel. Detection is accomplished by multiplying the received signal by $a_1(t - \tau_1) \cos(\omega_c t + \theta_1)$ and integrating over the data bit period. Since the time delays and phase angles are referenced to the desired channel, we can let $\tau_1 = \theta_1 = 0$. The integrator output at $t = T_d$ is

$$X = \sqrt{P_1/2} \int_0^{T_d} dt a_1^2(t) b_1(t) f_1(t) + \sum_{i=2}^{K} \sqrt{P_i/2} \cdot \int_0^{T_d} dt a_1(t) a_i(t-\tau_i) b_i(t-\tau_i) f_i(t-\tau_i) \cos\theta_i + \int_0^{T_d} dt a_1(t) n(t) \cos\omega_c t. \tag{4}$$

The first term in (4) is the desired signal, S, which can be simplified by noting that $a_1^2(t)$ is identically one and $b_1(t) = b_1$ is constant over the integration period.

$$S = \sqrt{P_1/2} b_1 \int_0^{T_d} dt f_1(t). \tag{5}$$

The second term in (4) is the interference from the undesired signals. To illustrate its effect on system performance, it is convenient to consider first the simple special case of one undesired signal.

$$I_i = \sqrt{P_i/2} \int_0^{T_d} dt a_1(t) a_i(t-\tau_i) b_i(t-\tau_i) f_i(t-\tau_i) \cos\theta_i. \tag{6}$$

The third term is noise

$$N = \int_0^{T_d} dt a_1(t) n(t) \cos\omega_c t. \tag{7}$$

Since $n(t)$ is a white zero-mean Gaussian process, N is also zero-mean Gaussian, and the variance is

$$\mathrm{Var}(N) = \sigma_N^2 = \eta_0 T_d/4 \tag{8}$$

where $\eta_o/2$ is the two-sided noise power spectral density at the receiver input.

In all that follows, the mean values of the fading envelopes f_1 and f_i are normalized to one. Differences in signal strength between the desired and interfering channels can be accommodated in the model by varying the effective received powers P_1 and P_i. In practice, fading is usually nonstationary. Depending on the mechanism which induces the fading, short-term statistics may vary drastically from long-term statistics. However, in many cases stationarity can be assumed for short-time periods on the order of a few seconds or longer.[5] If we assume the fading process is at least short-term stationary, the conditional mean and variance of the desired signal given the data bit are

$$E(S \mid b_1) = \sqrt{P_1/2} b_1 T_d$$

$$\mathrm{Var}(S \mid b_1) = P_1 T_d \sigma_f^2 \int_0^{T_d} d\tau (1 - \tau/T_d) \rho_f(\tau) \tag{9}$$

where σ_f^2 is the fading envelope variance and ρ_f is the normalized fading envelope autocovariance function. In the absence of fading, the conditional signal variance is zero because σ_f is zero.

To calculate the statistics of the interfering signal, we assume that the random phase angle θ_i is independent of the fading envelope f_i. Since θ_i is a random process which is uniformly distributed on $[0, 2\pi]$, the expected value of the interfering signal is zero. The variance is given by

$$\mathrm{Var}(I_i) = P_i/2 \int_0^{T_d} dt_1 \int_0^{T_d} dt_2 R_z(t_1 - t_2) a_1(t_1) a_1(t_2) \cdot \frac{1}{T_d} \int_0^{T_d} d\tau_i a_i(t_1 - \tau_i) a_i(t_2 - \tau_i) \cdot E[b_i(t_1 - \tau_i) b_i(t_2 - \tau_i)] \tag{10}$$

where R_z is the autocorrelation function associated with the random process

$$z(t) = f_i(t) \cos[\theta_i(t)]. \tag{11}$$

Equation (10) can be simplified by first evaluating the integral over the random delay τ_i. The expectation over the data sequence b_i is nonzero only when the arguments $(t_1 - \tau_i)$ and $(t_2 - \tau_i)$ correspond to the same data bit, i.e.,

$$E[b_i(t_1 - \tau_i)b_i(t_2 - \tau_i)] = \begin{cases} 1 & t_1 < \tau_i, t_2 < \tau_i \text{ or } t_1 > \tau_i, t_2 > \tau_i \\ 0 & \text{otherwise.} \end{cases} \quad (12)$$

Using (12) the integral over τ_i in (10) can be written

$$\gamma = 1/T_d \int_0^{\min(t_1,t_2)} d\tau_i a_i(t_1 - \tau_i)a_i(t_2 - \tau_i) + 1/T_d \int_{\max(t_1,t_2)}^{T_d} d\tau_i a_i(t_1 - \tau_i)a_i(t_2 - \tau_i). \quad (13)$$

After making the change of variables

$$t = t_1 - \tau_i$$

$$\tau = t_1 - t_2$$

γ becomes

$$\gamma = 1/T_d \int_{\max(0,\tau)}^{t_1} dt a_i(t)a_i(t - \tau) + 1/T_d \int_{t_1 - T_d}^{\min(0,\tau)} dt a_i(t)a_i(t - \tau). \quad (14)$$

Since the code sequence period is equal to the data bit period, T_d may be added to both the upper and lower integration limits of the second integral in (14). This gives

$$\gamma = 1/T_d \int_{\max(0,\tau)}^{T_d + \min(0,\tau)} dt a_i(t)a_i(t - \tau). \quad (15)$$

It is not difficult to show that γ is an even function of τ and can be written

$$\gamma = \frac{1}{T_d}\hat{R}_{ii}(\tau) = \frac{1}{T_d}\int_\tau^{T_d} dt a_i(t)a_i(t - \tau) \quad (16)$$

where we have used Pursley's notation for the partial autocorrelation function $\hat{R}_{ii}$.[6]

The expression for $\text{Var}(I_i)$ can be simplified further by substituting (16) into (10) and making the change of variables

$$\tau = t_1 - t_2$$

$$t = t_1.$$

The resultant integrand is an even function of τ which can be simplified to give

$$\text{Var}(I_i) = P_i/T_d \int_0^{T_d} d\tau \hat{R}_{11}(\tau)\hat{R}_{ii}(\tau)R_z(\tau). \quad (17)$$

It is convenient to express $\text{Var}(I_i)$ in terms of the normalized autocorrelation functions

$$\text{Var}(I_i) = P_i T_d(\sigma_f^2 + 1)/2 \int_0^{T_d} d\tau \hat{\rho}_{11}(\tau)\hat{\rho}_{ii}(\tau)\rho_z(\tau) \quad (18)$$

where

$$\hat{\rho}_{ii}(\tau) = \frac{1}{T_d}\hat{R}_{ii}(\tau)$$

$$\rho_z(\tau) = 2R_z(\tau)/(\sigma_f^2 + 1). \quad (19)$$

In the absence of fading, σ_f is zero and the normalized fading autocorrelation ρ_z is identically one. In this case, Eq. (18) becomes

$$\text{Var}(I_i) = P_i T_d/2 \int_0^{T_d} d\tau \hat{\rho}_{11}(\tau)\hat{\rho}_{ii}(\tau). \quad (20)$$

Equation (20) is consistent with the more recent results of Pursley and Sarwate.[7]

Our analysis can easily be extended to include K users if we assume the data bit sequences on each channel are statistically independent. The interfering signals are therefore mutually uncorrelated so that the variance of the total interfering signal is simply the sum of the variances of each interfering signal.

$$\text{Var}(I) = \sum_{i=2}^{K} P_i/T_d \int_0^{T_d} d\tau \hat{R}_{11}(\tau)R_{ii}(\tau)\hat{R}_z(\tau). \quad (21)$$

A useful parameter for assessing the reliability of SSMA systems is the signal-to-interference-plus-noise ratio (SNR). Although in general the SNR cannot be directly related to error rate, it does provide a useful measure of the relative system performance. We define the system SNR as the ratio of the expected signal energy to the sum of the noise and interference energies. Using Eqs. (8), (9), (19) and (21), we obtain

$$\text{SNR} = \frac{E(S^2)}{\text{Var}(N) + \text{Var}(I)} = \frac{\left[1 + 2\sigma_f^2/T_d \int_0^{T_d} d\tau(1 - \tau/T_d)\rho_f(\tau)\right]}{\text{SNR}_G^{-1} + \sum_{i=2}^{K} \frac{P_i(\sigma_f^2 + 1)}{P_1 T_d} \int_0^{T_d} d\tau \hat{\rho}_{11}(\tau)\hat{\rho}_{ii}(\tau)\rho_z(\tau)} \quad (22)$$

where

$$\text{SNR}_G = \frac{S^2}{\text{Var}(N)} = \frac{2P_1 T_d}{\eta_0}. \quad (23)$$

SNR_G is the equivalent Gaussian SNR in the absence of fading and interference (i.e., when $\sigma_f = P_i = 0$).

III. NUMERICAL RESULTS

The results derived in the previous section are fairly general, applying to most stationary (short-term) fading processes. It is clear from Eqs. (9) and (18) that the statistics of the signal and interference depend on the variance and autocorrelation properties of the fading and on the structure of the codes used in the system. In order to obtain some numerical results on the fading effects, we will use Gold codes[8,9] and consider the special case of Rayleigh fading.

The Rayleigh probability density for the fading envelopes of the desired and interfering signals is given by

$$p_f(f) = \frac{\pi}{2} \frac{f}{\langle f \rangle^2} \exp\left[-\frac{\pi f^2}{4\langle f \rangle^2}\right], \qquad f \geqslant 0 \tag{24}$$

where $\langle f \rangle$ denotes the mean of f. Because the Rayleigh density has a single parameter, the mean and variance of f are related by the equation

$$\sigma_f{}^2 = \left(\frac{4}{\pi} - 1\right) \langle f \rangle^2. \tag{25}$$

In the previous section and in this section $\langle f \rangle$ is assumed to be one. The average received strength of the desired and interfering signals is adjusted by varying the received powers P_1 and P_i.

To evaluate the system SNR given by Eq. (22), it is necessary to determine the relationship between the two covariance functions ρ_z and ρ_f.** The fading envelopes for each channel are assumed to be mutually independent and identically distributed Rayleigh processes. When f is Rayleigh distributed and θ is uniformly distributed on $[0, 2\pi]$, the process $z(t) = f(t) \cos \theta(t)$ is Gaussian distributed. In this case it can be shown[11]

$$\rho_f = \frac{{}_2F_1(-1/2, -1/2, 1, \rho_z{}^2) - 1}{\left(\frac{4}{\pi} - 1\right)} \cong \rho_z{}^2 \tag{26}$$

where ${}_2F_1$ is a hypergeometric function.

Fading correlation effects on the variance of the interfering signal are illustrated in Fig. 3 where the normalized interference variance is plotted versus correlation time for several values of the code sequence length $l = T_d/T_c$. The results were computed using a Gaussian shaped autocorrelation function for the fading process $z = f \cos \theta$

$$\rho_z(\tau) = \exp(-\tau^2/2T^2). \tag{27}$$

T is the correlation time. The Gaussian shape was chosen simply for mathematical convenience. The normalization factor used in Fig. 3 is the variance for infinite correlation time. The data

** Earlier we defined ρ_f as a covariance and ρ_z as a correlation. But, since $f \cos \theta$ is a zero mean process, its covariance and correlation functions are identical.

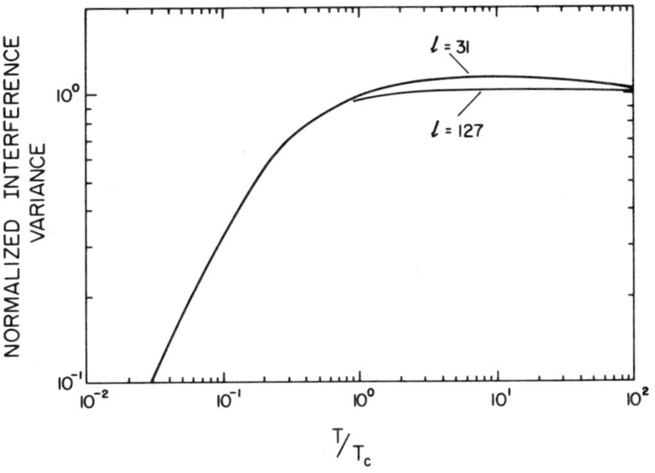

Figure 3. Normalized variance of the interfering signal. T is the fading correlation time and T_c is the spreading code chip duration.

for each code length represent the average variance calculated over ten randomly selected Gold code pairs.

The correlation time has little effect on the interference variance for $T > T_c$. Although this result is somewhat surprising, it can be explained by considering Eq. (17). For well-designed codes the partial autocorrelation functions $\hat{R}_{ii}(\tau)$ are highly peaked about $\tau = 0$. They have a maximum value of T_d at $\tau = 0$ and are nearly zero for $\tau \geqslant T_c$. Consequently, if the fading correlation time is greater than the chip time T_c, the major contribution to the integral in Eq. (17) occurs for $\tau \geqslant T_c$. Under these conditions the fading correlation function will be essentially constant and independent of T over the important range of integration. When the correlation time is small compared to the chip time, the fading correlation function limits the important range of integration to $\tau \leqslant T$. In this case the interference variance is roughly proportional to T. Under these rapidly fading conditions, the detection process averages the fading fluctuations over the chip duration thereby reducing the interference variance.

The system SNR given by Eq. (22) was evaluated using Eqs. (25) through (27) for the case of one interfering signal whose power is equal to the desired signal power. The results are plotted in Fig. 4 versus $\mathrm{SNR_G}$ for several values of the fading correlation time. The SNR saturates for large values of $\mathrm{SNR_G}$. Under these conditions the system noise is dominated by the interfering signal. As expected, the correlation time has little effect on the SNR for $T \geqslant T_c$.

The system error rate is a function of the signal, noise and interference probability densities. The signal and interference probability densities are strongly dependent on the fading correlation properties. For example, if the fading correlation time is long compared to the data bit time, the fading envelope in Eq. (5) will be essentially constant over the interval in $[0, T_d]$ and can be removed from under the integral. In this case the signal will be Rayleigh distributed. However, if the fading correlation time is short compared to T_d, the signal statistics will be approximately Gaussian. To illustrate the effects of fading on the system error rate, we will consider the case where the correlation time is long compared to the data bit time. This situation is most frequently encountered at HF and micro-

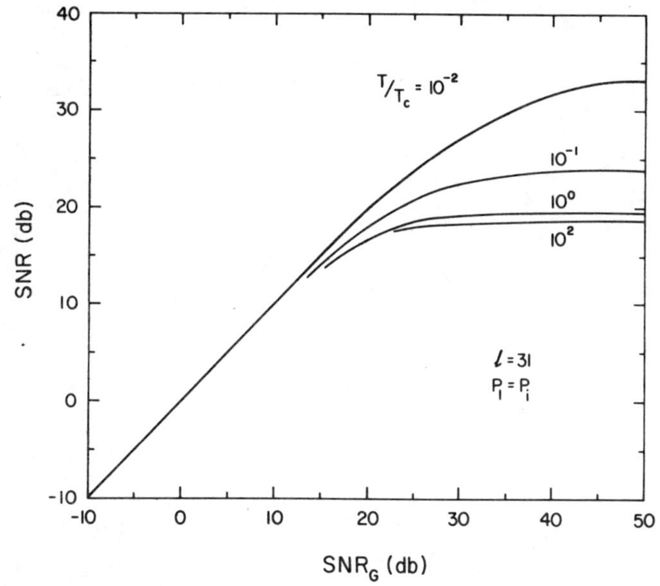

Figure 4. System SNR given by Equation (22) versus the equivalent Gaussian SNR (SNR_G).

wave frequencies for ionospheric multipath and turbulence induced fading where correlation times are typically a few milliseconds or longer.[5] Under these slow fading conditions the factor f_1 in the expression for the signal (Eq. (5)) and the factor $f_i \cos\theta_i$ in the expression for the interference (Eq. (6)) can be removed from under the integrals

$$S \cong \sqrt{P_1/2}\,T_d f_1 b_1 \tag{28}$$

$$I_i \cong \sqrt{P_i/2}\,f_i \cos\theta_i \int_0^{T_d} dt\,a_1(t)a_i(t-\tau_1)b_i(t-\tau_i)$$

$$= \sqrt{P_i/2}\,f_i \cos\theta_i C_i. \tag{29}$$

C_i is the cross-correlation between a_1 and a_i for a given time delay τ_i and data sequence b_i. Because $f_i \cos\theta_i$ is a zero mean Gaussian process, the conditional density of I_i given C_i will also be Gaussian. The unconditioned density will depend on the probability distribution of the cross-correlation values and in general will not be Gaussian.

If we consider one interfering signal, the conditional probability of error can be expressed in terms of complementary error functions

$$\Pr(E \mid f_1, f_i, \theta_i, C_i)$$

$$= \frac{1}{4}\,\mathrm{erfc}\left[\frac{\sqrt{P_1/2}\,T_d f_1 + \sqrt{P_i/2}\,f_i \cos\theta_i C_i}{\sqrt{2}(\eta_0 T_d/4)^{1/2}}\right]$$

$$+ \frac{1}{4}\,\mathrm{erfc}\left[\frac{\sqrt{P_1/2}\,T_d f_1 - \sqrt{P_i/2}\,f_i \cos\theta_i C_i}{\sqrt{2}(\eta_0 T_d/4)^{1/2}}\right]. \tag{30}$$

Notice that in the absence of fading and interference ($f_1 = f_i = 1$, $C_i = 0$) Eq. (30) reduces to the well-known error expression for signals in Gaussian noise

$$\Pr(E) = \frac{1}{2}\,\mathrm{erfc}\left(\sqrt{\frac{P_1 T_d}{\eta_0}}\right) = \frac{1}{2}\,\mathrm{erfc}\left(\sqrt{\frac{SNR_G}{2}}\right). \tag{31}$$

The conditional error probability given C_i is calculated by first integrating (30) over the probability density of $f_i \cos\theta_i$ which is zero mean Gaussian

$$\Pr(E \mid f_1, C_i) = \frac{1}{2}\,\mathrm{erfc}\left[\frac{\sqrt{P_1/2}\,T_d f_1}{\sqrt{2}\left(\frac{\eta_0 T_d}{4} + \frac{P_i C_i^2}{\pi}\right)^{1/2}}\right] \tag{32}$$

and then integrating (32) over the Rayleigh density of f_1

$$\Pr(E \mid C_i) = \frac{1}{2}\left\{1 - \left[1 + \frac{\pi/2}{SNR_G} + \frac{P_i}{P_1}\left(\frac{C_i}{T_d}\right)^2\right]^{-1/2}\right\}. \tag{33}$$

For comparison, the conditional error probability given C_i and θ_i in the absence of fading is obtained from (30) by setting $f_1 = f_i = 1$

$$\Pr(E \mid C_i, \theta_i)$$

$$= \frac{1}{4}\,\mathrm{erfc}\left[\sqrt{\frac{SNR_G}{2}}\left(1 + \sqrt{\frac{P_i}{P_1}}\frac{C_i}{T_d}\cos\theta_i\right)\right]$$

$$+ \frac{1}{4}\,\mathrm{erfc}\left[\sqrt{\frac{SNR_G}{2}}\left(1 - \sqrt{\frac{P_i}{P_1}}\frac{C_i}{T_d}\cos\theta_i\right)\right]. \tag{34}$$

The unconditioned error probabilities can be evaluated by multiplying $\Pr(E \mid C_i)$ times the probability density of C_i and integrating. $p(C_i)$ is computed by tabulating the frequency of occurrence of the cross-correlation values for all possible time delays, data sequences and code pairs. However, only time delays which are integer multiples of T_c need be considered since $\hat{R}_{ii}$ varies linearly over the intervals $\{[(k-1)T_c, kT_c], k = 1, \cdots, l\}$. Although this simplifies the problem somewhat, the computation of $p(C_i)$ is still very cumbersome, particularly for large code lengths. As an alternative, Yao has developed a moment space technique for bounding the error probabilities.[4] This approach requires the evaluation of various moments of the interfering signal which are generally easier to compute than $p(C_i)$. Yao's technique is an attractive alternative for use in systems analyses for which calculation of exact error probabilities is untractable.

$p(C_i)$ was computed for Gold codes of length 31 chips and used with Eq. (33) to calculate the unconditioned error probability for slow Rayleigh fading. The results are plotted versus SNR_G in Fig. 5 for the case where $P_i = P_1$. For comparison we have also plotted the corresponding error probabilities for the no interference and/or no fading conditions. The error probability for no fading with interference was computed using Eq. (34) for the worst case where $\cos\theta_i = 1$. As expected the best

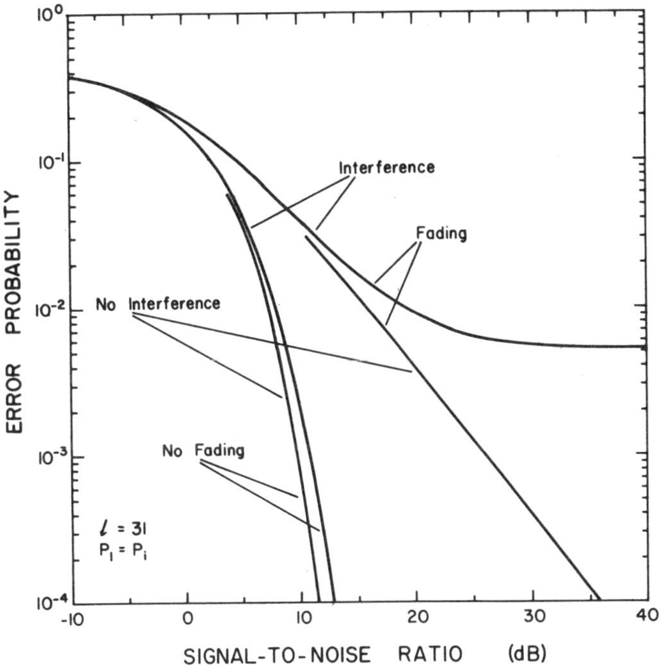

Figure 5. System error probability versus the equivalent Gaussian SNR (SNR_G). The data were calculated for slow Rayleigh fading and Gold codes of length $l = 31$.

performance occurs in the absence of interference and fading while the worst performance occurs when fading and interference are both present. The most interesting feature in these plots is the fact that the error probability for fading and interference saturates at large values of SNR_G. Under these conditions, performance is apparently limited by interference, not Gaussian noise. This is in contrast to the behavior for no fading where interference increases the error rate but does not cause it to level off at high values of SNR_G (see Eq. (34)). When there is no fading and $P_i \leqslant P_1$, the magnitude of the interference term I_i is always smaller than the magnitude of the signal term S, since $(C_i/T_d) \ll 1$ for well-designed multiple access systems. Consequently, there will be no errors when the Gaussian noise is zero, because the magnitude of the interference term will never be large enough to cause the polarity of $S + I_i$ to differ from the polarity of S. However, when fading is present on the interfering signal, desired signal or both, the magnitude of I_i could exceed the magnitude of S sometimes causing the polarity of $S + I_i$ to differ from the polarity of S, resulting in errors even in the absence of Gaussian noise. Using this simple argument it is clear that the saturating error rate effect is not unique to Rayleigh fading. It will occur whenever the fading process is such that $\Pr[|I_i| > |S|] > 0$. In fact, Orr's simulation results have indicated that the saturation effect also occurs with lognormal fading.[10] Notice that the limiting error rate in Figure 5 is equal to the error rate for the case of fading and no interference when SNR_G is approximately 18 dB. This is consistent with the plot of the system SNR versus SNR_G in Figure 4. For slow fading $T/T_c \gg 1$ and in this case the system SNR saturates at about 18 dB.

The limiting error probability can be estimated by setting SNR_G equal to infinity in Eq. (32), expanding the result in a power series and integrating over $p(C_i)$

$$\lim_{SNR_G \to \infty} \Pr(E) = \frac{1}{2} \sum_{n=1}^{\infty} (-1)^{n+1} \binom{2n}{n} \left(\frac{P_i}{4P_1}\right)^n \frac{\langle C_i^{2n} \rangle}{T_d^{2n}} \cong \frac{1}{4} \frac{P_i}{P_1} \frac{\langle C_i^2 \rangle}{T_d^2} \tag{35}$$

where the angle brackets denote ensemble average. The single term approximation in Eq. (35) is valid whenever the normalized moments $(P_i/P_1)^n \, (\langle C_i^{2n} \rangle / T_d^{2n})$ are small compared to one. This will be true for any well-designed multiple access codes provided P_i is less than or on the order of P_1.

Our simulation results have indicated that the statistical properties of Gold sequences are similar to those of completely random sequences particularly for the longer code lengths.[10] For example, the normalized second moment $\langle C_i^2 \rangle / T_d^2$ for completely random sequences is approximately $2/3l$, a value which compares favorably with the results for Gold codes. Apparently the data bit sequence in Eq. (30) helps to randomize the Gold sequences. Therefore if $P_i = P_1$, a good estimate for the limiting error probability is

$$\lim_{SNR_G \to \infty} \Pr(E) \cong \frac{1}{6l} . \tag{36}$$

When $l = 31$, Eq. (36) gives a value of 5.38×10^{-3} while the actual error probability is 5.22×10^{-3}. The accuracy of this approximation improves as l increases. When there are K equal energy signals using the channel and $K \ll l$, the limiting error probability is on the order of $(K-1)/6l$. These results suggest that under conditons of slow Rayleigh fading and interference, relatively long code sequences may be required to maintain low error rates.

IV. CONCLUSIONS

In this paper we have considered the effects of fading on the performance of SSMA communication systems. The analysis is applicable to systems employing direct sequence modulation. Expressions for the system signal-to-interference-plus-noise ratio and for the mean and variance of the desired and interfering signals were derived. The results depend on the variance and correlation properties of the fading and on the structure of the codes used in the system. The interference variance and signal-to-interference-plus-noise ratio were evaluated for systems using Gold codes and operating under Rayleigh fading conditions. The results are plotted in Figures 3 and 4.

The system error rate was also evaluated for the case of one interfering signal and compared with the corresponding error probabilities for no interference and/or no fading. The results are plotted in Figure 5 and apply to slow fading conditions where the correlation time is long compared to the data bit time. If fading and interference are both present, the error probability saturates when the equivalent Gaussian signal-to-noise ratio (SNR_G) becomes large. In this case the system performance is apparently limited by interference, not Gaussian noise. This is in contrast to the behavior without fading where interference increases the error rate but does not cause it to saturate at high values of SNR_G. The limiting error probability was estimated to be approximately $(6l)^{-1}$ where l is the Gold sequence length. This result suggests that under conditions of slow Rayleigh fading and interference, relatively long code sequences may be required to maintain low error rates.

REFERENCES

1. R. C. Dixon, *Spread Spectrum Systems,* John Wiley and Sons, 1976.
2. L. Gerhardt and R. C. Dixon, eds. "Special Issue on Spread Spectrum Communications," *IEEE Trans. Comm.,* COM-25, No. 8, August, 1977.
3. J. L. Massey and J. J. Uhran, "Sub-baud coding," *Proc. Thirteenth Annual Allerton Conference on Circuits and Systems Theory,* pp. 539-547, October, 1975.
4. K. Yao, "Error Probability of Asynchronous Spread Spectrum Multiple Access Communication Systems," *IEEE Trans. Comm.,* COM-25, No. 8, pp. 803-809, August, 1977.
5. M. Schwartz, W. Bennett and S. Stein, *Communication Systems and Techniques,* McGraw-Hill, 1966.
6. M. B. Pursley, "Evaulating Performance of Codes for Spread Spectrum Multiple Access Communications," *Proc. Twelfth Annual Allerton Conference on Circuits and Systems Theory,* pp. 765-774, October, 1974.
7. M. Pursley and D. Sarwate, "Performance Evaluation for Phase-Coded Spread-Spectrum Multiple-Access Communication–Part II: Code Sequence Analysis," *IEEE Trans. Comm,* COM-25, No. 8, pp. 800-803, August, 1977.
8. R. Gold, "Optimal Binary Sequences for Spread Spectrum Multiplexing," *IEEE Trans. Inform. Theory,* Vol. IT-13, pp. 619-621, October, 1967.
9. R. Gold, "Maximal Recursive Sequences with 3-Valued Recursive Cross-correlation Functions," *IEEE Trans. Inform. Theory,* Vol. IT-14, pp. 154-156, January, 1968.
10. J. A. Orr, "Effects of Fading on a Multi-User Spread Spectrum System Using Direct Sequence Modulation," Ph.D. dissertation, University of Illinois, Urbana, Illinois, 1977.
11. D. Middleton, *An Introduction to Statistical Communication Theory,* McGraw-Hill, 1960.

Part III
Interference Reduction/Cancelling

IN ADDITION to various modulation schemes that have been used to combat interference, direct interference reduction and cancelling has also received considerable attention over the years. The main approaches that have received special attention are "nullsteering," adaptive echo suppression, and antenna sidelobe reduction.

The subject of null-steering, where adaptive algorithms cause the nulls of the antenna directivity pattern to track noise sources, is studied by B. Widrow *et al.* It is shown that adaptive filtering techniques can be applied to processing the output of the individual elements in a receiving antenna array to reduce the sensitivity of the array to an interfering noise source whose characteristics may be unknown *a priori*. In the second paper, by A. M. Kowalski, the performance of an adaptive tapped delay line filter, based on a least mean square algorithm, is described for interference suppression in conjunction with a wide-band receiver. Finally, the third and fourth papers by P. S. Lubell *et al.* and W. F. Gerst, respectively, describe a rather popular adaptive technique for cancellation of interference by coherent comparison of the interference reference with a sample of the main line output.

Reflected signals in telephone systems create interference in the form of echo which is considered unacceptable when the round trip delay exceeds a few milliseconds. The general characteristics of the echo suppressors which are designed to eliminate or minimize this form of interference are analyzed in the paper by J. E. Unrue, Jr. O. A. Horna, the sixth paper, describes a novel approach by which the use of an adaptive filter enhances the echo return loss sufficiently while the residual echo is removed by an adaptive center clipper. A pseudo-logarithmic analog-to-digital conversion is used to reduce the memory size and the complexity of the processor used.

The analysis of sidelobes for small aperture earth stations and the implementation of a technique to reduce sidelobe excitations is studied in the seventh and eighth papers by J. Perini and J. M. Janky *et al.*, respectively.

Adaptive Antenna Systems

B. WIDROW, MEMBER, IEEE, P. E. MANTEY, MEMBER, IEEE, L. J. GRIFFITHS, STUDENT MEMBER, IEEE, AND B. B. GOODE, STUDENT MEMBER, IEEE

Abstract—A system consisting of an antenna array and an adaptive processor can perform filtering in both the space and the frequency domains, thus reducing the sensitivity of the signal-receiving system to interfering directional noise sources.

Variable weights of a signal processor can be automatically adjusted by a simple adaptive technique based on the least-mean-squares (LMS) algorithm. During the adaptive process an injected pilot signal simulates a received signal from a desired "look" direction. This allows the array to be "trained" so that its directivity pattern has a main lobe in the previously specified look direction. At the same time, the array processing system can reject any incident noises, whose directions of propagation are different from the desired look direction, by forming appropriate nulls in the antenna directivity pattern. The array adapts itself to form a main lobe, with its direction and bandwidth determined by the pilot signal, and to reject signals or noises occurring outside the main lobe as well as possible in the minimum mean-square error sense.

Several examples illustrate the convergence of the LMS adaptation procedure toward the corresponding Wiener optimum solutions. Rates of adaptation and misadjustments of the solutions are predicted theoretically and checked experimentally. Substantial reductions in noise reception are demonstrated in computer-simulated experiments. The techniques described are applicable to signal-receiving arrays for use over a wide range of frequencies.

Introduction

THE SENSITIVITY of a signal-receiving array to interfering noise sources can be reduced by suitable processing of the outputs of the individual array elements. The combination of array and processing acts as a filter in both space and frequency. This paper describes a method of applying the techniques of adaptive filtering[1] to the design of a receiving antenna system which can extract directional signals from the medium with minimum distortion due to noise. This system will be called an *adaptive array*. The adaptation process is based on minimization of mean-square error by the LMS algorithm.[2]–[4] The system operates with knowledge of the direction of arrival and spectrum of the signal, but with no knowledge of the noise field. The adaptive array promises to be useful whenever there is interference that possesses some degree of spatial correlation; such conditions manifest themselves over the entire spectrum, from seismic to radar frequencies.

Manuscript received May 29, 1967; revised September 5, 1967.

B. Widrow and L. J. Griffiths are with the Department of Electrical Engineering, Stanford University, Stanford, Calif.

P. E. Mantey was formerly with the Department of Electrical Engineering, Stanford University. He is presently with the Control and Dynamical Systems Group, IBM Research Laboratories, San Jose, Calif.

B. B. Goode is with the Department of Electrical Engineering, Stanford University, Stanford, Calif., and the Navy Electronics Laboratory, San Diego, Calif.

The term "adaptive antenna" has previously been used by Van Atta[5] and others[6] to describe a self-phasing antenna system which reradiates a signal in the direction from which it was received. This type of system is called adaptive because it performs without any prior knowledge of the direction in which it is to transmit. For clarity, such a system might be called an adaptive *transmitting* array; whereas the system described in this paper might be called an adaptive *receiving* array.

The term "adaptive filter" has been used by Jakowatz, Shuey, and White[7] to describe a system which extracts an unknown signal from noise, where the signal waveform recurs frequently at random intervals. Davisson[8] has described a method for estimating an unknown signal waveform in the presence of white noise of unknown variance. Glaser[9] has described an adaptive system suitable for the detection of a pulse signal of fixed but unknown waveform.

Previous work on array signal processing directly related to the present paper was done by Bryn, Mermoz, and Shor. The problem of detecting Gaussian signals in additive Gaussian noise fields was studied by Bryn,[10] who showed that, assuming K antenna elements in the array, the Bayes optimum detector could be implemented by either K^2 linear filters followed by "conventional" beam-forming for each possible signal direction, or by K linear filters for each possible signal direction. In either case, the measurement and inversion of a $2K$ by $2K$ correlation matrix was required at a large number of frequencies in the band of the signal. Mermoz[11] proposed a similar scheme for narrowband known signals, using the signal-to-noise ratio as a performance criterion. Shor[12] also used a signal-to-noise-ratio criterion to detect narrowband pulse signals. He proposed that the sensors be switched off when the signal was known to be absent, and a pilot signal injected as if it were a noise-free signal impinging on the array from a specified direction. The need for specific matrix inversion was circumvented by calculating the gradient of the ratio between the output power due to pilot signal and the output power due to noise, and using the method of steepest descent. At the same time, the number of correlation measurements required was reduced, by Shor's procedure, to $4K$ at each step in the adjustment of the processor. Both Mermoz and Shor have suggested the possibility of real-time adaptation.

This paper presents a potentially simpler scheme for obtaining the desired array processing improvement in real time. The performance criterion used is minimum mean-square error. The statistics of the signal are assumed

Reprinted from *Proc. IEEE*, vol. 55, pp. 2143-2159, Dec. 1967.

to be known, but no prior knowledge or direct measurements of the noise field are required in this scheme. The adaptive array processor considered in the study may be automatically adjusted (adapted) according to a simple iterative algorithm, and the procedure does not directly involve the computation of any correlation coefficients or the inversion of matrices. The input signals are used only once, as they occur, in the adaptation process. There is no need to store past input data; but there is a need to store the processor adjustment values, i.e., the processor weighting coefficients ("weights"). Methods of adaptation are presented here, which may be implemented with either analog or digital adaptive circuits, or by digital-computer realization.

Directional and Spatial Filtering

An example of a linear-array receiving antenna is shown in Fig. 1(a) and (b). The antenna of Fig. 1(a) consists of seven isotropic elements spaced $\lambda_0/2$ apart along a straight line, where λ_0 is the wavelength of the center frequency f_0 of the array. The received signals are summed to produce an array output signal. The directivity pattern, i.e., the relative sensitivity of response to signals from various directions, is plotted in this figure in a plane over an angular range of $-\pi/2<\theta<\pi/2$ for frequency f_0. This pattern is symmetric about the vertical line $\theta=0$. The main lobe is centered at $\theta=0$. The largest-amplitude side lobe, at $\theta=24°$, has a maximum sensitivity which is 12.5 dB below the maximum main-lobe sensitivity. This pattern would be different if it were plotted at frequencies other than f_0.

The same array configuration is shown in Fig. 1(b); however, in this case the output of each element is delayed in time before being summed. The resulting directivity pattern now has its main lobe at an angle of ψ radians, where

$$\psi = \sin^{-1}\left(\frac{\lambda_0 \delta f_0}{d}\right) = \sin^{-1}\left(\frac{c\delta}{d}\right) \tag{1}$$

in which

f_0 = frequency of received signal
λ_0 = wavelength at frequency f_0
δ = time-delay difference between neighboring-element outputs
d = spacing between antenna elements
c = signal propagation velocity $=\lambda_0 f_0$.

The sensitivity is maximum at angle ψ because signals received from a plane wave source incident at this angle, and delayed as in Fig. 1(b), are in phase with one another and produce the maximum output signal. For the example illustrated in the figure, $d=\lambda_0/2$, $\delta=(0.12941/f_0)$, and therefore $\psi=\sin^{-1}(2\delta f_0)=15°$.

There are many possible configurations for phased arrays. Fig. 2(a) shows one such configuration where each of the antenna-element outputs is weighted by two weights in parallel, one being preceded by a time delay of a quarter of a cycle at frequency f_0 (i.e., a 90° phase shift), denoted by $1/(4f_0)$. The output signal is the sum of all the weighted signals, and since all weights are set to unit values, the directivity pattern at frequency f_0 is by symmetry the same as that of Fig. 1(a). For purposes of illustration, an interfering directional sinusoidal "noise" of frequency f_0 incident on the array is shown in Fig. 2(a), indicated by the dotted arrow. The angle of incidence (45.5°) of this noise is such that it would be received on one of the side lobes of the directivity pattern with a sensitivity only 17 dB less than that of the main lobe at $\theta=0°$.

If the weights are now set as indicated in Fig. 2(b), the directivity pattern at frequency f_0 becomes as shown in that figure. In this case, the main lobe is almost unchanged from that shown in Figs. 1(a) and 2(a), while the particular side lobe that previously intercepted the sinusoidal noise in Fig. 2(a) has been shifted so that a null is now placed in the direction of that noise. The sensitivity in the noise direction is 77 dB below the main lobe sensitivity, improving the noise rejection by 60 dB.

A simple example follows which illustrates the existence and calculation of a set of weights which will cause a signal from a desired direction to be accepted while a "noise" from a different direction is rejected. Such an example is illustrated in Fig. 3. Let the signal arriving from the desired direction $\theta=0°$ be called the "pilot" signal $p(t)=P \sin \omega_0 t$, where $\omega_0 \triangleq 2\pi f_0$, and let the other signal, the noise, be chosen as $n(t)=N \sin \omega_0 t$, incident to the receiving array at an angle $\theta=\pi/6$ radians. Both the pilot signal and the noise signal are assumed for this example to be at exactly the same frequency f_0. At a point in space midway between the antenna array elements, the signal and the noise are assumed to be in phase. In the example shown, there are two identical omnidirectional array elements, spaced $\lambda_0/2$ apart. The signals received by each element are fed to two variable weights, one weight being preceded by a quarter-wave time delay of $1/(4f_0)$. The four weighted signals are then summed to form the array output.

The problem of obtaining a set of weights to accept $p(t)$ and reject $n(t)$ can now be studied. Note that with any set of nonzero weights, the output is of the form $A \sin(\omega_0 t+\phi)$, and a number of solutions exist which will make the output be $p(t)$. However, the output of the array must be independent of the amplitude and phase of the noise signal if the array is to be regarded as rejecting the noise. Satisfaction of this constraint leads to a unique set of weights determined as follows.

The array output due to the pilot signal is

$$P[(w_1 + w_3)\sin \omega_0 t + (w_2 + w_4)\sin(\omega_0 t - \pi/2)]. \tag{2}$$

For this output to be equal to the desired output of $p(t)=P \sin \omega_0 t$ (which is the pilot signal itself), it is necessary that

$$\left.\begin{aligned} w_1 + w_3 &= 1 \\ w_2 + w_4 &= 0 \end{aligned}\right\}. \tag{3}$$

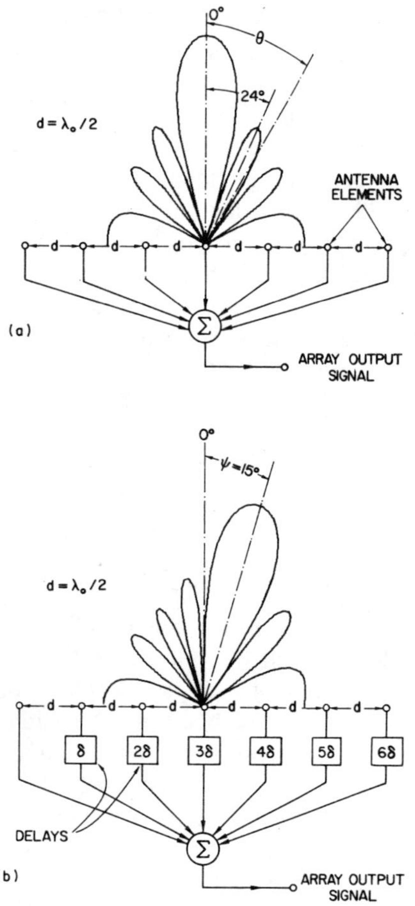

Fig. 1. Directivity pattern for a linear array. (a) Simple array. (b) Delays added.

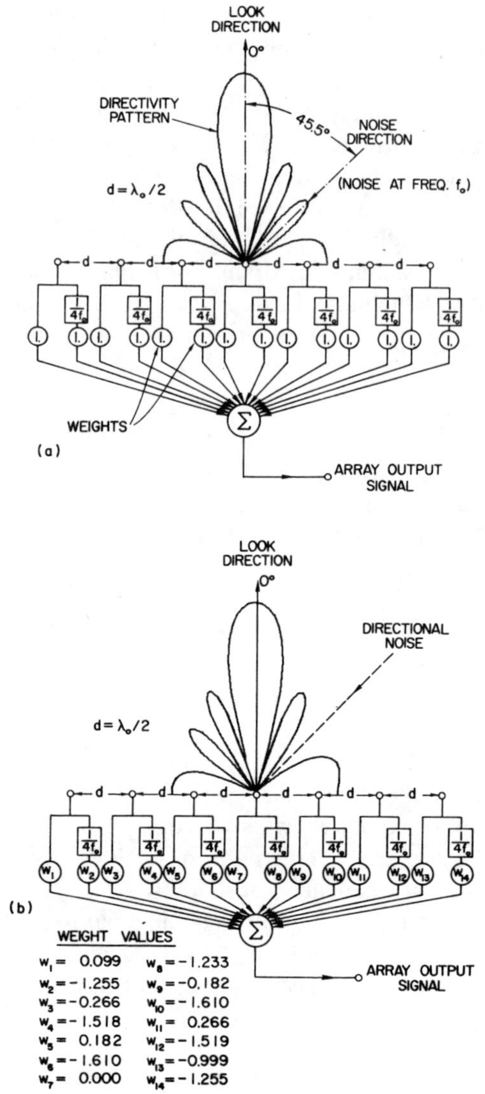

Fig. 2. Directivity pattern of linear array. (a) With equal weighting. (b) With weighting for noise elimination.

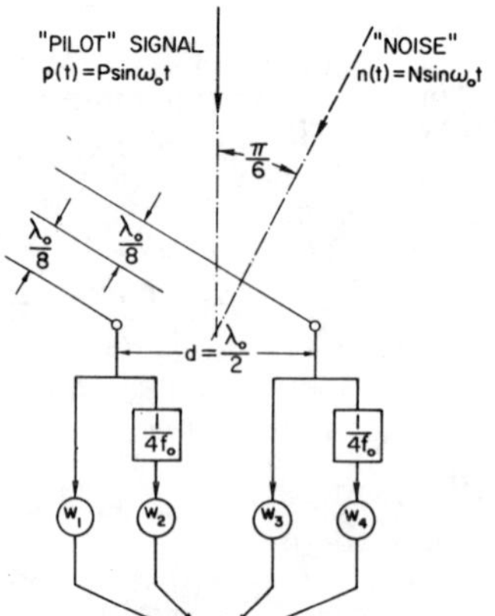

Fig. 3. Array configuration for noise elimination example.

With respect to the midpoint between the antenna elements, the relative time delays of the noise at the two antenna elements are $\pm[1/(4f_0)]\sin\pi/6 = \pm 1/(8f_0) = \pm\lambda_0/(8c)$, which corresponds to phase shifts of $\pm\pi/4$ at frequency f_0. The array output due to the incident noise at $\theta=\pi/6$ is then

$$N\left[w_1 \sin\left(\omega_0 t - \frac{\pi}{4}\right) + w_2 \sin\left(\omega_0 t - \frac{3\pi}{4}\right) + w_3 \sin\left(\omega_0 t + \frac{\pi}{4}\right) + w_4 \sin\left(\omega_0 t - \frac{\pi}{4}\right)\right]. \quad (4)$$

For this response to equal zero, it is necessary that

$$\left.\begin{aligned} w_1 + w_4 &= 0 \\ w_2 - w_3 &= 0 \end{aligned}\right\}. \quad (5)$$

Thus the set of weights that satisfies the signal and noise response requirements can be found by solving (3) and (5) simultaneously. The solution is

$$w_1 = \tfrac{1}{2},\ w_2 = \tfrac{1}{2},\ w_3 = \tfrac{1}{2},\ w_4 = -\tfrac{1}{2}. \quad (6)$$

With these weights, the array will have the desired properties in that it will accept a signal from the desired direction, while rejecting a noise, even a noise which is at the same frequency f_0 as the signal, because the noise comes from a different direction than does the signal.

The foregoing method of calculating the weights is more illustrative than practical. This method is usable when there are only a small number of directional noise sources, when the noises are monochromatic, and when the directions of the noises are known *a priori*. A practical processor should not require detailed information about the number and the nature of the noises. The adaptive processor described in the following meets this requirement. It recursively solves a sequence of simultaneous equations, which are generally overspecified, and it finds solutions which minimize the mean-square error between the pilot signal and the total array output.

Configurations of Adaptive Arrays

Before discussing methods of adaptive filtering and signal processing to be used in the adaptive array, various spatial and electrical configurations of antenna arrays will be considered. An adaptive array configuration for processing narrowband signals is shown in Fig. 4. Each individual antenna element is shown connected to a variable weight and to a quarter-period time delay whose output is in turn connected to another variable weight. The weighted signals are summed, as shown in the figure. The signal, assumed to be either monochromatic or narrowband, is received by the antenna element and is thus weighted by a complex gain factor $Ae^{j\phi}$. Any phase angle $\phi = -\tan^{-1}(w_2/w_1)$ can be chosen by setting the two weight values, and the magnitude of this complex gain factor $A=\sqrt{w_1^2+w_2^2}$ can take on a wide range of values limited only by the range limitations of the two individual weights. The latter can assume a continuum of both positive and negative values.

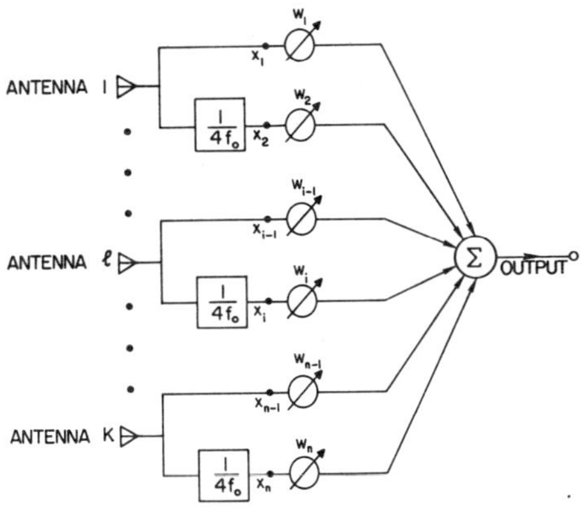

Fig. 4. Adaptive array configuration for receiving narrowband signals.

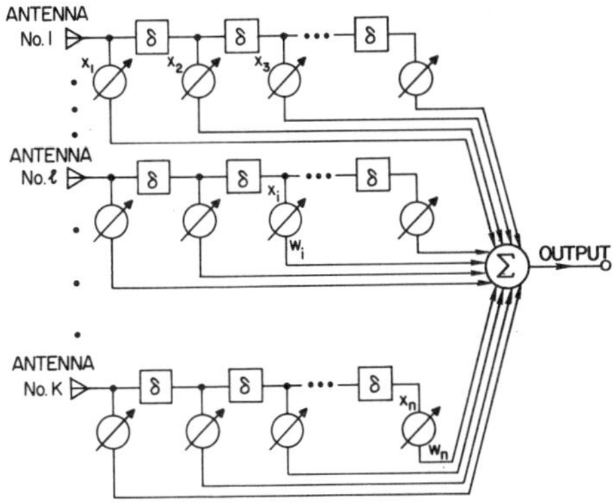

Fig. 5. Adaptive array configuration for receiving broadband signals.

Thus the two weights and the $1/(4f_0)$ time delay provide completely adjustable linear processing for narrowband signals received by each individual antenna element.

The full array of Fig. 4 represents a completely general way of combining the antenna-element signals in an adjustable linear structure when the received signals and noises are narrowband. It should be realized that the same generality (for narrowband signals) can be achieved even when the time delays do not result in a phase shift of exactly $\pi/2$ at the center frequency f_0. Keeping the phase shifts close to $\pi/2$ is desirable for keeping required weight values small, but is not necessary in principle.

When one is interested in receiving signals over a wide band of frequencies, each of the phase shifters in Fig. 4 can be replaced by a tapped-delay-line network as shown in Fig. 5. This tapped delay line permits adjustment of gain and phase as desired at a number of frequencies over the band of interest. If the tap spacing is sufficiently close, this network approximates the ideal filter which would allow complete control of the gain and phase at each frequency in the passband.

Adaptive Signal Processors

Once the form of network connected to each antenna element has been chosen, as shown for example in Fig. 4 or Fig. 5, the next step is to develop an adaptation procedure which can be used to adjust automatically the multiplying weights to achieve the desired spatial and frequency filtering.

The procedure should produce a given array gain in the specified look direction while simultaneously nulling out interfering noise sources.

Fig. 6 shows an adaptive signal-processing element. If this element were combined with an output-signal quantizer, it would then comprise an adaptive threshold logic unit. Such an element has been called an "Adaline"[13] or a threshold logic unit (TLU).[14] Applications of the adaptive threshold element have been made in pattern-recognition systems and in experimental adaptive control systems.[2],[3],[14]–[17]

In Fig. 6 the input signals $x_1(t), \cdots, x_i(t), \cdots, x_n(t)$ are the same signals that are applied to the multiplying weights $w_1, \cdots, w_i, \cdots, w_n$ shown in Fig. 4 or Fig. 5. The heavy lines show the paths of signal flow; the lighter lines show functions related to weight-changing or adaptation processes.

The output signal $s(t)$ in Fig. 6 is the weighted sum

$$s(t) = \sum_{i=1}^{n} x_i(t) w_i \tag{7}$$

where n is the number of weights; or, using vector notation

$$s(t) = \boldsymbol{W}^T \boldsymbol{X}(t) \tag{8}$$

where $\boldsymbol{W}^T$ is the transpose of the weight vector

$$\boldsymbol{W} \triangleq \begin{bmatrix} w_1 \\ \vdots \\ w_i \\ \vdots \\ w_n \end{bmatrix}$$

and the signal-input vector is

$$\boldsymbol{X}(t) \triangleq \begin{bmatrix} x_1(t) \\ \vdots \\ x_i(t) \\ \vdots \\ x_n(t) \end{bmatrix}.$$

For digital systems, the input signals are in discrete-time sampled-data form and the output is written

$$s(j) = \boldsymbol{W}^T \boldsymbol{X}(j) \tag{9}$$

where the index j indicates the jth sampling instant.

In order that adaptation take place, a "desired response" signal, $d(t)$ when continuous or $d(j)$ when sampled, must be supplied to the adaptive element. A method for obtaining this signal for adaptive antenna array processing will be discussed in a following section.

The difference between the desired response and the output response forms the error signal $\varepsilon(j)$:

$$\varepsilon(j) = d(j) - \boldsymbol{W}^T \boldsymbol{X}(j). \tag{10}$$

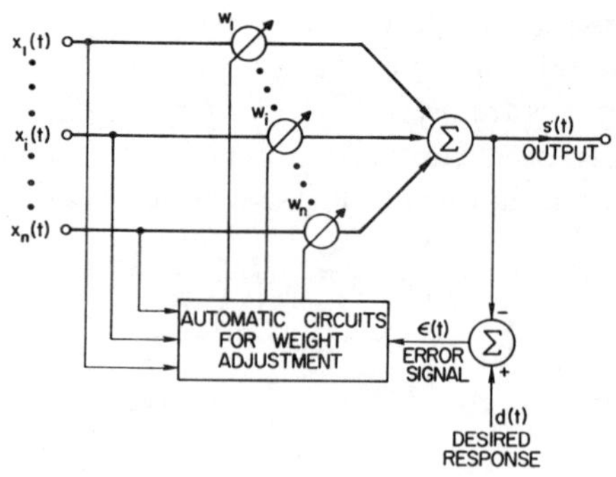

Fig. 6. Basic adaptive element.

This signal is used as a control signal for the "weight adjustment circuits" of Fig. 6.

Solving Simultaneous Equations

The purpose of the adaptation or weight-changing processes is to find a set of weights that will permit the output response of the adaptive element at each instant of time to be equal to or as close as possible to the desired response. For each input-signal vector $\boldsymbol{X}(j)$, the error $\varepsilon(j)$ of (10) should be made as small as possible.

Consider the finite set of linear simultaneous equations

$$\begin{cases} \boldsymbol{W}^T \boldsymbol{X}(1) = d(1) \\ \boldsymbol{W}^T \boldsymbol{X}(2) = d(2) \\ \quad \vdots \qquad \vdots \\ \boldsymbol{W}^T \boldsymbol{X}(j) = d(j) \\ \quad \vdots \qquad \vdots \\ \boldsymbol{W}^T \boldsymbol{X}(N) = d(N) \end{cases} \tag{11}$$

where N is the total number of input-signal vectors; each vector is a measurement of an underlying n-dimensional random process. There are N equations, corresponding to N instants of time at which the output response values are of concern; there are n "unknowns," the n weight values which form the components of $\boldsymbol{W}$. The set of equations (11) will usually be overspecified and inconsistent, since in the present application, with an ample supply of input data, it is usual that $N \gg n$. [These equations did have a solution in the simple example represented in Fig. 3. The solution is given in (6). Although the simultaneous equations (3) in that example appear to be different from (11), they are really the same, since those in (3) are in a specialized form for the case when all inputs are deterministic sinusoids which can be easily specified over all time in terms of amplitudes, phases, and frequencies.]

When N is very large compared to n, one is generally interested in obtaining a solution of a set of N equations [each equation in the form of (10)] which minimizes the sum of the squares of the errors. That is, a set of weights W is found to minimize

$$\sum_{j=1}^{N} \varepsilon^2(j). \tag{12}$$

When the input signals can be regarded as stationary stochastic variables, one is usually interested in finding a set of weights to minimize mean-square error. The quantity of interest is then the expected value of the square of the error, i.e., the mean-square error, given by

$$E[\varepsilon^2(j)] \triangleq \overline{\varepsilon^2}. \tag{13}$$

The set of weights that minimizes mean-square error can be calculated by squaring both sides of (10) which yields

$$\varepsilon^2(j) = d^2(j) + \boldsymbol{W}^T\boldsymbol{X}(j)\boldsymbol{X}(j)^T\boldsymbol{W} - 2d(j)\boldsymbol{W}^T\boldsymbol{X}(j) \tag{14}$$

and then taking the expected value of both sides of (14)

$$\begin{aligned} E[\varepsilon^2(j)] &= E[d^2 + \boldsymbol{W}^T\boldsymbol{X}(j)\boldsymbol{X}^T(j)\boldsymbol{W} - 2\boldsymbol{W}^T d(j)\boldsymbol{X}(j)] \\ &= E[d^2] + \boldsymbol{W}^T\boldsymbol{\Phi}(x, x)\boldsymbol{W} - 2\boldsymbol{W}^T\boldsymbol{\Phi}(x, d) \end{aligned} \tag{15}$$

where

$$\boldsymbol{\Phi}(x, x) \triangleq E[\boldsymbol{X}(j)\boldsymbol{X}^T(j)] \triangleq E\begin{bmatrix} x_1x_1 & x_1x_2 \cdots x_1x_n \\ x_2x_1 & \cdots x_2x_n \\ \vdots & \\ x_nx_1 & x_nx_n \end{bmatrix} \tag{16}$$

and

$$\boldsymbol{\Phi}(x, d) \triangleq E[\boldsymbol{X}(j)d(j)] \triangleq E\begin{bmatrix} x_1d \\ x_2d \\ \vdots \\ x_id \\ \vdots \\ x_nd \end{bmatrix}. \tag{17}$$

The symmetric matrix $\boldsymbol{\Phi}(x, x)$ is a matrix of cross correlations and autocorrelations of the input signals to the adaptive element, and the column matrix $\boldsymbol{\Phi}(x, d)$ is the set of cross correlations betweeen the n input signals and the desired response signal.

The mean-square error defined in (15) is a quadratic function of the weight values. The components of the gradient of the mean-square-error function are the partial derivatives of the mean-square error with respect to the weight values. Differentiating (15) with respect to $\boldsymbol{W}$ yields the gradient $\nabla E[\varepsilon^2]$, a linear function of the weights,

$$\nabla E[\varepsilon^2] = 2\boldsymbol{\Phi}(x, x)\boldsymbol{W} - 2\boldsymbol{\Phi}(x, d). \tag{18}$$

When the choice of the weights is optimized, the gradient is zero. Then

$$\begin{aligned} \boldsymbol{\Phi}(x, x)\boldsymbol{W}_{\text{LMS}} &= \boldsymbol{\Phi}(x, d) \\ \boldsymbol{W}_{\text{LMS}} &= \boldsymbol{\Phi}^{-1}(x, x)\boldsymbol{\Phi}(x, d). \end{aligned} \tag{19}$$

The optimum weight vector $\boldsymbol{W}_{\text{LMS}}$ is the one that gives the least mean-square error. Equation (19) is the Wiener-Hopf equation, and is the equation for the multichannel least-squares filter used by Burg[18] and Claerbout[19] in the processing of digital seismic array data.

One way of finding the optimum set of weight values is to solve (19). This solution is generally straightforward, but presents serious computational problems when the number of weights n is large and when data rates are high. In addition to the necessity of inverting an $n \times n$ matrix, this method may require as many as $n(n+1)/2$ autocorrelation and cross-correlation measurements to obtain the elements of $\boldsymbol{\Phi}(x, x)$. Furthermore, this process generally needs to be continually repeated in most practical situations where the input signal statistics change slowly. No perfect solution of (19) is possible in practice because of the fact that an infinite statistical sample would be required to estimate perfectly the elements of the correlation matrices.

Two methods for finding approximate solutions to (19) will be presented in the following. Their accuracy is limited by statistical sample size, since they find weight values based on finite-time measurements of input-data signals. These methods do not require explicit measurements of correlation functions or matrix inversion. They are based on gradient-search techniques applied to mean-square-error functions. One of these methods, the LMS algorithm, does not even require squaring, averaging, or differentiation in order to make use of gradients of mean-square-error functions. The second method, a relaxation method, will be discussed later.

The LMS Algorithm

A number of weight-adjustment procedures or algorithms exist which minimize the mean-square error. Minimization is usually accomplished by gradient-search techniques. One method that has proven to be very useful is the LMS algorithm.[1]–[3],[17] This algorithm is based on the method of steepest descent. Changes in the weight vector are made along the direction of the estimated gradient vector. Accordingly,

$$\boldsymbol{W}(j+1) = \boldsymbol{W}(j) + k_s\hat{\nabla}(j) \tag{20}$$

where

$\boldsymbol{W}(j) \triangleq$ weight vector before adaptation
$\boldsymbol{W}(j+1) \triangleq$ weight vector after adaptation
$k_s \triangleq$ scalar constant controlling rate of convergence and stability ($k_s<0$)
$\hat{\nabla}(j) \triangleq$ estimated gradient vector of $\overline{\varepsilon^2}$ with respect to $\boldsymbol{W}$.

One method for obtaining the estimated gradient of the mean-square-error function is to take the gradient of a single time sample of the squared error

$$\hat{\nabla}(j) = \nabla[\varepsilon^2(j)] = 2\varepsilon(j)\nabla[\varepsilon(j)].$$

From (10)

$$\begin{aligned} \nabla[\varepsilon(j)] &= \nabla[d(j) - \boldsymbol{W}^T(j)X(j)] \\ &= -\boldsymbol{X}(j). \end{aligned}$$

Thus

$$\hat{\nabla}(j) = -2\varepsilon(j)\boldsymbol{X}(j). \tag{21}$$

The gradient estimate of (21) is unbiased, as will be shown by the following argument. For a given weight vector $\boldsymbol{W}(j)$, the expected value of the gradient estimate is

$$E[\hat{\nabla}(j)] = -2E[\{d(j) - W^T(j)X(j)\}X(j)]$$
$$= -2[\Phi(x, d) - W^T(j)\Phi(x, x)]. \quad (22)$$

Comparing (18) and (22), we see that

$$E[\hat{\nabla}(j)] = \nabla E[\varepsilon^2]$$

and therefore, for a given weight vector, the expected value of the estimate equals the true value.

Using the gradient estimation formula given in (21), the weight iteration rule (20) becomes

$$W(j + 1) = W(j) - 2k_s\varepsilon(j)X(j) \quad (23)$$

and the next weight vector is obtained by adding to the present weight vector the input vector scaled by the value of the error.

The LMS algorithm is given by (23). It is directly usable as a weight-adaptation formula for digital systems. Fig. 7(a) shows a block-diagram representation of this equation in terms of one component w_i of the weight vector W. An equivalent differential equation which can be used in analog implementation of continuous systems (see Fig. 7(b)) is given by

$$\frac{d}{dt} W(t) = -2k_s\varepsilon(t)X(t).$$

This equation can also be written as

$$W(t) = -2k_s \int_0^t \varepsilon(\xi)X(\xi)\, d\xi.$$

Fig. 8 shows how circuitry of the type indicated in Fig. 7(a) or (b) might be incorporated into the implementation of the basic adaptive element of Fig. 6.

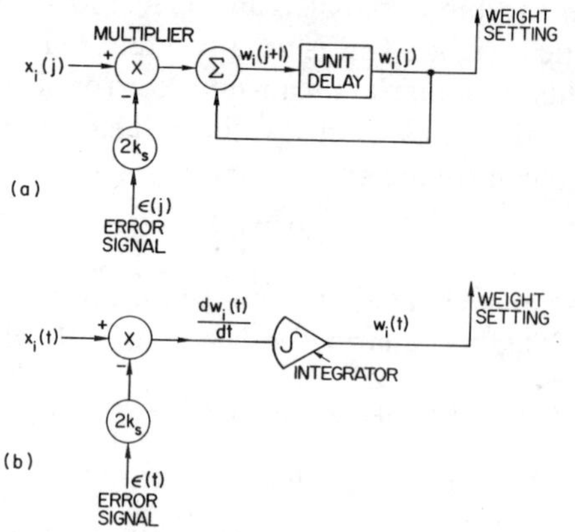

Fig. 7. Block diagram representation of LMS algorithm. (a) Digital realization. (b) Analog realization.

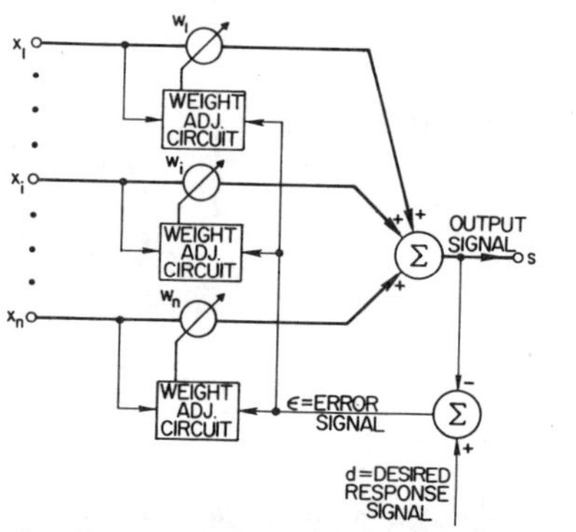

Fig. 8. Analog/digital implementation of LMS weight-adjustment algorithm.

Convergence of the Mean of the Weight Vector

For the purpose of the following discussion, we assume that the time between successive iterations of the LMS algorithm is sufficiently long so that the sample input vectors $X(j)$ and $X(j+1)$ are uncorrelated. This assumption is common in the field of stochastic approximation.[20]–[22]

Because the weight vector $W(j)$ is a function *only* of the input vectors $X(j-1), X(j-2), \cdots, X(0)$ [see (23)] and because the successive input vectors are uncorrelated, $W(j)$ is independent of $X(j)$. For stationary input processes meeting this condition, the expected value $E[W(j)]$ of the weight vector after a large number of iterations can then be shown to converge to the Wiener solution given by (19). Taking the expected value of both sides of (23), we obtain a difference equation in the expected value of the weight vector

$$E[W(j + 1)] = E[W(j)] - 2k_sE[\{d(j) - W^T(j)X(j)\}X(j)]$$
$$= [I + 2k_s\Phi(x, x)]E[W(j)] - 2k_s\Phi(x, d) \quad (24)$$

where I is the identity matrix. With an initial weight vector $W(0)$, $j+1$ iterations of (24) yield

$$E[W(j + 1)] = [I + 2k_s\Phi(x, x)]^{j+1}W(0) - 2k_s \sum_{i=0}^{j} [I + 2k_s\Phi(x, x)]^i\Phi(x, d). \quad (25)$$

Equation (25) may be put in diagonal form by using the appropriate similarity transformation Q for the matrix $\Phi(x, x)$, that is,

$$\Phi(x, x) = Q^{-1}EQ$$

where

$$E \triangleq \begin{bmatrix} e_1 & 0 & \cdots & 0 \\ 0 & e_2 & \cdots & 0 \\ \cdot & & \cdot & \cdot \\ \cdot & & & \cdot \\ \cdot & & & \cdot \\ 0 & 0 & \cdots & e_n \end{bmatrix}$$

is the diagonal matrix of eigenvalues. The eigenvalues are all positive, since $\Phi(x, x)$ is positive definite [see (16)]. Equation (25) may now be expressed as

$$E[W(j + 1)] = [I + 2k_sQ^{-1}EQ]^{j+1}W(0) - 2k_s \sum_{i=0}^{j} [I + 2k_sQ^{-1}EQ]^i\Phi(x, d)$$
$$= Q^{-1}[I + 2k_sE]^{j+1}QW(0) - 2k_sQ^{-1} \sum_{i=0}^{j} [I + 2k_sE]^iQ\Phi(x, d). \quad (26)$$

Consider the diagonal matrix $[\boldsymbol{I}+2k_s\boldsymbol{E}]$. As long as its diagonal terms are all of magnitude less than unity

$$\lim_{j\to\infty}[\boldsymbol{I}+2k_s\boldsymbol{E}]^{j+1}\to 0$$

and the first term of (26) vanishes as the number of iterations increases. The second term in (26) generally converges to a nonzero limit. The summation factor $\sum_{i=0}^{j}[\boldsymbol{I}+2k_s\boldsymbol{E}]^i$ becomes

$$\lim_{j\to\infty}\sum_{i=0}^{j}[\boldsymbol{I}+2k_s\boldsymbol{E}]^i=-\frac{1}{2k_s}\boldsymbol{E}^{-1}$$

where the formula for the sum of a geometric series has been used, that is,

$$\sum_{i=0}^{\infty}(1+2k_se_p)^i=\frac{1}{1-(1+2k_se_p)}=\frac{-1}{2k_se_p}.$$

Thus, in the limit, (26) becomes

$$\lim_{j\to\infty}E[\boldsymbol{W}(j+1)]=\boldsymbol{Q}^{-1}\boldsymbol{E}^{-1}\boldsymbol{Q}\boldsymbol{\Phi}(x,d)$$
$$=\boldsymbol{\Phi}^{-1}(x,x)\boldsymbol{\Phi}(x,d).$$

Comparison of this result with (19) shows that as the number of iterations increases without limit, the expected value of the weight vector converges to the Wiener solution.

Convergence of the mean of the weight vector to the Wiener solution is insured if and only if the proportionality constant k_s is set within certain bounds. Since the diagonal terms of $[\boldsymbol{I}+2k_s\boldsymbol{E}]$ must all have magnitude less than unity, and since all eigenvalues in $\boldsymbol{E}$ are positive, the bounds on k_s are given by

$$|1+2k_se_{\max}|<1$$

or

$$\frac{-1}{e_{\max}}<k_s<0 \tag{27}$$

where $e_{\max}$ is the maximum eigenvalue of $\boldsymbol{\Phi}(x, x)$. This convergence condition on k_s can be related to the total input power as follows.

Since

$$e_{\max}\leqq \operatorname{trace}[\boldsymbol{\Phi}(x,x)] \tag{28}$$

where

$$\operatorname{trace}[\boldsymbol{\Phi}(x,x)]\triangleq E[\boldsymbol{X}^T(j)\boldsymbol{X}(j)]$$
$$=\sum_{i=1}^{n}E[x_i^2]\triangleq \text{total input power},$$

it follows that satisfactory convergence can be obtained with

$$\frac{-1}{\sum_{i=1}^{n}E[x_i^2]}<k_s<0.$$

In practice, when slow, precise adaptation is desired, k_s is usually chosen such that

$$\frac{-1}{\sum_{i=1}^{n}E[x_i^2]}\ll k_s<0. \tag{29}$$

It is the opinion of the authors that the assumption of independent successive input samples used in the foregoing convergence proof is overly restrictive. That is, convergence of the mean of the weight vector to the LMS solution can be achieved under conditions of highly correlated input samples. In fact, the computer-simulation experiments described in this paper *do not* satisfy the condition of independence.

Time Constants and Learning Curve with LMS Adaptation

State-variable methods, which are widely used in modern control theory, have been applied by Widrow[1] and Koford and Groner[2] to the analysis of stability and time constants (related to rate of convergence) of the LMS algorithm. Considerable simplifications in the analysis have been realized by expressing transient phenomena of the system adjustments (which take place during the adaptation process) in terms of the normal coordinates of the system. As shown by Widrow,[1] the weight values undergo transients during adaptation. The transients consist of sums of exponentials with time constants given

$$\tau_p=\frac{1}{2(-k_s)e_p},\ p=1,2,\cdots,n \tag{30}$$

where e_p is the pth eigenvalue of the input-signal correlation matrix $\boldsymbol{\Phi}(x, x)$.

In the special case when all eigenvalues are equal, all time constants are equal. Accordingly,

$$\tau=\frac{1}{2(-k_s)e}.$$

One very useful way to monitor the progress of an adaptive process is to plot or display its "learning curve." When mean-square error is the performance criterion being used, one can plot the expected mean-square error at each stage of the learning process as a function of the number of adaptation cycles. Since the underlying relaxation phenomenon which takes place in the weight values is of exponential nature, and since from (15) the mean-square error is a quadratic form in the weight values, the transients in the mean-square-error function must also be exponential in nature.

When all the time constants are equal, the mean-square-error learning curve is a pure exponential with a time constant

$$\tau_{\text{mse}}=\frac{\tau}{2}=\frac{1}{4(-k_s)e}.$$

The basic reason for this is that the square of an exponential function is an exponential with half the time constant.

Estimation of the rate of adaptation is more complex when the eigenvalues are unequal.

When actual experimental learning curves are plotted, they are generally of the form of noisy exponentials because of the inherent noise in the adaptation process. The slower the adaptation, the smaller will be the amplitude of the noise apparent in the learning curve.

Misadjustment with LMS Adaptation

All adaptive or learning systems capable of adapting at real-time rates experience losses in performance because their system adjustments are based on statistical averages taken with limited sample sizes. The faster a system adapts, in general, the poorer will be its expected performance.

When the LMS algorithm is used with the basic adaptive element of Fig. 8, the expected level of mean-square error will be greater than that of the Wiener optimum system whose weights are set in accordance with (19). The longer the time constants of adaptation, however, the closer the expected performance comes to the Wiener optimum performance. To get the Wiener performance, i.e., to achieve the minimum mean-square error, one would have to know the input statistics *a priori*, or, if (as is usual) these statistics are unknown, they would have to be measured with an arbitrarily large statistical sample.

When the LMS adaptation algorithm is used, an excess mean-square error therefore develops. A measure of the extent to which the adaptive system is misadjusted as compared to the Wiener optimum system is determined in a performance sense by the ratio of the excess mean-square error to the minimum mean-square error. This dimensionless measure of the loss in performance is defined as the "misadjustment" M. For LMS adaptation of the basic adaptive element, it is shown by Widrow[1] that

$$\text{Misadjustment } M = \frac{1}{2} \sum_{p=1}^{n} \frac{1}{\tau_p}. \tag{31}$$

The value of the misadjustment depends on the time constants (settling times) of the filter adjustment weights. Again, in the special case when all the time constants are equal, *M is proportional to the number of weights and inversely proportional to the time constant.* That is,

$$\begin{aligned} M &= \frac{n}{2\tau} \\ &= \frac{n}{4\tau_{\text{mse}}}. \end{aligned} \tag{32}$$

Although the foregoing results specifically apply to statistically stationary processes, the LMS algorithm can also be used with nonstationary processes. It is shown by Widrow[23] that, under certain assumed conditions, the rate of adaptation is optimized when the loss of performance resulting from adapting too rapidly equals twice the loss in performance resulting from adapting too slowly.

Adaptive Spatial Filtering

If the radiated signals received by the elements of an adaptive antenna array were to consist of signal components plus undesired noise, the signal would be reproduced (and noise eliminated) as best possible in the least-squares sense if the desired response of the adaptive processor were made to be the signal itself. This signal is not generally available for adaptation purposes, however. If it were available, there would be no need for a receiver and a receiving array.

In the adaptive antenna systems to be described here, the desired response signal is provided through the use of an artificially injected signal, the "pilot signal", which is completely known at the receiver and usually generated there. The pilot signal is constructed to have spectral and directional characteristics similar to those of the incoming signal of interest. These characteristics may, in some cases, be known *a priori* but, in general, represent estimates of the parameters of the signal of interest.

Adaptation with the pilot signal causes the array to form a beam in the pilot-signal direction having essentially flat spectral response and linear phase shift within the passband of the pilot signal. Moreover, directional noises impinging on the antenna array will cause reduced array response (nulling) in their directions within their passbands. These notions are demonstrated by experiments which will be described in the following.

Injection of the pilot signal could block the receiver and render useless its output. To circumvent this difficulty, two adaptation algorithms have been devised, the "one-mode" and the "two-mode." The two-mode process alternately adapts on the pilot signal to form the beam and then adapts on the natural inputs with the pilot signal off to eliminate noise. The array output is usable during the second mode, while the pilot signal is off. The one-mode algorithm permits listening at all times, but requires more equipment for its implementation.

The Two-Mode Adaptation Algorithm

Fig. 9 illustrates a method for providing the pilot signal wherein the latter is actually transmitted by an antenna located some distance from the array in the desired look direction. Fig. 10 shows a more practical method for providing the pilot signal. The inputs to the processor are connected either to the actual antenna element outputs (during "mode II"), or to a set of delayed signals derived from the pilot-signal generator (during "mode I"). The filters $\delta_1, \cdots, \delta_K$ (ideal time-delays if the array elements are identical) are chosen to result in a set of input signals identical with those that would appear if the array were actually receiving a radiated plane-wave pilot signal from the desired "look" direction, the direction intended for the main lobe of the antenna directivity pattern.

During adaptation in mode I, the input signals to the adaptive processor derive from the pilot signal, and the desired response of the adaptive processor is the pilot signal

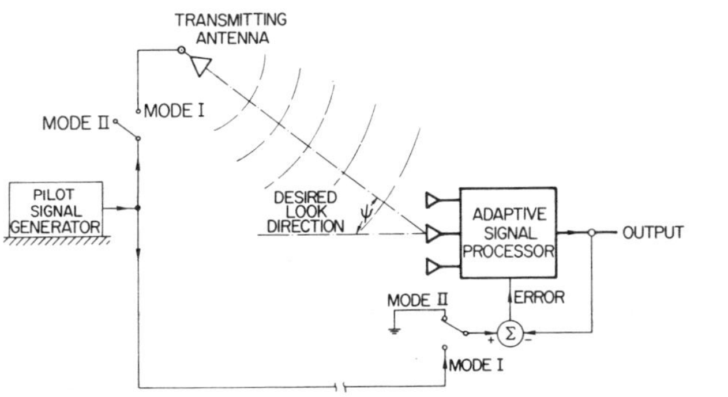

Fig. 9. Adaptation with external pilot-signal generator. Mode I: adaptation with pilot signal present; Mode II: adaptation with pilot signal absent.

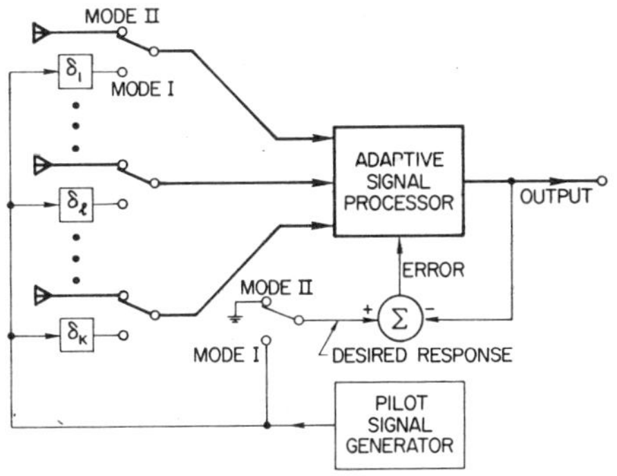

Fig. 10. Two-mode adaptation with internal pilot-signal generator. Mode I: adaptation with pilot signal present; Mode II: adaptation with pilot signal absent.

itself. If a sinusoidal pilot signal at frequency f_0 is used, for example, adapting the weights to minimize mean-square error will force the gain of the antenna array in the look direction to have a specific amplitude and a specific phase shift at frequency f_0.

During adaptation in mode II, all signals applied to the adaptive processor are received by the antenna elements from the actual noise field. In this mode, the adaptation process proceeds to eliminate all received signals, since the desired response is set to zero. Continuous operation in mode II would cause all the weight values to tend to zero, and the system would shut itself off. However, by alternating frequently between mode I and mode II and causing only small changes in the weight vector during each mode of adaptation, it is possible to maintain a beam in the desired look direction and, in addition, to minimize the reception of incident-noise power.

The pilot signal can be chosen as the sum of several sinusoids of differing frequencies. Then adaptation in mode I will constrain the antenna gain and phase in the look direction to have specific values at each of the pilot-signal frequencies. Furthermore, if several pilot signals of different simulated directions are added together, it will be possible to constrain the array gain simultaneously at various frequencies and angles when adapting in mode I. This feature affords some control of the bandwidth and beamwidth in the look direction. The two-mode adaptive process essentially minimizes the mean-square value (the total power) of all signals received by the antenna elements which are uncorrelated with the pilot signals, subject to the constraint that the gain and phase in the beam approximate predetermined values at the frequencies and angles dictated by the pilot-signal components.

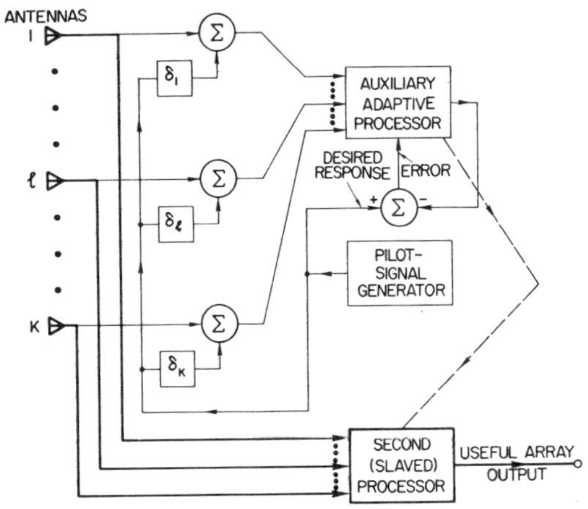

Fig. 11. Single-mode adaptation with pilot signal.

The One-Mode Adaptation Algorithm

In the two-mode adaptation algorithm the beam is formed during mode I, and the noises are eliminated in the least-squares sense (subject to the pilot-signal constraints) in mode II. Signal reception during mode I is impossible because the processor is connected to the pilot-signal generator. Reception can therefore take place only during mode II. This difficulty is eliminated in the system of Fig. 11, in which the actions of both mode I and mode II can be accomplished simultaneously. The pilot signals and the received signals enter into an auxiliary, adaptive processor, just as described previously. For this processor, the desired response is the pilot signal $p(t)$. A second weighted processor (linear element) generates the actual array output signal, but it performs no adaptation. Its input signals do not contain the pilot signal. It is slaved to the adaptive processor in such a way that its weights track the corresponding weights of the adapting system, so that it never needs to receive the pilot signal.

In the single-mode system of Fig. 11, the pilot signal is on continuously. Adaptation to minimize mean-square error will force the adaptive processor to reproduce the pilot signal as closely as possible, and, at the same time, to reject as well as possible (in the mean-square sense) all signals received by the antenna elements which are uncorrelated with the pilot signal. Thus the adaptive process forces a directivity pattern having the proper main lobe in the look direction in the passband of the pilot signal (satisfying the pilot signal constraints), and it forces nulls in the directions of the noises and in their frequency bands. Usually, the stronger the noises, the deeper are the corresponding nulls.

Computer Simulation of Adaptive Antenna Systems

To demonstrate the performance characteristics of adaptive antenna systems, many simulation experiments, involving a wide variety of array geometries and signal-

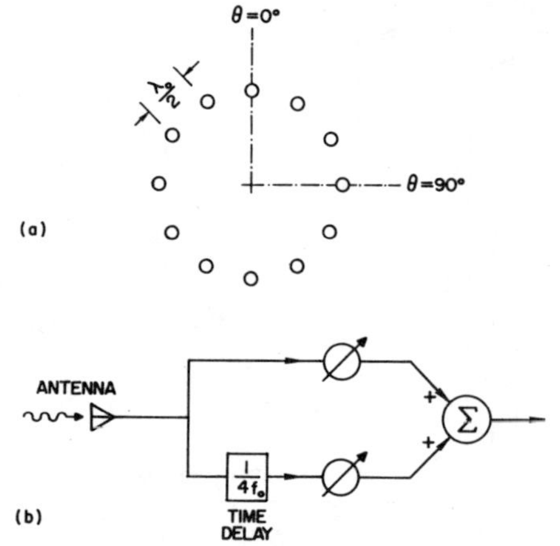

Fig. 12. Array configuration and processing for narrowband experiments.

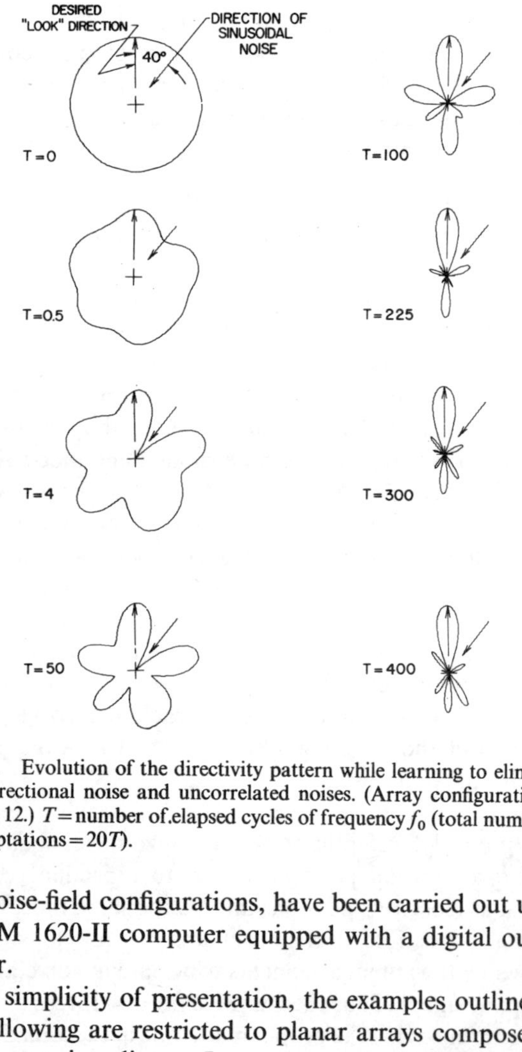

Fig. 13. Evolution of the directivity pattern while learning to eliminate a directional noise and uncorrelated noises. (Array configuration of Fig. 12.) T=number of elapsed cycles of frequency f_0 (total number of adaptations=$20T$).

and noise-field configurations, have been carried out using an IBM 1620-II computer equipped with a digital output plotter.

For simplicity of presentation, the examples outlined in the following are restricted to planar arrays composed of ideal isotropic radiators. In every case, the LMS adaptation algorithm was used. All experiments were begun with the initial condition that all weight values were equal.

Narrowband Processor Experiments

Fig. 12 shows a twelve-element circular array and signal processor which was used to demonstrate the performance of the narrowband system shown in Fig. 4. In the first computer simulation, the two-mode adaptation algorithm was used. The pilot signal was a unit-amplitude sine wave (power=0.5, frequency f_0) which was used to train the array to look in the $\theta=0°$ direction. The noise field consisted of a sinusoidal noise signal (of the same frequency and power as the pilot signal) incident at angle $\theta=40°$, and a small amount of random, uncorrelated, zero-mean, "white" Gaussian noise of variance (power)=0.1 at each antenna element. In this simulation, the weights were adapted using the LMS two-mode algorithm.

Fig. 13 shows the sequence of directivity patterns which evolved during the "learning" process. These computer-plotted patterns represent the decibel sensitivity of the array at frequency f_0. Each directivity pattern is computed from the set of weights resulting at various stages of adaptation. The solid arrow indicates the direction of arrival of the interfering sine-wave noise source. Notice that the initial directivity pattern is essentially circular. This is due to the symmetry of the antenna array elements and of the initial weight values. A timing indicator T, the number of elapsed cycles of frequency f_0, is presented with each directivity pattern. The total number of adaptations equals $20T$ in these experiments. Note that if $f_0=1$ kHz, $T=1$ corresponds to 1 ms real time; if $f_0=1$ MHz, $T=1$ corresponds to 1 μs, etc.

Several observations can be made from the series of directivity patterns of Fig. 13. Notice that the sensitivity of the array in the look direction is essentially constant during the adaptation process. Also notice that the array sensitivity drops very rapidly in the direction of the sinusoidal noise source; a deep notch in the directivity pattern forms in the noise direction as the adaptation process progresses. After the adaptive transients died out, the array sensitivity in the noise direction was 27 dB below that of the array in the desired look direction.

The total noise power in the array output is the sum of the sinusoidal noise power due to the directional noise source plus the power due to the "white" Gaussian, mutually uncorrelated noise-input signals. The total noise power generally drops as the adaptation process commences, until it reaches an irreducible level.

A plot of the total received noise power as a function of T is shown in Fig. 14. This curve may be called a "learning curve." Starting with the initial weights, the total output noise power was 0.65, as shown in the figure. After adaptation, the total output noise power was 0.01. In this noise field, the signal-to-noise ratio of the array[1] after adaptation was better than that of a single isotropic receiving element by a factor of about 60.

A second experiment using the same array configuration and the two-mode adaptive process was performed to investigate adaptive array performance in the presence of several interfering directional noise sources. In this example, the noise field was composed of five directional sinus-

[1] Signal-to-noise ratio is defined as

$$\text{SNR} = \frac{\text{array output power due to signal}}{\text{array output power due to noise}}.$$

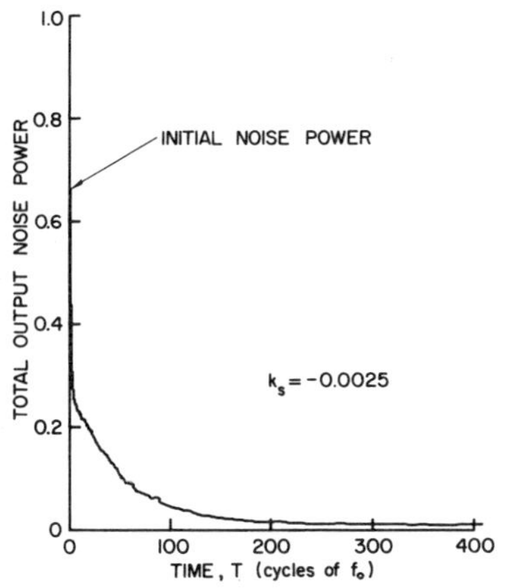

Fig. 14. Learning curve for narrowband system of Fig. 12, with noise from one direction only.

TABLE I

SENSITIVITIES OF ARRAY IN DIRECTIONS OF THE FIVE NOISE SOURCES OF FIG. 15, AFTER ADAPTATION

Noise Direction (degrees)	Noise Frequency (times f_0)	Array Sensitivity in Noise Direction, Relative to Sensitivity in Desired Look Direction (dB)
67	1.10	−26
134	0.95	−30
191	1.00	−28
236	0.90	−30
338	1.05	−38

oidal noises, each of amplitude 0.5 and power 0.125, acting simultaneously, and, in addition, superposed uncorrelated "white" Gaussian noises of power 0.5 at each of the antenna elements. The frequencies of the five directional noises are shown in Table I.

Fig. 15(a) shows the evolution of the directivity pattern, plotted at frequency f_0, from the initial conditions to the finally converged (adapted) state. The latter was achieved after 682 cycles of the frequency f_0. The learning curve for this experiment is shown in Fig. 15(b). The final array sensitivities in the five noise directions relative to the array sensitivity in the desired look direction are shown in Table I. The signal-to-noise ratio was improved by a factor of about 15 over that of a single isotropic radiator. In Fig. 15(b), one can roughly discern a time constant approximately equal to 70 cycles of the frequency f_0. Since there were 20 adaptations per cycle of f_0, the learning curve time constant was approximately $\tau_{\text{mse}} = 1400$ adaptations. Within about 400 cycles of f_0, the adaptive process virtually converges to steady state. If f_0 were 1 MHz, 400 μs would be the real-time settling time. The misadjustment for this process can be roughly estimated by using (32), although actually all eigenvalues were not equal as required by this equation:

$$M = \frac{n}{4\tau_{\text{mse}}} = \frac{24}{4\tau_{\text{mse}}} = \frac{6}{1400} = 0.43 \text{ percent.}$$

This is a very low value of misadjustment, indicating a very slow, precise adaptive process. This is evidenced by the learning curve Fig. 15(b) for this experiment, which is very smooth and noise-free.

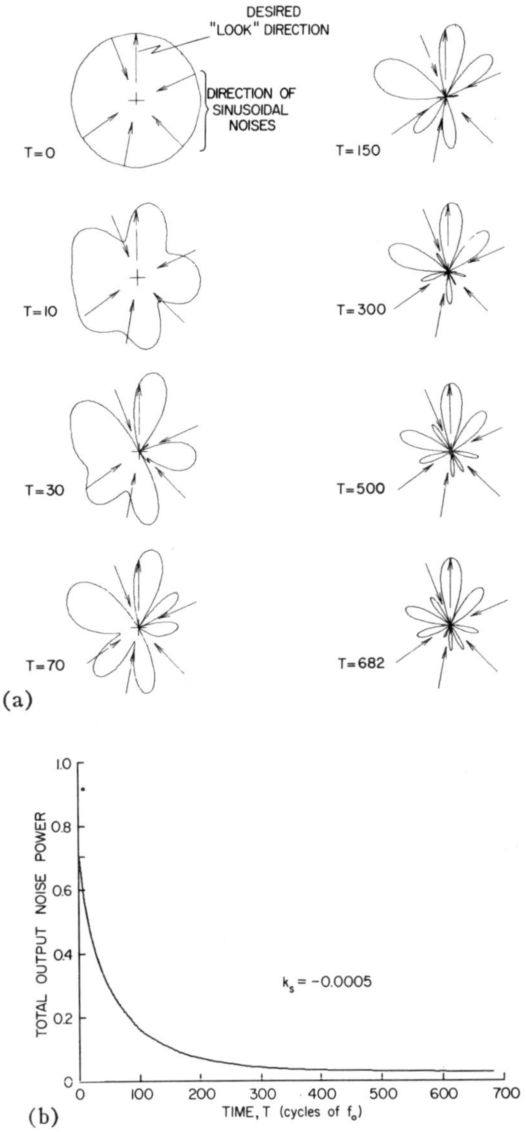

Fig. 15. Evolution of the directivity pattern while learning to eliminate five directional noises and uncorrelated noises. (Array configuration of Fig. 12.) (a) Sequence of directivity patterns during adaptation. (b) Learning curve (total number of adaptations = $20T$).

Broadband Processor Experiments

Fig. 16 shows the antenna array configuration and signal processor used in a series of computer-simulated broadband experiments. In these experiments, the one-mode or simultaneous adaptation process was used to adjust the weights. Each antenna or element in a five-element circular array was connected to a tapped delay line having five variable weights, as shown in the figure. A broadband pilot signal was used, and the desired look direction was chosen (arbitrarily, for purposes of example) to be $\theta = -13°$. The frequency spectrum of the pilot signal is shown in Fig. 17(a). This spectrum is approximately one octave wide and is centered at frequency f_0. A time-delay increment of $1/(4f_0)$ was used in the tapped delay line, thus providing a delay between adjacent weights of a quarter cycle at fre-

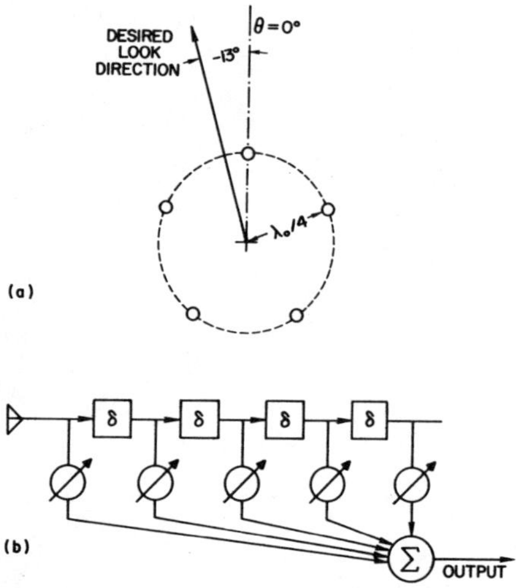

Fig. 16. Array configuration and processing for broadband experiments. (a) Array geometry. (b) Individual element signal processor.

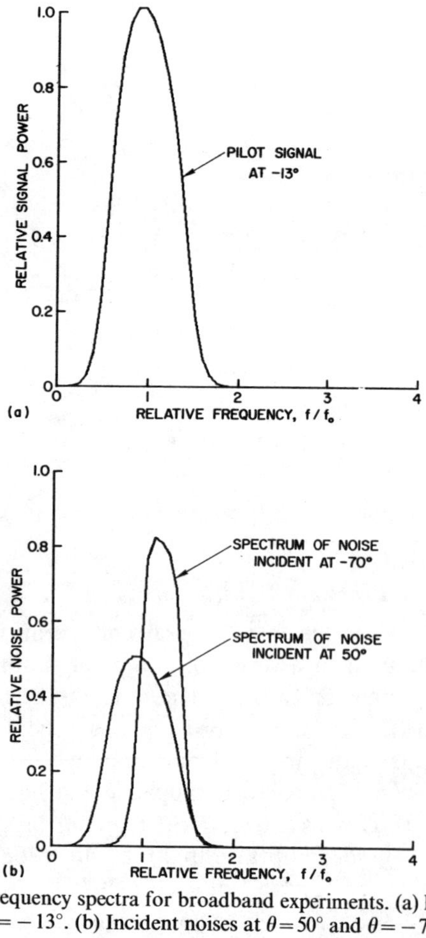

Fig. 17. Frequency spectra for broadband experiments. (a) Pilot signal at $\theta = -13°$. (b) Incident noises at $\theta = 50°$ and $\theta = -70°$.

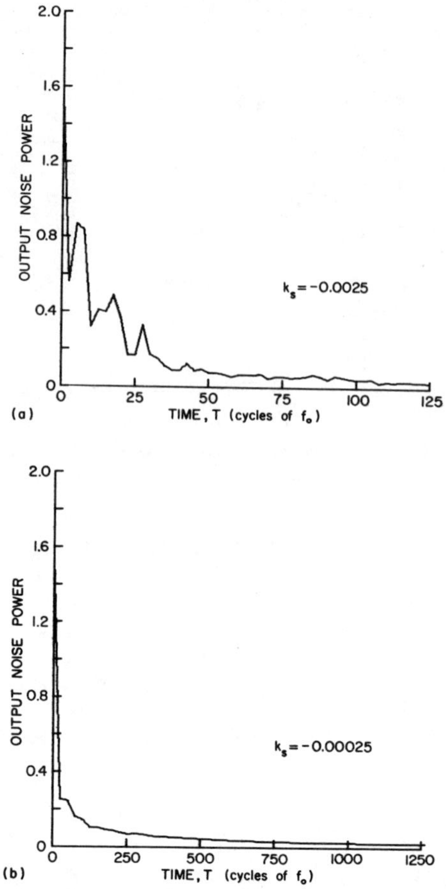

Fig. 18. Learning curves for broadband experiments. (a) Rapid learning ($M = 13$ percent). (b) Slow learning ($M = 1.3$ percent).

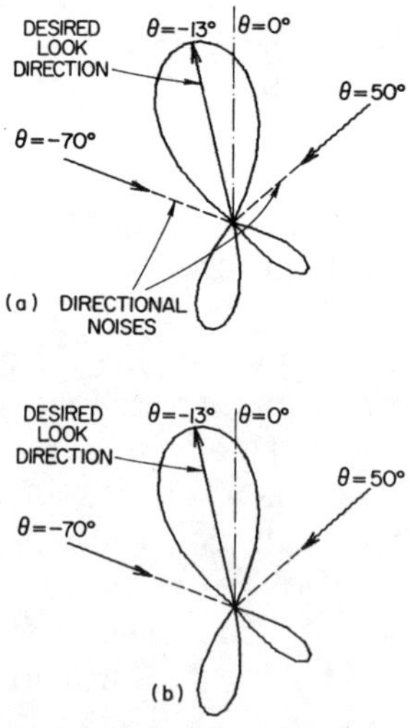

Fig. 19. Comparison of optimum broadband directivity pattern with experimental pattern after former has been adapted during 625 cycles of f_0. (Plotted at frequency f_0.) (a) Optimum pattern. (b) Adapted with $k_s = -0.00025$.

quency f_0, and a total delay-line length of one wavelength at this frequency.

The computer-simulated noise field consisted of two wideband directional noise sources[2] incident on the array at angles $\theta = 50°$ and $\theta = -70°$. Each source of noise had power 0.5. The noise at $\theta = 50°$ had the same frequency spectrum as the pilot signal (though with reduced power); while the noise at $\theta = -70°$ was narrower and centered at a slightly higher frequency. The noise sources were uncorrelated with the pilot signal. Fig. 17(b) shows these frequency spectra. Additive "white" Gaussian noises (mutually uncorrelated) of power 0.0625 were also present in each of the antenna-element signals.

To demonstrate the effects of adaptation rate, the experiments were performed twice, using two different values (-0.0025 and -0.00025) for k_s, the scalar constant in (23). Fig. 18(a) and (b) shows the learning curves obtained under these conditions. The abscissa of each curve is expressed in cycles of f_0, the array center frequency; and, as before, the array was adapted at a rate of twenty times per cycle of f_0. Note that the faster learning curve is a much more noisy one.

Since the statistics of the pilot signal and directional noises in this example are known (having been generated in the computer simulation), it is possible to check measured values of misadjustment against theoretical values. Thus the $\mathbf{\Phi}(x, x)$ matrix is known, and its eigenvalues have been computed.[3]

Using (30) and (31) and the known eigenvalues, the misadjustment for the two values of k_s is calculated to give the following values:

k_s	Theoretical Value of M	Experimental Value of M
-0.0025	0.1288	0.134
-0.00025	0.0129	0.0170

The theoretical values of misadjustment check quite well with corresponding measured values.

From the known statistics the optimum (in the least-squares sense) weight vector W_{LMS} can be computed, using (19). The antenna directivity pattern for this optimum weight vector W_{LMS} is shown in Fig. 19(a). This is a broadband directivity pattern, in which the relative sensitivity of the array versus angle of incidence θ is plotted for a broadband received signal having the same frequency spectrum as the pilot signal. This form of directivity pattern has few side lobes, and nulls which are generally not very deep. In Fig. 19(b), the broadband directivity pattern which resulted from adaptation (after 625 cycles of f_0, with $k_s = -0.0025$) is plotted for comparison with the optimum broadband pattern. Note that the patterns are almost indistinguishable from each other.

The learning curves of Fig. 18(a) and (b) are composed of decaying exponentials of various time constants. When k_s is set to -0.00025, in Fig. 18(b), the misadjustment is about 1.3 percent, which is a quite small, but practical value. With this rate of adaptation, it can be seen from Fig. 18(b) that adapting transients are essentially finished after about 500 cycles of f_0. If f_0 is 1 MHz, for example, adaptation could be completed (if the adaptation circuitry is fast enough) in about 500 μs. If f_0 is 1 kHz, adaptation could be completed in about one-half second. Faster adaptation is possible, but there will be more misadjustment. These figures are typical for an adaptive antenna with broadband noise inputs with 25 adaptive weights. For the same level of misadjustment, convergence times increase approximately linearly with the number of weights.[1]

The ability of this adaptive antenna array to obtain "frequency tuning" is shown in Fig. 20. This figure gives the sensitivities of the adapted array (after 1250 cycles of f_0 at $k_s = -0.00025$) as a function of frequency for the desired look direction, Fig. 20(a), and for the two noise directions, Fig. 20(b) and (c). The spectra of the pilot signal and noises are also shown in the figures.

In Fig. 20(a), the adaptive process tends to make the sensitivity of this simple array configuration as close as possible to unity over the band of frequencies where the pilot signal has finite power density. Improved performance might be attained by adding antenna elements and by adding more taps to each delay line; or, more simply, by band-limiting the output to the passband of the pilot signal. Fig. 20(b) and (c) shows the sensitivities of the array in the directions of the noises. Illustrated in this figure is the very striking reduction of the array sensitivity in the directions of the noises, within their specific passbands. The same idea is illustrated by the nulls in the broadband directivity patterns which occur in the noise directions, as shown in Fig. 19. After the adaptive transients subsided in this experiment, the signal-to-noise ratio was improved by the array over that of a single isotropic sensor by a factor of 56.

Implementation

The discrete adaptive processor shown in Figs. 7(a) and 8 could be realized by either a special-purpose digital apparatus or a suitably programmed general-purpose machine. The antenna signals would need analog-to-digital conversion, and then they would be applied to shift registers or computer memory to realize the effects of the tapped delay lines as illustrated in Fig. 5. If the narrowband scheme shown in Fig. 4 is to be realized, the time delays can be implemented either digitally or by analog means (phase shifters) before the analog-to-digital conversion process.

The analog adaptive processor shown in Figs. 7(b) and 8 could be realized by using conventional analog-computer

[2] Broadband directional noises were computer-simulated by first generating a series of uncorrelated ("white") pseudorandom numbers, applying them to an appropriate sampled-data (discrete, digital) filter to achieve the proper spectral characteristics, and then applying the resulting correlated noise waveform to each of the simulated antenna elements with the appropriate delays to simulate the effect of a propagating wavefront.

[3] They are: 10.65, 9.83, 5.65, 5.43, 3.59, 3.44, 2.68, 2.13, 1.45, 1.35, 1.20, 0.99, 0.66, 0.60, 0.46, 0.29, 0.24, 0.20, 0.16, 0.12, 0.01, 0.087, 0.083, 0.075, 0.069.

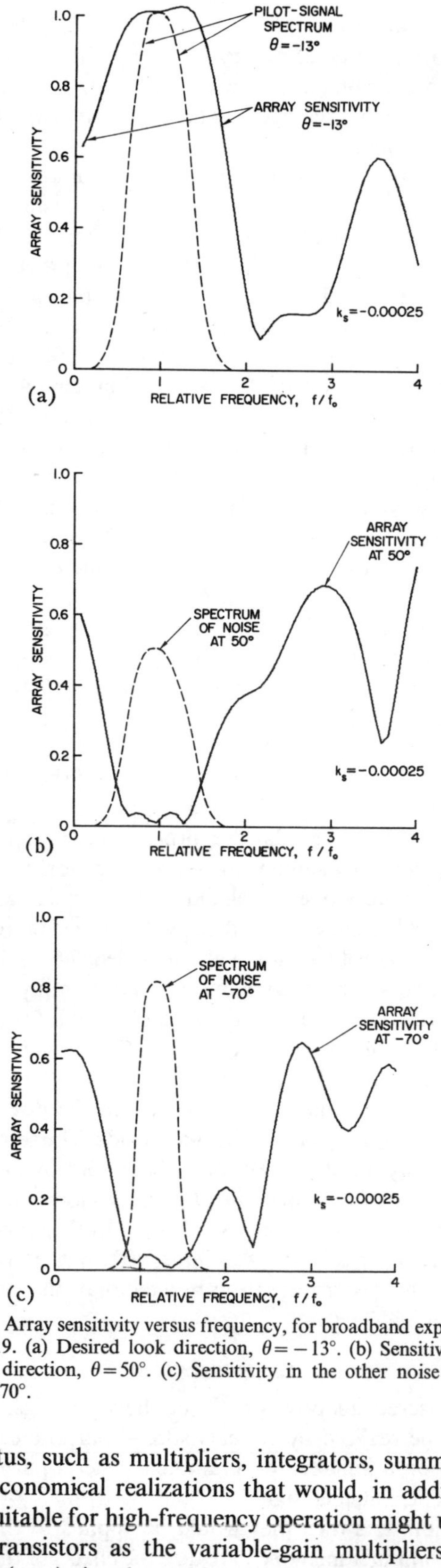

Fig. 20. Array sensitivity versus frequency, for broadband experiment of Fig. 19. (a) Desired look direction, $\theta=-13°$. (b) Sensitivity in one noise direction, $\theta=50°$. (c) Sensitivity in the other noise direction, $\theta=-70°$.

apparatus, such as multipliers, integrators, summers, etc. More economical realizations that would, in addition, be more suitable for high-frequency operation might use field-effect transistors as the variable-gain multipliers, whose control (gate) signals could come from capacitors used as integrators to form and store the weight values. On the other hand, instead of using a variable resistance structure to form the vector dot products, the same function could be achieved using variable-voltage capacitors, with ordinary capacitors again storing the weight values. The resulting structure would be a capacitive voltage divider rather than a resistive one. Other possible realizations of analog weights include the use of a Hall-effect multiplier combiner with magnetic storage[24] and also the electrochemical memistor of Widrow and Hoff.[25]

Further efforts will be required to improve existing weighting elements and to develop new ones which are simple, cheap, and adaptable according to the requirements of the various adaptation algorithms. The realization of the processor ultimately found to be useful in certain applications may be composed of a combination of analog and digital techniques.

Relaxation Algorithms and Their Implementation

Algorithms other than the LMS procedure described in the foregoing exist that may permit considerable decrease in complexity with specific adaptive circuit implementations. One method of adaptation which may be easy to implement electronically is based on a relaxation algorithm described by Southwell.[26] This algorithm uses the same error signal as used in the LMS technique. An estimated mean-square error formed by squaring and averaging this error signal over a finite time interval is used in determining the proper weight adjustment. The relaxation algorithm adjusts one weight at a time in a cyclic sequence. Each weight in its turn is adjusted to minimize the measured mean-square error. This method is in contrast to the simultaneous adjustment procedure of the LMS steepest-descent algorithm. The relaxation procedure can be shown to produce a misadjustment that increases with the *square* of the number of weights, as opposed to the LMS algorithm whose misadjustment increases only linearly with the number of weights. For a given level of misadjustment, the adaptation settling time of the relaxation process increases with the square of the number of weights.

For implementation of the Southwell relaxation algorithm, the configurations of the array and adaptive processor remain the same, as does the use of the pilot signal. The relaxation algorithm will work with either the two-mode or the one-mode adaptation process. Savings in circuitry may result, in that changes in the adjustments of the weight values depend only upon error measurements and not upon configurations of error measurements and simultaneous input-signal measurements. Circuitry for implementing the LMS systems as shown in Fig. 7(a) and (b) may be more complicated.

The relaxation method may be applicable in cases where the adjustments are not obvious "weight" settings. For example, in a microwave system, the adjustments might be a system of motor-driven apertures or tuning stubs in a waveguide or a network of waveguides feeding an antenna. Or the adjustments may be in the antenna geometry itself. In such cases, the mean-square error can still be measured, but it is likely that it would not be a simple quadratic function of the adjustment parameters. In any event, some very interesting possibilities in automatic optimization are presented by relaxation adaptation methods.

Other Applications and Further Work on Adaptive Antennas

Work is continuing on the proper choice of pilot signals to achieve the best trade-off between response in the desired look direction and rejection of noises. The subject of "null-steering," where the adaptive algorithm causes the nulls of the directivity pattern to track moving noise sources, is also being studied.

The LMS criterion used as the performance measure in this paper minimizes the mean-square error between the array output and the pilot signal waveform. It is a useful performance measure for signal *extraction* purposes. For signal *detection*, however, maximization of array output signal-to-noise ratio is desirable. Algorithms which achieve the maximum SNR solution are also being studied. Goode[27] has described a method for synthesizing the optimal Bayes detector for continuous waveforms using Wiener (LMS) filters. A third criterion under investigation has been discussed by Kelley and Levin[28] and, more recently, applied by Capon *et al.*[29] to the processing of large aperture seismic array (LASA) data. This filter, the maximum-likelihood array processor, is constrained to provide a *distortionless* signal estimate and simultaneously minimize output noise power. Griffiths[30] has discussed the relationship between the maximum likelihood array processor and the Wiener filter for discrete systems.

The examples given have illustrated the ability of the adaptive antenna system to counteract directional interfering noises, whether they are monochromatic, narrowband, or broadband. Although adaptation processes have been applied here exclusively to receiving arrays, they may also be applied to transmitting systems. Consider, for example, an application to aid a low-power transmitter. If a fixed amplitude and frequency pilot signal is transmitted from the receiving site on a slightly different frequency than that of the carrier of the low-power information transmitter, the transmitter array could be adapted (in a receiving mode) to place a beam in the direction of this pilot signal, and, therefore, by reciprocity the transmitting beam would be directed toward the receiving site. The performance of such a system would be very similar to that of the retrodirective antenna systems,[5],[6] although the methods of achieving such performance would be quite different. These systems may be useful in satellite communications.

An additional application of interest is that of "signal seeking." The problem is to find a coherent signal of unknown direction in space, and to find this signal by adapting the weights so that the array directivity pattern receives this signal while rejecting all other noise sources. The desired response or pilot signal for this application is the received signal itself processed through a narrowband filter. The use of the output signal of the adaptive processor to provide its own desired response is a form of unsupervised learning that has been referred to as "bootstrap learning."[31] Use of this adaptation algorithm yields a set of weights which accepts all correlated signals (in the desired passband) and rejects all other received signals. This system has been computer simulated and shown to operate as expected. However, much work of a theoretical and experimental nature needs to be done on capture and rejection phenomena in such systems before they can be reported in detail.

Conclusion

It has been shown that the techniques of adaptive filtering can be applied to processing the output of the individual elements in a receiving antenna array. This processing results in reduced sensitivity of the array to interfering noise sources whose characteristics may be unknown *a priori*. The combination of array and processor has been shown to act as an automatically tunable filter in both space and frequency.

Acknowledgment

The authors are indebted to Dr. M. E. Hoff, Jr., for a number of useful discussions in the early development of these ideas, and to Mrs. Mabel Rockwell who edited the manuscript.

References

[1] B. Widrow, "Adaptive filters I: Fundamentals," Stanford Electronics Labs., Stanford, Calif., Rept. SEL-66-126 (Tech. Rept. 6764-6), December 1966.

[2] J. S. Koford and G. F. Groner, "The use of an adaptive threshold element to design a linear optimal pattern classifier," *IEEE Trans. Information Theory*, vol. IT-12, pp. 42–50, January 1966.

[3] K. Steinbuch and B. Widrow, "A critical comparison of two kinds of adaptive classification networks," *IEEE Trans. Electronic Computers* (*Short Notes*), vol. EC-14, pp. 737–740, October 1965.

[4] C. H. Mays, "The relationship of algorithms used with adjustable threshold elements to differential equations," *IEEE Trans. Electronic Computers* (*Short Notes*), vol. EC-14, pp. 62–63, February 1965.

[5] L. C. Van Atta, "Electromagnetic reflection," U.S. Patent 2 908 002, October 6, 1959.

[6] "Special Issue on Active and Adaptive Antennas," *IEEE Trans. Antennas and Propagation*, vol. AP-12, March 1964.

[7] C. V. Jakowatz, R. L. Shuey, and G. M. White, "Adaptive waveform recognition," *4th London Symp. on Information Theory*. London: Butterworths, September 1960, pp. 317–326.

[8] L. D. Davisson, "A theory of adaptive filtering," *IEEE Trans. Information Theory*, vol. IT-12, pp. 97–102, April 1966.

[9] E. M. Glaser, "Signal detection by adaptive filters," *IRE Trans. Information Theory*, vol. IT-7, pp. 87–98, April 1961.

[10] F. Bryn, "Optimum signal processing of three-dimensional arrays operating on gaussian signals and noise," *J. Acoust. Soc. Am.*, vol. 34, pp. 289–297, March 1962.

[11] H. Mermoz, "Adaptive filtering and optimal utilization of an antenna," U. S. Navy Bureau of Ships (translation 903 of Ph.D. thesis, Institut Polytechnique, Grenoble, France), October 4, 1965.

[12] S. W. W. Shor, "Adaptive technique to discriminate against coherent noise in a narrow-band system," *J. Acoust. Soc. Am.*, vol. 39, pp. 74–78, January 1966.

[13] B. Widrow and M. E. Hoff, Jr., "Adaptive switching circuits," *IRE WESCON Conv. Rec.*, pt. 4, pp. 96–104, 1960.

[14] N. G. Nilsson, *Learning Machines*. New York: McGraw-Hill, 1965.

[15] B. Widrow and F. W. Smith, "Pattern-recognizing control systems," *1963 Computer and Information Sciences* (*COINS*) *Symp. Proc.* Washington, D.C.: Spartan, 1964.

[16] L. R. Talbert *et al.*, "A real-time adaptive speech-recognition system," Stanford Electronics Labs., Stanford University, Stanford, Calif., Rept. SEL 63-064 (Tech. Rept. 6760-1), May 1963.

[17] F. W. Smith, "Design of quasi-optimal minimum time controllers," *IEEE Trans. Automatic Control*, vol. AC-11, pp. 71–77, January 1966.

[18] J. P. Burg, "Three-dimensional filtering with an array of seismome-

ters," *Geophysics*, vol. 29, pp. 693–713, October 1964.

[19] J. F. Claerbout, "Detection of *P* waves from weak sources at great distances," *Geophysics*, vol. 29, pp. 197–211, April 1964.

[20] H. Robbins and S. Monro, "A stochastic approximation method," *Ann. Math. Stat.*, vol. 22, pp. 400–407, March 1951.

[21] J. Kiefer and J. Wolfowitz, "Stochastic estimation of the maximum of a regression function," *Ann. Math. Stat.*, vol. 23, pp. 462–466, March 1952.

[22] A. Dvoretzky, "On stochastic approximation," *Proc. 3rd Berkeley Symp. on Math. Stat. and Prob.*, J. Neyman, Ed. Berkeley, Calif.: University of California Press, 1956, pp. 39–55.

[23] B. Widrow, "Adaptive sampled-data systems," *Proc. 1st Internat'l Congress of the Internat'l Federation of Automatic Control* (Moscow, 1960). London: Butterworths, 1960.

[24] D. Gabor, W. P. L. Wilby, and R. Woodcock, "A universal nonlinear filter predictor and simulator which optimizes itself by a learning process," *Proc. IEE* (London), vol. 108 B, July 1960.

[25] B. Widrow and M. E. Hoff, Jr., "Generalization and information storage in networks of adaline 'neurons'," in *Self Organizing Systems 1962*, M. C. Yovits, G. T. Jacobi, and G. D. Goldstein, Eds. Washington, D. C.: Spartan, 1962, pp. 435–461.

[26] R. V. Southwell, *Relaxation Methods in Engineering Science*. London: Oxford University Press, 1940.

[27] B. B. Goode, "Synthesis of a nonlinear Bayes detector for Gaussian signal and noise fields using Wiener filters," *IEEE Trans. Information Theory* (*Correspondence*), vol. IT-13, pp. 116–118, January 1967.

[28] E. J. Kelley and M. J. Levin, "Signal parameter estimation for seismometer arrays," M.I.T. Lincoln Lab., Lexington, Mass., Tech. Rept. 339, January 8, 1964.

[29] J. Capon, R. J. Greenfield, and R. J. Kolker, "Multidimensional maximum-likelihood processing of a large aperture seismic array," *Proc. IEEE*, vol. 55, pp. 192–211, February 1967.

[30] L. J. Griffiths, "A comparison of multidimensional Wiener and maximum-likelihood filters for antenna arrays," *Proc. IEEE* (*Letters*), vol. 55, pp. 2045–2047, November 1967.

[31] B. Widrow, "Bootstrap learning in threshold logic systems," presented at the American Automatic Control Council (Theory Committee), IFAC Meeting, London, England, June 1966.

ADAPTIVE FILTER FOR INTERFERENCE SUPPRESSION

A. M. Kowalski
Hazeltine Corporation
Greenlawn, New York 11740

SUMMARY

This paper describes the performance of an adaptive tapped delay line filter based on the LMS algorithm and used for interference suppression in conjunction with a wideband receiver. Theoretical analysis as well as experimental results are included demonstrating the effectiveness of the approach when wideband information is subjected to multiple narrowband interferers. The discriminant used to force the system to work on narrowband interference signals is the relative correlation times of the signals involved. The delay in the tapped line is selected to permit correlation of the narrowband components of the composite input signal by the system LMS controller. This delay is selected large enough to minimize correlation of the desired signal terms.

A breadboard model of the technique was implemented with two adaptively-controlled weights. Substantial receiver sensitivity improvement was obtained for all amplitude and spectral combinations examined when the number of weights equaled or exceeded the number of interferers. When the number of interferers was greater than the number of weights, the improvement was a function of the amplitude and spectral distribution of the interferers. The processor, however, did not degrade performance when subjected to any of these amplitude and spectral combinations.

ADAPTIVE FILTER PERFORMANCE ANALYSIS

A block diagram of the adaptive filter is shown in Figure 1. The configuration uses two delay line taps and weights, although the number that can be cascaded is arbitrary. The discriminant used in system operation is the fact that the desired wideband signal is relatively decorrelated after τ seconds, while the interference shows some correlation from tap-to-tap. This discriminant is equivalent to assuming that the interference is narrowband compared to the signal. For a phase coded desired signal the correlation constraint implies that a single bit should be less than or equal to τ.

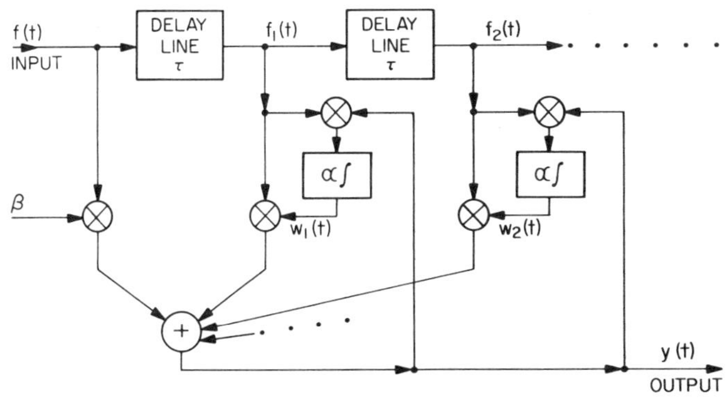

Figure 1. Adaptive Filter Block Diagram.

The following complex signals can be defined in Figure 1.

$s(t)$ = input signal
$j(t)$ = input interference
$f(t)$ = total input waveform
$f_i(t)$ = output of ith delay line tap
β = local oscillator phasor
$\underline{F}(t)$ = vector whose components are $f_i(t)$
$w_i(t)$ = weight of ith tap multiplier
$\underline{W}(t)$ = vector whose components are $w_i(t)$
$y(t)$ = output waveform

From the block diagram it follows that:

$$f(t) = s(t) + j(t)$$
$$f_i(t) = f(t-i\tau)$$

If the lower sideband is retained in all multiplications, then[1]:

$$\frac{d(w_i(t))}{dt} = \alpha f_i(t) y^*(t)$$

$$y(t) = \beta f(t) + \underline{F}_T(t) \cdot \underline{W}^*(t)$$

$$\frac{d(\underline{W}(t))}{dt} = \alpha[\beta^* f^*(t)\underline{F}(t) + \underline{F}(t)F^*_T(t) \cdot \underline{W}(t)]$$

[1]An underbar denotes a vector, a subscript T denotes transpose, an overbar denotes a matrix, and * denotes complex conjugate.

If α is assumed to be sufficiently small and the implementation of the integrator is sufficiently long time constant, then the system will be stable and reach a steady state condition where the average value of $\underline{W}(t)$ will be time invariant; $\underline{W}(t)$ will exhibit small fluctuations which, after long term averaging, will be essentially uncorrelated with the present values of $\underline{F}(t)$ and $f(t)$. Making these assumptions and taking the expected value of $\frac{d(\underline{W}(t))}{dt}$ at steady state yields:

$$E\left\{\frac{d(\underline{W}(t))}{dt}\right\} = 0 =$$
$$\alpha[\beta^* E\{f^*(t)\underline{F}(t)\} + E\{\underline{F}(t) \cdot \underline{F}^*_T(t)\} E\underline{W}(t)\}]_{t\to\infty}$$

Let

$$E\{\underline{F}(t) \cdot \underline{F}^*_T(t)\} = \bar{A}$$

$$E\{f^*(t)\underline{F}(t)\} = \underline{B}$$

$$E\{\underline{W}(t)|_{t\to\infty}\} = \underline{W}' \text{ a constant}$$

Then the steady state weight vector is given by:

$$\underline{W}' = -\beta^* \bar{A}^{-1} \underline{B}$$

Reprinted from *National Telecommun. Conf. Record*, vol. 1, Dec. 5-7, 1977, pp. 04:5-1-04-5-6.

In order to develop more insight into the functioning of this adaptive system it is of interest to consider what criterion applied to the filtering structure in Figure 1 would lead to the same weights in a non-adaptive case. One can expand the total system steady state output power, P_o, in terms of the weight vector.

$$P_o = E\{y(t)\cdot y^*(t)\}$$

$$P_o = E\{\beta\beta^* f(t)f^*(t)+\beta^* f^*(t)\underline{F}_T(t)\underline{W}^*(t) + \beta f(t)\underline{F}_T^*(t)\underline{W}(t)+\underline{W}_T^*(t)\underline{F}(t)\underline{F}_T^*(t)\underline{W}(t)\}$$

$$P_o = |\beta|^2 E\{f(t)f^*(t)\}+\beta^*\underline{B}_T E\{\underline{W}^*(t)\} + \beta\ \underline{B}^*_T E\{\underline{W}(t)\} + E\{\underline{W}_T^*\}\overline{A}E\{\underline{W}(t)\} \quad t\to\infty$$

To determine the minimum of P_o one can take the gradient, or vector of partial derivatives with respect to the $w_i(t)$, and set it equal to zero for large t.

$$\Delta_{\underline{W}} P_o = 0 = 2\beta^*\underline{B}+2\overline{A}\underline{W}(t)\Big|_{t\to\infty}$$

The weight vector $\underline{W}''$ which minimizes the average output power is, therefore, given by:

$$\underline{W}'' = -\beta^*\overline{A}^{-1}\ \underline{B}$$

which is the same as the expected value of the weights obtained through adaptation.

This LMS system attempts to arrive at weights which minimize the total output power. However, the system cannot shut off and eliminate the signal because (1) the weight β on the input is fixed, and (2) the signal correlation time is shorter than the delay line tap spacing. As a result the signal terms at different taps are independent and on the average cannot cancel the input term weighted by β. The narrowband interference is somewhat correlated from tap-to-tap and the weights can provide cancellation of this component. Excessive weight variation in performing the cancellation is bounded since the uncorrelated signals at the various taps multiplied by their weights generate terms which add to the total output power.

One measure of steady state system performance is obtained by evaluating the rms difference (ε) between the input signal and the system output and then comparing this to the input interference power. The parameter β can be chosen as real for convenience.

$$\varepsilon^2 = E\{|y(t) - s(t)|^2\}$$

Using the following identities and some moderately complex matrix algebra manipulations, an expression for ε^2 can be generated.

$E\{s(t)s^*(t)\} = P_S$ or average input signal power

$E\{j(t)j^*(t)\} = P_J$ or average input interference power

$$E\{s(t)j^*(t)\} = E\{s(t)\underline{F}^*_T(t)\} = 0$$

$$E\{\underline{F}\cdot\underline{F}^*_T\} = \overline{A} = \overline{A}^*_T$$

yielding;

$$\varepsilon^2 = (\beta-1)^2 P_S+\beta^2 P_J-\beta^2\underline{B}^*_T\overline{A}^{-1}\ \underline{B}$$

The system gain against interference can be defined as the ratio of the input interference power to ε^2

$$G = \frac{P_J}{\varepsilon^2} = \frac{P_J}{(\beta-1)^2 P_S+\beta^2 P_J-\beta^2\underline{B}^*_T\overline{A}^{-1}\underline{B}}$$

For the case where the reference oscillator has unit amplitude $\beta=1$

$$G = \frac{P_J}{P_J - \underline{B}^*_T\ \overline{A}^{-1}\ \underline{B}} = \frac{1}{1-\left[\dfrac{\underline{B}^*_T\ \overline{A}^{-1}\ \underline{B}}{P_J}\right]}$$

Performance represented by this expression was determined for a number of examples by programming a general purpose digital computer to evaluate the correlation vector $\underline{B}$ and the correlation matrix $\overline{A}$ for assumed interference and signal models and then using these quantities and the derived expressions to compute the steady state weight vector, $\underline{W}'$, and the system gain, G. In addition, the weight vector was used to compute the system transfer function vs. frequency. The program allowed control over the number of adaptive loops, number of interferers, power levels, bandwidths, and frequency separation between interferers. Both line interferers and ones of finite bandwidth were considered. For the finite bandwidth cases flat spectra were used.

A typical set of results is included in Figure 2. In all cases, the value of β was arbitrarily set equal to 1. For convenience the bandwidths and center frequencies were normalized to units of $1/\tau$, the reciprocal of the delay line tap spacings.

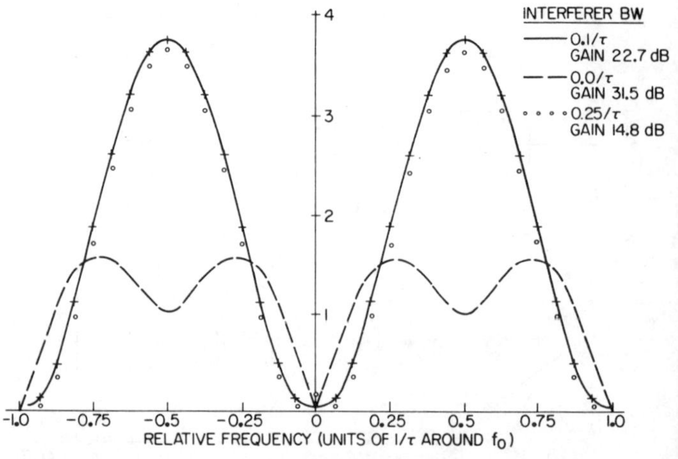

Figure 2. One Interferer Transfer Functions

Initial examples using one interferer were evaluated. Figure 2 shows the voltage transfer function vs. normalized frequency of the adaptive filter at steady state for interferers of three different spectra; a line of 0.0 bandwidth, and flat spectra of bandwidths $0.1/\tau$ and $0.25/\tau$. The processing gains realized were 31.5 dB, 22.7 dB and 14.8 dB respectively. This variation is to be expected since the narrower the interferer bandwidth the more effective the cancellation obtained by using nulls.

The figure shows, however, that the adaptive null width increases with the interferer bandwidth in an attempt to maintain cancellation. In these examples the interferer was assumed centered at frequency f_o. The results are independent of center frequency since the phase angle of the weights change so as to shift the transfer function null to the interferer location. The transfer functions shown are magnitude vs. frequency. Phase vs. frequency has not been plotted. If the adaptive filter is to be used for detection the conjugate filter should also be generated from the adapted tap weights so as to "match filter" the signal after cancellation.

EQUIPMENT CONFIGURATION

Figure 3 describes the equipment configuration used to evaluate the two loop adaptive filter. A wideband receiver was integrated with the adaptive filter; a four way hybrid at the RF input to the receiver was used to add interferers to the desired wideband signal. The desired signal was adjustable in power and simulated by a pseudo-noise code running at a 10 Mbs rate. The 20 MHz wide IF output spectrum was truncated to 10 MHz at the 60 MHz output of the receiver where an IF implementation of the adaptive filter was used. Performance of the processor was monitored with a spectrum analyzer before the receiver AGC and matched filter functions. An oscilloscope was used to measure system performance at the output of the matched filter.

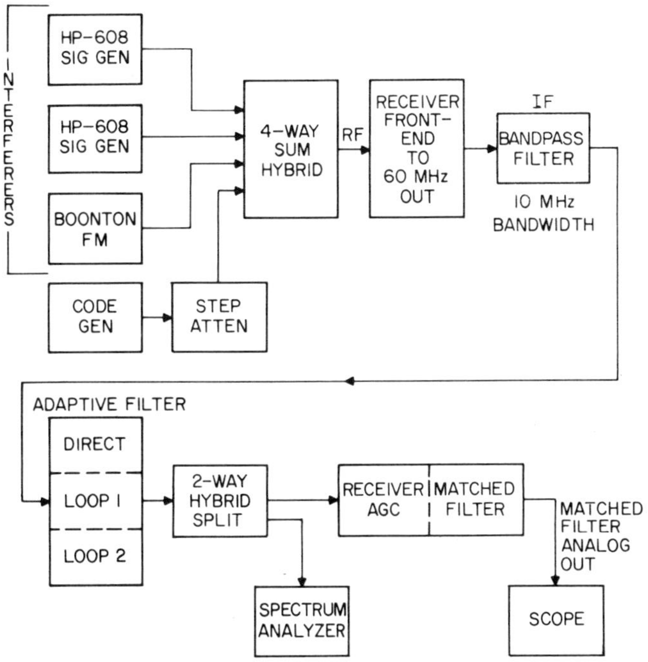

Figure 3. Equipment Configuration.

PERFORMANCE AGAINST TWO CW-INTERFERERS

Measurements were made to determine the performance of the adaptive filter against equal amplitude interferers as a function of the magnitude of the interference powers. Performance as a function of in-band frequency separation was also monitored for the equal power case. Interferers of different power levels were also used as test signals.

The receiver sensitivity was the measure of system performance used to determine the effects of the adaptive filter. When the receiver/matched filter system was subjected to interference and the adaptive filter was not used, the RF level of the coded signal had to be increased to maintain a given signal-to-interference-plus-noise ratio. When the adaptive filter was activated, the RF level of the coded signal could be decreased implying an increase in receiver sensitivity.

Figure 4 indicates receiver sensitivity as a function of the level of two equal CW interferers within the RF passband of the receiver. The frequency separation between the two interferers is 3.4 MHz. With the adaptive filter disabled, sensitivity is seen to be a linear function of interference power, degrading on a one-to-one basis with interference level. Operation with either one of the two adaptive loops provided at most a 3 dB improvement. With both loops operational as much as a 31 dB improvement in sensitivity was obtained.

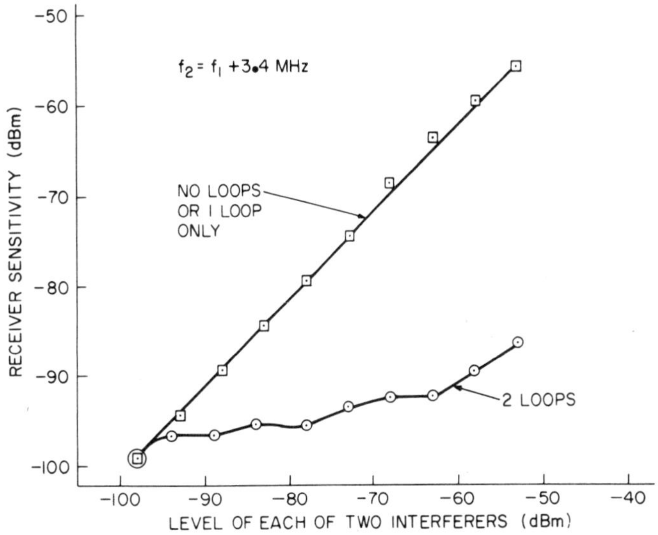

Figure 4. Performance Against Two Equal Amplitude Interferers.

The performance achieved in Figure 4 may be considered "typical" for the 3.4 MHz frequency difference examined. Figure 5 describes system performance by plotting the improvement in receiver sensitivity vs. the frequency separation of the two equal power CW interferers for various nominal in-band locations. On the average the processor provided between 20 and 40 dB of improvement for various frequency combinations.

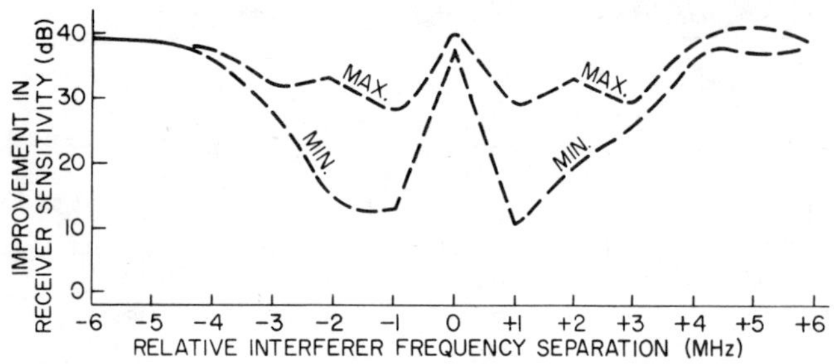

Figure 5. Performance Against Two CW Interferers vs. Frequency Separation.

Performance was generally good when both interferers were close in frequency near band center. For this case both are essentially attenuated by the same spectral notch. Good performance was also achieved with relatively large frequency separations for which the build-up of spurious mixer products was minimal. Raw data indicates that the poorest performance (10 dB) occurred with interferers having relatively small separations (+1 MHz) near the center of the receiver bandwidth. Loss of performance was caused by the build-up of spurious frequency components in the selected implementation. A choice of different IF frequencies within the breadboard would probably reduce this performance loss.

Adaptive filter performance is also a function of the differential amplitude between the two interferers. Figure 6 indicates receiver sensitivity as the ratio of interference powers is varied with the second interferer held at a fixed amplitude.

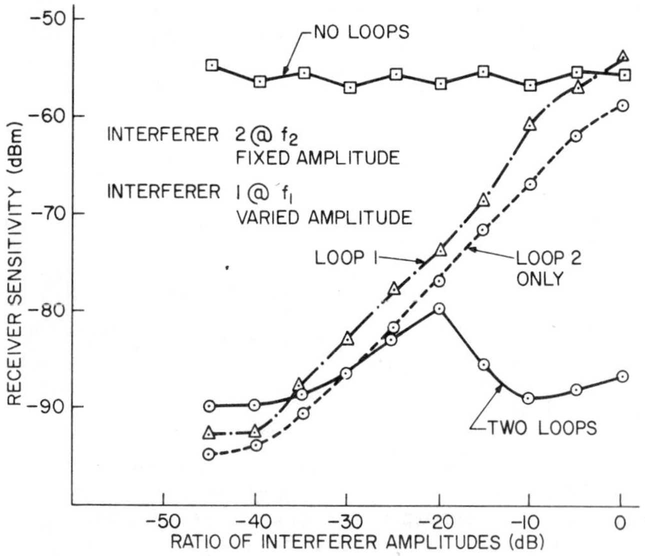

Figure 6. Two CW Interferers of Different Amplitudes.

For equal amplitudes (0 dB ratio) two loops provided a 30 dB improvement over a system without the adaptive filter. At low ratio values (-45 dB) where essentially only one interferer existed, each loop by itself provided between 35 and 40 dB of improvement. At intermediate ratio values (-20 dB) performance degraded from the equal power or 0 dB ratio case. This phenomenon is typical of LMS configurations where control loop gain is a strong function of interferer power level. For the intermediate ratios, both control loops are dominated by the larger interferer causing a decrease in system gain and cancellation capability against the weaker interferer.

PERFORMANCE AGAINST INTERFERERS EXCEEDING THE NUMBER OF LOOPS

Two Loops Against Three Interferers

When the number of interference signals exceed the number of control loops there are several specialized cases that exist. The important considerations, however, are those of relative interference power levels and frequency separation. If three CW interferers of equal amplitude occupy the receiver bandwidth, two-loop performance is dependent upon the relative spectral position of the interferers.

Figure 7 shows that as the frequency of a third interferer approaches that of either of two other equal amplitude interferers, the two adaptive loops can provide a significant improvement in receiver performance (up to 40 dB for coincident frequencies). At more distant frequencies, such as band center relative to two interferers at ± 3 MHz, the interferers do not fall within a common null of the filtering characteristic. On a statistical basis it is apparent that three equal interferers can be combatted to some degree by only two loops, but there can also be many combinations for which only minor improvements are obtained.

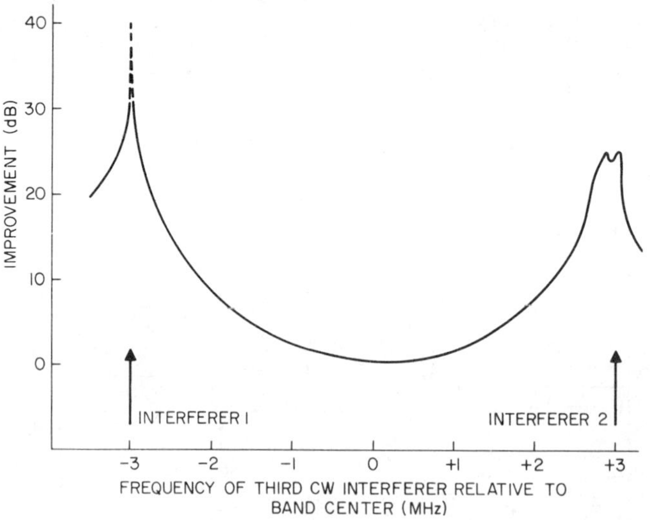

Figure 7. Performance Against Three Equal Amplitude Interferers.

One may then examine the effect of amplitude differentials between the three interferers. Tests were conducted under the spectral condition that provided no improvement in receiver sensitivity for three equal amplitude interferers (i.e., midband and ± 3 MHz). Figure 8 shows that as the amplitude of the midband interferer is decreased the two adaptive loops provide increasing improvement in receiver sensitivity. Alternatively, the sensitivity improvement is seen to be proportional to the ratio of the lower amplitude mid-band interferer to the combined power of the two high amplitude interferers.

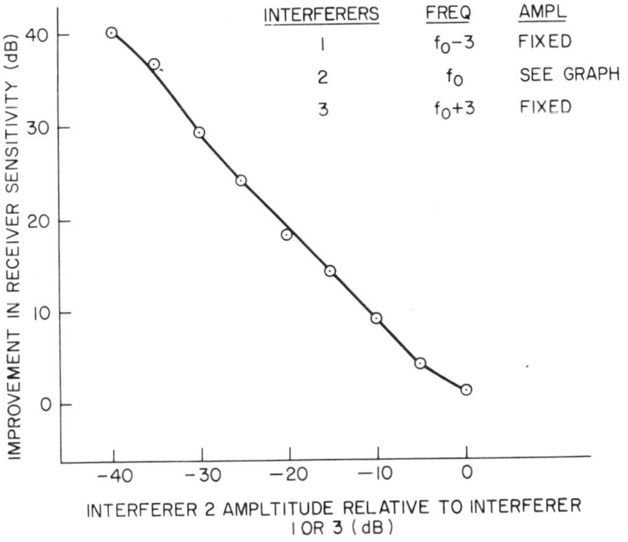

Figure 8. Performance Against Three Interferers (Two Fixed/ Equal Amplitude, One Varied Amplitude).

Figure 9 shows the related case of a high amplitude midband interferer and two equal but lower amplitude interferers at ± 3 MHz. Again, the two loops provide improvement proportional to the ratio between the interferers. Examination of the IF spectrum after the adaptive loops helps to explain this performance. It was observed that the two loops attack the higher amplitude single interferer and have no significant effect on the two equal but lower amplitude signals. Similarly for the previous case of two equally high amplitude signals and one weaker signal, the two loops attack the two equal amplitude interferers but do not significantly affect the weaker interference signal.

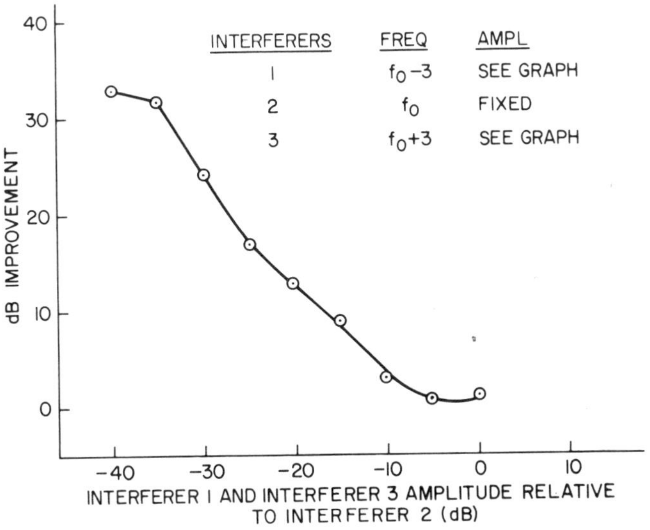

Figure 9. Performance Against Three Interferers (One Fixed Amplitude, Two Varied But Equal Amplitudes).

The term "no significant effect" deserves some comment, however. Depending upon how the two loops suppress the IF bandpass, there will in fact be a gain reinforcement in certain spectral regions, causing the "unattacked" interferer to increase by several dB at the loop output.

The performance of the two adaptive loops was then evaluated against three CW interferers (midband and ± 3 MHz) with a set of linear (in dB) relative amplitude distributions from the strongest one on down. Table 1 summarizes the results of the experiment, where it is seen that the two loops provide an improvement in performance that may be approximated by the amplitude differential between the two strongest interferers (or two weakest) for the relative amplitude distributions tested. The spectrum of the system output again explains these results by showing that the two adaptive loops attack the highest amplitude interferer, with no significant effect upon the lower amplitude signals.

Table 1

PERFORMANCE OF TWO ADAPTIVE LOOPS AGAINST THREE CW INTERFERERS

I_1 (RELATIVE AMPLITUDES)	I_2	I_3	IMPROVEMENT OVER NO LOOPS
0 dB	0 dB	0 dB	1 dB
0 dB	-2 dB	-4 dB	2 dB
0 dB	-4 dB	-8 dB	5 dB
0 dB	-6 dB	-12 dB	7 dB
0 dB	-8 dB	-16 dB	9 dB
0 dB	-10 dB	-20 dB	12 dB
0 dB	-12 dB	-24 dB	15 dB
0 dB	-14 dB	-28 dB	16 dB
0 dB	-16 dB	-32 dB	18 dB
0 dB	-18 dB	-36 dB	19 dB
0 dB	-20 dB	-40 dB	21 dB

$f_1 = f_0 - 3$

$f_2 = f_0$

$f_3 = f_0 + 3$

Two Loops Against Four Interferers

The performance of two loops against three interferers suggested that these results might be extended to additional loop overload conditions where there are a greater number of interferers than adaptive loops. Table 2 indicates that under a 4 unequal amplitude interferer "overload" of two loops, the improvement provided by the loops is again related to the differential between the two highest amplitude interferers. A similar format of a linear relative amplitude distribution was used. Spectral analysis again confirmed that the highest amplitude interferer is attacked by the two loops, with no significant influence on the other interferers.

On a statistical basis the loops still offer improvement under this "overload" condition.

Table 2

PERFORMANCE OF TWO ADAPTIVE LOOPS AGAINST FOUR CW INTERFERERS

I_1	I_2	I_3	I_4	IMPROVEMENT OVER NO LOOPS
RELATIVE AMPLITUDES				
0 dB	0 dB	0 dB	0 dB	1 dB
0 dB	-2 dB	-4 dB	-6 dB	1 dB
0 dB	-4 dB	-8 dB	-12 dB	4 dB
0 dB	-6 dB	-12 dB	-18 dB	6 dB
0 dB	-8 dB	-16 dB	-24 dB	9 dB
0 dB	-10 dB	-20 dB	-30 dB	11 dB
0 dB	-12 dB	-24 dB	-36 dB	13 dB

$$f_1 = f_0 - 3$$
$$f_2 = f_0 - 1$$
$$f_3 = f_0 + 1$$
$$f_4 = f_0 + 3$$

PERFORMANCE AGAINST FREQUENCY MODULATED INTERFERERS

When subjected to equal amplitude, CW and frequency modulated interferers, performance is found to be essentially equivalent to the case of two equal amplitude CW interferers. Figure 10 shows the frequency domain at IF of one CW and one FM interferer in the presence of the wideband signal. With the loops operating, (Figure 10(b)), the interferers are both seen to be suppressed greater than 40 dB. The FM interferer was deviated ± 210 kHz at a modulation rate of 1 kHz.

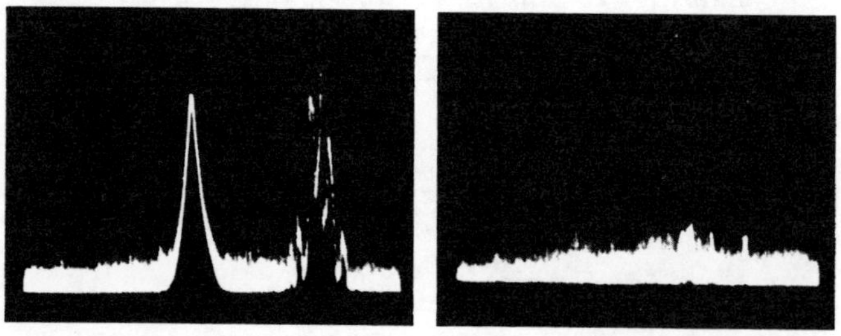

(a) Open Loop (b) Closed Loop

Figure 10. Performance Against One CW and One FM Interferer (1 MHz/CM; 10 dB/CM).

Figure 11 shows the receiver output in the time domain after matching filter processing both in the absence of interference, (Figure 11(a)), and in the presence of interference, (Figures 11(b) and 11(c)). The two interferers capture the system AGC preventing proper operation of the system matched filter (Figure 11(b)). Detectability is restored in Figure 11(c). The second pulse in Figure 11(c) is due to the delayed replica of the wideband pseudo noise signal. This signal along with its undelayed counterpart become inputs to the system matched filter. Since the interference has been rejected both terms can now be "matched filtered". In many applications the presence of the second pulse is tolerable. Conjugate filtering techniques can be used to remove the delayed replica of the desired signal depending upon overall system requirements.

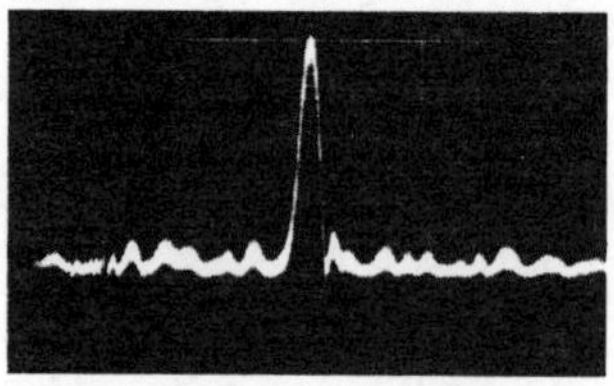

(a) Closed Loop - No Interferers.

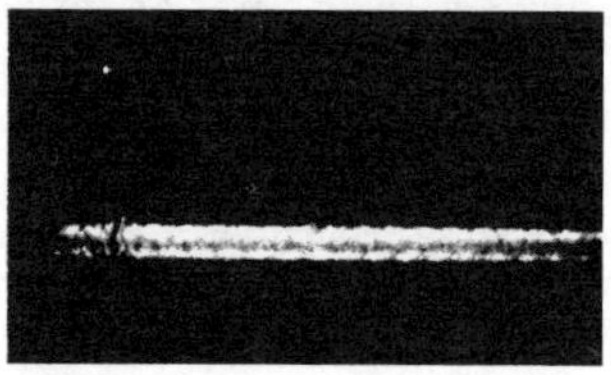

(b) Open Loop - 2 CW or 1 CW and 1 FM Interferer.

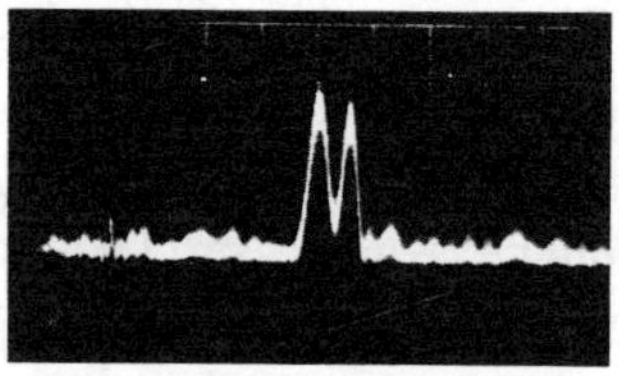

(c) Closed Loop - 2 CW or 1 CW and 1 FM Interferer.

Figure 11. Adaptive Filter Performance (Time Domain).

ACKNOWLEDGEMENTS

The work reported in this paper was performed on internal study programs at the Hazeltine Research Laboratories. The system concept is credited to R. J. Masak and is described in his U.S. Patent (No. 3,932,818). Dr. M. Somin performed the theoretical analysis of system behavior. A. M. Kowalski and M. Brody designed, constructed and tested the breadboard model.

SUPPRESSION OF CO-CHANNEL INTERFERENCE WITH ADAPTIVE CANCELLATION DEVICES AT COMMUNICATIONS SATELLITE EARTH STATIONS

P. D. Lubell F. D. Rebhun
AIL, a division of CUTLER-HAMMER
Melville, New York

SUMMARY

The widespread use of domestic satellite communication systems with dedicated earth stations has focused renewed attention on co-channel interference reduction. Traditional methods involving site selection, physical barriers, and frequency coordination are not always sufficient when economic viability is determined by proximity to the individual user and when station costs are about the same as those of a data transmission link which could permit remote location of the station. This paper discusses an alternative technique where co-channel interference is suppressed by an adaptive interference cancellation network (AICN). Included in this discussion are: an explanation of AICN operation, a review of recent activities in adaptive cancellers, and a description of theoretical considerations and practical operational factors which are requisite to a satisfactory application.

INTRODUCTION

Current commercial communication satellite systems operate with uplink (earth-to-satellite) transmissions in the 5925- to 6425-MHz band and downlink transmissions in the 3700- to 4200-MHz band. These frequency allocations are shared with those assigned to terrestrial microwave communications systems. The result is that a high degree of possible interference exists between these two independent sources, particularly in the downlink portion of the satellite system. Such interferences are derived from several possible causes, such as direct reception on an earth station antenna side lobe, multipath reflection into a side lobe, and introduction into the antenna through the ducting action of anomalous propagation. Domestic communication satellite systems are more likely to be faced with such interference situations than are their international counterparts. INTELSAT stations can usually be sited in relatively undeveloped "quiet" areas, while DOMSAT stations are economically viable when located close to the user, i.e., near large metropolitan areas. The same areas already abound in terrestrial microwave systems for much the same economic reasons. These urban locations also militate against the application of such solutions as geographic shielding and radiation fences which are described elsewhere in this session. This susceptibility problem is further accentuated by recent technological advances in space and ground equipment design, which encourage the use of smaller, less costly earth station antennas which are more susceptible to side-lobe-induced interference.[1]

The AICN provides an attractive solution since it deals directly with the problem at the desired location and does so in a relatively inexpensive and practical manner. In general, it offers an alternative solution to any ground station interference problem caused by one or two distinct interferers, particularly in those situations which are not amenable to siting, construction, and frequency coordination solutions. It is this application of electronic adaptive cancelling which is described in this paper.

CO-CHANNEL CANCELLER SYSTEM DESCRIPTION

AICN networks are attractive because they automatically compensate for variations in interference signals, supressing them with almost no effect on the desired signal. A block diagram of an AICN is given in Figure 1.

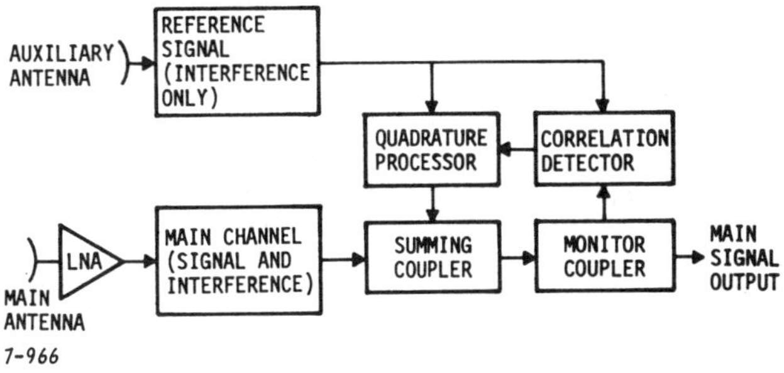

Fig. 1. AICN block diagram

The output signal of the main channel low-noise amplifier contains the desired signal and the undesired interference, both received by the main station antenna. An independent sample of the interference signal is obtained from a separate auxiliary antenna which is directed toward this source. This interference reference is split into two paths. One path, through the quadrature processor, produces an interference signal equal in amplitude and antiphase to that existing in the main line, including the effects of the summing coupler. This nulling signal is combined with the main channel composite signal in the feedforward summing coupler to theoretically produce complete cancellation of the interference signal.

The second path for the interference reference is to the correlation detector which measures the degree of cancellation at the summer output. This measurement is made by coherent comparison of the interference reference with a sample of the mainline output obtained from the monitor coupler. This sample contains both the desired signal and the residual interference. However, only the residual contributes

Reprinted from *Proc. 1977 Int. Conf. Commun.*, vol. 3, June 12-15, 1977, pp. 284-289.

to the detector error signal output since the desired signal is uncorrelated with the interference reference.

The error signal is used to make null seeking adjustments of both the phase and amplitude of the interference sample which passes through the quadrature processor. This adaptive process is required to adjust for phase and amplitude variations due to internally and externally contributed fluctuations in the two paths.

Adaptive cancellers can be realized entirely at either RF or IF, or through some hybrid combination of the two. The hybrid approach usually involves detection at IF with RF processing of the interference reference. It is more complex than the RF approach. The IF implementation requires direct interaction with the station receiver which often is impractical. In addition, it is still more complex (than the others) and may also be band limited due to the lower frequency of operation. The RF approach is the simplest and most promising of three alternatives.

RECENT ACTIVITIES IN ADAPTIVE CANCELLERS

This review describes a currently operational English system as well as two United States developments which are being translated into field units.

Early in 1971, the INTELSAT site at Goonhilly Downs, England began to experience severe interference problems which were traced to an LOS link located 300 km away in France. The duration of the interference often exceeded one-half hour with signal-to-interference (S/I) power ratios as low as 0 dB. A solution in the form of an adaptive tunable canceller was developed by the British Post Office working with Plessey Electronics. This hybrid unit (IF detection) is described by N. White.[2] It monitors both horizontal and vertical polarizations and is tunable because the French link is switched frequently within its six-channel (three on each of two polarizations) allocation. During the first year of operation (commencing March 1975) only four significant conditions were recorded with an S/I of 3 to 8 dB existing for periods of one-half to 2-1/2 hours. In all cases, the canceller system was able to adequately correct the S/I ratio.[3]

A more recent (1976) canceller activity is described by E. D. Horton of Electrospace.[4]. This is an all RF unit employing a four-phase interference processor driven by four outputs from a correlation detector. A laboratory evaluation was conducted based on a small-aperture 4-GHz earth station model employing a 10-m main antenna and a 1-m auxiliary dish. Data was taken for S/I power ratios of -10 to +30 dB at the main antenna output. A maximum suppression of 26 dB (at S/I = -10 dB) was achieved for a 10-MHz-wide interference signal and a 40-MHz-wide desired signal. This paper also contains a theoretical analysis of transient response. Field test data has not been reported.

Another current canceller development has been carried out by the authors at AIL. The unit is an RF implementation of the AICN that employs a two-phase quadrature processor, driven by the two outputs from a correlation detector. Each output is bipolar with the result that only two channels are required, a significant hardware reduction over previously built four-channel systems and with a corresponding increase in reliability. In bench tests, using a carrier frequency modulated by sine waves, the AICN demonstrated at least 25-dB cancellation at the edge of a 35-MHz spectrum and more than 30 dB at band center. These results, and a photo of the unit, are shown in Figure 2. Additional testing was performed to determine its performance with TV signals.

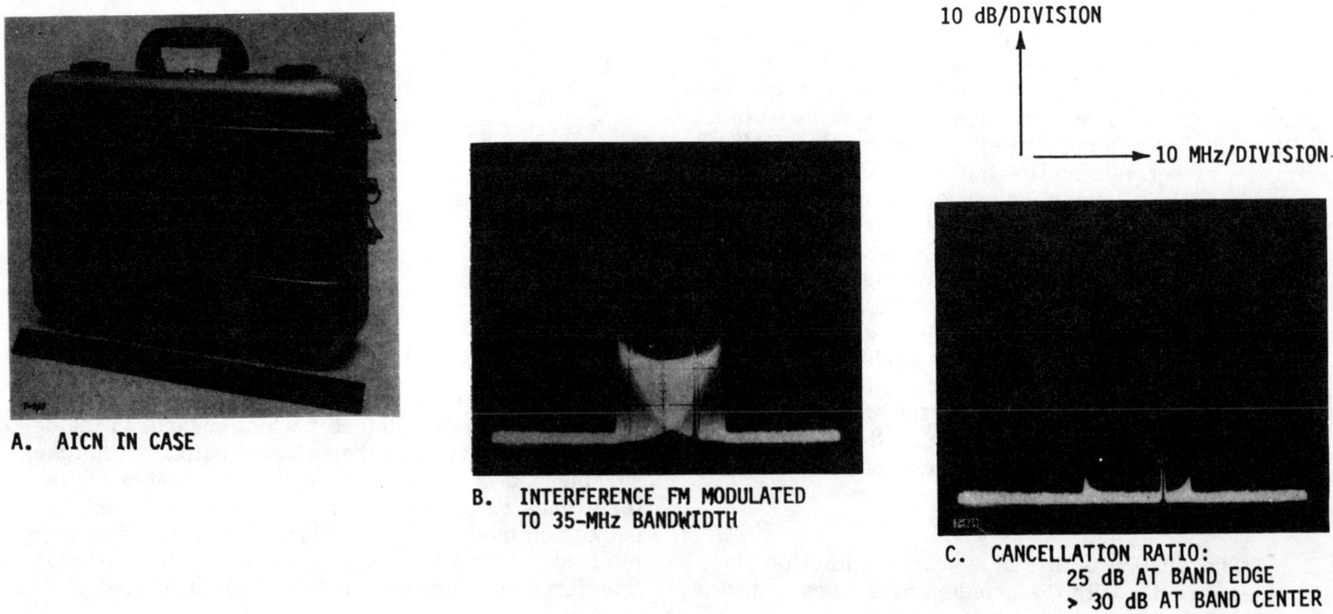

Fig. 2. AIL's automatic interference cancelling network

In the lab two independently controlled TV generator/FM generator signal sources were set up to represent the desired and interfering signals. Each was adjustable in frequency, modulation width, and amplitude. With the AICN cancelling the undesired TV interference signal, the impairments to the desired TV signal were measured using NTSC-type tests. These tests were performed at an S/I of 0 dB and a frequency separation between carriers of 10 MHz (resulting in an overlapping spectrum). The results were that either no distortion or tolerable impairments remained after cancellation. Spot checks were made at different frequencies in the 3.7- to 4.2-GHz band with similar results after optimizing the AICN (two adjustments needed). Finally, the performance as a function of S/I level was checked and again performance was acceptable. The specific tests were conducted per NTC Report No. 7 and included the following: insertion gain, field-time waveform distortion, line-time waveform distortion, short-time waveform distortion, chrominance-luminance gain inequality, gain frequency distortion, chrominance nonlinear gain distortion, dynamic gain distortion, differential phase, and random noise-weighted. Field tests were performed at the Collins Radio facility in Dallas, Texas, using signals from the Westar II satellite transponder. The area was relatively interference free so a controlled interference signal was generated and space coupled to the main antenna. Several transponder channels were operating which enabled evaluation to be made across the entire 3.7- to 4.2-GHz frequency band. The test results showed adequate cancellation at some frequencies and inadequate performance at others. This is discussed further under "Operational Factors."

THEORETICAL CONSIDERATIONS

Loop Analysis

Figure 3 is a simplified diagram of the AICN loop. The correlator is a quadrature coherent detector that generates two orthogonal components of the correlation between the two input signals. The vector generator similarly splits its input signal into two orthogonal components, each controlled independently in phase and amplitude so that, when recombined, the resultant signal can be rotated through 360 degrees of phase and up to 40 dB in amplitude. In Figure 3 the system is modeled as a feedback loop containing two multipliers, the correlator, and the vector generator. Assuming that the correlator satisfactorily measures the phase and the vector generator responds accurately, we can assume the antiphase requirement is established and treat the problem from the amplitude viewpoint.

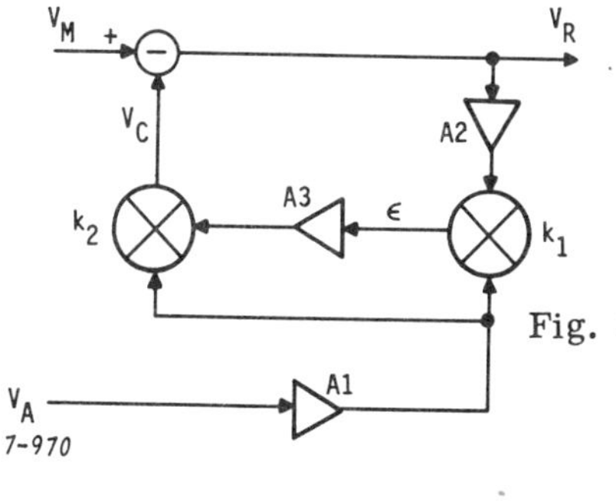

Fig. 3. Simplified loop model

We can express the residual uncancelled interference as:

$$V_R = \frac{V_M}{(1 + GV_A^2)} \tag{1}$$

where $G = k_1k_2A_1A_2A_3$ is the loop gain constant; V_M is the main-line interference voltage; V_A is the auxiliary channel interference voltage; k_1 and k_2 are the correlator and quadrature processor gain constants respectively; and amplifiers A_1, A_2, and A_3 are used to obtain the proper operating levels.

The loop gain constant G has the units $(\text{volts})^{-2}$. The loop gain itself is given by GV_A^2 and is a function of the interference reference level. For very low levels of V_A, corresponding to weak interference reference levels, there is no suppression since $GV_A^2 << 1$ and $V_R = V_M$. As the interference increases and the loop gain increases, the loop begins to suppress the interference. When $GV_A^2 >> 1$, $V_R = V_M/GV_A^2$ and the loop efficiency and suppression both increase. Although the loop is theoretically unconditionally stable, high interference reference levels will suppress the interference more than necessary (below the noise level, for example). The result is that nonlinearities set in and can cause the loop to be unstable. To overcome this, either an AGC or limiter can be employed.

If we denote the limited or AGC'd level as V_L, we can rewrite equation 1 as follows:

$$V_R = \frac{V_M}{1 + G_L V_A}; \quad G_L = G\frac{V_L}{A_1} \tag{2}$$

where G_L is the new loop gain constant. The performance changes so that a constant residual voltage remains as the interference level increases.

Part A of Figure 4 illustrates the performance of the AICN. In region (1), the loop gain is so low that no suppression occurs and the residual level follows the main-channel interference level with unity gain. In region (2), the interference level is high enough to increase the loop gain and start to reduce the residual level. In region (3), limiting or AGC action is effective and the residual level is constant at a level determined by the loop gain constant.

The frequency response is determined as follows. Amplifier A3 actually contains a low-pass filter of bandwidth W_0 which was neglected in the previous analysis because we were interested only in the steady-state performance. The transfer function of the system, including this filter, is:

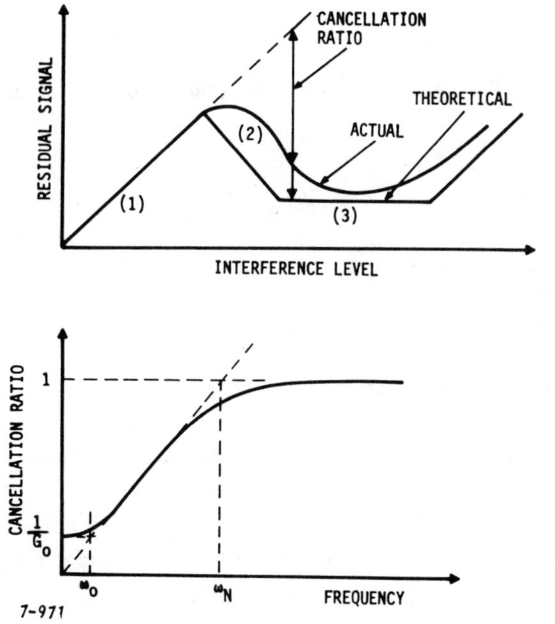

Fig. 4. Loop Characteristics

$$\frac{V_R}{V_M} = \frac{\frac{1}{G_o} + \frac{s}{G_o W_o}}{1 + \frac{s}{G_o W_o}} \tag{3}$$

where $G_o = 1 + G_L V_A$. The ratio V_R/V_M is called the cancellation ratio (CR). Equation 3 is plotted in part B of Figure 4. Assuming that the loop gain G_o is much greater than unity, the CR is $1/G_o$ for frequencies below W_o radians per second and approaches unity at $G_o W_o = W_N$ radian per second. As expected, the closed-loop 3-dB bandwidth is equal to the low-pass filter bandwidth multiplied by the loop gain and the closed-loop gain is equal to the reciprocal open-loop gain.

Although this analysis would predict very high CR's as the loop gain increases, other factors militate against this theoretical performance. Such factors are discussed next and include the affects of VSWR, phase, amplitude, and line equalization errors.

Sources of Error

Cancellation of the interference occurs because a replica of the main channel interference is vectorially added equal in amplitude and in phase opposition. An estimate of the degree of amplitude and phase match required over the bandwidth of the interference can be obtained from a simple sinusoidal analysis. The correlator and vector generator will be adaptively centered on the interference spectrum but the circuitry must maintain the phase linearity and amplitude uniformity to insure cancellation over the requisite bandwidth.

Figure 5 relates null depth to amplitude and phase errors. Note that the amplitude error scale is logarithmic in dB. A 30 dB null with equal amplitude and phase errors requires an amplitude variation under 0.2 dB and a phase error less than 1.3 degrees.

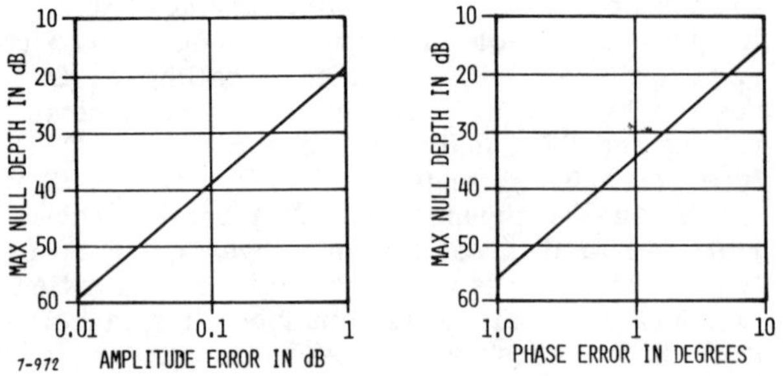

Fig. 5. Effect of amplitude and phase errors

Another source of error is unequal path lengths from the source of interference through the two paths to the point of cancellation. If the path lengths differ, there will be a phase divergence over any finite bandwidth. Figure 6 relates null depth to RF bandwidth as a function of differential line length errors. These curves do not include the effects of dispersive structures (waveguides) or anomalies in the intervening paths or antennas. The path lengths must be within one wavelength of each other to achieve at least 30-dB cancellation over a 20-MHz bandwidth.

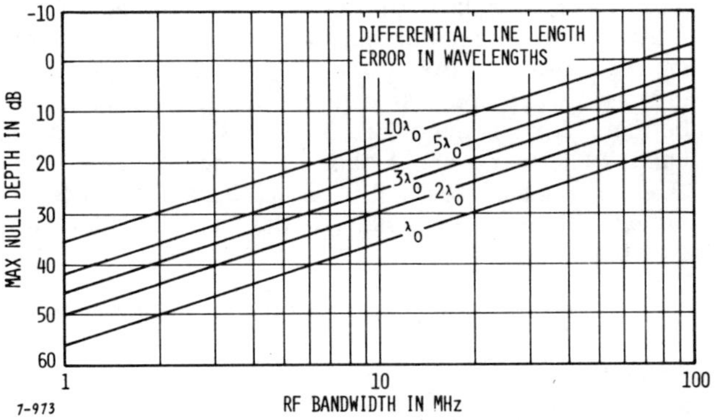

Fig. 6. Effect of line length matching errors

An additional source of error is caused by mismatches which result in phase and amplitude errors. This arises when two devices with standing wave ratios other than unity are interconnected. The multiple reflections add vectorially with the incident wave to form a resultant with phase and amplitude perturbations. Figure 7 depicts the maximum null depth achievable when either the phase or the amplitude is at a peak. If both reach a peak simultaneously, the error could be 3 dB worse; otherwise, the two errors would add in an rms fashion. The phase and amplitude ripple have a period determined by the line length connecting them. For a line length of 8.5 in., corresponding to a two-way delay of about 2 ns, there is a periodicity of 500 MHz. This means that a complete cycle of ripple occurs from one end of the frequency band to the other (3.7 to 4.2 GHz). This effect alone may require some tuning across the frequency band.

A further source of error arises because the interference reference signal is either limited or AGC'd, as discussed in the loop analysis. As shown in Figure 3, it is necessary to take the vector generator input from the same point as the correlator reference input in order to cancel phase errors associated with changes in the voltage controlled attenuator (VCA) setting in response to input fading variation. Had the vector generator input been derived before the VCA, the resulting phase errors would reduce the cancellation ratio. The same effects are present with a limiter through the mechanism of AM/PM conversion.

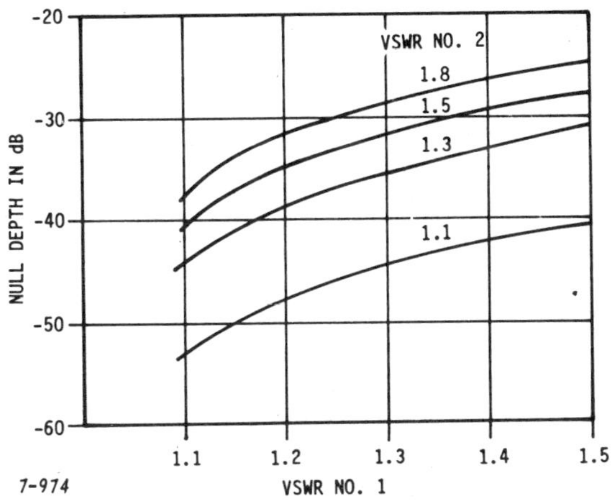

Fig. 7. Effect of mismatch errors

The final source of error considered is caused by dc offset effects and its variation with temperature. In Figure 3, the coherent detector drives the quadrature processor (also known as a vector generator, weighting network, and adaptive filter) through a dc amplifier-filter combination. The dc amplifier is subject to dc offset changes, although this effect can be minimized. The correlator employs double-balanced mixers (DBM) in its structure with one input to the DBM at approximately +5 dBm. Slight differences in the RF structure and the diodes themselves cause an imbalance at the correlator output. Typically, the offset is in the range of ±5 mV, although it is frequency sensitive, varying about 2 mV from 3.7 to 4.2 GHz. In addition, it exhibits a temperature variation of a similar amount. These offset variations, if not compensated, would severely restrict AICN performance, reducing cancellation by as much as 10 dB.

To overcome this problem, AIL employs a PIN diode switch to chop the RF signal before it gets to the correlator. The output of the correlator is a square-waveform signal whose amplitude is determined by the residual error and whose frequency is the switching frequency, approximately 1000 Hz. The effects of the offset are eliminated, as well as similar effects of the dc amplifier following the correlator. With the AIL switching correlator, a 1-kHz amplifier is required, followed by a detector-filter to reconvert to dc. The only temperature effects to contend with are variations of gain, much less serious a problem.

OPERATIONAL FACTORS

The AIL canceller was tested in November 1976 at the Collins Radio facility in Dallas, Texas where a prototype earth station has been set up as part of the PBS network. These tests would not have been possible without the able cooperation of Mr. E. Devletoglou of Collins Radio. For the field tests, an Andrews 10-meter antenna with AIL paramp was used to receive signals from the Westar II satellite transponder. A separate antenna was set up and fed from a TV generator/FM generator to represent the interference. This antenna was placed about 50 to 75 feet from the large dish. A small horn was suspended from the Andrews antenna to pick up the interference. The test results showed acceptable cancellation at some frequencies and inadequate performance at others in the frequency band.

A more dramatic effect of the canceller performance is shown in Figure 8 where the interference is equal to the signal. The signal was "live" from Westar II; the interference was a locally generated TV test pattern.

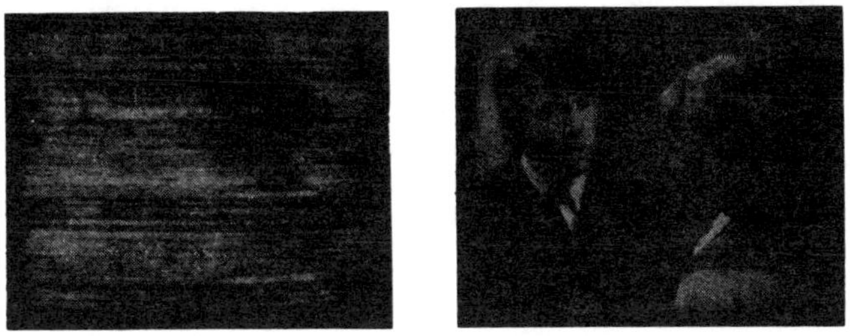

A. AICN OFF   B. AICN ON

Fig. 8. AICN performance at C/I = 0 dB

One of the problems discovered during the field tests is that the amplitude/phase response of the auxiliary antenna (picking up the interference reference signal) in its main lobe can be substantially different than the main antenna in its side lobes. This was confirmed by setting up a network analyzer to measure differential gain and phase between the two paths. The first path was from the interference antenna through the pickup horn and LNA to the network analyzer. The second path was from the interference antenna through the main antenna and paramp to the network analyzer. Figure 9 shows the differential amplitude and phase response over 50-MHz bands centered at 3860 MHz and at 4120 MHz. Over a 20-MHz band there is about a 2-dB amplitude and 25-degree phase variation at 3860 MHz, whereas at 4120 the amplitude and phase curves are relatively flat. This explains why the AICN performed adequately at some frequencies and not at others.

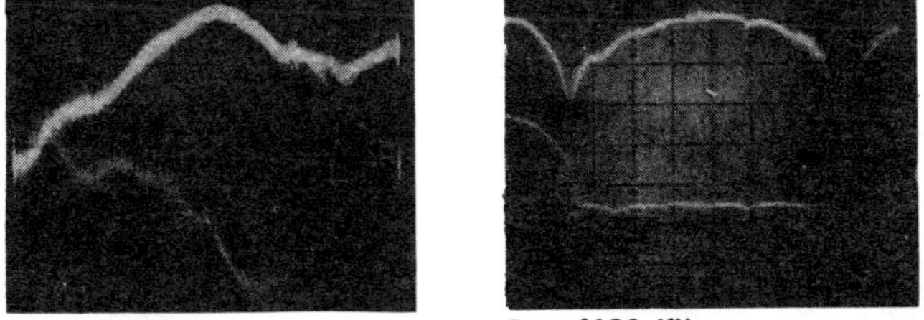

A. 3860 MHz   B. 4120 MHz

Fig. 9. Differential amplitude and phase response

This phenomenon has been mentioned in reference 2 where it is observed that side lobes which are natural lobes of the large dish are highly frequency dependent whereas side lobes caused by aperture blocking or feed spillover are less sensitive. The success of the Goonhilly Downs experiment confirms their prediction that the interference arrived on a side lobe caused by the broader aperture blocking pattern.[2,3]

Several solutions are possible when this frequency variation is a problem. One approach is to increase the AICN loop bandwidth so that it tracks the instantaneous frequency of the interference. Unfortunately, the increased noise generated could adversely effect the desired signal. A second approach is to build an equalizer to predistort the reference signal so that cancellation is effective. AIL is currently pursuing this method. The approach we selected is based on a microwave transversal equalizer[5]. It employs a multitapped delay line with each of the N taps independently adjusted in gain and phase prior to being summed with the input signal. It is equivalent to dividing the spectrum of interest into N parts and using a piecemeal approximation to the required response. The summation process could be incorporated in a closed loop but a separate loop is required for each tap.

AIL is convinced that the AICN will prove indispensable as the trend toward smaller diameter earth stations accelerates. The lower antenna gain and broader beam width combine to reduce signal strength and increase interference levels. With narrow bandwidth signals, the AICN alone is sufficient; in broad bandwidth applications, an equalizer may be required.

REFERENCES

1. Janky, J. M., et al., "New Sidelobe Envelopes for Small Aperture Earth Stations," IEEE Trans on Broadcasting, Vol BC-22, pp 39-44, June 1976.

2. White, N., Brandwood, D., and Raymond, G., "The Application of Interference Cancellation to an Earth Station," IEE Publication 126, "Satellite Communications Systems Technology," pp 233-238, April 1975.

3. Private Correspondence; White, N./Lubell, P., May 1976.

4. Horton, E. D., "An Adaptive Co-Channel Interference Suppression System to Suppress High Level Interference in Satellite Communication Earth Terminals," National Telecommunications Conference Record, Dallas, Texas, Section 13.4, pp 1-5, 30 November to 2 December 1976.

5. Taub, J. J. and Kurpis, G. P., "Microwave Transversal Equalizers," Microwave Journal, January 1969.

AN ADAPTIVE INTERFERENCE CANCELLATION SYSTEM FOR ELIMINATION OF CO-LOCATED INTERFERENCE SIGNALS

W. F. Geist

Hazeltine Corporation
Greenlawn, New York

Abstract

Many communication sites must receive and transmit information simultaneously within a given band of frequencies. To avoid interference between co-located transmitters and receivers, careful placement of receive and transmit antennas is required. An adaptive-interference cancellation system which permits a substantial reduction in the frequency and spatial separation of co-located receive and transmit frequencies and antennas is described in this paper.

The cancellation system protects VHF voice communication receivers from multiple AM, FM, or CW interference signals. The canceller employs a direct sample of the unwanted signal which is correlated with the output from the receiving antenna. Variable weights in the canceller are automatically adjusted in phase and amplitude on a least-mean-squares basis to null the unwanted signal prior to processing by the receiver.

A single loop adaptive canceller system has been built. Its experimental performance for several types of jamming signals, its effects upon the desired signal, the minimum frequency separation between desired and undesired signals, and its speed of response are described. Photographs of the canceller outputs with cancelled and uncancelled signals are presented on both a time domain and frequency domain and frequency domain basis.

The experimental results show that cancellation of AM and CW jammers is in excess of 55 dB and that rejection levels as high as 75 dB can be obtained. Convergence times measured from 0 dB rejection level (uncancelled) to a 55 dB rejection level were less than 20 ms. A frequency separation between desired and undesired signals of only 14 kHz was achieved without degradation of the desired signal. This data clearly demonstrates the effectiveness of this technique in eliminating co-located interference.

Introduction

A problem that exists at many communication sites which operate full duplex communication transceivers in a given frequency band is the interference of co-located transmitters and receivers. This problem is solved, or at least minimized, by physically separating receive and transmit antennas and by maximizing the separation of receive and transmit channel frequencies. For example, the Army VRC-12 field manual recommends an antenna separation of five feet and a channel frequency separation of at least 10 MHz for co-located voice communication receivers and transmitters. The FAA has similar problems at their air traffic control sites which operate VHF (118 MHz to 136 MHz) voice communications. As communication needs expand, it becomes more difficult and costly to implement the solutions outlined above. This paper describes an alternate approach which employs adaptive signal processing techniques for the elimination of co-located interference.

Adaptive signal processing can be categorized as either spatial domain processing which relies on the angular separation of desired or undesired signals or as frequency domain processing which relies on the spectral difference. However, when a sample of the undesired signal is available it may be used directly as the reference signal to achieve cancellation. This direct approach does not require multiple antennas and nulls only the specific undesired signal used for the reference.

To evaluate the effectiveness of the direct cancellation approach, Hazeltine developed and tested under an internally funded program a single loop interference canceller with the following performance objectives:

- o Frequency Range: 118 MHz to 136 MHz
- o Cancellation: >55 dB null depth
- o Response Time: <20 ms

Description of the Interference Cancellation System

The Interference Cancellation System (ICS) is designed for application to a multiple interference environment as shown in figure 1. Four transmitting antennas are interconnected to the receiver via the ICS. Samples of each transmitter output are adaptively processed (i.e. correlated with) the composite signal at the receiving antenna and are removed from the composite signal prior to processing by the receiver. Except for the addition of a directional coupler which is inserted in each transmit line to extract a small amount of power for the direct sample, no modification to the existing terminal receivers, transmitters or antennas is required.

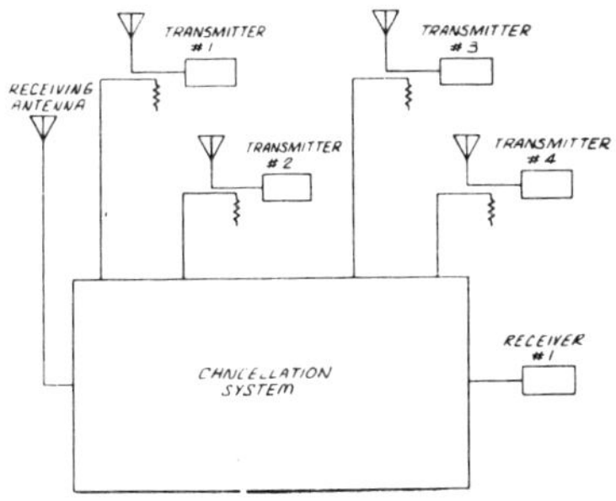

Figure 1. ICS Interface with Communications Sets

Each sample of the interfering transmitter signal is processed in a correlator loop which is based on the Least Mean Square (LMS) algorithm (Ref. 1) as shown in figure 2. Use of the LMS algorithm ensures rapid response to permit nulling of the transmitted signal in a few milliseconds. This avoids annoying clicks in the demodulated output. The composite signals from the receiving antenna pass through two, low-loss directional couplers. At coupler #1, the weighting circuits provide a replica of the transmit signal which is of equal amplitude, but of opposite phase to the signal from the receiving antenna. Time delay matching is required to

Reprinted from *Proc. 1977 Int. Conf. Commun.*, vol. 3, June 12-15, 1977, pp. 273-276.

ensure that the signal from the weight circuit and the interference signal from the receiving antenna arrive in time coincidence at coupler #1. Coupler #2 provides the feedback signal which is correlated with the sample from the interfering transmitter.

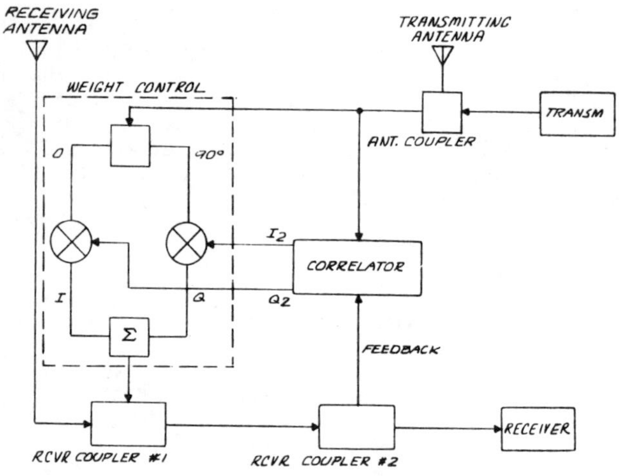

Figure 2. Basic Correlator Loop

The feedback signal contains the desired signal, residue of the signal being cancelled, and other signals from the receiving antenna. The residue of the cancelled signal controls the cancellation process. It is extracted from the composite signal by correlation with the reference signal received directly from the transmitter by coaxial cable. It is converted to baseband, resolved into in-phase and quadrature components of the signals from the transmitter. These components are combined in a hybrid transformer and fed to the directional coupler which couples it into the main line where it cancels the interfering transmission received via the antenna. High feedback gain in the control loops provides fast, accurate response, handles drift, and eliminates the need for adjustments. When the transmitter is turned on, cancellation occurs in less than 20 milliseconds.

Experimental Test Set-Up

The correlator loop described above was assembled and experimentally evaluated using the test set-up diagrammed in figure 3. The upper grouping contains the test signal generators and modulators used to simulate various combinations of desired and unwanted transmitted signals, while the lower grouping contains the essential elements of the ICS. The ICS evaluated has a receiver and a transmitter channel input, each of which receives signals from the test signal generator. The loop can be opened via a switch to evaluate the cancellation ratio achieved.

"Multiple-loop" ICS processors would possess a single receiver channel input and a sequence of transmit channel inputs each associated on a one-to-one basis with a known interference source. The basic ICS performance, however, can still be adequately demonstrated with the "single-loop" configuration of figure 3.

Test Results

A sequence of tests was conducted to evaluate the depth of null for AM and CW interference sources, the nulling capability vs the frequency separation between desired and undesired signals and transient response time.

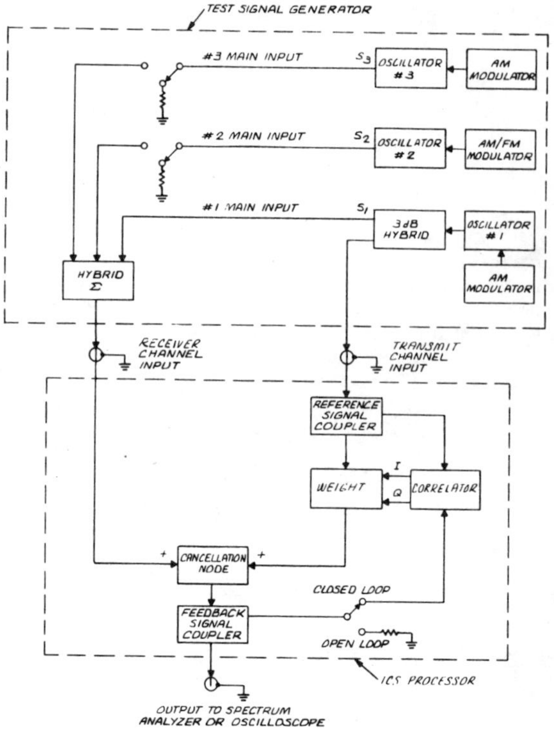

Figure 3. Experimental Test Set Up

Single Interference Cancellation as a Function of Modulation Rate

The purpose of this sequence of tests is to determine the basic cancellation ability of the ICS against a single AM interfering signal as a function of the AM modulation rate. Modulation rates of 1000, 3000, 6000 and 20,000 Hz were evaluated. A 95% modulation depth was used for all tests performed and the signals were of equal amplitude.

Figure 4 illustrates the performance of the ICS processor under the described conditions at a modulation rate of 1000 Hz. Part (a) is a photograph of the open loop spectral output of the ICS. A comparison

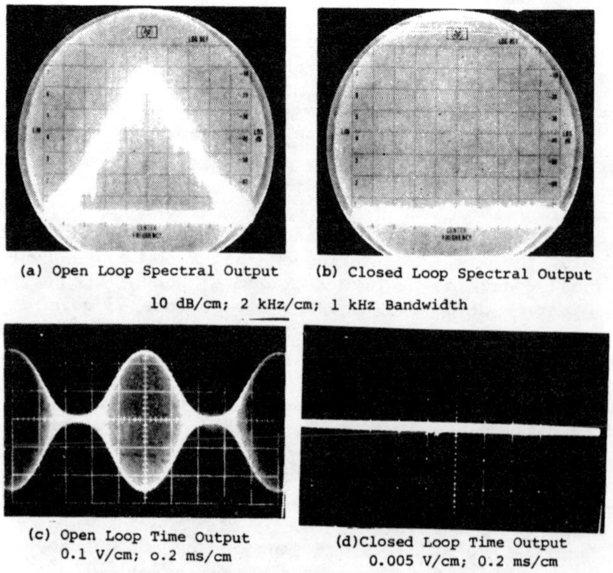

(a) Open Loop Spectral Output (b) Closed Loop Spectral Output

10 dB/cm; 2 kHz/cm; 1 kHz Bandwidth

(c) Open Loop Time Output 0.1 V/cm; o.2 ms/cm

(d)Closed Loop Time Output 0.005 V/cm; 0.2 ms/cm

Figure 4. Single AM Jammer Modulated 95% at 1000 Hz Rate

with the corresponding closed loop spectral output of (b) indicates that a cancellation in excess of 60 dB has been achieved. Parts (c) and (d) contain photographs of the open and closed loop time domain outputs of the ICS. Because of the high cancellation achieved, the closed loop time domain system output is essentially unmeasurable, even at the 0.005 v/cm oscilloscope sensitivity of (d).

Cancellation in excess of 60 dB is achieved at the modulation rates of 3000 and 6000 Hz. Small degradations began to occur at modulation rate in excess of 20,000 to 30,000 Hz. The conclusion is, therefore, that the ICS yields substantial AM interference rejection at a 95% depth of modulation at rates of at least up to 20,000 Hz.

Single Interference Cancellation in the Presence of a Desired Signal

This sequence of tests describes the cancellation capability of the ICS in the presence of a low level AM desired signal. Both CW and AM interference signals are considered.

Figure 5 describes CW interference cancellation in the presence of a low-level desired AM signal separated from the jammer in frequency by only 14 kHz. The AM signal is modulated 95 percent at a 1 kHz rate.

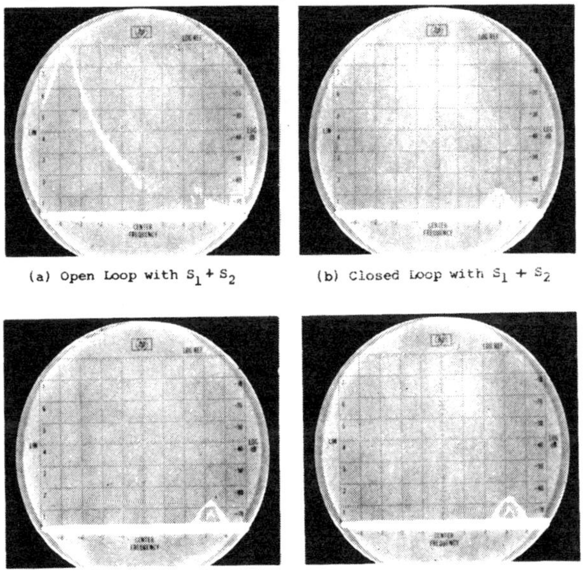

(a) Open Loop with $S_1 + S_2$ (b) Closed Loop with $S_1 + S_2$

(c) Open Loop with S_2 only (d) Closed Loop with S_2 only

Figure 5. CW Jammer with a Low Level Desired Signal Separated by 14 kHz (All oscillographs at 10 dB/cm; 2 kHz/cm; 1 kHz Bandwidth)

Figure 6 describes AM interference cancellation in the presence of a low-level desired signal. The desired signal is again modulated 95% at a 1000 Hz rate. The interference signal is now also 95% modulated at a 1000 Hz rate. The frequency separation for this test is approximately 200 kHz. A comparison of the open and closed loop spectral outputs of parts (a) and (b) of figure 6 shows that the AM interference signal has been cancelled by about 70 dB.

When the 200 kHz frequency difference of figure 6 is reduced to about 16 kHz, cancellation of the undesired signal in excess of 70 dB is maintained.

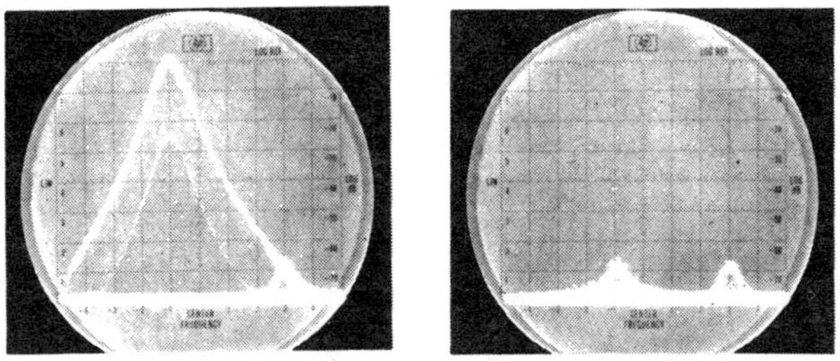

(a) Open Loop with $S_1 + S_2$ (b) Closed Loop with $S_1 + S_2$

Figure 6. AM Jammer with a Low-Level Desired Signal Separated by 200 kHz (10 dB/cm; 50 kHz/cm; 30 kHz Bandwidth)

Cancellation in the Presence of a Second Interference Signal

The following sequence of tests describes the cancellation performance in a multiple interference environment. For these tests, modifications were made to the error sensing circuitry involving the use of additional internal filtering which enables the processor to perform optimally in a multiple signal environment where strong uncancelled signals are present at frequencies close to interference which is to be cancelled.

Figure 7 demonstrates interference reception of a desired low-level signal in the presence of two strong signals. Unmodulated S1 and S3 signals (figure 2) are used to simulate the two equal, high level waveforms. The desired signal is simulated by S2 which is 95% modulated at a 1000 Hz rate.

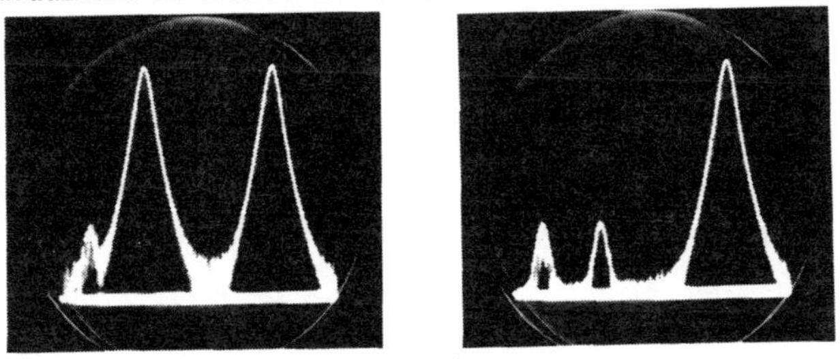

(a) Open Loop - $S_1 + S_2 + S_3$ (b) Closed Loop - $S_1 + S_2 + S_3$

10 dB/cm; 10 kHz/cm; 3 kHz Bandwidth

Figure 7. CW Jammer with a Desired Signal and an Equal Level Uncancelled Signal Differing in Frequency by 50 kHz

Part (a) of figure 7 shows the open loop system spectral output with the three input signals as described above. The AM modulated desired signal is 20 kHz removed from the interfering signal S1 against which cancellation is desired. The second signal (S3) is about 50 kHz removed from the first jamming signal, and at the same power level. The closed loop operation shown in part (b) demonstrates 55 dB of cancellation against the interference which is only 20 kHz removed from the desired signal. Cancellation occurs even though another high-level signal is present only 50 kHz away from cancelled interference.

Transient Response Time

A single square wave modulated CW interference signal was used to determine the response time of the ICS. The oscillator was alternately turned on and off through the use of an external square wave generator which controlled an in-line rf switch. In this manner periodic bursts of interference lasting for about one second were forced to enter both the receive and transmit inputs of the ICS. The burst was repeated every two seconds.

All Oscillograms at 50 mV/cm

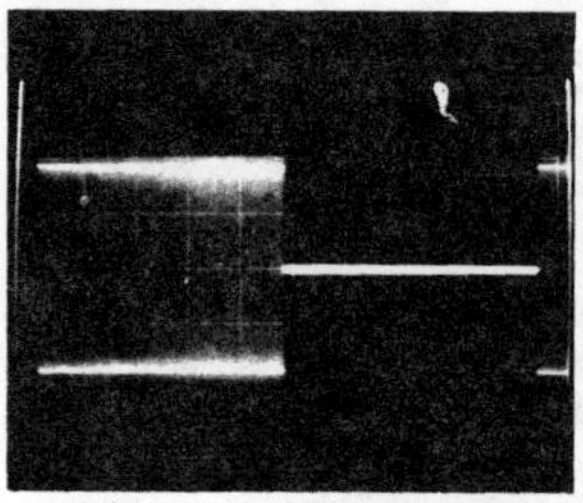

(a) Open Loop Time Output
200 ms/cm

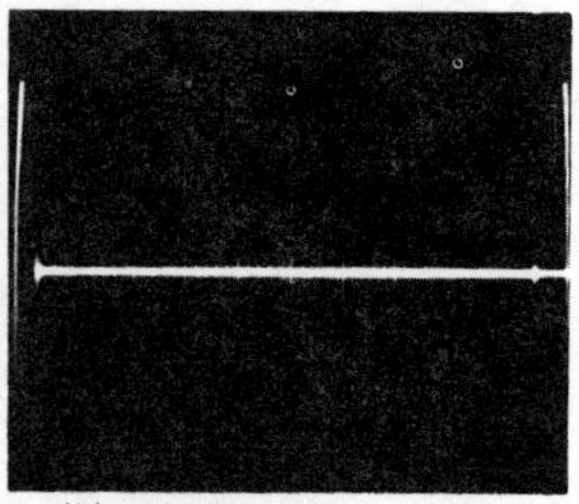

(b) Closed Loop Time Output
200 ms/cm

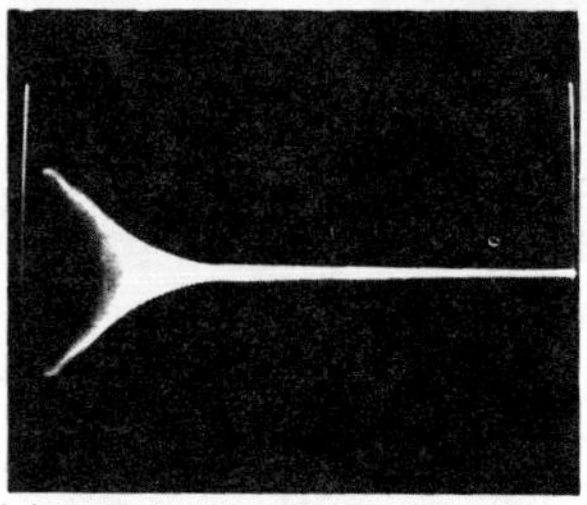

(c) Expanded Closed Loop Time Output
2 ms/cm

Figure 8. System Response Time

Figure 8 contains a time expanded photograph of the first output transient spike. The horizontal scale factor is 2 ms/cm. The full horizontal extent of the photograph is 20 ms. It is apparent that the system has attained its full cancellation in less than 20 ms.

Conclusions

The adaptive ICS offers a suitable alternative to currently used methods of eliminating co-located interference signals. The experimental data shows that rejection levels exceeding 55 dB can be achieved with a frequency separation between receive and transmit frequencies of only 14 kHz. Since the ICS requires minimal alterations for installation, the impact on existing physical terminal facilities is negligible.

Acknowledgments

The author wishes to acknowledge the contributions of Mr. Anthony Kowalski and Mr. William Frost of Hazeltine Corporation who formulated the design approach and obtained the experimental data upon which this paper is based.

References

1. B. Widrow et al, Adaptive Antenna Systems, Proceedings of the IEEE, Vol. 55, No. 12, Dec. 1967 pp. 2143-2158

Echo Suppressor Design Considerations

JOHN E. UNRUE, JR., MEMBER, IEEE

REFERENCE: Unrue, J. E., Jr.: ECHO SUPPRESSOR DESIGN CONSIDERATIONS, Bell Telephone Laboratories, Inc., Holmdel, N. J. Rec'd 12/12/67; revised 4/11/68. Paper 67TP1375-COM, approved by the IEEE Wire Communication Committee for publication after sponsored presentation at the 1967 National Electronics Conference, Chicago, Ill. IEEE TRANS. ON COMMUNICATION TECHNOLOGY, 16-4, August 1968, pp. 616–624.

ABSTRACT: The growth of telephone circuits with long propagation times, particularly those using synchronous satellite facilities, has renewed interest in the design of echo suppressors. The increase in delay intensifies potential degradations associated with the use of echo suppressors and demands echo suppressor performance far superior to that of older designs. Several organizations, including the Bell System, have designed long delay echo suppressors in the last few years and others are likely to become involved in the future as the growth of long delay facilities increases. Echo suppressors of different design will often be used in the same connection and must be compatible. To assure this compatibility a working party of the International Telephone and Telegraph Consultative Committee (CCITT) Study Group XV, with the cooperation of many of the organizations presently concerned, has recently completed a recommendation on the operating characteristics of long delay echo suppressors. This paper will present the operating characteristics of the newest Bell System echo suppressors, the reasons for them, and the relevant provisions of the CCITT working party recommendation. It is believed that all echo suppressors having these characteristics will perform sufficiently well and be compatible with any other so designed.

I. Introduction

Most telephone connections involving long distances use four-wire facilities between exchange areas and two-wire facilities within the exchange area. Impedance irregularities in the two-wire facilities and at the two- to four-wire junctions result in reflected signals. Such an "echo" becomes disturbing when there are more than a few milliseconds of round-trip delay, and steps must be taken to reduce its volume. The loss required to reduce it to acceptable limits is an increasing function of delay [1]. When the round-trip delay is less than about 45 ms, sufficient echo reduction can be obtained by operating the trunks at a loss (called *via net loss*) proportional to the length of the trunk [2]. For longer connections the loss required to render the echo unobjectionable would excessively degrade transmission if it were incorporated in the connection in a static fashion. On such connections echo suppressors are used to insert the loss, on a voice switched basis, in that direction of the four-wire facility that is not being used at the time for direct transmission. Echo suppressors can introduce transmission degradations, primarily in the form of speech mutilation, due to the losses that are switched in and out of the transmission paths. The potential disturbance of this mutilation is also found to be an increasing function of delay [3].

The first echo suppressors were designed almost 40 years ago, but there was little motivation for design improvement until early in this decade. Then the increasing demand for very long circuits, particularly the anticipated use of satellite facilities, inspired new interest. A synchronous satellite results in about 600 ms of round-trip delay compared with about 100 ms for a submarine cable connection between New York and London. The performance of the older echo suppressors was inadequate for the increased delay.

During the past several years, studies undertaken by Bell Telephone Laboratories led to the development of an echo suppressor designed for long delay circuits [4]. Other organizations in several countries have also been active in this area and have produced several new echo suppressors. It quickly became evident that echo suppressors of different design would often be used in the same circuit and that the compatibility of one echo suppressor with another would have to be assured. Toward this end most of the organizations involved have worked together in attempting to determine and standardize the desirable operating characteristics of echo suppressors. Ideas have been shared in informal discussions, several Ad Hoc committees, and Study Group XV of the International Telephone and Telegraph Consultative Committee (CCITT). In March 1968, a Study Group XV working party completed a draft recommendation on specifications for echo suppressors, incorporating the characteristics of those which have good performance and have proved to be compatible [5].[1] It is expected that this draft recommendation will be adopted and become the recommendation of the CCITT for echo suppressor characteristics.

The Bell System's echo suppressor specifications will be presented with the relevant provisions of the draft recommendation of the working party. We will first briefly review the operation of an echo suppressor, pointing out the environment in which it must function and the situations it must handle. This will be followed by the detailed specifications that have evolved, with explanations and supporting data. We will conclude with some general remarks concerning echo suppressor design and compatibility.

II. General Operation of Echo Suppressors

Fig. 1 indicates the important aspects of the environment of an echo suppressor. Speakers *A* and *B* are connected by a combination of two- and four-wire facilities.

[1] Modifications to this document were made at the March 1968 meeting of the Working Party. At the time of this writing the paper documenting these changes had not been issued by the CCITT.

Reprinted from *IEEE Trans. Comm. Tech.*, vol. COM-16, pp. 616-624, Aug. 1968.

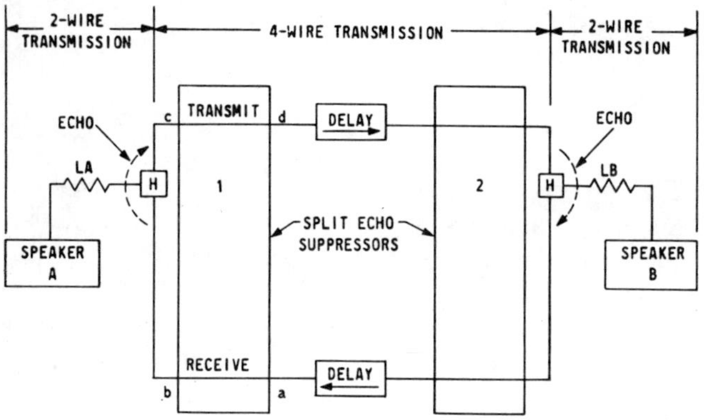

Fig. 1. Echo suppressor environment.

The conversion is accomplished by hybrid transformers labeled H. The major delay in the connection is associated with the four-wire facility (which may be a satellite link) although some delay is also present in the two-wire facilities. On long delay circuits, echo suppressors are located in the four-wire facility at each end. Each unit provides for suppressing echoes generated in the end circuit nearest it. When used in this manner the units are referred to as *split* or *half* echo suppressors in order to distinguish the configuration from one used on short delay circuits. In the latter case, the functions of two split echo suppressors are often combined in a single unit, called a *full* echo suppressor, which is located at one end of the four-wire facility. Echo suppressors must be split on long delay circuits for reasons that will become evident.

The four-wire facility is often operated with 0-dB net loss between echo suppressor locations. The two end sections generally have losses associated with them, indicated by L_A and L_B in Fig. 1. Although these two losses can be different, there is no consistent bias favoring one or the other. Speech from B, arriving at a will, on the average, be equivalent in power to speech from A arriving at c. Signal powers will always be referred to a 0-dB transmission level point (0TL). At echo suppressor locations in the Bell System, the transmit path is normally adjusted to a level 16 dB less than 0TL, while the receive path is adjusted to a level 7 dB greater than 0TL. In other telephone networks they may be at different level points. Referring all powers to 0TL avoids confusion resulting from these differences.

When B is speaking, the signal will be transmitted through the lower path of the four-wire facility, eventually reaching point b and the two-wire facility at A's end. A portion will be reflected and coupled to point c, due to impedance irregularities in the two-wire facility. It would eventually reach B as an echo if it were not blocked by the echo suppressor. The echo signal will suffer a net loss between b and c, the amount of which is usually a function of frequency. The average steady-state loss between 500 and 2500 Hz is defined as *echo return loss.*

In addition to undergoing a loss, the signal at c may be delayed relative to that at b. This is referred to as *end delay.* Both the return loss and end delay vary from one

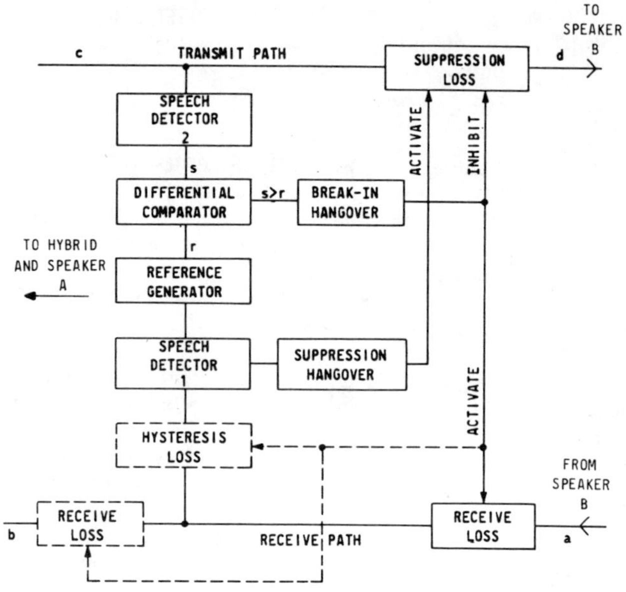

Fig. 2. Functional diagram of split echo suppressor.

connection to another. The echo suppressor must be designed to operate properly for the worst case of minimum return loss and maximum end delay. These extremes, of course, depend on the network in which the echo suppressor is used. Considering present applications a split echo suppressor may, on rare occasions, encounter a return loss as low as 6 dB and an end delay as long as 25 ms.

Echo suppressor 1 blocks (or suppresses) B's echo in the transmit path between c and d. Echo suppressor 2 suppresses A's echo in a similar way. The suppression is accomplished by inserting a large loss in the path, such as that obtained from a short circuit. The suppression must be applied rapidly at the onset of speech arriving at a and must remain in for a period of time after the cessation of speech at a in order to suppress echoes stored in the end delay. These periods are referred to as *suppression pickup* and *suppression hangover* times, respectively. If B is speaking, causing echo suppressor 1 to suppress, and A interrupts, A's speech would tend to be blocked. In these *double talking* situations the suppression must be removed, i.e., A must be able to *break in* on B. Break-in should occur rapidly at the onset of A's interruption. The time lag is called *break-in pick-up* time. The echo suppressor should not resuppress as long as A and B both continue to talk.

The latter requirement is facilitated by forcing the break-in condition to continue for a period of time after the echo suppressor ceases to recognize that double talking exists, so as to bridge the intersyllabic intervals in A's speech. This period is called *break-in hangover* time.

Echo is allowed to return unsuppressed when an echo suppressor is in the break-in mode. Although this echo is usually masked by the interrupting talkers speech, it can be disturbing on long delay circuits, especially during the break-in hangover period when the masking speech is absent. Thus, it is desirable to insert some loss in one of the transmission paths during double talking, even though it will result in reduced received volume for the direct speech. This is found to be less disturbing than the echo, as long as the loss is not too large. The loss is inserted between *a* and *b* of Fig. 1 for reasons which will become evident.

An echo suppressor's ability to correctly distinguish between one party talking and double talking is crucial. When it fails to perform this function well, degradations in the form of clipped or chopped speech and annoying echo result. The detection of double talking is not a simple task, as can be appreciated by considering the information available to the echo suppressor. It is assumed that only *B*'s speech appears at point *a*, i.e., echo suppressor 2 does a perfect job of removing *A*'s echo. However, a signal at point *c* can be due to *A*'s speech, *B*'s echo, or both. That portion due to *B*'s echo will be correlated with the signal at *a*, but will be delayed and attenuated by unknown amounts. The detection of double talking requires a comparison of the two signals in a way that accounts for the possible return loss and end delay that may be present. Double talking is assumed if the echo suppressor concludes that the signal at *c* consists of more than the echo of *a* alone.

It will prove helpful to have a functional model of an echo suppressor in mind in the ensuing discussion. Fig. 2 shows echo suppressor 1 of Fig. 1 in more detail. This model is not intended to limit the possible implementations of an echo suppressor, nor does it accurately represent some echo suppressors existing today. However, it is believed that all implementations should perform functions essentially the same as those obtained from the model.

If speaker *B* alone is talking, his speech in the receive path is recognized by speech detector 1, and the suppression loss in the transmit path is activated through the suppression hangover circuit. The suppression hangover circuit will maintain the suppression as long as speech persists at *a* and for the hangover period afterwards, provided that break-in does not occur. Signals at point *c* are detected by speech detector 2, whose output *s* is one input to a differential comparator. The signal from speech detector 1 is operated on by a reference generator, whose output *r* is the second input to the differential comparator. The reference generator integrates its input in such a way that the output *r* will always exceed *s* provided that only *B*'s echo is present at *c*, and the return loss and end delay bounds are not exceeded [6]. Under these conditions there will be no output from the differential comparator. However, if speech from speaker *A* is also present *s* will, at times, exceed *r*. Then the differential comparator will have an output which, in conjunction with the break-in hangover circuit, will cause the suppression loss to be inhibited (without waiting for the suppression hangover to expire) and the receive loss to be inserted. Full duplex transmission is thereby restored, with some echo reduction as a result of the receive loss, and with some reduction in volume of *B*'s speech as heard by *A*. The break-in hangover circuit maintains the break-in state for the hangover period during those times when *r* exceeds *s*. Break-in will be continuously maintained for the entire duration of double talking provided that *s* exceeds *r* at least once in every hangover period.

Inserting the echo-reducing loss in the receive path aids the maintenance of the break-in state by reducing the signal at speech detector 1 and, hence, *r* relative to *s*. This is referred to as the *break-in hysteresis advantage*. It should be noted that some echo suppressors insert the receive loss after the input to speech detector 1, as shown by the dotted box. In order to obtain the hysteresis advantage, it is necessary to simultaneously switch an *equivalent* loss in the control path as indicated by the dotted box labeled *hysteresis loss*. The hysteresis loss must not exceed the receive loss or echo can maintain the break-in state after double talking when the return loss is small.

The functions that must be performed by a full echo suppressor differ only slightly from those applicable to a split suppressor. Essentially a full echo suppressor is equivalent to two split echo suppressors connected together, with one inverted to insert suppression in the receive path. However, the end delay of one side now includes the round-trip delay of the four-wire facility and may be considerably longer than 25 ms. To compensate for the longer delay, the memory time of the reference generator must be increased, thus resulting in a generally larger reference signal for the break-in talker to overcome. This tends to increase the amount of chopping during double talking, so the allowable end-delay protection is limited. This, in turn, limits the application of full echo suppressors to circuits where the total delay from speaker *A* to speaker *B* is less than about 100 ms. In all other respects (with the possible exception of the receive loss, discussed later), the specifications for full and split echo suppressors are the same. We will now cover these specifications in detail from the point of view of a split suppressor, pointing out the differences in a full suppressor when differences exist.

III. Specifications

The most significant measure of the performance of an echo suppressor is its acceptability by the customer using the circuit for ordinary voice communications. However, to be useful for design, the specifications must be in terms of characteristics measurable in an objective way. Generally, these specifications are in terms of the state of the echo suppressor, or the transition between states resulting from signals applied to its input ports. These signals simulate conditions that can arise during a telephone conversation. The word signal means a 1000-Hz sinusoid unless otherwise noted.

It will be convenient to have a "map" of the possible states of an echo suppressor as a function of signal powers at its two inputs. Such a map is provided by Fig. 3.[2] It shows the condition of the echo suppressor as a function of signals appearing at the transmit and receive inputs.

[2] This map was conceived by D. L. Richards of the British Post Office.

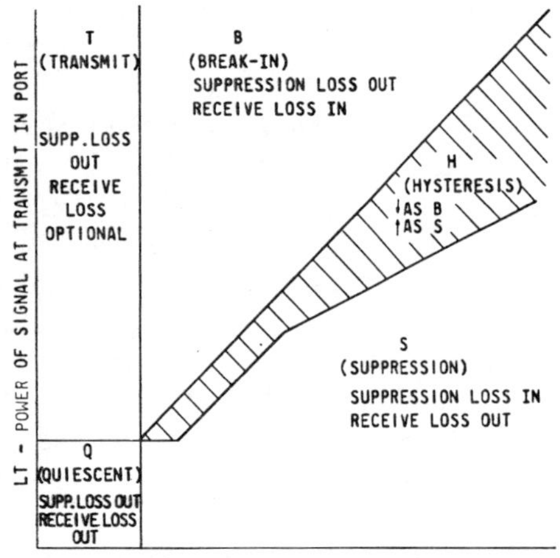

Fig. 3. Echo suppressor state map.

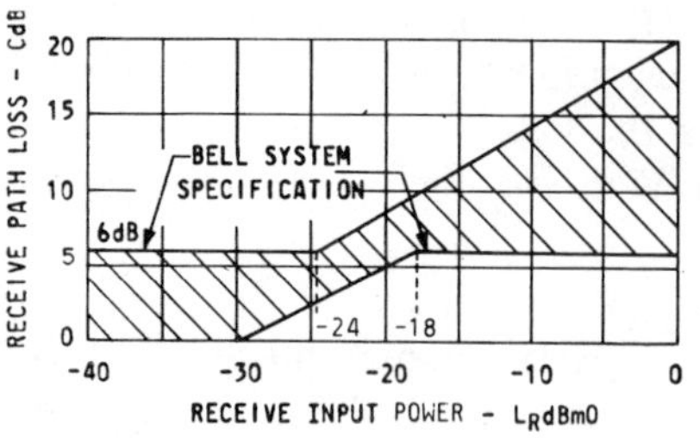

Fig. 4. Receive loss recommendation.

These are denoted by L_T and L_R and correspond to signal power in dBm0 at points c and a of Fig. 2, respectively. The hysteresis region describes the transitions between the suppression and break-in states and depends upon the direction of the transition. The peculiar shape of the lower boundary is due to possible receive loss characteristics. The "transmit state" implies that the near-end party alone is talking. Some echo suppressors treat this the same as break-in with the receive loss inserted, while others distinguish the state from break-in and keep all losses out. Since no speech is present in the receive path, it seems to make very little difference whether the receive loss is in or out and is considered optional. If the loss is inserted, the circuit noise will be attenuated during the transmit state, but experience indicates that this is not disturbing. For brevity, we will often refer to the states by the letter names indicated, which is the first letter of the word description of the state, i.e., H for hysteresis.

In conjunction with some of the specifications to follow, supporting data referred to as degradation counts will be presented [5]. These are the results of a device that monitors several points in a test circuit equipped with echo suppressors. The presence or absence of signals at these points allows a determination of whether chopping or echo degradations are occurring. The device is used in conjunction with a standard tape-recorded conversation and counts the number of occurrences of each of these degradations.

A. *Loss Specifications*

Several methods of applying suppression loss were considered, including transistor or relay switches that insert a large loss instantaneously, and one that inserts an increasing amount of loss smoothly over a short period of time. Subjective tests indicated that the required suppression loss at long delays is very large and must be inserted quickly, so all present echo suppressors use fast operating transistor or relay switches. The suppression loss in the transmit path should exceed 50 dB.

The loss inserted in the receive path during double talking should be large enough to reduce echo to tolerable limits (in conjunction with the masking effect of the near party's speech), but not so large as to seriously reduce the receive speech volume. Some echo suppressors use a speech compressor to obtain the receive loss. This results in an increasing amount of loss as L_R increases, and has the advantage of yielding more echo reduction when the echo would tend to be loud. Also, this loss is inserted and removed in a smooth fashion over a period of time to make it less noticeable to the near-end party. Other echo suppressors have used a fixed loss somewhere between 6 and 12 dB. Subjective tests have not indicated that either means of obtaining receive loss has a significant subjective advantage over the other. Of course, the fixed loss is usually more economical. These same subjective tests did not reveal any compatability problems when echo suppressors using both types of receive loss were used in the same circuit. For these reasons it appears that the type of receive loss should be optional. The draft recommendation of Study Group XV calls for a receive loss characteristic which falls within the shaded region of Fig. 4. In future echo suppressor designs the Bell System will use a 6-dB fixed loss. The receive loss corresponding to a particular receive input power is denoted by C.

It should be mentioned that some consideration was given to inserting the double talking loss in the transmit path or splitting it between both paths. If this is done, full advantage of the hysteresis effect as an aid toward maintaining the break-in state is not realized. Also, some tests indicate that compatibility problems can arise if the loss is not applied in a similar location in the two echo suppressors in the trunk [5]. Operation in this manner tends to bias the connection in favor of one party of the conversation at the expense of chopping the other party's speech.

Since a full echo suppressor is intended to be used in connections where the total delay is less than about 100 ms, it is not essential that double talking loss be provided and is considered optional.

B. *Sensitivity and Bandwidth*

A fundamental requirement of an echo suppressor is to reliably detect the presence of speech signals in the two transmission paths. It must do this in the presence of normal circuit noise without mistaking the noise for a speech signal. Older echo suppressors used a speech detector sharply tuned to 1000 Hz. These detectors will fail to detect many speech sounds unless the sensitivity is made

TABLE I

1000-Hz Transition Sensitivities (Bell System Specifications and Those of the Draft Recommendation of Study Group XV Are the Same)

Transition	Symbol	Sensitivity, dBm0	Notes
Suppression			
Q to S	T_{QS}	$L_R = -31 \pm 2$	Variation with frequency as shown in Fig. 5 (Bell System specification as shown in Fig. 6)
S to Q	T_{SQ}	$L_R = T_{QS} \begin{smallmatrix}+0\\-3\end{smallmatrix}$	
Transmit			
Q to T	T_{QT}	$L_T = -31 \begin{smallmatrix}+2\\-5\end{smallmatrix}$	Variation with frequency as shown in Fig. 6
T to Q	T_{TQ}	$L_T = T_{QT} \begin{smallmatrix}+0\\-3\end{smallmatrix}$	
Break-in			
H to B	T_{HB}	$L_T = L_R \pm 1.5$ $-26.5 \leq L_R \leq +4.5$	Variation with frequency as shown in Fig. 6. For T_{HB} tolerance of ± 1.5 dB applies throughout 500- to 3000-Hz bandwidth. For T_{BH} the nominal sensitivity is $L_T = L_R - C$. The vertical extent of the H area should not exceed C by more than 3 dB.
B to H	T_{BH}	$L_T = L_R - C$ -3 minimum $-26.5 \leq L_R \leq +4.5$	

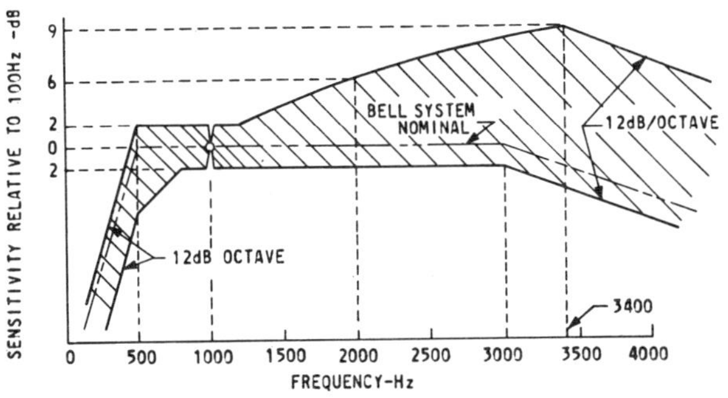

Fig. 5. Recommended frequency characteristic of suppression control path.

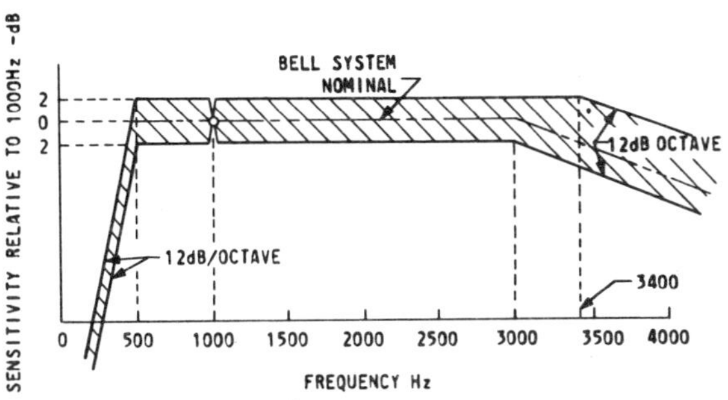

Fig. 6. Recommended frequency characteristic of break-in control path.

very high. It is desirable to keep the sensitivity reasonably low to reduce the dynamic range that the detectors must handle. Studies made in conjunction with the development of time assignment speech interpolation (TASI) indicated that the wider the bandwidth of the detector the less sensitivity is required for the same performance. Furthermore, the advantage gained in sensitivity is greater than the added noise due to the increased bandwidth [7].

For transitions such as between S and B (Fig. 3), the differential comparator will be comparing signals in the two transmission paths, and there should be no frequency bias in the comparison. Thus, the control circuitry to the differential comparator should be essentially flat across the voice band. For purposes of operating the suppression loss, i.e., for transitions from Q to S, some are of the opinion that the best performance is obtained when the sensitivity of the detector increases slightly with increasing frequency. Subjective tests conducted during actual service conditions have not indicated any significant differences in performance of such a detector over one with a flat sensitivity characteristic. Since the control circuit must be flat for the differential comparator, a different characteristic for the Q to S transition could complicate the design of the echo suppressor without yielding any significant benefits. We believe that this characteristic should also be flat and the same as that for the control path of the differential comparator. Experience has not indicated any compatibility problems when an echo suppressor with a flat characteristic is used in the same trunk with one using an increasing characteristic. Thus, as shown in Fig. 5, the draft recommendation of Study Group XV allows a large tolerance with regard to this characteristic. The specifications applicable to the control paths of the differential comparator, including all tolerances, are shown in Fig. 6.

The 1000-Hz sensitivity for all transitions from the Q state should nominally be -31 dBm0. Transitions from S to B should occur when the signals in the transmit and receive path are equal. The term *differential sensitivity* is used to measure the bias in this transition. It is defined as the difference, in decibels, between the power in the transmit path L_T and the power in the receive path L_R when break-in occurs. As the differential sensitivity increases, break-in becomes more difficult leading to excessive chopping. Thus, the differential sensitivity should nominally be 0 dB.

The transition from B to S should occur when s, in Fig. 2, becomes less than r. Due to hysteresis, this corresponds to

$$L_T = L_R - C.$$

TABLE II
MAXIMUM PICK-UP TIMES[a]

Transition		Initial Signals L_T, dBm0	Initial Signals L_R, dBm0	Final Signals L_T, dBm0	Final Signals L_R, dBm0	Study Group XV Recommendation	Bell System Specifications
Q to S		≤ -40	≤ -40	≤ -40	-25	5 ms	5 ms
		≤ -40	≤ -40	≤ -40	-11	5 ms	5 ms
H to B	L_R Constant	≤ -40	-25	-19	-25	30 ms[b]	10 ms
		≤ -40	-15	-9	-15	30 ms[b]	10 ms
	L_T Constant	-25	-22	-25	≤ -40	40 ms[c]	30 ms
		-25	-19	-25	≤ -40	60 ms[c]	50 ms
		-25	-16	-25	≤ -40	70 ms[c]	70 ms
		-25	-9	-25	≤ -40	75 ms[c]	75 ms

Notes:
a) Operation of echo suppressors by signals of short duration, such as circuit impulse noise, is undesirable.
b) From theoretical considerations it is desirable that this time be short. With conversation in the English language, no deleterious effects have been observed for pick-up times of 30 ms.
c) It is desirable that these values be as low as possible consistent with the need to protect against false break-in on echo for the case of maximum end delay and minimum return loss.

In this case we would say that the "break-in hysteresis" is C dB. The 1000-Hz sensitivities for all transitions, including tolerances, are summarized in Table I.

C. Dynamic Requirements

The transition time from the Q to S state (suppression pick-up time) should be short to prevent an initial burst of echo from being returned. The transition time from the S to B state (break-in pick-up time) should also be short to avoid clipping the initial portion of the interrupting talker's speech. However, the break-in pick-up time requirement is not as stringent as that for suppression pick-up time. A long suppression pick-up time would result in an initial burst of echo being returned for every new utterance of the talker. A long break-in pick-up time will result in initial clipping of the interrupting party's speech during break-in, but this is a relatively rare occurrence in a conversation.

The break-in pick-up time will usually depend upon the angle in which the H–B boundary is crossed (see Fig. 3). The most important case is when a receive path signal has been established and the transmit signal suddenly increases. This condition (L_R constant) simulates the near party interrupting the far party. It corresponds to a vertical crossing of the H–B boundary in Fig. 3. For this condition the pick-up time should be as short as possible, limited only by the speed of the differential comparator and break-in control circuit. Another possible case arises when a transmit signal is established in the presence of a stronger receive signal and then the receive signal is removed. This condition (L_T constant) is approximated when the near-end speaker starts talking, at a low volume, just as the other speaker ceases. The near speaker must break in because of the suppression hangover time. Here the H–B boundary is crossed horizontally. The pick-up time is likely to be longer since the reference signal at the differential comparator must first decay to the point where the transmit signal exceeds it. The reference signal lags the receive signal because of the end delay compensation.

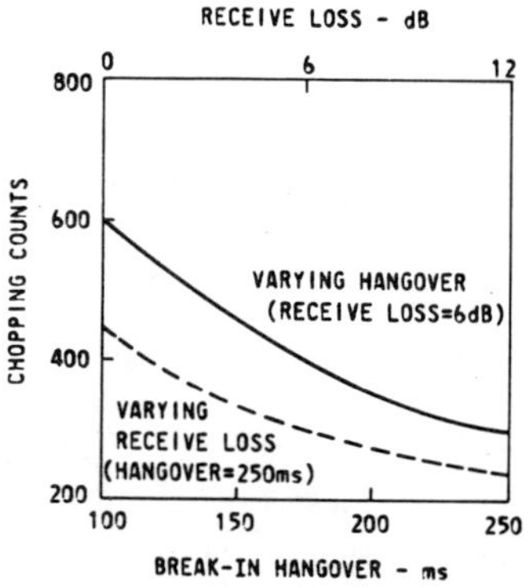

Fig. 7. Chopping degradation as a function of receive loss and break-in hangover.

The echo suppressor should not be unduly susceptible to operation on noise impulses encountered on a telephone trunk. One means of preventing this is to add some smoothing in the control paths so that the pick-up time will bridge most noise impulses. Also, time measurements will generally depend on the power difference in the signals before and after the transition between states, so the specifications are stated for several initial and final signal powers. The pick-up time specifications are summarized in Table II.

Transition times in the reverse directions must also be controlled. The time required to switch from S to Q (suppression hangover) should be at least as long as the maximum end delay in order to suppress the echo stored in the near-end circuit. It should actually be somewhat longer since the ending of a speech signal often tends to be weak and prematurely drops below the sensitivity level. The maximum time limit is not critical since the echo suppressor has the ability to break in without waiting for the suppression hangover to expire.

The transition from B to S (break-in hangover) has a large influence on the double talking performance of the echo suppressor. It should be long enough to bridge those time periods, during double talking, when the transmit signal drops below the reference signal. The optimum period

TABLE III

HANGOVER TIMES

Transition[c]	Initial Signals L_T, dBm0	Initial Signals L_R, dBm0	Final Signals L_T, dBm0	Final Signals L_R, dBm0	Study Group XV Recommendation	Bell System Specifications
S to Q	≤ -40 ≤ -40	-25 -11	≤ -40 ≤ -40	≤ -40 ≤ -40	40 to 75 ms[a]	200 ± 40 ms
H to S	-19 -9	-25 -15	≤ -40 ≤ -40	-25 -15	150 to 350 ms[b]	200 ± 40 ms

Notes:

a) The upper limit for the S to Q transition may be as great as 240 ms *provided* the requirements for the H to B transition with L_T constant are met. (See Table II.)

b) The amount of break-in hangover necessary depends upon the values of loss inserted in the receive path under break-in conditions. For low fixed loss for C, the hangover will tend to be toward the upper end of the range. For higher values of loss, as inserted by a compressor, the hangover time will tend to be at the lower end of the range.

c) It is not considered necessary to measure the T to Q hangover time. However, it is desirable that the echo suppressor should be so designed that this hangover is not unnecessarily long, for example, not greatly in excess of that applicable for the H to S transition.

TABLE IV

CHOPPING DEGRADATION VERSUS END DELAY PROTECTION

End Delay[a] Protection, ms	Chopping Counts
25	563
35	620
45	717
55	890
75	1057
100	7123

Note:

a) With 6-dB return loss, no false break-in occurs for the end delay stated. A 5-ms increase results in false break-in.

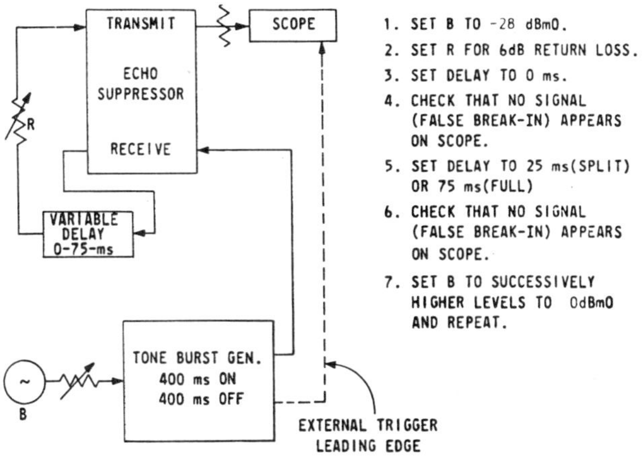

Fig. 8. Test for false break-in due to echo.

depends on the differential sensitivity and hysteresis in addition to the statistical properties of speech. Fig. 7 shows the variation of chopping degradation counts as functions of hysteresis and break-in hangover time for an experimental echo suppressor. For purposes of reducing chopping it is desirable to make both the hysteresis and break-in hangover large. However, if the receive loss is too large, the near-end party would experience difficulty in hearing. If the hangover is too large, the echo during the hangover period after double talking ceases would be disturbing. Thus, a compromise is required, based on judgement. We believe that a nominal break-in hangover period of about 200 ms, in conjunction with a 6-dB receive loss is near optimum. The hangover time specifications are summarized in Table III.

Sometimes significant economies can result by making both the suppression and break-in hangovers the same. Since the upper limit of the suppression hangover is not critical, this is allowed as indicated in note a) of Table III. It has also been suggested that further economies could result by changing the echo suppressor operation so that the suppression loss would be inserted for both the Q and S states and would simply be removed for the T and B states. We feel that this should be discouraged for several reasons. Operation in this manner would result in blocked transmission paths if a failure occurs in the echo suppressor. Also it would result in initial clipping for each transition from the Q to T state. Operation as assumed in Fig. 3 results in initial clipping only for the rare transitions from S to Q to T within the suppression hangover time.

The preceding specifications are applicable when the return loss is large so that any echo appearing at point c of Fig. 2 can be ignored. However, proper behavior of the echo suppressor must also be assured for the most adverse echoes that may reasonably be expected. When the return loss is small and the end delay is zero, false break-in, lasting for the break-in hangover period, may result if the response of the transmit control circuit is faster than that of the receive control circuit. This would tend to occur when a signal is suddenly applied to the receive in port (point a of Fig. 2). Similarly, if the end delay compensation of the reference generator were inadequate, false break-in would result when the return loss was small and the end delay large, and would occur when a signal was suddenly removed from the receive in port. However, an excessive amount of end delay compensation will degrade the break-in performance of the echo suppressor. It is axiomatic that the end delay compensation biases the differential comparator in favor of receive path signals, making break-in more difficult. Increasing the compensation increases the bias. Table IV indicates the increase in chopping degradation counts for one echo suppressor as the end delay compensation is increased.

The end delay compensation of a full echo suppressor results from a compromise involving both economic and performance considerations. A considerable cost savings results from the use of a single full echo suppressor on a trunk compared to two split echo suppressors. In the Bell System DDD network a full echo suppressor could encounter end delays of about 75 ms on the longest trunks. The resulting chopping performance would be acceptable provided the echo suppressor met the specifications described here. Split echo suppressors should provide end delay compensation of about 25 ms in order to minimize chopping (which is more disturbing at long delays).

With regard to the preceding items, the performance can be measured by applying a suitable sequence of signals to the echo suppressor while varying the end circuit parameters. One such test procedure is shown in Fig. 8.

D. *Other Specifications*

In addition, an echo suppressor must satisfy all of the transmission requirements such as structural return loss overload, envelope delay distortion, etc., applicable to all devices used in the transmission paths. These requirements will not be discussed.

Any transients induced in the transmission paths, due to operation of the suppression and receive losses, should be inaudible to the customers using the circuit and should not lead to the false operation of any other echo suppressor in the connection. This can be assured by measuring the peak voltage of the transient, after bandpass filtering it to a 500- to 3000-Hz band. The peak should not exceed 20 mV at 0TL (−34 dBv0).

An echo suppressor should include means for controlling its operation by the presence or absence of a ground supplied by the associated trunk circuit. This *enabling feature* is used in some networks to control the assignment of echo suppressors in a multilink connection. Upon the receipt of a ground, the echo suppressor should be enabled to operate normally. Otherwise the suppression and receive losses should be inhibited so that the echo suppressor has no influence on transmission. In some cases, the enabling is initiated by the absence of a ground, so both possibilities should be allowed as options.

An echo suppressor should also be equipped with a *tone disabler* when used on trunks that may be used for data services. This auxiliary circuit, upon detection of a signal from the data set, disables the suppression and receive losses during data transmission. The requirements for Bell System tone disablers are given elsewhere [8], and the CCITT is in the process of preparing a recommendation.

IV. Conclusion

It is believed that an echo suppressor incorporating the features specified previously will result in performance which is as good as presently feasible at a reasonable cost. Cheaper devices could be made by giving up certain features such as the receive loss and hysteresis. If this were done, an increase in degradation would certainly result. At moderate delays some increase can be accepted. Thus, as pointed out, the specifications for a full echo suppressor are more lenient. At long delays even the most sophisticated echo suppressors are only marginally acceptable. In tests conducted during the initial operation of the Early Bird satellite, a significant increase in conversational difficulty was experienced by customers using the satellite compared with those using the submarine cable, even though the satellite circuits were equipped with the best modern echo suppressors available [9]. Therefore, we are not in a position to sacrifice performance in order to achieve a smaller echo suppression cost. The increased cost of long delay echo suppressors should properly be assigned to the service need or facilities which make them necessary.

Even the best performing echo suppressor is of small value if it is not compatible with other echo suppressors used in the same circuit. This compatibility is crucial since the fact that long delays are involved implies that often the circuit will transcend the boundaries of any one telephone administration, and terminal equipment of different designs will usually be used. It is not difficult to imagine ways in which certain deviations from the preceding specifications could result in incompatibility. Differences in suppression sensitivity, for example, could result in an echo signal at the receive in port of the more sensitive echo suppressor, due to the failure of the other to suppress. The echo signal would tend to cause the echo suppressor to insert suppression, possibly leading to excessive chopping. Large differences in pick-up and hangover times could cause similar difficulties.

The compatibility of a new echo suppressor can be verified only by observing its behavior when it is used in a connection which includes echo suppressors of different design. It is not practical to make such tests for every new echo suppressor. The objective of the draft recommendation of Study Group XV is to eliminate the need for much of this testing. It is presumed that an echo suppressor meeting the specifications will be compatible with another so designed.

Acknowledgment

The author is indebted to his colleagues at Bell Telephone Laboratories, particularly P. T. Brady and G. K. Helder who jointly developed many of the concepts and specifications reported, and to the many persons in other organizations concerned with echo suppressors by whose efforts the draft recommendation of CCITT Study Group XV for echo suppressors was established.

References

[1] J. W. Emling and D. Mitchell, "The effects of time delay and echoes on telephone conversations," *Bell Sys. Tech. J.*, vol. 42, pp. 2869–2891, November 1963.

[2] H. R. Huntley, "Transmission design of intertoll telephone trunks," *Bell Sys. Tech. J.*, vol. 32, pp. 1019–1036, September 1953.

[3] R. R. Riesz and E. T. Klemmer, "Subjective evaluation of delay and echo suppressors in telephone communications," *Bell Sys. Tech. J.*, vol. 42, pp. 2919–2924, November 1963.

[4] E. W. Holman and V. Suhocki, "A new echo suppressor," *Bell Lab. Rec.*, pp. 139–142, April 1966.

[5] Report to CCITT, Study Group XV of Working Party on Echo Suppressors (Question 10/XV), December 1, 1966, COM XV-No. 93, COM XII-No. 52.

[6] P. T. Brady and G. K. Helder, "Echo suppressor design in telephone communications," *Bell Sys. Tech. J.*, vol. 42, pp. 2893–2917, November 1963.

[7] H. Miedema and M. G. Schachtman, "TASI quality—effect of speech detectors and interpolation," *Bell Sys. Tech. J.*, vol. 41, pp. 1455–1473, July 1962.

[8] L. F. Bugbee, Jr., "A tone disabler for Bell System 1A echo suppressors," *AIEE Trans. (Communication and Electronics)*, vol. 80, pp. 596–600, 1961 (January 1962 sect.).

[9] G. K. Helder, "Customer evaluation of telephone circuits with delay," *Bell Sys. Tech. J.*, vol. 45, pp. 1157–1191, September 1966.

ECHO CANCELLER UTILIZING PSEUDO-LOGARITHMIC CODING*

Otakar A. Horna
COMSAT Laboratories
Clarksburg, Maryland 20734

Abstract

This paper describes a novel digital echo canceller (DEC) with an adaptive transversal filter which is capable of enhancing the echo return loss by more than 22 dB, while the residual echo is removed by an adaptive center clipper.

Pseudo-logarithmic analog-to-digital conversion of the signal reduces the memory size and simplifies the convolution processor. A new convergence algorithm using two dither signals and nonlinear stabilization of the correction loop substantially reduces the cross-correlator complexity and shortens the convergence time of the transversal filter. This DEC performs well during simultaneous double talk and phase roll, which are the most adverse circuit conditions for an echo canceller. The center clipper with an adaptive threshold and distortion reduction circuit eliminates residual echo.

Introduction

In long distance telephone connections, an echo signal may be generated primarily at the hybrid coil that converts a 4-wire link to a 2-wire circuit. It has been determined[1,2] that the echo return loss (ERL), measured directly at the hybrid, has a mean, μ_{ERL}, of 11-15 dB and a standard deviation, σ_{ERL}, between 3 and 5 dB; thus, the ERL can be 5 dB or less in as many as 10-12 percent of the cases.

The subjective disturbing influence of echo increases with the round trip delay, t_{RD}, of the connection.[3,4] With round trip delays encountered in long distance international terrestrial connections ($t_{RD} > 50$ ms) or satellite communications ($t_{RD} \approx 500$ ms), μ_{ERL} must be greater than 30 dB; therefore, an echo control device is necessary. Considerable effort has been directed[5,6] toward improving echo suppressors, i.e., voice-activated switches interrupting the return path in the 4-wire section. These devices fail to provide adequate quality for 2-way communications mainly in the double talk condition, where ERL < 15 dB and $t_{RD} > 100$ ms.

Various approaches have been investigated. E. G. Sondhi and Flanagen et al.[7] synthesized a transversal filter approximating the impulse response of the echo path. The output of this filter, i.e., the simulated echo signal, is subtracted from the actual echo to cancel it. This idea was improved by Kelly, who used self-adaptive tap gain settings of the filter.

Campanella et al.[10] derived the linear theory of convergence of the adaptive filter loop, and in cooperation with the Nippon Electric Company, developed a DEC which underwent an extensive field trial to demonstrate that the performance of a satellite link equipped with echo cancellers is nearly equivalent to that of a short 2-wire connection. This has been confirmed by recent subjective field tests performed by AT&T. However, in former designs, the limits of economic viability were exceeded[11] because the equipment was nearly two orders of magnitude more complex than an echo suppressor.[6]

This paper presents the design philosophy of an echo canceller whose complexity, size, and cost are comparable to those of echo suppressors. The end result can be judged from the composite photographs in Figure 1, where the latest model of the new echo canceller is compared with the early version used in worldwide tests.

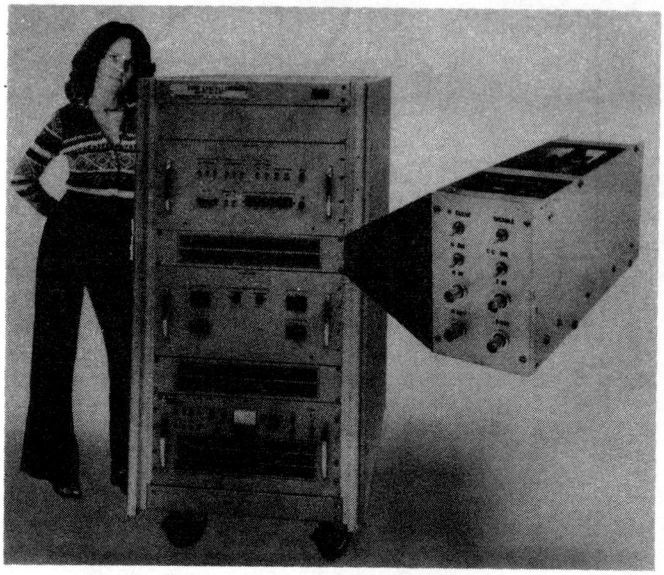

Figure 1. Comparison of the Echo Canceller Used in the Worldwide Tests with the New Model

Digital Echo Canceller Design

Figure 2 shows the echo canceller inserted in the 4-wire circuit some distance from the hybrid coil, i.e., with some delay t_E. In the H-register $\underline{n}$ samples, h_i, of the unit impulse response of the echo path are stored. The X-register stores the $\underline{n}$ most recent speech samples, x_{i+j}.

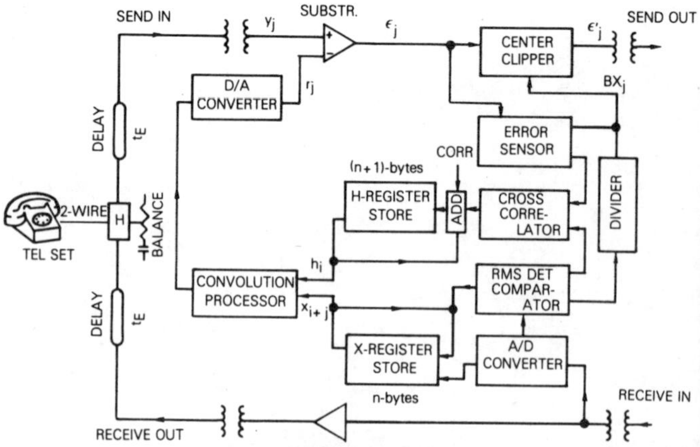

Figure 2. Block Diagram of the DEC with Logarithmic Encoding

*This paper is based upon work performed at COMSAT Laboratories under the sponsorship of the International Telecommunications Satellite Organization (INTELSAT). Views expressed in this paper are not necessarily those of INTELSAT.

Reprinted from *National Telecommun. Conf. Record*, vol. 1, Dec. 5-7, 1977, pp. 04:7-1-04:7-8.

During the jth sampling period the jth estimate of the echo, r_j, is computed in the convolution processor:

$$r_j = \sum_{i=1}^{m} h_i \times x_{i+j} = \underline{H}(i)*\underline{X}(j) \qquad (1)$$

This is subtracted from the true echo, y_j, to result in the residual echo signal sample,

$$\varepsilon_j = y_j - r_j \qquad (2)$$

at the output of the subtractor. This error signal is used to compute a correction, $\Delta h_{i,j}$, for every impulse response sample h_i with a modified steepest descent method.[8,9] Hence, a new value is obtained

$$h_i' = h_i + \Delta h_{i,j} \qquad (3)$$

This is repeated to compute the next (j + 1)th estimate, r_{j+1}, for a new set of x_{i+j+1} samples of speech and a new error value, ε_{j+1}, etc., leading ultimately to a replica of the impulse response in the H-register which is continuously updated whenever changes occur.

Because values of ERL > 30 dB are required, the echo canceller must attenuate the echo signal power by more than 26 dB; i.e., the echo return loss enhancement (ERLE) must be

$$\text{ERLE} = -10 \log_{10} \frac{\overline{\varepsilon_j^2}}{\overline{y_j^2}} \geq 26 \text{ dB}$$

The dynamic range of speech is about 40 dB and the possible amplitude range of the different impulse responses exceeds 24 dB[2,12]; therefore, the quantization of speech samples x_{i+j} requires 12 bits, and that of the impulse response samples h_i at least 9 bits. For the end delay, $t_E \leq 15$ ms, the impulse response can be delayed by 30 ms. Approximately m $\geq$ 30 samples h_i are required to emulate the echo response; therefore, for a sampling period $T_S \leq 125$ µs at least $m + 2t_E/T_S \geq 250$ samples must be stored in the H- and X-registers.

To compute the echo estimate r_j according to equation (1), 250 multiplications of 12 by 9 bits and 250 additions of these products must be performed every 125 µs. The speed of the convolution processor is limited mainly by the access time of available memories; therefore, a parallel multiplier seems to be the only means of avoiding custom-built circuits. This multiplier and memories represent most of the hardware cost.[10,11]

DEC with Logarithmic Coding

The following steps have been taken to reduce the hardware complexity. Speech samples x_{i+j} are encoded in a sign plus magnitude 13-segment pseudo-logarithmic 7-bit format which has a linear segment around zero (similar to A-law) and the dynamic range of an 11-bit linear code. Impulse response samples h_i are encoded in a similar 11-segment 8-bit format. This coding reduces the memory size by more than 28 percent. Multiplication in the convolution processor [see equation (1)] is therefore performed as a simple addition of the logarithms of x_{i+j} and h_i.

Another advantage of logarithmic format is that the signal-to-quantization noise ratio (S/N_q) is nearly constant over a dynamic range of 30 dB. Therefore, the S/N_q ratio of the echo estimate r_j [equation (1)] is less dependent on the number (m) of h_i samples required to yield a response H(t) with a given accuracy.

The logarithmic code permits a simple implementation of a new fast adaptive algorithm. The impulse response samples, h_j, are updated by multiplying by constants $(1 \pm \Delta h_{i,j})$, rather than by addition of $\Delta h_{i,j}$ according to equation (3). The convergence for large values of h_i is nearly one order of magnitude faster than that of other algorithms.[9,10,12] Therefore, the DEC is relatively insensitive to the so-called "phase roll" in the echo path without requiring complicated double convolution processing.[12,13] The fast convergence permits the use of the simplest sign correlation algorithm. The two required dither signals Φ_{i+j} and $F(\varepsilon_j)$ are combined with other DEC internal operational functions.

The residual echo signal is suppressed by an adaptive center clipper (to be shown in Figure 7) which is disabled during double talk. The DEC attenuates the echo signal by about ≈22 dB even without the center clipper. The average convergence time is <200 ms over the input signal dynamic range of >28 dB, even with a phase roll of 6 radians/second.

Convolution Processor

The input signal sample code has a sign bit, a 3-bit exponent $0 \leq e_x \leq 7$, and a 3-bit mantissa, which is by definition $0 \leq m < 1$. A similar code is used for the impulse response sample; mantissa m_h is four bits wide, but only three bits enter the convolution processor.

Assume that the digital-to-analog scaling factor is unity. When the exponents $e_x = e_h = 0$, the amplitudes of samples $\underline{s}$ are in the linear segment of the code; i.e., their equivalent absolute values are

$$|s| = 2m \qquad (5)$$

and the resolution is four bits. For $e_x > 0$ and $e_h > 0$, samples $\underline{s}$ are encoded quasi-logarithmically; their values are

$$|s| = 2^e(1 + m) \qquad (6)$$

The resolution is equivalent to that of a linear code with B = 5 bits.

The processor computes the jth echo estimate, r_j, according to equation (1). That is, it accumulates $\underline{n}$ products $P_i = h_i \times x_{i+j}$ of the samples stored in the registers and at the end of the cycle stores the result r_j, in another register which drives a digital-to-analog converter. The analog value of r_j is then subtracted from the echo signal, y_j (see Figure 2). There are four possible combinations of |x| and |h| formats entering the convoluiton processor's multiplier.

If $e_x = e_h = 0$ and |x| and |h| are only three bits wide [see equation (5)], the product $|P_i|$ rounded off to four bits is found in a 256-bit ROM table. If either $e_x = 0$, $e_h \geq 1$, or $e_x \geq 1$, $e_h = 0$, then for the latter case the following multiplication formula is used:

$$|P_i| = 2^{e_x+1}(m_h + m_x \times m_h) \qquad (7)$$

The 4-bit product $m_x \times m_h$ is read from the ROM and added to the mantissa with the zero exponent. If both $e_x \geq 1$ and $e_h \geq 1$, then approximate expressions for the binary logarithm and antilogarithm are used:

$$\log_2(1 + m) \doteq m \qquad (8)$$

$$\text{antilog}_2\, m = 2^m \pm 1 + m \qquad (9)$$

The $\log_2$ of product $|P_i|$ is therefore

$$\log_2 |P_i| = e_x + e_h + m_x + m_h = E_i + M_i \qquad (10)$$

where exponent E_i is the whole part of the sum of equation (10) and mantissa M_i is the remainder < 1. Product $|P_i|$ is thus

$$|P_i| = \text{antilog}_2\,(E_i + M_i) = 2^{E_i}(1 + M_i + \varepsilon_M) \qquad (11)$$

where ε_M is a correction which is a function of m_x, m_h, and M_i.

Equation (10) can obviously be implemented by a 8-bit adder; multiplication by 2^{E_i} according to equation (14), where E_i is an integer $E_i \in \{2, . . ., 14\}$, is performed by shifting the partial products by E_i binary places before they enter the accumulator (see Figure 3).

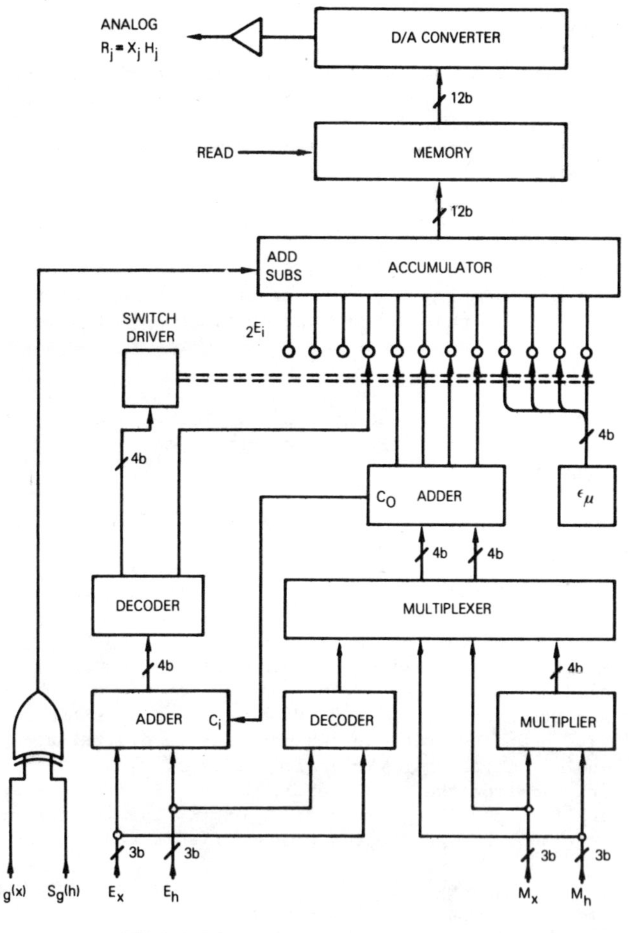

Figure 3. Block Diagram of the Convolution Processor

In addition to circuit simplification, there are two other reasons for using this seemingly complicated scheme. First, this multiplication algorithm provides a simple solution to the problem of zero values associated with logarithmic encoding. Secondly, since the multiplication is performed by one read command from the ROM and by a single addition, the entire operation of forming product $|P_i|$ and adding it into the accumulator can be done in one clock period, which may be as short as 150 ns with current T^2L logic.

Correction Algorithm

If the exponent of the h_i samples stored in the H-register is $e_h = 0$, i.e., h_i is in the linear segment of the code, the correction formula is the same as that given by equation (3). For exponents $e_h \geq 1$ the sample h_i is in pseudo-logarithmic format if $\log_2$ is

$$\log_2 |h_i| = \log_2 2^{e_h} (1 + m_h) = e_h + m_h \qquad (12)$$

according to equations (6) and (8). To establish the corrected value, h_i', the increment $|\Delta_{i,j}| << 1$ is added to mantissa m_h of $\log_2 |h_i|$. In linear format this is equivalent to

$$|h_i'| = |h_i| \; (1 \pm \Delta_{i,j}) = |h_i| \pm |h_i| \; \Delta_{i,j} \qquad (13)$$

according to equation (9). Thus, sample h_i is multiplied by $(1 + \Delta_{i,j}) > 1$ or $(1 - \Delta_{i,j}) < 1$ according to the sign of $\Delta_{i,j}$. In effect, correction step $\Delta h_{i,j}$ is directly proportional to the amplitude of sample h_i:

$$|\Delta h_{i,j}| = |h_i| \; \Delta_{i,j} \qquad (14)$$

With this correction algorithm, the convergence time, τ_{cn}, is less dependent on the amplitude of input response H(i).[8,10] For large values of h_i with the same correction size, $\Delta h_{i,j}$, it is nearly eight times faster than the correction method of equatipn (3). The speed of convergence resulting from this adaptive algorithm permits the use of only one fixed value of correction increment, $\Delta_{i,j} = \pm 2^{-4}$.

Because correction $\Delta_{i,j}$ is added to the fourth bit of the mantissa, which does not enter the convolution processor, a "digital inert zone" is created in the correlator's loop. This inert zone contributes to the stability of the H-response during extreme and transient conditions and in the presence of uncorrelated noise on the send-in path.

Error Analysis

Four main contributors to the total energy of the error signal $P(\varepsilon_j) = k\, E[\varepsilon_j^2]$ limit the degree of cancellation: uncorrelated noise appearing at the send side, truncation of impulse response H(i), quantization noise, and error due to the $\log_2$ approximation. During normal operation the level of the send side noise is usually <-50 dBm0 and the m impulse response bytes h_i usually occupy only a small fraction of the capacity (n >> m) of the memory.

Under these conditions the degree of cancellation is determined mainly by the size of the quantization steps Δ_h and Δ_x of samples h_i and x_{i+j}, respectively.[10,12] As shown in equation (1), the jth estimate of echo r_j is

$$r_j = \sum_{i=1}^{m} P_i = \sum_{i=1}^{m} h_i \cdot x_{i+j} \qquad (15)$$

Suppose that the process is stationary and that samples h_i of the stored response H(i) converge to their optimal values. Therefore, every product P_i contains only a quantization error, i.e.,

$$P_i + \Delta P_i = (h_i + \Delta_h)(x_{i+j} + \Delta_x)$$
$$= h_i x_{i+j} + \Delta h^{x_{i+j}} + \Delta_x h_i + \Delta_x \Delta_h \qquad (16)$$

Neglecting $\Delta_x \Delta_h$ in a first approximation leads to the quantization error

$$\Delta P_i = \Delta_h x_{i+j} + \Delta_x h_i \qquad (17)$$

Because Δ_h and Δ_x are uncorrelated, the variance σ_i^2 is

$$\sigma_i^2 = E[(\Delta P_i)^2] = \frac{\Delta_h^2}{12} x_{i+j}^2 + \frac{\Delta_x^2}{12} h_i^2 \tag{18}$$

If both samples x_{i+j} and h_i are encoded in a pseudo-logarithmic format [see equation (6)], quantization steps Δ_h and Δ_x are approximately a constant fraction ($\delta << 1$) of sample h_i or x_{i+j}; e.g.,

$$\delta = \frac{\Delta_h}{|h_i|} \doteq \text{const} \tag{19}$$

Substituting δ into equation (18) yields

$$\sigma_i^2 = \frac{\delta^2}{6} h_i^2 x_{i+j}^2 \tag{20}$$

and the expected value of the error ε_j^2 of all products P_i of convolution (15) is

$$E[\varepsilon_j^2] = \sum_{i=1}^{m} \sigma_i^2 = \frac{\sigma^2}{6} \sum_{i=1}^{m} h_i^2 x_{i+j}^2 \tag{21}$$

which is obviously proportional to the total power of the estimated echo signal, $P(r_j) = r_j^2/R$. At the end of the convergence process the estimated and true echo signals are nearly equal so that

$$y_j^2 \doteq r_j^2 = \sum_{i=1}^{m} h_i^2 x_{i+j}^2 \tag{22}$$

If equations (21) and (22) are substituted into equation (4) and $\delta^2 = 2^{-B}$, ERLE is a function of the number of quantization steps, 2^B, of the logarithmically encoded signals:

$$\text{ERLE} \leq -10 \log_{10} \frac{\delta^2}{6} = 3B + 7.8 \text{ dB} \tag{23}$$

For the code used in the digital echo canceller, the resolution of B is five bits; i.e., the expected ERLE is 22.8 dB, which is in good agreement with measured values (see the section entitled "Measurements").

For small amplitudes $|h_i|$ and $|x_{i+j}|$ the encoding is linear and equation (19) does not hold; the quantization error increases with decreasing $|P_i|$. In the present digital echo canceller design this is the main reason for decreasing ERLE for low-level receive-in signals. Figure 4 shows the error, ε_ℓ, of the $\log_2$ approximation according to equation (8) as a function

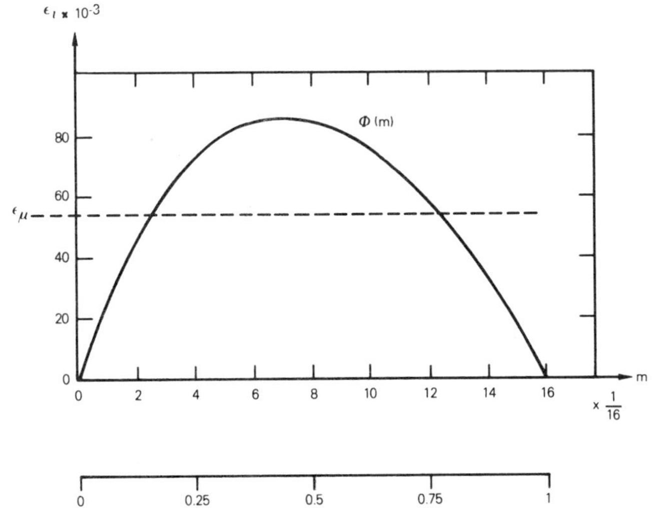

Figure 4. Error ε_1 of the $\log_2$ Approximation as a Function $\Phi(m)$ of Mantissa $0 \leq m < 1$

$\Phi(m)$ of the values $0 < m < 1$. The peak error value is $\varepsilon_p = 0.0086$. The error of the antilog$_2$ approximation according to equation (9) is approximately $\varepsilon_a = -\Phi(m)$. Dill has suggested[14] that these errors can be halved by adding $\varepsilon_p/2 = 0.043$ to or subtracting it from mantissa m when computing $\log_2$ and antilog$_2$.

For the convolution process [equation (1)] it is obviously more advantageous to decrease the average error, $\varepsilon_\mu = 0.0573$, to zero, i.e., to add to mantissa M_i [equation (11)] a correction $\varepsilon_M = \varepsilon_\mu$ which in binary representation is $\varepsilon_M = 0.0000111$. This correction can be directly added to the output of the 2^{Ei} multiplier, as shown in Figure 3, because mantissa M_i is represented by only three bits. The rms value of the relative error of this $\log_2$ or antilog$_2$ approximation is $\varepsilon_{rms} = 0.027$, which is also equal to the standard deviation, σ, because the average has been adjusted for zero.

If it is assumed that all three errors in computing the product according to equations (10) and (11) are uncorrelated and the central limit theorem is applied, the rms error of $|P_i|$ is $\sigma_p = \sigma\sqrt{3} \doteq 0.0316 \doteq 2^{-5}$. Because a sufficiently low level of quantization noise is achieved with a 3-bit mantissa, the rms and probable errors of the logarithmic approximation with the correction $\varepsilon_M = \varepsilon_\mu$ are sufficiently below this noise.

Stability of the Correction Loop

The convergence, or more exactly, the nondivergence, of the correction process is assured when absolute values of errors ε_j and ε_j' before and after the corrections of the response H(j), respectively, fulfill the inequality

$$|\varepsilon_j'| < |\varepsilon_j| \tag{24}$$

The stability criterion has been derived under different limiting assumptions.[8-10,12,13,15] From 1951 to 1953 Svoboda[16] and Pokorna[17] designed and tested a rapidly converging least mean square (lms) algorithm for matrix inversion. When applied to the input signal symmetric matrix, which is composed of an ordered set of input signal row vectors $X(j) = [x_{1+j}, \ldots, x_{i+j}, \ldots, x_{n+j}]$, this computational method is as follows. From the set of products $P_i = h_i \cdot x_{i+j}$ used to compute the jth estimate r_j [equation (1)] one of the highest absolute values, $|P_k| = |h_k \cdot x_{k+j}|$, is selected. The correction $\Delta h_{k,j}$ is then computed from

$$\Delta h_{k,j} = \varepsilon_j \frac{x_{k+j}}{\sum_i (x_{i+j})^2} \tag{25}$$

and this value is used to update the selected coefficient h_k:

$$h_k' = h_k + \Delta h_{k,j} \tag{26}$$

This correction reduces the error to

$$\varepsilon_j' = \varepsilon_j \left[1 - \frac{(x_{k+j})^2}{\sum_i (x_{i+j})^2} \right] \tag{27}$$

Then a second product with the next highest absolute value of $|P_i|$ is chosen and the correction process repeated with the new resulting values of ε_j'. Since under all conditions

$$0 < \frac{(x_{k+j})^2}{\sum_i (x_{i+j})^2} \leq 1 \tag{28}$$

nondivergence is assured. This algorithm ensures fast convergence and also minimizes the probability of "wrong" corrections.[16-18] A similar result was obtained by Suyderhoud[10] using the pseudo-inverse theory of non-square matrices.

In echo cancellers several hundred samples x_{i+j} are stored and only 10-30 significantly participate in the computation of r_j. Therefore, the square of the length of the vector $||X(j)||^2 = \Sigma_i (x_{i+j})^2$, which is proportional to the total signal power entering the adaptive filter, changes relatively slowly and can be approximated by the square of the input signal rms value $\overline{X}_j^2$. Substituting further from equation (14) into equation (25) makes it possible to express the stability criterion as

$$\Delta h_{i,j} = \Delta_{i,j}|h_i| = \frac{\varepsilon_j x_{k+j}}{X_j^2}$$

$$= \alpha\beta \, \mathrm{Sgn}(\varepsilon_j) \, \mathrm{Sgn}(x_{i+j}) \, \Phi_{i+j} \, F(\varepsilon_j) \qquad (29)$$

where Φ_{i+j} and $F(\varepsilon_j)$ are 2-valued functions defined as follows:

$$\Phi_{i+j} = 1 \text{ for } \left|\frac{x_{i+j}}{\alpha X_j}\right| > 1$$

$$\Phi_{i+j} = 0 \text{ for } \left|\frac{x_{i+j}}{\alpha X_j}\right| < 1 \qquad (30)$$

where α is a dimensionless constant.

Function Φ_{i+j} selects the samples x_{i+j} with the highest amplitudes. Thus, in accordance with the lms correction strategy outlined above, the correction step which is proportional to the previous amplitude h_i will be largest for the largest values of h_i and hence for the largest product $|P_i|$. The value of Φ_{i+j} is stored in the memory as a part of the digital signal x_{i+j} and fed into the correlation processor to compute the H-response corrections according to equation (29). The other function,

$$F(\varepsilon_j) = 1 \text{ for } \left|\frac{\varepsilon_j}{\beta X_j}\right| \geq 1$$

$$= 0 \text{ for } \left|\frac{\varepsilon_j}{\beta X_j}\right| < 1 \qquad (31)$$

will initiate the correction process only if error $|\varepsilon_j|$ is greater than a fraction (β) of the rms value X_j of the input signal. $F(\varepsilon_j)$ is implemented by making the error sensor reference voltage proportional to X_j so that the detector is less sensitive in the presence of higher level receive-in signals. Multiplier Δ_{i+j} has a fixed value of 2^{-4}; the value of α is 1.5 (see the following subsection), and the condition for correction is $F(\varepsilon_j) \cdot \Phi_{i+j} = 1$. Therefore, the coefficient β, which determines the sensitivity of the error detector must be equal to

$$\beta \geq \frac{|h_i|}{1.5} 2^{-4} \qquad (32)$$

to satisfy equation (29).

For stable operation β should be adjusted according to the value of $|h_i|$. Although this adjustment is easy to implement, it has been found that a compromise value for β which satisfies condition (29) for ERLE = 14 dB can be used instead. This will cause a burst oscillation of the stored response H(i) in circuits with ERL < 10 dB, but at the frequency $F_s/2 = 4$ kHz. This causes no problem since the "reconstruction" filter high-frequency limit is 3.2 kHz. The influence of this instability on response r_j is further attenuated because the fourth bit of mantissa m_h does not enter the convolution processor and the adaptive loop tends to adjust the values of H(i) to minimize the "overcorrection" so that it does not affect the ERLE.

For values of ERL > 20 dB, the probability, P_ε, of signal peaks, $\varepsilon_j > \beta X_j$, is $<10^{-3}$, but in this case no more than 16 correction steps are required to create the impulse response and the center clipper removes most of the residual echo. This heuristically explains the good performance of the DEC over the 40-dB range of input signal levels.

Sign Correlation

Under idealized conditions the echo canceller in the circuit shown in Figure 2 can be represented by an adaptive filter which identifies an unknown system (see Figure 5). Discrete coefficients h_i must be adjusted so that they emulate the unit impulse response H(t) of the unknown system with minimum mean square error $E[\varepsilon^2]$.

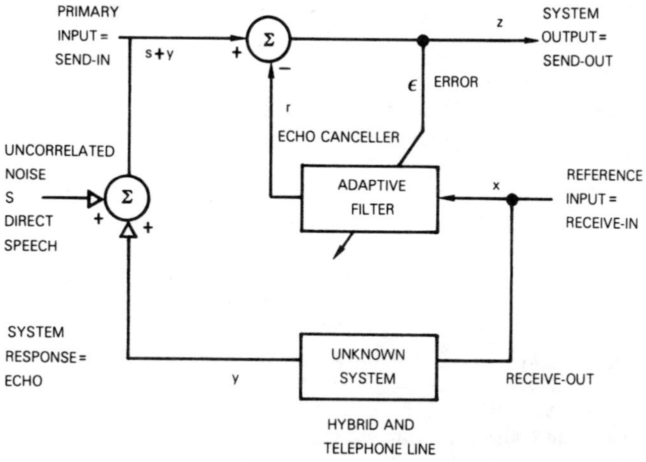

Figure 5. Equivalent Diagram of the Digital Echo Canceller

This process is based on the well-known[9] relationship between the input/output signal crosscorrelation, $R_{xy}(\tau)$, and the system's response, $H(\nu)$:

$$R_{xy}(\tau) = E[\underline{x}(\nu) \; y(\tau - \nu)]$$

$$= \int_{-\infty}^{+\infty} H(\nu) \, R_{xx}(\tau - \nu) \, d\nu \qquad (33)$$

where $R_{xx}(\tau - \nu)$ is the autocorrelation function of the input signal $\underline{x}$ and ν the normalized time. Without an additive noise signal (whose effect will be analyzed later), $H(\nu)$ can be unambiguously determined from equation (33) by using an input signal $\underline{x}$ whose R_{xx} = constant for $\tau = 0$ and $R_{xx} = 0$ elsewhere (e.g., a Poisson wave). Then

$$H(\nu) = E[x(\nu) \; y(\tau - \nu) \qquad (34)$$

Speech, however, is a highly correlated signal; for example, successive speech samples x_{i+j} and x_{i+1+j} taken at intervals of 125 μs have an autocorrelation coefficient $\rho_{xx} > 0.8$.[19] Therefore, a dither signal[20] must be added to $\underline{x}$. Function Φ_{i+j} defined in equation (37) serves this purpose; as the coefficient α increases, the sequence of pulses $|x_{i+j}| > \alpha \overline{X}_j$ more closely approximates the statistical properties of the Poisson wave, but the number of corrections [equation (29)] which can be made every convolution cycle is reduced. To prevent the canceller from correlating on a continuous sine wave signal, e.g., signaling tone, which would create a completely distorted response H(i), α must be greater than the peak-to-rms ratio, i.e., $\alpha > \sqrt{2}$. It has been determined experimentally that, for continuous speech and for the quantization code used herein, the optimum compromise is α between $\sqrt{2} < \alpha < 1.6$; for $\alpha = 1.5$ the probability that the sample $|x_{i+j}| > 1.5\ \bar{X}_j$ is $p_x \le 0.15$.[7]

The true crosscorrelation [equation (34)] can be approximated by the discrete sign correlation $B_{xy}(i)$. That is, according to Reference 20,

$$H(\nu) \doteq H(i) \doteq B_{xy}(i)$$
$$= E[\mathrm{Sgn}(x_{i+j}\Phi_{i+j})\ \mathrm{Sgn}(y_j\theta_j)] \qquad (35)$$

where θ_j is another random or pseudo-random process statistically independent from Φ_{i+j}. Function $F(\varepsilon_j)$ defined by equation (31) fulfills this condition because it changes its value once while Φ_{i+j} changes n times in one convolution cycle. During the H-response buildup process $|y_j| > |r_j|$; therefore, according to equation (2)

$$\mathrm{Sgn}(\varepsilon_j) = \mathrm{Sgn}(y_j) \qquad (36)$$

Substituting into equation (35) and separating both dither functions makes it possible to compute the unit impulse response $H(\nu)$:

$$H(\nu) \doteq H(i) \doteq B_{xy}(i)$$
$$= E[\Phi_{i+j}\ F(\varepsilon_j)\ \mathrm{Sgn}(x_{i+j})\ \mathrm{Sgn}(\varepsilon_j)] \qquad (37)$$

which proves that the correction algorithm [equation (29)] used in the digital echo canceller will identify the unknown system response H(t).

Double Talk Detector

In Figure 5, which is an equivalent diagram of the DEC, $\underline{y}$ is the echo signal as a result of a signal $\underline{x}$, and $\underline{s}$ is an uncorrelated signal, e.g., the second party's speech signal plus noise of the tail circuit. Assume that $\underline{s}$, $\underline{y}$, $\underline{x}$, and $\underline{r}$ have zero means. The send-out signal is $z = s + y - r$. Because $\underline{s}$ is uncorrelated with $\underline{y}$ and $\underline{r}$, the variance $E[z^2]$ is

$$E[z^2] = E[s^2] + E[(y - r)^2] \qquad (38)$$

The adaptive rms algorithm minimizes the square of errors $\varepsilon_j^{\hat{2}}$, i.e., $E[z^2]$. Minimizing $E[z^2]$ minimizes only the echo output power $E[(y - r)^2] \to 0$ without regard to the presence of the uncorrelated signal $\underline{s}$. However, this is true only when echo $\underline{y}$ is approximately equal to or larger than signal $\underline{s}$, e.g., in circuits with very low ERL and a low level of $\underline{s}$ at the hybrid end of 2-wire sections.

In the double talk condition signal $\underline{s}$ is usually substantially larger than echo $\underline{y}$. Therefore, the correction algorithm cannot work properly because the minimum of $E[z^2]$ remains relatively constant as a function of h_i. Hence, the gradient $\nabla(\varepsilon_j)$ of the error function is below the resolution of the error sensor (Figure 2), whose sensitivity is limited by the stability criterion [equation (32)]. In this condition, the correction estimate [equation (29)] can lead to a solution which introduces distortion of signal $\underline{s}$.[9]

Because of the short convergence time, τ_{cn}, the correction process must be inhibited immediately when a send-out signal $\underline{s}$ larger than echo $\underline{y}$ is detected. However, this fast convergence ensures that any distortion of the impulse response is corrected, usually during one syllable. A relatively simple double talk detector as shown in Figure 6 can give satisfactory results even when the echo path is unstable or has some phase roll.

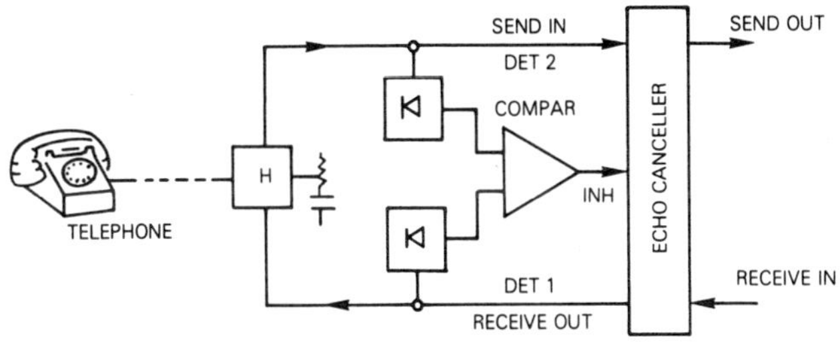

Figure 6. Block Diagram of the Double Talk Detector

Both receive-in and send-out rectified signals are compared in a voltage comparator. When the send-in signal is 0.5 dB stronger than the strongest possible echo signal directly at hybrid H, the output of the comparator inhibits the crosscorrelator's adaptation process.

Adaptive Center Clipper

The DEC described herein can enhance the ERL only by 20 to 23 dB. With ERL in the range of 6-20 dB, only a very weak, mostly incoherent echo can be heard in the circuit during loud speech passages. To eliminate this residual echo signal, an adaptive center clipper (CC) has been included in the send-out path[21] of the DEC (see Figure 7).

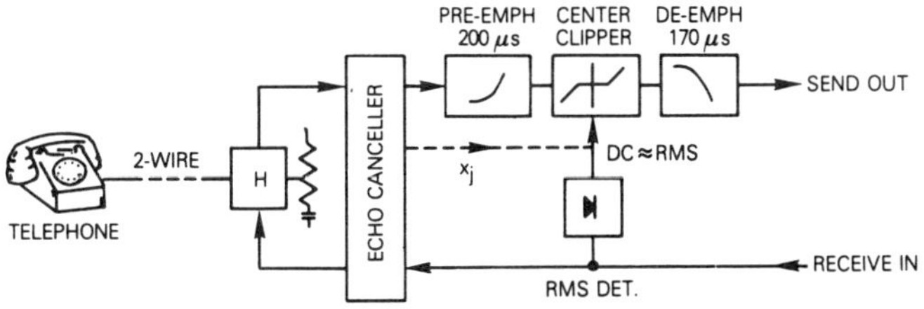

Figure 7. Connection of the Center Clipper in the 4-Wire Circuit with the Echo Canceller

The nonlinear transfer characteristic of the center clipper suppresses signals with instantaneous amplitudes less than the threshold a, as shown in Figure 8. Tests have shown that best results can be obtained if the clipping level $|d|$ is held approximately proportional to the rms value $\bar{X}_j$ of the speech integrated over a "window" of 50 to 150 ms. When properly adjusted, this device can suppress the residual echo in the whole dynamic range of the telephone signal by another 8 to 50 dB without noticeably affecting the direct signal in the send-out path, which is usually much stronger than the residual echo.

Theoretically distortion is only by odd harmonics (Figure 8) which are in the frequency band above 800 Hz, where the speech power is falling more than 12 dB/octave. The signal can therefore be pre-emphasized by

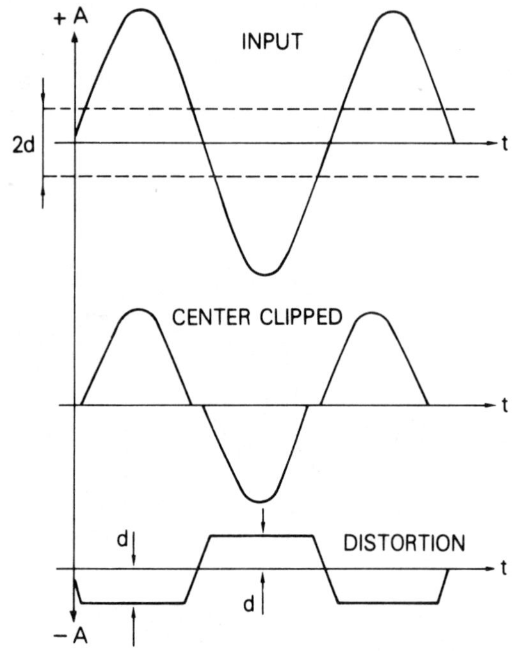

Figure 8. Effect of Center Clipping on a Sinusoidal Waveform

6 dB/octave before clipping and de-emphasized by 6 dB/octave after clipping (Figure 6) to attenuate the harmonics. Subjective testing has shown that, for continuous speech with the same rms value as a 600-Hz sine wave used for calibration, a clipping level, $d \leq 0.2$ cannot be detected even by immediate comparison of the clipped and unclipped signals when pre-emphasis/de-emphasis is used. It is usually advantageous to put the pre-emphasis filters in the send and receive paths before the DEC (Figure 2). This reduces the correlation of the successive $\underline{x}$ and $\underline{y}$ samples hence improving the ERLE by 2 to 3 dB and shortening the convergence time, τ_{cn}, by 15 to 20 percent.

Measurements

The performance of DEC has been evaluated by a series of measurements. For example, Figure 9 shows ERLE as a function of receive-in signal level with the center clipper switched off. A white noise bandwidth limited by a 0.3- to 3.2-kHz filter is used as the test (receive-in) signal and the send-out output is measured with C-message weighting. The center clipper suppresses the echo signal below the system's noise level. The convergence time (τ_{cn}) with a -10-dBm0 signal and an enabled center clipper is less than 100 ms and is slightly longer for speech. One word, e.g., "yeah", can correlate the response in the H-register. Figure 10 shows the impairment in ERLE as a result of phase roll with a disabled center clipper. With center clipping, for example, the 6-dB impairment is not detectable with the speech signal.

DEC performance was also tested in a circuit used for data transmission at a 2400-bit/s rate to verify that the device does not interfere with this application. As expected, no measurable changes in data transmission performance were detected.

Subjective Tests

Subjective tests performed by COMSAT used a simulated satellite circuit in which the DEC was compared with two commercially available echo suppressors.[22]

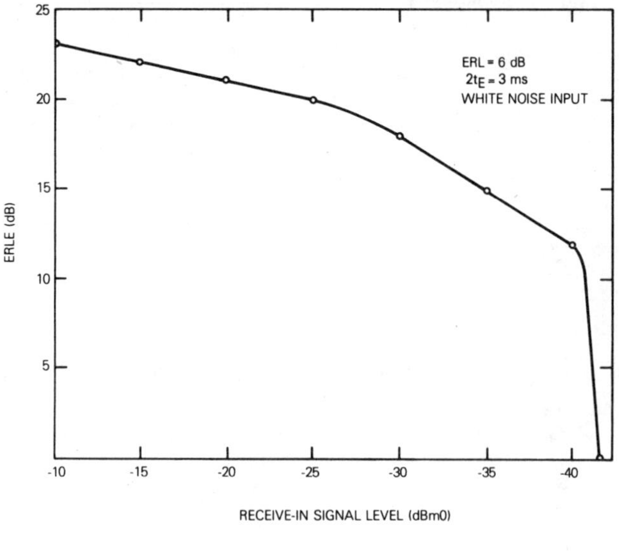

Figure 9. ERLE as a Function of the Receive-in Signal Level

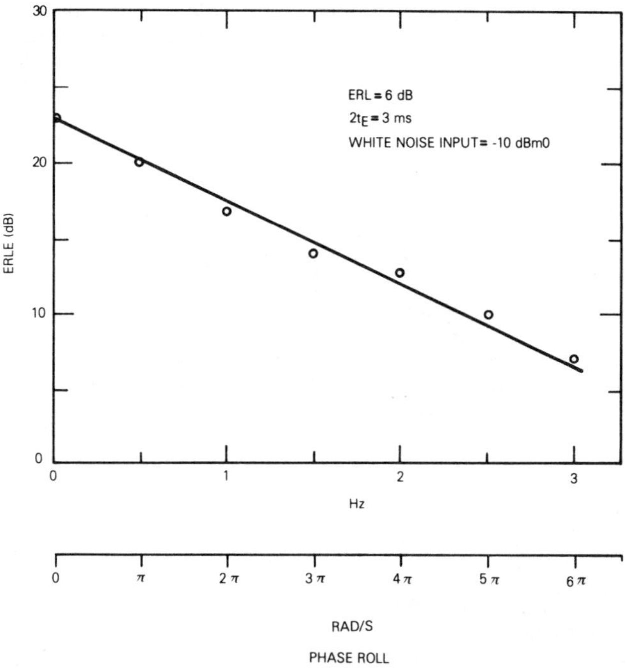

Figure 10. DEC Performance in a Circuit with Phase Roll

Sixteen listeners indicated their relative preference for these three echo controlling devices. In all tests the digital echo canceller was considered superior to both echo suppressors with a confidence level greater than 97.5 percent. These results were confirmed by extensive subjective tests performed by AT&T on an actual satellite circuit. Statistical evaluation of these tests[23] has shown that the quality of a satellite connection equipped with echo cancellers equals that of a short-distance domestic terrestrial connection.

Acknowledgment

The author wishes to thank S. J. Campanella, H. G. Suyderhoud, and M. Onufry for their many valuable suggestions concerning the work and the manuscript, and P. R. Culley, who carried out most of the experiments and implementations of several models of the digital echo canceller.

References

1. L. B. Bogan, "Impedance and Return Loss Performance of Telephone Plant in Metropolitan Areas," AIEE Transactions, Vol. 77, pt. 1, July 1958, pp. 257-261.

2. F. P. Duffy et al., "Echo Performance of Toll Telephone Connections in the United States," Bell System Technical Journal, Vol. 54, February 1975, pp. 209-243.

3. Transmission Systems for Communications, E. F. O'Neil, ed., Fourth Edition, Winston-Salem, N.C.: Bell Telephone Laboratories, 1970, Chapter 4.

4. H. R. Huntley, "Transmission Design of Intertoll Telephone Trunks," Bell System Technical Journal, Vol. 52, September 1953, pp. 1019-1036.

5. P. T. Brady and G. K. Helder, "Echo Suppressor Design in Telephone Communications," Bell System Technical Journal, Vol. 62, November 1963, pp. 2893-2917.

6. S. J. Campanella, M. Onufry, and H. G. Suyderhoud, "Digital Controlled Echo Suppressor," Patent Application, COMSAT Disclosure 13-E-24.

7. J. L. Flanagan, Speech Analysis, Synthesis, and Perception, New York: Academic Press, 1965.

8. M. M. Sondhi, "An Adaptive Echo Canceller," Bell System Technical Journal, Vol. 46, March 1967, pp. 497-511.

9. B. Widrow et al., "Adaptive Noise Cancelling," Proc. IEEE, Vol. 63, No. 12, December 1975, pp. 1692-1716.

10. S. J. Campanella, H. G. Suyderhoud, and M. Onufry, "Analysis of an Adaptive Impulse Response Echo Canceller," COMSAT Technical Review, Vol. 2, No. 1, Spring 1972, pp. 1-36.

11. H. G. Suyderhoud, S. Campanella, and M. Onufry, "Results and Analysis of Worldwide Echo Canceller Field Trial," COMSAT Technical Review, Vol. 5, No. 2, Fall 1975, pp. 253-273.

12. N. Demytko and L. Mackechnie, "A High Speed Digital Adaptive Echo Canceller," ART, Vol. 7, No. 1, May 1973, pp. 20-28.

13. N. Demytko et al., "Study of Echo Canceller Under the Conditions of Phase Roll," INTELSAT Contract IS-668, Phase II.b Telecom Australia, November 1975.

14. G. D. Dill, Private Communication, July 1, 1959.

15. M. M. Sondhi and D. Mitra, "New Results on the Performance of a Well-Known Class of Adaptive Filters," Proc. IEEE, Vol. 64, No. 11, November 1976, pp. 1586-1597.

16. A Svoboda, "Linear Analyser in Czechoslovakia," (in Czech), Czechoslovak Journal for Physics, Vol. 1, No. 10, October 1951.

17. O. Pokorna, "Solution of a System of Algebraic Equations by Minimization of Squares of Residues," (in Czech), Information Processing Machines, Prague: Academia, 1954, Vol. 2, pp. 111-116.

18. V. Cerny and A. Marek, "Application of the Method of Minimization of the Sum of Absolute Values..." (in Czech), Information Processing Machines, Prague: Academia, 1958, Vol. VI, pp. 209-225.

19. K. Virupaksha, Private Communication, September 1975.

20. H. Berndt, "Correlation Function Estimation by a Polarity Method Using Stochastic Reference Signals," IEEE Transactions on Information Theory, Vol. IT-14, November 1968, pp. 796-801.

21. O. A. Horna, "Adaptive Center Clipper," Patent Application, COMSAT Disclosure 15-E-8.

22. M. Onufry, Private Correspondence, August 1975.

23. G. K. Helder and P. C. Lopiparo, "Improving Transmission in Domestic Satellite Circuits," Bell Laboratories Record, Vol. 46, No. 9, September 1977.

An Unusually Simple Technique for Sidelobe Reduction

JOSÉ PERINI, MEMBER, IEEE

Abstract—The technique presented here consists of superposing three constant amplitude current excitations on a uniformly spaced array. The first one has constant phase and the other two have equal progressive phase but of opposite signs. The amplitude as well as the progressive phase are chosen in such a way that a high degree of sidelobe reduction is achieved. Results are comparable to those of Chebyshev arrays with the advantage that the current distribution is much simpler to compute. Because of this simplicity the designer can much more easily guess what corrections to introduce in the array excitation in case the measured patterns do not agree with the computed ones.

I. Introduction

When point-to-point communication is attempted it is advantageous to use highly directional antennas to avoid radiation of electromagnetic energy in directions other than the line connecting both stations. One way to achieve this directivity is by using phased arrays. If the elements of the array are excited with uniform currents the radiation pattern will be a sharp beam with the first sidelobes only 13.6 dB down. Dolph [1] in 1946 proposed a technique of controlling the excitation of the array elements in order to suppress the sidelobes. This led to the well-known Chebyshev arrays. Element currents can be computed in such a way that sidelobe reduction to any extent can be achieved at least theoretically. The current excitations of such arrays are cumbersome to compute but extensive tables have been published on the subject [2]. This cumbersomeness completely obscures the contribution of each element to the total result such that when an array is constructed, based on this technique and does not perform as predicted, the designer has no way to guess what currents to modify, in order to achieve the proper results. These modifications are almost inevitably necessary because of the mutual coupling effects among the array elements. Since Dolph's paper many other techniques of sidelobe reduction have been devised a few of which are listed in the references [3]–[6]. These methods exhibit to a lesser or greater extent the same difficulty of Dolph's method. The designer is at a loss when discrepancies occur.

The method introduced here will achieve sidelobe reduction by a very simple technique which will allow the designer to guess which currents to modify when deviations occur in the measured patterns.

II. Summary of Equations from Array Theory

Assume a linear array of N identical equally spaced elements as shown in Fig. 1(a). Each element is excited by a current represented by the phasor $I_n = |I_n|e^{-j\alpha_n}$, where $|I_n|$ is the amplitude and α_n is the initial phase referred to some point in the array. All currents are varying with time as $e^{j\omega t}$.

It can be easily shown [7] that the normalized array factor (AF)[1] of an array of N antennas with constant current magnitudes ($I_n = 1$) and progressive phase ($\alpha_n = n\alpha$ referred to I_0) can be written as

$$e(u) = \frac{1}{N} \sum_{0}^{N-1} ne^{jn(u-\alpha)} \tag{1}$$

where

$$u = \frac{2\pi d \sin\theta}{\lambda} \tag{2}$$

λ is the wavelength at the frequency of operation and d the element spacing.

Equation (1) can also be written in a closed form as

$$e(u) = \frac{1}{N} \frac{\sin \frac{N}{2}(u-\alpha)}{\sin \frac{1}{2}(u-\alpha)} e^{j[(N-1)/2](u-\alpha)}. \tag{3}$$

Manuscript received July 20, 1968. This work was supported in part by Rome Air Development Center, Griffiss AFB, N. Y., Contract AF30(602) 2636, and pursued under Rome Air Development Center Postdoctoral Program in cooperation with Syracuse University under Contract F30602-8-C-0086.

The author is with the Department of Electrical Engineering, Syracuse University, Syracuse, N. Y., and Rome Air Development Center, Griffiss Air Force Base, Rome, N. Y.

[1] The name array factor (AF) is given to the radiation pattern of an array with isotropic sources as elements. If the elements are other than isotropic and as long as they are all identical and identically oriented, the resulting pattern is the product of the AF by the element pattern. This is the so-called principle of pattern multiplication [7]. Therefore, it is advantageous to talk about AF instead of actual radiation patterns in array theory since AF is a more general concept.

Reprinted from *IEEE Trans. Electromagn. Compat.*, vol. EMC-11, pp. 29-34, Feb. 1969.

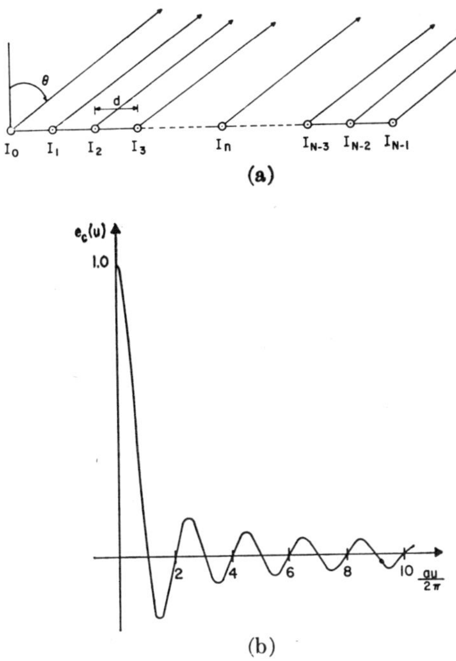

Fig. 1. Discrete array. (a) Array elements. (b) Array factor $N = 10$.

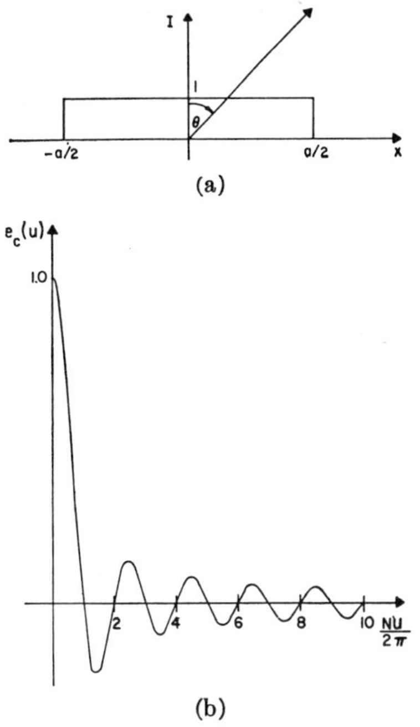

Fig. 2. Continuous Array. (a) Aperture illumination. (b) Diffraction field.

If the phase reference is the center of the array, instead of I_0 (3) becomes [7]

$$e_c(u) = \frac{1}{N} \frac{\sin \frac{N}{2}(u - \alpha)}{\sin \frac{1}{2}(u - \alpha)}. \tag{4}$$

Examining this equation we see that $e_c(u)$ is a periodic function of u with period 2π or 4π depending on whether N is odd or even. Let us for a moment concentrate in the behavior of this function for just one period (or half if N is even.)

We see that at $u = \alpha$, $e_c(u) = 1$ which is also its absolute maximum. To simplify our further considerations let $\alpha = 0$. In this case $e_c(u)$ has a maximum at $u = 0$ and $N - 1$ zeros at

$$\frac{Nu}{2} = n\pi, \quad n = 1,2,3, \cdots, N - 1 \tag{5}$$

or

$$u = n \frac{2\pi}{N}. \tag{6}$$

In general, if the normalized coordinate

$$u' = \frac{Nu}{2\pi} \tag{7}$$

is used with α still zero, the absolute maxima of $e_c(u)$ (or absolute maxima and minima for N even) occur at[2]

$$u_m' = pN, \quad p = 0,1,2, \cdots \tag{8}$$

and the zeros at

$$u_0' = n, \quad n \neq pN. \tag{9}$$

The general behavior of $e_c(u)$ is shown in Fig. 1(b) for $\alpha = 0$ with normalized coordinates. Note that if $\alpha \neq 0$ then $e_c(u)$ will be displaced to the right or left by $N\alpha/2\pi$ depending on whether α is positive or negative.

For the case of continuously illuminated apertures, as shown in Fig. 2(a), the corresponding equations are [7]

$$e_c(u) = \frac{1}{2} \int_{-a/2}^{a/2} e^{j(u-\alpha)x}\, dx \tag{10}$$

$$e_c(u) = \frac{\sin \frac{a}{2}(u - \alpha)}{\frac{a}{2}(u - \alpha)} \tag{11}$$

where

$$u = \frac{2\pi \sin \theta}{\lambda}. \tag{12}$$

A graph of $e_c(u)$ is shown in Fig. 2(b) for $\alpha = 0$. Note that in contrast with (4), (11) is not a periodic function.

[2] The parenthetical note about N even is needed because in this case $e_c(u_m')$ equals $+1$ for p even and -1 for p odd. For N odd $e_c(u_m')$ is always $+1$.

This completes all the necessary preliminary work to carry out the sidelobe reduction.

III. Sidelobe Reduction—Discrete Arrays

Suppose the current distribution of the array of Fig. 1 is the superposition of the following three current distributions:

a) constant current of magnitude 1 and constant phase assumed equal to zero (Fig. 3, curve *a*),
b) constant current of magnitude k and progressive phase $-\alpha$ (Fig. 3, curve *b*),
c) constant current of magnitude k and progressive phase α (Fig. 3, curve *c*).

The progressive phase α is so adjusted that the main peaks of the AFs of b) and c) coincide with each of the nulls adjacent to the main peak of the AF of a). This implies that α has to be equal to $2\pi/N$ and $-2\pi/N$, respectively. This situation is shown in Fig. 3 for $N = 20$. It is seen there that the sidelobes of b) and c) always have the same sign but oppose those of a). Therefore, if the three AFs are added, there will be a cancellation of most of the sidelobes. The cancellation depends in this case on the choice of k. This being a linear system, the resulting current distribution for the sum of the three AFs is the sum of those of a), b), and c). Let us look at some possible choices of k.

First choose k such that a null is produced at the maximum of the second sidelobe of a). The value of k is obtained by equating the maximum of the second sidelobe of a) to k times the sum of the maxima of the first sidelobe of c) and the third of b). The result is $k = 0.426$.[3] The AF and the current distribution for this value of k are shown in Fig. 4. It is seen that the first sidelobe is at -44.5 dB and the highest one is at -42 dB. In this figure, as well as in the subsequent ones, the sidelobe region is drawn in a scale tenfold that of the main beam so that the sidelobe reduction can be more fully appreciated. A comparison of the main beam with that of the corresponding Chebyshev array at -3 dB, -10 dB, and first zero is also shown. The gains of the present and the Chebyshev array are given for comparison. It can be seen that the AFs as well as the gains are essentially the same.

If $k = 0.416$, then the two sidelobes near the main beam become equal (-48 dB) but the fourth goes up to -40 dB, as shown in Fig. 5.

If $k = 0.480$, then a null is produced at the peak of the third sidelobe of a). The resulting AF is shown in Fig. 6. The first sidelobe goes up to -36 dB, the second is at -48 dB, but from the third on they are all below -60 dB.

[3] The value of these local maxima (or minima) can be obtained from (4) assuming that they occur midway between the two adjacent nulls.

A common property of all the AFs presented up to now is that the main beam is double that of the AF with uniform illumination. If, however, instead of making $\alpha = 2\pi/N$, a smaller value is used, then a narrower main beam results. This is shown in Fig. 7 where α is 96 percent of the value used so far. It is seen that a narrower beam is obtained at the price of increasing the first sidelobe levels to -38 dB. All others remain below -40 dB.

The resulting AF as well as the corresponding current distributions for these or any other cases are easily computed by writing down the AF of (1) or (4) for the three current distributions a), b), and c) and adding them up. If (1) is used, the AF for current distribution a) is

$$e^a(u) = \frac{1}{N} \sum_{n\,0}^{N-1} e^{jnu}. \tag{13}$$

The AF for current distribution b) is

$$e^b(u) = \frac{k}{N} \sum_{n\,0}^{N-1} e^{jn(u+\alpha)} \tag{14}$$

and the AF for current distribution c) is

$$e^c(u) = \frac{k}{N} \sum_{n\,0}^{N-1} e^{jn(u-\alpha)}. \tag{15}$$

The AF of the sum of these three current distributions is the sum of the above three AFs. Therefore:

$$e(u) = \frac{1}{N} \sum_{n\,0}^{N-1} (1 + ke^{jn\alpha} + ke^{-jn\alpha})e^{jnu} \tag{16}$$

or

$$e(u) = \frac{1}{N} \sum_{n\,0}^{N-1} [1 + 2k \cos (n\alpha)]e^{jnu}. \tag{17}$$

Equation (17) gives the resulting AF and the expression in brackets is the current distribution

$$I_n = 1 + 2k \cos (n\alpha) \tag{18}$$

where α is the progressive phase. Observe that the resulting current distribution is equiphase and symmetrical about the center of the array.

Another way of writing the AF of (17) would be to use (4) for each of the current distributions. The result is

$$e(u) = \frac{1}{N}\left[\frac{\sin \frac{Nu}{2}}{\sin \frac{u}{2}} + k\left(\frac{\sin \frac{N}{2}(u-\alpha)}{\sin \frac{1}{2}(u-\alpha)} + \frac{\sin \frac{N}{2}(u+\alpha)}{\sin \frac{1}{2}(u+\alpha)}\right)\right]. \tag{19}$$

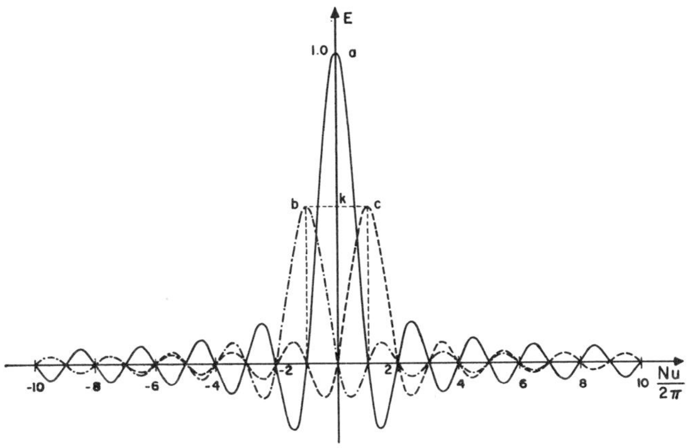

Fig. 3. Array factors of current distributions, a)–c), $N = 20$.

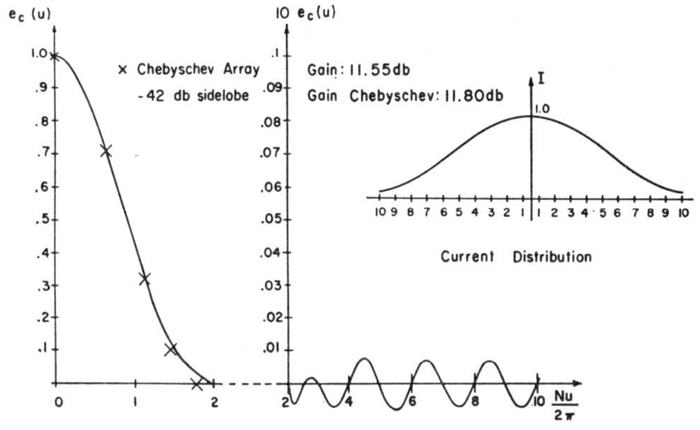

Fig. 4. Resulting array factor for $k = 0.426$, $N = 20$.

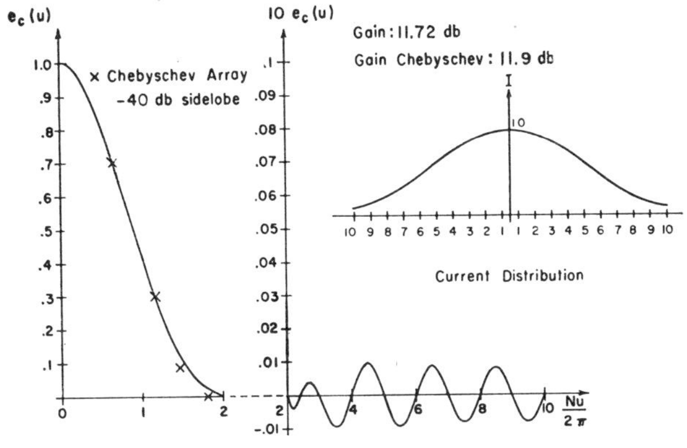

Fig. 5. Resulting array factor for $k = 0.416$, $N = 20$.

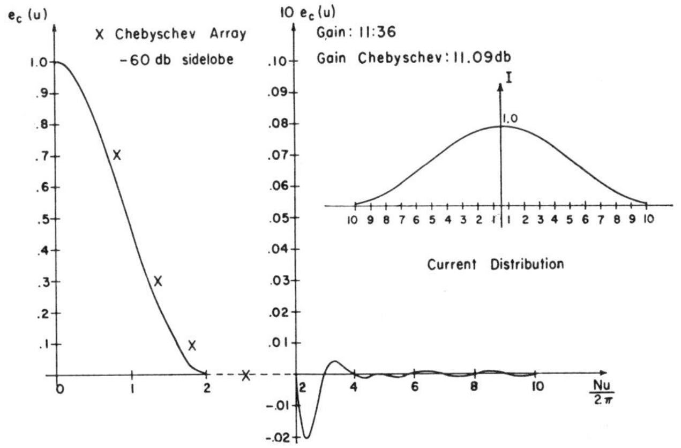

Fig. 6. Resulting array factor for $k = 0.480$, $N = 20$.

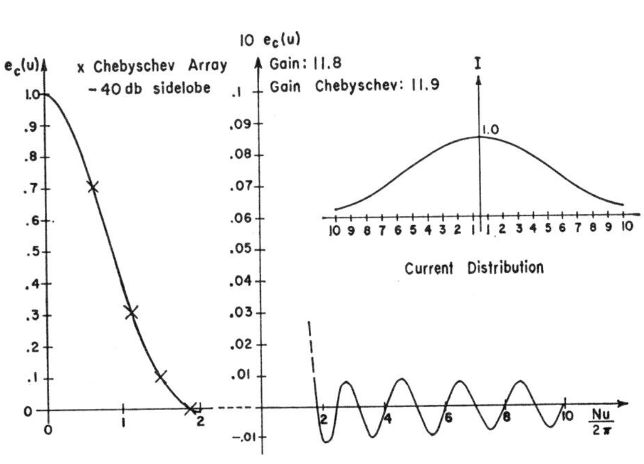

Fig. 7. Beam reduction $\alpha = 0.96 \times 2\pi/N$, $N = 20$, $k = 0.426$.

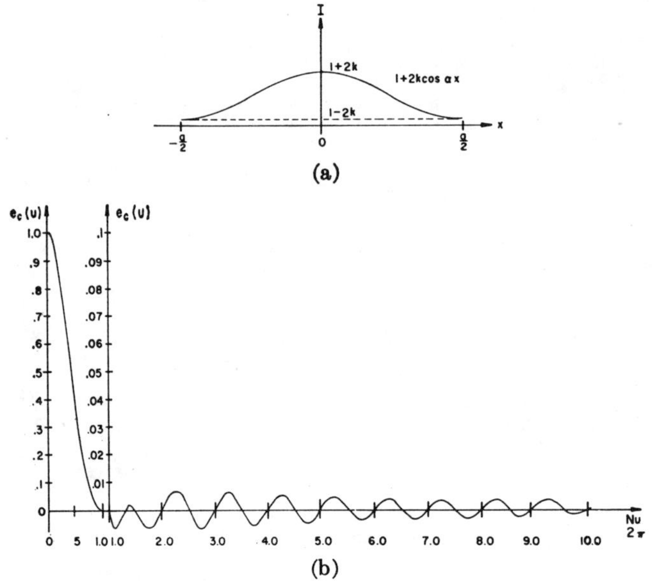

Fig. 8. Continuous array. (a) Aperture illumination. (b) Diffraction field for $k = 0.426$.

IV. Sidelobe Reduction—Continuously Illuminated Apertures

Exactly the same procedure as in Section III can be followed in the case of continuously illuminated apertures. The three following excitations are superposed:

a) constant illumination of magnitude 1 and zero phase,

b) constant illumination of magnitude k and progressive phase $-\alpha$,

c) constant illumination of magnitude k and progressive phase α.

α is set equal to $2\pi/a$, the value of u for the first null of Fig. 2. This displaces the peak of the beams of b) and c) such that they coincide with each of the first nulls of a). Fig. 8 shows the result of the superposition of a), b), and c) for k equal to 0.426. This produces a zero at the second sidelobe of a). The sidelobes are reduced to 43 dB.

In order to obtain the resulting pattern and the corresponding aperture illumination it is only necessary to superpose the patterns of the three aperture illuminations as the system is linear. If (10) is used, the result is

$$e_c(u) = \frac{1}{a} \int_{-a/2}^{a/2} [1 + 2k \cos(\alpha x)]e^{jux}]\, dx. \quad (20)$$

We recognize from (20) that the resulting aperture illumination is

$$1 = 1 + 2k \cos(\alpha x). \quad (21)$$

The total pattern can be obtained by carrying out the integration in (20) or, more simply, by summing the patterns of the aperture illumination a), b), and c) as given by (11). We therefore have

$$e_c(u) = \frac{1}{a}\left[\frac{\sin u}{u} + k\left(\frac{\sin(u + \alpha)}{u + \alpha} + \frac{\sin(u - \alpha)}{u - \alpha}\right)\right]. \quad (22)$$

V. Correcting Deviations Between Computed and Measured Patterns

As was mentioned in the Introduction, corrections in the excitation of arrays which do not perform as predicted can be done quite easily as follows.The array can initially be excited with uniform current and phase and small corrections introduced to compensate for mutual coupling. This is not too hard to do since the pattern of the uniform array is known and the current excitations should not be too different from each other. After this is obtained, then a proper progressive phase is superposed to the corrected one and the three component current distributions obtained. The sum of them will be the desired array excitation.

Another approach would be to excite the elements with a $1/N$ power divider and measure the resulting pattern. Apply now the method outlined in the paper using this measured pattern instead of the AF.

VI. Effect of the Element Spacing d

A look at (2) shows that if $d = \lambda$ then u will vary between $\pm 2\pi$ as θ varies between $\pm 90°$. In this case (8) shows that the AF will have two peaks of magnitude 1 (or $+1$ and -1 if N is even) at $\theta = 0°$ and $\theta = 90°$ (for $\alpha = 0$). This occurs because of the periodic nature of (4). Sidelobe reduction will be impaired here by the appearance of the peak at 90° (called grating lobe). If $d = n\lambda$ then n such grating lobes will occur. This phenomena takes particular importance when harmonics of the main frequency are considered. Even if at the fundamental $d = \lambda/2$, which implies inexistence of any grating lobes, at all other harmonics grating lobes will exist. This undesired feature can be eliminated to a great extent by properly using directional antennas as elements of the array. The directional antennas should have low radiation in the grating lobes directions. Consideration should be given to the radiation patterns of the elements at all harmonic frequencies since they very probably are not the same as the one at the fundamental frequency and the grating lobes will point at different directions for different frequencies.

VII. Conclusions

The sidelobe reduction technique presented here leads to results almost indistinguishable from those of more complicated methods, such as the Chebyschev arrays, using a much more direct and simple route. They allow the designer to correct for discrepancies between theoretcal and experimental data.

References

[1] C. L. Dolph, "A current distribution for broadside arrays which optimizes the relationship between beam width and side-lobe level," *Proc. IRE*, vol. 34, pp. 335–447, June 1946.
H. J. Riblet and C. L. Dolph, "Discussion on 'A current distribution for broadside arrays which optimizes the relationship between beam width and side-lobe level'," *Proc. IRE*, vol. 35, pp. 489–492, May 1947.

[2] L. B. Brown and G. A. Scharp, "Tschebyscheff antenna distribution beam-width and gain tables," NAVORD Rept. 4629 (NOLC Rept. 383), February 28, 1958.

[3] M. T. Ma and D. K. Cheng, "A critical study of linear arrays with equal sidelobes," *1961 IRE Natl. Conv. Rec.*, vol. 9, pp. 110–122.

[4] D. K. Cheng and B. J. Strait, "Sidelobe reduction and gain characteristics of partially uniform antenna arrays," *Proc. NEC*, vol. 20, p. 38, October 1964.

[5] A. Ishimaru, "Theory of unequally-spaced arrays," *IRE Trans. Antennas and Propagation*, vol. AP-10, pp. 691–702, November 1962.

[6] R. F. Harrington, "Sidelobe reduction by nonuniform element spacing," *IRE Trans. Antennas and Propagation*, vol. AP-9, pp. 187–192, March 1961.

[7] J. D. Krauss, *Antennas*. New York: McGraw-Hill, 1950, pp. 76, 110–114.

New Sidelobe Envelopes for Small Aperture Earth Stations

JAMES M. JANKY, MEMBER, IEEE, BRUCE B. LUSIGNAN, LIN-SHAN LEE, ENG CHONG HA, AND EDWARD F. REINHART

Abstract—The advent of domestic satellite systems and technological improvements in both space and ground equipment makes the use of small aperture earth stations economically feasible and attractive for broadcast satellite applications. With increasing use of the orbit spectrum, there is some concern that broadcast satellite systems may not be able to obtain satisfactory allocations of this orbit-spectrum resource because of interference considerations. Specifically, interference calculations based on the present envelope of $9 + 20 \log (\theta/\theta_0)$ result in comparatively large orbit spacings for broadcast satellites. Theoretical and experimental investigations of simple sidelobe suppression techniques, plus an analysis of experimental data on small aperture antennas (from data over a range of $40 < D/\lambda < 100$), indicate that much better sidelobe performance can be obtained with a very small cost or performance penalty.

I. Introduction

Analysis of the present trends in satellite communications technology indicates that higher satellite EIRP's and smaller aperture earth station antennas are cost effective. With increasing use of the orbit-spectrum resource there is some concern that broadcasting satellite systems may not be able to utilize this resource effectively in the face of intraservice and interservice interference. Specifically, interference calculations based on the reference antenna pattern envelope suggested in Report 215-3 for individual reception result in comparatively large orbit spacings for broadcasting satellites.

An examination of experimental pattern data on moderate aperture antennas ($40 < D/\lambda < 100$), supplemented by theoretical and experimental investigations of simple sidelobe suppression techniques, indicates that much better sidelobe performance can be obtained with a very small cost or performance penalty. On the basis of this examination, two new reference patterns, each applicable to both community and individual reception, are proposed to represent ground receiving antennas in the broadcasting satellite service depending on whether or not sidelobe suppression techniques are used. As with other reference patterns, these are intended for use in interference calculations when data on actual patterns are not available.

Manuscript received March 1, 1976.

J. M. Janky, B. B. Lusignan, L. S. Lee, and E. C. Ha are with the Communication Satellite Planning Center, Stanford University, Stanford, CA 94305.

E. F. Reinhart is with the Jet Propulsion Laboratory, Pasadena, CA.

II. The Current Reference Patterns

The current reference patterns for receiving antennas for the broadcasting satellite service are given in Report 215-3 [8] as follows. For individual reception

$$d(\theta) = \min \begin{cases} 9 + 20 \log (\theta/\theta_0) \\ G_m \\ 30 \end{cases} \tag{1}$$

For community reception

$$d(\theta) = \min \begin{cases} 10.5 + 25 \log (\theta/\theta_0) \\ G_m \end{cases} \tag{2}$$

where

$d(\theta)$ $G_m - G(\theta)$ = angular discrimination in decibels;
$G(\theta)$ gain in decibels at off-axis angle θ;
G_m $G(0)$ = maximum or on-axis gain in decibels;
θ_0 full half-power beamwidth.

Reprinted from *IEEE Trans. Broadcast.*, vol. BC-22, pp. 39-44, June 1976.

These patterns are determined entirely by the slope chosen to represent the angular dependence of the pattern envelope (-2 for individual reception and -2.5 for community reception), and the somewhat irrelevant condition that the envelope passes through the half-power point on the main lobe, i.e., the condition that

$$d(\theta_0/2) = 3 \text{ dB}.$$

As noted in the Introduction, the pattern for individual reception is unnecessarily conservative and leads to excessively large satellite spacings when used in interference calculations. More accurate models, based on both theoretical and practical considerations, are derived in the following sections.

III. Factors in the Formulation of a New Model

A. Method

The analytic determination of sidelobe levels is well known [5], and for circular apertures involves the use of a Hankel (Fourier) transform. This approach is reasonably accurate only for unblocked apertures, such as with Cassegrain horns [3]. However, most antennas use a primary radiator which incurs some blockage of the aperture due to either the radiator itself (feed) or to the supporting structure for the radiator. The effect of this blockage is to superimpose a second transform on the primary transform pattern, which results in an increase in sidelobe levels and a slight decrease in main beam gain. Since the exact transform depends on the shape and size of the blockage, it is very difficult to model the effects accurately. To be of any use, however, a sidelobe model must take these effects due to blockage into account. This has been done with the following approach.

To a first approximation, the slope of the sidelobe levels ($d(\theta)$ versus θ/θ_0) is dependent only on the illumination function. Using an illumination function which is relatively easy to achieve in practice, one can calculate the slope of the sidelobe envelope. By examining data available for a wide range of apertures, one can determine how well an envelope with this slope could actually represent measured results. This curve-fitting procedure was done for antennas which had no special consideration taken to ensure low sidelobes, nor was any information on the actual primary radiator patterns available. The reference pattern obtained by this method is discussed in Section IV.

The next step is to consider what level of sidelobe suppression could be obtained with inexpensive techniques. Theoretical and experimental results are described in Section V, and a second reference pattern is derived which takes these techniques into account.

B. Theoretical Estimation of the Slope

A theoretical radiation pattern was examined by using the Fourier-Bessel or Hankel transform for a circular aperture.

$$E_c(u) = 2\pi \int_0^{D/2} F_c(\rho) J_0(2\pi u \rho)\rho \, d\rho \tag{3}$$

where

- $F_c(\rho)$ the illumination function;
- $E_c(u)$ the radiation pattern;
- D the aperture diameter;
- u $\sin\theta/\lambda$;
- θ the off-axis angle;
- λ the wavelength.

A Bessel-on-10-dB-pedestal illumination was assumed because this is a reasonable approximation for patterns available from simple feeds:

$$F_c(\rho) = 0.3162 + 0.6838 J_0\left(4.81 \frac{\rho}{D}\right).$$

With this substitution, (3) yields

$$E_c(u) = 0.3162\lambda_1(\pi Du) + \frac{1.710 J_0(\pi Du)}{5.784 - (\pi Du)^2}$$

where

$$\lambda_1(\pi x) = \frac{2J_1(\pi x)}{\pi x}$$

is the so-called λ function.

If the maximum values of the radiation pattern $E_c(u)$ are plotted versus log (Du), it will be found that all points except first and second sidelobes can be fitted quite well by a straight line given by 21 + 29.8 log (Du) dB, whereas the first and second sidelobes are a few decibels above it. Since

$$Du = \frac{D}{\lambda}\sin\theta$$

and the 3-dB beamwidth

$$\theta_0 = K \cdot \frac{\lambda}{D}$$

it is reasonable to write

$$\frac{\theta}{\theta_0} = \frac{\sin^{-1}[Du \cdot \lambda/D]}{K(\lambda/D)} \sim \frac{Du \cdot \lambda/D}{K(\lambda/D)} = \frac{Du}{K} \text{ (for small } \theta).$$

Thus the scale Du will be proportional to θ/θ_0 when $\sin\theta$ can be approximated by θ. With Du converted into θ/θ_0 for some typical D/λ values, ranging from 400 (for 10-m antenna at 12 GHz) down to 13 (for 1-m antenna at 4 GHz), the sidelobe levels relative to the maximum gain are plotted versus θ/θ_0 in Fig. 1. Note that for the outer sidelobes, where $\sin\theta$ cannot be approximated by θ, there are multiple values of θ/θ_0 for specific values of Du. However, the spread of points over the range of D/λ is considered very small. Therefore the points can be approximated by the straight line

$$d(\theta) = 22 + 30 \log(\theta/\theta_0)(\text{dB}). \tag{4}$$

However, a better model would actually be

$$d(\theta) = 16 + 30 \log(\theta/\theta_0)(\text{dB}) \tag{5}$$

so that the first two sidelobes are also under the curve.

This model does not yet account for the effects of blockage. To a first approximation, the effects of blockage will result in

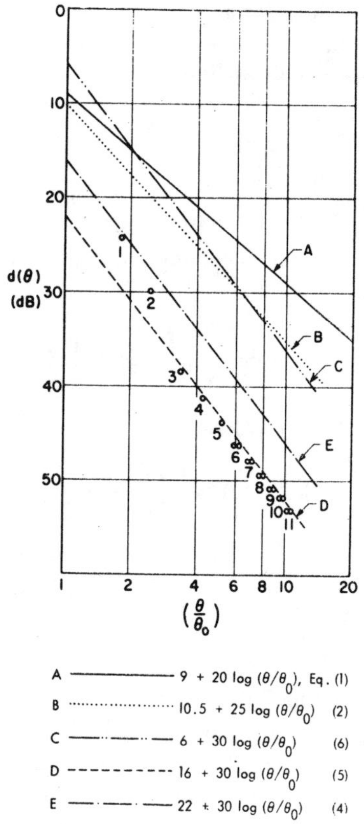

Fig. 1. Calculated peak sidelobe levels and various reference antenna pattern envelopes.

A ——— $9 + 20 \log(\theta/\theta_0)$, Eq. (1)
B ········· $10.5 + 25 \log(\theta/\theta_0)$ (2)
C —·—·— $6 + 30 \log(\theta/\theta_0)$ (6)
D – – – – $16 + 30 \log(\theta/\theta_0)$ (5)
E —··—··— $22 + 30 \log(\theta/\theta_0)$ (4)

a secondary pattern (superimposed on the primary pattern) which has the same general character as the primary pattern; i.e., the slope of the secondary pattern (which increases the sidelobe levels) will essentially follow the same $30 \log(\theta/\theta_0)$ dependence. Therefore a good model for the sidelobe envelope will be of the form

$$d(\theta) = x + 30 \log(\theta/\theta_0)\,(\text{dB}) \qquad (6)$$

where the "slope" 30 was obtained from the theoretical model, and the intercept x should be determined by examining measured data.

Fig. 1 also includes plots of the current reference patterns for individual and community reception given by (1) and (2), the "theoretical" patterns given by (4) and (5) and, for subsequent reference, the pattern given by (6) with $x = 6$ dB. This value for the intercept x will be justified in Section IV.

IV. EXAMINATION OF MEASURED DATA

Data extracted from measured antenna patterns are shown in Fig. 2. The list of antennas from which the data were taken is given in Table I. The data are presented in groups. Each group is represented by a vertical bar spanning the range of gain variation of the sample of data points in that group. Such partitioning into groups is done with due caution to ensure that sufficient data are encompassed by each group. The upper circle on each vertical bar is an upper 20-percent cutoff point above which 20 percent of the data lie. The lower circle is the corresponding lower 20-percent cutoff point. The median is shown as an open circle.

| | upper limit
• | upper 20% point
○ | median point
• | lower 20% point
| | lower limit

A ——— $9 + 20 \log(\theta/\theta_0)$
B ········· $10.5 + 25 \log(\theta/\theta_0)$
C —·—·— $6 + 30 \log(\theta/\theta_0)$
D – – – – 25

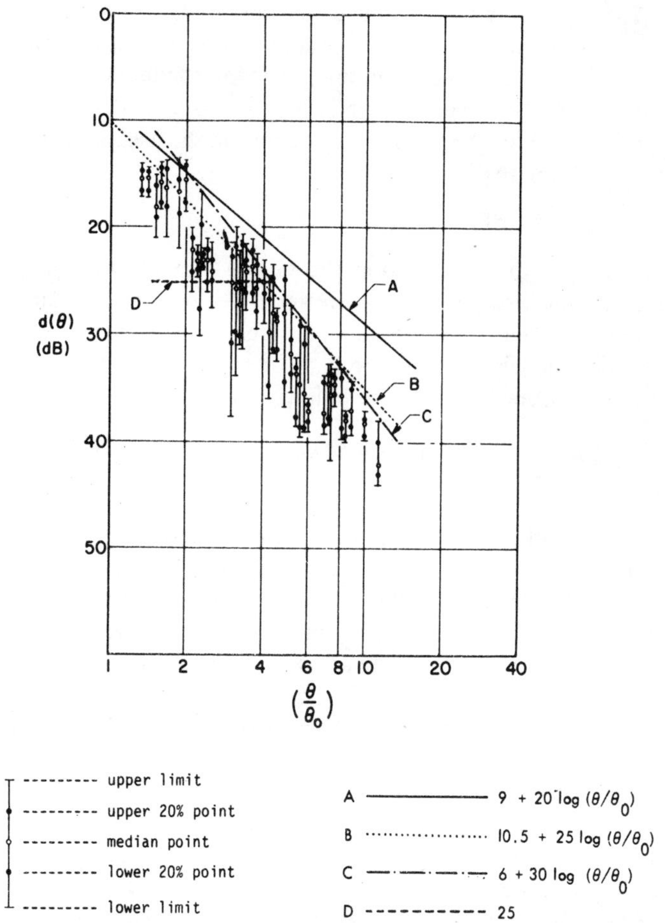

Fig. 2. Measured peak sidelobe levels and various reference antenna pattern envelopes.

TABLE I

Company	Diameter	Frequency (GHz)	D/λ	Remarks	Date
Andrew Corporation	5 m	3.7–4.2	~61		10/74 to 3/75
Andrew Corporation	5 m	5.925–6.425	~92		
Andrew Corporation	4 m	3.7–4.2	~49		
Andrew Corporation	4 m	5.925–6.425	~73		
Prodelin	3 m	2.7–2.7	~66	from NASA Goddard	1975
Prodelin	1.2 m	10.5	~43	test in PhilcoFord	1972
Scientific Atlanta	3 m	3.7–4.2	~41	Prime Focus	1975
Scientific Atlanta	3 m	5.925–6.425	~61	Prime Focus	1975
RCA Ltd. of Canada	2.1 m	11.7–11.9 12.0–12.3 14.0–14.5	~84 ~85 ~100	Prime Focus	1975
Sumitomo Electric Industries	1.0 m	11.7–12.2	~40	Prime Focus	1975
Sumitomo Electric Industries	1.6 m	11.7–12.2	~64	Prime Focus	1975

In addition to the measured data, Fig. 2 also includes plots of three of the reference antenna patterns shown in Fig. 1. The current pattern for individual reception [(1)] given by the solid line is seen to be a reasonable fit for θ/θ_0 less than 5, but is much too conservative for larger off-axis angles. The current pattern for community reception [(2)] given by the dotted line is a reasonable fit for θ/θ_0 greater than 5, but badly underestimates peak sidelobe levels for smaller off-axis angles. The third reference pattern is the one obtained by setting $x = 6$ dB in (6). As a measure of how well it fits the data, it may be noted that it lies above all of the median values and above 95.5 percent of all the points plotted. Even the unnecessarily conservative pattern for individual reception failed to include all of the points.

In compiling the data, no correlation was found between sidelobe performance and antenna cost on the one hand or between sidelobe performance and antenna size (in wavelengths) on the other. Therefore it was concluded that separate patterns for individual and community reception are not justified, and that when sidelobe reduction techniques are not used, the pattern

$$d(\theta) = 6 + 30 \log (\theta/\theta_0) (\text{dB}) \quad (7)$$

with a suitable maximum value for large off-axis angles can represent antennas for both types of reception. For antennas in the size and cost range considered suitable for broadcasting satellite applications, it is unlikely that gains below isotropic will be consistently achieved in the far sidelobes and backlobes. Therefore the limiting value of angular discrimination is taken to be

$$d(\theta) = G_m (\text{dB}).$$

V. Sidelobe Suppression Techniques

There are numerous ways to achieve the goal of reduced sidelobes, ranging from extremely simple to extremely complicated [2]. The simple approaches rely on the choice of feed illumination pattern and the use of strategically located microwave absorbers. More complicated techniques, such as active adaptive cancellation feeds [7], will not be discussed because of their relatively high cost in comparison to the first two methods.

A reduction in sidelobes can be achieved by increasing the taper of the feed pattern across the aperture of the reflector [2], [6]. The penalty paid for this reduction in sidelobe level is a loss of on-axis gain. However, overall efficiencies of 50 percent are still achievable. With simple under-illumination, the sidelobe levels are reduced in all planes of rotation about the boresight. For this application in the broadcasting satellite service, a reduction is only necessary in an azimuthal plane corresponding to the aperture view of the equatorial arc. This is because most broadcasting satellites will be in a geostationary orbit, and consequently the interference from adjacent satellites will be confined to a particular azimuthal plane of the aperture. Such a selective reduction in sidelobes can be obtained by the use of microwave absorber material attached at appropriate points on the surface of the reflector [1], [2]. This method introduces some loss of gain in the main beam also, and increases the noise temperature of the antenna in proportion to ratio of the area covered by the absorber material to the remaining area of the reflector. But for selective reduction in a particular sector of the azimuthal plane of the aperture of the dish, these absorbers are relatively small. Reduction of the first few sidelobes in the azimuthal plane, $\phi = 0°$ (the plane containing the tangent to the geostationary orbit at the satellite) for example, can be achieved by placing

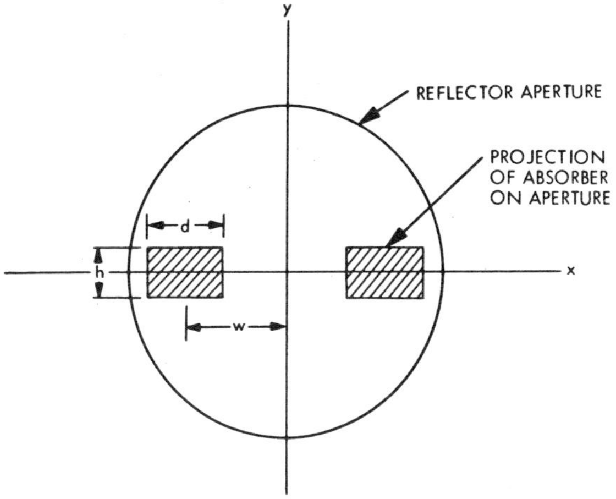

Fig. 3. Absorber array at the aperture of reflector.

rectangular absorbers in the positions shown in Fig. 3. Note that here the x axis corresponds to the plane $\phi = 0°$.

In theoretical and experimental work by Albernaz [1] and Han [2], a 1.27-m reflector was equipped with rectangular absorbers designed to provide a maximum suppression of the first and second sidelobes, without regard to the others. The predicted and the measured patterns are shown in Figs. 4-7. The feed pattern had an 8-dB taper. In the $\phi = 0°$ plane the highest sidelobe level is reduced by 7 dB, while the beam center gain is reduced by 0.7 dB. From Fig. 5 it can be seen that the radiated energy in the sidelobes has been diverted to the $\phi = 90°$ plane. The sidelobe levels relative to the maximum beam center gain stay below -27 dB up to the plane $\phi = 30°$. The relative insensitivity of this sidelobe suppression technique to azimuthal variation makes alignment and adjustment in the field quite simple and noncritical.

The experimental results are presented in Figs. 6 and 7, corresponding to measurements in the planes $\phi = 0°$ and $\phi = 2.5°$, respectively. The maximum sidelobe level reduction is over 12 dB with a beam center gain reduction of 1.6 dB.

These experimental data indicate the validity of the method. But instead of trying for the maximum suppression possible, it is suggested that suppression to a 25-27-dB level for the first two sidelobes be sought with either this technique or with a highly tapered feed pattern. This would reduce the loss of gain in the main beam and would reduce the magnitude of the increase in the remaining sidelobes.

The extra costs associated with the use of either of these techniques are primarily for nonrecurring engineering and testing to obtain a production design; the additional production costs are minimal. Therefore, if a fairly large quantity of antennas is required, the sales price increase could be fairly small. The real impact on an overall system cost cannot be estimated without due consideration for the tradeoffs between the ground segment cost performance and the space segment cost performance [4]. However, the net reduction in receiver figure of merit G/T for a 1.5-m antenna is estimated to be in the range of 1 dB, which could be made up by using a 1.7-m reflector with commensurate sidelobe performance. The cost increase associated with the slightly larger antenna is not a significant one, and so the overall system cost penalty is not likely to be severe.

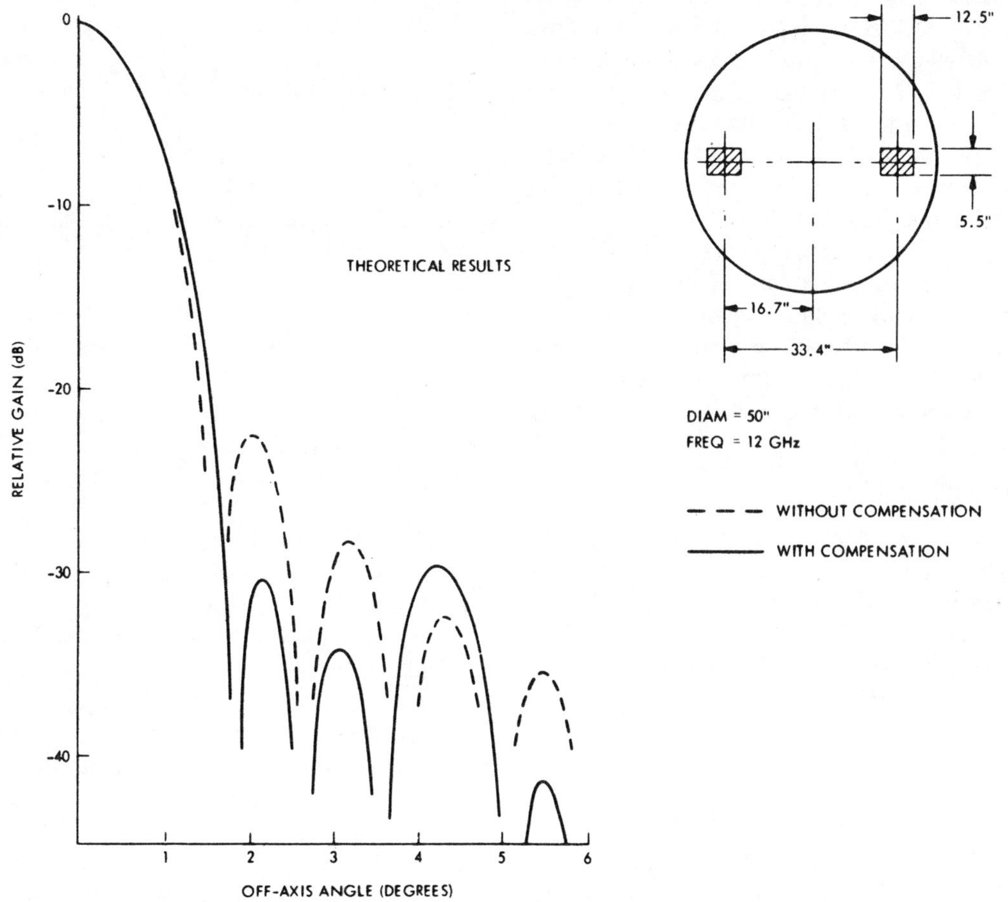

Fig. 4. Radiation pattern with and without the pair of absorbers (8-dB taper illumination) [2].

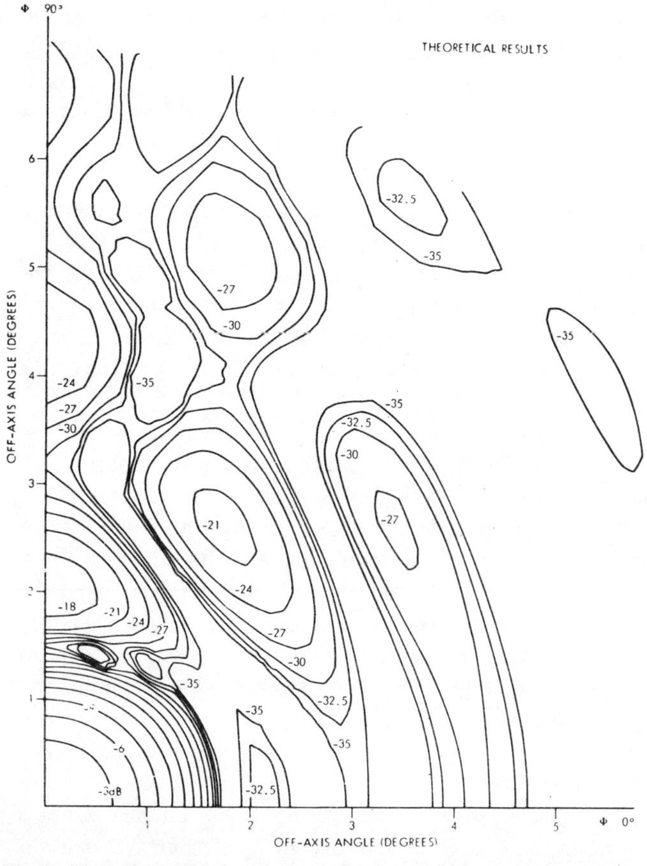

Fig. 5. Two-dimensional radiation contours with compensation on the aperture field distribution (8-dB taper illumination) [2].

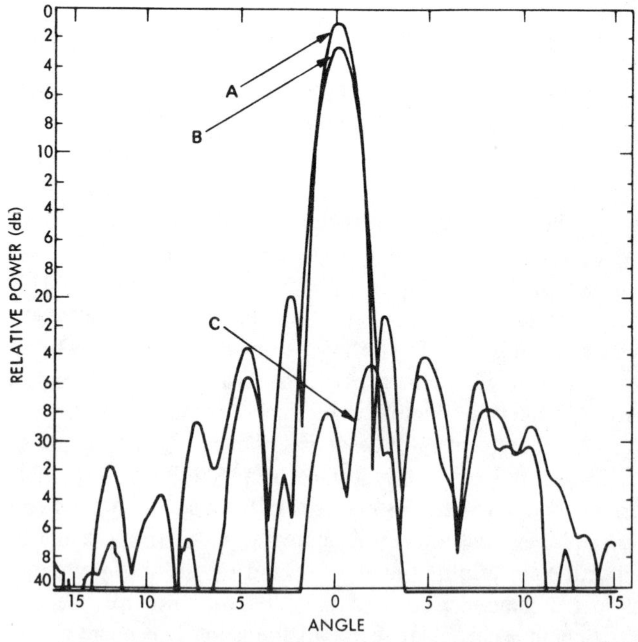

Fig. 6. Measured radiation patterns of a reflector—with and without absorber [2]. A: without absorbers. B: with absorber. C: cross polarization (with absorbers). Frequency 10.5 GHz. Distance between absorber = 26 in. Absorber size = 9 × 13 in. Absorber location: *H* plane. Pattern plane: Phi = 0°. Reflector diameter = 4 ft. *F/D* = 0.4.

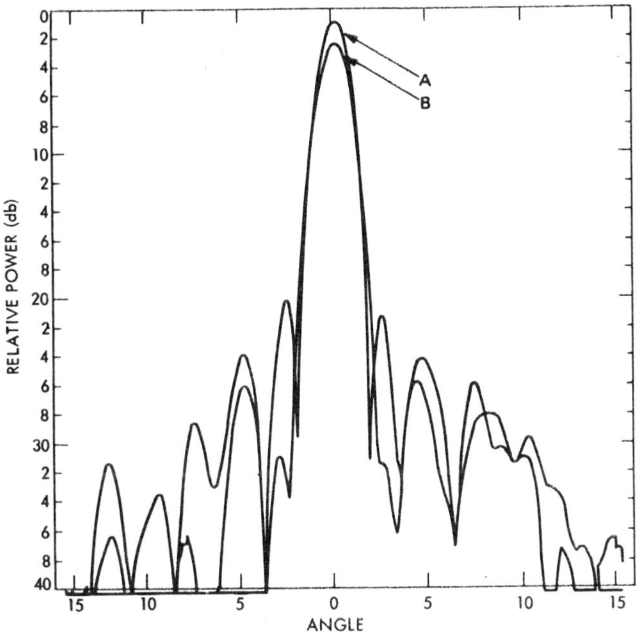

Fig. 7. Measured radiation patterns of a reflector—with and without absorber 7. A: without absorber. B: with absorbers. Frequency = 10.5 GHz. Distance between absorber = 26 in. Absorber size = 9 X 13 in. Absorber location: *H* plane. Pattern plane: Phi = 2.5°. Reflector diameter = 4 ft. $F/D = 0.4$.

VI. Conclusions

An examination of experimental measurements of the co-polarized patterns of antennas with diameters in the range from 40 to 100 wavelengths at frequencies from 2.5 to 12.2 GHz leads to the following conclusions.

Separate reference patterns for individual and community reception are not justified by the measured data. Moreover, neither of the reference patterns suggested in Report 215-3 provides a good fit to these data.

Based on both theoretical considerations and the measured pattern data for antennas which do not employ sidelobe suppression techniques, the following reference pattern is recommended for such antennas:

$$d(\theta) = G_m - G(\theta) = \begin{cases} 6 + 30 \log (\theta/\theta_0), & 1.5\theta_0 \leqslant \theta < \theta_1 \\ G_m & \theta \geqslant \theta_1 \end{cases}$$

As indicated, the quantity $d(\theta)$ is the relative gain or angular discrimination of the antenna in decibels, and θ_1 is the value of θ for which $G(\theta) = 0$.

When sidelobe suppression techniques are employed to reduce the level of the first two sidelobes, the recommended reference pattern is given by

$$d(\theta) = \begin{cases} 25 & 1.5\theta_0 \leqslant \theta < 4.5\theta_0 \\ 6 + 30 \log (\theta/\theta_0), & 4.5\theta_0 \leqslant \theta < \theta_1 \\ G_m & \theta > \theta_1 \end{cases}$$

References

[1] J. C. F. Albernaz, [November, 1972] "Side-lobe control in antennas for an efficient use of the geostationary orbit." Ph.D. dissertation, Stanford University.

[2] C. C. Han, [June, 1972] "Optimized earth terminal antenna systems for broadcast satellites." Ph.D. dissertation, Department of Electrical Engineering, Stanford University.

[3] J. N. Hines, T. Li, and R. H. Turrin, [1963] "The electrical characteristics of the conical horn-reflector antenna. The Telstar Experiment, Part 2," *Bell System Technical Journal*, 42, pp. 1187–1212.

[4] W. C. Mitchell, [June, 1975] "The use of satellites in meeting the telecommunications needs of developing nations." Technical Report No. 1 of Communication Satellite Planning Center, NASA-Ames Grant NGR 05-020-659, Stanford University.

[5] J. F. Ramsey, [June, 1967] "Lambda functions describe antenna/diffraction patterns." *Microwaves*, p. 71.

[6] S. Silver, [1949] *Microwave Antenna Theory and Design*. M.I.T. Radiation Laboratory Series, vol. 12, McGraw-Hill, New York, p. 195.

[7] W. F. Williams, [January, 1972] "Reduction of near-in sidelobes using phase reversal aperture rings." *JPL Quarterly Technical Review*, vol. 1, no. 4, p. 34.

[8] "Systems for sound and television broadcasting from satellites," CCIR Rep. 215-3 [Rev. 1974].

Part IV
Computer Simulation

IN ADDITION to the widespread use of hardware simulation, computer simulations have also become an important adjunct to the design of many types of communication systems. Recent developments in satellite communications gave significant thrust to the almost essential use of such tools in modern communication system designs. The basic goal of a software simulation is to investigate the influence of particular sources of degradation in the link.

Simulation techniques use mathematical models to represent the function of the devices that perform the signal processing in a communications path. As such, the results are, of course, dependent on the accuracy of this modeling to reflect the most significant factors of design and degradation. There are limitations and cost considerations that play an important role in the achievement of this goal. First, an accurate model of a particular device might involve the use of hardware simulations which may be expensive and time consuming. Second, there are limitations on the complexity of the mathematical model that can be used in such cases and still serve its purpose. These limitations stem from the requirement of manageability, acceptable accuracy, and computational speed.

In the four papers in this part, examples of these types of software simulations are presented. In the paper by C. S. Lorens a parametrized analysis of distortion of an FDM system due to clipping is presented. The cases of digital systems are treated in the following two papers. A Monte-Carlo simulation method for the determination of BER for digital satellite links is presented by M. C. Jeruchim. Primary emphasis is placed on modeling considerations. Distortions examined include nonlinearities (modulator, receiver, filter), thermal noise, phase noise, and bit timing jitter. Estimation of BER is performed by error-counting or by the use of extreme value statistical theory. In the paper by D. L. Hedderly and L. Lundquist, a similar simulation technique is extended to include modeling of adjacent channel interference. Finally, the description of an *interactive* simulation program as an important experimental tool and as an aid to the system design of quite general communication systems is given, in the fourth paper, by W. L. Cook.

MULTI-CHANNEL FREQUENCY DIVISION MULTIPLEX SIMULATION

Charles S. Lorens
Defense Communications Agency
Reston, Virginia 22091

ABSTRACT

This paper presents the simulation of one to twelve voice channels frequency multiplexed into a signal which is distorted by clipping. The distortion produces a channel noise which is calculated by individually demodulating each channel and comparing the demodulated sample with the modulating sample. The clipping level is characterized by the peak factor defined as the ratio of the clipping level power to the average multiplexed power level. The simulation results are parameterized in terms of demultiplexed signal to noise ratio, activity factor, peak factor, sample peaking percentage, number of channels, and voice statistics.

The results indicate that exponential voice and Gaussian statistics produce essentially identical results for six or more fully active multiplexed channels, and substantially different results for twelve or less one quarter active multiplexed channels.

FREQUENCY DIVISION MULTIPLEXED CHANNELS

A classic technique of achieving multi-communication channel multiplexing is with the use of frequency division multiplexing (FDM). This technique has been in wide-spread use for many years and will continue to be utilized as long as channels are multiplexed in the analog domain.

The principal use of channel multiplexing is for long distance communication systems where channel multiplexing significantly reduces the number of transmission channels at the off-setting expense of added complexity of the individual transmission channels.

Frequency Division Multiplexing as depicted in Figure 1 translates each communication channel to a unique band of the frequency spectrum so that upon addition the translated channels are non-overlapping in frequency. The recovery or de-multiplexing of the individual channels is accomplished by filtering and translating the channels back to baseband. Each translation is accomplished by multiplying by a sinusoid having a frequency of the translation.

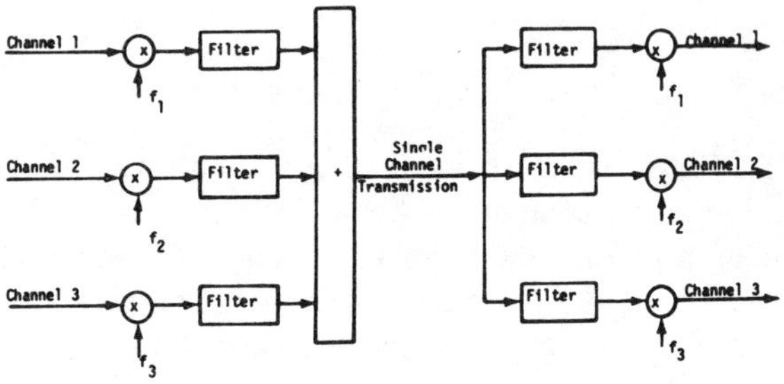

Figure 1. Communication Channel Multiplexing and De-multiplexing through the use of Frequency Translation and Filtering To Achieve Single Channel Transmission

Where the multiplexing and transmission are distortion free, the channel modulation and demodulation result in distortion free channels. Due to limited output power, the transmission amplifiers introduce an element of distortion through a process of clipping peaks of the multiplexed signal. This clipping can be minimized by reducing the power level of the multiplexed signal at the expense of increased system background noise.

The purpose of the simulation reported in this paper is to explore the relationship between the clipping level power and the multiplexed signal power level as a function of the channel performance in terms of signal to noise ratio, the number of channels, the channel statistics, and the activity of the channels.

EXPONENTIAL VOICE MODEL

Two models of voice are generally used: (1) an exponential voice model developed experimentally by Davenport [1] and reproduce in the experimental work of Sciulli, [2] and (2) a Gaussian model developed for testing multiplexed systems.

The exponential voice model is a random combination of zero samples corresponding to the natural speech pauses and of active samples selected from an exponential distribution. The amplitude density function for the exponential voice model with activity "a" and average power $S = 2a/\lambda^2$ is

$$P(y) = (1-a)\delta(y) + a\frac{\lambda}{2}e^{-\lambda|y|}$$

Reprinted with permission from *Proc. Fifth Annual Pittsburgh Conf. Modeling and Simulation*, vol. 5, Apr. 1974, pp. 873-877.

where $\delta(y)$ is the Kronecker delta function. This voice model is non-Gaussian in distribution and has many high peak amplitude values relative to its standard deviation.

Clipping the peak amplitude values of non-multiplexed speech results in a distortion having a noise power

$$N = 2 \int_B^{\infty} (y - B)^2 P(y)\, dy$$

$$= \frac{2a}{\lambda^2} e^{-\lambda B}$$

The signal to noise ratio is then

$$\frac{S}{N} = \frac{1}{a} e^{\lambda B} = \frac{1}{a} e^{\sqrt{2aP_k}}$$

where P_k is the peak factor defined as the ratio of the clipping power B^2 to the fully active average power $S = 2/\lambda^2$. It follows that a peak factor of 15.11 db is required to produce a 35 db signal to noise ratio for fully active non-multiplexed voice signal. A peak factor of 15.11 db corresponds to a clipping level of 5.6 standard deviations, far beyond the common practice of considering density functions to two to three times the standard deviations.

MULTIPLEX CHANNEL SIMULATION

A combination of models has led to a simulation of the frequency division multiplexed (FDM) system described in first section.

Due to the bandwidth limited characteristics of voice, the voice signals of a set of n channels can be represented by a sequence of samples where each sample is a corresponding set of n randomly generated samples $\{x_i\}$, $i = 1,\ldots,n$ representing each voice signal.

Since multiplexing in frequency is carried out at a frequency substantially above the highest frequency component in the voice signal, a sample in time $x(t)$ of the multiplexed signal is a sum of modulated sinusoids at frequencies $m + 1, \ldots,$ $m + n$ times an artifical multiplier such that the highest frequency is sampled k times per cycle and the lowest sinusoid has $m + 1$ complete cycles.

$$x(t) = \sum_{i=1}^{n} x_i \cos \frac{2\pi (i + m)}{k\ (m + n)} t$$

The sample in time is in turn sampled at times $t = 1, \ldots, k\ (m+n)$.

Figure 2 is an example of a sample in time of six fully active voice samples selected with the exponential voice density distribution with variance $\sigma^2_{x_i} = 1/3$ and multiplexed at relative frequencies one through six. The two dominate channels in this example are channels 2 and 6.

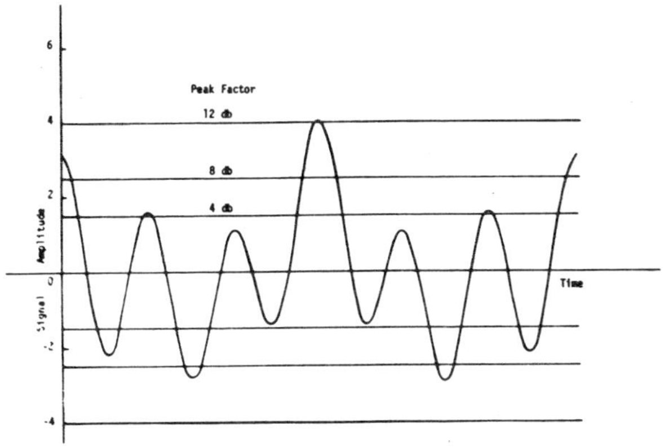

Figure 2. A Sample in Time of Six Voice Channels Multiplexed in Frequency

The average power P_a of the sample in time is

$$P_a = \frac{1}{2} \sum_{i=1}^{n} \sigma^2_{x_i}$$

assuming that the channels are independent.

Clipping is introduced into the sample in time $x(t)$ to produce a clipped signal $x_b(t)$. Clipping occurs at an amplitude B such as to result in a peak factor P_k which is the ratio of the peak power B^2 and the fully active average power P_a

$$P_k = \frac{B^2}{P_a}$$

Each channel is demodulated by again multiplying a sinusoid at the frequency of the channel, filtering in time, and amplifying the result so as to regain the original voice sample when there is no clipping. The filter impulse response is a constant amplitude function for the sample interval. The demodulated channels are thus

$$y_j(t) = \frac{2}{k(m+n)} \sum_{t=1}^{k(m+n)} x_b(t) \cos \frac{2\pi (j + m)}{k\ (m + n)} t$$

In the limiting case where there is no clipping ($x_b(t) = x(t)$), there is no distortion leakage or cross talk between channels.

The noise in each channel is defined as the difference between the demodulated output and the input modulating sample.

For each sample $\{x_i\}$ and each peak factor of interest, the noise in each channel was calculated and averaged in power over the number of samples and channels. Reasonably consistent results at the lower noise levels required 10,000 samples. Though it is known analytically that the center channels in an FDM system experience more intermodulation products (and thus more noise) than the side channels, this variation in noise level over the channels was not readily apparent. The channels utilized were 1 through 12 (m=0). Sampling of the highest multiplexed frequency was at the lowest level (8 samples per cycle), which appeared to produce asymtolic results. Clipping the multiplexed signal normally produced a suppression of the output upon demodulation. This suppression was particulary true of the larger valued inputs. Figure 3 is a plot of the demodulated amplitudes corresponding to the example sample in time of Figure 2. For each peak factor there exists the possibility of amplifying the demodulated signals to counteract the suppression and thus obtain a higher signal **to** noise ratio than calculated in this simulation.

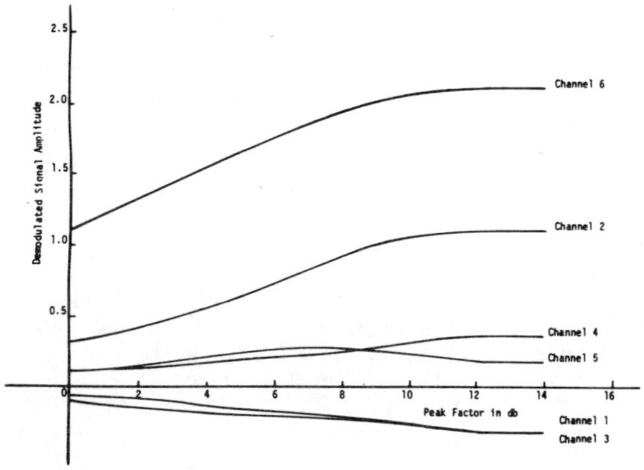

Figure 3. Demodulated Signal Amplitude

The simulation was run for both the exponential voice and Gaussian models and full and one-quarter active channels. In the one-quarter active simulation, the noise in the inactive channels was 10 to 15 decibels below that calculated in the active channels and essentially independent of the number of channels, three and greater. The simulation results indicate that inactive channel noise is considerably below that which is experienced in the active channels, and as such, is non-representative of the noise in the active channels.

The simulation results are plotted for constant signal to noise ratio in Figures 4 and 5. From Figure 4, representing the results of the fully active channel simulation, it can be seen that the results with exponential voice statistics approach those limits obtained with Gaussian statistics. It appears that for fully active channels, a 35 db signal to noise ratio is obtained with a peak factor of 9.8 db. A 20 db signal to noise ratio requires a peak factor of 6.0 db. The fully active results for the exponential voice and Gaussian models are essentially identical for six or more channels. The results of Figures 5 indicate that twelve one quarter active channels is a relatively small number of channels in relation to the system reaching some limiting value independent of the number of channels.

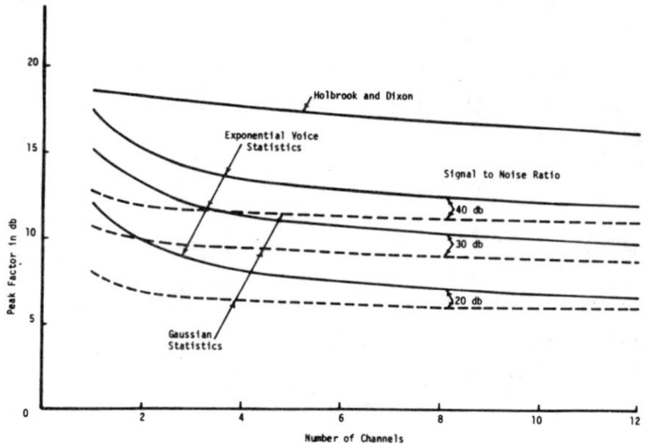

Figure 4. Peak Factor
Fully Active Channel

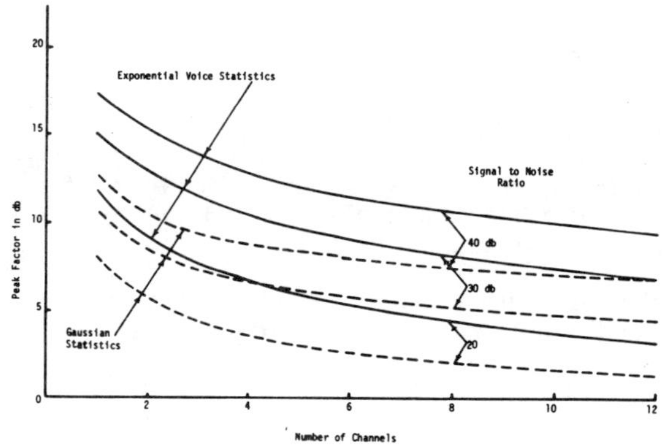

Figure 5. Peak Factor
One-Quarter Active Channel

A further result of the multiplex simulation was the sample peaking percentage plotted in Figures 6 and 7, for exponential voice statistics with full and one-quarter active channels. Similar results were obtained for Gaussian statistics but with higher signal-to-noise ratios and sample peaking percentages. The results are of particular interest in that the sample peaking percentage required to produce a 35 db signal to noise ratio with six or more fully active multiplexed voice channels appears to be greater than two percent. With one multiplexed channel the sample peaking percentage must be one-tenth of a percent to produce a 35 db signal to noise ratio.

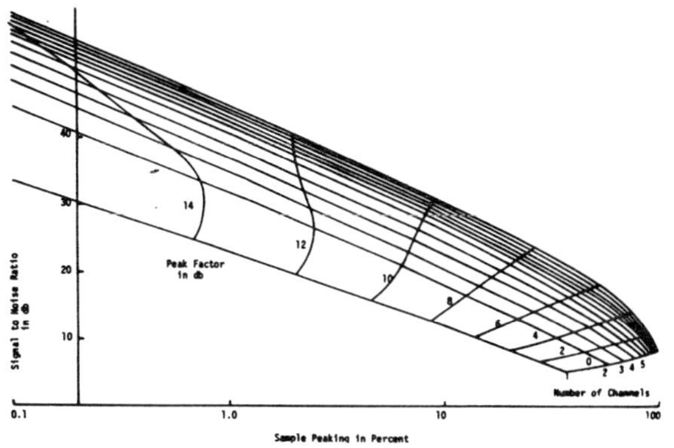

Figure 6. FDM Demultiplexed Signal to Noise Ratio
Fully Active Voice Channels

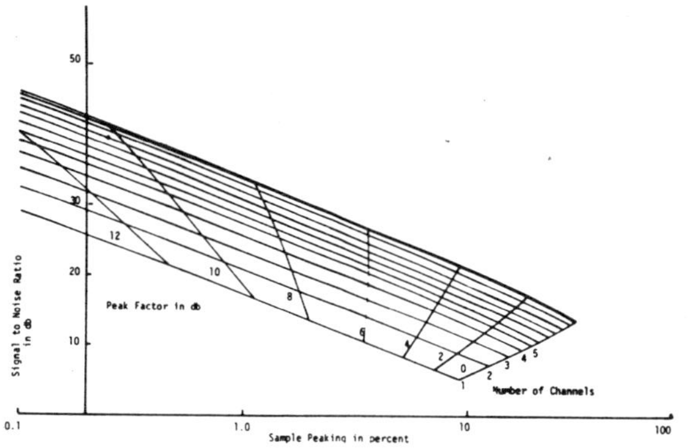

Figure 7. FDM Demultiplexed Signal to Noise Ratio
One-Quarter Active Voice Channels

CURRENT PRACTICE

The current practice and standards for the loading of FDM systems are based principally upon the work of Holbrook and Dixon.[3] A dominate results of their work is the multi-channel peak factor (reproduced in Figure 4) for fully active toll circuit frequency division multiplex communications. The one quarter activity factor appears to originate from their work.

Both the CCITT[4] and CCIR [5] standards utilize the work of Holbrook and Dixon in developing a slightly adjusted peak factor.

Experimentally it is known that a minimum reasonable voice signal-to-noise ratio is between 20 and 30 decibels. Systems operating at 10 decibels are intelligible but noticeably noisy. The threshold for intelligibility seems to be at about 0 decibels signal to-noise ratio. The CCITT and CCIR establish a 35 decible signal to noise ratio for a hypothetical reference circuit of 2500 km.

LIMITATIONS

As with any analysis or experimental work, there are a number of areas which were not explored under the supposition and assumption that the area is irrelevant to the results. The purpose of this section is to set down those areas which have been identified as potential limitations to the results.

The areas in which the simulation of this paper were limited are principally in the use of the signal to noise ratio as a criteria, identical voice channels, sample size, random phase, peak factor, number of channels, filtering, and modulation. These limitations were principally the result of the computational speed limitation. Though the present day computers are fast, they are still not fast enough to compete effectively with experimental simulations.

It was impractical to develop a continuous modulation/demodulation simulation where the individual channels were translated to a particular frequency, filtered to a single side band, multipled, clipped, and demultiplexed. Such a model would have been at least an order of magnitude more complicated than the one utilized in the reported simulation.

The sample size of 10,000 was marginally adequate to sort out the relevant results and resolve apparent inconsistancies in the results which consistantly occured at the higher signal to noise ratios.

The computing machine speed also limited the number of channels to at most twelve. The number of samples per cycle in the highest modulated sinusoid was limited to eight based upon a number of preliminary test runs.

The modulated sinusoids were held invariant in phase throughout the simulation. The manner in which this was done could only produce conservative results in comparison to those results which would be obtained with random phase.

The computation of the clipping amplitude was based upon the peak factor and the theoretical average power of a set of fully active channels. The use of the simulated power level would have produced slightly different results in the fully active case and significantly different results in the partially active case due to a change in the definition of peak factor. An appropriate definition of peak factor is not absolute and depends upon the simplicity of the concept and the ease with which it could be used.

Though there exists a relatively good model of an active voice channel which is based upon experimental results, there exist only a very poor model of channel activity and distribution of the voice model over an ensemble of users.

A principal criteria used in this report was that of signal to noise ratio. A more appropriate criteria is that of subjective acceptance and intelligibility. Though signal to noise ratio is a well respected criteria, it is admittedly a weak one which is used for its precision of definition and wide acceptance.

The important aspect of these limitations is that as a result of this simulation, there exists a departure point from which further investigation can proceed as time becomes available for further exploration. Examination of the limitations may produce further understanding of the requirements of multiplexed channels.

One of the principal objectives of further work is the examination of frequency modulated systems where the underlying modulation is frequency division multiplexed voice. An analysis [6] has been developed which has potential application to such an examination. This analysis is particulary useful in that it treats not only the distribution of the modulating signal, but also the threshold effects of frequency modulation/demodulation systems.

REFERENCES

1. Davenport, W. B., "An Experimental Study of Speech-Wave Probability Distributions," Journal of the Acoustical Society of America, Volume 24, Number 4, July 1952, pp 390-399.

2. Sciulli, J. A., "A Statistical Characterization of Digital Voice Signals for Compression System Applications," COMSAT Corp., 1971.

3. Holbrook, B. D., Dixon, J. T., "Load Rating Theory for Multi-Channel Amplifiers," The Bell System Technical Journal, Volume 18, Number 4, October 1939, pp 624-644.

4. CCITT, Fourth Plenary Assembly, White Book, Volume III, Lone Transmission, ITU, 1969.

5. CCIR, Twelveth Plenary Assembly, Volume IV, Part 1, Radio-Relay Systems, and Part 2, communication Satellites, ITU, 1970.

6. Lorens, C. S., "Threshold Frequency Demodulation," Aerospace Corporation Report No. TDR-699(9990)-2, 16 September 1965.

DIGITAL COMPUTER SIMULATION
OF
SATELLITE QUADRATURE DATA COMMUNICATION SYSTEMS

Michel C. Jeruchim
General Electric Company, Space Division
Philadelphia, Pa. 19101

SUMMARY

This paper describes a monte-carlo simulation developed to predict the bit error rate (BER) for digital satellite communication links using quadrature signaling (QPSK or QAM). Primary emphasis is placed on modeling considerations. The model contains two nonlinear amplifiers and thermal noise, phase noise, and bit timing jitter. Provision is made to either insert arbitrary transfer functions or to specify the filtering through only a few parameters. Two types of equalizers are included, a "preset" and an adaptive type. Estimation of the BER is performed by error-counting or by the use of extreme-value statistical theory.

INTRODUCTION

Simulation has become an important adjunct to the design of many types of satellite communication systems. This paper considers in particular, data communication systems using quadrature signaling. A simulation is the synthesis of three distinct disciplines: communication theory, computational methods, and modeling techniques. These aspects are inextricably linked within the simulation itself, but for the purpose of discussion their identities can be kept separate. This paper, thus, is a review of modeling considerations, which have received relatively little attention in previous literature. The discussion uses as a point of reference a simulation developed to predict the BER for satellite quadrature links.

The basic purpose of a simulation determines the manner in which it is structured. Practical constraints will inevitably limit its capabilities. Therefore a simulation must be designed to solve a restricted range of problems. Thus, for example, some simulations are intended to investigate the influence of a particular source of degradation and tend to model that effect with accuracy while idealizing other aspects. On the other hand, the type of simulation discussed here is intended to provide a reasonably accurate estimate of actual performance (BER). Thus, many practical factors have to be modeled with relatively good correspondence to reality. The number and complexity of these factors, and computing limitations, preclude the extensive modeling that might be appropriate if only one of them were being investigated. Fortunately, for the purpose at hand what is really needed is a phenomenological model rather than an explicit one. That is, a device needs to be modeled only in the aspects of its behavior that affect the BER. A case in point is clock jitter. An actual bit synchronizer is a complex device but its attribute of relevance to the BER is the lack of regularity of the sampling pulses. Thus, to a reasonable approximation, we can replace the complex dynamical model of a bit synchronizer with an appropriate random process that represents the jitter in the sampling pulses.

Another major consideration in the modeling arises from the practical requirement that the simulation be able to estimate the BER of: (a) actual systems, and (b) proposed systems. These two capabilities may not lead to compatible constraints. The need for the first capability may arise in a number of ways. For example, it may not be economical to assemble the equipment for the transmitter, the transponder, and the receiver, all in one location to perform a BER test. Measured characteristics for these equipments, however, can be "assembled" in the simulation which can then substitute as a composite test facility. For the simulation to perform such a function the model must be sufficiently detailed to reflect the more significant factors of design and degradation.

Simulating hypothetical systems, on the other hand, requires a sufficient level of flexibility to be applicable to a relatively broad range of possible designs. For this purpose, in fact, it may be counter-productive to have too specific a model, for at the planning stages a system is usually specified by a small set of broadly descriptive parameters. The evolution of the same system from conception to implementation will successively call on capabilities (a) and (b). In the beginning, one performs trade-off studies on the system level parameters, and arrives at a set of specifications. As the design begins to be realized, one would like to be able to substitute measured for specified characteristics. Thus, a balance between specificity and generality has to be designed into a simulation in order to fulfill both functions reasonably well.

With the preceding points in mind we developed a simulation with the following features and capabilities:

- Source/Modulator
  - Data bits are modeled as trapezoids with independently specifiable rise and fall times.
  - The two data streams can have arbitrary relative timing.
  - The modulated carrier output can exhibit amplitude unbalance.
  - The modulated carrier output can exhibit phase unbalance.
- Filtering
  - Two filtering models are included; one permits use of arbitrary amplitude and phase characteristics when actual measurements are available. The second model permits filtering to be specified by a small set of parameters.
- Power Amplifier
  - Gain characteristic
    - Can use actual amplitude saturation curve
    - Can simulate hard limiter
    - Can be linear
  - AM/PM conversion

Reprinted from *National Telecommun. Conf. Record*, vol. 2, Dec. 1-3, 1975, pp. 33-16-33-26.

Can use actual phase vs. amplitude characteristic

Can use "worst case" constant coefficient (°/dB).

Thermal Noise

- Independently generated at satellite input and ground station input.
- Generated as Gaussian but no other assumptions made.

Interference

- CW interference can be input on both uplink and downlink.

Receiver

- Independent static phase error can be used on the two channels
- Carrier tracking phase jitter can be simulated
- Clock instability can be simulated via bit sync sampling jitter
- BER can be obtained for either a single bit synchronizer or for independent bit synchronizers in both channels
- Sampling threshold error (dc) offset can be specified
- Data filters
    - Integrate and dump (I&D)
    - 2-Pole Butterworth
        - Variable bandwidth
        - Variable sampling instant
- Equalization
    - Preset equalizer
    - Adaptive tapped-delay line (zero-forcing)
- BER estimation
    - The estimation philosophy is based on Monte Carlo simulation; two statistical methods are available -
        - Classical (error-counting); binomial distribution
        - Generalized extreme-value statistics; Gumbel/Weinstein distribution.

The specific nature of the models for the preceding items is elaborated in subsequent paragraphs.

MODELING CONSIDERATIONS

Signal Definition

The signal transmitted is referred to as quadrature because its representation is a set of four phasors nominally of equal amplitude and successively displaced by $\pi/2$. Such a representation applies to 4-level phase modulation (QPSK) and 4-level quadrature amplitude modulation (QAM). When these methods are ideally implemented, the definitions coalesce. In practice, whether it is QPSK or QAM that is intended, the actual representation will in general consist of non-orthogonal and unequal amplitude phasors. The degree of non-orthogonality (phase unbalance) and amplitude variation (amplitude unbalance) is a function both of the information source and the modulator. A particular way of implementing QPSK may result in a signal that is more naturally modeled as QAM. For example, two basic methods of biphase modulator implementation use diode switches as the state-changing mechanism [1]. In one method, using a resistive diode phase shifter, the carrier phase can be considered ideally switched but the carrier envelope has deep nulls at the transition, while in the reactive diode phase shifter method the envelope is nearly constant and the switching transient appears in the RF phase. Clearly, what is significant is the envelope and phase variation. The terms QPSK or QAM can be used as a matter of taste though one may be more appropriate than the other in any given case. The complex envelope is the unifying concept that can represent any desired phasor configuration for which the term quadrature is appropriate.

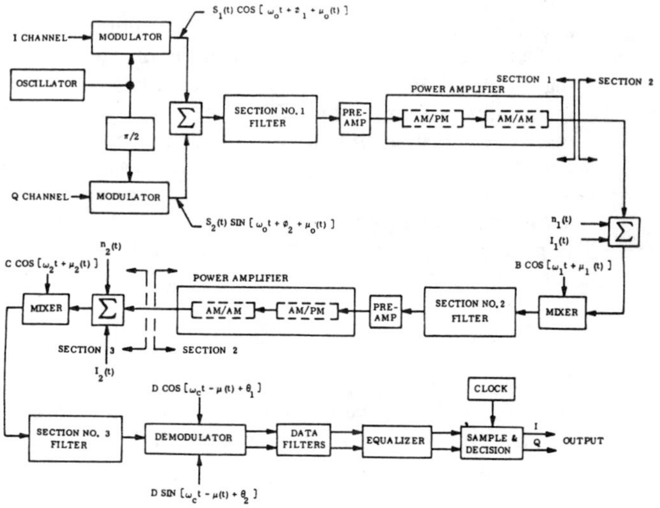

Figure 1. Simulation Block Diagram

System Description

The model is shown in Fig 1. Our motivation lies in space applications for which the model represents a standard system, comprising the ground transmitter, translating repeater, and receiving earth station; the separation between these sections is indicated in Fig 1. A second possible application would be to space-space-earth systems, (e.g., the NASA tracking and data relay satellite system). In such cases, contrary to the usual one, the uplink may well be weaker than the downlink. It is with such applications in mind that we separately generate noise at the satellite input. In much previous work; e.g., [2] the total noise is input directly into the ground receiver. This method is appropriate for a linear system or where the uplink is strong, but needs to be re-examined for the case under discussion. We have made runs where the total noise was either input to the receiver or split between the uplink and downlink and found appreciable differences in degradation. Incidentally, the model can also be used for antipodal systems by suppressing one of the two data streams.

Data Source/Modulator

As discussed above the generated signal is the combined result of the source and the modulator. For convenience we have chosen to view the signal as arising from two amplitude-modulated carriers. The modulators are taken to be ideal product modulators but the two carriers issuing from them may not be in phase quadrature. The data source consists of a pair of nominally two-state waveforms (the I and Q channels). The gain in the two channels may be different so that the amplitude in each can be independently specified. The waveform of either channel is a trapezoidal sequence with fall time (FT) independently specifiable from the rise time (RT). The BER will differ for ones and zeros whenever FT ≠ RT. A convenient indicator of this property is the data asymmetry, α, defined as

$$\alpha = \frac{T_b + 0.5FT - 0.5RT}{T_b + 0.5RT - 0.5FT}, \quad FT \geq RT \tag{1}$$

where T_b is the bit duration, and in (1) FT and RT are interchanged when FT ≤ RT. In actuality the transients may arise from imperfections in the source generator or the modulator, or both. These transients may be significant in high data rate applications [3]. Thus the transmitted signal exhibits two features that ultimately lead to degradation. First, the four symbols may not all have the same magnitude. This amplitude unbalance creates an envelope variation which, because of AM/PM conversion, subsequently induces degradation. Secondly, the four phasors may not be orthogonal. This phase unbalance leads to crosstalk. The binary sequences can either be user-specified or pseudo-randomly generated. Finally, the two bit streams need not be in synchronism. This may arise unintentionally or deliberately. When the channels are staggered by one-half bit this has been termed "offset"[4]. In the general case we refer to the relative time slip between the channels as the skew. An appreciation of the possible influence of skew may be had from Fig 2 which shows a computer generated plot of the carrier envelope with zero skew. The severe fluctuations give some indication of the potential benefit of skewing when significant AM/PM conversion is present.

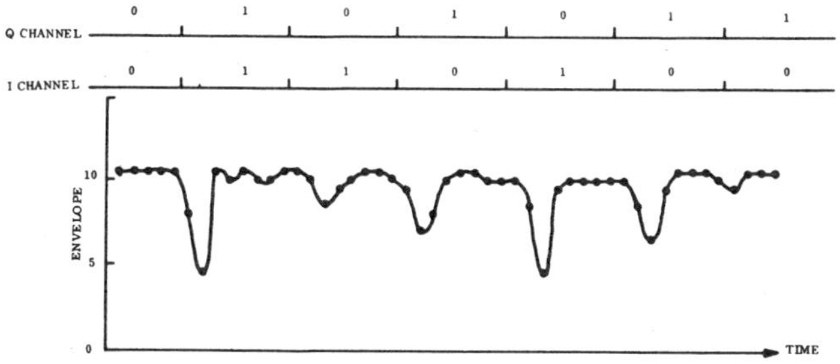

Figure 2. Illustrative Envelope Waveform - Zero Skew

In connection with the capability mentioned earlier of studying hypothetical implementations, data asymmetry, skew, amplitude and phase unbalance are all simple parameters that can be specified by the system designer to control performance irrespective of the particular realization.

Section (Bandpass) Filtering

In the typical system, a section will have several filters. There will also be unintentional filtering. Because filtering is the most time consuming operation to simulate, it is desirable to model the section filtering as a single "complex" filter. In most cases this is a valid representation. But it is not practical to devise a general model, particularization of which could reduce to any desired filtering characteristic. Consequently, the filter model was designed so that it could accommodate conveniently either of the two basic functions of the simulation, analysis and design. The model is indicated in Fig 3. By "analysis" we mean the determination of performance given actual (measured) characteristics. In this option, the user has complete freedom to specify any section's amplitude and phase transfer function by means of look-up tables. In the "design" option we wish to provide a model that has the flexibility to approximate

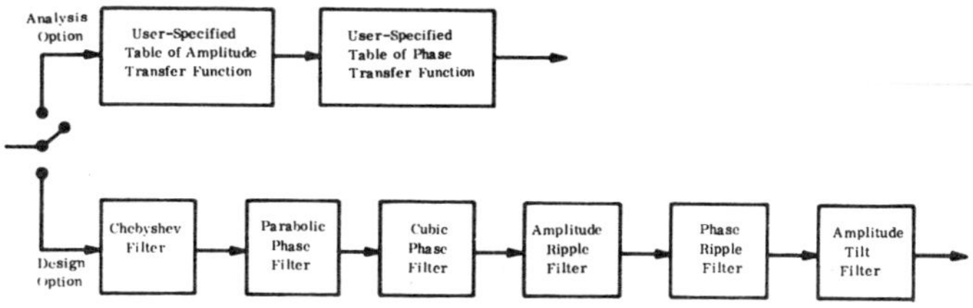

Figure 3. Section Filtering Model

reasonable transfer functions by specifying a small set of parameters under the system designer's control. One common procedure, e.g., is to expand the transfer function into some series and set limits on the first few coefficients with the implicit assumption that the higher-order terms are negligible. The "design" filter model is based on similar reasoning, i.e., that specifying the parabolic and cubic phase components, the phase ripple, the amplitude tilt and ripple, and the nominal bandwidth, is sufficient to control the essential effects of filtering. The specifications are actually made on the lowpass equivalents.

The first component is a lowpass prototype Chebyshev filter. The purpose of this filter is to provide the desired bandlimiting. A fit of the phase characteristic provides a linear and a cubic component and a residual ripple. The "parabolic phase filter" provides the parabolic phase component of the section filter. The "cubic phase filter" provides the flexibility to simulate any amount of cubic phase. This capability is desirable since the cubic phase of the Chebyshev filter cannot be chosen independently of its other parameters. Since the amplitude and phase of the Chebyshev filter are interdependent, a given filter may not have the desired amount of amplitude ripple. The "amplitude ripple filter" is included to adjust the total ripple to that amount. The phase ripple in the Chebyshev filter may not be the amount it is desired to simulate for the section. The phase ripple filter fills this gap. Finally, the "amplitude tilt filter" simulates first-order frequency selective effects by inserting a slope (or tilt) of arbitrary magnitude in dB/Hz. The slope can be independently specified on either side of center frequency.

Originally we sought a physically realizable filter model. It turns out to be rather difficult to find a realizable filter that: (1) over a given bandwidth closely approximates a specified characteristic; (2) adds no characteristic other than the desired one; (3) is capable of variation as large as desired in the simulated variable where the latter is simply related to the filter parameters. A case in point is parabolic phase; i.e., linear time delay (LTD). It is a complicated problem to relate LTD of any desired magnitude to the number and type of network and their individual parameters, with a minimum of residual error. One is naturally led to the alternative of assuming a parabolic phase characteristic over all frequencies. Although such a network is not casual, it is in cascade with a band-limiting filter. Hence, outside a given bandwidth, the phase characteristic is actually immaterial. Thus, the cascade of the Chebyshev and all-pass parabolic phase filter is very close to the cascade of the Chebyshev and some realizable filter with (nearly) parabolic phase over the band of interest. This artifice permits the use of simple mathematical models for the blocks of Fig 3 with acceptable approximation to the behavior of a real device. In the end there is no compelling reason for physical realizability so long as the model is reasonable, just as the ideal lowpass filter, though unrealizable, is a frequent and useful model for real lowpass filters.

Power Amplifier

Most power amplifiers (PA) exhibit amplitude saturation (AM/AM) and amplitude-to-phase (AM/PM) conversion [5]. However, there is little available on phenomenological models. For example, should AM/AM follow AM/PM, or is it even meaningful for a given device to think of it in such terms? For the traveling-wave tube (TWT), the most common PA for space applications, a proper model is AM/PM followed by AM/AM [6]. For this reason, the PA is so modeled in Fig 1.

Actually, even for the TWT, the PA model may be too crude for all applications. In this model the AM/AM and AM/PM blocks have zero-memory. For high data rates, however, the signal bandwidth may be of the order of the PA bandwidth and then at least two other aspects of PA behavior need to be considered: frequency selectivity, and frequency dependence of AM/PM conversion. Reflection indicates that to first-order approximation linear filtering effects precede AM/PM and AM/AM effects and thus can be combined with the section filtering. But the question of PA modeling needs further investigation.

a. AM/AM Characteristic. The AM/AM block relates output amplitude (i.e., envelope) to input amplitude. In practice this information is embodied in the conventional single-carrier input-output power measurement, which can be meaningfully interpreted in terms of the instantaneous time-discrete input-output envelope transfer function. Since we inevitably deal with samples, this enables us to approximate the true but unknown instantaneous envelope transfer function using a standard measurement. Optionally, the gain characteristic can be made linear or hard-limiting. It is necessary to control the input to the PA so that it is varying over some predetermined portion of the transfer characteristic. This control is symbolized by the preamplifiers in Fig 1.

b. AM/PM Characteristic. The AM/PM characteristic is a function relating output phase to input envelope. Here too we normally wish to use measured characteristics. As before, we can interpret standard measurements on an instantaneous time-discrete basis. The user can also optionally use an equation which assumes a constant AM/PM coefficient. For the purpose of observing the waveform the average phase shift $\bar{\psi}$ is subtracted from every sample value $\psi(i\Delta t)$. In most cases studied using typical parameters for high rate applications we have found AM/PM conversion to be a more serious cause of degradation than saturation. An illustrative example is given in Fig 4 which shows BER vs. SNR when AM/PM is absent (in both PA's) or present.

Oscillators/Mixers/Phase Noise

There are groups of devices in a system that act as a unit insofar as the system performance is concerned. A case in point is the set of items that produce a "coherent" carrier reference. It is not the spectral purity of oscillators, per se, that matters, nor the characteristics of frequency-converters, nor the specific design of the PLL, but it is rather the combined effect that is of consequence, namely the phase noise. Clearly, if the

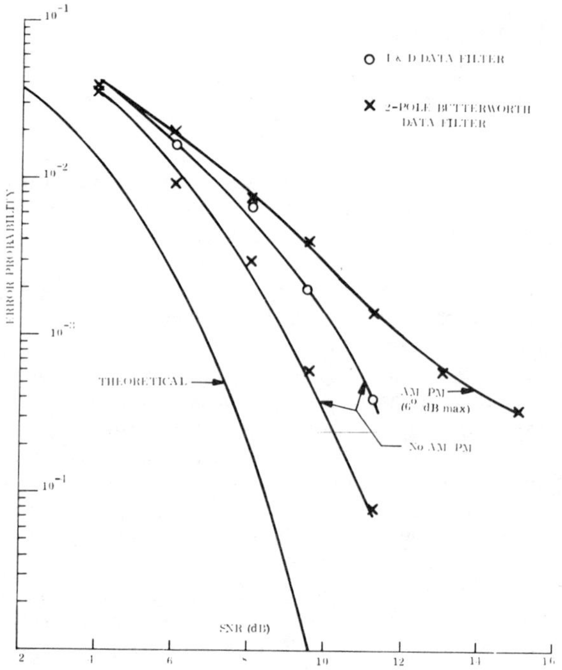

Figure 4. Typical Simulation Results Showing Effect of AM/PM Conversion and Comparison of Data Filters

latter can be modeled in a reasonable fashion it is more desirable to do so than to explicitly simulate all the causes of its existence. Even if the system model were sufficiently flexible, simulating every single device would be impractical. Furthermore simulation of the PLL itself is quite complex because of the transient dynamics, which are not of concern here, while the steady-state behavior of the loop, which is of interest, can be essentially represented by a filter in the phase domain. The evolution of the phase-noise model rests of this last aspect of PLL behavior and on the excellent assumption that imperfect frequency sources can be modeled as randomly phase-modulated carriers [7]. Thus, at the demodulator input the signal can be represented as

$$s(t) = \rho(t) \cos [\omega_c t + \theta(t) + \mu(t)] \quad (2)$$

where $\rho(t)$ is the envelope, $f_c = \omega_c/2\pi$ is the carrier frequency, $\theta(t)$ is the (distorted) signal phase, and $\mu(t)$ is the accumulated phase error. The PLL output can be written as

$$f(t) = 2A \cos [\omega_p t + \beta(t) + \theta_1] \quad (3)$$

where $\beta(t)$ is a filtered version of the carrier phase, $f_p = \omega_p/2\pi$ is the PLL output frequency and θ_1 is a static phase error. Demodulation is represented by multiplication of (2) and (3), and by multiplication of (2) and a $\pi/2$ shifted version of (3) for the I and Q channels, respectively, and with $\omega_p = \omega_c$. A static phase error in the Q channel, θ_2, possibly different from θ_1, is permitted. Rejecting sum-frequency terms we get

$$I(t) = A\rho(t) \cos [\theta(t) + \epsilon(t) - \theta_1] \quad (4)$$

$$Q(t) = A\rho(t) \sin [\theta_2 - \theta(t) - \epsilon(t)] \quad (5)$$

where $\epsilon(t) \triangleq \mu(t) - \beta(t)$ is defined as the phase noise. This is what is directly modeled in the simulation. Our model for $\epsilon(t)$ is given by the autocorrelation function

$$R_\epsilon(\tau) = P_\epsilon e^{-\alpha |\tau|}. \quad (6)$$

The performance degradation is limited through specification of the rms phase error $\sqrt{P_\epsilon}$. The main motivating factor behind (6) lies in the desire to use an efficient algorithm to generate correlated noise samples. The corresponding spectral density is peaked at f = 0. While this may not be the expected shape for all disturbances; e.g., untracked oscillator noise, this should have little effect since phase noise can usually be regarded as lowpass with respect to the signal spectrum.

Thermal Noise and Interference

As indicated in Fig 1 we simulate thermal noise at the front end of two sections. Interference may also be present. It is easy to incorporate these disturbances by means of the complex envelope which can be written

$$z(t) = \{ S_c(t) + n_c(t) + I_c(t) \} + j \{ S_s(t) + n_s(t) + I_s(t) \} \quad (7)$$

where S, n, and I stand for signal, noise, and interference, respectively, and where the subscripts c and s represent "in-phase" and "quadrature" components. I(t) is taken to be CW interference with random phase. The noise is generated by two normal number generators. Moreover, a rectangular spectrum with bandwidth B and single-sided PSD n_o has associated autocorrelation functions $R_c(\tau)$ and $R_s(\tau)$ that have zeros at all $\tau = n/B$, $n = \pm 1, \pm 2, \ldots$ Thus we chose $1/B = \Delta t$ where Δt is the sampling interval. This permits us at each sampling instant to independently generate a random sample both for $n_s(t)$ and $n_c(t)$, and in effect simulate white noise of density n_o over the effective bandwidth in the simulation.

Data Filters

Following the demodulator the data streams are processed through "data filters" that attempt to "match" the signal. The waveforms are usually such complicated functions that a priori design of a matched filter is not feasible. Consequently, two types of filters are included to provide a comparison: an integrate and dump (I&D) filter and a two-pole Butterworth filter.

Integrate and Dump Filter. As is well known the I&D filter performs a repeated integration of its input every bit duration T_b. The signal suffers an effective delay, D_I, which we term decision delay. Thus the I&D filter computes the decision variable for the kth bit in the I channel as follows:

$$V_I(k) = \int_{t_k}^{t_{k+1}} S_1(t)\, dt; \quad t_k \triangleq (k-1)\, T_b + D_I \quad (8)$$

where $S_1(t)$ is the I channel demodulator output. In the simulation $S_1(t)$ is available only as a sequence of samples which are joined to form a trapezoidal approximation. For the Q channel an equation identical to (8) is used but delayed by the skew. We actually implement a modified version of (8) to take into account timing jitter. That is, the upper and lower limits of integration are $t_{k+1} + J_{k+1}$ and $t_k + J_k$ where J is taken as a Gaussian random process with autocorrelation function

$$R(J) = (SDJ)^2\, e^{-|\tau|(ALP/T_b)} \quad (9)$$

The jitter is generally a slow process and is thus correlated from bit to bit. The jitter in the Q channel is an identical but independent process. The standard deviation SDJ and the decorrelation parameter ALP are user inputs.

Butterworth Filter. A two-pole Butterworth filter provides a reasonable match to a variety of distorted pulses by varying the 3 dB bandwidth f_3 [8]. Denote the Butterworth filter output (say in the I channel) by $S_{1B}(t)$. Once every bit interval $S_{1B}(t)$ is sampled to produce the decision variable $V_B(k)$. Here too we must account for a decision delay D_B. Thus the decision variable for the kth bit is given by

$$V_B(k) = S_{1B}\,(\,(k-1)\,T_b + D_B) = S_{1B}\,(t_k). \quad (10)$$

Once again, because of jitter we replace t_k by $t_k + J_k$ where J_k is a normal random variable with autocorrelation function having the form (9). The Q channel is handled the same way except for the delay due to skew. If the decision instant is not a sampling instant, interpolation on the straddling samples is performed. In most cases studied the Butterworth filter with proper f_3 performed better than the I&D filter. Fig 4 shows an example. For the Butterworth filter itself, performance can vary significantly with f_3, or it can be relatively insensitive, depending on the specifics. In some instances, e.g., we found over a 2:1 improvement in going from $f_3 = 0.5/T_b$ to $f_3 = 0.6/T_b$. In any event the optimum f_3 for a given case can be determined since the simulation can provide in a single run the BER corresponding to 30 values of f_3.

Equalization

There are many types of equalizers and it is not practical to pre-build every one into the simulation. An approach was developed, however, utilizing auxiliary programs, that provides much flexibility. First, to standardize program structure the (non-restrictive) assumption was made that the equalizer would be implemented at baseband as a tapped delay-line. Next, two categories were defined as "preset" and "adaptive". Preset equalization connotes the type traditionally implemented with passive filters. We use the term preset because there is no self-adjusting mechanism. The preset equalizer structure in general requires four sets of taps, two of these being cross-coupled to the other channel. The tap gains are anti-symmetric. The tap spacing τ is constrained to be a multiple of the sampling interval Δt so as to avoid interpolation. This is hardly a constraint, however, for $1/\tau$ represents the effective bandwidth over which equalization takes place and it can be made slightly different to accommodate this requirement with negligible effect. The preset equalizer subroutine is thus easy to implement and very fast.

Adaptive equalizers are more complex. In estimating the BER it is the steady-state value we want; thus some means for separating the transient from the settled behavior is required. Since the duration of the transient state is usually not known before-hand it is difficult to estimate the required length of a run. Inefficiency in computer utilization would result if the time required for the tap gains to settle exceeds that required purely from statistical consideration. Besides, in an adaptive equalizer there are usually a number of design parameters to be adjusted for optimum performance. This optimization must be done before applying the equalizer in the simulation. Consequently it was decided to implement adaptive equalizers as preset equalizers whose tap gains would be the steady-state gains observed in an auxiliary "training" program that dynamically simulates the equalizer. The observed BER in the simulation would then be the desired long-term performance. The training signal is a sequence of samples generated by the simulation itself so that the equalizer trains on the actual signal it would encounter. This approach was tested with a generalized version of Lucky's zero-forcing equalizer [9], applicable to a pair of skewed channels. This equalizer is simple to program and converges rapidly to a steady set of tap gains. This approach to simulating the effect of adaptive equalization was thus verified as both accurate and efficient.

Bit Synchronization

In accordance with the point of view expressed earlier, we do not simulate a bit synchronizer in the sense that particular circuitry is assumed. For our purposes this would be both unnecessarily complicated and overly specific. The BER is affected, in the end, only by the behavior of the sampling pulses themselves. For practical purposes this behavior can be decomposed into two categories: static (timing bias) and dynamic (timing jitter). The latter is characterized by a two-dimensional pdf and is incorporated as described in Section 2.8. The actual distribution may be either difficult to obtain or difficult to simulate efficiently, but an adequate model is provided by the normal distribution when one uses the true rms value and decorrelation time. The jitter takes place about a nominal sampling instant which itself may not occur at the optimum time for a variety of reasons. Timing bias is not amenable to mathematical modeling as such. However it is simple to investigate the effect of this bias in relation to optimum sampling, simply by sampling the final waveform at a number of instants. In the simulation, provision is made for obtaining the BER for up to 30 values of decision delay. These 30 values of delay, however, must be clustered reasonably near the true delay. Thus we first need to determine a nominal delay about which the decision instants can be varied. This determination is made by the computer, preliminary to a BER run, using an algorithm designed for the purpose.

Bit Detection and Decision and BER Estimate

Two methods provide an estimate of the BER. The first is the traditional "error-counting" technique in which the output signal is regenerated and compared to the input bit stream. Specifically, for the I&D filter decide "1" was sent for the kth bit if $V_I(k) > TSTI$, or a "0" was sent otherwise, where the threshold (or "test" number) TSTI is normally zero but different values can be used; e.g., to investigate the effect of DC bias. Similarly, for the Butterworth filter, decide "1" was sent if $V_B(k) > TSTB$, and a "0" was sent otherwise, where the threshold value TSTB can be input as any desired number. The BER is simply the number of errors divided by the total number of bits transmitted. The confidence intervals are obtained from classical parameter estimation theory for a binomial distribution. The second BER estimation method, available as an option, is based on the statistics of extremes as generalized by Weinstein [10]. For given confidence interval and level the error-counting method implies an increasing number of bits as the BER decreases until, at some point, the required computer time becomes excessive. Extreme-value theory is a tail-estimation method that uses the analog waveform before decision, which has greater informational content than that possessed by a hard decision. In effect the method permits an accurate extrapolation of the BER about two orders of magnitude below that provided by error-counting.

CONCLUDING REMARKS

This paper has discussed in a broad way some of the device and system modeling considerations involved in simulation of quadrature data systems. As mentioned in the introduction, analytical and computational considerations play an equal role in the final product. Frequently these three facets are in conflict. Judicious approximations and compromises are always required in trading off cost and accuracy, even if complete models are available. In a complex system, it is generally not possible to predict the loss in accuracy incurred by these compromises in rigor. Thus the accuracy of the simulation must ultimately be tested against known results. We have done this by simulating a number of cases for which experimental results were available, one such case being shown in Fig 5. The correspondence between simulated and actual results that has been obtained to date is encouraging of a more general agreement.

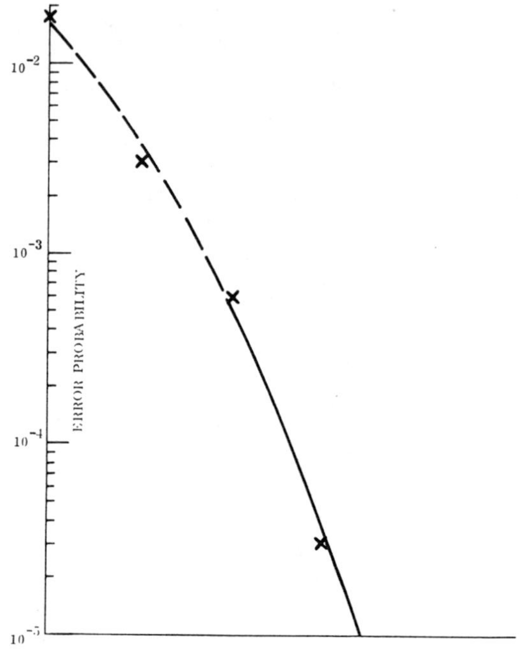

Figure 5. Comparison of Simulated (Crosses) vs Experimental (Solid Line) Results.

REFERENCES

1. C.L. Cuccia, "Study of Efficient Transmission and Reception of Image-Type Data Using Millimeter Waves", Philco-Ford Corp. TR-DA 2180, June 15, 1970; Prepared for NASA on Contract NAS 12-2186.
2. D.L. Hedderly and L. Lundquist, "Computer Simulation of a Digital Satellite Communications Link", IEEE Trans. Comm. Vol. COM-21, No. 4, April 1973.
3. C.L. Cuccia, J.L. Spilker and D.T. Magill, "Digital Communications at Gigahertz Data Rates", The Microwave Journal, Part I, January 1970; Part II, February 1970; Part III, April 1970.
4. S.A. Rhodes, "Effects of Hardlimiting on Bandlimited Transmission with Conventional and Offset QPSK Modulation", 1st IEEE Nat. Conf. Telecomm, December 1972.
5. D. Chakraborty and D. Geden, "Measurements of AM-PM Conversion in Low Noise TWT's, TDA's and Parametric Amplifier", Proc. IEEE (Letters), Vol. 56, No. 11, November 1968.
6. A.L. Berman and C.E. Mahle, "Nonlinear Phase Shift in Traveling-Wave Tubes as Applied to Multiple Access Communications Satellites", IEEE Trans. Com. Tech., Vol. COM-18, No. 1, February 1970.
7. L.S. Culter and C.L. Searle, "Some Aspects of the Theory and Measurement of Frequency Fluctuations in Frequency Standards", Proc. IEEE, Vol. 54, No. 2, February 1966.
8. J.J. Jones, "Filter Distortion and Intersymbol Interference Effects on PSK Signals", IEEE Trans. Com. Tech. Vol. COM-19, No. 2, April 1971.
9. R.W. Lucky, "Techniques for Adaptive Equalization of Digital Communication Systems", Bell Sys. Tech. J., February 1966.
10. S.B. Weinstein, "Theory and Application of Some Classical and Generalized Asymptotic Distributions of Extreme Values", IEEE Trans. Information Theory, Vol. IT-19, No. 2, March 1973.

Computer Simulation of a Digital Satellite Communications Link

D. L. HEDDERLY AND L. LUNDQUIST

Abstract—**This paper describes a computer program for simulating 4-phase digital transmission through a multitransponder satellite link. It takes account of intersymbol and adjacent channel interference and TWT nonlinearity. Results presented include the effects of different TWT characteristics, bit rates, and filter characteristics.**

SYSTEM MODEL

The system model is shown in Fig. 1, where the heavy line indicates the path of the simulated signal and the dotted lines X and Y indicate the paths of the adjacent channel interferences. The program operates by applying a particular "statistically balanced" pseudorandom data sequence to the modulator. Simulation parameters may be varied, but the bulk of the work reported here has been carried out with a 256-bit message and 16 samples per symbol, giving a matrix size of 2048 complex numbers. Thus the total frequency interval simulated is $\pm 4x$ (symbol frequency) from the carrier frequency.

The modulator is a variable-envelope four-phase quadrature modulator with a signal space diagram as shown in Fig. 2. Its two quadrature inputs, I and Q, are fed at the same instant with alternate bits from the data sequence. These data bits are

Paper approved by the Space Communications Committee of the IEEE Communications Society for publication after presentation at the 1972 International Conference on Communications, Philadelphia, Pa., June 16–19. Manuscript received August 14, 1972; revised December 11, 1972.

D. L. Hedderly is with Plessey Telecommunications Research, Maidenhead, England.

L. Lundquist is with the European Space Research Organization, European Space Research and Technology Center, Noordwijk, The Netherlands.

Reprinted from *IEEE Trans. Comm.*, vol. COM-21, pp. 321-325, Apr. 1973.

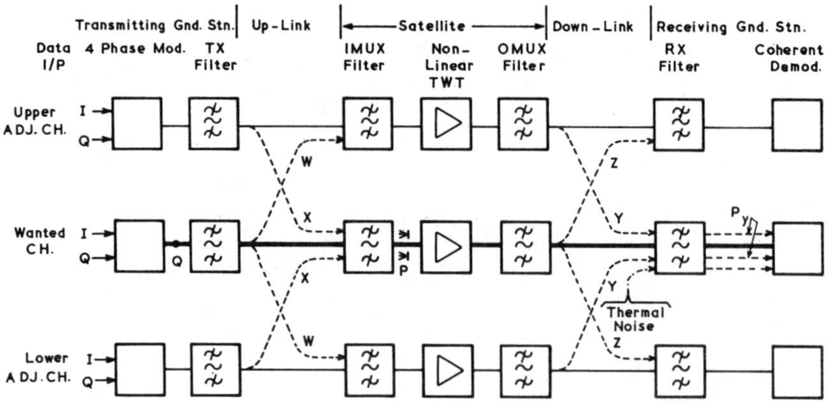

Fig. 1. Model for simulation of a digital satellite communications link.

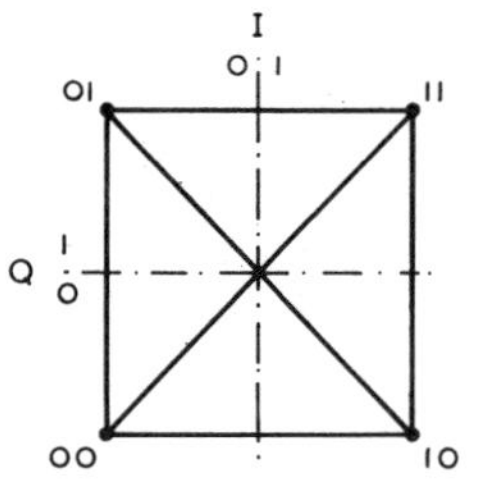

Fig. 2. Signal space for diagram for absolutely encoded 4-phase modulator.

converted to narrow pulses in the time domain and a complex fast Fourier transform (FFT) is then used to convert this signal to the complex frequency plane at the output of the modulator. The signal spectrum associated with this pulse shape is corrected by a compensating filter to give a level signal spectrum within the region of interest.

All filters are simulated in the complex frequency plane and conceived as bandpass filters, although their passband characteristics may be arbitrary, including assymmetry if required.

The signal, before being applied to the input of the nonlinear TWT, is transformed to a time series of complex values from which signal envelope amplitude and phase with respect to a (virtual) carrier are obtained. The nonlinear TWT characteristic is stored as a set of point data of AM to AM and AM to PM values, between which the computer performs an interpolation to provide output values of envelope amplitude and phase. The input backoff, defined as the difference (in decibels) between the actual mean input power of the modulated signal and the CW input to the TWT required to saturate it, must be specified; it is taken as positive for input values less than that required for saturation. The program delivers as output a value for the TWT output backoff analogously defined. The signal is then transformed back to the frequency plane (and also a spectral plot of the signal at the TWT output if provided) together with histograms of signal amplitude at the TWT output and output.

Adjacent channel interference is, for the simulations reported here, not directly simulated. Instead the spectrum of the adjacent channel interference is computed from the spectrum of the wanted channel.

The total interference power is computed and "multipath" signals that enter an adjacent transponder by path W and return to interfere with the wanted signal by path Y are not directly simulated.

This method of course assumes the total interference to have the same distribution as the thermal noise, i.e., Gaussian. For the simulations reported here we have later made comparisons with simulations actually simulating the interferences. There appears to be no significant error in the results.

Demodulation and Timing Recovery

Demodulation is carried out by converting the signal to two baseband signals, representing the in-phase and quadrature components, and taking a time sample of each component for every symbol period (after having determined the optimum sampling instant as explained later).

Formulas are then used to determine the probability of error P_b for each bit, for an initial assumed signal to noise ratio; the average bit error probability for the message being then found by averaging over all the P_b. If the probability of error for the message does not equal the required error rate to within some margin, the signal to noise ratio is altered and the whole process recycled until the signal to noise ratio for the required error rate is found. The thermal noise is represented at the input of the receiving filter by a signal of constant spectral power density. This is equivalent to setting the value of Boltzmann's constant within the program and is taken account of in calculating the final output value of E_{sat}/N_0.

Due to its passage through the filters in the system, the signal becomes delayed in time, and due to its passage through the TWT it suffers a (message dependent) rotation of average phase and envelope compression. Prior to demodulation, it is necessary to resolve the signal into its two quadrature channels correctly and to choose the optimum sampling instant. Both of these operations are done by carrying out a complex correlation between the transmitted data signal and the received signals at the demodulation input. The modulus of the correlation coefficient goes through a maximum around the sampling time and the argument of the correlation coefficient at this optimum sampling instant indicates net phase rotation ϕ of the signal. Accordingly, the phase of the received signal is rotated by $-\phi$, and the I and Q components at the optimum sampling instant are fed to the detector subroutine.

During early trials of the program we found that, as the binary eye on both channels closed due to increasing intersymbol interference, the criterion of greatest correlation did not always correspond to the best E_{sat}/N_0. Accordingly, the detection subroutine was modified to find E_{sat}/N_0 at adjacent sampling instants until a minimum E_{sat}/N_0 was found. This procedure not only yields the best E_{sat}/N_0, but also yields a measure of the sensitivity to incorrect bit timing recovery, since the values of E_{sat}/N_0 are found at the neighboring sampling instants.

The program also prints out an ordered list of the 15 bits with the largest error probabilities and gives the proportion of

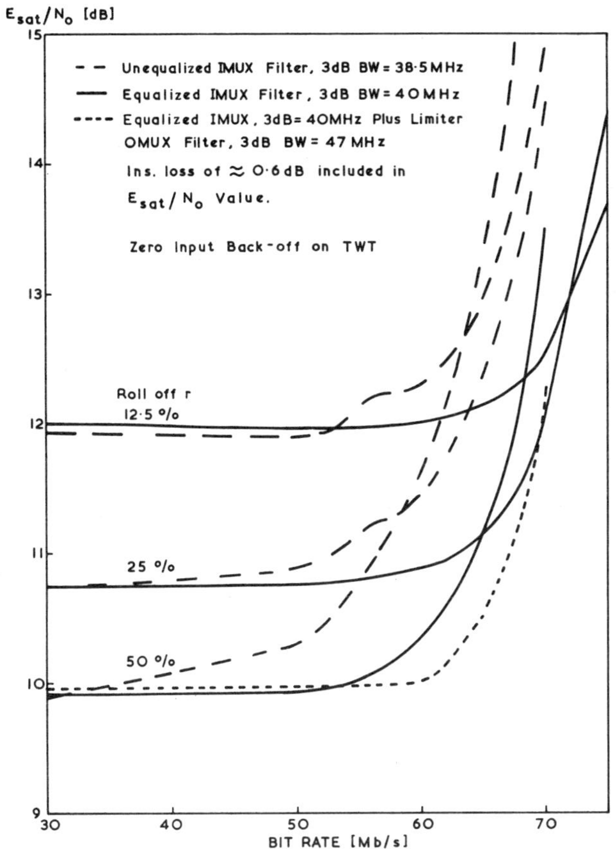

Fig. 3. Variation of E_{sat}/N_0 with bit rate for equalized and unequalized satellite input (IMUX) filters.

the total error rate that each bit contributes. This has been found to be a very sensitive test of the correct resolution of the signal into the I and Q channels, because, with correct resolution of the signal, there should be roughly equal number of errors on both channels.

Calculation of E_{sat}/N_0

The output of the program and all the results presented here are given in terms of the system E_{sat}/N_0 where E_{sat}/N_0 is referred to the saturated CW output power of the nonlinear TWT in the satellite. Thus the results for E_{sat}/N_0 presented here are related to other system parameters by

$$\frac{E_{sat}}{N_0} = \frac{P_{sat} \cdot G}{L \cdot K \cdot T \cdot F_b} M,$$

where

P_{sat} transmitter EIRP at TWT saturation;
L path loss;
G/T gain/temperature for the earth station;
K Boltzmann's constant;
F_b bit rate;
M available margin.

(For an ideal 4-phase PSK system with coherent detection $P_b = 10^{-4}$ corresponds to $E_{sat}/N_0 = 8.4$ dB.)

The simulation includes, within the value of E_{sat}/N_0 calculated, the effects of intersymbol interference, adjacent channel interference, down-link noise, TWT nonlinearity in amplitude and phase, TWT power backoff, and the insertion loss of the satellite output filter. It excludes up-link noise, implementation, rain, and operating margins, together with imperfections in carrier and bit timing recovery; all of these being contained within the available margin M. In addition to E_{sat}/N_0 values, the program provides a breakdown of the causes of degradation and additional line printer graphs of spectra, eye diagrams, and signal amplitude histograms.

Results

The program has been used for degradation studies of a digital satellite channel. The European Space Research Organization (ESRO) satellite system was first thought of as a frequency-division multiple-access (FDMA) system, and hardware work was begun based on this. During the system studies it became clear that a time-division multiple-access (TDMA) system and, in particular, one using 4-phase PSK with sampling detection was a more desirable system. The decision was therefore made to go to TDMA. Unfortunately, the filter studies were well advanced and we therefore had to first find the degradation of the digital signal in an analog channel. Therefore, the results presented here are primarily concerned with optimizing the 4-phase signal itself to get through various FDM channels with a minimum of distortion. All of the results presented here are for a fixed frequency separation of 40 MHz between repeaters on adjacent channels. In all results shown E_{sat}/N_0 values are given for $P_b = 10^{-4}$.

Degradation in FDM Filters

Before starting the simulation study we had to make a choice of earth station filter. We decided to pick an ideal shaping for the transmitter and receiver back to back. We chose to study various degrees of cosine roll-off in the filters in order to see the effect of amplitude variation. Furthermore, we divided up the channel shaping equally between transmitter and receiver.

Figs. 3 and 4 show E_{sat}/N_0 as a function of bit rate for three

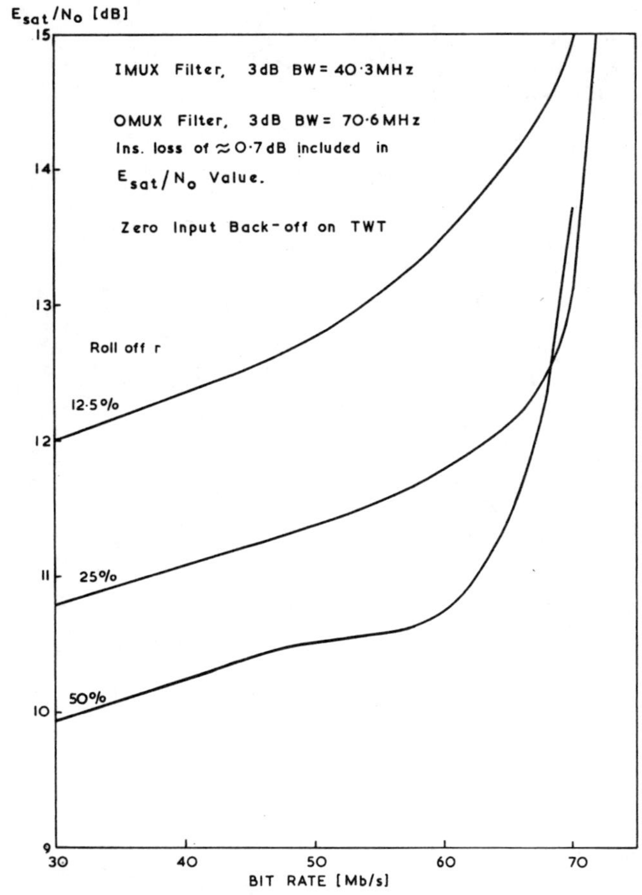

Fig. 4. Variation of E_{sat}/N_0 with bit rate for equalized satellite input filter (IMUX) and a broad satellite output filter (OMUX).

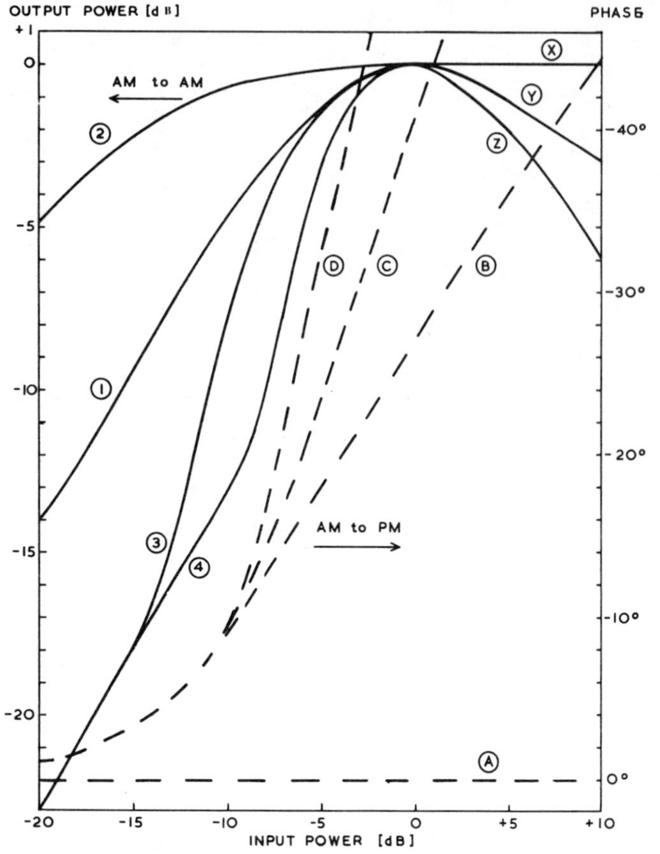

Fig. 5. TWT AM to AM and AM to PM characteristics. For results of Table I.

different combinations of filters, each for various degrees of roll-off in the earth station pulse-shaping filters. First of all we see in both figures that 50 percent cosine roll-off gives the best performance. We have run checks with 100 percent and found it to be worse than 50 percent. There is a tradeoff here because a gentle roll-off such as 100 percent allows too much adjacent channel interference in the receiving earth station. A sharp roll-off, on the other hand, produces long tails on the pulses, which makes the RF signal have large envelope variations. This in turn causes much distortion in the satellite TWT. We believe the optimum is around 50 percent at the bit rate of interest (60 Mbits/s). At high bit rates, the adjacent channel interference is large and it pays to go to sharper filters.

Fig. 3 shows results (in full line) for an equalized satellite input filter that is very flat in the passband and cuts off very fast outside. The satellite output filter is wider but still sharp. The dashed curves are for the same output filter but with an unequalized input filter. In comparing the two sets of curves in Fig. 3 we would conclude that at 60 Mbits/s it is perhaps necessary to have equalized filters since the difference between equalized and unequalized is over 1 dB. This conclusion is somewhat misleading because the input filters are a bit narrow and we are working in a steep region of the curve where any change may cause a large degradation. Of course, we would like to save size and weight and equalizers are quite large and heavy.

Fig. 4 is for a satellite input filter with larger bandwidth and with a much smoother attenuation characteristic. This makes both weight and size more reasonable. The output filter is also wider and this causes the rise in the E_{sat}/N_0 values in the 30–60-MHz region, due to adjacent channel interference.

The Q of the Satellite Input Filter

One way to save weight in the repeater is to have filters with low Q values. We have run several cases with Q varying from 1200 to 600. In all these cases the difference in E_{sat}/N_0 is negligible; a typical difference being 0.5 dB. Thus the obvious conclusion is that we can save weight and size by reducing the Q.

TWT Contribution

We have run several cases with different TWT characteristics. Fig. 5 shows the amplitude and phase characteristics used and Table I shows results from runs with various combinations of these characteristics. We conclude the main cause of degradation is the AM to AM, whereas the AM to PM is not very important unless the signal envelope varies a great deal from decision sample to decision sample, as is the case for sharp roll-off of the earth station filters.

This conclusion appears reasonable. If we look at the signal in the time domain, we have a set of two-dimensional time samples that, after the satellite input filter, have some added intersymbol interference, e.g., a sample in say the I channel now has some contribution from all other samples in the I channel. We can minimize the contribution from samples distant in time by using a pulse with tails falling off very fast, i.e., increasing the roll-off. Going to 100 percent roll-off, however, is too far because of the increased adjacent channel interference. At the TWT output, the main effect is loss of power, because the sample input amplitudes do not all fall at the point of saturation and, therefore, the average of all output samples is less than the saturated value.

In the satellite output filter the main effect is a loss of power due to the insertion loss. The receiver filter finally determines

TABLE I
RESULTS OF TWT STUDY

TWT Characteristic See Fig. 5	E_{sat}/N_0 (dB) Without Adjacent Channel Interference	Increase in E_{sat}/N_0 (dB) Due to Adjacent Channel Interference
1 - Y - A	10.08	0.10
1 - Y - D	10.50	0.16
1 - X - A	9.93	0.09
1 - X - D	10.34	0.15
1 - Z - A	10.23	0.11
1 - Z - D	10.65	0.18
2 - Y - A	9.98	0.14
2 - Y - D	10.41	0.21
2 - X - A	9.83	0.12
2 - X - D	10.26	0.20
2 - Z - A	10.11	0.15
2 - Z - D	10.54	0.23
3 - Y - A	10.12	0.09
3 - Y - D	10.53	0.15
3 - X - A	9.97	0.08
3 - X - D	10.37	0.14
3 - Z - A	10.26	0.10
3 - Z - D	10.67	0.17
4 - Y - A	10.23	0.08
4 - Y - D	10.63	0.14
4 - X - A	10.08	0.07
4 - X - D	10.46	0.13
4 - Z - A	10.38	0.09
4 - Z - D	10.78	0.15

Note: 1) Satellite I/P filter (IMUX) is equalized and is used for results in Fig. 3, 3 dB bandwidth = 40 MHz.
2) Satellite O/P filter (OMUX) has insertion loss of 0.6 dB, which is included in E_{sat}/N_0.
3) Bit rate = 60 Mbits/s. Channel separation—40 MHz.
4) Zero input backoff used on TWT.
5) 50 percent roll-off factor used for earth station filters.
6) $P_b = 10^{-4}$.

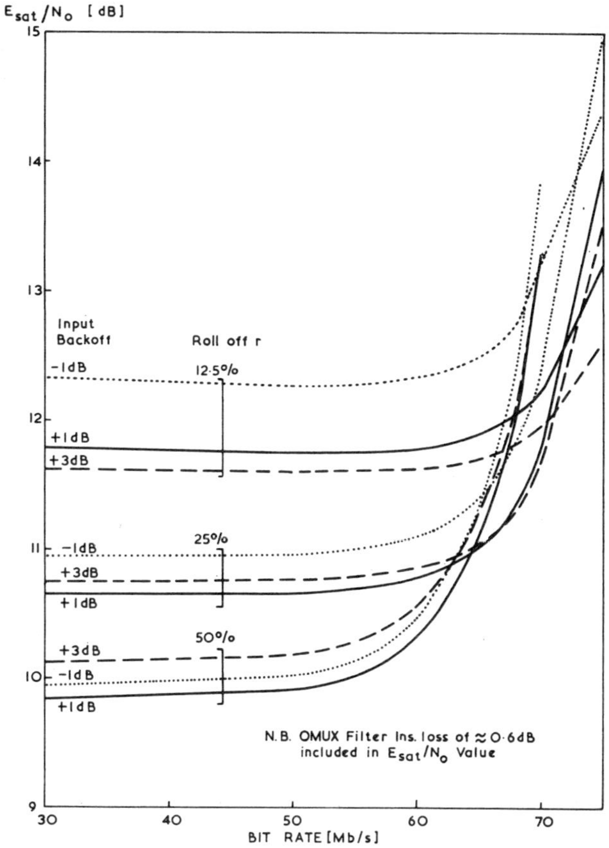

Fig. 6. Effect of TWT backoff on E_{sat}/N_0 (filters as for Fig. 3—solid curves).

the noise bandwidth and, with 50 percent of the shaping in the receiver, the noise bandwidth is equal to the signaling rate.

TWT Operating Point

Fig. 6 shows E_{sat}/N_0 as a function of bit rate for various operating points of the TWT. The filter/TWT combination is that of Fig. 3 with one exception. The conclusion is that backoff does not pay for 50 percent roll-off. The distortion in the TWT is small and backing off means losing more in power than we gain in reduction of distortion. As we decrease the roll-off, however, the distortion increases and it pays to operate the TWT in a more linear region. This is shown in Fig. 6 where we see that, with 12.5 percent roll-off, it pays to have as much as 3-dB backoff. The 12.5 and 25 percent roll-off cases are, of course, not very interesting since the 50 percent case falls far below both for the bit rates of interest.

FURTHER WORK

Channel Optimization of a Nonlinear Channel for Digital Signals

At this point we have not done enough work on optimization. Several suggestions can be made however. First, if there were no nonlinear element in the channel we would optimize the channel by choosing filters satisfying Nyquist's first criterion. Ideally we would get no intersymbol interference. The tradeoff would be between filter realization, receiver noise bandwidth, and adjacent channel interference. Based on the simulation runs so far, introducing a nonlinear element into the channel using the variable envelope form of modulation would add very little intersymbol interference because, at sampling, the input envelope is approximately constant.

Another suggestion would be to put a limiter before the TWT. Intersymbol interference in the I and Q channels, would, at the limiter input, cause amplitude variations at the sampling instants. By using a limiter, these amplitude variations would not be translated to phase variations or power reductions. In Fig. 3 there is one curve showing E_{sat}/N_0 as a function of bit rate assuming a limiter without AM/PM before the TWT. It appears we can improve the system by using a limiter.

INTERACTIVE COMPUTER SIMULATION OF SATELLITE TRANSMISSION SYSTEMS

William L. Cook
Staff Scientist
COMSAT Laboratories
Clarksburg, Maryland 20734

ABSTRACT

A comprehensive satellite transmission system simulation program has been developed for evaluating signal impairments in an interactive mode. The essential features of the software design are discussed, and the analysis capabilities are summarized. The provision of an on-line graphical display capability, together with interactive control over the execution of the program, has been found to greatly enhance the usefulness of the program as a system design tool.

INTRODUCTION

Digital computer simulation of signals transmitted through a communications link has been shown to provide accurate estimates of the actual performance of communications systems.[1,2] A problem of particular concern has been the calculation of error rates in digital signals due to the nonlinear and dispersive properties of the link and the effects of digital receiver performance.[3] The simulation techniques required for these and similar studies have relied heavily on recent developments in digital filter design and the application of fast Fourier transform techniques.[4]

The simulation model of a satellite transmission link has traditionally assumed the form of a single signal passing through a sequence of modules representing modulators, bandpass filters, nonlinear power amplifiers, and receivers. In most cases, only signals of a single type (e.g., M-ary PSK) have been considered.

In the satellite communications system proposed for the late 1970's, such a simplistic model may no longer be adequate. A typical complication is the introduction of frequency reuse, in which several carriers may share the same frequency band. Separation between signals is provided by nonintersecting "spot" and "hemispheric" antenna beams, and by using orthogonal polarization modes.[5] Neither of these techniques provides the degree of isolation found in present systems, which rely principally on frequency (FDMA) and time separation (TDMA). Consequently, the mutual interference between signals will most often be the determining factor in assessing system performance. Simulation tools developed for studying such systems must be sufficiently general to accommodate simultaneously a number of signals having arbitrary characteristics and to account for the primary modes of interaction between signals. Specifically, the model must include mutual interference phenomena such as intermodulation caused by nonlinear devices, adjacent and cochannel interference, and multipath effects.

SOFTWARE DESIGN REQUIREMENTS

A satellite transmission system simulation program has been developed for evaluating signal impairments in an interactive mode. The software operates on an IBM 360/65 computer under the TSO time-sharing system. An on-line graphics capability is provided by a Tektronic 4010 display, which has been found to be an important element in using the software as an experimental tool and as an aid to interactive system design. The design objectives which guided the development of the software may be summarized briefly as follows:

a. Generality of the system model. The user is allowed complete freedom in specifying the system configuration, including such factors as the number and type of components, the number of signals sharing a common link, and the number of parallel paths a signal may follow.

This paper is based upon work performed in COMSAT Laboratories under the sponsorship of the Communications Satellite Corporation.

Reprinted with permission from *Proc. Fifth Annual Pittsburgh Conf. Modeling and Simulation*, vol. 5, Apr. 1974, pp. 867-872.

b. Generality of signal types. The program is capable of accommodating all types of angle and/or amplitude modulated bandpass signals. No restrictions are placed on sampling rate, signal bandwidth, or length of the signals which are simulated.
c. Ease of use. The user communicates with the program by means of a mnemonic-oriented free-field command language. A wide variety of input options are available for defining each signal or component in the system, and checks for consistency are provided to minimize the chance of input errors.
d. Graphics capability. The user may request the immediate display of graphical and numerical data relating to components and signals at any point in the system. This capability is useful in assessing the validity and accuracy of the simulation model, as well as in providing direct measures of system performance.

The program structure is based on a few simple principles. The first is that all signals are stored on disk files and processed in a sequential manner, one record (256 samples) at a time. This approach not only removes any restrictions on the length of signals which may be processed, but also allows for storage of signals at various points in the system for subsequent display or reprocessing. Second, all operations are performed entirely in the time domain. This restriction simplifies the interface between system blocks, eliminates the computation time required for transformations between the time and frequency domains, and eliminates the need for large in-core arrays.

SIGNAL GENERATION

The simulation program is designed to accommodate narrowband bandpass signals, i.e., signals whose bandwidth is small compared to the carrier frequency. The general form of such a signal is

$$a(t) = \rho(t) \cos [2\pi f_c t + \theta(t) + \phi] \quad (1)$$

where f_c and ϕ are the carrier frequency and phase, and $\rho(t)$ and $\theta(t)$ are the amplitude and phase modulation, respectively. The modulation of the signal may be equally well characterized in terms of the complex envelope,[6] $u(t)$, defined by,

$$u(t) = \rho(t)\, e^{j\theta(t)} \quad (2)$$

In the program, all signals are represented by time-sampled values of the complex envelope.

The modulation functions $\rho(t)$ and $\theta(t)$ may be continuous functions of time, or may take on only a finite number of discrete values. An example of the latter is an M-ary phase-shift keyed (PSK) signal. In this case the amplitude modulation $\rho(t)$ is constant, and the angle modulation $\theta(t)$ is limited to M discrete values.

The complex envelope of a periodic 4-phase PSK signal is shown in Figure 1. The signal is composed of a sequence of symbols, each of constant duration. If the phase of the signal were sampled at the mid-point of each symbol, only four distinct values should occur, each representing a pair of bits (00, 01, 10, or 11) in the binary sequence which is being transmitted.

The bit sequences to be simulated may be specified by the user or generated in a random manner. In many cases it may be desirable to generate a maximal length pseudorandom sequence, which has the advantage of guaranteeing certain statistical properties usually found only in random sequences of infinite length.[7] In addition, the shape of each symbol may be specified by the user. This option has been used to generate the pulse train whose complex envelope is shown in Figure 2.

A variety of continuous signal generators are available, including Gaussian noise, summations of sinusoids equally spaced in frequency, or user-defined signals. When used in conjunction with amplitude, phase, and frequency modulation modules, these signals may simulate most commonly encountered signal types, including FM telephony and television test pattern signals.

SIGNAL PROCESSING

A library of basic operations is provided to allow the user to simulate the effects of each component in the transmission path. The most essential components are nonlinear amplifiers and bandpass filters, which are the principal contributors to signal distortion. A summation module is provided to account for the effects of frequency multiplexing, multipath, cochannel interference, and additive noise. Additional operations, used primarily to account for receiver performance, include squaring of a signal or finding the product of two signals at either the sum or difference frequency, time delays of arbitrary duration, simple phase-lock loops, limiters, rectifiers, and demodulators.

Bandpass filter characteristics may be defined in a number of ways:

a. by specifying the filter type (e.g., Butterworth, Chebychev, inverse-Chebychev, elliptic, or Bessel) and appropriate parameters, including filter center frequency and bandwidth, in-band and out-of-band attenuation, dissipation loss factor, and type of bandpass transformation;

b. by specifying the symmetric or nonsymmetric pole and zero locations of the low-pass equivalent filter;
c. by specifying a set of measured values for attenuation and group delay as functions of frequency; or
d. by specifying a set of measured values for the real and imaginary components of the impulse response as functions of time.

Two distinct approaches to bandpass filter simulation may be used, depending on the filter definition. If the complex poles and zeros of the filter transfer function are known (as in the first two cases above), a digital filtering approach is used. The poles and zeros are transformed to the time-sampled domain by means of either the standard or bilinear Z transform.[8]

If the filter is defined in terms of measured transfer curves, the output signal is obtained by repetitive convolutions of the filter impulse response with overlapping segments of the input signal using the discrete Fourier transform technique.[9]

In either case, transformation of the problem from the continuous to the discrete-time domain may introduce modeling errors which are influenced by a number of factors including the sampling rate, degree of out-of-band attenuation, and type of transformation used. Consequently, after the user has entered data defining a bandpass filter, the original transfer curves and those corresponding to the simulation model are immediately displayed to provide him with an estimate of the modeling errors which are introduced.

The transfer curves displayed for both a Butterworth and inverse-Chebychev filter are shown in Figures 3 and 4, respectively. In each case the dashed curves represent the desired attenuation and group-delay characteristics, while the solid curves represent those of the discrete-time model. The discrepancy between curves in Figure 4 demonstrates the effects of frequency distortion introduced by the bilinear z-transform.

The model of the nonlinear amplifier is defined by specifying values of output power level and phase as functions of input power level.

As in the case of filters, the amplifier transfer characteristics generated from the user-defined data points are immediately displayed to allow the user to verify the accuracy of the input data, as well as the curve-fitting procedure. The curves shown in Figure 5 are for the INTELSAT IV traveling wave tube amplifier.

EVALUATION OPTIONS

The on-line graphical display capability allows the user to determine the validity of the system model and to obtain measures of system performance by examining the characteristics of signals at any point in the system. Output options available to the user include the following:

a. complex envelope (real and imaginary components),
b. complex envelope (amplitude and phase),
c. power spectral density,
d. signal space diagram,
e. eye pattern,
f. scatter diagram, and
g. error probability versus carrier-to-noise ratio.

Power spectral density calculations are based on a technique for averaging modified periodograms of the signal, which are obtained by taking the discrete Fourier transform of consecutive records of the signal after weighting by an appropriate data window.[10] The final three options listed above are used primarily to investigate the performance of M-ary PSK signals. In each case, symbol timing intervals are recovered by averaging the zero crossings of the complex envelope components. The decision process is assumed to be based on an instantaneous sampling at the mid-point of each symbol, or alternatively at some point specified by the user. A reference phase angle is estimated by averaging the phase at each sampling instant over M equally spaced sectors in the complex plane. (e.g., for 4-phase PSK, the average phase is calculated for each of the four quadrants and averaged).

Plots may be generated showing the expected probability of error as a function of carrier-to-noise ratio. The noise is assumed to be Gaussian and to have been added to the signal immediately prior to the receiver filters (e.g., in the down-link of the satellite). If these conditions are satisfied, and sufficiently low error rates are assumed, the probability of error for 4-phase PSK is simply

$$P_e = \frac{1}{2} \sum_{i=1}^{n} \left[\operatorname{erfc}\left(\frac{x_i}{\sqrt{2}\sigma}\right) + \operatorname{erfc}\left(\frac{y_i}{\sqrt{2}\sigma}\right) \right] \qquad (3)$$

where x_i and y_i are the real and imaginary values of the complex envelope at the sampling instants, respectively, σ^2 is the noise power, and the complementary error function is defined by

$$\operatorname{erfc}(\xi) = \frac{2}{\sqrt{\pi}} \int_{\xi}^{\infty} e^{-u^2}\, du \qquad (4)$$

Equation (3) is extended to PSK signals having an arbitrary number of phases by redefining x_i and y_i as the minimum distances between the sampled point and the straight lines bounding the region of a decision error. The accuracy of this approximation improves as the probability of error becomes small. If desired, the effects on the error probability of various types of convolutional and block coding schemes may be calculated.

The on-line display options available to a user of the program may be demonstrated by means of a simple example. Consider the block diagram of a simplified communications system shown in Figure 6. Two 4-phase pseudo-random PSK signals are generated. The signals have identical complex envelopes, and differ only in carrier frequency. Each signal is passed through a bandpass filter, and the output signals are frequency multiplexed.

The complex envelope of the combined signal is shown in Figure 7. (Note the high-frequency component, which corresponds to the difference in frequency between the two carriers.) This signal is now passed through a nonlinear amplifier. The power spectral density of the amplifier's signal output is shown in Figure 8. In addition to the two original signals appearing on the right-hand side of the figure, an intermodulation product has been generated due to the nonlinear characteristics of the amplifier. The power levels and frequencies of intermodulation products generated in this manner are in close agreement with predicted results.[11]

The output of the amplifier is filtered to recover the individual signals. The distortion in each signal due to the nonlinear and dispersive properties of the link is evident in the eye pattern shown in Figure 9. The eye pattern, obtained by superimposing the complex envelopes of all transmitted symbols, provides a rough estimate of the performance of the system as well as the optimum receiver sampling instant.

Once a sampling instant has been selected (in this case the vertical line in Figure 9), the actual sampled values may be plotted in the signal-space plane. Such a plot is commonly called a scatter diagram (Figure 10). Based on these sampled values, the bit-error probability may be calculated as a function of carrier-to-noise ratio and displayed as shown in Figure 11. An additional perspective may be obtained from a plot of the trajectory of the signal in the signal-space plane (Figure 12), which often provides useful information on amplitude and delay distortions.[12]

REFERENCES

(1) Rappeport, M. A., "Digital Computer Simulation of a Four-Phase Data Transmission System," Bell System Technical Journal, May 1964.

(2) Hedderly, D. L., and Lundquist, L., "Computer Simulation of a Digital Satellite Communications Link," Proc. of ICC Conference, Philadelphia, June 19-21, 1972.

(3) Rhodes, S. A., and Palmer, L. C., "Computer Simulation of a Digital Satellite Communications System Utilizing TDMA and Coherent Quadriphase Signalling," Proc. of ICC Conference, Philadelphia, June 19-21, 1972.

(4) Digital Signal Processing, edited by L. R. Rabiner and C. M. Rader, New York: IEEE Press, 1972.

(5) Kreutel, R. W., "The orthogonalization of polarized fields in dual-polarized radio transmission systems," COMSAT Technical Review, Vol. 3, No. 2, Fall 1973.

(6) Stein, S., and Jones, J., Modern Communication Principles, McGraw-Hill, 1967.

(7) Peterson, W. W., Error Correcting Codes, Cambridge, Mass.: M.I.T. Press, 1961.

(8) Kaiser, J. F., "Design Methods for Sampled Data Filters," Proc. of the 1st Annual Allerton Conference on Circuit System Theory, 1963.

(9) Helms, H. D., "Fast Fourier Transform Method of Computing Difference Equations and Simulating Filters," IEEE Trans. on Audio and Electroacoustics, Vol. AU-15, No. 2, June 1967.

(10) Welch, "The Use of Fast Fourier Transform for the Estimation of Power Spectral Density: A Method Based on Time Averaging Over Short, Modified Periodograms," IEEE Trans. on Audio and Electroacoustics, Vol. AU-15, No. 2, June 1967.

(11) Fuenzalida, J. C., Shimbo, O., and Cook, W. L., "Time domain analysis of intermodulation effects caused by nonlinear amplifiers," COMSAT Technical Review, Vol. 3, No. 1, Spring 1973.

(12) Davey, J. R., "Digital Data Signal Space Diagrams," Bell System Technical Journal, November 1964.

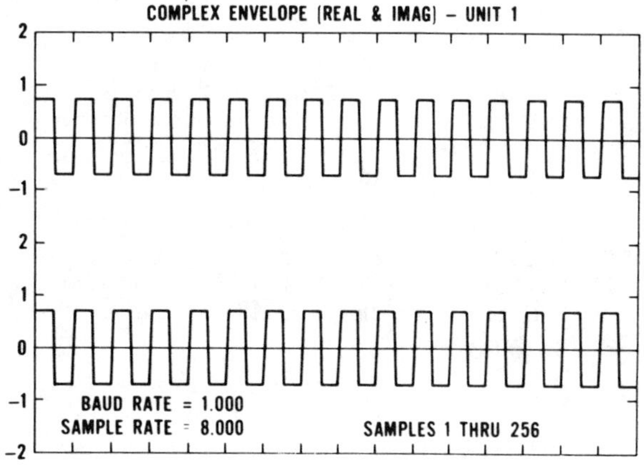

Figure 1. Four-Phase PSK

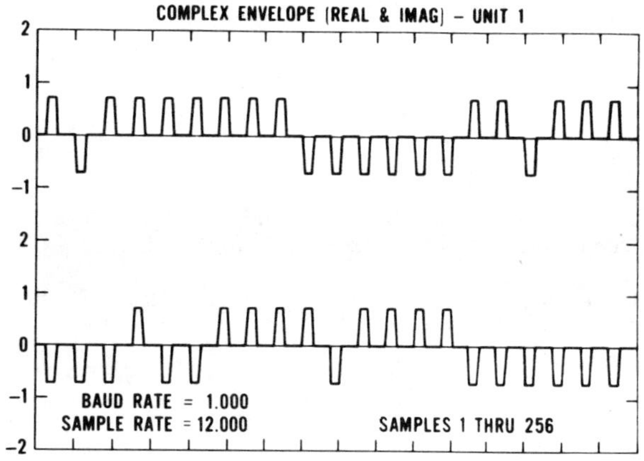

Figure 2. Random Pulse Train

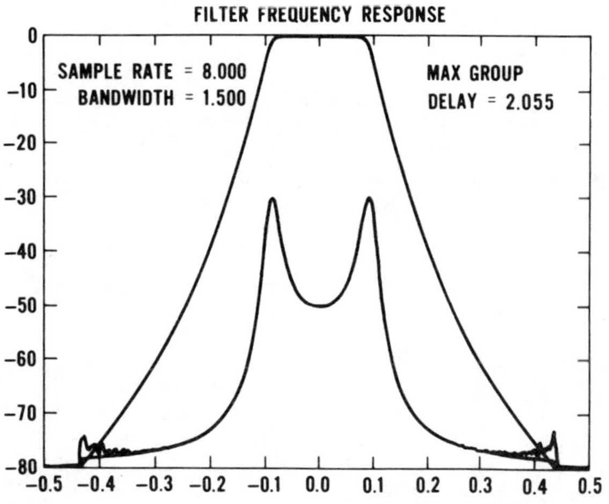

Figure 3. Butterworth Filter

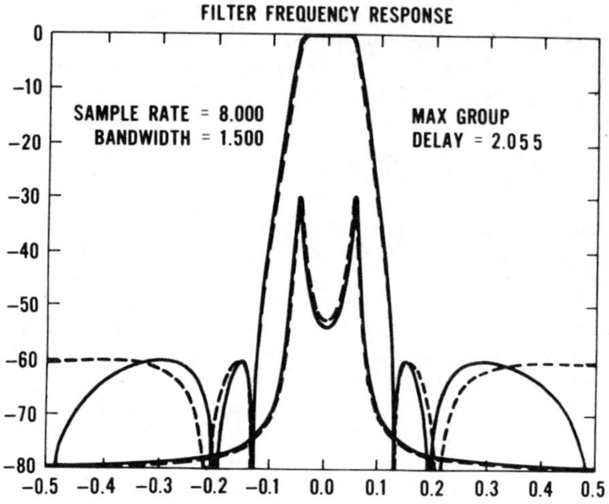

Figure 4. Inverse Chebychev Filter

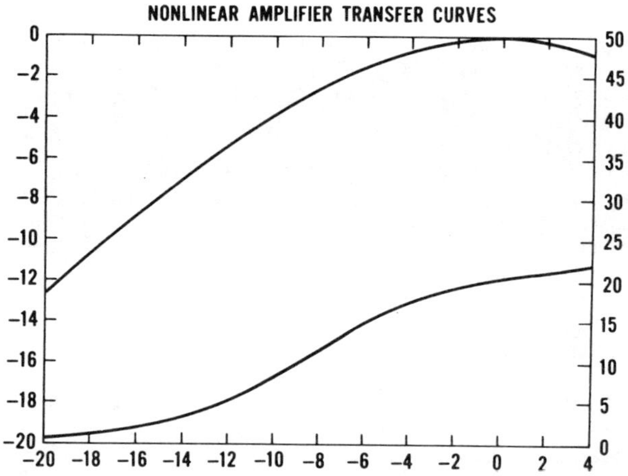

Figure 5. Amplifier Transfer Characteristic

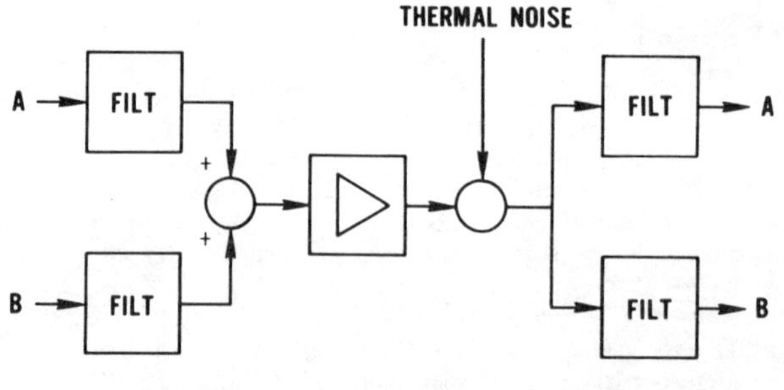

Figure 6. System Block Diagram

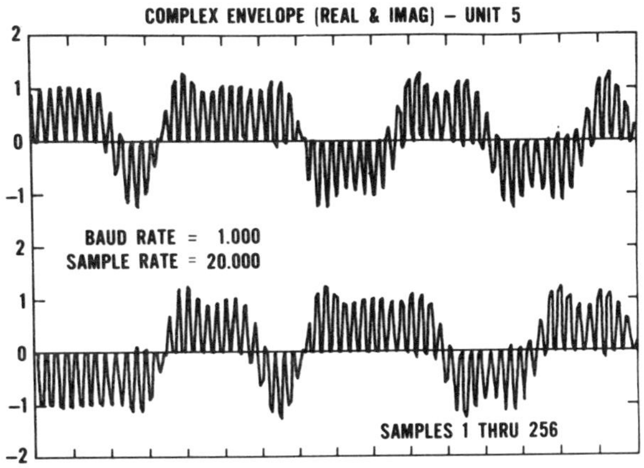

Figure 7. Complex Envelope of Combined Signal

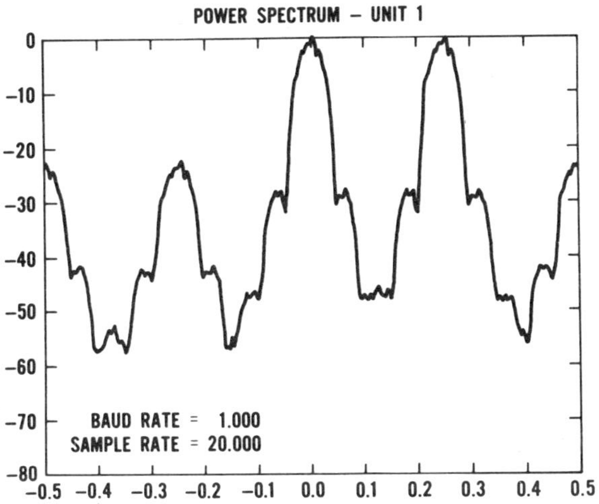

Figure 8. Power Spectrum of Amplifier Output

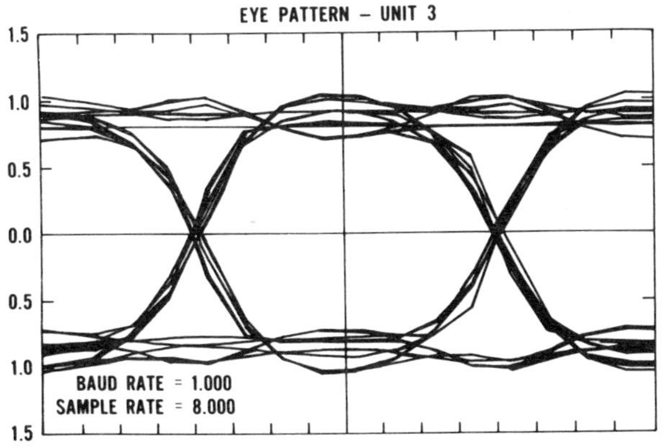

Figure 9. Eye Pattern

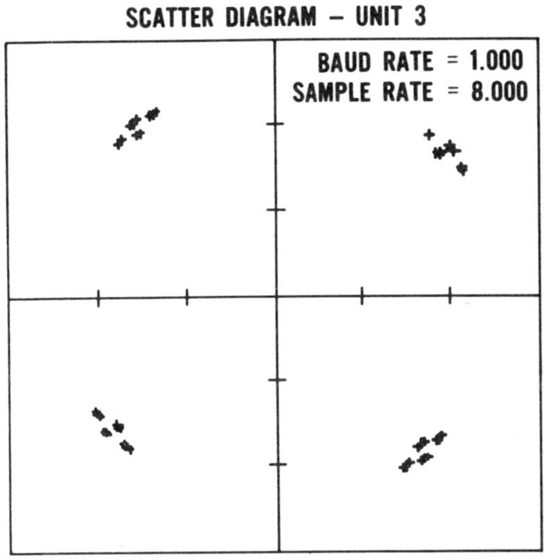

Figure 10. Scatter Diagram

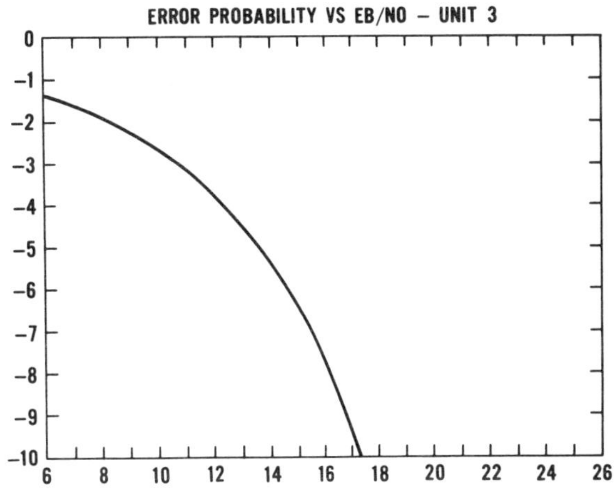

Figure 11. Error Probability

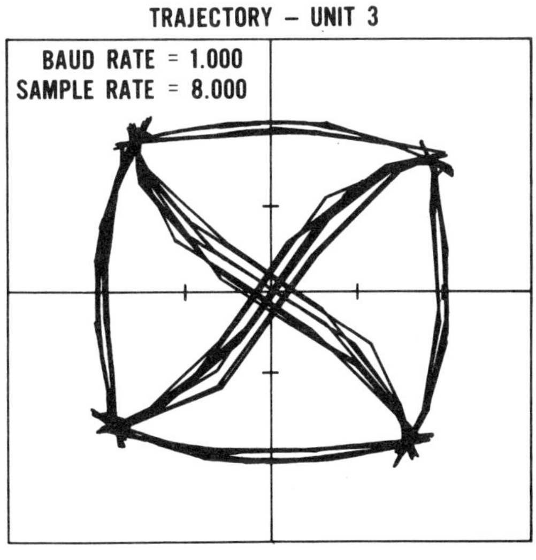

Figure 12. Signal Space Trajectory

Author Index

Subject Index

Editor's Biography

Peter Stavroulakis received the Ph.D. degree in 1973 from New York University.

From 1973 to 1979 he was with Bell Laboratories in New Jersey working on various projects involving technoeconomic studies of small and large telephone telecommunications systems and interference studies of satellite systems. During the academic year 1975–1976 he was a visiting Assistant Professor at the University of Athens, Greece. Since August 15, 1979 he has been at Oakland University, Rochester, MI, as an Associate Professor of Engineering teaching and doing research in communications and systems engineering. He is the author of two monographs in Greek on control and communication systems. He has published a substantial number of papers in the field of communication and control.

Dr. Stavroulakis is a member of Tau Beta Pi, Eta Kappa Nu, and he is mentioned in the *Who's Who of Intellectuals.*